TULLIO REGGE

An Eclectic Genius

From Quantum Gravity to Computer Play

TULLIO REGGE

An Eclectic Genius

From Quantum Gravity to Computer Play

Editors

Leonardo Castellani
University of Eastern Piedmont, Italy

Anna Ceresole
National Institute for Nuclear Physics, Turin, Italy

Riccardo D'Auria
Polytechnic University of Turin, Italy

Pietro Fré
University of Turin, Italy

World Scientific

NEW JERSEY • LONDON • SINGAPORE • BEIJING • SHANGHAI • HONG KONG • TAIPEI • CHENNAI • TOKYO

Published by

World Scientific Publishing Co. Pte. Ltd.
5 Toh Tuck Link, Singapore 596224
USA office: 27 Warren Street, Suite 401-402, Hackensack, NJ 07601
UK office: 57 Shelton Street, Covent Garden, London WC2H 9HE

Library of Congress Control Number: 2019044511

British Library Cataloguing-in-Publication Data
A catalogue record for this book is available from the British Library.

TULLIO REGGE: AN ECLECTIC GENIUS
From Quantum Gravity to Computer Play

ISBN 978-981-121-343-4

For any available supplementary material, please visit
https://www.worldscientific.com/worldscibooks/10.1142/11643#t=suppl

To Rosanna Cester

Computer art by T. Regge, Rosanna's dream (see http://www.arnoldreggecenter.com/about/bio-regge/ for colour version and more drawings).

Preface

This volume of original articles, collected papers and commentaries by contemporary scientists and intellectuals is devoted to celebrating the work of Tullio Regge: his outstanding mathematical insight in the fundamental properties of physical theories, his unique creativity and unusual scientific breadth have made him a giant in the panorama of Theoretical Physics in the second half of the 20th century. His contributions to quantum theory and to general relativity have marked significant turning points in the development of scientific knowledge, and they still have a profound impact today on the work of a large theoretical community.

Born in 1931, of Piedmontese origin, he graduated in 1952 from the University of Turin under the supervision of Mario Verde and Gleb Wataghin. In those times, academic life in the Physics Department was quite effervescent, and many important international figures in quantum mechanics and quantum field theory, including Wolfgang Pauli, were passing through Turin: this surely ignited Regge's imagination. In 1957, he got his PhD in Physics from Rochester University (USA) under Robert Marshak, and then moved to the Max Planck Institute for Physics in Munich till 1959, where he interacted with Werner Heisenberg. In 1961, he earned the chair of Relativity at the University of Turin, and collaborated with various members of the Physics department: beside the contributors to this volume, one can mention Alessandro Bottino, Vittorio De Alfaro, Ferdinando Gliozzi, Stefano Sciuto, Alfredo Molinari among others.

Upon a call by Oppenheimer, he then became a member of the Institute for Advanced Studies in Princeton in the early sixties, and there he continued his work till 1979. Besides, the international prestige of IAS, he was attracted by a centre for frontier scientific investigations of theoretical and multidisciplinary character, with pure knowledge *per se* as its mission. These were for Regge golden years where, while living in Einstein Drive, he could interact scientifically and socially with an amazing crowd of past and future Nobel prize winners, and with highly innovative thinkers such as Freeman Dyson and Kurt Gödel. In 1983, back in Turin, he founded with Mario Rasetti from Turin Polytechnic University, the Institute for Scientific Interchange, inspired by the spirit of the IAS, and by the belief that societal crises must be fought by promoting basic science as a driver of profound structural changes. He spent several periods as a visiting scientist at CERN in Geneva and then returned as an emeritus professor to Turin Polytechnic, symbolically closing the circle to the beginnings of his scientific career. He suffered from a muscular degenerative illness that began to impair his walking in the '80s, and it soon forced him to a wheelchair for the rest of his life. After some experience as a politician in the European Parliament and a long period of work as a popular science communicator, he died in Turin on October 23, 2014. He was married to Rosanna Cester, an experimental physicist who shared his daily life, his passion for physics and his cosmopolitan attitude. They had three children: Daniele, Marta and Anna, and eight grandchildren. We are grateful to all of them for their contributions to this volume, the article by Daniele (a radiology specialist and professor at U. Turin) in "Reminiscences", the portraits and the computer drawings provided by Rosanna, Marta and Anna. Regge wrote an autobiography in collaboration with Stefano Sandrelli, published in Italian by Einaudi in 2012 with the title "L'infinito cercare — Autobiografia di un curioso".

During his early years in Princeton, in a collaboration with John Archibald Wheeler, Regge created the basis for the theory of black hole perturbations. It was Wheeler who induced him to examine spherical solutions of Einstein field equations in general relativity, the Schwarzschild metric, even before those objects were named "black holes". Regge was only a PhD student at the time: it is impressive that Wheeler trusted him to sustain his physical intuition with mathematical rigour, and that Regge succeeded (as reported in Wheeler's autobiography, "Regge rose to the occasion").

Their paper of 1957 is still the basis for studies in black hole stability, and some later improvements on the Regge–Wheeler equation by Frank Zerilli are still relevant for analytic studies of black hole collisions, and for gravitational waves detection at large interferometers. This topic is discussed in Part I, with a detailed contribution by Kip Thorne, the Feynman Professor of Theoretical Physics, Emeritus at the California Institute of Technology (Caltech), who claims that "Over the sixty years since the paper's publication, it has been a foundation for an enormous edifice of ideas, insights, and quantitative results". The reprinted paper by Regge and Wheeler of 1957 from Physical Review follows the commentary.

One of Regge's first very impressive scientific contributions in 1959 at the Max Planck, in investigating mathematical properties of high energy potential scattering in the Schrödinger equation, was the idea to study the scattering amplitudes of elementary particles as functions of a complex angular momentum. He thus discovered the intimate relation between the poles in the angular momentum — the famous *Regge poles* — the bound states of the system and the power-law growth of the amplitudes. Regge was only 27 years old at the time, and it was due to the encouragement of Kurt Symanzik and then Werner Heisenberg that convinced him to pursue his "poles" as something beyond an intriguing mathematical result. Indeed this idea immediately generated an enormous interest in the study of strong interactions, and it was soon extended to the scattering amplitudes in field theory. Thus followed a massive production of results and extensions, which in the end, opened the way to the Veneziano amplitudes and the first formulation of dual resonance models. These were later reinterpreted as string theories, which, together with supersymmetry and supergravity, have produced a tremendous impact on modern ideas in theoretical physics and on the goal of unification of fundamental interactions including gravity. This topic is discussed in Part II, starting with a contribution by one of the founding fathers of String Theory, Gabriele Veneziano (CERN and Collège de France), commenting on its deep roots into Regge theory and on the "almost unimaginable consequences in fundamental Physics" of Regge's bold idea of making angular momentum complex, with "still very many to be uncovered". There follows an article by two of Regge's students during their B.S. degrees in Turin, Vittorio Del Duca (ETH, Zurich) and Lorenzo Magnea (Turin University), today active researchers in particle theory, who provide a panorama on today's relevance of the Regge limit of gauge-theory amplitudes and cross-sections as a powerful tool for the study of fundamental interactions. Then, the section includes an article on the deep mathematical path between the S-matrix theory of the '70s and modern developments in the computations of Feynman amplitudes by Giampiero Passarino, professor of Theoretical Physics at the University of Turin, a world leading figure in perturbative Quantum Field Theories and in bridging Standard Model Physics towards current collider experiments. Three reprinted papers by Regge are included (see table of contents), two of them reproduced from the *Nuovo Cimento* by courtesy of the Società Italiana di Fisica. As contributions from his former colleagues in Turin, one could mention on this topic also a recent article by Alessandro Bottino, `https://arxiv.org/pdf/1807.02456.pdf`, reviving his work with Regge and Anna Longoni of 1962, and the classic book of 1965 on Potential Scattering written with Vittorio De Alfaro, who also wrote some personal reminiscences `http://inspirehep.net/record/1330474/files/dealfaro.pdf`.

Slightly later, in the early 1960s, Regge created a completely new and revolutionary approach to general relativity based on a discretisation of space-time, since then universally known as *Regge Calculus* (1961). In the same way, a curved surface can be tessellated by flat triangles as closely as we like, he introduced a "triangulation" of the space-time with four-dimensional polyhedra, the curvature residing at the vertex (or more properly, at the hinges) where the simplices meet. This curvature is measured by the "deficit angles", analogous to the deficit angle that arises when trying to flatten a tetrahedron, since the edges meeting at the vertex cannot fit together in the flat plane. Remarkably, he also gave a discrete action principle in terms of such deficit angles and edge lengths which in the continuum limit exactly reproduces the Einstein action. Even today, Regge calculus, which is considered a precursor of lattice gauge theory, is widely studied and applied to solve numerically Einstein equations when the absence of any symmetry would make it impossible to use analytic methods. In 1968, Regge and Giorgio Ponzano developed a quantum variant of Regge calculus in three space-time dimensions, which today is known as the *Ponzano–Regge model.* This evolved into a series of attempts to formally describe (discrete) quantum theories of gravity known as spin foam models. It also developed into mathematical models, such as the Turaev–Viro model, which in the context of topological quantum field theories provides an example of a quantum invariant. These topics are discussed in Part III, starting with a review by Annalisa Marzuoli from Pavia University, on the Ponzano–Regge model, illustrating the fundamental breakthrough for 3D discretised quantum gravity historically introduced by viewing the 6-j symbol geometrically as a tetrahedron, or *magic brick* of the Racah–Wigner tensor algebra. Since then, the elegant mathematical apparatus of Regge "skeletons" and Regge calculus has reached beyond regularizations of quantum gravity, with many diverse applications ranging from topological quantum field theories to statistical lattice models and quantum computing. The reprint of the paper by Regge and Ponzano "Semiclassical Limit of Racah Coefficients" originally published in a Proceedings by North Holland in 1968 is included thereafter. Reproductions from the original manuscript of the Ponzano–Regge paper are also included.

The chapter continues with the seminal article of Regge "General Relativity Without Coordinates", that is reprinted in its original version, courtesy again of *Nuovo Cimento.* As a celebration of this paper, there follows a recent application of simplicial gravity by Alessandro D'Adda, from the Turin Division of the National Institute for Nuclear Physics (INFN), where he explores the role of space-time coordinate systems defined on each simplex and of a discrete version of general coordinate invariance in discretized gravities.

Discrete structures in classical and quantum gravity with emphasis on modern applications are also illustrated in Part IX, with a contribution by John Barrett (University of Nottingham), Daniele Oriti (Potsdam, Max Planck Institute) and Ruth Williams from Cambridge, followed by the original article of 2000 by Regge and Williams that is presented as a reprint. The authors provide specific focus on the recent developments in spin foam models, Barrett–Crane models and group field theories that have emerged, inspired by Regge's discretizations, in relation with the loop approach to quantum gravity.

Regge's interest in general relativity included its Hamiltonian formulation, and prompted him to adapt the Dirac theory of constrained Hamiltonians to the important problems of surface integrals and asymptotically defined symmetries, which is the subject of the top cited article that Regge wrote with Claudio Teitelboim (now Claudio Bunster) while at Princeton University, in 1974. This article opens as a reprint in Part IV, which continues with a contribution by Claudio Bunster, now at the CECs in Valdivia, Andrès Gomberoff (U. Adolfo Ibàñez, Santiago) and Alfredo Perez (CECs, Valdivia) on a modern evolution of this subject. They examine asymptotically null space-like surfaces that turn out to be particularly suitable to describe radiation and allow an Hamiltonian description of Bondi, van der Burg, Metzner and Sachs for both electromagnetism and gravitation, displaying asymptotic electro-magnetic duality. We should also mention the classical reference on constrained Hamiltonian systems by Hanson, Regge and Teitelboim, published in the *Rendiconti of the Accademia dei Lincei* in 1976, and the less known but noteworthy covariant Hamiltonian formulation of 1985 for gravity and supergravity theories (with Alessandro D'Adda and Jeanette Nelson in `https://doi.org/10.1016/0003-4916(86)90057-6`, and later with Alberto Lerda from University of Eastern Piedmont at Alessandria on `https://doi.org/10.1016/0370-2693(85)90764-6`).

The lack of space-time does not allow us to properly cover many other important contributions in the physics of many-body systems explored with the late Alfredo Molinari, quantum vortices in liquid helium and his very elegant paper with Mario Rasetti (emeritus professor at the Turin Polytechnic University and president of the ISI Foundation after Regge, which has continued to promote a distinctive borderless attitude to science, particularly focussed on complex systems) on the two-dimensional Ising model having a dodecahedral group of symmetry (`https://link.springer.com/article/10.1007%2FBF02740736`), which was later recognized as the symmetry of the fullerene molecule. As a sample, Part V contains a contribution by Mario Rasetti on quantum vortices showing the intertwining of physics, topology and algebra that was a signature of Regge's scientific approach, and a reprinted paper by Fernando Lund (at the time Regge's PhD student, presently at the University of Chile in Santiago) and Regge of 1976 on strings and quantum vortices in soliton solutions.

Leaving Princeton in 1979, Regge came back to Torino, where he founded an approach to the construction of unified theories of (super)gravity inspired by his work with Yuval Ne'eman (Tel Aviv University) done in 1978, based on the simple realization that gravity and its generalizations are all gauge theories of the Poincarè group extended to other groups and supergroups. The spirit was very much that of Élie Cartan, who was the first to worry about gauge theories from a geometrical viewpoint, formalizing them in terms of connections and torsion. Since it was heavily based on group-theoretical methods, it became known as group-manifold or geometric approach among his early collaborators Alessandro D'Adda, Riccardo D'Auria and Pietro Fré at Turin University, or else dubbed "rheonomic approach" by some of them. This work largely expanded its applications to more geometric formulations of supergravity theories, that became a signature of the Turin school on unified gravity. This topic is discussed in Part VI with an extended review by Riccardo D'Auria (Professor Emeritus at Turin Polytechnic, who was also a student of Regge and his assiduous collaborator), a contribution on more recent developments by Leonardo Castellani (University of Eastern Piedmont in Alessandria) about integration on supermanifolds applied to supergravity, and two of the original reprinted papers by Ne'eman and Regge. We like to include an image taken from the Les Houches lectures of 1983 on "The group manifold approach for unified gravity", depicting the "spinor clock" that Regge devised to explain the properties of spinors in any dimension, an example of his unique mixture of playfulness and mental clarity. The geometric approach to supergravity is the topic of the three-volume book by Castellani, D'Auria and Frè "Supergravity and Superstrings, a geometric perspective", published by World Scientific in 1991.

In the last decade before his retirement, Regge became deeply interested in a broader variety of geometry-inspired research themes. Firstly, since 1989 he studied (2+1)-dimensional gravity extensively in a long collaboration with Jeanette Nelson from Turin University, that explored the connection with Chern–Simons theories, AdS-CFT correspondence and Liouville theory, with fascinating topological aspects. This is covered in Part VII by two reprints by Nelson, Regge and Federico Zertuche (from U. Mexico and Sissa at the time), and by Nelson and Regge, with an introduction by Nelson, emphasizing the connection with braid groups, quantum groups, knot theory and recalling Regge's dream of a "swiss cheese universe" modeled on higher genus Riemann surfaces.

Then, in the early '90s, Regge got involved in some developments in algebro-geometric aspects of string theory and Calabi–Yau manifolds that are covered in Part VIII, containing a contribution by Anna Ceresole (Turin division of INFN) and Riccardo D'Auria, reviewing some geometric aspects of string compactifications, followed by one of their articles with Regge. The techniques used heavily hinged on his early work on the monodromy of Feynman integrals in scattering theory, as illustrated by Passarino in this volume.

As a final challenge, Regge tackled the three-dimensional Ising model by topological and combinatorial methods. His basic observation was that a three-dimensional Ising model can be considered as a two-dimensional one whose graph has a very high genus. Because of its very high symmetry, one can in many cases perform the evaluation of the partition function, at least in principle. These were the same kind of considerations that were applied to three-dimensional gravity, which, despite having no propagating degrees of freedom, has topological non-trivial observables related to non-trivial topology. Many of these ideas have been shared with Mario Rasetti, Jeanette Nelson and Riccardo Zecchina (Bocconi University). In the article https://iopscience.iop.org/article/10.1088/0305-4470/33/4/308 published in *Journal of Physics* A33 (2000), Regge and Zecchina proposed a combinatorial/topological formalism for the study of the Ising problem over lattices of arbitrarily high topological genus which generalized Kasteleyn's approach. They managed to write the partition function as a sum over Pfaffians with a topological signature and examined the 3D cubic Ising problem, reaching some preliminary assessment on the expansion in high and low temperature ranges. The method was also further explored for its potential physical and algorithmic relevance and was an ideal setting where Regge could express his mastering of symbolic manipulations with Mathematica.

It is amusing that these lines of thinking naturally originated from an important idea introduced in the Ponzano–Regge model of 1969, reviewed by Marzuoli in Part III, concerning the asymptotic limit of the 6-j symbols. This seems unrelated to previous work, but in fact, since a 6-j symbol can be described by a tetrahedron, it was recognized as a model for discretization of three-dimensional quantum gravity. Indeed the search for invariants of three-dimensional manifolds brought the mathematicians Turayev and Viro to a regularized version of the Ponzano–Regge model. Edward Witten (IAS, Princeton) conjectured that this theory could be a Feynman path integral formulation of a Chern–Simons action for a quantum group. In the course of the many developments of this pioneering work we may quote spin networks, Ooguri model, Archer Crane–Yetter and Roberts theory, BF theory etc., most of them aiming at a formulation of a four-dimensional theory of quantum gravity.

For many of the above outstanding contributions, Regge received various recognitions: the Dannie Heineman Prize for Mathematical Physics in 1964, the Città di Como prize in 1968, the Albert Einstein Award in 1979, and the Cecil Powell Medal in 1987. In 1996, he also received the Dirac medal by ICTP, the Marcel Grossmann Award in 1997, and the Pomeranchuk Prize in 2001. The asteroid 3778 Regge has been named after him, to honour his immense passion for astronomy: he used to say "The Universe is my hobby", before turning to his beloved amateur telescope.

Finally, a separate but fundamental part must cover Regge's efforts to devising innovative teaching techniques and to the dissemination of scientific knowledge within the general public, which was one of the main missions in his late career. After his return from Princeton, he deliberately devoted time, talent and passion to this activity, he took part in public events and TV shows and produced a large number of popular articles for daily newspapers and science magazines. He certainly was a forerunner of multimedia outreach. All this was very important for stimulating and bringing many young people, including some of us, to the world of physics. Part IX contains three contributions devoted to this important component of Regge's production and philosophy, starting with an article by the renowned Italian journalist and scientific communicator Piero Bianucci from La Stampa, who initiated Regge to his public role on television and many daily newspapers, and accompanied him throughout his career as a popular character on the Italian scientific outreach. Then, an article by Giulio Peruzzi, science historian at the University of Padua, describes their interaction when writing a popular book on space, time and the universe, an occasion for Peruzzi to get in touch with Regge's eclectic culture while discussing relativity and quantum mechanics.

To close the chapter, an article by the physicist and scientific writer Federico Tibone narrates the story of their collaboration on the very first examples of computer applications and multimedia works on CD-ROM, prepared for the teaching of Physics and to popularize scientific ideas: Astrophysics, Relativity and the Physics behind music (done in collaboration with Luciano Berio) were the topics of the CD-ROM produced by Regge and Tibone for the broad diffusion in Italian newspapers, and precursors of an app that today you can use to play with Einstein's relativity on your smartphone (please see the link in Tibone's article).

A section on Reminiscences contains some brief recollections by Sergio Ferrara (CERN), recipient of the 2019 Special Breakthrough Prize in Fundamental Physics for his co-discovery of Supergravity, and a contribution by Daniele Regge on the perspective of a son on Regge's multifaceted and towering figure.

Regge's polyhedral personality attracted him towards many other interests outside his passion for physics: firstly, he was a pioneer in computer art and design, at a time when computer graphics was taking its first steps. He wrote: "It is well known that those structures enjoying a profound geometrical meaning are also aesthetically very beautiful".

The personal computer, at the time called an *electronic calculator*, gave him the chance to play with shapes and colors in an extremely sophisticated way, since behind each of his apparently innocent drawings were complicated mathematical equations and the whole lore of his geometric vision. He used to explain that for him, any hand drawing would be a challenge, while with the computer it is easy to give cartesian coordinates and to derive an image. This resulted in a series of drawings mixing mathematical abstraction, computer graphics and an amazing humor, always associated with hilarious titles. In 2017/2018, an exhibition in Torino at the Accademia delle Scienze, "L'infinita curiosità", was put up to celebrate not only his scientific and human figure (nicely illustrated in the book in Italian `https://www.edizionidedalo.it/collane/l-infinita-curiosita.html` curated by Vincenzo Barone of the University of Eastern Piedmont at Alessandria and by Piero Bianucci), but also his creative vein. Next to impressive scientific installations about Regge's impact on contemporary physics, was a collection of his drawings, together with a reconstruction of his desk with the computer that he used to compose his works of art, often inspired by his research. Some samples of his art are displayed in the book cover.

Interestingly, the Detecma armchair (whose name is an acronym for design, technology and mathematics) that he designed in 1967 based on the fourth order equation of a Dupin's cyclic, is still produced and sold as an object of design for slightly over $1500.

Regge had a solid culture in music, as an amateur piano player (mostly Bach and Mozart) and in literature. Great interest was raised by the publication of his dialogues with Primo Levi in 1984 (republished in 2005 by Einaudi), containing conversations between the writer and the scientist on topics ranging from the Talmud to the limits of the human language and Gödel's incompleteness theorem. Also memorable are his literary interactions with Jorge Luis Borges and the comparison between the library of Babel and the universe contained in one of his popular articles, mentioned by Bianucci in his contribution.

Among his public engagements, Regge was also an active politician. In 1989, he was elected to the European Parliament in the European Socialism Group as an independent candidate for the Italian Communist Party. He served until 1994, focussing on energy policies and fighting fiercely for the rights of the disabled.

Regge served as president of the Turin division of the Association for Research in Handicap Prevention (AIRH) and as a co-founder of CICAP, the Italian Committee for the fight against pseudosciences and superstition https://www.cicap.org/n/index.php celebrating in 2019 its 25 years of activity.

Today, a Google search of the world Regge in the titles of current scientific literature in any of the modern preprint repositories gives thousands of hits. We currently hear in scientific discourse mentions to Regge theory of strong interaction phenomenology at high energies, of Regge behavior in field theory and of Regge calculus in general relativity. The main lecture hall at the University Physics Department in Turin is dedicated to his memory.

Regge's work significantly incarnates the interplay between mathematics and theoretical physics that has always been at the basis of scientific discovery, and that more recently has prompted the birth of a new discipline called Physical Mathematics (see the article by Gregory Moore (Yale) on https://arxiv.org/pdf).

This has also stimulated in Torino the founding in 2017 of the inter-university Arnold Regge Center for Algebra, Geometry and Theoretical Physics http://www.arnoldreggecenter.com, a research center devoted to fundamental research at the interface between mathematics and physics. The center, promoted by P. Fré after a long period in Russia as Scientific attaché, has also been named after the famous mathematician Vladimir Arnold, who shared with Regge much of his scientific vision, and his unconventional originality and charisma in communicating science to general audiences. The ARCenter supports various scientific activities of common interest to mathematicians and physicists ranging from extended visits, seminars, colloquia, workshops and lecture series aimed at faculty and advanced graduate students, together with some public outreach initiatives.

Tullio Regge, in his personal recollections http://inspirehep.net/record/1330473, repeatedly emphasized the importance of one's family and of finding good and wise mentors and advisors as key secrets for a bright scientific career, but also the fire of a maniacal passion: his was really contagious.

Everybody who had the chance to interact with him in scientific discussions and collaborations (and we are among those lucky ones), has always been impressed by the outstanding power of his reasoning and his non-conservative way of handling problems. An eclectic genius with pungent humour, a unique character endowed with enormous curiosity and charisma, he has indelibly marked the fundamental physics scene and also ourselves, both as intellectuals and as human beings.

Leonardo Castellani (UPO, Alessandria), *Anna Ceresole* (*INFN*, Turin),
Riccardo D'Auria (Turin Polytechnic) and *Pietro Fré* (Turin University),
for the Arnold–Regge Center for Algebra, Geometry and Theoretical Physics.

The logo of the Arnold Regge Center for Algebra, Geometry and Theoretical Physics (http://www.arnoldreggecenter.com).

Acknowledgments

The Editors are grateful to Rosanna, Daniele, Marta and Anna Regge for their friendship and support during this project, and for providing useful material. They also gladly thank Lucette Ponzano for sharing the manuscript of the Ponzano Regge paper.

The Editors and Publishers are indebted to the following publishers of the various journals and books for their assistance and permission to reproduce the reprinted papers and images found in this volume: Societa Italiana di Fisica, Annals of Physics, North Holland Publishing Co., Publications of the Research Institute for Mathematical Sciences Kyoto, Zanichelli Editore, Bologna. Wherever requested or possible, Copyrights of articles have been obtained through Copyright Clearance Center.

" I think I can be quite satisfied with myself. Not only for what I have discovered and learnt, but for the constant fun I had: guided by curiosity, I have opened many doors to snoop inside. And every time I got bored, I easily found another one that let itself be opened. I also had the good fortune of finding myself in stimulating environments, with skilled collaborators, who proved able to autonomously deepen many ideas. Today, I think I did my part well. As best as I could."

from Regge's autobiography "L'infinito cercare. Autobiografia di un curioso", written with S. Sandrelli (Einaudi, 2012)

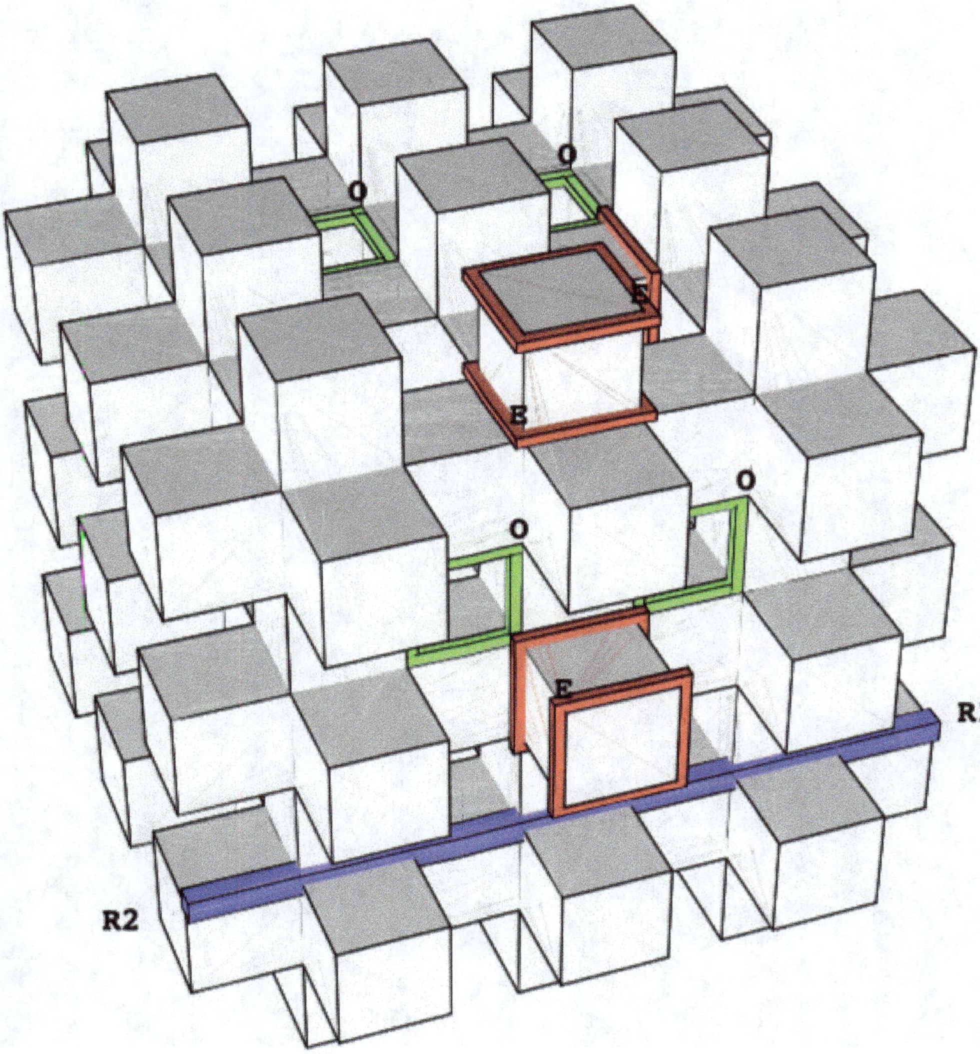

The Ising 3D lattice. Orientable surface of genus $g = N = 4 + 1$ containing all sites and bonds of the 3D cubic lattice Λ. Some examples of the even (E), odd (O) and exceptional (R) cycles are also shown. (Permission to republish from
`https://iopscience.iop.org/article/10.1088/0305-4470/33/4/308kindlygrantedbybyIOPPublishing`.)

Tullio Regge at the blackboard in Torino, in the late '70s.

Mikhail Gorbachov with Anna and Tullio Regge.

Contents

PART I

Black Hole Stability under Pertubations

Introduction to Regge and Wheeler "Stability of a Schwarzschild Singularity"

Kip S. Thorne

Theoretical Astrophysics, California Institute of Technology, Pasadena, CA 91125 USA

1. Introduction

The paper "Stability of a Schwarzschild Singularity" by Tullio Regge and John Archibald Wheeler[1] was the pioneering formulation of the theory of weak (linearized) gravitational perturbations of black holes. This paper is remarkable because black holes were extremely poorly understood in 1957, when it was published, and the very name *black hole* would not be introduced into physics (by Wheeler) until eleven years later.

Over the sixty years since the paper's publication, it has been a foundation for an enormous edifice of ideas, insights, and quantitative results, including, for example:

- Proofs that black holes are stable against linear perturbations — both non-spinning black holes (as treated in this paper) and spinning ones.[1–4]
- Demonstrations that all vacuum perturbations of a black hole, except changes of its mass and spin, get radiated away as gravitational waves, leaving behind a quiescent black hole described by the Schwarzschild metric (if non-spinning) or the Kerr metric (if spinning).[5] This is a dynamical, evolutionary version of Wheeler's 1970 aphorism "a black hole has no hair"; it explains how the "hair" gets lost.
- Formulations of the concept of *quasinormal modes* of a black hole with complex eigenfrequencies,[6] and extensive explorations of quasinormal-mode spectra as functions of the black hole's mass and spin.[7]
- Excitations of the quasinormal modes of the final black hole when two black holes collide and merge, and expectations that gravitational waves from those modes will be observed and used to measure the mass and spin of the final black hole and test general relativity's predictions of the black hole's properties.[8]
- Insights into physical structures, made from curved spacetime, that stick out of black holes: so-called *frame-drag vortices* and *tidal tendices* — and their remarkable dynamical evolutions when a black hole is perturbed.[9]

- The discovery of *superradiant scattering* of gravitational waves by a spinning black hole, in which the waves extract spin energy from the hole, so the outgoing, scattered wave is more energetic than the incoming wave.[10,11]
- The discovery, by Zel'dovich, of a corresponding superradiant scattering of gravitational vacuum fluctuations, which manifests itself as spontaneous generation of gravitational waves that feed off the spin energy of a black hole.[10,12]
- Zel'dovich's thereby triggering Stephen Hawking to discover *Hawking radiation* from non spinning as well as spinning black holes.[13]
- The exploitation of quasinormal modes to elucidate the *AdS/CFT correspondence* and issues in quantum gravity.[14]

2. The context in which this paper was written

In 1952 John Wheeler began a transition to relativity from nuclear physics (where, among other things, he had formulated the theory of nuclear fission with Niels Bohr).

His plunge into relativity was triggered, at least in part (see Chap. 10 of his autobiography[15]), by reading the classic 1939 paper by Robert Oppenheimer and Hartland Snyder[16] on the collapse of a highly idealized, spherical, pressure-free star. The spacetime geometry outside the star was the Schwarzschild solution to Einstein's field equations,

$$ds^2 = -(1 - 2m^*/r)dT^2 + (1 - 2m^*/r)^{-1}dr^2 + r^2(d\theta^2 + \sin^2\theta d\varphi^2) = g_{\mu\nu}dx^\mu dx^\nu \ , \quad (1)$$

where $m^* = 2G\mathcal{M}/c^2$ with $\mathcal{M}$ the star's mass, and G and c Newton's gravitational constant and the speed of light.

As the star's surface shrank, more and more of the Schwarzschild spacetime was opened up to view. Oppenheimer and Snyder showed, analytically, that as measured by a distant observer (for whom T is proper time), the star's surface required infinite time T to reach $r = 2m^*$, but as measured by an observer on the star's surface, it took only finite proper time τ; and they showed analytically that it would be impossible for an external observer to see what happened to the star thereafter. The star would be "cut off" from external view (Oppenheimer's words). Remarkably, Oppenheimer and Snyder did not explore, mathematically, what happened thereafter as seen by observers inside the star or on its surface; but their limited exploration was enough to sow confusion in the minds of even the world's best physicists. As Yevgeny Lifshitz (of Landau and Lifshitz fame) described it to me, many years later, "You cannot appreciate how difficult it was for the human mind to understand how both viewpoints [distant, and stellar-surface] can be true simultaneously."

Not until the early 1960s did the full physics of the collapse's endpoint become clear: the formation of an *event horizon* (also called *future horizon* below) and what Wheeler, in 1968, would call a black hole. Until the early 1960s, confusion reigned.

In their 1957 paper, Regge and Wheeler did not mention the Oppenheimer–Snyder analysis of stellar collapse, but it was likely on their minds to some degree, since that collapse analysis did pique Wheeler's interest in general relativity five years before the Regge–Wheeler paper was submitted, and Wheeler would soon initiate his research project with Kent Harrison and Masami Wakano on the fates of stars with large masses, resulting in a momentous 1958 confrontation between Wheeler and Oppenheimer over the physical outcome of stellar collapse.[17]

In 1957, with collapse and black holes a domain of confusion, Regge and Wheeler chose to focus on a different physical venue for the Schwarzschild metric (1) and its perturbations: A bridge between two asymptotically flat spaces or between two widely separated regions of our own universe, first explored in detail by Albert Einstein and Nathan Rosen in 1935.[18] Wheeler had recently coined the name *wormhole* for this bridge, and that name — like so many of Wheeler's coinages — was quickly embraced by the physics community. In parallel with the Regge–Wheeler research on perturbations of a wormhole, Wheeler and his student Charles Misner were exploring in great detail possible roles of wormholes in physics. In fact, the Misner–Wheeler paper[19] was received at Annals of Physics just four days before the Regge–Wheeler paper,[1] at Physical Review.

In the Regge–Wheeler paper, in addition to *wormhole*, we also see the name *Schwarzschild singularity* used for the object that is perturbed, but we see no explanation of how this name is related to wormholes (or to the Oppenheimer–Snyder analysis of stellar collapse). The Misner–Wheeler paper contains the answer: In their footnote 18, Misner and Wheeler refer to a wormhole as a "bridge across the Schwarzschild singularity", i.e. across the sphere $r = 2m^*$ where the Schwarzschild metric coefficient g_{rr} becomes singular and g_{TT} becomes zero. Although Regge and Wheeler were aware that this was a mere coordinate singularity (after all, a wormhole can extend smoothly through it), they and other relativists of that era were unaware of the past and future horizons that reside at $r = 2m^*$, and unaware of the dynamical evolution that the wormhole must undergo near there. These would get lucidly exposed in 1960 by the work of Wheeler's Princeton colleague Martin Kruskal, in a paper that Wheeler would ghost write for Kruskal,[20] and in related 1958 work of David Finkelstein.[21]

Although the Regge–Wheeler analysis focuses on the stability of a Schwarzschild wormhole, a few years later, when black holes and their horizons were correctly understood, it became clear that the analysis is also applicable, without change, to a Schwarzschild black hole. This is because outside $r = 2m^*$ and at the future horizon $\{r = 2m^*, T = +\infty\}$ the spacetime geometries are identically the same for the black hole and the wormhole.

3. The Regge–Wheeler collaboration

In February 1955, at the 5th Rochester Conferece on High Energy Nuclear Physics, Tullio Regge (then in the second year of his PhD study) was introduced to John Wheeler by Robert Marshak, Regge's PhD advisor and the chair of the University of Rochester Physics Department. Wheeler describes that introduction and subsequent events in his autobiography[15] as follows:

"Marshak described Regge to me as a brilliant mathematical physicist with some interest in general relativity. My conversation with him confirmed Marshak's assessment. It happened, at the time, that I was trying to come to grips with the problem of the stability of what we then called a 'Schwarzschild singularity'. My intuition told me that the Schwarzschild singularity should be stable, but I had not yet been able to prove it. It was important to find out. There is no use looking for such entities in nature if they are not stable. They wouldn't last long enough to be seen. Regge, with his mathematical power, seemed to be just what I needed. I outlined the problem and my thinking about it to him, and he agreed to work with me. ...

"Since I had a vision of how the whole problem should be tackled and how it would work itself out, I sat down and wrote a paper with spaces left for equations, and sent it to Regge. He rose to the occasion, and filled in the blanks. Then, early in 1956, I was able to round up travel funds ... to bring Regge to Leiden [Netherlands, where I was on an 8 month sabbatical, January to September]. We spent ten days together and made the pieces of our work fit smoothly together. Indeed, the Schwarzschild singularity is stable."

Regge, in his short autobiography,[22] remembers the collaboration as beginning in Leiden at the end of his PhD study, which would be in spring 1956, so it appears likely that there was much left to do when Regge arrived in Leiden. Indeed, their paper was not submitted to Physical Review until more than a year later, on July 15, 1957 (though the delay may have been due in large measure to the enormous number of other research projects that Wheeler was pursuing in parallel; see Chap. 12 of Wheeler's autobiography[15]).

4. Overview of the Regge–Wheeler paper

Although the Regge–Wheeler paper was a foundation for the major insights that I list in Sec. 1, and although the mathematical analysis in the paper is correct (aside from typographical errors found by Misner's students Lester Edelstein and C. V. Vishveshwara[23]), and was the template for the analyses in many subsequent papers, nevertheless there are some important conceptual misunderstandings in the Regge–Wheeler paper. These misunderstandings arose from the paper being written in the early era of confusion about the physical nature of the Schwarzschild spacetime.

In this section, I shall walk the reader through the paper, highlighting key, fully correct aspects of the analysis and discussion as well as the conceptual misunderstandings.

4.1. *Section I. Introduction and summary*

In Section I, Regge and Wheeler (R&W) explain that they are focusing on the Schwarzschild spacetime, interpreted physically as a wormhole connecting two regions of our universe that are arbitrarily far apart. They also motivate the idea of a wormhole by describing the closely related Reissner–Nordstrom solution of the Einstein equations, which can be interpreted as a Schwarzschild wormhole with electric field lines threading through it and, via the electrical stresses and energy, modifying the metric. These electrical side remarks give us a sense of why wormholes were of great interest to Wheeler and his students in 1957, and provide a link to the Misner–Wheeler paper[19] that was being submitted for publication almost simultaneously with the Regge–Wheeler paper.[1]

In this Section I, R&W begin refering to the perturbations of the Schwarzschild metric that they will analyze as waves — obviously *gravitational waves*, though they use that full phrase only late in their paper's last section. Although they have waves very much on their minds, they do discover, in their analysis in Section III, that for wavelengths comparable to the size of the Schwarzschild radius $2m^*$ and longer, the perturbations are wavelike only very near $r = 2m^*$ and far from there, $r \gg 2m^*$.

4.2. *Section II. Differential equations in polar coordinates for small first-order changes away from the Schwarzschild metric*

In Section II, R&W develop their mathematical formalism for first-order perturbations $h_{\mu\nu}$ of the Schwarzschild metric. As one might expect, their formalism begins with the separation of variables: writing each perturbed component of the metric as a product of a spherical-harmonic function of angles θ and φ; a monotonic, oscillatory function of time $\exp(-ikT)$; and a to-be-computed function of radius r.

The harmonics that are needed are scalar spherical harmonics for h_{TT}, $h_{Tr} = h_{rT}$ and h_{rr}; vector spherical harmonics on a 2-sphere for $\{h_{T\theta}, h_{T\varphi}\} = \{h_{\theta T}, h_{\varphi T}\}$ and $\{h_{r\theta}, h_{r\varphi}\} = \{h_{\theta r}, h_{\varphi r}\}$; and tensor spherical harmonics on a 2-sphere for $\{h_{\theta\theta}, h_{\theta\varphi} = h_{\varphi\theta}, h_{\varphi\varphi}\}$. R&W build their vector and tensor spherical harmonics in the simplest, most straightforward way possible: by applying 2-sphere gradients once or twice to the scalar spherical harmonics and taking cross products with the Levi–Civita tensor $\epsilon_\mu{}^\nu$ [Eqs. (RW 6)–(RW 11); i.e. Eqs. (6)–(11) in their paper]. The relativity community quickly gave these the name "Regge–Wheeler spherical harmonics", and they have been widely used ever since — though by 1980 a large number of alternative approaches to vector and tensor spherical harmonics were also in use in the relativity community, each with its own powers in special situations.[24]

With their spherical harmonics in hand, R&W focus on perturbations with fixed spherical-harmonic orders L, M and fixed parity — i.e. symmetry + or − under reflection through the origin. They introduce slightly peculiar names for the parity: *even* if the parity is $(-1)^L$ and *odd* if it is $(-1)^{L+1}$. These names were quickly adopted by the relativity community and are widely used even today, though the alternative names *electric-type parity* in place of even and *magnetic-type parity* in place of odd have also gained currency.

R&W write down the most general mathematical form for the metric perturbations of fixed L, M and parity [Eqs. (RW 12) and (RW 13)]. Then, to simplify their subsequent analysis of stability, they specialize to $M = 0$, knowing from standard group theoretic arguments that the eigenfrequencies k of modes must be independent of the harmonic order M. R&W then simplify the mathematical forms of the metric perturbations (RW 12), (RW 13) by means of carefully chosen infinitesimal coordinate transformations (for which they adopt the electromagnetic-like name *gauge transformations* — a name forever thereafter adopted by the relativity community).

For odd parity, they find a gauge (forever thereafter called Regge–Wheeler gauge) in which the perturbed Einstein field equations *decouple* in the following sense: (i) one of the metric perturbations' two unknown radial functions $h_0(r)$ can be expressed in terms of the other $h_1(r)$, and (ii) $h_1(r)$ — or equivalently the quantity $Q(r) = (1 - 2m^*/r)h_1/r$ — satisfies a single ordinary differential equation (RW 24): the radial wave equation, which can be rewritten as

$$d^2Q/dr^{*2} + \left[k^2 - V_{\text{eff}}(r)\right] Q = 0\ , \tag{2}$$

where

$$V_{\text{eff}}(r) = (1 - 2m^*/r)[L(L+1)/r^2 - 6m^*/r^3] \tag{3}$$

is the effective potential that R&W discuss in Sec. III; and where $r^* = r + 2m^* \ln(r/2m^* - 1)$ is a new radial coordinate that Wheeler later dubs the "tortoise coordinate". From Eq. (2)

(which quickly came to be called the *Regge–Wheeler equation*), it follows immediately that $(\partial^2/\partial r^{*2} - \partial^2/\partial T^2)(Qe^{-ikT}) = V_{\rm eff}(r)$, whence near the Schwarzschild singularity $r = 2m^*$ and far away at $r \gg 2m^*$, where the effective potential is vanishingly small, the perturbations do behave like waves, and propagate along radial rays $r^* = \pm T + \text{constant}$ [cf. Eq. (RW 39)].

Once the Regge–Wheeler equation (2) has been solved, all other aspects of the black hole's perturbations can be computed from the resulting $Q(r)$, or equivalently $h_1(r)$.

For even parity, R&W are not able to decouple the field equations. The best they can do is find a gauge in which there remain three radial functions $H(r)$, $K(r)$ and $H_1(r)$. They can express H_1 in terms of H and K via their Eq. (RW 28), but they then wind up with two coupled ordinary differential equations for H and K. As a result, the even parity analysis becomes much more difficult than the odd parity analysis.

[Remarkably, 12 years later, in 1969, another student of Wheeler's named Frank Zerilli[25] succeeded in decoupling the even parity equations and thereby obtaining a radial wave equation identical to Eq. (2) but with a different effective potential, for a single radial function $Z(r)$, from which all properties of the perturbations could be deduced. Even more remarkably, in 1975 S. Chandraskhar and S. Detweiler[26] discovered a one-to-one mapping between the even parity perturbations and the odd parity perturbations: For even parity there is a single radial function $Q_e(r) = A(k,r)Z(r) + B(k,r)dZ/dr^*$ (for specific functions A and B) that satisfies the Regge–Wheeler equation (2) with the *same* effective potential as for odd parity. From any solution of this equation, one can construct all the properties of the corresponding even parity perturbations as well as of the odd parity perturbations, and because the effective potential is the same as for odd parity, the spectrum of frequencies of the even parity modes is identical to that of the odd parity modes! The mathematical route to this amazing result was so complex that it is no wonder R&W, in their pioneering foray into this problem, failed to find it.]

With the even parity and odd parity radial equations in hand, R&W focus on static (time-independent) perturbations. They easily show, as one might expect, that for the even parity perturbations the static, spherically symmetric $M = 0$ perturbation represents a simple change of mass of the wormhole; and the static, dipolar $M = 1$ perturbation represents a simple physical displacement of the wormhole from its original position. For the odd parity perturbation, they miss the fact that the static, dipolar $M = 1$ perturbation represents adding a small spin angular momentum to the wormhole.[2]

For $L \geq 2$, R&W note that one class of static perturbations represents deformations of the wormhole by the gravity of external masses, but they miss a second class of static perturbations, which represent *singular* multipolar pertiurbations of the Schwarzschild singularity $r = 2m^*$ (the wormhole's throat and horizon). Again, missing this is not at all surprising, considering relativists' confusion about the physical nature of $r = 2m^*$ in 1957.

4.3. *Section III. Boundary conditions and stability*

In Sec. III, R&W focus on dynamical perturbations of a wormhole. Dynamical perturbations are possible only for spherical harmonics of quadrupole order or higher, $L \geq 2$.

R&W require, of course, that at some initial moment of time the perturbations be regular at both $r = 2m^*$ and $r = \infty$, but they do not discuss what "regularity" means, physically. Instead, they examine the solutions to their odd parity radial wave equation (2) and see that for real frequencies k the solutions $Q(r^*)$ are all oscillatory with asymptotically finite amplitude at both $r = 2m^*$ ($r^* = -\infty$) and $r = +\infty$; and they assume (correctly it turns out) that these solutions are physically regular. A proof of physical regularity would entail, for example, examining the components of the Riemann curvature tensor as measured by observers far from the Schwarzschild singularity and by observers falling into the Schwarzschild singularity (falling through the future horizon).

Continuing to focus on odd parity and real frequencies k, R&W discuss three types of solutions of their radial wave equation:

- Case 1: High frequency solutions, $k \gtrsim 1/m^*$, which are oscillatory at all radii. These represent waves that propagate freely, little affected by the effective potential (by the wormhole's spacetime curvature) from $r = 2m^*$ to $r = \infty$ and the reverse. Today we recognize them as waves that propagate nearly freely up from the wormhole's past horizon and on outward to $r = \infty$, or from $r = \infty$ inward to and down the wormhole's future horizon.
- Case 2a: Low frequecy solutions, $k \ll 1/m^*$, which have large oscillatory amplitudes near $r = 2m^*$ ($r^* \to -\infty$), die out inside the effective potential, and have tiny oscillatory amplitudes at $r \to \infty$. R&W identify these as waves that are trapped near the Schwarzschild singularity much like the electromagnetic waves trapped in *geons* that Wheeler had recently conceived and explored.[27] Today we recognize them as waves that come up from the wormhole's past horizon, and reflect off the Schwarzschild spacetime curvature where the effective potential is large, so almost all of the wave energy goes down into the wormhole's future horizon but a tiny portion leaks out and travels toward $r = \infty$; or the time reversal of this.
- Case 2b: Low frequency solutions, $k \ll 1/m^*$, which have large oscillatory amplitudes at large radii $r \gg 2m^*$, die out inside the effective potential, and have tiny oscillatory amplitudes near $r = 2m^*$ ($r^* \to -\infty$). Today we recognize these as waves that impinge on the wormhole from $r = \infty$ and are largely reflected by the wormhole's spacetime curvature, with a tiny amount of transmission toward and down the future horizon; or the time reversal of this.

R&W then discuss waves that "go through the wormhole", and infer, correctly of course, that the phases of these waves are arbitrary. Throughout this discussion, and throughout their paper, R&W assume that the wormhole is static. But it is not. As Wheeler himself would explain with great clarity in 1960, in a paper he would ghostwrite for Martin Kruskal,[20] the Schwarzschild wormhole is dynamical. The waves that R&W think of as going into and through the wormhole actually go into and down the future horizon and get caught and destroyed in the wormhole's interior, future singularity. And those that R&W think of as coming to us from the wormhole's other mouth actually originate in the wormhole's past singularity, and travel up through the wormhole's past horizon and onward toward $r = \infty$ (with some reflection off the spacetime curvature).

In their last subsection, R&W address the problem of the wormhole's stability.

Focusing again on the odd parity case, they easily show that for imaginary frequencies k

all solutions $Q(r^*)$ of their odd parity wave equation grow without bound either as $r \to 2m^*$ (the wormhole throat) or as $r \to \infty$, and so are unphysical. For the even parity case, they state without proof that a detailed but more complex analysis shows that all solutions with imaginary frequency are unphysical. Based in these results, they assert that a wormhole (or Schwarzschild singularity) is stable against weak perturbations.

What R&W do not say but could have, to make this more firm, is that their odd parity radial wave equation (2) with boundary conditions of regularity on $Q(r^*)$ at $r = 2m^*$ and $r = \infty$ is a self-adjoint eigenvalue problem for k^2; and by self adjointness k^2 must be real, so any instability must show up as an imaginary frequency k with regular $Q(r^*)$. The self adjointness further implies that the modes with real frequency form a complete set, in terms of which any regular initial perturbation can be expanded and evolved; and their stability implies stability of evolution of that regular initial perturbation.

Even after taking account of self adjointness, there are several holes in the R&W stability proof — holes that were filled in by others over the ensuing years and decades:

- R&W did not prove that *physical* regularity at $r = 2m^*$ and at $r = \infty$ implies regularity of $Q(r^*)$ there. In 1970, with the past and future horizons well understood, Vishveshwara fixed this hole.[2]
- R&W did not succeed in constructing a self-adjoint eigenvalue problem for k^2 in the even parity case. In 1970 Zerilli fixed this hole.[25]
- The problem of stability for a black hole that is formed by gravitational collapse of a star is actually very slightly different from that for the pure vacuum Schwarzschild spacetime, which is truly a wormhole: the past horizon is replaced by the world tube of the star that collapsed long long ago. For ordinary theoretical physicists like me, that difference seems exceedingly unlikely to be important, and I glossed over it above; but in 2015 Kay and Wald gave a rigorous proof of stability for such a black hole, taking that difference into account.[28]

This level of rigor is enough to satisfy almost all physicists. However mathematicians and the most mathematical of physicists are focusing today on attempts to prove stability of a Schwarzschild wormhole in stronger senses, what they call "linear and nonlinear stability". For a detailed discussion and progress report, see a recent monumental paper by Klainerman and Szeftel.[29]

5. Quasinormal modes and their stability

In 1970 Misner's student Vishveshwara simulated[30] the scattering of wave packets of gravitational waves off a Schwarzschild black hole by solving numerically the odd parity Regge–Wheeler radial wave equation (2). For wave packets with widths comparable to the black hole's size, the scattered waves oscillated sinusoidally with rapidly decaying amplitude. The following year my student William Press performed other numerical experiments showing similar results and interpreted the scattered waves as coming from what he called *quasinormal modes* of vibration of the black hole, which were excited by the incoming waves.[6]

These quasinormal modes have boundary conditions of no upcoming waves at $r = 2m^*$

and no incoming waves at $r = \infty$. The waves are purely downgoing at the horizons and outgoing far from the black hole.

Any quasinormal mode can be constructed from a superposition of modes with real frequencies k, and so is guaranteed to be stable.

In 1975 Chandrasekhar and Detweiler[26] carried out the first numerical computations of the complex eigenfrequencies of a Schwarzschild black hole's (or wormhole's) quasinormal modes. For subsequent history, in extreme brevity, see Sec. I of this paper; for greater detail see the 2015 review article by Teukolsky[4] and references therein.

6. A personal conclusion

I conclude with some personal remarks. As a graduate student of John Wheeler in 1962–65, and then when building my own research group at Caltech in the late 1960s and early 1970s, I was greatly influenced by the Regge–Wheeler paper. I pushed my students to generalize the R&W analysis to spinning (Kerr) black holes, and they did so: Teukolsky separated variables and decoupled the resulting perturbation equations to obtain a radial wave equation for a variable $R(r)$ from which all properties of the perturbations could be deduced;[31] and Press and Teukolsky proved (with a mixture of numerical and analytical manipulations of the Teukolsky equation) that a Kerr black hole is stable.[3] I myself, with colleague Alfonso Campolattaro and student Richard Price, generalized the R&W analysis to study perturbations and quasinormal modes of nonrotating relativistic stars.[32,33] I helped my friend and colleague James Hartle extend the R&W analysis of time-independent perturbations to stars, which we set into rotation and explored to second order in their rotational angular velocity.[34,35] And a later generation of my students, beginning with Fernando Echeverria,[36] focused on the gravitational waves from black-hole normal modes as targets for gravitational wave detectors.

I therefore owe a huge debt of personal gratitude to Tullio Regge and John Archibald Wheeler for their pioneering work on perturbations of the Schwarzschild spacetime.

References

1. T. Regge and J. A. Wheeler, Phys. Rev. **108**, 1063 (1957).
2. C. V. Vishveshwara, Phys. Rev. D **1**, 2870 (1970).
3. W. H. Press and S. A. Teukolsky, Astrophys. J. **185**, 649.
4. For a review see S. A. Teukolsky, Class. Quantum Grav. **32**, 124006 (2015).
5. R. H. Price, Phys. Rev. D **5**, 2419 and 2439 (1972).
6. W. H. Press, Astrophys. J. **170**, L105 (1971).
7. For example, H. Yang, A. Zimmerman, A. Zenginoglu, F. Zhang, E. Berti and Y. Chen, Phys. Rev. D **88**, 044047 (2013).
8. E. Berti, K. Yagi, H. Yang, and N. Yunes, Gen. Relativ. Gravit. **50**, 49 (2018) and earlier references therein.
9. D. A. Nichols *et al.*, Phys. Rev. D **86**, 104028 (2012).
10. Ya. B. Zel'dovich, Sov. Phys.—JETP Lett. **14**, 180 (1971).
11. C. W. Misner, Phys. Rev. Lett. **28**, 994 (1972).
12. A. A. Starobinsky, Sov. Phys.—JETP **37**, 28 (1973).

13. S. W. Hawking, Nature **248**, 30 (1974).
14. G. T. Horowitz and V. E. Hubeny, Phys. Rev. D **62**, 024027 (2000); also R. A. Konoplya and A. Zhidenko, Rev. Mod. Phys. **83**, 793 (2011) and papers cited therein.
15. J. A. Wheeler with K. Ford, *Geons, Black Holes, and Quantum Foam: A Life in Physics* (Norton, 2000).
16. J. R. Oppenheimer and H. Snyder, Phys. Rev. **56**, 455 (1939).
17. K. Harrison, M. Wakano and J. A. Wheeler, in Onzième Conseil de Physique Solvay, *La Structure et l'évolution de l'univers* (Stoops, 1958), p. 124 and especially p. 147.
18. A. Einstein and N. Rosen, Phys. Rev **48**, 73 (1935).
19. C. W. Misner and J. A. Wheeler, Ann. Phys. (USA) **2**, 525 (1957).
20. M. D. Kruskal, Rev. Mod. Phys. **119**, 1743 (1960).
21. D. Finkelstein, Phys. Rev. **110**, 965 (1958).
22. T. Regge, `http://inspirehep.net/record/1330473/files/regge.pdf`.
23. L. A. Edelstein and C. V. Vishveshwara, Phys. Rev. D **1**, 3514 (1970).
24. K. S. Thorne, Rev. Mod. Phys. **52**, 299 (1980).
25. F. J. Zerilli, Phys. Rev. Lett. **24**, 737 (1970).
26. S. Chandrasekhar and S. Detweiler, Proc. Roy. Soc. **344**, 441 (1975).
27. J. A. Wheeler, Phys. Rev. **97**, 511 (1955).
28. B. S. Kay and R. M. Wald, Class. Quantum Grav. **4**, 893 (2015).
29. S. Klainerman and J. Szeftel, `https://arxiv.org/pdf/1711.07597.pdf` (2018).
30. C. V. Vishveshwara, Nature **277**, 936 (1970).
31. S. A. Teulolsky, Phys. Rev. Lett. **29**, 1114 (1972).
32. K. S. Thorne and A. Campolattaro, Astrophys. J. **149**, 591 (1967).
33. R. H. Price and K. S. Thorne, Astrophys. J. **155**, 163 (1969).
34. J. B. Hartle, Astrophys. J. **150**, 1005 (1967).
35. J. B. Hartle and K. S. Thorne, Astrophys. J. **153**, 802 (1968).
36. F. Echeverria, Phys. Rev. D **40**, 3194 (1989).

PHYSICAL REVIEW VOLUME 108, NUMBER 4 NOVEMBER 15, 1957

Stability of a Schwarzschild Singularity

TULLIO REGGE, *Istituto di Fisica della Università di Torino, Torino, Italy*

AND

JOHN A. WHEELER, *Palmer Physical Laboratory, Princeton University, Princeton, New Jersey*

(Received July 15, 1957)

It is shown that a Schwarzschild singularity, spherically symmetrical and endowed with mass, will undergo small vibrations about the spherical form and will therefore remain stable if subjected to a small nonspherical perturbation.

I. INTRODUCTION AND SUMMARY

SCHWARZSCHILD found long ago the solution of the Einstein equations for the metric around a fixed spherically symmetrical center-of-mass:

$$ds^2=-(1-2m^*/r)dT^2+(1-2m^*/r)^{-1}dr^2 +r^2(d\theta^2+\sin^2\theta d\varphi^2)=g_{\mu\nu}dx^\mu dx^\nu, \quad (1)$$

with $x^0=T$, $x^1=r$, $x^2=\theta$, $x^3=\varphi$. Here the quantity

$$m^*(\text{cm})=Gm/c^2=(0.74\times10^{-28}\text{ cm/g})m(\text{g}) \quad (2)$$

denotes the value of the mass as expressed in the geometric units of length. A similar but more complicated expression was found by Reissner and Nordström[1] for the case where the system appears at a distance not only as a spherically symmetrical center of gravitational attraction but also as a spherically symmetrical center of electric lines of force. Such a charge-like solution permits interpretation in terms of a multiply connected space.[2] The lines of force can be considered to emerge from one mouth of a wormhole, the other end of which is located somewhere else in space. This interpretation has the following features: (1) There is no real charge present anywhere in space. Lines of force never end. The Maxwell field is free of singularity. However, the lines of force are trapped in the topology of space so that their number cannot change and the charge remains constant. (2) The other wormhole mouth can be supposed as far away as one pleases. We shall assume it to be infinitely far away, so that it does not disturb the spherical symmetry and dynamics of the mouth under consideration. Then we can use the Reissner-Nordström solution as an idealized representation of the metric down to the throat of the wormhole. (3) The electric field, being divergence-free, has a strength q/r^2 and an energy density $q^2/8\pi r^4$. Translated to the units of length, the charge has the value

$$q^*(\text{cm})=(G^{\frac{1}{2}}/c^2)q(\text{electrostatic units}). \quad (3)$$

(4) The stress energy density of this electric field acts as a source of gravitational field in Einstein's field equations. There is no other source of the gravitational field in the usual way of writing these equations. However, the equations can be rearranged[3] in such a way as to bring into evidence an additional production of gravitational field by the stress energy tensor *of* the gravitational field. On this account the geometrized mass, m^*, is not uniquely determined by the geometrized charge, q^*; it only follows that m^* is no less than q^*. (5) One can therefore think of the field energy—or the mass and stress that goes with it—as in equilibrium under its own gravitational attraction.

We have equilibrium, but is it stable? A sphere of water held together by gravitational forces is stable against small departures from sphericity. A sphere of water surrounded by a spherical shell of liquid mercury is also an equilibrium configuration for gravitational forces, but a situation of unstable equilibrium. Initial small departures from sphericity at the water-mercury interface will grow exponentially, and the mercury will concentrate with a rush at the center of the sphere. Which situation will more closely correspond to the behavior of a Schwarzschild singularity subjected to a small initial perturbation?

We have investigated this question here up to terms of the first order in the departures from sphericity. In this approximation, as in every other kind of stability problem in physics, the equations are linear and it is possible to analyze the disturbance into proper modes and find for each its frequency, real (stability) or imaginary (instability). We have determined these proper frequencies and find that the Schwarzschild singularity is essentially stable because imaginary frequencies would require a clearly unrealistic spatial behavior of the initial perturbation.

We therefore conclude that the object in question, built out of the mass-free Einstein field, is stable against small departures from sphericity. A typical disturbance from the equilibrium configuration will not grow with time but will oscillate around equilibrium.

The analysis proceeds in the following way: In Sec. II we write down, following Komar and Eisenhart, the *linear* differential equations for small first-order departures from the Schwarzschild metric. These equations ensure that the perturbed field will satisfy the Einstein equations and represent a mass-free space

[1] H. Reissner, Ann. Physik **50**, 106 (1916); G. Nordström, Proc. Acad. Sci. Amsterdam **20**, 1238 (1918).

[2] J. A. Wheeler, Phys. Rev. **97**, 511 (1955); C. W. Misner and J. A. Wheeler, Ann. Phys. (to be published); also A. Einstein and N. Rosen, Phys. Rev. **48**, 73 (1935).

[3] S. N. Gupta, Phys. Rev. **96**, 1683 (1954).

of nearly spherical symmetry. We look for those solutions of these differential equations which can be expressed in the form of a product of four factors, each depending upon a single one of the quantities T, r, θ, φ.

We find mathematical expressions for a complete set of functions of this kind, by superposition of which one can represent any arbitrary small first-order perturbation that satisfies the appropriate boundary conditions.

The typical mode of disturbance of the Einstein field has the circular frequency ω. Here the quantity ω is an eigenvalue parameter to be chosen so that the disturbance satisfies the radial wave equation with the appropriate boundary conditions for small and large r.

In Sec. III we formulate the boundary conditions for this eigenvalue problem and analyze the radial dependence of the functions involved in the problem by way of the JWKB procedure. In this manner we can avoid the problem of securing an exact solution of the radial wave equation and still conclude that the proper frequencies are real.

II. DIFFERENTIAL EQUATIONS IN POLAR COORDINATES FOR SMALL FIRST-ORDER CHANGES AWAY FROM THE SCHWARZSCHILD METRIC

General Equations

We shall indicate here the background metric with $g_{\mu\nu}$ and the perturbation in it with $h_{\mu\nu}$. The quantity $g_{\mu\nu}$ will be later specialized to be the ordinary Schwarzschild metric. The perturbations $h_{\mu\nu}$ are supposed to be very small as compared with $g_{\mu\nu}$. The contracted Ricci tensor will be, as usual, called $R_{\mu\nu}$ if calculated from $g_{\mu\nu}$ and $R_{\mu\nu}+\delta R_{\mu\nu}$ if calculated from $g_{\mu\nu}+h_{\mu\nu}$. It is not difficult to derive an expression for $\delta R_{\mu\nu}$. The calculation has been made by Eisenhart and independently by Komar.[4] We shall follow here Eisenhart's point of view as being more suitable for our kind of calculation. His result can be expressed in the form

$$\delta R_{\mu\nu}=-\delta\Gamma^{\beta}{}_{\mu\nu;\beta}+\delta\Gamma^{\beta}{}_{\mu\beta;\nu}, \tag{4}$$

where the semicolons indicate covariant differentiation and where we use the symbol

$$\delta\Gamma^{\alpha}{}_{\beta\gamma}=\tfrac{1}{2}g^{\alpha\nu}(h_{\beta\nu;\gamma}+h_{\gamma\nu;\beta}-h_{\beta\gamma;\nu}). \tag{5}$$

Although $\Gamma^{\alpha}{}_{\beta\gamma}$ is not a tensor, its variation is a tensor. Putting (5) into (4), we get a second-order differential equation on the $h_{\mu\nu}$ from the condition $\delta R_{\mu\nu}=0$. This equation is a generalization in a curved space of the known Schrödinger equation for a massless particle of spin 2 in a flat space.[5] When the background metric is given by (1), then of course $R_{\mu\nu}=0$. Then the equation $\delta R_{\mu\nu}=0$ means that the perturbed space is also empty of matter or distributed energy. These equations we shall analyze and separate in polar coordinates.

Analysis into Spherical Harmonics

We first undertake to separate the solution into a product of four factors, each a function of a single coordinate. This separation is best achieved by generalizing to tensors the well-known development in spherical harmonics, already firmly established for vectors, scalars, and spinors. For all four cases the symmetry of the metric allows angular momentum to be defined. The angular momentum is investigated by studying rotations on the 2-dimensional manifold $x^0=T=$constant, $x^1=r=$constant. Under a rotation of the frame around the origin, the ten components of the perturbing tensor transform like 3 scalars (h_{00}, h_{01}, h_{11}), 2 vectors (h_{02}, h_{03}; h_{12}, h_{13}), and a second-order tensor, when considered as covariant quantities on the sphere. For the scalar and vector parts we know already how to develop into spherical harmonics. A typical scalar function has the form

$$\phi_L{}^M=\text{const}\,Y_L{}^M(x_2,x_3)=\text{const}\,Y_L{}^M(\theta,\varphi). \tag{6}$$

This term belongs to a wave of parity $(-)^L$ and of angular momentum L, whose projection on the z axis is M. For vectors we have two distinct types with opposite parity:

$$\psi_L{}^M{}_{,\mu}=\text{const}\frac{\partial}{\partial x^\mu}Y_L{}^M(x_2x_3),\qquad \text{parity } (-)^L; \tag{7}$$

$$\phi_L{}^M{}_{,\mu}=\text{const}\,\epsilon_\mu{}^\nu\frac{\partial}{\partial x_\nu}Y_L{}^M(x_2x_3),\quad \text{parity } (-)^{L+1}. \tag{8}$$

Here the labels μ and ν run over the values 2 and 3, when $x^2=\theta$, $x^3=\varphi$; and the $\epsilon_\mu{}^\nu$ represent the quantities $\epsilon_2{}^2=\epsilon_3{}^3=0$; $\epsilon_2{}^3=-1/\sin x^2$, $\epsilon_3{}^2=\sin x^2$. Similarly we shall later introduce the quantities $\gamma_{22}=1$, $\gamma_{23}=0=\gamma_{32}$; $\gamma_{33}=\sin^2 x^2$. Both types of vector carry an angular momentum L. For tensors similar properties hold. We outline here only the results. There are three fundamental types of tensor of angular momentum L:

$$\psi_L{}^M{}_{\mu\nu}=\text{const}\,Y_L{}^M{}_{;\mu\nu}(\text{covariant derivatives}),\qquad \text{parity } (-)^L; \tag{9}$$

$$\phi_L{}^M{}_{\mu\nu}=\text{const}\,\gamma_{\mu\nu}Y_L{}^M,\quad \text{parity } (-)^L; \tag{10}$$

$$\chi_L{}^M{}_{\mu\nu}=\tfrac{1}{2}\,\text{const}[\epsilon_\mu{}^\lambda\psi_L{}^M{}_{\lambda\nu}+\epsilon_\nu{}^\lambda\psi_L{}^M{}_{\lambda\mu}],\qquad \text{parity } (-)^{L+1}. \tag{11}$$

Any one of these terms can obviously be multiplied by an arbitrary function of r and T, without changing its transformation properties under a rotation. It is clear that term (10) is a simple combination of a scalar with the metric tensor on the sphere, $\gamma_{\mu\nu}=g_{\mu\nu}/r^2$. Term (8) can be considered as a "pseudogradient" of $Y_L{}^M(x_2x_3)$. Similarly (9) and (11) can be obtained by operating twice with gradients and pseudogradients. The trace

[4] L. P. Eisenhart, *Riemannian Geometry* (Princeton University Press, Princeton, 1926), Chap. VI; A. Komar, Ph.D. thesis, Princeton, 1956 (unpublished).

[5] See for example T. Regge, Nuovo cimento **5**, 325 (1957).

of (11) is identically 0. The trace of (9) has the value

$$g^{\mu\nu}\psi_L{}^M{}_{\mu\nu} = -L(L+1)Y_L{}^M(\theta,\varphi)\times\text{const.}$$

From the results that we have just outlined, it follows that the most general perturbation belonging to a given L, M, and parity $(-)^{L+1}$ is of the form:

$$h_{\mu\nu} = \left|\begin{array}{llll} 0 & 0 & -h_0(T,r)(\partial/\sin\theta\partial\varphi)Y_L{}^M & h_0(T,r)(\sin\theta\partial/\partial\theta)Y_L{}^M \\ 0 & 0 & -h_1(T,r)(\partial/\sin\theta\partial\varphi)Y_L{}^M & h_1(T,r)(\sin\theta\partial/\partial\theta)Y_L{}^M \\ \text{Sym} & \text{Sym} & h_2(T,r)(\partial^2/\sin\theta\partial\theta\partial\varphi - \cos\theta\partial/\sin^2\theta\partial\varphi)Y_L{}^M & \text{Sym} \\ \text{Sym} & \text{Sym} & \tfrac{1}{2}h_2(T,r)(\partial^2/\sin\theta\partial\varphi\partial\varphi + \cos\theta\partial/\partial\theta - \sin\theta\partial^2/\partial\theta\partial\theta)Y_L{}^M & -h_2(T,r)(\sin\theta\partial^2/\partial\theta\partial\varphi - \cos\theta\partial/\partial\varphi)Y_L{}^M \end{array}\right|. \tag{12}$$

Here the rows and columns are numbered in the order 0, 1, 2, 3 (T, r, θ, φ). The symbol "Sym" indicates that the missing components of $h_{\mu\nu}$ are to be found from the symmetry $h_{\mu\nu} = h_{\nu\mu}$. We shall refer to (12) as the "odd" type of perturbation.

Even Waves

The grouping of the terms of even parity yields the "even" perturbation:

$$h_{\mu\nu} = \left|\begin{array}{llll} (1-2m^*/r)H_0(T,r)Y_L{}^M & H_1(T,r)Y_L{}^M & h_0(T,r)(\partial/\partial\theta)Y_L{}^M & h_0(T,r)(\partial/\partial\varphi)Y_L{}^M \\ H_1(T,r)Y_L{}^M & (1-2m^*/r)^{-1}H_2(T,r)Y_L{}^M & h_1(T,r)(\partial/\partial\theta)Y_L{}^M & h_1(T,r)(\partial/\partial\varphi)Y_L{}^M \\ \text{Sym} & \text{Sym} & r^2[K(T,r) + G(T,r)(\partial^2/\partial\theta^2)]Y_L{}^M & \text{Sym} \\ \text{Sym} & \text{Sym} & r^2G(T,r)(\partial^2/\partial\theta\partial\varphi - \cos\theta\partial/\sin\theta\partial\varphi)Y_L{}^M & r^2[K(T,r)\sin^2\theta + G(T,r)(\partial^2/\partial\varphi\partial\varphi + \sin\theta\cos\theta\partial/\partial\theta)]Y_L{}^M \end{array}\right|. \tag{13}$$

Frequency Analysis; Specialization to $M=0$

Owing to the spherical symmetry of the background metric, Eqs. (4) and (5) cannot mix terms belonging to different L and parity. To apply quantum language to a classical problem, we can say that L, M, and the parity are constants of the motion. The existence of still another constant follows from the circumstance that the background metric (1) is independent of the cotime, $T=ct$. On this account we can consider a perturbation of a definite frequency, $\omega=kc$, so that every component of the perturbation $h_{\mu\nu}$ will have a time dependence of the form $\exp(-i\omega t)=\exp(-ikT)$. We therefore proceed to determine completely the form of the individual solution of specified parity, L and M values, and frequency. The general solution will be a superposition of these individual solutions with coefficients adjusted to fit the appropriate boundary conditions and initial values.

In working out the typical individual solution, we need not occupy ourselves with the angular dependence, which is completely specified by (12) and (13). Moreover, there is no need to work with an arbitrary M. For any specified choice of L, k, and partly all values of M $(M=-L, -L+1, \cdots L)$ will lead to the same radial equation. We prefer to take $M=0$ with the advantage that φ will completely disappear from the calculations. We are still left with a considerable amount of labor. An odd wave contains three unknown functions of r. Worse, an even wave contains seven unknown functions. Here comes a fact which greatly simplifies and illuminates the calculation.

Gauge or Coordinate Transformations and the Simplifications They Can Bring

Different waves can represent the same physical phenomena viewed in different systems of coordinates. Consider an infinitesimal coordinate transformation:

$$x'^{\alpha} = x^{\alpha} + \xi^{\alpha} \quad (\xi^{\alpha} \ll x^{\alpha}). \tag{14}$$

The infinitesimal displacements ξ^{α} transform like a vector. In the new frame we shall have:

$$g_{\mu\nu}' + h_{\mu\nu}' = g_{\mu\nu} + \xi_{\mu;\nu} + \xi_{\nu;\mu} + h_{\mu\nu}. \tag{15}$$

Now $h_{\mu\nu}$ is defined as the difference between the perturbed metric and the Schwarzschild metric *written in spherical coordinates.* According to this definition, the difference in the new frame will have the value

$$h_{\mu\nu}{}^{\text{new}} = h_{\mu\nu}{}^{\text{old}} + \xi_{\mu;\nu} + \xi_{\nu;\mu}. \tag{16}$$

This result can be interpreted by saying that for infinitesimal changes in the coordinates the $h_{\mu\nu}$ undergo a "gauge transformation" quite similar to the well-known gauge transformation for the electromagnetic field. We can now use this important circumstance in order to simplify the description of the perturbation and to make it unique.

The gauge transformation (16) can be performed on any individual partial wave. Obviously no real simplification will result unless the resulting wave still belongs to the original eigenvalues. This requirement limits the possible choices for ξ^α. This vector turns out to be a spherical harmonic of the same L and parity as the partial wave in consideration. Such a gauge transformation allows us to impose additional simplifying conditions on the perturbation, $h_{\mu\nu}$. We have therefore chosen to eliminate those terms which contain the derivatives of the highest order with respect to the angles. Then the final radial equations simplify. Moreover, the desired gauge transformation ξ^α can then be found by the use of finite operations only, without arbitrary constants and boundary conditions.

The gauge vector ξ^α that simplifies the general odd wave (12) must have the form

$$\xi^0=0,\quad \xi^1=0,\quad \xi^\mu=\Lambda(T,r)\epsilon^{\mu\nu}(\partial/\partial x^\nu)Y_L{}^M(\theta,\varphi),\quad (\mu,\nu=2,3),\tag{17}$$

according to the foregoing arguments. Moreover, the radial function Λ can be adjusted to annul the radial factor $h_2(T,r)$ or that component of the perturbation $h_{\mu\nu}$ which has the form (11).

Canonical Form of Odd and Even Waves

The final canonical form for an odd wave of total angular momentum L and projection $M=0$ is then

$$h_{\mu\nu}=\begin{vmatrix}0 & 0 & 0 & h_0(r)\\ 0 & 0 & 0 & h_1(r)\\ 0 & 0 & 0 & 0\\ \mathrm{Sym} & \mathrm{Sym} & 0 & 0\end{vmatrix}\times\exp(-ikT)(\sin\theta\partial/\partial\theta)P_L(\cos\theta).\tag{18}$$

In a similar way we write the gauge transformation that simplifies even waves in the form

$$\begin{aligned}&\xi^0=M_0(T,r)Y_L{}^M(\theta,\varphi);\quad \xi^1=M_1(T,r)Y_L{}^M(\theta,\varphi);\\ &\xi^2=M(T,r)(\partial/\partial\theta)Y_L{}^M(\theta,\varphi);\\ &\xi^3=M(T,r)(\partial/\sin^2\theta\partial\varphi)Y_L{}^M(\theta,\varphi).\end{aligned}\tag{19}$$

We adjust the factors M_0, M_1, and M to annul the factors G, h_0, and h_1 in expression (13), thereby obtaining the even wave in the canonical form

$$h_{\mu\nu}=\exp(-ikT)P_L(\cos\theta)\times\begin{vmatrix}H_0(1-2m^*/r) & H_1 & 0 & 0\\ H_1 & H_2(1-2m^*/r)^{-1} & 0 & 0\\ 0 & 0 & r^2K & 0\\ 0 & 0 & 0 & r^2K\sin^2\theta\end{vmatrix}\tag{20}$$

There are therefore two unknown functions of r in the odd case (h_0 and h_1) and four unknowns for the even case (H_0, H_1, H_2, and K).

Radial Wave Equations

We now substitute the expressions (18) and (20) for the first-order perturbation into the first variation of the Einstein field equations,

$$\delta\Gamma_{\mu\nu}{}^\beta{}_{;\beta}-\delta\Gamma_{\mu\beta}{}^\beta{}_{;\nu}=0.\tag{21}$$

The resulting equations for the radial factors may be derived and discussed separately for odd and even waves.

For odd waves the variation $\delta\Gamma_{\mu\beta}{}^\beta$ vanishes identically. Out of the 10 Einstein equations only 3 nontrivial ones can be obtained:

$$\begin{aligned}&(1-2m^*/r)^{-1}kh_0+(d/dr)(1-2m^*/r)h_1=0, &&\text{from}\quad \delta R_{23}=0;\\ &(1-2m^*/r)^{-1}k(dh_0/dr-kh_1-2h_0/r)\\ &\quad+(L-1)(L+2)h_1/r^2=0, &&\text{from}\quad \delta R_{13}=0;\\ &(d/dr)(kh_1-dh_0/dr)+2kh_1/r=r^{-2}(1-2m^*/r)^{-1}\\ &\quad\times(4m^*h_0/r-L(L+1)h_0), &&\text{from}\quad \delta R_{03}=0.\end{aligned}\tag{22}$$

The last equation is a consequence of the other two. Define

$$Q=(1-2m^*/r)h_1/r,\tag{23}$$

eliminate h_0, and find for Q the second-order wave equation,

$$d^2Q/dr^{*2}+k_{\mathrm{eff}}{}^2(r)Q=0.\tag{24}$$

Here we have made the abbreviations

$$dr^*=\exp(\tfrac{1}{2}\lambda-\tfrac{1}{2}\nu)dr,$$

and

$$k_{\mathrm{eff}}{}^2=k^2-L(L+1)e^\nu/r^2+6m^*e^\nu/r^3,\tag{25}$$

where λ and ν are defined by the expression for the metric,

$$ds^2=-e^\nu dT^2+e^\lambda dr^2+r^2(d\theta^2+\sin^2\theta d\varphi^2);\tag{26a}$$

thus,

$$e^\nu=e^{-\lambda}=1-2m^*/r.\tag{26b}$$

For even waves the 10 Einstein field equations give 7 nontrivial conditions: one algebraic relation; 3 first-order differential equations; and 3 of the second order. We use one relation,

$$H_2=H_0\equiv H, \tag{27a}$$

to eliminate H_0 from the other 6 equations:

$$dK/dr+r^{-1}(K+H)-L(L+1)H_1/2kr^2 -(1-2m^*/r)^{-1}m^*K/r^2=0, \tag{27b}$$

$$(d/dr)(1-2m^*/r)H_1=k(K-H), \tag{27c}$$

$$kH_1-(1-2m^*/r)(d/dr)(K+H)-(2m^*/r^2)H=0; \tag{27d}$$

$$(d/dr)(1-2m^*/r)(2rH+(d/dr)r^2K)-L(L+1)K -2krH_1+k^2r^2(1-2m^*/r)^{-1}K=0; \tag{27e}$$

$$\begin{aligned}&2k(1-2m^*/r)(dH_1/dr)+2km^*H_1/r^2+k^2H\\&-(1-2m^*/r)^2[(d/r^2dr)(r^2dH/dr)\\&+2(d/r^2dr)(r^2dK/dr)]-(1-2m^*/r)[L(L+1)H/r^2\\&+4m^*(d/r^2dr)H+2m^*(d/r^2dr)K]=0;\end{aligned} \tag{27f}$$

$$\begin{aligned}&(1-2m^*/r)^2d^2H/dr^2+(2/r)(1-2m^*/r)dH/dr-k^2H\\&-L(L+1)(1-2m^*/r)(H/r^2)-2km^*H_1/r^2\\&-2k(1-2m^*/r)(d/r^2dr)r^2H_1+2k^2K\\&+2(1-2m^*/r)(m^*/r^2)dK/dr=0.\end{aligned} \tag{27g}$$

We are facing a system of 6 equations in 3 unknowns. One will expect that the 3 first-order Eqs. (27b,c,d) will suffice to determine the solutions, apart from the boundary conditions. Actually the second-order Eqs. (27e,f,g) contain an additional nontrivial piece of information about the solution that is consistent with (27b,c,d). Specifically, a rather elaborate investigation shows that (27e,f,g) can all be deduced from (27b,c,d) plus the algebraic relation

$$\begin{aligned}&[6m^*/r+(L-1)(L+2)]H+[2kr-L(L+1)m^*/r^2k]H_1\\&+[(2m^*/r)+(L-1)(L+2)\\&-2(1-2m^*/r)^{-1}(m^{*2}/r^2+k^2r^2)]K\equiv F=0;\end{aligned} \tag{28}$$

and conversely, (27b,c,d) and (28) follow from (27e,f,g), provided that the frequency k is nonzero. The static case will be discussed separately.

It is remarkable that (28) is an *algebraic* relation consistent with (27b,c,d). It can properly be called a "first integral." Thus, let $K^\dagger$, $H^\dagger$, $H_1^\dagger$ be any solution of (27b,c,d) and let $F^\dagger$ be the corresponding function constructed from (28). We have the following identity:

$$dF^\dagger/dr+(m^*/r^2)(1-2m^*/r)^{-1}F^\dagger=0. \tag{29}$$

If now $F^\dagger$ vanishes at one point—as is possible by an appropriate choice of the arbitrary constants—then $F^\dagger=0$ everywhere.

Static Case and Its Physical Interpretation

In the static case where $k=0$ it follows from (27b,c) that H_1 vanishes. The other two radial factors can be shown to satisfy the relations

$$\begin{aligned}&(d/dr)(K+H)+(2m^*/r^2)(1-2m^*/r)^{-1}H=0,\\&(dH/dr)+(2m^*/r^2)(1-2m^*/r)^{-1}H+2H/r\\&\qquad+(L-1)(L+2)(K+H)/2m^*=0.\end{aligned} \tag{30}$$

For $L=0$ the solution is trivial. We get the difference between the Schwarzschild metrics for the two reduced masses m^* and $m^*+\delta m^*$. For $L=1$ we expect to find a solution that corresponds to a displacement of the center of attraction by the amount δz:

$$\begin{aligned}&x^\mu\to x^\mu+h^\mu; \qquad \theta\to\theta-\sin\theta\delta z/r;\\&r\to r+\cos\theta\delta z; \quad \varphi\to\varphi; \quad T\to T.\end{aligned} \tag{31}$$

This transformation is not acceptable (1) because it assumes Euclidean rather than Schwarzschild geometry for the displacement and (2) because it leads to a change $h_{\mu\nu}$ in the metric which does not have the diagonal form of (20). The correct infinitesimal coordinate transformation has the form

$$\begin{aligned}&r\to r+\cos\theta\delta z,\\&\theta\to\theta+(\sin\theta\delta z/2m^*)\ln(1-2m^*/r),\end{aligned} \tag{32}$$

and the two radial factors that satisfy (30) have the values

$$\begin{aligned}&H_2=H_0=H=-(2m^*/r^2)\delta z/(1-2m^*/r);\\&K=(\delta z/m^*)[(2m^*/r)+\ln(1-2m^*/r)].\end{aligned} \tag{33}$$

For $L>1$ we eliminate $(K+H)$ from Eqs. (30) and obtain an equation for the quantity $M=r(r-2m^*)H$:

$$(d^2M/dr^2)-[r^{-1}+(r-2m^*)^{-1}](dM/dr) -(L-1)(L+2)M/r(r-2m^*)=0. \tag{34}$$

The general solution of this equation can be expressed in terms of the associated Legendre functions:

$$M=\alpha P_L^{(2)}(1-r/m^*)+\beta Q_L^{(2)}(1-r/m^*), \tag{35}$$

where

$$\begin{aligned}&P_2^{(2)}(x)=3(1-x^2);\quad P_3^{(2)}(x)=15x(1-x^2);\quad\cdots;\\&Q_2^{(2)}(x)=(1-x^2)(d^2/dx^2)[\tfrac{1}{2}(3x^2-1)\,\mathrm{arc\,tanh}x-(3x/2)],\\&Q_3^{(2)}(x)=(1-x^2)(d^2/dx^2)[\tfrac{1}{2}(5x^3-3x)\,\mathrm{arc\,tanh}x\\&\qquad-(5x^2/2)+(2/3)],\\&\cdots.\end{aligned}$$

By way of comparison with the solution (32), consider the expression,

$$V=Q/r+\sum(\alpha_{LM}r^L+\beta_{LM}r^{-L-1})Y_{LM}(\theta,\varphi), \tag{36}$$

for electrostatic potential in flat space. Here the α terms represent the asymmetries in the potential due to remote sources and the β terms are due to asymmetries in the distribution of internal sources. The origin of the two kinds of terms in (35) is a little more subtle. We interpret them *both* as asymmetries in the metric arising from remote masses, as follows: (1) masses in the present region of space, and (2) masses in the region of space which unfolds from the other end of the wormhole. Naturally motions of both kinds of sources will be demanded by the field equations themselves. In contrast, the field equations of electrostatics are linear and are not sufficient in and by themselves to give the equations of motion. On this account, in the expression (36) for the electrostatic potential one is content to consider the α's and β's as constants. In contrast, one might well ask how the α and β in (35) can be constant if they represent the effect of remote masses that will inevitably be set in motion. However, we deal always with the metric *interior* to these remote masses. Therefore there is no obvious reason for a change with time in these coefficients. Moreover, the equations themselves say that the solution (35) is static.

Static solutions of odd type also exist. To analyze these solutions, we annul the frequency parameter k in Eqs. (22), note that the radial factor h_1 must vanish, and find for the other radial factor the second-order equation

$$d^2h_0/dr^2+(1-2m^*/r)^{-1}(4m^*/r^3 - L(L+1)/r^2)h_0=0. \quad (37)$$

Here again the general solution is a linear combination of two terms which reduce at large r to the form r^{-L} and r^{L+1}. These terms are interpreted as before as the effect of asymmetrically distributed masses which are very remote and located on the one side or the other of the wormhole. That solution which behaves asymptotically as r^{L+1} allows itself to be expressed as a hypergeometric function that reduces to a polynomial in $z=r/2m^*$:

$$\begin{aligned} h_0&=z^2F(L+2,1-L,4,z)\\ &=z^2+[(L+2)(1-L)/4]z^3\\ &\quad+[(L+2)(L+3)(1-L)(2-L)/2!4\times5]z^4+\cdots\\ &\quad+[(L+2)(L+3)\cdots(2L)\cdot(1-L)(2-L)\cdots\\ &\quad\times(-1)/(L-1)!4\times5\cdots(L+2)]z^{L+1}. \end{aligned} \quad (38)$$

We shall not investigate further the behavior of static perturbations. We interpret them as the effect of distant masses, producing inhomogeneous external gravitational fields, and thereby deforming the mouth of the wormhole that would otherwise be spherically symmetric. These perturbations have nothing directly to do with the dynamics or stability of the Schwarzschild metric itself.

III. BOUNDARY CONDITIONS AND STABILITY OF THE SCHWARZSCHILD SINGULARITY

Regularity Conditions; Odd Waves Analyzed Via the Concept of Effective Potential

A physically acceptable dynamic variation away from the background metric must be represented by functions which have at the starting instant a reasonably regular behavior both at the point $r=2m^*$ and at infinity.

It is simplest to analyze first the odd dynamic disturbances, as they are governed by the single relatively simple differential Eq. (24). The general solution behaves so:

$$\begin{aligned} Q&\sim c_1e^{i\delta}(r/2m^*-1)^{2ikm^*}+c_1e^{-i\delta}(r/2m^*-1)^{-2ikm^*} &&\text{for } r\to 2m^*,\\ Q&\sim c_2\sin(kr+\eta) &&\text{for } r\to\infty. \end{aligned} \quad (39)$$

Between these two extreme values of r the behavior of Q can be found by interpreting k_{eff}^2 in (24) in terms of an effective potential or effective refractive index encountered by gravitational waves. There is little difference in equations between the analysis here and that given previously for electromagnetic waves[6] and for neutrinos[7] moving under the influence of a spherically symmetric metric of the form (26b):

$$k_{\text{eff}}^2(\text{neutrinos})=k^2-\kappa^2e^\nu/r^2\pm(\kappa r^{-2}e^{\nu-\frac{1}{2}\lambda} - \kappa r^{-1}e^{\frac{1}{2}(\nu-\lambda)}de^{\frac{1}{2}\nu}/dr), \quad (40a)$$

$$k_{\text{eff}}^2(\text{light})=k^2-L(L+1)e^\nu/r^2, \quad (40b)$$

$$k_{\text{eff}}^2(\text{gravitational waves}) =k^2-L(L+1)e^\nu/r^2+6m^*e^\nu/r^3. \quad (40c)$$

The k^2 factor on the right-hand side of each expression plays the part of the energy in the Schrödinger wave equation, and the quantity subtracted from k^2 can be given the name of effective potential.

The effective potential starts from 0 at the Schwarzschild radius, rises to a maximum and then falls off again to zero at very large r. Therefore there are three regimes to be considered. In case 1, the quantity k^2 exceeds the height of the barrier and the solution is everywhere oscillatory. In case 2, the quantity k^2 is still positive, but less than the height of the barrier. The critical barrier height, k_{crit}^2, is given by the equation

$$\begin{aligned} 2m^*r_{\text{crit}}&=\tfrac{3}{2}+\tfrac{1}{2}(L+\tfrac{1}{2})^{-2}+\cdots;\\ k_{\text{crit}}&=(2.598\times 2m^*)^{-1}[(L+\tfrac{1}{2}) - (9/8)(L+\tfrac{1}{2})^{-1}+\cdots], \end{aligned} \quad (41)$$

when L is large. This case 2 recalls the problem of binding photons in the gravitational field of a geon. The interesting wave either falls off or rises exponentially in the barrier region. The one case, case 2a, corresponds to gravitational waves which never escape to large r. Such waves are of interest for the theory of gravitational geons: geons which derive all their mass and energy from gravitational waves trapped in the metric. Case 2b corresponds to waves which are large outside the barrier. These solutions like the case 1 solutions represent freely running gravitational waves disturbed to a greater or lesser extent by the gravitational field of the mass that they encounter.

[6] J. A. Wheeler, Phys. Rev. **97**, 511 (1955).
[7] D. Brill and J. A. Wheeler, Revs. Modern Phys. **29**, 465(1957).

Waves that Go through the Wormhole

The behavior of the waves will be affected by the conditions they encounter at small r. The metric acting on the gravitational wave in the present problem is multiply connected, whereas the metric acting on the neutrinos and electromagnetic waves in the published examples was topologically equivalent to Euclidean space. In those examples there was a definite origin at which the wave amplitude had to go to zero. By that condition the phase of the wave was uniquely determined, as showed up most readily in the JWKB approximate representation of the solution. The contrary will be the case for the present metric. The phase is not determined as it was for electromagnetic and neutrino geons, and as it would be for a gravitational geon. The ambiguity in phase has a clear physical significance. The ingoing wave need not equal in strength the outgoing wave. The Schwarzschild space is to be visualized as not inwardly bounded.[2] Instead, it can be considered as one mouth of a wormhole, the other mouth of which emerges elsewhere. Gravitational waves must be able to propagate through this tunnel as they propagate anywhere in space. The undetermined phase of this wave gives one the freedom that one can and must demand for a complete description of such "waves through the wormhole."

Space Behavior Unacceptable for Waves of Imaginary Frequency

Regime 3 corresponds to negative values of k^2. The analysis of the space dependence of such solutions shows that the radial part is uniquely determined by the requirement that it shall not go to infinity at large r. The solution that is acceptable because it falls off for large r proves, however, to fall off also at the Schwarzschild radius. Therefore there is no possibility to join it smoothly to a solution "in the other half of the tunnel" which will be acceptable. Consequently we conclude that there are no unstable solutions for odd waves.

There remains the discussion for the even waves. Here we have to examine the system (27b,c,d) supplemented by the condition (28). Unfortunately, owing to the complication of the equations involved, we were not able to establish a convenient "effective potential" picture. However, as far as stability is concerned there is no difficulty in recognizing that the same arguments used for odd waves are still valid.

We cannot avoid pointing out the essential importance of condition (28) in this analysis. It eliminates spurious solutions which do not have the correct behavior on the singular sphere. This can be clearly shown by a power series analysis near $r=2m^*$. Indeed to have the correct "wave through the tunnel" behavior it proves essential that the radial factor K be small for $r\sim 2m^*$ as compared with the other unknowns. This is plainly a consequence of (28). Curiously enough the same equation (28) also demands that the same factor K has to be negligible for large r, insuring the correct behavior at ∞. The discussion is then the same as for the odd waves. Consequently we conclude that Schwarzschild's solution of the gravitational field equations is stable.

ACKNOWLEDGMENTS

We wish to express our appreciation to Professor S. de Groot for the hospitality of the Lorentz Institute of the University of Leiden, where we enjoyed a collaboration of ten days. Also one of us (T.R.) is indebted to the Friends of Elementary Particle Physics of Princeton University for a travel grant. Finally, we are grateful to Mr. M. Wakano for checking our derivations and equations.

PART II

Regge Poles

Regge Poles and Their Unexpected Offsprings

Gabriele Veneziano

Theory Division, CERN, CH-1211 Geneva 23, Switzerland
Collège de France, 11 place M. Berthelot, 75005 Paris, France

We are all very accustomed in theoretical physics to extending real variables to complex ones. Since in any case quantum mechanics deals with complex numbers (wave functions, scattering amplitudes), considering such quantities as functions of complex arguments looks extremely natural or maybe even trivial.

Before Tullio Regge's seminal paper[1] it was common practice to extend the domain of energy values from the real line to the complex plane. Poles in the complex-E plane would correspond to bound states — if the poles were at real negative values — or resonances — if the poles were off the real axis — the distance from the real axis being inversely proportional to the resonance's lifetime.

For this viewpoint, Regge's bold idea of making angular momentum complex, also looks extremely natural. It was however already revolutionary in that is extended to complex values a variable which physically, and unlike energy, takes only discrete values according to quantum mechanics. But the consequences of that observation turned out to be far greater than expected. In the following I will try to summarize my own personal view on how this happened.

1. Organizing the hadronic zoo

The first, immediate use of Regge poles, both in their original context in non-relativistic potential scattering theory and in a relativistic context, is as a classification tool. Before Regge's work we knew about how to group different elementary particles in multiplets thanks to "internal" symmetries such as isospin or $SU(3)$. Such symmetries connect particles having the same spin (proton to neutron, charged and neutral pions etc.). An attempt was also made in the sixties to group together particles with spins differing by one unit, but, in spite of some phenomenological success, such symmetries could never be put on solid grounds.

Regge poles, by contrast, assemble together particles with the same internal quantum numbers (charge, strangeness etc.) but with spins that differ by (in principle) an arbitrary integer. Such particles get connected by the fact that they lie on the *same* Regge trajectory. The combination of internal symmetries and Regge poles managed to put some (although perhaps not enough) order in the ever increasing zoo of strongly-interacting elementary particles in the sixties.

2. The double role of Regge poles

The big boost to Regge's idea came with the work of Froissart, Gribov, Chew, Frautchi and Mandelstam.[2] They considered Regge poles for the scattering amplitude of relativistic particles using a crucial property of that situation — crossing. Crossing (a more general property than crossing symmetry) is a property (believed to be exact in any relativistic quantum theory) by which one and the same analytic function describes different processes in different regions of its kinematic (Mandelstam) variables.

One was then able to show that, under some reasonable assumptions, the *same* Regge poles that describe the spectrum of bounds states and resonances in one channel's physical region also control the high energy behavior in some other channel's physical region. In the latter case the relevant Regge poles are to be considered in the unphysical region for the channel in which the poles appear. In other words, the same Regge trajectory, say in the t channel, has two complementary roles: it describes particles in the t-channel at $t > 0$, and high energy behavior in the s and u channels for $t < 0$.

But up to here those two roles are distinct and complementary.

3. The duality bootstrap

In 1967 Dolen, Horn and Schmit made a fundamental empirical observation.[3] They found that in pion-nucleon charge-exchange scattering the resonance description (which can be based on Regge poles in the pion-nucleon s-channel) is on average equivalent to the description based on Regge poles exchanged in the crossed (i.e. nucleon-antinucleon to two pions) channel. This was largely unexpected and certainly not a consequence of crossing alone. This observation took the name of DHS (for Dolen–Horn–Schmit) duality.

From DHS duality to the duality bootstrap the step is not a long one: consider, instead of the original pion-nucleon scattering, a more symmetric — even if less feasible — process, like pion-pion scattering. For such a process the same Regge poles appear in all three channels and the statement of DHS duality becomes a self-consistency condition on the mesonic Regge poles themselves i.e. a bootstrap.

That program went on for almost a year (particularly by a group of four people: Marco Ademollo, Hector Rubinstein, Miguel Virasoro and myself) with the outside encouragement of Sergio Fubini, and obtained very encouraging results i.e. accurate, though not exact, solutions to the duality Bootstrap (see Ref. 4 for an account of those results). Then, in the summer of 1968 I managed to find[5] a simple solution satisfying exact crossing symmetry as well as duality in the form of Euler's Beta-function. For the process at hand, $\pi\pi \to \pi\omega$, it reads:

$$A(s,t) = g^2\,[B(1-\alpha(s), 1-\alpha(t)) + B(1-\alpha(s), 1-\alpha(u)) + B(1-B(1-\alpha(t), 1-\alpha(u))]$$
$$\alpha(s) = \alpha_0 + \alpha' s \;\; ; \;\; B(x,y) = \frac{\Gamma(x)\Gamma(y)}{\Gamma(x+y)} \tag{1}$$

where α_0 and α' are the Regge intercept and slope, respectively, and g^2 is an overall coupling for strong interactions.

Actually, DHS duality turned out to be a direct consequence of Regge behavior (with sufficiently low Regge trajectories at sufficiently negative t) and of the amplitude having only poles (i.e. only resonances contributions) in the Mandelstam variables. In other words Regge behavior (which gives branch points in the complex Mandelstam variable plane) was only holding on average and a bit off the real axis. This was the small prize to pay in order to have all the desired features.

That exact solution was further generalized and studied (becoming known as the dual resonance model (DRM)) until, a few years later, it became clear that it described the scattering of string-like (instead of point-like) objects. Unfortunately, in spite of its wonderful mathematical properties and physical consistency, the string of the DRM did not fit well certain features of strong interactions and was eventually abandoned in favor of a better theory, QCD, the theory of quarks and gluons interacting according to a non-abelian gauge theory based on the group $SU(3)$, still today an integral part of the celebrated standard model of elementary particles and fundamental forces. The two main killers were the wrong Regge intercept ($\alpha_0 = 1$ the famous Virasoro condition[6] for the absence of ghosts) and string theory's softness, not to talk about the necessity of additional spatial dimensions.

Why were people led to discover string theory in strong interactions if the correct theory is a standard quantum field theory? The answer is quite simple: because QCD is not like any old-fashioned QFT. The fundamental quanta of QCD are not really observable since they are permanently confined inside the hadrons; under certain circumstances the confinement mechanism appears through the formation of long and thin tubes of chromo-electric field that look like strings but are not as elementary as the string discovered through the DRM. An even closer connection can be found through an appropriate generalization of QCD to be discussed next.

4. Strings and large-N QCD

If the gauge group of QCD is generalized from its actual realization in Nature, $SU(3)$, to $SU(N)$ and one considers[7] the large-N limit, the theory is supposed to simplify enormously because its bound states will become stable. The corresponding scattering amplitudes will become meromorphic functions of their Mandelstam variables precisely like in the DRM. A big open question is whether, in the same large-N limit, QCD also has Regge-pole behavior at high energy and, if yes, how its Regge trajectories behave as a function of t. Arguments have been recently given[8] that large-N QCD and the DRM string should agree at large positive t and s i.e. in an unphysical region. Drastic differences are instead expected at negative t, positive s (e.g. fixed angle scattering). However, if the scattering amplitude goes to zero at large s and sufficiently negative t, DHS duality would follow irrespectively from whether such asymptotics is completely described by Regge poles. Given the data that were at the origin of DHS duality, it looks quite possible that this will turn out to be the case.

The most likely possibility, in my opinion, is that Regge behavior is indeed realized in large-N QCD, but with Regge trajectories that grow linearly at large positive t but bend, probably going to some negative constant, at large negative t in order to fulfill asymptotic freedom.

5. String-unification of all forces and the ubiquity of Regge poles

In 1974 Scherk and Schwarz[9] made the daring proposal that string theory should be reinterpreted as a theory of all interactions upon a huge rescaling of the characteristic energy-scale of string theory. In this new context, both the integer Regge intercepts ($\alpha_0 = 1$ and $\alpha_0 = 2$ for open and closed strings, respectively) and softness become advantages rather than obstacles.

If this would be the case, Regge theory would become very relevant for the totality of particle spectra. Known elementary particles would just represent the lightest sector of an infinity of particles lying on infinitely rising Regge trajectories. At the same time Regge poles would control the high-energy behavior of all processes and assure the absence of dangerous ultraviolet divergences.

Although the birth of string theory through the DRM was deeply rooted, as we have seen, on Regge theory, the attention of the string community after the Green–Schwarz revolution[10] shifted very quickly on its low-energy spectrum and interactions. This was clearly motivated by the extremely high new string energy scale, pretty close numerically to the Planck mass. Therefore, present tests of string theory appear to be necessarily related to its effective low energy theory.

Nevertheless, and to conclude, there are interesting aspects of string theory that call for going away from the low-energy approximation into the high energy/short distance (both the Regge and the fixed angle) regimes. Let me just mention a few:

- Very early cosmology. During a regime of extremely high temperatures (close to the Planck scale) string theory may have left interesting imprints on cosmological observables. The big bang singularity itself may be avoided by the soft behavior of string theory and by the existence of a limiting (Hagedorn–Fubini–Veneziano) temperature. This led to the original pre-big bang scenario[11] which has been followed by many similar "big bounce" cosmologies.
- Mini black holes and/or the late stage of BH evaporation information paradox. The Hawking temperature of a BH is inversely proportional to its mass and therefore blows up when the BH mass tends to zero, i.e. in the latest stages of its evaporation. Before reaching zero mass a BH is supposed to stop being a black hole thus avoiding a potential singularity problem.[12]
- Trans-planckian energy collisions are controlled[13] by the exchange of many gravi-Reggeons, the Regge pole replacing the old Pomeron of strong interactions. Regge behavior plays an important role in the study of transplanckian string collisions. For instance, the fact that gravi-Reggeon exchange replaced the graviton exchange of ordinary QFT implies that the former is related to the production of heavy strings in the direct channel (via DHS duality) and these lead to some precocious black-hole behavior[14] even before the threshold of BH production is reached.
- Uncover genuine stringy effects like the Generalized Uncertainty Principle[15] constraining $\Delta x > l_s$ where l_s is the fundamental length parameter of string theory.
- Causality problems that can be generically present in higher-derivative gravity theories[16] are neatly avoided thanks to Regge behavior.[17]

- Going off shell via Regge? It is notoriously difficult to define off-shell correlation functions in string theory as opposed to QFT. This may be well related to the Regge-behavior of string theory. Perhaps the concept of "mass shell" should be changed, in string theory in that of "Regge shell" i.e. going simultaneously way from both the physical mass and the physical angular momentum while staying on the Regge trajectory $J = \alpha(t)$. Indeed the so-called Pomeron vertex operator[18] (which should have been called gravi-Reggeon operator) is an example of an "on the Regge shell" quantity.

In conclusion, Regge's original idea had almost unimaginable consequences in fundamental physics and I'm sure there are still very many to be uncovered.

Acknowledgments

I would like to express my deep gratitude to the theory group of the University of Turin which, starting with Sergio Fubini and Tullio Regge, has always been for me a source of knowledge, inspiration and support for over 50 years of research.

References

1. T. Regge, "Introduction to complex orbital momenta", Nuovo Cim. **14**, 951 (1959).
2. V. N. Gribov, "Asymptotic behaviour of the scattering amplitude at high energies", Nucl. Phys. **22**, 249 (1961). M. Froissart, "Asymptotic behavior and subtractions in the Mandelstam representation", Phys. Rev. **123**, 1053 (1961). G. F. Chew and S. C. Frautschi, "Regge Trajectories and the Principle of Maximum Strength for Strong Interactions", Phys. Rev. Lett. **8**, 41 (1962). G. F. Chew, S. C. Frautschi and S. Mandelstam, "Regge poles in pi pi scattering", Phys. Rev. **126**, 1202 (1961).
3. R. Dolen, D. Horn and C. Schmid, "Prediction of Regge Parameters of rho Poles from Low-Energy pi N Data", Phys. Rev. Lett. **19**, 402 (1967); "Finite energy sum rules and their application to pi N charge exchange", Phys. Rev. **166**, 1768 (1968).
4. M. Ademollo, H. R. Rubinstein, G. Veneziano and M. A. Virasoro, "Bootstrap of meson trajectories from superconvergence", Phys. Rev. **176**, 1904 (1968).
5. G. Veneziano, "Construction of a crossing - symmetric, Regge behaved amplitude for linearly rising trajectories", Nuovo Cim. A **57**, 190 (1968).
6. M. S. Virasoro, "Subsidiary conditions and ghosts in dual resonance models", Phys. Rev. D **1**, 2933 (1970).
7. G. 't Hooft, "A Planar Diagram Theory for Strong Interactions", Nucl. Phys. B **72**, 461 (1974).
8. S. Caron-Huot, Z. Komargodski, A. Sever and A. Zhiboedov, "Strings from Massive Higher Spins: The Asymptotic Uniqueness of the Veneziano Amplitude", JHEP **1710**, 026 (2017).
9. J. Scherk and J. H. Schwarz, "Dual Models for Nonhadrons", Nucl. Phys. B **81**, 118 (1974).
10. M. B. Green and J. H. Schwarz, "Anomaly Cancellation in Supersymmetric D=10 Gauge Theory and Superstring Theory", Phys. Lett. **149B**, 117 (1984).
11. M. Gasperini and G. Veneziano, "Pre - big bang in string cosmology", Astropart. Phys. **1**, 317 (1993).
12. M. J. Bowick, L. Smolin and L. C. R. Wijewardhana, "Does String Theory Solve the Puzzles of Black Hole Evaporation?", Gen. Rel. Grav. **19**, 113 (1987).
13. D. Amati, M. Ciafaloni and G. Veneziano, "Classical and Quantum Gravity Effects from Planckian Energy Superstring Collisions", Int. J. Mod. Phys. A **3**, 1615 (1988).

14. G. Veneziano, "String-theoretic unitary S-matrix at the threshold of black-hole production", JHEP **0411**, 001 (2004).
15. D. Amati, M. Ciafaloni and G. Veneziano, "Can Space-Time Be Probed Below the String Size?", Phys. Lett. B **216**, 41 (1989).
16. X. O. Camanho, J. D. Edelstein, J. Maldacena and A. Zhiboedov, "Causality Constraints on Corrections to the Graviton Three-Point Coupling", JHEP **1602**, 020 (2016).
17. G. D'Appollonio, P. Di Vecchia, R. Russo and G. Veneziano, "Regge behavior saves String Theory from causality violations", JHEP **1505**, 144 (2015).
18. M. Ademollo, A. Bellini and M. Ciafaloni, "Superstring Regge Amplitudes and Emission Vertices", Phys. Lett. B **223**, 318 (1989). R. C. Brower, J. Polchinski, M. J. Strassler and C. I. Tan, "The Pomeron and gauge/string duality", JHEP **0712**, 005 (2007).

The Long Road from Regge Poles to the LHC

Vittorio Del Duca

Institute for Theoretical Physics, ETH Zürich, 8093 Zürich, Switzerland[a]

Lorenzo Magnea

Dipartimento di Fisica and Arnold–Regge Center, Università di Torino, and INFN, Sezione di Torino, Via P. Giuria 1, I-10125 Torino, Italy

The Regge limit of gauge-theory amplitudes and cross sections is a powerful theory tool for the study of fundamental interactions. It is a vast field of research, encompassing perturbative and non-perturbative dynamics, and ranging from purely theoretical developments to detailed phenomenological applications. It traces its origins to the proposal of Tullio Regge, almost sixty years ago, to study scattering phenomena in the complex angular momentum plane. In this very brief contribution, we look back to the early days of Regge theory, and follow a few of the many strands of its development, reaching to present day applications to scattering amplitudes in non-abelian gauge theories.

1. Introduction

The beginnings of Tullio Regge's career were characterised by deep and early forays into new and fast-developing fields of theoretical physics, where his extraordinary mastery of mathematical techniques never failed to leave a lasting mark. This was the case for his paper with John Wheeler on the stability of the Schwarzschild solution in general relativity,[34] and certainly for Regge Calculus,[1] but it is probably fair to say that his most prophetic contribution in those years was his proposal to study scattering amplitudes by means of their analiticity properties in the complex angular momentum plane.[3]

Regge's original work was in the context of non-relativistic potential scattering, and it

[a]On leave from Laboratori Nazionali di Frascati, INFN, Italy.

succeeded in uncovering a remarkable connection between the bound states of a quantum mechanical potential and the behaviour of the corresponding scattering amplitudes for complex values of the scattering angle. It was however quickly realised that these ideas could be generalised to relativistic quantum field theory (for a historical account, see the classic textbooks[4,5]), where the constraints of Lorentz invariance and crossing symmetry would make the formalism even more powerful. In particular, singularities in the complex angular momentum plane are tied to the high-energy behaviour of scattering amplitudes in the crossed channel, which is of great interest for collider applications.

The generalisation to quantum field theory was spearheaded by Vladimir Gribov,[6] and the resulting body of knowledge is often appropriately called *Gribov–Regge Theory*: it has turned out to be one of the most far-reaching and fruitful set of ideas in high-energy physics in the past half century. To name just three sets of applications and consequences of Gribov–Regge theory, we first note that it played a crucial role in the origins of string theory: indeed, the Veneziano model[7] was developed in search of a crossing-symmetric and unitary realization of Regge's ideas; next, in the context of the theory of strong interactions, at the non-perturbative level, linearly rising Regge trajectories provide excellent fits to mass spectra (see for example Ref. 8), and give simple and elegant parametrisations for total hadronic cross sections, across more than three orders of magnitude in energy;[9,10] finally, it has been known since the seminal work of Lev Lipatov[11] that Regge behaviour is realised at the perturbative level in the scattering amplitudes and cross sections of non-abelian gauge theories, in the high-energy limit. On the one hand, this leads to improved perturbative predictions of great relevance for collider phenomenology (see for example Ref. 12); on the other hand, it provides a laboratory for the most advanced studies of the properties of scattering amplitudes, in particular in the interesting case of $N = 4$ Super–Yang–Mills (SYM) theory (see for example Refs. 13 and 14).

The above set of examples makes it abundantly clear that any attempt to review, even superficially, the developments that flowed from Regge's original idea is doomed to fail: literally tens of thousands of papers have been written in the various fields of high-energy physics tracing their inspiration to Ref. 3. Our goal here will be much more limited: in Section 2 we will briefly illustrate the concept of a 'Regge Trajectory', and we will sketch the implications of Regge's approach for the high-energy limit of scattering amplitudes. In Section 3, we will attempt to provide a bird's eye view of the manifold implications of Regge's ideas for scattering amplitudes in perturbative non-abelian gauge theories. Finally, in Section 4, we will briefly present three short examples illustrating the vast phenomenological relevance of Regge's ideas for the modern understanding of strong interactions. Inevitably, our own history and viewpoint tilts our account towards the issues that we understand best and we have been actively working on: we apologise in advance for the omission of many very relevant topics from these notes. We hope that nonetheless the remarkable breadth of the impact of Regge Theory on contemporary high energy physics will be apparent.

2. From potential scattering to field theory

The basic intuition concerning the connection between poles of the scattering amplitude in angular momentum and bound states can be gathered by considering the best-known quantum-mechanical system: the hydrogen atom.[15] The textbook solution for energy levels in the Coulomb potential (neglecting spin) is

$$E = E_n = -\frac{\mu e^4}{2\hbar^2}\frac{1}{N^2}, \qquad N = n + l + 1, \tag{1}$$

where μ is the reduced mass, and the orbital angular momentum quantum number l is a non-negative integer. It is clearly possible to invert this expression and express the angular momentum as a function of the energy E. One finds

$$l(E) = -n - 1 + \mathrm{i}\frac{e^2}{\hbar}\sqrt{\frac{\mu}{2E}} \equiv -n + g(E). \tag{2}$$

This is an example of a *Regge Trajectory*: as a function of the energy, even for real E, the angular momentum quantum number l sweeps a curve in the complex plane, crossing the real axis at non-negative integer points when the energy equals a bound state energy E_n. For the Coulomb problem, one can also compute the S-matrix as a function of energy and angular momentum. The answer can be written as[16]

$$S(E, l) = \frac{\Gamma\left[l - g(E)\right]}{\Gamma\left[l + g(E)\right]}\, \mathrm{e}^{\mathrm{i}\pi l}. \tag{3}$$

As expected, the S-matrix has simple poles, due to the numerator Γ function, where l and E satisfy the bound state quantisation condition.

In order to generalise this observation to other potentials, and ultimately to relativistic field theory, a good starting point is the well-known partial-wave expansion of the scattering amplitude in a series of Legendre polynomials. For a 2-particle scattering process we write

$$A(s, \theta) = 16\pi \sum_{l=0}^{\infty} (2l + 1)\, a_l(s) P_l(\cos\theta), \tag{4}$$

where s is squared center-of-mass energy and θ the scattering angle in the center-of-mass frame in the s-channel. For equal masses, the scattering angle θ is related to the squared momentum transfer t by

$$t = -\frac{1}{2}\left(s - 4m^2\right)(1 - \cos\theta), \tag{5}$$

so that $t \le 0$ above threshold. Eq. (4) can be inverted to yield the partial-wave amplitude $a_l(s)$ according to

$$a_l(s) = \frac{1}{2}\int_{-1}^{1} dz\, A(s, z)\, P_l(z). \tag{6}$$

The crucial step taken by Regge[3,17] was to propose that the partial-wave amplitude in Eq. (6), defined for integer l, could be usefully generalised to an analytic function $a(l, s)$ depending on the complex variables l and s. Under suitable conditions on the asymptotic behaviour for large l, which can be shown to be met for potential scattering, the definition of the 'interpolating function' $a(l, s)$ is unique. One can then derive an expression for the scattering amplitude in terms of a contour integral in the l plane, performing a Sommerfeld–Watson transform.[b] For example, assume $a(l, s)$ to be analytic in l for $\mathbf{Re}\, l > l_0$ and not growing too fast for $l \to \infty$, and let n be the smallest integer $n > l_0$: then one can write

$$A(s, \theta) = 16\pi \sum_{l=0}^{n-1} (2l + 1)\, a_l(s) P_l(\cos\theta) + 8\pi\mathrm{i}\int_C dl\, \frac{(2l + 1)\, a(l, s)\, P_l(-\cos\theta)}{\sin(\pi l)}, \tag{7}$$

[b] In this brief account of the consequences of treating l as a complex variable we sketchily follow Ref. 4. For a recent historical perspective, see also Ref. 18.

with the integration contour $\mathcal{C}$ closely encircling the real axis for $l > l_0$, enclosing the poles induced by the denominator for $\mathbf{Re}\, l \geq n$. One may now attempt to open up the integration contour, making it parallel to the imaginary axis, and then move it to smaller values of $\mathbf{Re}\, l$. In the absence of complex singularities in $a(l,s)$, this just reconstructs the original partial-wave expansion; if however $a(l,s)$ has, for example, isolated poles in the l plane away from the real axis, in the process of moving the integration contour one picks up the corresponding residues. The location of these poles will in general depend on s, so that, varying s, the poles will follow trajectories such as Eq. (2). Denoting the trajectory of the i-th pole by $l_i = \alpha_i(s)$, and the residue of that pole by $r_i(s)$, the contribution of the poles to the scattering amplitude will be proportional to

$$\sum_i \frac{(2\alpha_i(s)+1)\; r_i(s)\, P_{\alpha_i(s)}(-\cos\theta)}{\sin\left(\pi\alpha_i(s)\right)} \,. \tag{8}$$

One can now use the knowledge of the asymptotic behaviour of Legendre polynomials

$$P_l(-z) \sim (-z)^l \,, \qquad z \to \infty \,, \tag{9}$$

to prove that in the (unphysical) limit $\cos\theta \to \infty$ (corresponding to $t \to \infty$ by Eq. (5)) the scattering amplitude is dominated by the contribution of the rightmost pole in the angular momentum plane, $l_{\rm max} = \alpha_{\rm max}(s)$, and behaves like an s-dependent power of t for large t.

Even though the large t region is not physical, the result just derived is of great interest: since Regge trajectories are related to bound states for the potential being considered, we now have a relation between the location of bound states and the asymptotic behaviour of the scattering amplitude; this allows, for example, to broadly extend the domain of applicability of dispersion relations. The extension of this result to relativistic field theory is however much more powerful: by Lorentz invariance and crossing symmetry, one can in fact transfer the s-channel argument to the t channel, and derive results for the asymptotic behaviour of the scattering amplitude for large s, which is physical, and of great phenomenological interest.

The main tool for achieving the relativistic generalisation is the use of dispersion relations to reconstruct the scattering amplitude in terms of its discontinuities along the cuts originating from normal thresholds. For example, for equal mass particles and up to rational contributions, one may write

$$A(s,t) \;=\; \frac{1}{\pi}\left[\int_{4m^2}^{\infty} dt'\, \frac{{\rm Disc}_t(A(s,t))}{t'-t} + \int_{4m^2}^{\infty} du'\, \frac{{\rm Disc}_u(A(s,t))}{u'-u}\right] \,. \tag{10}$$

Using Eq. (10) in Eq. (6), one may perform the z integration by means of Neumann's formula

$$\frac{1}{2}\int_{-1}^{1} dw\, \frac{P_l(w)}{z-w} \;=\; Q_l(z) \,, \tag{11}$$

where $Q_l(z)$ are the Legendre functions of the second kind. This leads to the Froissart–Gribov representation of the partial wave amplitude

$$a_l(s) \;=\; \frac{1}{\pi}\left[\int_{z_s}^{\infty} dz\, {\rm Disc}_t[A(s,t(z))]\; Q_l(z) - \int_{z_s}^{\infty} dz\, {\rm Disc}_u[A(s,t(z))]\; Q_l(-z)\right] \,, \tag{12}$$

where $z_s = 1 + 2m^2/s$ in the equal-mass case. The Sommerfeld–Watson transform is not directly applicable to Eq. (12) because of the bad convergence properties of the second term for large l. This leads to the introduction of the concept of *signature*: one needs to consider the (anti)symmetric combinations of t- and u-channel contributions defined by

$$a_l^{\pm}(s) = \frac{1}{\pi} \int_{z_s}^{\infty} dz \Big[\mathrm{Disc}_t[A(s,t(z))] \pm \mathrm{Disc}_u[A(s,t(z))] \Big] Q_l(z) \,. \tag{13}$$

Applying Regge's reasoning to partial waves of definite signature constrains the asymptotic behaviour of relativistic scattering amplitudes. In particular, and most interestingly, one can use crossing symmetry to study partial waves in the t-channel physical region, and for complex l: the presence of a 'Regge' pole in these partial wave amplitudes leads to a t-dependent power behaviour of the scattering amplitude at high energy, according to

$$a_l^{\pm}(t) \simeq \frac{1}{l - \alpha(t)} \qquad \longrightarrow \qquad A(s,t) \overset{s \to \infty}{\longrightarrow} f(t)\, s^{\alpha(t)} \,. \tag{14}$$

Clearly, Eq. (14) is only the simplest example of connection between singularities for complex l and asymptotic behaviours. Indeed, it was clear from the early days that one should in general expect contributions from 'Regge cuts', which are required at the non-perturbative level by unitarity arguments, and can be seen to arise perturbatively from the analysis of high-order Feynman diagrams. In the decades following Regge's original contribution, the manifold consequences of these subtle analytic properties for scattering amplitudes have been steadily explored, with important consequences both for non-perturbative studies of strong interactions, and for the exploration of perturbative gauge-theory scattering amplitudes.

3. The Regge limit for perturbative gauge theories

The discussion in Section 2 is general, and, for quantum field theories, it is based only on Lorentz invariance and analiticity: as a consequence, it is expected to apply in both the perturbative and non-perturbative regimes. When it became clear that strong interactions are described by QCD, and that the theory admits a rich and very relevant perturbative sector, a natural question was how Regge theory results would emerge from the perturbative expansion of a non-abelian gauge theory.

In the Regge limit, in which the squared centre-of-mass energy s is much larger than the momentum transfer $|t|$, $s \gg |t|$, any scattering process is dominated by the exchange in the t channel of the highest-spin particle. In the case of QCD, or $\mathcal{N} = 4$ Super–Yang–Mills (SYM), that entails the exchange of a gluon in the t channel. The study of this limit for QCD was pioneered by Lev Lipatov,[11] and the results are described by the Balitsky–Fadin–Kuraev–Lipatov (BFKL) theory, which models strong-interaction processes with two large and disparate scales (s and t) by resumming radiative corrections to parton-parton scattering. This is achieved, at leading logarithmic (LL) accuracy in $\ln(s/|t|)$, through the BFKL equation,[19–22] an integral equation for the evolution of t-channel gluon exchange in transverse-momentum space and Mellin moment space, which can be written as

$$\omega\, f_\omega(q_1, q_2) = \frac{1}{2}\, \delta^{(2)}(q_1 - q_2) + (\mathcal{K} \star f_\omega)\,(q_1, q_2) \,. \tag{15}$$

In Eq. (15), f_ω is related by Laplace–Mellin transform to the discontinuity of the four-point amplitude in the Regge limit, and depends on the momenta flowing in the t-channel on either side of the cut; the convolution is defined by

$$(\mathcal{K} \star f_\omega)(q_1, q_2) \equiv \int d^2k \, K(q_1, k) \, f_\omega(k, q_2) \,, \tag{16}$$

and $K(q_1, q_2)$ the BFKL kernel. The kernel is real and symmetric, $K(q_1, q_2) = K(q_2, q_1)$, so that the integral operator $\mathcal{K}$ is hermitian and its eigenvalues are real. At LL accuracy in $\ln(s/|t|)$, the building blocks of the BFKL kernel are a real correction (the emission of a gluon along the ladder, *a.k.a.* the central emission vertex, displayed in Fig. 1(a)), and a virtual correction (the one-loop Regge trajectory,[11] displayed in Fig. 1(b)). The BFKL equation is then obtained by iterating those one-loop corrections to all orders in the strong coupling α_s, resumming the terms of $\mathcal{O}(\alpha_s^n \ln^n(s/|t|))$.

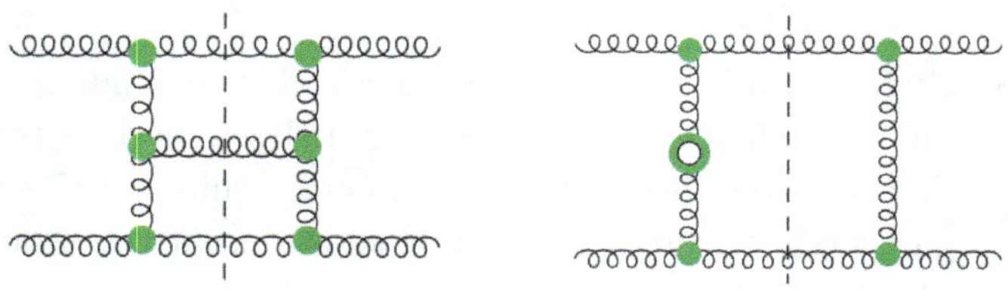

Figure 1. (a) Emission of a gluon along the ladder; (b) One-loop Regge trajectory.

Underpinning BFKL theory is the diagrammatic ladder structure of scattering amplitudes in the Regge limit. For instance, in the case of gluon-gluon scattering in QCD, the gluon ladder is described by the $8 \otimes 8$ colour representation, which is decomposed as $8 \otimes 8 = \{1 \oplus 8_s \oplus 27\} \oplus [8_a \oplus 10 \oplus \overline{10}]$, where the term in curly brackets in the direct sum is even under $s \leftrightarrow u$ exchange, while the term in square brackets is odd. To LL accuracy in $\ln(s/|t|)$, however, parton-parton scattering amplitudes are real and only the antisymmetric octet contributes: this implies gluon Reggeisation at LL accuracy.[11] For phenomenological purposes, the solution of the BFKL equation must then to be endowed with process-dependent impact factors describing the evolution of the emitting partons: the simplest cases are the quark and gluon impact factors,[20,23] by means of which one can obtain, for example, the di-jet cross section at large rapidities.[24]

The kernel of the BFKL equation (15) admits the perturbative expansion,

$$K(q_1, q_2) = \overline{\alpha}_\mu \sum_{l=0}^{\infty} \overline{\alpha}_\mu^l \, K^{(l)}(q_1, q_2) \,, \tag{17}$$

where $\overline{\alpha}_\mu = N_C \, \alpha_s(\mu^2)/\pi$ is the renormalised strong coupling constant evaluated at an arbitrary scale μ^2. $K^{(0)}$ is the leading-order BFKL kernel, which leads to the resummation of terms of $\mathcal{O}(\alpha_s^n \ln^n(s/|t|))$, corresponding to the LL accuracy discussed above; similarly, the NLO kernel $K^{(1)}$ resums terms at next-to-leading-logarithmic (NLL) accuracy, *i.e.* of $\mathcal{O}(\alpha_s^n \ln^{n-1}(s/|t|))$, and so forth.

The computation of the NLO BFKL kernel[25–28] requires the real corrections to the central emission vertex, corresponding to the emission of two gluons, or of a quark-antiquark pair, along the ladder, as depicted in Fig. 2(a);[29–32] furthermore, one needs the one-loop correction to the central emission vertex (Fig. 2(b)),[33–37] as well as the two-loop Regge

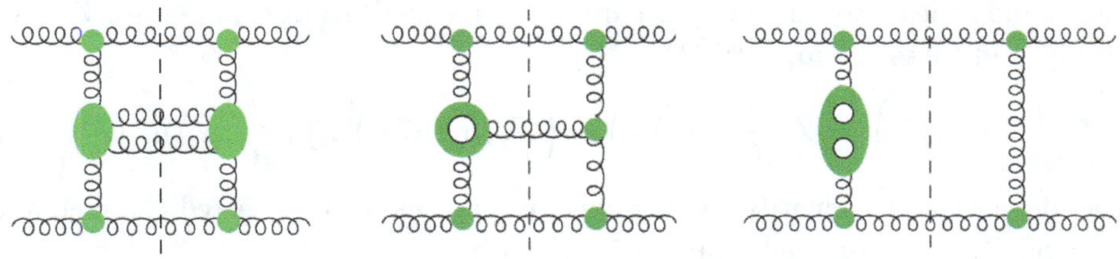

Figure 2. (a) Emission of two gluons (or of a quark-antiquark pair) along the ladder; (b) one-loop corrections to the central emission vertex; (c) two-loop Regge trajectory.

trajectory (Fig. 2(c)).[38–42] Underpinning BFKL theory at NLL accuracy is the fact that, while parton-parton scattering amplitudes develop an imaginary part at one loop, the BFKL equation is computed through the real part of the amplitudes, which receives contributions only from the antisymmetric octet representation; this, again, implies gluon Reggeisation at NLL accuracy.[43,44]

At LL accuracy, the kernel of the BFKL equation is conformally invariant, and thus the leading-order eigenfunctions of the BFKL equation are fixed[45] by conformal symmetry. NLL corrections[25–28] to the BFKL equation were computed by acting with the next-to-leading-order (NLO) kernel of the equation on the leading-order eigenfunctions. This procedure is not fully consistent, and it was already clear to Fadin and Lipatov[25] that the terms which make the procedure inconsistent are related to the running of the coupling. Consistent NLO eigenfunctions were later constructed by Chirilli and Kovchegov,[46,47] who found that indeed the additional pieces which occur at NLO are proportional to the QCD beta function.

3.1. *The Regge limit and infrared factorisation*

The backbone of the BFKL equation are the quark and gluon scattering amplitudes, which at high energy are dominated by the exchange of a gluon in the t channel. Because loop-level scattering amplitudes of massless partons are infrared divergent, and so are the impact factors, the Regge trajectory, and the central emission vertex, which build up the amplitudes in the Regge limit, the study of scattering amplitudes at high energy has benefited from a cross breeding with infrared factorisation.[13,48–57] In particular infrared factorisation leads to the exponentiation of infrared singularites: in the high-energy limit, this leads to all-order expressions for singular contributions to Regge-theory quantities such as the Regge trajectory and the impact factors.

Infrared (soft and collinear) divergences in a multi-parton amplitude $\mathcal{A}$ are known to factorise according to

$$\mathcal{A}\left(\frac{p_i}{\mu}, \alpha_s(\mu), \epsilon\right) = \mathbf{Z}\left(\frac{p_i}{\mu}, \alpha_s(\mu), \epsilon\right) \mathcal{H}\left(\frac{p_i}{\mu}, \alpha_s(\mu), \epsilon\right), \tag{18}$$

where the scattering amplitude $\mathcal{A}$ and the finite hard factor $\mathcal{H}$ should be seen as vectors in the space of available color configurations, while IR divergences are generated by the color operator $\mathbf{Z}$. Using renormalisation-group arguments, $\mathbf{Z}$ can be expressed in terms of an anomalous dimension matrix, known up to three loops in the massless case.[58–67] In the

Regge limit, and at least to three-loop accuracy, the infrared-divergent operator $\mathbf{Z}$ simplifies considerably, and takes the form[13,55,56]

$$\mathbf{Z}\left(\frac{p_i}{\mu},\alpha_s(\mu),\epsilon\right) = \widetilde{\mathbf{Z}}\left(\frac{s}{t},\alpha_s(\mu),\epsilon\right) Z_i\left(\frac{t}{\mu^2},\alpha_s(\mu),\epsilon\right) Z_j\left(\frac{t}{\mu^2},\alpha_s(\mu),\epsilon\right), \tag{19}$$

where the indices i and j identify the high-energy partons in the selected channel, and the factors Z_i and Z_j are color-singlet quantities given by

$$Z_i\left(\frac{t}{\mu^2},\alpha_s(\mu),\epsilon\right) = \mathrm{P}\exp\left\{-\int_0^{\mu^2}\frac{d\lambda^2}{\lambda^2}\left[\frac{\gamma_K\left(\alpha_s(\lambda)\right)}{4}\ln\frac{-t}{\lambda^2}\right]+\gamma_i\left(\alpha_s(\lambda)\right)\right\}. \tag{20}$$

Equation (20) naturally defines infrared divergent contributions to the impact factors in terms of the cusp γ_K and collinear γ_i anomalous dimensions. Colour and energy dependencies, on the other hand, are confined to the operator $\widetilde{\mathbf{Z}}$, defined by

$$\widetilde{\mathbf{Z}}\left(\frac{s}{t},\alpha_s(\mu),\epsilon\right) = \exp\left\{K(\alpha_s(\mu),\epsilon)\left[\left(\ln\left(\frac{s}{-t}\right)-\mathrm{i}\frac{\pi}{2}\right)\mathbf{T}_t^2+\mathrm{i}\pi\mathbf{T}_{s-u}^2\right]+Q_{\mathbf{\Delta}}^{(3)}(\alpha_s(\mu),\epsilon)\right\}. \tag{21}$$

In Eq. (21) we have introduced the colour operators

$$\mathbf{T}_s = \mathbf{T}_1+\mathbf{T}_2\,, \quad \mathbf{T}_u = \mathbf{T}_1+\mathbf{T}_3\,, \quad \mathbf{T}_t = \mathbf{T}_1+\mathbf{T}_4\,, \quad \mathbf{T}_{s-u}^2 = \frac{1}{2}(\mathbf{T}_s^2-\mathbf{T}_u^2)\,, \tag{22}$$

where $\mathbf{T}_i$ is a gluon insertion operator acting on parton i. Up to two loops, the exponent in Eq. (21) is completely determined by dipole colour correlations, and all singularities are generated by the integral

$$K(\alpha_s(\mu),\epsilon) = -\frac{1}{4}\int_0^{\mu^2}\frac{d\lambda^2}{\lambda^2}\widehat{\gamma}_K\left(\alpha_s(\lambda)\right), \tag{23}$$

which relates the infrared-divergent part of the Regge trajectory to the cusp anomalous dimension,[c] to all orders in perturbation theory. At three loops, quadrupole contributions arise for the first time, and, in the high-energy, limit, they contribute through the integral

$$Q_{\mathbf{\Delta}}^{(3)}\left(\alpha_s(\mu),\epsilon\right) = -\frac{\mathbf{\Delta}^{(3)}}{2}\int_0^{\mu^2}\frac{d\lambda^2}{\lambda^2}\left(\frac{\alpha_s(\lambda)}{\pi}\right)^3, \tag{24}$$

where $\mathbf{\Delta}^{(3)}$ is a single-logarithmic contribution with an intricate colour structure.[56]

Using Eqs. (19)–(24) one can, for example, determine the singular contributions to one-loop impact factors in terms of the one-loop cusp anomalous dimension and the one-loop quark and gluon collinear anomalous dimensions;[54,55] similarly, singular contributions to the two-loop trajectory emerge from the two-loop cusp anomalous dimension.[51] Furthermore, we know that the picture of Regge-pole factorisation, based on gluon Reggeisation, breaks down at NNLO accuracy.[42] The origin of the violation can be explained by means of infrared factorisation, by showing that the real part of the amplitudes becomes non-diagonal in the t-channel-exchange basis.[13,52] Accordingly, it is possible to predict how the violation propagates to higher loops, and the three-loop prediction,[54,55] which has NNLL accuracy, has been confirmed by the explicit computation of the three-loop four-point function of full $\mathcal{N}=4$ SYM in Ref. 71. In the Regge-pole factorisation picture, the violation is due

[c]In Eq. (23), $\widehat{\gamma}_K(\alpha_s)$ is the cusp anomalous dimension, rescaled by the quadratic Casimir eigenvalue of the relevant parton representation.

to the exchange of three Reggeised gluons.[56,72] Thus, although a full study of the BFKL ladder at NNLL accuracy has yet to be undertaken (the three-loop Regge trajectory in a specific scheme,[56] and the emission of three partons along the gluon ladder being the only building blocks computed so far[73–75]), we already have a clear picture of how the factorisation violations occur at NNLL accuracy.

While we have a precise knowledge of how infrared poles occur in loop corrections to impact factors and to the Regge trajectory, the finite parts of those corrections are not constrained by either Regge or infrared factorisation, and must be directly computed. Recently, however, a relation has been exposed[76] between the $\mathcal{O}(\varepsilon)$ terms of the one-loop gluon impact factor[37] and the $\mathcal{O}(\varepsilon^0)$ terms of the two-loop Regge trajectory:[38–42] this hints at a connection between the Regge limit and infrared factorisation extending beyond the infrared poles.

3.2. *The Regge limit in $\mathcal{N} = 4$ SYM and the analytic structure of scattering amplitudes*

In the last few years, it has been realised that the Regge limit of QCD and of $\mathcal{N} = 4$ SYM, and thus the BFKL equation, are endowed with a rich mathematical structure: the Regge limit has thus become a laboratory for advanced studies of gauge-theory scattering amplitudes at high orders.

It is well-known that $\mathcal{N} = 4$ SYM is a super-conformal theory. In the planar limit, the scattering amplitudes of the theory also exhibit a dual conformal symmetry,[77–80] where the conformal group acts on dual variables x_i defined in terms of (cyclically ordered) particle momenta by

$$p_i = x_i - x_{i-1} \, . \tag{25}$$

Dual conformal invariance is broken by infrared divergences,[81,82] but it is possible to construct finite functions of dual-conformal invariant ratios of x_i's, in which all infrared divergences cancel. As a consequence, the analytic structure of scattering amplitudes in planar $\mathcal{N} = 4$ SYM is highly constrained. In particular, the four- and five-point amplitudes are fixed to all loop orders by their symmetries, in terms of the one-loop amplitudes and of the cusp anomalous dimension.[82,83] The first occurrence of non-trivial corrections is then for amplitudes with at least six legs.[14,82,84–86] The ordinary and dual conformal symmetries are also at the heart of a surprising identity between scattering amplitudes and Wilson loops computed along light-like polygonal contours[78–82,85] in the dual space parametrised by the x_i coordinates. Thus, in what follows, any statement about amplitudes applies also to the corresponding polygonal Wilson loops.

The relevance of the Regge limit for studies of these amplitudes became apparent when it was observed that the four-point amplitude is Regge exact,[79,87] *i.e.* it is not modified by the Regge limit. Regge exactness extends to the five-point amplitude,[88,89] if computed in multi-Regge kinematics (MRK),[11] characterised by the central-emission vertex; similarly, it extends to the six-point amplitude,[14] if computed in quasi-multi-Regge kinematics,[29] characterised by the emission of two gluons along the ladder. In fact, Regge exactness extends in general to n-point MHV amplitudes, if computed in quasi-multi-Regge kinematics for the emission of $(n-4)$ gluons along the ladder.[86]

The Regge limit provides further insights in the mathematical structure of amplitudes in planar $\mathcal{N} = 4$ SYM, specifically on the functional spaces where the amplitudes are defined. In general, it is expected that n-point amplitudes in this theory should be expressed as iterated integrals of differential one-forms[90] defined on the space of configurations of points in three-dimensional projective space, $\mathrm{Conf}_N(\mathbb{CP}^3)$.[91,92] The simplest instance of iterated integrals that one encounters in these amplitudes are multiple polylogarithms (MPL),[93,94] which correspond to iterated integrals over rational functions. It is believed that all maximally helicity violating (MHV) and next-to-MHV amplitudes in planar $\mathcal{N} = 4$ SYM should be expressed in terms of MPLs having uniform transcendental weight.[95] The Regge limit provides evidence and abundant data to test these ideas. In fact, in the Euclidean region, where all Mandelstam invariants are negative, scattering amplitudes in planar $\mathcal{N} = 4$ SYM in multi-Regge kinematics factorise to all orders in perturbation theory into building blocks: the impact factor, the central-emission vertex, and the Regge trajectory, describing the Reggeised gluons exchanged in the t-channel, and the emission of gluons along the t-channel ladder. The building blocks are determined to all orders by the four- and five-point amplitudes, hence scattering amplitudes in multi-Regge kinematics are trivial in the Euclidean region.[88,89,96–99] In particular, the Regge trajectory is fixed, to all orders in perturbation theory, by the cusp anomalous dimension and the gluon collinear anomalous dimension.[96] Starting from six legs, however, scattering amplitudes exhibit Regge cuts, if the multi-Regge limit is taken after analytic continuation to a specific Minkowski region.[89,97] The discontinuity across these cuts is described, to all orders, by a dispersion relation closely related to the BFKL equation, which can be expressed in terms of single-valued functions.[100,101] In particular, the discontinuity of the six-point amplitude[102] in multi-Regge kinematics can be expressed in terms of single-valued harmonic polylogarithms (SVHPL).[103] Further, it was later realised that n-point amplitudes in multi-Regge kinematics are described by the geometry of the moduli space $\mathfrak{M}_{0,n-2}$ of Riemann spheres with $(n-2)$ marked points.[104] It can be shown that the algebra of iterated integrals on $\mathfrak{M}_{0,n}$ factors through certain hyperlogarithm algebras,[94] so that the iterated integrals on $\mathfrak{M}_{0,n}$ can always be expressed in terms of MPLs and rational functions, with singularities when two marked points coincide. To be more precise, one can introduce dual coordinates in transverse space, according to

$$\mathbf{q}_i = \mathbf{x}_{i+2} - \mathbf{x}_1 \,, \qquad \mathbf{k}_i = \mathbf{x}_{i+2} - \mathbf{x}_{i+1} \,, \tag{26}$$

as illustrated in Fig. 3. The discontinuity of the n-point amplitude in multi-Regge kinematics is then parametrised by $(n-5)$ conformally invariant cross-ratios, given by

$$z_i = \frac{(\mathbf{x}_1 - \mathbf{x}_{i+3})(\mathbf{x}_{i+2} - \mathbf{x}_{i+1})}{(\mathbf{x}_1 - \mathbf{x}_{i+1})(\mathbf{x}_{i+2} - \mathbf{x}_{i+3})} = -\frac{\mathbf{q}_{i+1}\mathbf{k}_i}{\mathbf{q}_{i-1}\mathbf{k}_{i+1}} \qquad i = 1, \ldots, N-5 \,. \tag{27}$$

It can be shown[104] that, in the soft limit for any of the $(n-5)$ transverse momenta of the gluons emitted along the ladder, one of the $(n-5)$ ratios z_i vanishes, and thus two marked points on the Riemann sphere coincide. The transverse momenta of the ladder gluons, however, never vanish. This requirement constrains the iterated integrals that can appear in multi-Regge kinematics to be single-valued functions: more precisely, linear combinations of products of iterated integrals on $\mathfrak{M}_{0,n}$ (and their complex conjugates) such that all branch cuts cancel. These are the single-valued multiple polylogarithms (SVMPL).[103,105,106] The statements of Ref. 104, which were made at LL accuracy, can be generalised at NLL accuracy and beyond.[107,108]

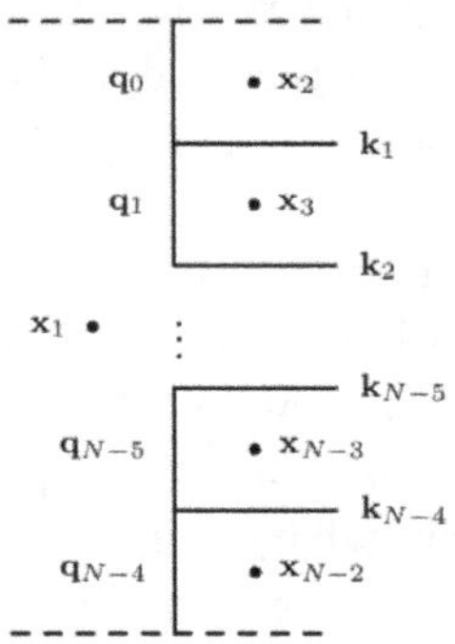

Figure 3. Dual coordinates in transverse space for the ladder of the n-point amplitude.

The foundation of the analysis of n-point amplitudes in planar $\mathcal{N} = 4$ SYM in multi-Regge kinematics is provided by Lipatov's picture[109] of the BFKL-like dispersion relation, which describes the discontinuity of the amplitude in terms of the Hamiltonian of an integrable open Heisenberg spin chain.[110,111] According to this picture, the discontinuity of the amplitude in multi-Regge kinematics and in specific Minkowski regions is described by a two-Reggeon cut; similarly, the double discontinuity is described by a three-Reggeon cut, and in general the m-fold discontinuity is described by an $(m+1)$-Reggeon cut. As outlined above, this picture has been fully explored at the level of the single discontinuity, while the analysis of the double discontinuity has just begun.[112]

Previously, Lipatov had characterised the BFKL equation, and, more generally, the exchange of a ladder of n Reggeised gluons in a singlet configuration in QCD at large N_c, in terms of the Hamiltonian[110,113,114] of a closed spin chain. It should then come as no surprise that, just like the case of the six-point amplitude of $\mathcal{N} = 4$ SYM in multi-Regge kinematics,[102] also the analytic structure of the BFKL ladder at LL accuracy in QCD can be described in terms of single-valued iterated integrals on the moduli space $\mathfrak{M}_{0,4}$ of Riemann spheres with four marked points, which are also given by SVHPLs.[115] In this case, the single-valuedness can be traced back to the fact that neither of the momenta at the ends of the gluon ladder can vanish. That analysis can be extended to NLL accuracy for the BFKL ladder in QCD, as well as in full $\mathcal{N} = 4$ SYM,[116] however it requires a generalisation of the SVHPLs, recently introduced by Schnetz.[117] The control of the analytic structure of the BFKL ladder at NLL accuracy, and the freedom in defining its matter content, allowed Ref. 116 to prove that there is no gauge theory of uniform and maximal transcendental weight which can match the maximal weight part of QCD, and to identify a set of conditions which allow to constrain the field content of theories for which the BFKL ladder has maximal weight.

4. Regge phenomenology: three highlights

As is the case for theoretical developments, the field of phenomenological applications of Regge theory is too vast to be even superficially overviewed. We will therefore confine ourselves to three examples, highlighting the fact that strong interactions display Regge behaviour both in their bound state structure and in the asymptotic behaviour of their cross sections; furthermore, high-energy logarithms may affect in a measurable way important multi-jet cross sections at LHC.

As a first example, in Fig. 4, we show two sample fits of hadron spectra, clearly displaying the alignment of meson and baryon states along linearly rising Regge trajectories. The first panel of Fig. 4, from Ref. 8, shows a set of isovector meson resonances (starting with ρ mesons), with the red diamonds representing predictions derived from a relativistic quark model. The second panel shows a baryon example, taken from the review,[118] displaying resonances of the Δ^* baryon; in this case the linearly rising Regge trajectory is not fit from data, but it is derived from a non-perturbative model based on the AdS/QCD correspondence.[119] The existence of linearly rising trajectories was an early inspiration for the development of string theory, and effective string theories for a non-perturbative description of strong interactions continue to be an active field of research.

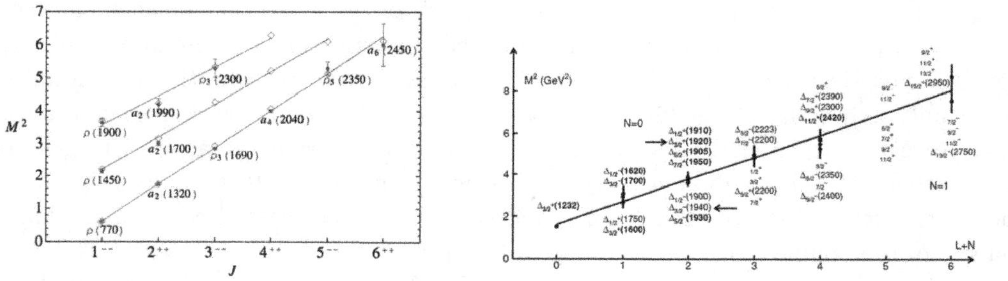

Figure 4. In the first panel, a sample fit to the light isovector meson spectrum, for the case of ρ resonances, interpolated with linear Regge trajectories, from Ref. 8; in the second panel, from Ref. 118, data for the Δ^* baryonic resonances, fit with a linear Regge trajectory with parameters derived in Ref. 119.

As a second example, we display in Fig. 5 the well-known fits of total cross sections for proton-proton and proton-antiproton collisions by Donnachie and Landshoff. To quote the short abstract of their 1992 paper,[9] the fits indeed show that "Regge theory provides a very simple and economical description of all total cross sections". Indeed, Regge theory suggests that the total hadronic cross section should be described at high energies by the exchange of Reggeons, corresponding to sets of particles belonging to Regge trajectories. Donnachie and Landshoff showed in Ref. 9 that the total cross sections data then available could be fit accordingly, using simple expressions of the form

$$\sigma_{ab}^{\rm tot} = X_{ab}\, s^{\alpha} + Y_{ab}\, s^{\beta} \,, \tag{28}$$

where a and b denote the selected hadrons, and X and Y the corresponding fit parameters, while $\alpha \simeq 0.08$ and $\beta \simeq -0.45$ are effective powers, roughly representing the contributions of the pomeron and ρ trajectories: remarkably, the same α and β can be used to describe many different processes, highlighting the fact that the success of the fit depends on properties of the particles being exchanged in the scattering, rather than the particles being scattered.

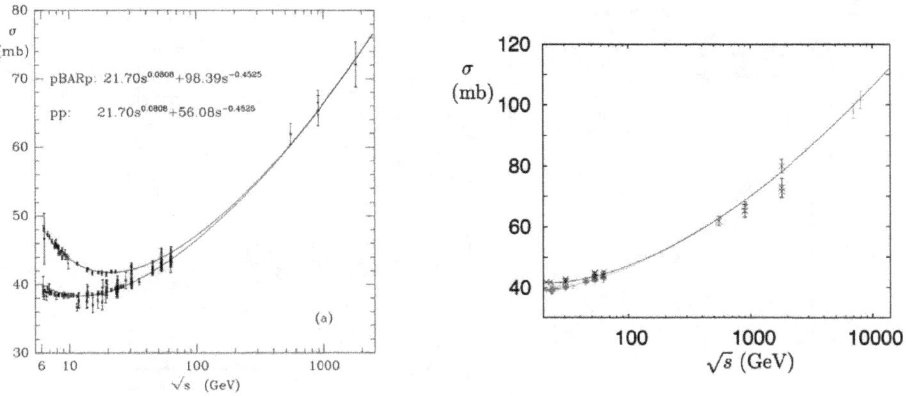

Figure 5. In the first panel, the original Donnachie-Landshoff fit of the total proton-proton and proton-antiproton cross sections with simple Regge parametrisations, from Ref. 9; in the second panel, the 2013 update for the proton-proton case, including early LHC data, from Ref. 10.

The second panel of Fig. 5 shows an update of the fit, from Ref. 10, which successfully includes the first LHC data.

To conclude, we note that Regge theory provides, in principle, an all-order understanding of perturbative rapidity logarithms that arise in multi-particle production cross sections, most notably in multi-jet cross sections currently measured at LHC, when the jets are produced at large rapidity intervals. Figure 6 shows a first successful attempt to fit such cross sections including Regge contributions at leading logarithmic accuracy. In this case, the plot shows the cross section for the production of W bosons in association with at least 2 jets, as a function of the di-jet invariant mass of the two leading jets. As shown in the second panel, the theoretical prediction using the High Energy Jet framework[12] successfully covers the high invariant mass region of the data. As future LHC measurements become more detailed and statistical errors are reduced as accumulated luminosity increases, it is very likely that the implementation of accurate high-energy resummations will become a necessary ingredient for data analysis. Many detailed theoretical results are already available in this direction (for a recent example, see Ref. 121).

As we have tried to show with these brief notes, the legacy of Tullio Regge and his work on complex angular momenta will continue to play a central role in high-energy physics for a long time to come.

Acknowledgments

We thank the Galileo Galilei Institute for Theoretical Physics in Firenze, and the Higgs Centre for Theoretical Physics in Edinburgh, for their kind hospitality.

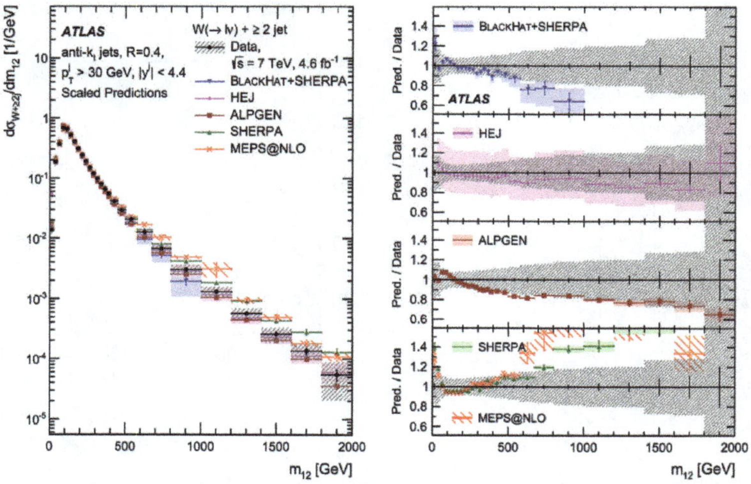

Figure 6. ATLAS data giving the cross section for the production of W bosons in association with at least 2 jets, as a function of the di-jet invariant mass of the two leading jets, compared with various theoretical predictions, from Ref. 120.

References

1. T. Regge and J. A. Wheeler, Phys. Rev. **108** (1957) 1063.
2. T. Regge, Nuovo Cim. **19** (1961) 558.
3. T. Regge, Nuovo Cim. **14** (1959) 951.
4. R.J. Eden, P.V. Landshoff, D.I. Olive and J.C. Polkinghorne, *The Analytic S-matrix*, Cambridge University Press, 1966.
5. P.D.B. Collins, *An Introduction to Regge Theory and High-Energy Physics*, Cambridge University Press, 1977.
6. V.N. Gribov, *The Theory of Complex Angular Momenta*, Cambridge University Press, 2003.
7. G. Veneziano, Nuovo Cim. A **57** (1968) 190.
8. D. Ebert, R. N. Faustov and V. O. Galkin, Phys. Rev. D **79** (2009) 114029.
9. A. Donnachie and P. V. Landshoff, Phys. Lett. B **296** (1992) 227, [hep-ph/9209205].
10. A. Donnachie and P. V. Landshoff, Phys. Lett. B **727** (2013) 500, Erratum: [Phys. Lett. B **750** (2015) 669], [arXiv:1309.1292 [hep-ph]].
11. L. N. Lipatov, Sov. J. Nucl. Phys. **23** (1976) 338–345. [Yad. Fiz.23,642(1976)].
12. J. R. Andersen and J. M. Smillie, JHEP **1106** (2011) 010, [arXiv:1101.5394 [hep-ph]].
13. V. Del Duca, C. Duhr, E. Gardi, L. Magnea, and C. D. White, JHEP **12** (2011) 021 [arXiv:1109.3581].
14. V. Del Duca, C. Duhr and V. A. Smirnov, JHEP **1003** (2010) 099, [arXiv:0911.5332 [hep-ph]].
15. `https://en.wikipedia.org/wiki/Regge_theory`
16. S. Weinberg, *Lectures on Quantum Mechanics*, Cambridge University Press, 2013.
17. A. Bottino, A. M. Longoni and T. Regge, Nuovo Cim. **23** (1962) 954.
18. A. Bottino, arXiv:1807.02456 [physics.hist-ph].
19. V. S. Fadin, E. A. Kuraev, and L. N. Lipatov, Phys. Lett. **B60** (1975) 50–52.
20. E. A. Kuraev, L. N. Lipatov, and V. S. Fadin, Sov. Phys. JETP **44** (1976) 443–450. [Zh. Eksp. Teor. Fiz.71,840(1976)].
21. E. A. Kuraev, L. N. Lipatov, and V. S. Fadin, Sov. Phys. JETP **45** (1977) 199–204. [Zh. Eksp. Teor. Fiz.72,377(1977)].

22. I. I. Balitsky and L. N. Lipatov, Sov. J. Nucl. Phys. **28** (1978) 822–829. [Yad. Fiz.28,1597(1978)].
23. V. Del Duca, Phys. Rev. D **52** (1995) 1527 [hep-ph/9503340].
24. A. H. Mueller and H. Navelet, Nucl. Phys. B **282** (1987) 727.
25. V. S. Fadin and L. N. Lipatov, Phys. Lett. B **429** (1998) 127–134 [hep-ph/9802290].
26. M. Ciafaloni and G. Camici, Phys. Lett. B **430** (1998) 349–354 [hep-ph/9803389].
27. A. V. Kotikov and L. N. Lipatov, Nucl. Phys. B **582** (2000) 19–43 [hep-ph/0004008].
28. A. V. Kotikov and L. N. Lipatov, Nucl. Phys. B **661** (2003) 19–61 [hep-ph/0208220]. [Erratum: Nucl. Phys. B685, 405 (2004)].
29. V. S. Fadin and L. N. Lipatov, JETP Lett. **49** (1989) 352 [Yad. Fiz. **50** (1989) 1141] [Sov. J. Nucl. Phys. **50** (1989) 712].
30. V. Del Duca, Phys. Rev. D **54** (1996) 989 [hep-ph/9601211].
31. V. S. Fadin and L. N. Lipatov, Nucl. Phys. B **477** (1996) 767 [hep-ph/9602287].
32. V. Del Duca, Phys. Rev. D **54** (1996) 4474 [hep-ph/9604250].
33. V. S. Fadin and L. N. Lipatov, Nucl. Phys. B **406** (1993) 259.
34. V. S. Fadin, R. Fiore and A. Quartarolo, Phys. Rev. D **50** (1994) 5893 [hep-th/9405127].
35. V. S. Fadin, R. Fiore and M. I. Kotsky, Phys. Lett. B **389** (1996) 737 [hep-ph/9608229].
36. V. Del Duca and C. R. Schmidt, Phys. Rev. D **59** (1999) 074004 [hep-ph/9810215].
37. Z. Bern, V. Del Duca and C. R. Schmidt, Phys. Lett. B **445** (1998) 168 [hep-ph/9810409].
38. V. S. Fadin, M. I. Kotsky and R. Fiore, Phys. Lett. B **359** (1995) 181.
39. V. S. Fadin, R. Fiore and M. I. Kotsky, Phys. Lett. B **387** (1996) 593 [hep-ph/9605357].
40. V. S. Fadin, R. Fiore and A. Quartarolo, Phys. Rev. D **53** (1996) 2729 [hep-ph/9506432].
41. J. Blumlein, V. Ravindran and W. L. van Neerven, Phys. Rev. D **58** (1998) 091502 [hep-ph/9806357].
42. V. Del Duca and E. W. N. Glover, JHEP **10** (2001) 035 [hep-ph/0109028].
43. V. S. Fadin, R. Fiore, M. G. Kozlov and A. V. Reznichenko, Phys. Lett. B **639** (2006) 74 [hep-ph/0602006].
44. V. S. Fadin, M. G. Kozlov and A. V. Reznichenko, Phys. Rev. D **92** (2015) no. 8, 085044 [arXiv:1507.00823 [hep-th]].
45. L. N. Lipatov, Sov. Phys. JETP **63** (1986) 904 [Zh. Eksp. Teor. Fiz. **90** (1986) 1536].
46. G. A. Chirilli and Y. V. Kovchegov, JHEP **1306** (2013) 055 [arXiv:1305.1924 [hep-ph]].
47. G. A. Chirilli and Y. V. Kovchegov, JHEP **1405** (2014) 099 Erratum: [JHEP **1508** (2015) 075] [arXiv:1403.3384 [hep-ph]].
48. M. G. Sotiropoulos and G. F. Sterman, Nucl. Phys. B **419** (1994) 59 [hep-ph/9310279].
49. G. P. Korchemsky, Phys. Lett. B **325** (1994) 459 [hep-ph/9311294].
50. I. A. Korchemskaya, G. P. Korchemsky, Nucl. Phys. B **437** (1995) 127 [hep-ph/9409446].
51. I. A. Korchemskaya, G. P. Korchemsky, Phys. Lett. B **387** (1996) 346 [hep-ph/9607229].
52. V. Del Duca, C. Duhr, E. Gardi, L. Magnea, and C. D. White, Phys. Rev. D **85** (2012) 071104 [arXiv:1108.5947].
53. S. Caron-Huot, JHEP **1505** (2015) 093 [arXiv:1309.6521].
54. V. Del Duca, G. Falcioni, L. Magnea, and L. Vernazza, Phys. Lett. B **732** (2014) 233–240 [arXiv:1311.0304 [hep-ph]].
55. V. Del Duca, G. Falcioni, L. Magnea, and L. Vernazza, JHEP **02** (2015) 029 [arXiv:1409.8330 [hep-ph]].
56. S. Caron-Huot, E. Gardi, and L. Vernazza, JHEP **06** (2017) 016 [arXiv:1701.05241 [hep-ph]].
57. S. Caron-Huot, E. Gardi, J. Reichel and L. Vernazza, JHEP **1803** (2018) 098 [arXiv:1711.04850 [hep-ph]].
58. S. M. Aybat, L. J. Dixon and G. F. Sterman, Phys. Rev. Lett. **97** (2006) 072001 [hep-ph/0606254].
59. S. M. Aybat, L. J. Dixon and G. F. Sterman, Phys. Rev. D **74** (2006) 074004 [hep-ph/0607309].
60. L. J. Dixon, L. Magnea and G. F. Sterman, JHEP **0808** (2008) 022 [arXiv:0805.3515 [hep-ph]].

61. T. Becher and M. Neubert, Phys. Rev. Lett. **102** (2009) 162001. Erratum: [Phys. Rev. Lett. **111** (2013) no. 19, 199905], [arXiv:0901.0722 [hep-ph]].
62. E. Gardi and L. Magnea, JHEP **0903** (2009) 079 [arXiv:0901.1091].
63. E. Gardi and L. Magnea, Nuovo Cim. C **32N5-6** (2009) 137 [Frascati Phys. Ser. **50** (2010)], [arXiv:0908.3273 [hep-ph]].
64. T. Becher and M. Neubert, JHEP **0906** (2009) 081. Erratum: [JHEP **1311** (2013) 024], [arXiv:0903.1126 [hep-ph]].
65. L. J. Dixon, E. Gardi and L. Magnea, JHEP **1002** (2010) 081, [arXiv:0910.3653 [hep-ph]].
66. Ø. Almelid, C. Duhr and E. Gardi, Phys. Rev. Lett. **117** (2016) 172002 [arXiv:1507.00047].
67. Ø. Almelid, C. Duhr, E. Gardi, A. McLeod and C. D. White, JHEP **1709** (2017) 073 [arXiv:1706.10162].
68. V. S. Fadin and R. Fiore, Phys. Lett. B **294** (1992) 286.
69. V. S. Fadin, R. Fiore and A. Quartarolo, Phys. Rev. D **50** (1994) 2265 [hep-ph/9310252].
70. V. Del Duca and C. R. Schmidt, Phys. Rev. D **57** (1998) 4069 [hep-ph/9711309].
71. J. M. Henn and B. Mistlberger, Phys. Rev. Lett. **117** (2016) 171601, [arXiv:1608.00850].
72. V. S. Fadin, AIP Conf. Proc. **1819** (2017), no. 1 060003 [arXiv:1612.04481].
73. V. Del Duca, A. Frizzo and F. Maltoni, Nucl. Phys. B **568** (2000) 211 [hep-ph/9909464].
74. E. N. Antonov, L. N. Lipatov, E. A. Kuraev and I. O. Cherednikov, Nucl. Phys. B **721** (2005) 111 [hep-ph/0411185].
75. C. Duhr, Ph.D. thesis, Universitè Catholique de Louvain, 2009.
76. V. Del Duca, JHEP **1802** (2018) 112 [arXiv:1712.07030 [hep-ph]].
77. J. M. Drummond, J. Henn, V. A. Smirnov and E. Sokatchev, JHEP **0701** (2007) 064 [hep-th/0607160].
78. L. F. Alday and J. M. Maldacena, JHEP **0706** (2007) 064 [arXiv:0705.0303 [hep-th]].
79. J. M. Drummond, G. P. Korchemsky and E. Sokatchev, Nucl. Phys. B **795** (2008) 385 [arXiv:0707.0243 [hep-th]].
80. A. Brandhuber, P. Heslop and G. Travaglini, Nucl. Phys. B **794** (2008) 231 [arXiv:0707.1153 [hep-th]].
81. J. M. Drummond, J. Henn, G. P. Korchemsky and E. Sokatchev, Nucl. Phys. B **795** (2008) 52 [arXiv:0709.2368 [hep-th]].
82. J. M. Drummond, J. Henn, G. P. Korchemsky and E. Sokatchev, Nucl. Phys. B **826** (2010) 337 [arXiv:0712.1223 [hep-th]].
83. Z. Bern, L. J. Dixon and V. A. Smirnov, Phys. Rev. D **72** (2005) 085001 [hep-th/0505205].
84. Z. Bern, L. J. Dixon, D. A. Kosower, R. Roiban, M. Spradlin, C. Vergu and A. Volovich, Phys. Rev. D **78** (2008) 045007 [arXiv:0803.1465 [hep-th]].
85. J. M. Drummond, J. Henn, G. P. Korchemsky and E. Sokatchev, Nucl. Phys. B **815** (2009) 142 [arXiv:0803.1466 [hep-th]].
86. V. Del Duca, C. Duhr and V. A. Smirnov, JHEP **1005** (2010) 084 [arXiv:1003.1702 [hep-th]].
87. S. G. Naculich and H. J. Schnitzer, Nucl. Phys. B **794** (2008) 189 [arXiv:0708.3069 [hep-th]].
88. R. C. Brower, H. Nastase, H. J. Schnitzer and C. I. Tan, Nucl. Phys. B **814** (2009) 293 [arXiv:0801.3891 [hep-th]].
89. J. Bartels, L. N. Lipatov and A. Sabio Vera, Phys. Rev. D **80** (2009) 045002 [arXiv:0802.2065 [hep-th]].
90. K. T. Chen, Bull. Am. Math. Soc. **83** (1977) 831.
91. J. Golden, A. B. Goncharov, M. Spradlin, C. Vergu and A. Volovich, JHEP **1401** (2014) 091 [arXiv:1305.1617 [hep-th]].
92. J. Golden, M. F. Paulos, M. Spradlin and A. Volovich, J. Phys. A **47** (2014) no. 47, 474005 [arXiv:1401.6446 [hep-th]].
93. A. B. Goncharov, [math/0103059].
94. F. C. S. Brown, Annales Sci. Ecole Norm. Sup. **42** (2009) 371 [math/0606419].
95. N. Arkani-Hamed, J. L. Bourjaily, F. Cachazo, A. B. Goncharov, A. Postnikov and J. Trnka, Cambridge University Press, Cambridge U.K. (2012).

96. V. Del Duca and E. W. N. Glover, JHEP **0805** (2008) 056 [arXiv:0802.4445 [hep-th]].
97. J. Bartels, L. N. Lipatov and A. Sabio Vera, Eur. Phys. J. C **65** (2010) 587 [arXiv:0807.0894 [hep-th]].
98. R. C. Brower, H. Nastase, H. J. Schnitzer and C. I. Tan, Nucl. Phys. B **822** (2009) 301 [arXiv:0809.1632 [hep-th]].
99. V. Del Duca, C. Duhr and E. W. N. Glover, JHEP **0812** (2008) 097 [arXiv:0809.1822 [hep-th]].
100. L. N. Lipatov and A. Prygarin, Phys. Rev. D **83** (2011) 125001 [arXiv:1011.2673 [hep-th]].
101. V. S. Fadin and L. N. Lipatov, Phys. Lett. B **706** (2012) 470 [arXiv:1111.0782 [hep-th]].
102. L. J. Dixon, C. Duhr and J. Pennington, JHEP **1210** (2012) 074 [arXiv:1207.0186].
103. F. C. S. Brown, C. R. Acad. Sci. Paris, Ser. I **338** (2004) 527.
104. V. Del Duca, S. Druc, J. Drummond, C. Duhr, F. Dulat, R. Marzucca, G. Papathanasiou and B. Verbeek, JHEP **1608** (2016) 152 [arXiv:1606.08807].
105. F. C. S. Brown, [www.ihes.fr/ brown/RHpaper5.pdf].
106. F. C. S. Brown, SIGMA **2** (2014) e25 [arXiv:1309.5309].
107. V. Del Duca, S. Druc, J. Drummond, C. Duhr, F. Dulat, R. Marzucca, G. Papathanasiou and B. Verbeek, JHEP **1806** (2018) 116 [arXiv:1801.10605 [hep-th]].
108. R. Marzucca and B. Verbeek, arXiv:1811.10570 [hep-th].
109. L. N. Lipatov, J. Phys. A **42** (2009) 304020 [arXiv:0902.1444 [hep-th]].
110. L. N. Lipatov, JETP Lett. **59** (1994) 596 [Pisma Zh. Eksp. Teor. Fiz. **59** (1994) 571] [hep-th/9311037].
111. L. D. Faddeev and G. P. Korchemsky, Phys. Lett. B **342** (1995) 311 [hep-th/9404173].
112. V. Del Duca, C. Duhr, F. Dulat and B. Penante, arXiv:1811.10398 [hep-th].
113. L. N. Lipatov, Phys. Lett. B **309** (1993) 394.
114. L. N. Lipatov, Nucl. Phys. B **548** (1999) 328 [hep-ph/9812336].
115. V. Del Duca, L. J. Dixon, C. Duhr, and J. Pennington, JHEP **1402** (2014) 086 [arXiv:1309.6647].
116. V. Del Duca, C. Duhr, R. Marzucca and B. Verbeek, JHEP **1710** (2017) 001 [arXiv:1705.10163].
117. O. Schnetz, Phys. Rev. D **97** (2018) 085018, [arXiv:1606.08598 [hep-th]].
118. E. Klempt and J. M. Richard, Rev. Mod. Phys. **82** (2010) 1095, [arXiv:0901.2055 [hep-ph]].
119. H. Forkel, M. Beyer and T. Frederico, JHEP **0707** (2007) 077, [arXiv:0705.1857 [hep-ph]].
120. G. Aad *et al.* [ATLAS Collaboration], Eur. Phys. J. C **75** (2015) no. 2, 82, [arXiv:1409.8639 [hep-ex]].
121. M. Bonvini and S. Marzani, Phys. Rev. Lett. **120** (2018) no. 20, 202003, [arXiv:1802.07758 [hep-ph]].

IL NUOVO CIMENTO Vol. XIV, N. 5 1° Dicembre 1959

Introduction to Complex Orbital Momenta.

T. Regge

Max-Planck-Institut für Physik und Astrophysik - München (*)

(ricevuto il 18 Luglio 1959)

Summary. — In this paper the orbital momentum j, until now considered as an integer discrete parameter in the radial Schrödinger wave equations, is allowed to take complex values. The purpose of such an enlargement is not purely academic but opens new possibilities in discussing the connection between potentials and scattering amplitudes. In particular it is shown that under reasonable assumptions, fulfilled by most field theoretical potentials, the scattering amplitude at some fixed energy determines the potential uniquely, when it exists. Moreover for special classes of potentials $V(x)$, which are analytically continuable into a function $V(z)$, $z=x+iy$, regular and suitable bounded in $x>0$, the scattering amplitude has the remarcable property of being continuable for arbitrary negative and large cosine of the scattering angle and therefore for arbitrary large real and positive transmitted momentum. The range of validity of the dispersion relations is therefore much enlarged.

---

1. - In the following we shall choose dimentionless variables, by putting $x=kr$, where r is the distance from the origin, k the wave number (fixed). We can write then Schrödinger's equation as follows:

$$\psi'' - \frac{\lambda^2-\frac{1}{4}}{x^2}\psi + \psi - U(x)\psi = 0\,. \tag{1.1}$$

Here λ is a generalized complex orbital momentum; when λ assumes positive half-integer values (hereafter referred to as the physical values) we shall write $\lambda = j+\frac{1}{2}$.

Eq. (1.1) is even in λ. For brevity we shall single out all quantities cor-

(*) Now at the Palmer Physical Laboratory - Princeton, N.J.

responding to $U(x)=0$ with the index ^{0}. We shall also suppose

$$(1.2)\qquad \int_0^\infty x|U(x)|\,\mathrm{d}x < Q < \infty\,, \qquad |x^2 U(x)| < M = \text{const.}$$

Although much of the theory here developed could be retained under a weaker condition, some interesting results would not hold or would require too lengthy proofs. We list here a set of solutions of (1.1) with the appropriate boundary conditions:

$$(1.3)\qquad \begin{cases} F(\lambda,\eta,x) & x\to\infty & \sim\cos\left(x-\eta-\dfrac{\pi}{4}\right) \\ G(\lambda,x) & x\to\infty & \sim\cos\left(x-\dfrac{\pi}{4}\right) \\ S(\lambda,x) & x\to\infty & \sim\sin\left(x-\dfrac{\pi}{4}\right) \\ \varphi(\lambda,x) & x\to 0 & \sim x^{\lambda+\frac{1}{2}}\,. \end{cases}$$

Clearly:

$$(1.4)\qquad \begin{cases} F(\lambda,0,x)=G(\lambda,x)\,; \qquad F\left(\lambda,\dfrac{\pi}{2},x\right)=S(\lambda,x)\,, \\ F(\lambda,\eta,x)=\cos\eta\, G(\lambda,x)+\sin\eta\, S(\lambda,x)\,, \\ F(\lambda,\eta,x)=F(-\lambda,\eta,x)\,. \end{cases}$$

A slight generalized form of a theorem of Poincaré states that $F(\lambda,\eta,x)$, for fixed η and x, is an entire function of λ. If $U(x)=0(x^{-2+2C})$, $C>0$ (*) small x, then $\varphi(\lambda,x)$ is analytic in $R(\lambda)>-C$ if $C<1$ and in $R(\lambda)>-1$ if $C>1$. Therefore $\varphi(\lambda,x)$ and $\varphi(-\lambda,x)$ have a common domain of analyticity in the strip: $|R(\lambda)|<C$, (<1).

For those values for which $\varphi(\lambda,x)$ exists we must have a relation of the kind:

$$(1.5)\qquad \varphi(\lambda,x)=2\lambda\,[C(\lambda)S(\lambda,x)-S(\lambda)C(\lambda,x)]\,,$$

$C(\lambda)$, $S(\lambda)$ depend on λ only and are analytic in $R(\lambda)>-C(-1)$ with the possible exception of $\lambda=0$, where there may be a simple pole. We shall not

(*) In the following we shall use, when there is no danger of confusion, the same letter C for quite different constants in different formulas.

reproduce here the details of the proofs of all these statements which are mostly by-products of more careful estimates contained in the Appendix.

We now observe that the Wronskian of any two solutions of (1.1) is a constant. We have for instance from the small x limit:

$$\varphi(\lambda, x)\varphi'(-\lambda, x) - \varphi(-\lambda, x)\varphi'(\lambda, x) = -2\lambda = W[\varphi(\lambda, x), \varphi(-\lambda, x)] . \tag{1.6}$$

Similarly:

$$W[C(\lambda, x), S(\lambda, x)] = 1 . \tag{1.7}$$

Introducing (1.5) into (1.6) and using (1.7) we find:

$$C(\lambda)\, S(-\lambda) - S(\lambda)\, C(-\lambda) = \frac{1}{2\lambda} . \tag{1.8}$$

This identity is of paramount importance in our theory. From it and (1.5) we can write $C(\lambda, x)$ and $S(\lambda, x)$ in terms of $\varphi(\lambda, x)$ and $\varphi(-\lambda, x)$:

$$\begin{cases} C(\lambda, x) = C(\lambda)\varphi(-\lambda, x) + C(-\lambda)\varphi(\lambda, x) , \\ S(x, \lambda) = S(\lambda)\varphi(-\lambda, x) + S(-\lambda)\varphi(\lambda, x) . \end{cases} \tag{1.9}$$

From (1.9) if $R(\lambda) > 0$ we deduce:

$$C(\lambda) = \lim_{x\to 0} x^{\lambda-\frac{1}{2}} C(\lambda, x) \qquad \text{and similarly for } S(\lambda) .$$

Finally we list here the unperturbed functions:

$$\begin{cases} \varphi^0(\lambda, x) = x^{\frac{1}{2}}\Gamma(\lambda+1)2^\lambda J_\lambda(x) , \\ C^0(\lambda, x) = \sqrt{\dfrac{\pi}{2}}\, x^{\frac{1}{2}} \dfrac{1}{2\cos(\pi\lambda/2)} [J_\lambda(x) + J_{-\lambda}(x)] , \\ S^0(\lambda, x) = \sqrt{\dfrac{\pi}{2}}\, x^{\frac{1}{2}} \dfrac{1}{2\sin(\pi\lambda/2)} [J_\lambda(x) - J_{-\lambda}(x)] , \\ C^0(\lambda) \quad = \sqrt{\dfrac{1}{2\pi}}\, 2^\lambda \sin\dfrac{\pi\lambda}{2}\, \Gamma(\lambda) , \\ S^0(\lambda) \quad = \sqrt{\dfrac{1}{2\pi}}\, 2^\lambda \cos\dfrac{\pi\lambda}{2}\, \Gamma(\lambda) . \end{cases} \tag{1.10}$$

2. - If $\lambda = j + \frac{1}{2}$, $\varphi(\lambda, x)$ is the only solution which satisfies the boundary conditions required for the physical interpretation. The phase δ is defined

then from its asymptotic behaviour at large distances:

$$(2.1)\qquad \begin{cases} \varphi(\lambda, x) \simeq 2\lambda T(\lambda)\cos\left[x - \dfrac{\pi\lambda}{2} + \delta(\lambda)\right], \\ T^2(\lambda) \;\;= C^2(\lambda) + S^2(\lambda)\,. \end{cases}$$

We define $\delta(\lambda)$ through (2.1) also when λ is generally complex. Comparing (2.1) with (1.5) we find:

$$(2.2)\qquad K(\lambda) = \frac{S(\lambda)}{C(\lambda)} = -\operatorname{ctg}\left[\frac{\pi\lambda}{2} - \delta(\lambda)\right].$$

The so defined $\delta(\lambda)$ will be hereafter referred to as the « interpolation » of the phase-shifts in the physical points. Not all analytic functions which interpolate the phase shifts can be generated by a potential. Some necessary conditions are:

a) $K(\lambda)$ is regular analytic in $R(\lambda) > 0$ with the exception of simple poles.

b) $K(\lambda)$ is an hermitian function, that is $K(\lambda) = [K(\lambda^*)]^* = K^\dagger(\lambda)$. (Clearly this follows from $C(\lambda)$, $S(\lambda)$ being also hermitian), if λ is real then $K(\lambda)$ is also real.

a) and *b*) are by no means sufficient. To see it let us evaluate the integral (see Appendix):

$$(2.3)\qquad P(\lambda) = \int_0^\infty |\varphi(\lambda, x)|^2 \frac{dx}{x^2} > 0 \qquad R(\lambda) > 0\,.$$

(Notice that $P(\lambda)$ is *not* an analytic function of λ!).
The result can be stated as follows:

$$(2.4)\qquad \begin{cases} P(\lambda) = 2\lambda[C(\lambda)S'(\lambda) - C'(\lambda)S(\lambda)] = 2\lambda T^2(\lambda)\left[\dfrac{\pi}{2} - \delta'\right] > 0\,, & \lambda \text{ real} > 0, \\ P(\lambda) = \dfrac{2\lambda^2}{R(\lambda)I(\lambda)}\cdot|T^2(\lambda)|\sinh[\pi I(\lambda) - 2I(\delta)] > 0\,, & \lambda \text{ complex } K(\lambda) > 0. \end{cases}$$

The following inequalities can be therefore derived:

$$(2.5)\qquad \begin{cases} c) \quad \dfrac{d\delta(\lambda)}{d\lambda} < \dfrac{\pi}{2} & \lambda > 0 \text{ real}, \\ d) \quad I(\lambda)[\pi I(\lambda) - 2I(\delta)] > 0 & I(\lambda) \neq 0\,. \end{cases}$$

Consider now the function $W(L) = K(\lambda)$, $L = \lambda^2$. The half-plane $R(\lambda) > 0$ is mapped into the whole L plane cut along the negative real axis (cut plane). *a*) ... *d*) can be then translated into properties of $W(L)$

a) $W(L)$ is regular analytic in the cut plane with the exception of simple poles.

b) $W(L) = W^{+}(L)$. In particular W is real if L is real > 0.

c) if L real > 0 then $\mathrm{d}W/\mathrm{d}L > 0$ if the derivative is defined.

d) $W(L)$ is a following function in the sense of WIGNER, that is $I(W)$ and $I(L)$ have always the same sign, moreover W is real only if L is real and it can have poles or zeros only on the real axis.

All these properties agree in characterizing $W(L)$ as Wigner's function of L (1), slightly generalized in the sense that continuous distributions of singularities are admitted on the cut while Wigner's function $R(E)$ was meromorphyc. Cond. *a*) ... *d*) restrict considerably the class of interpolations but they are not yet sufficient since some limitation has still to be imposed on the growth of W or K for large λ. We shall derive them in the next part. We only observe that *c*) implies a limitation also on the physical phase shifts. Indeed let us integrate (2.5) *c*) between $\lambda = j + \frac{1}{2}$ and $\lambda = j + \frac{3}{2}$; we find:

$$\delta_{j+1} - \delta_j < \frac{\pi}{2} . \tag{2.6}$$

This is a weak but simple condition on the phase shifts in order that they may be produced by a potential.

3. – If λ is large we expect the perturbation created by $U(x)$ to decrease so that perturbed functions will eventually approach the unperturbed at ∞. Although this statement is in general true there is a number of cases in which it is grossly false so that great care has to be taken in deriving results along this line. For instance it is not generally true that $\delta(\lambda)$ vanishes in the limit of large λ. We shall see that this holds for real λ only, if no condition is imposed on $U(x)$, being valid along any ray in $\pi/2 \geqslant \arg \lambda \geqslant -\pi/2$ for a very special class of potentials. We shall not bother here with a detailed explanation of how these results are derived. The general procedure is the following. We transform (1.1) into an equivalent Green's integral equation of the Volterra type in which the boundary conditions as already included. Suitable upper bounds are then placed upon the sum of the perturbative expansion using a Titchmarsh's lemma. In the Appendix we show the essential points in our procedure. It must be noted that sometimes the lack of simple formulae for

(1) E. WIGNER and J. NEUMANN: *Annals of Math.*, **59**, 418 (1954).

the Bessel functions of general complex order makes it impossible to derive some limit directly. It is then much better to derive the result, usually a condition on the growth of some analytic function in an angle, on the boundary of the domain and then to extend it in the inside using the Phragmen-Lindeloef theorem or its by-products. Using this technique it is possible to derive some more conditions on the interpolation which we state as follows:

$$\exp[-i\pi\lambda](\exp[2i\,\delta(\lambda)]-1)\to 0\,,\qquad -\frac{\pi}{2}<\arg\lambda\leqslant 0\,. \tag{3.1}$$

We can write it into the equivalent form

$$e):\quad \begin{cases} \lim\limits_{\lambda\to\infty}[K(\lambda)-K^0(\lambda)]=0\,, & \arg\lambda\neq 0,\\[2ex] \delta(\lambda)=O\left(\dfrac{1}{\lambda}\right). & \arg\lambda=0, \end{cases}$$

If $U(x)$ us specialized then we have stronger results.

For instance if $U(x)$ admits some bound of the kind: $(A>0,\ C>0,\ B>1)$:

$$U(x)<C\exp[-Ax^B]\,,\qquad x>x_0=\text{const.} \tag{3.2}$$

Then

$$|\ln(K(\lambda)-K^0(\lambda))|<-\left(1-\frac{1}{B}\right)2|\lambda|\ln|\lambda|+0(|\lambda|)\,.$$

This bound is essentially the one derived by CARTER [2] for $\lambda=j+\frac{1}{2}$. If $B=1$, similar bounds of some use can be derived. A second way of restricting $U(x)$, of remarkable interest in connection with dispersion relations, is provided by the theorem:

« Let $U(x)$ admit an analytic continuation $U(z)$, $z=x+iy$, regular in the sector $\arg z<\sigma+\varepsilon<\pi/2$, and such that:

$$\int\limits_0^{\infty e^{i\xi}}|z\|dz\|U(z)|<\infty\,,\qquad |\xi|\leqslant\sigma$$

is uniformly bounded within the sector, and moreover $U(z)=0(z^{-2+2C})$, $C>0$,

---

(2) D. S. CARTER: *Thesis*, Princeton; N. KHURI: *Phys. Rev.* **107** 1148 (1957). We have had no opportunity of examining directly CARTER's work and we know of his result only through KHURI's paper.

$z \to 0$, in the sector, then:

$$|K(\lambda) - K^0(\lambda)| = O\left(\exp\left[-2\sigma I(\lambda)\right]\right). \tag{3.3}$$

Potentials satisfying the conditions of this theorem will be named σ-potentials those which do not 0-potentials. »

The proof of this theorem is based upon the WKB method. Indeed if one takes for granted the first WKB approximation then the above results already follow.

(3.2) and (3.3) are in a way complementary conditions on $U(x)$, indeed the first implies an upper bound on $\delta(\lambda)$ along $\arg\lambda = 0$ and nothing along $\arg\lambda = \pm\pi/2$, the second one works the opposite way. In the derivation of (3.3) it is essential to have $\sigma < \pi/2$, correspondingly it can be proved that a higher value of σ cannot improve the bound. Actually no potential attains:

$$|K(\lambda) - K^0(\lambda)| = O(\exp[-(\pi+\varepsilon)|\lambda|), \quad \varepsilon > 0, \qquad \arg\lambda = \pm\frac{\pi}{2}. \tag{3.4}$$

This can be seen by applying Carleman's theorem to both sides of (3.4) or using Carlson's theorem on the function $F^+(\lambda)$ defined in Appendix IV. One can include $\pi/2$ in our discussion using the weaker statement that if some potentials is a σ-potential, where σ is any angle $< \pi/2$, then:

$$|K(\lambda) - K^0(\lambda)| = O(\exp[-(\pi-\varepsilon)I(\lambda)]), \qquad \varepsilon > 0$$

ε can be taken small at will.

Finally we report some limit:

$$\lim_{\lambda\to\infty}\frac{S(\lambda)}{S^0(\lambda)} = \lim_{\lambda\to\infty}\frac{C(\lambda)}{C^0(\lambda)} = 1 \qquad \arg\lambda \neq 0, \qquad R(\lambda) > 0. \tag{3.5}$$

4. – We shall give here a method of construction of $U(x)$ from the interpolation $K(\lambda)$. We sketch here the essential points only without giving detailed proofs of our statements. We wish to point out also that some more research will be needed in order to cast the theory into a fully satisfactory and rigorous form. We believe, however, that this is more properly a mathematician's task and that what we are going to show here is already enough for the physicist's needs.

The starting point is, as in Gel'fand and Levitan's theory [3-5] an integral

(3) I. M. GEL'FAND and LEVITAN: *Amer. Math. Soc. Trans.*, Sec. 2.1.250 (1955).
(4) R. JOST and W. KOHN: *Math. Phys. Medd.*, **27**, (9) (1953).
(5) L. D. FADDEEV: *Soviet Phys. Dokl. Transl.*, **3**, 747 (1959).

equation of the Volterra type which relates $\varphi(\lambda, x)$ to $\varphi^0(\lambda, x)$ through a kernel which does not depend on λ:

$$\varphi(\lambda, x) = \varphi^0(\lambda, x) + \int_0^x \frac{K(x, y)\,\varphi^0(\lambda, y)}{y^2}\,dy\,. \tag{4.1}$$

One can justify (4.1) from two points of view. The first is that by virtue of the asymptotic expansion (see Appendix):

$$\varphi(\lambda, x) = x^{\lambda+\frac{1}{2}}(1 + D) \qquad \text{where} \qquad D < \frac{Qx^2}{C|\lambda|}\,. \tag{4.2}$$

we have the Mellin representation:

$$\varphi(\lambda, x) = x^{\lambda+\frac{1}{2}} + \int_0^x H(x, y)\,y^{\lambda+\frac{1}{2}}\,\frac{dy}{y^2}\,, \tag{4.3}$$

and of course a similar for $\varphi^0(\lambda, x)$. It can be shown that the latter can be solved in $x^{\lambda+\frac{1}{2}}$ and if this function is fed into (4.3) then (4.1) follows. The second way will be apparent later. The y^{-2} factor is merely inserted for symmetry.

In considering the functions $\varphi(\lambda, x)$ or $\varphi^0(\lambda, x)$ the natural problem arises whether they can be considered a complete and orthogonal set in the interval $0 \dots \infty$. This problem has been considered since long by H. WEIL and several other mathematicians. The reader will find a fully satisfactory overall view on this subject in (6). We shall report here the results found with the help of (6). The $\varphi(\lambda, x)$ are eigenfunctions of the linear operator: ($R(\lambda) \geqslant 0$!)

$$\Omega_x = x^2\left(\frac{d^2}{dx^2} + 1 - U(x)\right), \tag{4.4}$$

subjected to the condition of being of class L^2 in $0 \dots \infty$ if we adopt the norm:

$$(\varphi, \psi) = \int_0^\infty \varphi(x)\,\psi^*(x)\,\frac{dx}{x^2}\,. \tag{4.5}$$

(It must be clear that under this norm all solutions of (1.1) are L^2 in any interval $H \dots \infty$). Using the standard terminology we are in the so called

(6) E. C. TITCHMARSH: *Eigenfunction Expansions*, associated with second order differential equations I, II. (Oxford, 1946).

limit circle case. If no other condition is implied our set is clearly overcomplete and not orthogonal. We must find a subset which satisfies these conditions. This can be accomplished by imposing another boundary condition at ∞, which cannot be square integrability as in quantum mechanics, because it is already satisfied. The new condition is that all eigenfunctions should have the same asymptotic phase μ or, in other words, they must be multiples of $F(\lambda, \mu, x)$. Now there are only two types of choices of λ for this to happen:

A) $R(\lambda)=0$, we refer to it as the continuum

B) $R(\lambda)>0$, and $F(\lambda)=C(\lambda)\cos\mu+S(\lambda)\sin\mu=0$.

In this case

$$F(\lambda, \mu, x) = F(-\lambda)\varphi(\lambda, x)$$

μ is here real and $0<\mu<\pi$; for each value of μ we have a different expansion. It can be checked that all functions in *A*, *B*) satisfy the orthogonality relations:

$$(4.6)\quad \begin{cases} (F(\lambda,\mu,x),\, F(\lambda',\mu,x)) = 4\pi\,\delta(\lambda'-\lambda'^2)\,F(\lambda)\,F(-\lambda)\lambda\,, \\ \qquad\qquad\qquad \text{if } R(\lambda)=0,\quad R(\lambda')=0\,, \\ (F(\lambda,\mu,x),\, F(\lambda_n,\mu,x)) = 0 \qquad F(\lambda_n)=0, \qquad F(\lambda_m)=0, \\ (F(\lambda_n,\mu,x),\, F(\lambda_m,\mu,x)) = -\,\delta_{nm}\,\dfrac{1}{2\lambda F(\lambda_n,\mu-(\pi/2))\,F'(\lambda_n,\mu)}\,. \end{cases}$$

Moreover there are infinitely many zeros of $F(\lambda)$, all real. We have correspondingly the expansion theorem:

$$(4.7)\qquad \delta(x-y)=\frac{1}{2i\pi}\int_P d\lambda\,\frac{\varphi(\lambda,x)}{x}\,\frac{F(\lambda,\mu,y)}{y\,F(\lambda,\mu)}\,.$$

The path P is shown in Fig. 1. This path can be deformed into $R(\lambda)=0$, and to loops which enclose the zeros of $F(\lambda,\mu)$. From $R(\lambda)=0$ we have the *A* terms, from the loops the *B* terms. Some remarks are needed for (4.7); it is clearly a symbolical formula and statements of the kind: the Dirac function in (4.7) is independent of μ, and so must be the second term, are in general meaningless or false since $\delta(x-y)$ is a distribution in a space whose definition depends on μ. However, under some

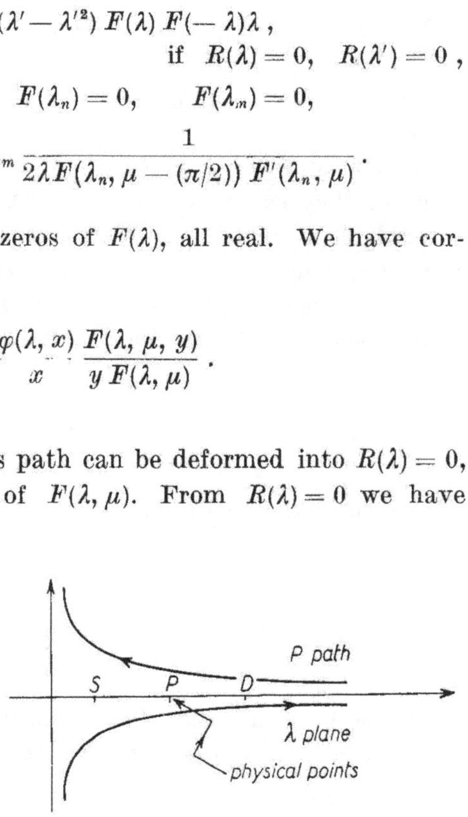

Fig. 1.

conditions, whose exact form will not be stated here, it is true that

$$(4.8)\qquad f(x)=\frac{1}{2\pi i}\int_P \mathrm{d}\lambda\, g(\lambda)\frac{F(\lambda,\mu,x)}{xF(\lambda,\mu)}\,,\qquad \text{if}\quad g(\lambda)=\int_0^\infty \mathrm{d}x\,\frac{1}{x}\,\varphi(\lambda,x)f(x)\,,$$

in particular (4.8) holds if $f(x)$ decreases fast enough. In (4.1) we have always $x>y$; we define $K(x,y)$ also when $y<x$ through the reversed equation:

$$(4.9)\qquad \varphi^0(\lambda,y)=\varphi(\lambda,x)-\int_0^y K(x,y)\,\varphi(\lambda,x)x^{-2}\,\mathrm{d}x\,.$$

Under these assumptions we find the following integral representation from (4.9), (4.1) and (4.7):

$$(4.10)\qquad K(x,y)=\frac{1}{2\pi i}\int_P \mathrm{d}\lambda\left[\frac{F^0(\lambda,\mu,x)}{F^0(\lambda,\mu)}\,\varphi(\lambda,x)-\frac{F(\lambda,\mu,x)}{F(\lambda,\mu)}\,\varphi^0(\lambda,y)\right].$$

From (4.10) one can verify the differential equation:

$$(4.11)\qquad \Omega_x K(x,y)=\Omega_y K(x,y)\,,$$

(4.1), (4.9) and (4.11) are consistent with (1.1) provided

$$(4.12)\qquad K(x,x)=\frac{x}{2}\int_0^x y\,U(y)\,\mathrm{d}y\,,\qquad K(x,y)=O(x)\,,\qquad O(y)\,,\qquad x,y\to 0.$$

Conversely (4.11), (4.12) can be used to define $K(x,y)$ as the solution of a certain differential partial equation with appropriate boundary conditions. The so defined $K(x,y)$, if fed into (4.1) generates a function which satisfies (1.1) with the appropriate boundary condition. This is a second way of justifying (4.1). By rewriting (4.10) as follows ($\mu=0$):

$$(4.13)\qquad K(x,y)=\frac{1}{2\pi i}\int_{-i\infty}^{+i\infty}\mathrm{d}\lambda\left[\frac{C^0(-\lambda)}{C^0(\lambda)}-\frac{C(-\lambda)}{C(\lambda)}\right]\varphi(\lambda,x)\,\varphi^0(\lambda,y)\,+$$
$$+\frac{1}{2\pi i}\int_{\text{loops}}\mathrm{d}\lambda\,\frac{1}{2\lambda}\left[\frac{1}{C(\lambda)\,S(\lambda)}-\frac{1}{C^0(\lambda)\,S^0(\lambda)}\right]\varphi(\lambda,x)\,\varphi^0(\lambda,y)\,,$$

where the « loops » integral encloses the zeros of $C(\lambda)$ and $C^0(\lambda)$ only, avoiding those of $S(\lambda)$, $S^0(\lambda)$, and defining $Q(x, y)$ through the same formula where $\varphi(\lambda, x)$ has been replaced everywhere by $\varphi^0(\lambda, x)$ the integral equation follows:

$$K(x, y) = Q(x, y) + \int_0^\infty K(x, z)\, Q(x, y)\, \frac{dz}{z^2}\,. \tag{4.14}$$

If $K(x, y)$ is regarded as an unknown in (4.14) this equation is of the Fredholm type. It can be shown that the homogeneous counterpart of (4.14) has no non-trivial solutions and we assume therefore that (4.14) can be always solved under very large conditions. From $Q(x, y)$ we can deduce therefore $K(x, y)$ and $K(x, x)$ in particular. (4.12) yields then $U(x)$. Moreover, $Q(x, y)$ involves $\varphi^0(\lambda, x)$, which is an elementary function, and essentially the ratios $C(-\lambda)/C(\lambda)$, $S(-\lambda)/S(\lambda)$, having noticed from (1.8) that

$$\frac{C(-\lambda)}{C(\lambda)} - \frac{S(-\lambda)}{S(\lambda)} = -\frac{1}{2\lambda\, C(\lambda)\, S(\lambda)}\,.$$

The knowledge of these ratios implies that of $U(x)$. We are left with the task of connecting them to the interpolation $K(\lambda)$. To carry it out we observe that in virtue of $C(ia) = C(-ia)^*$ and $S(ia) = S(-ia)^*$ these ratios are simply exponential functions of $\arg C(ia)$ and $\arg S(ia)$. At the same time from (1.8) we know that:

$$K(\lambda) - K(-\lambda) = \frac{1}{2\lambda\, C(\lambda)\, C(-\lambda)}\,.$$

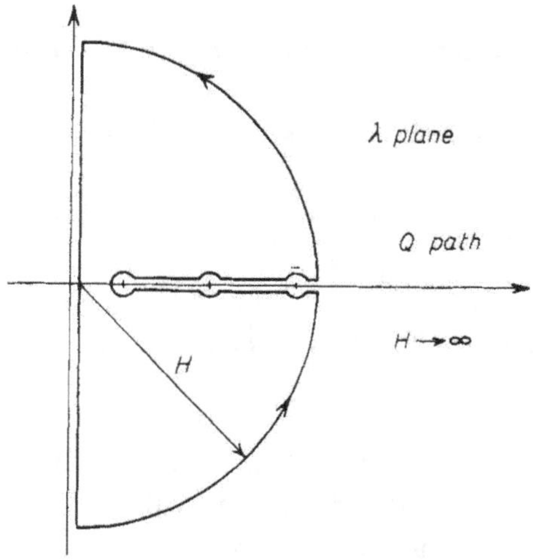

Fig. 2.

We can calculate $\ln|C(\lambda)|$ and similarly $\ln|S(\lambda)|$ from the interpolation. However, owing to (3.5) $\arg C(\lambda)/C^0(\lambda)$ and $\ln|C(\lambda)/C^0(\lambda)|$ if $\lambda = ia$ are in some sense conjugate functions if one forgets about the zeros of $C(\lambda)$ and $C^0(\lambda)$. We can still relate them by taking the real and imaginary part of Cauchy's formula:

$$\frac{1}{2\pi i}\int_Q \frac{1}{\lambda - ia - \varepsilon} \ln\frac{C(\lambda)}{C^0(\lambda)}\, d\lambda = \ln\frac{C(ia)}{C^0(ia)}\,, \qquad \varepsilon > 0 \text{ small.} \tag{4.15}$$

Q means here a path which avoids the branch points of the integrand, as shown in Fig. 2. The result expresses $\arg C/C_0$ as an Hilbert transform of $\ln|C/C_0|$

with an infinite series of additional terms which depend on the location of the branch points, which in turn are perfectly known from $K(\lambda)$. This series converges because the zeros of $C(\lambda)$ and those of $C^0(\lambda)$ tend to be very close when λ is large and to cancel each other. Similarly one proceeds with $S(\lambda)$. We have completed the chain from $K(\lambda)$ up to $U(x)$. The only problem left now is the actual construction of $K(\lambda)$ from the values that it takes at $\lambda = j+\frac{1}{2}$. This problem is still unsolved although we are well on the way to do it. We shall discuss it in the next part.

5. – We shall give here some heuristic arguments on the interpolation problem. Our starting point is Carleman's theorem (5):

$$(5.1)\quad \sum_n \left(\frac{1}{h_n} - \frac{h_n}{H^2}\right)\sin\theta_n - \sum_m \left(\frac{1}{k_m} - \frac{k_m}{H^2}\right)\sin\theta_m = \frac{1}{\pi H}\int_{-\pi/2}^{\pi/2} \ln|L(He^{i\theta})|\cos\theta\, d\theta + \frac{1}{2\pi}\int_0^H \left(\frac{1}{a^2} - \frac{1}{H^2}\right)\ln|L(ia)L(-ia)|\, da + \frac{1}{2}R[L'(0)],$$

$L(\lambda)$ is in this theorem a general analytic function of λ; regular in $R(\lambda) > 0$ with the exception of poles in $k_m \exp[i\theta_m]$ and having zeros at $h_n \exp[i\theta_n]$. Suppose now that we have two interpolations of the same set of phase shifts: $K(\lambda)$ and $H(\lambda)$. The difference $K(\lambda) - H(\lambda)$ must have zeros at least in $\lambda = j+\frac{1}{2}$ and poles at most at the poles of $K(\lambda)$ and $H(\lambda)$. (We cannot exclude that $K(\lambda)$ and $H(\lambda)$ coincide elsewhere and that either a new zero arises or two poles coalesce). The poles are distributed with the density 1. Supposing $L(\lambda) = K(\lambda) - H(\lambda)$ the contributions of zeros and poles in (5.1) are nearly opposite and of the order of $\ln H$ + const. Their algebrical sum will tend to a constant limit. What can be said about the contribution of the boundary? For the sake of simplicity let us restrict ourselves to $H(\lambda) = K^0(\lambda)$ so that the problem now is to find a potential (in the following $V^0(x)$) which produces no scattering at a given energy. We have now some very useful estimate of the decrease of $L(\lambda)$ on the boundary. Quite generally $L(\lambda)$ will decrease in such a fashion as to make $Y(H)$ eventually negative and $J(H)$ decreasing. If, however, condition (3.2) is used we see that $Y(H)$ can be made arbitrarily large and negative if H is chosen sufficiently large, so that unless $K(\lambda) = K^0(\lambda)$ we face a contradiction in (5.1). Similarly, if $V^0(x)$ were a σ-potential $J(H)$ could be made arbitrarily large and negative and (5.1) would be again an absurdity. We see therefore that $V^0(x)$ must be a σ-potential which does not decrease faster than any exponential. The usual field theoretical potentials are therefore excluded. These results can be generalized and it can be shown that there is at most one σ-potential which yields a given set of phase shifts at a given energy. In some sense we find unicity in the inversion problem under rather broad conditions, that is, analiticity requirements in an arbitrarily small angle.

A specific example of $V^0(x)$ can be constructed as follows. We take $H^0(\lambda)=(K^0(\lambda)-1)/(K^0(\lambda)+1)$, $H^0(\lambda)$ is again a Wigner's function of λ^2 and it satisfies therefore all the same requirements of $K^0(\lambda)$. It has alternatively zeros and poles in the physical points. If $H(\lambda)=(K(\lambda)-1)/(K(\lambda)+1)$ then the ansatz $H(\lambda)=H^0(\lambda)\cdot[1+(C/(\lambda^2-\lambda_0^2))]$, where C and λ_0 are constants, still satisfies the correct properties and yields vanishing phases since both $H(\lambda)$ and $H^0(\lambda)$ take the same values at the physical points. λ_0 and C are not entirely arbitrary otherwise $H(\lambda)$ may be somewhere a non-increasing function. Since it is enough here to show the existence of at least two different interpolations we shall be satisfied with one example only which is provided by taking $H^0(\lambda)<0$ and C small as readily checked. The general problem is still unsolved and we hope to tackle it in a forthcoming paper. In analogy to Bargman's potentials which are solvable for all energies but for a single partial wave it is possible to give potentials which are solvable for all values of the angular momentum at a fixed energy. Since these potentials are a rather academic case we shall deal with them in some future work.

6. – In this part we shall establish some results in the field of dispersion relations. As well-known these relations are statements of analiticity of the scattering amplitude as function of the energy and the transmitted momentum. Although here the energy is kept fixed it is still possible to derive for special classes of potentials enough properties as to guarantee for the existance of such relations. We restrict ourselves to σ-potentials ($\sigma\neq 0$) and moreover we impose some additional condition on $U(x)$ as follows:

$$|U(x)|<C\,\frac{\exp[-\alpha x]}{x}\,,\qquad x>x_0\,. \tag{6.1}$$

Correspondingly, the phase shift will decrease like $\exp[-\alpha'\lambda]$ along the real λ, where $\alpha'=\ln(1+(\alpha^2/2))<\alpha$.

Under these conditions we have the following asymptotic estimate:

$$\exp[2i\,\delta(\lambda)]-1\sim O(\exp[i(\pi-2\sigma)\lambda-\alpha'\lambda]),\qquad -\frac{\pi}{2}\leqslant\arg\lambda\leqslant 0\,. \tag{6.2}$$

In these rough extimates we have neglected powers of λ which are not essential in our discussion. (6.2) is very useful if the Legendre expansion of the scattering amplitude:

$$f(z)=f(\cos\theta)=\frac{1}{2i}\sum_{j=0}^{\infty}(2j+1)(\exp[2i\delta_j]-1)\,P_j(z)\,,\qquad z=\cos\theta, \tag{6.3}$$

is transformed into the integral:

$$f(z)=\frac{1}{2}\int\lambda\,\mathrm{d}\lambda\,(\exp[2i\,\delta(\lambda)]-1)\,\frac{1}{\cos\pi\lambda}\,P_{\lambda-\frac{1}{2}}(-z)\,. \tag{6.4}$$

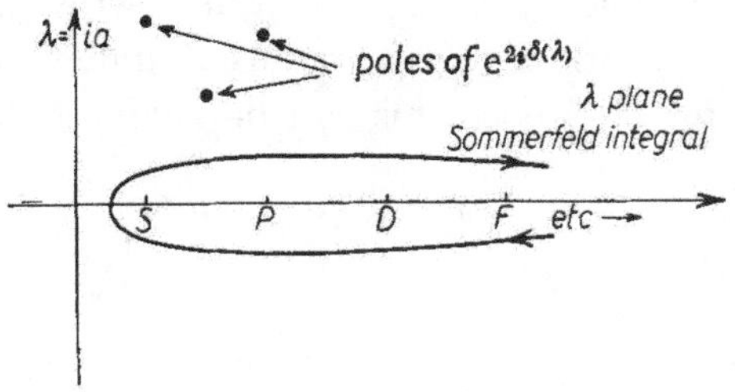

Fig. 3.

This artifice is due to WATSON and it was used by SOMMERFELD (7) in some wave propagation problems. The integration path is shown in Fig. 3. It encloses the zeros of $\cos\pi\lambda$ and avoids the poles of the integrand in the upper quadrant. The important fact about it is that it may converge outside the customary Legendre ellipse. To prove it, one needs an asymptotic expansion of Legendre functions of large order:

$$P_{\lambda-\frac{1}{2}}(z)\sim\frac{1}{\sqrt{2\pi\sin\theta\cdot\lambda}}\exp[\pm i\lambda(\theta-\pi)]\,. \tag{6.5}$$

One must choose the sign which yields the larger result. From (6.5) and (6.2) one sees that it is possible to deform the path in the lower quadrant into $\lambda=ia$, $a<0$, provided one has $(\theta=\theta_0+i\theta_1)$

$$\theta_1<\alpha' \qquad \text{and} \qquad \theta_0>\pi-2\sigma\,. \tag{6.6}$$

If the first condition is dropped the lower part of the deformed integral still makes sense but the original expansion diverges. We have to deal now with the upper part where we did not apply immediately the same artifice because of the poles in the integrand. From (6.2) we see that:

$$\exp[2i\,\delta(\lambda)]\to 1 \qquad \text{if} \qquad 0<-\arg\lambda<\xi_0\,,$$

where

$$\operatorname{tg}\xi_0=\frac{\alpha'}{\pi-2\sigma}\,.$$

Correspondingly, the same limit holds in $0<\arg\lambda<\xi_0$.

It follows that within this sector there is at most a finite number of poles

(7) A. SOMMERFELD: *Partielle Differential-Gleichungen der Physik* (Leipzig, 1947), p. 285.

of $\exp[2i\delta(\lambda)]$. The upper part of the path can be deformed along $\arg\lambda = \xi_0$, and the poles in the sector yield a finite sum of residues which are analytic functions (Legendre functions) of z, in the z plane cut along z real > 1.

If we now try to shift the upper path along $\lambda = ia$, $a > 0$ we find that, apart from the contribution of poles, (6.4) still makes sense provided $\theta_0 > 0$ which is included in (6.6). The infinitely many poles yield a converging series if the values of the integral along an arc of arbitrarily large radius A indented in $\lambda = iA$ and $\lambda = A\exp[i\xi_0]$, vanishes for $A \to \infty$. This can be achieved if the condition

$$\theta_1 < \frac{\alpha'}{\pi - 2\sigma}[4(\pi - \sigma) - \theta_0], \tag{6.7}$$

is satisfied. Now particularly interesting are the values $\theta = \pi + i\theta_1$ which correspond to z real < -1 or in the dispersion relation language to $\tau > 2k$, τ being the transmitted momentum and k the wave number, $(x = kr)$. If σ can be taken arbitrarily near to $\pi/2$ then θ_1 may be arbitrarily large and correspondingly there is analyticity in z along the whole negative axis and, for arbitrary large real values of τ.

This result holds in particular for all those potentials which can be included into the formula

$$x\,U(x) = \int_{\alpha}^{\infty} \exp[-px]\,f(p)\,dp\,, \tag{6.8}$$

Now from the previous literature (see [2,8]) one knows that $f(k, \tau)$ is, for τ fixed real, and an analytic function of k in the whole upper plane $I(k) > 0$ (*) approaching $f_B(\tau)$ (Born approximation) for large k, if $\tau < 2\alpha$. If the latter condition is not satisfied, a domain D arises where analyticity cannot be proved with the usual methods. This domain $D = D(\tau)$ however can be enclosed in a sufficiently large semicircle with center in $k = 0$. Moreover, $f(k, \tau)$ still approaches $f_B(\tau)$ when k is large. We know that $f(k, \tau)$, if k is real, is analytic in a region which in terms of $z = 1 - (\tau^2/2k^2)$ is the sum of the interior of the ellipse of convergence and of the domain defined in (6.7) and the second of (6.6). In particular if k is kept fixed real and $\neq 0$, $f(k, \tau)$ is an analitic function of τ for real arbitrary positive values of τ and in a suitable neighbourhood of them. The same result can be proved with a limiting procedure when $k = 0$. It is remarkable then that analyti-

(8) S. Gasiorowicz and H. P. Noyes: *Nuovo Cimento*, **10**, 78 (1958).

(*) Bound states will not be considered here. However these can be easily included into the theory.

city follows in the whole $I(k) > 0$ plane. Indeed one can take a semicircle enclosing $D(\tau)$, where $\tau < T = \text{constant}$. Applying Cauchy's theorem to this semicircle we get

$$(6.9)\qquad f(k, T) = \frac{1}{2\pi i}\int_{\bigcirc} \frac{f(k, T)}{h-k}\, dh\,.$$

If $\tau > 2\alpha$ this formula still defines a function, analytic in k and τ, the latter analyticity follows from the fact that all values on the boundary are analytic functions of τ. (6.9) defines a function also within D which must be the analytic continuation of the original function because it coincides with it outside D where is was defined. Take namely k outside $D(\tau)$ but inside the semicircle and split the semicircle into two closed loops, the first enclosing D but not k, the latter k but not $D(\tau)$. The contribution of the first is an analytic function of τ which vanishes if $\tau < 2\alpha$ and therefore always the last just yields the original function. The result follows. The above arguments are standard ideas from the theory of several complex variables. For special potentials one gets still stronger results. Indeed suppose that $|V(z)| < H/z^2$ within and along the boundary of $|\arg z| \leqslant \pi/2$. Let λ^0 be a pole of $\exp[2i\,\delta(\lambda)]$. From the discussion above we know that $R(\lambda^0) > 0$, $I(\lambda^0) > 0$. Moreover $\varphi(\lambda^0, z) \sim C\exp[iz]$ if $z \to \infty$. Also

$$(6.10)\qquad \varphi''(\lambda^0, z) + \varphi(\lambda^0, z) - \frac{\lambda_0^2 - \frac{1}{4}}{z^2}\varphi(\lambda^0, z) - U(z)\varphi(\lambda^0, z) = 0\,.$$

Take now $z = iy$; y real. Then $\varphi \sim C\exp[-y]$ and

$$(6.11)\qquad \ddot{\varphi} - \varphi - \frac{\lambda^{02} - \frac{1}{4}}{y^2}\varphi - U(iy)\varphi = 0\,,$$

the dots now refer to y derivatives. The conjugate reads

$$(6.12)\qquad \ddot{\varphi}^* - \varphi^* - \frac{\lambda^{0*2} - \frac{1}{4}}{y^2}\varphi^* - U(-iy)\varphi^* = 0\,.$$

We multiply now (6.11) by φ^* and subtract (6.12) multiplied by φ. The resulting equation is then integrated between $0 \dots \infty$. We obtain

$$(6.13)\qquad (\lambda^{02} - \lambda^{0*2})\int_0^\infty \frac{|\varphi|^2}{y^2}\, dy = -2i\int_0^\infty I(U(iy))\,|\varphi|^2\, dy\,.$$

If $I(U(z))$ is replaced by M/y^2 we obtain

$$(6.14)\qquad R(\lambda^0)\, I(\lambda^0) < \frac{M}{2}\,.$$

This is a restriction on the position on poles: this result enables us to say that only a finite number of poles have $R(\lambda^0) > m+\frac{1}{2}$ where $m+\frac{1}{2}$ is some positive constant. In the Sommerfeld integral one can use $R(\lambda)= m+\frac{1}{2}$ as path, the contribution of the extra poles can be evaluated and yields a finite sum which diverges at most like some power of $-\cos\theta$ or τ when those quantities are large. The contribution of the $R(\lambda)= m+\frac{1}{2}$ path can be best evaluated by using (see App. V) asymptotic expansions of Legendre functions:

$$(6.15)\quad \begin{cases} f(\cos\theta) = \dfrac{1}{2\sqrt{\pi}}\displaystyle\int_{-\infty}^{\infty}\dfrac{\Gamma(\lambda)}{\Gamma(\lambda+\frac{1}{2})\cos\pi\lambda}\,\zeta\,d\zeta\,\{\exp[2i\,\delta(\lambda)]-1\}\cdot \\ \qquad\qquad\cdot\exp[i\zeta\theta_1+\zeta(\theta_0-\pi)]\cdot(-2\cos\theta)^m\,, \\ \lambda = m+i\zeta+\frac{1}{2}+\text{contr. extra poles.} \end{cases}$$

This contribution is therefore seen to be growing at most like some power of $-\cos\theta$ or τ when these variables are large in modulus (the cut $\cos\theta$ real > 1 being excluded). The Mandelstam representation can then be derived.

* * *

The author wishes to thank Dr. SYMANZIK for many suggestions and constant advice. This work was partially supported by the American Air Force.

APPENDIX I

In order to calculate $P(\lambda)$ we notice that $\varphi(\lambda, x)^* = \varphi(\lambda^*, x)$ and that :

$$(A.1)\qquad \varphi(\lambda, x)\,\varphi''(\lambda^*, x) - \varphi''(\lambda, x)\,\varphi^*(\lambda, x) = \frac{1}{x^2}(\lambda^{*2}-\lambda^2)\,|\varphi(\lambda, x)|^2\,,$$

or if λ is real:

$$(A.2)\qquad \varphi(\lambda, x)\frac{\partial}{\partial\lambda}\varphi''(\lambda. x) - \varphi''(\lambda, x)\frac{\partial}{\partial\lambda}\varphi(\lambda, x) = \frac{2\lambda}{x^2}\varphi(\lambda, x)^2\,.$$

Integration of these indentities between $0 \ldots \infty$ yields the desired formulas. The asymptotic behaviour of $\varphi(\lambda, x)$ for large x can be obtained through (15) and (1.3).

APPENDIX II

Most of the bounds used in this paper can be derived from suitable integral equations of the Volterra type and from the following lemma (5):

Lemma: Let $f(x) \geqslant 0$, $g(x) \geqslant 0$ and let $f(x)$ be continuous $g(x)$ integrable in $0 \leqslant x \leqslant X$. Let $f(x) < C + \int_0^x f(t)\, g(t)\, dt$, $0 \leqslant x \leqslant X$. Then $f(x) < C \exp\left(\int_0^x g(t) dt\right)$ $0 \leqslant x \leqslant X$.

This lemma will be used also in the interval $X \ldots \infty$ with obvious modifications. We list here the integral equations which we have used and the functions which in our derivation correspond to $f(x)$ and $g(x)$.

$$\text{(I)} \qquad \varphi(\lambda, x) = x^{\lambda+1} + \frac{1}{2\lambda}\int_0^x (U(y) - 1)\, \varphi(\lambda, y) \left[\frac{x^{\lambda+1}}{y^\lambda} - \frac{y^{\lambda+1}}{x^\lambda}\right] dy\,.$$

We have if $R(\lambda) > 0$: $(y < x)$

$$\left|\frac{x^{\lambda+1}}{y^\lambda} - \frac{y^{\lambda+1}}{x^\lambda}\right| < 2\left|\frac{y}{x}\right|^\lambda |x|\,;$$

we take $f(x) = |\varphi(\lambda, x) x^{-\lambda-1}|$ and $g(x) = |U(x) - 1|\, |x| (1/|\lambda|)$: the result is essentially (4.2). A similar bound holds for $\varphi'(\lambda, x)$

$$\text{(II)} \qquad C(\lambda, x) = \cos\left(x - \frac{\pi}{4}\right) + \int_x^\infty \left[\frac{\lambda^2 - \frac{1}{4}}{y^2} + U(y)\right] C(\lambda, y) \sin(y - x)\, dy\,;$$

here we replace the sines and cosines with 1 and take $C = 1$, $f(x) = |C(\lambda, x)|$, $g(x) = |((\lambda^2 - \frac{1}{4})/x^2) + U(x)|$. We obtain a result of the form:

$$|C(\lambda, x)| < C \exp\left[\frac{\lambda^2}{x}\right], \qquad \lambda \text{ large.}$$

Similarly we find a corresponding bound on $C'(\lambda, x)$ and on $S(\lambda, x)$ and $S'(\lambda, x)$. Introducing these bounds in the identity:

$$2\lambda\, C(\lambda) = C(\lambda, x)\, \varphi(\lambda, x) - C'(\lambda, x)\, \varphi(\lambda, x)$$

and taking $x = \lambda$ we find

$$\text{(A.3)} \qquad |C(\lambda)| < |\lambda^\lambda C^\lambda|\,,$$

where C is some constant. This bound, although very rough, can be sharpened with the help of Phragmen-Lindeloef's theorem [9]:

$$\text{(III)}\qquad C(\lambda, x) = C^0(\lambda, x) + \int_x^\infty U(y)\, C(\lambda, y)[C^0(\lambda, x)\, S^0(\lambda, y) - C^0(\lambda, y)\, S^0(\lambda, x)]\, dy\,.$$

We consider this equation for $\lambda = ia$ (a real) only, since otherwise no reliable bound on the unperturbed solutions is available. We know from the Poisson integral representation of Bessel functions that:

$$C^0(\lambda, x)\,,\quad S^0(\lambda, x) < E\sqrt{\pi x}\,,\qquad \lambda = ia,$$

where $E > 1$ is a suitable constant.

The bound holds uniformly in x, λ since E does not depend on them. We take $C^0(\lambda, x) = f(x)E\sqrt{\pi x}$, $g(x) = 2xU(x)\,\pi E^2$ and since $\int_0^\infty x\,|U(x)| = Q < \infty$, we have:

$$\text{(A.4)}\qquad C(\lambda, x) < E\sqrt{\pi x}\,\exp[2\pi E^2 Q] = E'\sqrt{\pi}\, x\,,$$

which is of the same type. The same results follows for $S(\lambda, x)$,

$$\text{(IV)}\quad \varphi(\lambda, x) = \varphi^0(\lambda, x) + \frac{1}{2\lambda}\int_0^x U(y)\,\varphi(\lambda, y)[\varphi^0(\lambda, x)\,\varphi^0(-\lambda, y) - \varphi^0(\lambda, -x)\,\varphi^0(\lambda, y)]\, dy.$$

As in (III) we suppose $\lambda = ia$. We have the preliminary inequality:

$$\varphi^0(\lambda, x) < F\sqrt{\lambda\pi x}\qquad F \text{ const.}$$

we take then $\varphi(\lambda, x) = f(x)F\sqrt{\lambda\pi}x$ and $g(x) = 2xU(x)\pi F^2$ and as before we derive a bound of the kind:

$$\text{(A.5)}\qquad \varphi(\lambda, x) < F'\sqrt{\lambda\pi x}\qquad F'' \text{ is a constant.}$$

From (A.5) and the formula:

$$\text{(A.6)}\qquad C^0(\lambda) + \frac{1}{2\lambda}\int_0^\infty U(x)\, C^0(\lambda, x)\,\varphi(\lambda, x)\, dx = C(\lambda)\,.$$

we can prove (3.5) along $R(\lambda) = 0$, $a \to \infty$. Indeed if we split the interval of integration in $0 \ldots \sqrt{a}$ and $\sqrt{a} \ldots \infty$ we find the following bound using (A.5):

$$\int_{\sqrt{a}}^\infty U(x)\, C^0(\lambda, x)\,\varphi(\lambda, x)\, dx < \pi E F' \sqrt{|\lambda|}\int_{\sqrt{|\lambda|}}^\infty x\,|U(x)|\, dx\,.$$

(9) R. P. Boas: *Entire Functions* (New York, 1955).

This part is $o(\sqrt{\lambda})$ and negligible $[C^0(\lambda)$ is $o(\sqrt{\lambda^{-1}})]$. In $0 \dots \sqrt{a}$ we expand $C^0(\lambda, x)$ following (1.9) and use (4.2). If $0 < x < \sqrt{a}$ from (4.2) we deduce:

$$\varphi(\lambda, x) = x^{\lambda+1}(1 + O(1)) \qquad \lambda \text{ large}.$$

Two terms arise from (A.6):

$$C^0(\lambda)\int_0^{\sqrt{a}} x U(x)(1 + O(1)),$$

and

$$C^0(-\lambda)\int_0^{\sqrt{\lambda}} x U(x) x^{2\lambda}(1 + O(1))\,\mathrm{d}x .$$

Both are negligible. The limit follows. We have shown here the simplest obtainable bounds. A more refined estimate can be derived using the WKB method but we deliver it to the next Appendix.

Appendix III

The WKB method gives fairly simple estimates of the wave function when λ is large but it is difficult to estimate the error when λ and x are simultaneously complex. Fortunately it is enough for our purposes to take $\lambda = ia$ and x complex. In this case, provided a is large enough, there are no turning points if $U(x)$ is a σ-potential. Indeed it is clear that in this case $U(z)$ admits the Cauchy's rapresentation:

$$zU(z) = \frac{1}{2\pi i}\int_{\infty \exp[-i\sigma']}^{\infty \exp[i\sigma']} \frac{z' U(z')}{z - z'}\,\mathrm{d}z' \qquad \sigma' = \sigma + \varepsilon.$$

The integration path is stretched along the two half-lines $+\infty \exp[-i\sigma']\dots 0$ and $0 \dots \infty \exp[i\sigma']$. Now within $|\arg z| \leqslant \sigma$ we have $|z - z'| \geqslant z \sin\varepsilon$ and clearly:

$$|zU(z)| < \frac{M}{|z|\sin\varepsilon} \qquad \text{or} \qquad U(z) < \frac{M}{z^2 \sin\varepsilon},$$

were $M = (1/\pi)\int_0^{\infty \exp[i\sigma]} |z|| \mathrm{d}zU||(z)|$. Similarly

$$U^{(n)}|z| < \frac{M_n}{z^{2+n}}; \qquad M_n \text{ const.}$$

In order to avoid turning points the function $a^2+z^2+z^2U(z)$ must have no zeros in the sector B $|\arg z|<\sigma$. Suppose now that there is Z such that $a^2+Z^2U(Z)=-Z^2$. We have $Z^2U(Z)<(M/\sin\varepsilon)=M'$. Now $Z^2=-a^2+pe^{i\xi}$, where $p<M'$ and if a is taken large enough

$$z\sim\pm\left(ia+\frac{p\exp[i\xi]}{2ia}\right)\quad\text{and}\quad \arg z\sim\pm\frac{\pi}{2}.$$

Hence Z will eventually fall out of the sector Σ which will be free of zeros. If we define $s^2(a,z)=a^2+z^2+z^2U(z)$ one can prove with similar arguments that there exists, for sufficiently large a, a constant C such that:

$$|s(z)|>Ca \qquad (C \text{ ind. of } a \text{ and } z). \tag{A.7}$$

It follows that if we defines $\xi(z)$ through the equations:

$$\frac{d\xi}{dz}=\frac{S(a,z)}{z},\qquad \lim_{z\to a}(\xi(z)-a\ln z)=0 \qquad z \text{ in } \Sigma,$$

$\xi(z)$ is a single valued regular analytic function of z in Σ. In particular

$$\xi^0(z)=a\ln z-a\ln(a+\sqrt{a^2+z^2})+\sqrt{a^2+z^2}-a+a\ln 2a\,.$$

From (A.7) one can prove then

$$|\xi(z)-\xi^0(z)|<\frac{H}{a}, \tag{A.8}$$

where H is independent of z and a. All these inequalities require explicitly $|\sigma|<\pi/z$ equality beeing excluded. Let us define also $w(\xi)=\sqrt{s(z)}\cdot\psi(z)$, where $\psi(z)$ is any solution of (1.1). (1.1) transforms then into:

$$\frac{d^2w}{d\xi^2}+w=-J(z)\,w, \tag{A.9}$$

where

$$J(z)=\frac{1}{4s^2}\left\{\frac{5}{4}\left[z\frac{d}{dz}s^2(z)\right]^2+s(z)^2\left[z^2\frac{d^2}{dz^2}s^2(z)+zs^2(z)\right]\right\}.$$

Putting then $X(z)=\sqrt{s(z)/az}\,\varphi(\lambda,z)$, we readily derive the integral equation:

$$X(z)=\exp[i\xi(z)]-\int_0^z\sin[\xi(z)-\xi(t)]\,s(t)\,X(t)\,s(t)\frac{dt}{t}. \tag{A.10}$$

Take now $X_1(z)=\exp[-i\xi]X(z)$. $X_1(z)$ satisfies:

$$X_1(z)=1-\int_R\frac{1-\exp[2i(\xi(z)-\xi(t))]}{2i}J(t)\,X_1(t)\,s(t)\frac{dt}{t}. \tag{A.11}$$

We choose a path R of integration such that along it $I(\xi^0(z)) < I(\xi^0(t))$. It can be proved from the formula for $\xi^0(z)$ that $I(\xi^0(t))$ increases along $\arg t = \text{const}$ and decreases along $I(\sqrt{a^2+t^2}) = \text{const}$. Suppose $I(\xi^0(z)) = a\sigma$. A suitable path is then R_1: $\arg t = \sigma$ and the line R_2:

$$I(\sqrt{t^2+a^2}) = I(\sqrt{z^2+a^2}) .$$

With this position

$$|\exp[2i(\xi^0(z) - \xi^0(t))]| < 1 .$$

Consequently from (A.8) we have when a is large:

$$\left|\frac{1-\exp[2i(\xi(z)-\xi(t))]}{2i}\right| < K = \text{const} .$$

R has a corner T where R_1 and R_2 meet. We have $T = O(a)$. We need now some simple bound on $J(t)$. If $0 < t < T$ one can use (A.7), if $t > T$ then $s(t) > C't$ is also needed. Omitting here the details we arrive at the conclusion that:

$$\int_R |J(t)\,s(t)|\frac{dt}{t} = O(a^{-1}) .$$

The above estimate holds uniformly for all z such that $I(\xi(z)) = a$. It follows from the lemma that on this line Q one has $X_1(z) = O(1)$ uniformly. Also on Q we have:

$$|\varphi(ia, z)| < C\sqrt{z_>}\exp[-a\sigma] ,$$

where C is some constant, $z_>$ the largest between z, a.

Finally from (A. 6) we derive:

$$C(\lambda)\,S^0(\lambda) - C^0(\lambda)\,S(\lambda) = -\frac{1}{4\lambda^2}\int_0^\infty U(z)\,\varphi^0(\lambda, z)\,\varphi(\lambda, z)\,dz . \tag{A.12}$$

In (A.12) the integration can be carried out along Q and the above found inequality introduced. Dividing both sides by $C(\lambda)\,C^0(\lambda)$ the result follows. More details on this technique can be found in ([6]).

Appendix IV

We suppose now that λ is large positive. Our starting point is eq. (IV) (Appendix III). We take firstly $x \leqslant \lambda$. From the general theory of Bessel functions it is easy to prove that $\varphi^0(\lambda, x) > 0$ and that for sufficiently large

λ, $\varphi^0(\lambda, x)x^{-M}$, $M > \frac{1}{2}$ is an increasing function. From the identity:

$$G(x, y, \lambda) = \varphi^0(\lambda, x)\varphi^0(-\lambda, y) - \varphi^0(\lambda, y)\varphi^0(-\lambda, x) =$$

$$= 2\lambda\varphi^0(\gamma, x)\varphi^0(\lambda, y)\int_y^x \frac{dz}{\varphi^0(\lambda, z)^2}, \qquad \infty \geqslant y,$$

replacing $\varphi^0(\lambda, z)$ with the smaller quantity $\varphi^0(\lambda, y)(z/y)^M$ we obtain the bound:

$$|G(\lambda, x, y)| < \frac{y}{M - \frac{1}{2}} \frac{\varphi^0(\lambda, x)}{\varphi^0(\lambda, y)}.$$

Putting then $\varphi(\lambda, x) = \zeta(\gamma, x)\varphi^0(\lambda, x)$ we find the inequality:

$$|\zeta(\lambda, x)| < 1 + \int_0^x |y\, U(y)|\, \zeta(\lambda, y) \frac{1}{M - \frac{1}{2}} dy.$$

From the lemma it follows then

$$\varphi(\lambda, x) \sim \varphi^0(\lambda, x)\, O(1)$$

uniformly in $0 < x < \lambda$. Take then the identity (A.6):

$$C(\lambda) = C^0(\lambda) + \frac{1}{2\lambda}\left[\int_0^\lambda U(x)\, C^0(\lambda, x)\varphi^0(\lambda, x) O(1)\, dx + \right.$$

$$\left. + \int_\lambda^\infty U(x)\, C^0(\lambda, x)[C(\lambda) S(\lambda, x) - S(\lambda)\, C(\lambda, x)]\, dx\right].$$

In this formula $C^0(\lambda, x)\varphi^0(\lambda, x) \sim C^0(\lambda)/(\sqrt{1 - (z/\lambda)^2})$ as it can be deduced from the power series or from the asymptotic expansions of Bessel functions. It can be moreover shown from the Schlaefli representation that:

$$\text{(A.14)} \qquad \left|\frac{C^0(\lambda, x)}{\sqrt{x}}\right|, \left|\frac{S^0(\lambda, x)}{\sqrt{x}}\right| < C = \text{constant}, \; x \geqslant \lambda.$$

From eq. (III) (Appendix II) it follows that the same kind of inequality holds for $C(\lambda, x)$ and $S(\lambda, x)$. We are now in condition to write:

$$C(\lambda)[1 + O(1)] + S(\lambda)\, O(1) = C^0(\lambda)[1 + O(1)]$$

and similarly

$$S(\lambda)[1 + O(1)] + C(\lambda)\, O(1) = S^0(\lambda)[1 + O(1)].$$

It follows:

$$C(\lambda) = C^0(\lambda)[1 + O(1)] + S^0(\lambda)\, O(1) \quad \text{etc.},$$

and

$$\lim_{\lambda\to\infty} \frac{C(\lambda)^2 + S(\lambda)^2}{C^0(\lambda)^2 + S^0(\lambda)^2} = 1\,.$$

Probably our estimate here can be ameliorated at the expense of additional complication. It would be also desirable to have a simpler proof. We turn our attention now to (A.12). Putting $C(\lambda)^2+S(\lambda)^2 = T(\lambda)^2$ we can write it as follows:

$$\sin \delta(\lambda) = \frac{1}{-4\lambda^2}\int_0^\infty U(x)\,\frac{\varphi^0(\lambda, x)}{T^0(\lambda)}\,\frac{\varphi(\lambda, x)}{T(\lambda)}\,\mathrm{d}x\,.$$

If $x > \lambda$ then from the corresponding bounds (A.14) we deduce:

$$|\varphi^0(\lambda, x)| < 2C\lambda T^0(\lambda)\cdot\sqrt{x}\,; \qquad |\varphi(\lambda, x)| < 2C'\,\lambda T(\lambda)\cdot\sqrt{x}\,,$$

C, C' being, as usual, two suitable constants. It follows that the contribution to $\sin\delta$ coming from the interval $\lambda \ldots \infty$ is of the order of

$$CC'\int_\lambda^\infty x\,|U(x)|\,\mathrm{d}x = O(1)\,.$$

Actually a more elaborate analysis shows that this bound is pessimistic and one has actually $O(\lambda^{-1})$. Between $0 \ldots \lambda$ one has

$$|\varphi(\lambda, x)\,T^{-1}(x)| < H\,|\varphi^0(\lambda, x)\,T^{0-1}(\lambda)|\,.$$

This part of the integral is certainly smaller than:

$$\frac{1}{4\lambda^2}\,H\int_0^\infty |U(x)|\,\frac{\varphi^0(\lambda, x)}{T^0(\lambda)^2}\,\mathrm{d}x\,.$$

Since $|U(x)| < Mx^{-2}$ from (2.4) we get the upper bound:

$$\frac{H\pi M}{4\lambda}\,.$$

If $U(x)$ is of the (3.2) type then the above technique yields easily the stated result. It is not clear to us if the above analysis is included in Carter's result in the sense that his bound may hold regardless of λ being not an half-integer.

A few words on the use of the Phragmen-Lindeloef theorem in extending these results to complex λ are also needed.

Take

$$F^+(\lambda) = \frac{C(\lambda) + iS(\lambda)}{C^0(\lambda) + iS^0(\lambda)} = \frac{T(\lambda)}{T^0(\lambda)}\exp[-i\delta(\lambda)]\,.$$

This function is easily seen to be regular in $R(\lambda) > 0$. We have also $F^+(\lambda) \to 1$ along $\arg\lambda = 0$ and $\arg\lambda = -(\pi/2)$. $F^+(\lambda)$ is also bounded by an exponential function in the intermediate angles and is therefore bounded by a constant. From Montel's theorem the limit holds in all

$$0 > \arg\lambda > -\frac{\pi}{2},$$

$F^+(\lambda)$ has also no zeros in this sector since if $F^-(\lambda)$ is the adjoint function, then:

$$\exp[-2i\delta] = \frac{F^+(\lambda)}{F^-(\lambda)}$$

and also this function would vanish since (F^+, F^-) cannot vanish simultaneously, for also $C(\lambda)$ and $S(\lambda)$ would vanish and also $\varphi(\lambda, x)$ because of (1.5). But

$$|\exp[-2i\,\delta(\lambda)]| = \exp[2I[\delta(\lambda)]] > \exp[\pi I(\lambda)] > 0$$

and the above claim is clearly impossible, Along $\arg\lambda = \pi/2$ one has $|F^+(\lambda)| = O(\exp[\pi(\lambda) - 2\sigma(\lambda)])$. If one applies Carleman's theorem then unless $\sigma = \pi/2$, one has always zeros of F^+ in this sector. The properties of $F^-(\lambda)$ are of course the same of $F^+(\lambda)$ provided one changes $\arg\lambda$ into $-\arg\lambda$. All extensions to complex λ can be carried out with the same technique.

Appendix V

We shall sketch here briefly the low energy limit. The potential will be supposed to decrease exponentially. In this case a simple generalization of known arguments yields the limit:

$$\text{(A.15)} \qquad \sin_{k\to 0}\delta(\lambda) \sim k^{2\lambda}\eta(\lambda),$$

where $\eta(\lambda)$ is finite.

This limit follows from (A.12) and from the similar indentity:

$$\text{(A.16)} \qquad C(\lambda)\,S^0(-\lambda) - C^0(-\lambda)\,S(\lambda) = \frac{1}{2\lambda} + \frac{1}{4\lambda^2}\int_0^\infty \underset{0}{U}(r)\,r^{-\lambda+\frac{1}{2}}\underset{0}{\varphi}(\lambda, r)\,\mathrm{d}r = \frac{\alpha(\lambda)}{2\lambda}.$$

$\underset{0}{U}(r)$ is here the usual potential and r the distance $(x = kr)$. Also $k^{-2}\underset{0}{U}(r) = U(x)$. $\underset{0}{\varphi}(\lambda, r)$ is that solution of:

$$\underset{0}{\varphi''} - \frac{\lambda^2 - \frac{1}{4}}{r^2}\underset{0}{\varphi} - \underset{0}{U}(r)\underset{0}{\varphi} = 0,$$

which behaves like $r^{\lambda+\frac{1}{2}}$ for small r.

(A.12) can be correspondingly written as (k small):

$$C(\lambda)\,S^0(\lambda) - S(\lambda)\,C^0(\lambda) = -\frac{1}{4\lambda^2}\int_0^\infty \underset{0}{U}(r)\,r^{\lambda+\frac{1}{2}}\varphi(\lambda, r)\,\mathrm{d}r\cdot k^{2\lambda} = \beta(\lambda)\,k^{2\lambda}\,.$$

It follows that if $R(\lambda) > 0$:

$$\tau^{0^2}(\lambda)\eta(\lambda) = \frac{\beta(\lambda)}{\alpha(\lambda)}\,, \qquad \frac{\tau(\lambda)}{\tau^0(\lambda)} = \alpha(\lambda)\,.$$

Therefore $\eta(\lambda)$ is a meromorphyc function of λ in $R(\lambda) > 0$. If λ is large $\alpha(x) \to 1$ and $\beta(\lambda)$ becomes in general large. A good approximation to $\varphi(\lambda, r)$ is then simply $r^{\lambda+\frac{1}{2}}$. Also one can show that if $\underset{0}{U}(r)$ is a σ potential then $T^0(\lambda)^2\eta(\lambda) = 0(\exp[-2\sigma|\lambda|])$ for λ large imaginary. Take now Sommerfeld's integral (6.4). In the low energy limit

$$-z = \frac{\tau^2}{2k^2} - 1 \sim \frac{\tau^2}{2k^2}\,,$$

if τ is kept constant. But if $-z$ is large positive then:

$$P_{\lambda-\frac{1}{2}}(-z) \sim \frac{\Gamma(\lambda)\,T^{2\lambda-1}k^{1-2\lambda}}{\sqrt{\pi}\,\Gamma(\lambda+\frac{1}{2})}\,.$$

A factor k^{-1} was omitted for simplicity in (6.4). Having care of this factor and others, Sommerfeld's representation takes the limiting form:

$$f(0, \tau) = \frac{1}{\tau}\int_{-i\infty}^{i\infty}\lambda\,\mathrm{d}\lambda\,\frac{\beta(\lambda)}{\alpha(\lambda)}\left(\frac{\tau}{2}\right)^{2\lambda}\frac{\sqrt{\pi}}{\Gamma(\lambda)\,\Gamma(\lambda+\frac{1}{2})\cos\pi\lambda}\,. \tag{A.17}$$

The discussion of this integral in no point is essentially different from the general case already treated in Section **6**, and it is actually simpler because it involves the theory of elementary Mellin transforms.

In (A.17) the path avoids the zeros of α. The resulting analiticity domain is the sector $|\arg\tau| < \sigma$ plus the interms of the circle $|\tau| < \alpha$.

RIASSUNTO

Nel presente lavoro viene definita l'interpolazione degli sfasamenti nello scattering da potenziale per valori generalmente complessi del momento orbitale. Tale definizione si presta particolarmente a discutere le proprietà analitiche (tra cui la rappresentazione di Mandelstam) dell'ampiezza di scattering giovandosi all'uopo di un metodo dovuto a Watson sueccessivamente perfezionato da Sommerfeld.

IL NUOVO CIMENTO VOL. XVIII, N. 5 1° Dicembre 1960

Bound States, Shadow States and Mandelstam Representation.

T. REGGE (*)

Palmer Physical Laboratory, Princeton University - Princeton, N. J.
Department of Physics, University of Rochester - Rochester, N. Y.

(ricevuto il 5 Agosto 1960)

Summary. — In a previous paper a technique involving complex angular momenta was used in order to prove the Mandelstam representation for potential scattering. One of the results was that the number of subtractions in the transmitted momentum depends critically on the location of the poles (shadow states) of the scattering matrix as a function of the complex orbital momentum. In this paper the study of the position of the shadow states is carried out in much greater detail. We give also related inequalities concerning bound states and resonances. The physical interpretation of the shadow states is then discussed.

1. – Introduction.

The validity of the Mandelstam representation for the potential scattering of two spinless parti les has now been firmly established for a class of generalized Yukawa potentials (1-4). Double dispersion relations can now be proved in a variety of methods each one adding an interesting angle to the overall picture of the analytic properties of the scattering amplitude.

In spite of these advances many details are missing and in particular little is known of the connection between bound states and subtraction terms in the transmitted momentum.

(*) Now at the University of Torino, Italy.

(1) T. REGGE: *Nuovo Cimento*, **14**, 951 (1959).

(2) R. BLANKENBECLER, M. L. GOLDBERGER, M. N. KHURI and S. B. TREIMAN: *Ann. of Phys.*, **10**, 62 (1960).

(3) A. KLEIN: *Journ. Math. Phys.*, **1**, 41 (1960).

(4) L. FONDA, L. A. RADICATI and T. REGGE: *Mandelstam representation for the non-relativistic many channel problem* (in proof).

BLANKENBECLER *et al.* [2] have shown how unitarity and Mandelstam representation can replace the Schrödinger equation in the construction of the scattering matrix under some simplifying hypothesis on bound states and subtraction terms. Their analysis would attain full generality and beauty if these restrictions could be removed and their statements rigorously backed.

In this respect part of the gap is filled in a previous paper [1] of the author where it is shown that, under some restrictive hypothesis on the potential, the number of needed subtractions is finite.

In the present work explicit bounds are placed on L and some restrictions on the potential are removed. Similar bounds are also derived for bound states. Finally a simple physical interpretation is given for the shadow states.

2. – Definitions and previous results.

In [1] complex angular momenta were employed in the proof of Mandelstam representation. We summarize here briefly the most important definitions and results needed for the proof. The starting point is the partial wave equation

$$D(E, \beta)\psi = 0 \,, \tag{2.1}$$

where

$$D(E, \beta) = \frac{\mathrm{d}^2}{\mathrm{d}x^2} + E + \left(\beta + \frac{1}{4}\right) x^{-2} - V(x) \,,$$

and

$$\beta = -(l + \tfrac{1}{2})^2 \,.$$

Eq. (2.1) will be considered for generally complex values of β. $l = \sqrt{-\beta} - \frac{1}{2}$ will be defined as that branch of the function which is one valued in the β plane with the cut $0 \leqslant \beta < \infty$ and such that $\operatorname{Re}(l + \frac{1}{2}) > 0$. The «regular» solution φ of eq. (2.1) is defined by the following boundary conditions:

$$\begin{cases} \varphi(E, \beta) \sim \cos\left[x\sqrt{E} - \dfrac{\pi\sqrt{-\beta}}{2} + \delta(\beta)\right], & x \to \infty \,. \\ \varphi(E, \beta) \sim C x^{l+1} \,, & x \to 0 \,, \end{cases} \tag{2.2}$$

φ defines therefore an asymptotic phase shift $\delta(\beta)$ which reduces to the physical values when l is integer. In [1] was shown that $S(\beta) = \exp[2i\delta(\beta)]$ has the following properties:

a) $S(\beta)$ is meromorphyic in the β plane with the exception of a discontinuity across the cut $0 \leqslant \beta < \infty$.

b) $S(\beta)S^+(\beta) = 1$ where $S^+(\beta) = (S(\beta^*))^*$. This is equivalent to $\delta(\beta) = \delta^+(\beta)$.

In particular $\delta(\beta)$ is real if β is negative.

c) If β is negative $\dfrac{\partial\delta(\beta)}{\partial\beta} > \dfrac{\pi}{2}\sqrt{-\beta}$.

d) $\operatorname{Im}\sqrt{-\beta}[\pi \operatorname{Im}\sqrt{-\beta} - 2 \operatorname{Im}\delta(\beta)] > 0$.

Condition d) implies that $S(\beta)$ can have poles in $\operatorname{Im}\beta < 0$ only. c) shows that if there is a pole in β there is a zero in β^*.

e) $\lim\limits_{|\beta|\to\infty} \exp[-i\pi\sqrt{-\beta}][S(\beta) - 1] = 0$, $\qquad \arg\beta \geqslant 0$.

If in addition to (2.1) we suppose that $V(x)$ is a generalized Yukawa potential

$$V(x) = \int_\mu^\infty \mathrm{d}m\,\sigma(m)\,\frac{\exp[-mx]}{x}, \tag{2.3}$$

we find that e) can be improved as follows:

$$e')\quad \begin{cases} S(\beta) - 1 = 0(\exp[i\varepsilon\sqrt{-\beta} - \alpha\sqrt{-\beta}])\,, & \arg\beta \geqslant 0, \\ \cosh\alpha = 1 + \dfrac{\mu^2}{2E}\,, & \varepsilon > 0 \text{ and small.} \end{cases}$$

Properties a) ... e') are then used in connection with the partial wave expansion of the scattering amplitude

$$f(z) = \frac{1}{2i\sqrt{E}} \sum_l (2l+1)(\exp[2i\delta_l] - 1)\,P_l(z); \qquad z = \cos\theta; \tag{2.4}$$

This expansion is known to converge in the small Lehmann ellipse $\operatorname{Im}\theta < \alpha$. According to WATSON we can transform it into the integral

$$f(z) = \frac{1}{4\sqrt{E}} \int_C \mathrm{d}\beta[S(\beta) - 1]\,P_{\sqrt{-\beta}-\frac{1}{2}}(-z)\,\frac{1}{\cos\pi\sqrt{-\beta}}\,. \tag{2.5}$$

The contour C loops around the poles of $(\cos\pi\sqrt{-\beta})^{-1}$ which occur on the physical values of β on the negative axis. C must not enclose any singularity of $S(\beta)$. It is possible to deform C around the positive axis and transform

(2.5) into

$$(2.6)\qquad \sqrt{E}f(z)=\frac{i\pi}{2}\sum_n S_n P_{l_n}(-z)(2l_n+1)\frac{1}{\sin\pi l_n}+$$

$$+\frac{1}{4}\int_0^\infty \mathrm{d}\beta\,[S(\beta+i\varepsilon)-S(\beta-i\varepsilon)]P_{\sqrt{-\beta}-\frac{1}{2}}\frac{1}{\cos\pi\sqrt{-\beta}},$$

where S is the residuum of $S(\beta)$ in the pole $\beta_n=-(l_n+\frac{1}{2})^2$. The summation is carried out on all poles of $S(\beta)$. The functions $P_{l_n}(-\cos\theta)$, where l_n is not an integer, are Legendre functions. It is apparent from (2.6) and (2.4) that $f(\cos\theta)$ is analytic in the $\cos\theta$ plane with the cut $\cosh\alpha<\cos\theta<\infty$.

The contribution of the integral can be shown to vanish when $\operatorname{Im}\theta\to\infty$ (or $\cos\theta\to\infty$). The function $P_{l_n}(-\cos\theta)$ instead of the order of $\cos\theta^{\operatorname{Re}(l_n)}$. If the set is bounded above by L' the behavior of $f(\cos\theta)$ will be dominated by $\cos\theta^{L'}$. If $L'=\infty$ $f(\cos\theta)$ will have an essential singularity at $\cos\theta=\infty$. In ([1]) we proved that L' is finite if $|y^2\operatorname{Im}V(iy)|<A<\infty$. In this paper we shall give an explicit estimate for L'. The number of subtractions needed in the Mandelstam representation is then the mallest integer $L>L'$. In (2) it is proved that unitarity requires $L>K$, where K is the largest angular momentum for which there is a bound state.

3. - Bound states.

A bound state is defined as a solution of the equation

$$(3.1)\qquad D(E,\beta)\psi=0 \qquad \text{where } E,\ \beta<0;\ (l \text{ integer})$$

which is L^2 in $0\ldots\infty$. We shall name bound states the solutions of (3.1) with the same boundary conditions also when l is any positive number. L^2 implies the asymptotic behaviour

$$(3.2)\qquad \begin{cases}\psi\sim cx^{l+1} & x\to 0\,,\\ \psi\sim\exp[-\sqrt{-E}x] & x\to\infty\,.\end{cases}$$

In order to have a bound state E and β must be related by some equation $g(E,\beta)=0$, where g depends on V. This equation, if solved in E, yields $E=f(\beta)$ where f is a multivalued function since there may exist several bound states with the same β.

For definiteness we choose the branch with the largest binding energy.

It is clear that in general $f(\beta)$ is differentiable. Write now the equation

$$\int_0^\infty \mathrm{d}x\, \psi D(E, \beta)\psi = 0 . \tag{3.3}$$

After an integration by parts and some algebra (3.3) becomes

$$\int_0^\infty \mathrm{d}x \left[\left(\psi' - \frac{\psi}{2x}\right)^2 + |E|\psi^2 + |\beta|\frac{\psi^2}{x^2}\right] = -\int_0^\infty V(x)\psi^2\, \mathrm{d}x . \tag{3.4}$$

A necessary condition for the existence of bound states is that

$$|E| + (|\beta|/x^2) + V(x)$$

is somewhere negative. This implies that somewhere

$$|E| + (|\beta|/x^2) \leqslant V(x).$$

This latter condition is of course the same as in classical mechanics if $l + \frac{1}{2}$ is the classical angular momentum.

If V is repulsive, *i.e.* $V \geqslant 0$, there are no bound states. Suppose now that V satisfies an inequality of the kind

$$|V(x)| \leqslant \frac{M_\alpha}{x^{2\alpha}} . \tag{3.5}$$

By derivation it can be checked that

$$P(\lambda) = \frac{|E|}{\lambda^\alpha} + \lambda^{\alpha-1}\frac{|\beta|}{x^2} , \tag{3.6}$$

satisfies the inequality

$$P(\lambda) \geqslant \alpha^{-\alpha}(1-\alpha)^{\alpha-1}|E|^{1-\alpha}\left|\frac{\beta}{x}\right|^{\alpha} ,$$

putting $\lambda = 1$ one finds

$$|E| + \frac{|\beta|}{x^2} \geqslant \left|\frac{E}{1-\alpha}\right|^{1-\alpha}\left|\frac{\beta}{\alpha}\right|^{\alpha} x^{-2\alpha} .$$

Comparing this result with (3.4) we get

$$\left|\frac{\beta}{\alpha}\right|^{\alpha}\left|\frac{E}{1-\alpha}\right|^{1-\alpha} \leqslant M_\alpha . \tag{3.7}$$

Some consequences of this formula are

$$(3.8)\qquad |E|< M_0; \qquad l+\frac{1}{2}\leqslant K+\frac{1}{2}\leqslant\sqrt{M_1}; \qquad |E|\leqslant\frac{M_{\frac{1}{2}}^2}{4(l+\frac{1}{2})^2}.$$

The latter can be compared with the exact result for the Coulomb potential $V=-(M_{\frac{1}{2}}/x)$

$$|E|\leqslant\frac{M_{\frac{1}{2}}^2}{4(l+1)^2}.$$

Finally by integration of the identity

$$\psi D(E,\beta)\frac{\partial\psi}{\partial\beta}-\frac{\partial\psi}{\partial\beta}D(E,\beta)\psi=\frac{\psi^2}{x^2}+\frac{dE}{d\beta}\psi^2,$$

n the interval $0\ldots\infty$ we find

$$(3.9)\qquad \frac{dE}{d\beta}\int_0^\infty\psi^2\,dx=-\int_0^\infty\frac{\psi^2}{x^2}\,dx.$$

Eq. (3.9) can be inserted in eq. (3.4) thus obtaining after some algebra

$$(3.10)\qquad P\left(\frac{d\beta}{dE}\right)\leqslant M_\alpha\frac{dE}{d\beta}.$$

If $\alpha=0$ or $\alpha=1$ this inequality can be integrated and the result is

$$(3.11)\qquad \begin{cases} |E(l')|<|E(l)|\dfrac{M_1-(l'+\frac{1}{2})^2}{M_1-(l+\frac{1}{2})^2}, & l'>l.\\[2ex] |E(l)|<M_0-\dfrac{(l+\frac{1}{2})^2}{(l'+\frac{1}{2})^2}[M_0-|E(l')|]. \end{cases}$$

4. – The shadow and resonance states.

We know from the results of ref. (1) that if β_0 is a pole of $S(\beta)$ there is a solution ψ of the Schrödinger equation

$$(4.1)\qquad D(E,\beta_0)\psi=0, \qquad E>0,\ \mathrm{Im}\,\beta_0<0,$$

with the boundary conditions:

$$(4.2)\qquad \begin{cases} \psi\sim cx^{l_0+1} & x\to 0, \qquad l_0+\frac{1}{2}=\sqrt{-\beta_0}.\\ \psi\sim\exp[i\sqrt{E}x] & x\to\infty. \end{cases}$$

If $V(x)$ is of the form (2.3) we can continue it for complex values of x in the domain $\operatorname{Re}(x) \geqslant 0$. According to a general theorem (see ref. (5)) ψ can also be continued in the same region. Let us consider the function $\psi(iy) = \chi(y)$. χ satisfies the differential equation

$$D(E, \beta)\chi = \ddot{\chi} - E\chi + (\beta + \tfrac{1}{4})y^{-2}\chi + V(iy)\chi = 0\,, \tag{4.3}$$

where dots mean y-derivatives. From $\int \chi^*(D(E, \beta_0)\chi)\,\mathrm{d}y = 0$ by using the same techniques of Section **3** we find

$$\operatorname{Im}\beta \int_0^\infty |\chi|^2 \frac{\mathrm{d}y}{y^2} + \int_0^\infty \mathrm{d}y\, |\chi|^2 \operatorname{Im} V(iy) = 0\,, \tag{4.4}$$

$$\int_0^\infty \left[\left|\dot{\chi} - \frac{\chi}{2y}\right|^2 + E|\chi|^2 - \operatorname{Re}\beta \frac{|\chi|^2}{y^2} - \operatorname{Re}\,(V(iy))\,|\chi|^2\right] \mathrm{d}y = 0\,, \tag{4.4'}$$

$E - \operatorname{Re}\beta - \operatorname{Re} V$ must be negative somewhere in $0 \ldots \infty$.

In discussing these integral we must remember that for large positive y we have $\chi \sim \exp[-\sqrt{E}y]$. χ is therefore L^2.

If we assume now that $V(iy)$ satisfies the set of inequalities

$$\begin{cases} y^{2\alpha}|\operatorname{Im} V(iy)| < N'_\alpha\,; & y^{2\alpha}|\operatorname{Re} V(iy)| < N''_\alpha\,, \\ y^{2\alpha}|V(iy)| < N_\alpha\,: & N_\alpha < |N'_\alpha + iN''_\alpha|\,, \end{cases} \tag{4.5}$$

we can draw the following conclusions by using the same methods of Section **3**:

$$|\operatorname{Im}\beta| < N'_1; \qquad \left(\frac{E}{1-\alpha}\right)^{1-\alpha}\left(\frac{\operatorname{Re}\beta}{\alpha}\right)^{\alpha} \leqslant N''_\alpha \qquad \text{if} \quad \operatorname{Re}\beta < 0\,, \tag{4.6}$$

$$\text{no restrictions} \qquad \text{if} \quad \operatorname{Re}\beta > 0\,.$$

Condition (4.6) provides some bounds on L if N is finite. We have for instance $L \leqslant \operatorname{Re}\sqrt{N''_1 + iN'_1} - \frac{1}{2}$. The interesting case of Yukawa potential does not admit a finite N'_1 and we must therefore use a more sophysticated technique. To this purpose let us multiply (4.4) by $\cos\eta$, (4.4') by $\sin\eta$, and add, where $0 < \eta < \pi/2$. The result to be used only when $(\cos\eta \operatorname{Im}\beta + \sin\eta \operatorname{Re}\beta) \leqslant 0$ is

$$\left(\frac{E}{1-\alpha}\right)^{1-\alpha} \sin\eta^{1-\alpha} \left(\frac{\cos\eta \operatorname{Im}\beta + \sin\eta \operatorname{Re}\beta}{\alpha}\right)^{\alpha} \leqslant N_\alpha\,. \tag{4.7}$$

---

(5) E. T. WHITTAKER and G. N. WATSON: *A Course of Modern Analysis* (Cambridge, 1952).

In the β plane cond. (4.5) is in general different from (4.6). If equality would hold in (4.5) this equation would represent a set of straight lines in the β plane. The permissible domain lies then below the real axis and above the envelope of these lines. For $\alpha = 0$ there is no envelope. For $\alpha = 1$ the envelope is the circle $|\beta| = N_1$, with $0 \leqslant \arg\sqrt{-\beta} \leqslant \pi/2$, and the line $\operatorname{Im}\beta = -N$. The corresponding conditions are $0 > \operatorname{Im}\beta > -N_1$ and $\operatorname{Re}\beta > -\sqrt{N^2 - \operatorname{Im}\beta^2}$. If $\alpha = \frac{1}{2}$ the envelope is most simply expressed in the plane as

$$\operatorname{Re} l < \frac{N_{\frac{1}{2}}}{2\sqrt{E}} - \frac{1}{2}. \tag{4.8}$$

This condition can be easily applied to the Yukawa potential. The other values of α require a much more involved algebra and provide a bound only if $\alpha > \frac{1}{2}$.

The same procedure can be applied to the resonance states. The resonance state can be defined as the solution of (4.1) with the boundary conditions (4.2) but with l real and integer and E complex. The branch of $\sqrt{E}$ has to be choosen in order to have $\operatorname{Im} E < 0$, this condition follows from the continuity equation. We find instead of (4.6)

$$\operatorname{Im} E < N_0'; \qquad \left|\frac{\operatorname{Re} E}{1-\alpha}\right|^{1-\alpha}\left|\frac{\beta}{\alpha}\right|^{\alpha} < N_\alpha'', \qquad \text{if} \quad \operatorname{Re} E > 0, \tag{4.9}$$

otherwise no restriction.

And instead of (4.7)

$$\left|\frac{\beta \sin\eta}{\alpha}\right|^{\alpha} |\cos\eta \operatorname{Im} E + \sin\eta \operatorname{Re} E|^{1-\alpha} < N_\alpha. \tag{4.10}$$

The allowed areas for the resonances in the E plane are for $\alpha = 0$ the union of the circle $|E| < N_0$ and of the strip $\operatorname{Re} E < 0$. If $\alpha = \frac{1}{2}$ and $k^2 = E$ we have $\operatorname{Re} k < N_{\frac{1}{2}}/(2l+1)$.

5. – The physics of the shadow states.

It can be easily seen that shadow states are closely connected with resonances. Take namely the formula

$$\frac{\mathrm{d}E}{\mathrm{d}\beta} = -\frac{\int_0^\infty \psi^2 x^{-2}\,\mathrm{d}x}{\int_0^\infty \psi^2\,\mathrm{d}x}. \tag{5.1}$$

Eq. (5.1) is a close analog of (3.9). Let we suppose now that $\text{Im}\,\beta$ is very small. A small variation $\Delta\beta$ in β produces a small change ΔE in E. We choose $\Delta\beta = -i\,\text{Im}\,\beta$ so that $\beta' = \Delta\beta + \beta$ is real. Correspondingly the energy will become complex $E + \Delta E = E + i\,\text{Im}\,\beta/(R^2)$, $1/R^2$ being the average value of $1/x^2$ in the shadow state. These rough arguments show that there should be a resonance at the energy E if $\text{Re}\,l$ is integer. The mean life of the resonance is given by $1/\Delta E$ or $R^2/\text{Im}\,\beta$ having adopted units such that $\hbar = c = 2m = 1$. What physical meaning can we associate with $\Delta\beta$? Classically a resonance is a scattering process where the path of the incoming particle winds up several times around the center of force and has therefore the opportunity of interacting for a long period with the potential. We can suppose that the radius of the orbit can be conveniently approximated by R. If l is the orbital momentum the angular speed $\mathrm{d}\varphi/\mathrm{d}t$ is $2l/R^2$.

During the interaction time the particle travels the angle $\Delta\varphi = \Delta t \cdot \mathrm{d}\varphi/\mathrm{d}t = 1/\text{Im}\,l$ before it is lost again in the scattering process. The above estimate can be also checked as follows. We write the total wave function as

$$\Psi = \frac{1}{r}\, Y_l^m(\theta, \varphi)\, \psi(r)\,. \tag{5.2}$$

Of course Ψ is not a true wave function since it is multivalued in the angles. However if we disregard these irregularities it provides a fairly accurate description of the scattering process when we are sufficiently far from the branch points of Y_l^m. Take now $m = l$. We have $Y_l^l = P_l^l(\theta)\exp[il\varphi]$. This function is multivalued in φ and $\text{Im}\,l$ provides a damping factor in the orbit which will allow the particle to travel by an angle $\Delta\varphi$ of the same order of magnitude as found before. We can confidently assert that a shadow state with a very small $\text{Im}\,l$ originates from a resonance in the energy. In writing up a Mandelstam representation of a given process the number of subtractions L has to be equal or larger than the angular momentum of the intermediate resonance. The above arguments lose their strength if $\text{Im}\,\beta$ is no longer small because then the life-time of the resonance becomes too short.

An interesting interpretation of the shadow states, from which the name derives, follows from the work of LEVY and KELLER on the theory of diffraction [6]. These authors have shown that in calculating the diffraction of a field against a smooth object it is possible to modify the ordinary geometrical optics, and obtain reliable results, by introducing new kind of rays, the diffracted rays, besides the optical ones.

The diffracted rays are produced by incident rays which are tangent to the surface of the body. A part of the incident ray, instead of continuing along

---

[6] B. LEVY and J. KELLER: *Diffraction on a smooth object.*

the geometrical path, follows a geodesics along the body, until it leaves eventually the body along a tangent. The amplitude of the part of the ray which undergoes this process decreases exponentially with the arc of geodesics travelled by the diffracted ray. Diffracted rays obviously penetrate into the geometrical shadow of the object. By analyzing the scattering amplitude in terms of complex angular momenta LEVY and KELLER prove that diffracted rays arise from the shadow states. In their case Im is essentially a angular penetration factor of the ray into the shadow, thereby justifying the name shadow state. Unfortunately there is no immediate generalization of Levy and Keller's theory of diffraction to the simple potential scattering. It has to be pointed out that the boundary condition for a hard core potential, corresponding to a mooth object, yield to somewhat simple analytic properties of $S(\beta)$, in fact in this case $S(\beta)$ is meromorphye in β, there being no cut in $0<\beta<\infty$.

In view of the considerable amount of work being done on phenomenological hard core potentials for nucleon-nucleon scattering it would be desirable to extend our results to more realistic cases.

* * *

The author wishes to thank Professor B. LEVY and Professor J. KELLER for an illuminating discussion. This work was partially supported by an AEC grant. The author is also particularly grateful to Prof. R. E. MARSHAK for his hospitality at the University of Rochester.

RIASSUNTO

In questo lavoro si estendono i risultati di una nota precedente dell'autore in cui si è fatto uso di momenti angolari complessi. In particolare vengono derivate ineguaglianze concernenti il numero di sottrazioni nelle regole di dispersione alla Mandelstam per vari tipi di potenziale tra cui il potenziale di Yukawa.

S-matrix and Feynman Amplitudes

Giampiero Passarino

Dipartimento di Fisica, Università di Torino,
and INFN, Sezione di Torino, Via P. Giuria 1, I-10125 Torino, Italy

1. S-matrix and Feynman amplitudes

In 1976 Tullio Regge joined the Oji International Seminar on Algebraic Analysis, where he synthesized his views on the problem of constructing a convenient set of axioms for the S-matrix directly in terms of the scattering variables (S-matrix theory was developed heavily by Geoffrey Chew, Steven Frautschi, Stanley Mandelstam, Vladimir Gribov, and Tullio Regge).

In his contribution[1] we also find a comparison with the conventional field theory (QFT) approach; in his conclusions Regge writes "My personal belief however, is that the possibility of different approaches originates from a basic inadequacy of the theory and that ultimately field theory and S-matrix theory will be unified".

We can summarize the argument by saying that in QFT one assumes that the singularity spectrum of the S-matrix is confined to the union of the singularity spectra of the Feynman integrals, and then proceed to construct the singularity spectra of the Feynman integrals. Therefore, the scattering amplitude appears as the sum of infinitely many diagrams of increasing complexity and each diagram in principle can be completely investigated. In principle there is no reason for the S-matrix to be the sum of the diagrams but QFT works under the assumption that the diagrams represent the local behavior of the amplitude and that the whole picture can be recovered by gluing together all these local behaviors.

Each Feynman diagram is a multivalued analytic function of the relevant variables whose branching locus is in general an extremely complicated reducible algebraic variety; however, the set of singularities is very well defined by the Landau rules: usually Landau–Nakanishi[2,3] equations are called Landau equations for short, see also Bjorken (thesis, Stanford Univ. 1959) and Refs. 4–9. For an interpretation of the Landau singularities as macroscopic causality see Refs. 10, 11.

One can say that Feynman diagrams provide a mathematical representation of the various contributions to the S-matrix in the context of perturbative QFT. However, Landau equations establish a link between Feynman graphs and singularities of the S-matrix, going beyond the concept of weakly-coupled quantum field theories. These considerations led to a variety of features that could be aimed at in model building.

In Ref. 1 we find other important comments concerning the difference between the Feynman diagram approach and the S-matrix theory. Feynman integrals can be analytically continued around a Landau singularity: as a consequence Feynman diagram integrals are clearly singular only on the Landau surfaces obtained from the so-called $+\alpha$ Landau equations.[12] In S-matrix theory there is, a-priori, no $i\epsilon$ prescription (unless it is adopted as an additional postulate). Furthermore, one needs to gain some control on the proliferating singularities. The main tool is based on the fact that the physical amplitude is singular only on a very limited real subset of the whole variety and that this subset has rather nice support properties. A Landau variety L is a reducible algebraic variety; the singular points of a generic plane section of L are expected to be transverse intersections, tacnodes or cusps.[13] Indeed, in a set of related papers[13–15] Regge and collaborators have developed an algebraic approach on Feynman diagrams which works on selected classes, e.g. one-loop graphs, two-vertex self-energy graphs, necklace graphs.

It remains the problem of classifying all the Landau varieties (the position of these depends on the masses of the interacting particles), a problem manageable in a local context. Landau equations for a given Feynman integral are a set of kinematic constraints that are necessary for the appearance of a pole or branch point in the integrated function (as a function of external kinematics and masses).

Landau equations admit many families of solutions which are naturally classified as leading Landau singularities (LLS), sub-leading Landau singularities (SLLS), sub-sub-leading (S^2LLS) etc. A description of the LLS is as follows: 3 and higher point functions can be cut in more than two pieces; putting all propagators on-shell corresponds to $\alpha_i \neq 0$ at the level of the Landau equations, i.e. the LLS goes beyond the concept of unitarity cuts.[16–18]

Leading Landau singularities

Important results on the LLS follow from the so-called Coleman–Norton theorem:[19] a Feynman amplitude has singularities on the physical boundary if and only if the relevant Feynman diagram can be interpreted as a picture of an energy- and momentum-conserving process occurring in space-time, with all internal particles real, on the mass shell, and moving forward in time. As a by-product, the Feynman parameter associated with an internal line is identified (within a proportionality factor) with the time the particle exists between collisions, divided by its mass.

Most of the modern developments are based on the so-called Kershaw theorem:[20] the singular part of a scattering amplitude around its leading Landau singularity may be written as an algebraic product of the scattering amplitudes for each vertex of the corresponding Landau graph times a certain explicitly determined singularity factor which depends only on the type of singularity (triangle graph, box graph, etc.) and on the masses and spins of the internal particles.

Finally, solutions to the Landau conditions, corresponding to kinematic configurations where the modified Cayley determinant[15,21] vanishes, are called singularities of the first type (singularities of the first type comprise all solutions to the Landau conditions for finite values of loop momentum); a discussion of LLS using the Cayley language can be found in Refs. 22–24.

To conclude we can say that the most elementary form of Landau singularities is connected to holomorphic functions defined by integrals of rational functions as they appear in the "perturbative" approach to quantum field theory. From a mathematical point of view these holomorphic structures have a deeper meaning and we learned from Regge that general questions should receive an answer in this context.

On the negative side, the singular locus, the fundamental group of its complement and the monodromy group could be calculated only for very restricted classes of Feynman diagrams.[25] Furthermore, one needs to specify the local Laurent expansion near the singular locus and not only the monodromy group; a partial solution follows from the Kershaw theorem,[23,24] at least for one-loop diagrams while multi-loops diagrams generally are out of reach.

The ingredients which went into these attempts are well-known: relativistic invariance, unitarity, and finally some degree of analyticity with the hope that by combining them according to some recipe one obtains the S-matrix. There is no mentioning of the phenomenological impact of the Landau singularities which requires additional ingredients: the complex-mass scheme[26–30] is essential in this context, given the non-integrable character of the LLS for boxes and pentagons (if internal masses are kept real).

Modern developments

It is worth noting that the consequences of the Kershaw theorem have been reinterpreted by various authors in terms of multiple cuts on Feynman diagrams,[31–38] the beginning of the "NLO revolution": Witten invented twistor string theory,[39] Britto, Cachazo, and Feng (and Witten) found the BCFW recursion relations.[40,41] At the amplitude level, a given branch cut is generically shared by several integrals while the leading singularities receive contributions from a single integral (for the most general scenario this is an assumption, see the discussion on equivalent diagrams in Refs. 42, 43); one-loop amplitudes in any massless theory are much simpler that they ought to be, unitary is all that is necessary.[32]

Landau equations and diagrams have been discussed in many contexts and interpretations: the theory of asymptotic operation,[44] a geometric interpretation in terms of volumes of polytopes,[45] an interpretation in terms of projective geometry and momentum twistors,[46] in the analyticity properties of amplitudes in theories with nonlocal vertices of the type occurring in string field theory[47] or in theories with gravity duals.[48] Furthermore, for amplitudes of generalized polylogarithm type there should be a close connection between symbol entries and solutions of the Landau equations.[48]

It is worth noting that any Feynman integral should satisfy a holonomic system of linear differential equations (see Ref. 49 for a modern approach), a conjecture given by Regge who first understood and emphasized the importance of the role of the differential equations in the investigation of Feynman integrals.

In conclusion, much of the lasting impact of the analytic S-matrix program has been on how we understand the results of Feynman diagram calculations and, recently, several authors have been digging up old concepts from those days and finding new applications. Nowadays, there is a boom in research devoted to understanding the analytic behavior and remarkable structures present in scattering amplitudes and then turning that knowledge into new methods. Unfortunately, Regge and collaborators did not have the tools we do now, e.g. a whole field of techniques out there for manipulating polylogarithms.

In his conclusions[1] Regge makes some remarks and puts questions forward that are still valid and worth asking today, e.g. "are there infinitely many parameters left into the theoretical frame, which need to be specified before we attempt comparison with experiment?". It may happen that a physical theory merely predicts a manifold of possible physical worlds; in this view the theory would merely provide us with a set of relations between masses, couplings and the like without specifying all of them. Further axioms would be needed in order to establish uniqueness of the predictions. Perhaps we have to decide whether or not to continue working with the existing rule of physics, after all we still talk about Feynman while many prophets have been forgotten.

References

1. T. Regge, Old Problems and New Hopes in S Matrix Theory, Publ. Res. Inst. Math. Sci. Kyoto 12 (1977) 367–375.
2. N. Nakanishi, Classical motion of particles and the physical-region singularity of the feynman integral, Prog. Theor. Phys. 39 (1968) 768–771.
3. N. Nakanishi, Ordinary and anomalous thresholds in perturbation theory, Progress of Theoretical Physics 22 (1) (1959) 128–144.
4. J. C. Taylor, Analytic Properties of Perturbation Expansions, Phys. Rev. 117 (1960) 261–265.
5. J. C. Polkinghorne, G. R. Screaton, The analytic properties of perturbation theory — i, Il Nuovo Cimento (1955-1965) 15 (2) (1960) 289–300.
6. M. J. W. Bloxham, D. I. Olive, J. C. Polkinghorne, S-matrix singularity structure in the physical region. 1. properties of multiple integrals, J. Math. Phys. 10 (1969) 494–502.
7. M. J. W. Bloxham, D. I. Olive, J. C. Polkinghorne, S-matrix singularity structure in the physical region. 2. unitarity integrals, J. Math. Phys. 10 (1969) 545–552.
8. M. J. W. Bloxham, D. I. Olive, J. C. Polkinghorne, S-matrix singularity structure in the physical region. 3. general discussion of simple landau singularities, J. Math. Phys. 10 (1969) 553–561.
9. Mathematical Society of Japan, K. Itō, N. Sūgakkai, no. v. 1 in Encyclopedic Dictionary of Mathematics, MIT Press, 1993.
10. C. Chandler, H. P. Stapp, Macroscopic causality conditions and properties of scattering amplitudes, J. Math. Phys. 10 (1969) 826–859.
11. D. Iagolnitzer, H. P. Stapp, Macroscopic causality and physical region analyticity in S-matrix theory, Commun. Math. Phys. 14 (1969) 15–55.
12. A. White, The Past and future of S matrix theory (2000) 1483–1504, http://arxiv.org/abs/hep-ph/0002303
13. G. Ponzano, T. Regge, E. R. Speer, M. J. Westwater, The monodromy rings of one loop feynman integrals, Commun. Math. Phys. 18 (1970) 1–64.
14. G. Ponzano, T. Regge, E. R. Speer, M. J. Westwater, The monodromy rings of a class of self-energy graphs, Commun. Math. Phys. 15 (1969) 83–132.

15. T. Regge, G. Barucchi, On the properties of Landau curves, Il Nuovo Cimento (1955-1965) 34 (1) (1964) 106–140.
16. R. E. Cutkosky, Singularities and discontinuities of Feynman amplitudes, J. Math. Phys. 1 (1960) 429–433.
17. S. Mandelstam, Unitarity Condition Below Physical Thresholds in the Normal and Anomalous Cases, Phys. Rev. Lett. 4 (1960) 84–87.
18. R. E. Cutkosky, Anomalous thresholds, Rev. Mod. Phys. 33 (1961) 448–455.
19. S. Coleman, R. E. Norton, Singularities in the physical region, Nuovo Cim. 38 (1965) 438–442.
20. D. Kershaw, Algebraic factorization of scattering amplitudes at physical Landau singularities, Phys. Rev. D5 (1972) 1976–1982.
21. D. B. Melrose, Reduction of Feynman diagrams, Nuovo Cim. 40 (1965) 181–213.
22. A. Ferroglia, M. Passera, G. Passarino, S. Uccirati, All purpose numerical evaluation of one loop multileg Feynman diagrams, Nucl. Phys. B650 (2003) 162–228.
23. S. Goria, G. Passarino, Anomalous Threshold as the Pivot of Feynman Amplitudes, Nucl. Phys. Proc. Suppl. 183 (2008) 320–325.
24. G. Passarino, Peaks and cusps: anomalous thresholds and LHC physics, http://arxiv.org/abs/1807.00503
25. T. Regge, The fundamental group of poincare and the analytic properties of Feynman relativistic amplitudes, Conf. Proc. C680519 (1968) 285–309.
26. M. J. G. Veltman, Unitarity and causality in a renormalizable field theory with unstable particles, Physica 29 (1963) 186–207.
27. H. Weldon, The Description of Unstable Particles in Quantum Field Theory, Phys. Rev. D14 (1976) 2030.
28. A. Denner, S. Dittmaier, The Complex-mass scheme for perturbative calculations with unstable particles, Nucl. Phys. Proc. Suppl. 160 (2006) 22–26, [22(2006)], http://arxiv.org/abs/hep-ph/0605312
29. S. Actis, G. Passarino, Two-Loop Renormalization in the Standard Model Part II: Renormalization Procedures and Computational Techniques, Nucl. Phys. B777 (2007) 35–99, http://arxiv.org/abs/hep-ph/0612123
30. A. Denner, J.-N. Lang, The Complex-Mass Scheme and Unitarity in perturbative Quantum Field Theory, Eur. Phys. J. C75 (8) (2015) 377, http://arxiv.org/abs/1406.6280
31. G. Ossola, C. G. Papadopoulos, R. Pittau, Reducing full one-loop amplitudes to scalar integrals at the integrand level, Nucl. Phys. B763 (2007) 147–169, http://arxiv.org/abs/hep-ph/0609007
32. Z. Bern, L. J. Dixon, D. C. Dunbar, D. A. Kosower, One loop n point gauge theory amplitudes, unitarity and collinear limits, Nucl. Phys. B425 (1994) 217–260, http://arxiv.org/abs/hep-ph/9403226
33. C. F. Berger, Z. Bern, L. J. Dixon, D. Forde, D. A. Kosower, Bootstrapping One-Loop QCD Amplitudes with General Helicities, Phys. Rev. D74 (2006) 036009, http://arxiv.org/abs/hep-ph/0604195
34. Z. Bern, N. E. J. Bjerrum-Bohr, D. C. Dunbar, H. Ita, Recursive calculation of one-loop QCD integral coefficients, JHEP 11 (2005) 027, http://arxiv.org/abs/hep-ph/0507019
35. R. Britto, B. Feng, P. Mastrolia, The Cut-constructible part of QCD amplitudes, Phys. Rev. D73 (2006) 105004, http://arxiv.org/abs/hep-ph/0602178
36. R. K. Ellis, W. T. Giele, G. Zanderighi, The One-loop amplitude for six-gluon scattering, JHEP 05 (2006) 027, http://arxiv.org/abs/hep-ph/0602185
37. C. Anastasiou, R. Britto, B. Feng, Z. Kunszt, P. Mastrolia, D-dimensional unitarity cut method, Phys. Lett. B645 (2007) 213–216, http://arxiv.org/abs/hep-ph/0609191
38. F. Cachazo, Sharpening The Leading Singularity, http://arxiv.org/abs/0803.1988
39. E. Witten, Perturbative gauge theory as a string theory in twistor space, Commun. Math. Phys. 252 (2004) 189–258, arXiv:hep-th/0312171
40. R. Britto, F. Cachazo, B. Feng, New recursion relations for tree amplitudes of gluons, Nucl. Phys. B715 (2005) 499–522, arXiv:hep-th/0412308

41. R. Britto, F. Cachazo, B. Feng, E. Witten, Direct proof of tree-level recursion relation in Yang-Mills theory, Phys. Rev. Lett. 94 (2005) 181602, arXiv:hep-th/0501052
42. J. Coster, H. P. Stapp, Physical-region discontinuity equations for many-particle scattering amplitudes. 1., J. Math. Phys. 10 (1969) 371–396.
43. J. Coster, H. P. Stapp, Physical-region discontinuity equations for many-particle scattering amplitudes. 2., J. Math. Phys. 11 (1970) 1441–1463.
44. F. V. Tkachov, Landau equations and asymptotic operation, Int. J. Mod. Phys. A14 (1999) 683–715, arXiv:hep-ph/9703423
45. S. Abreu, R. Britto, C. Duhr, E. Gardi, Cuts from residues: the one-loop case, JHEP 06 (2017) 114, arXiv:1702.03163
46. T. Dennen, M. Spradlin, A. Volovich, Landau Singularities and Symbology: One- and Two-loop MHV Amplitudes in SYM Theory, JHEP 03 (2016) 069, arXiv:1512.07909
47. P. Chin, E. T. Tomboulis, Nonlocal vertices and analyticity: Landau equations and general Cutkosky rule, arXiv:1803.08899
48. J. Maldacena, D. Simmons-Duffin, A. Zhiboedov, Looking for a bulk point, JHEP 01 (2017) 013, arXiv:1509.03612
49. J. M. Henn, Lectures on differential equations for Feynman integrals, J. Phys. A48 (2015) 153001, arXiv:1412.2296

Publ. RIMS, Kyoto Univ.
12 Suppl. (1977), 367-375.

Old Problems and New Hopes in S-Matrix Theory

by

Tullio REGGE*

1) The problem of constructing a convenient ser of axioms for the S-matrix directly in terms of the scattering variables and without recourse to conventional field theory is still receiving wide attention by theoreticians in recent years. The ingredients which go into these attempts are well-known. There is relativistic invariance, unitarity, and finally some degree of analyticity. The hope is that by combining them according to some recipe one obtains the S-matrix. The trouble with this program is that really no one seems to know how to combine the ingredients in the proper order; moreover there is wide disagreement as to what analyticity really means. It is on this last point that I wish to concentrate the discussion. Obviously Lorentz invariance and unitarity have quite an ambiguous meaning. Every attempt toward S-matrix theory usually starts with the rather ambitious goal of doing away with conventional field theory. The basic philosophy is opposite to that of constructive field theory whose remarkable progress in recent years has overshadowed any alternative approach.

My personal belief however, is that the possibility of different approaches originates from a basic inadequacy of the theory and that ultimately field theory and S-matrix theory will be unified. Secondly the constructive approach is still facing tremendous difficulties and is still far removed from any physical application; but basically I feel that committing ourselves to any particular philosophy at this early stage is an essentially unsound strategy.

2) Analyticity statements usually follow from causality through the

Received September 22, 1976.
* Institute for Advanced Study, Princeton, New Jersey 08540.
Research sponsored by the National Science Foundation Grant No. GP-40768X.

well-oiled machinery of dispersion relations. However this limited kind of analyticity is usually unsatisfactory to the S-matrix physicist. For two basic reasons. 1) It bears the clear imprint of the original sin of field theory. 2) It is too limited and it remains meaningful really only at the lowet thresholds. In order to obviate these shortcomings one looks at individual contributions of Feynman relativistic amplitudes (FRA) in the conventional perturbative approach. Here the scattering amplitude appears as the sum of infinitely many FRA of increasing complexity. Each FRA in principle can be completely investigated.

There is no reason however, for the S-matrix to be the sum of the FRA and in fact, this is likely not the case. The basic assumption is however that the FRA represents the local behaviour of the amplitude and that the whole picture can be recovered by gluing together all these local behaviours. Each FRA is a multivalued analytic function of the relevant variables whose branching locus is in general an extremely complicated reducible algebraic variety. Luckily the set of singularities is very well defined by the Landau rules. By introducing two scattering amplitudes into the unitarity relation we see that if each of the factors has a given singularity locus then the resulting amplitude must be singular on a variety defined by some algebraic reduction procedure from those of the factors. In other words unitarity tells us that given a Landau variety we can form a third one through unitarity. The set of all singularities must be closed under this composition rule. One sees quickly that starting from the simplest polar singularities at the lowest perturbative orders the set is indefinitely enlarged and includes eventually all Landau singularities of every permissible FRA. This set is probably denumerably infinite and the algebraic degree of the variety grows exponentially with the perturbative order.

At this point one needs to gain some control on these proliferating singularities. The main tool is the pleasant discovery that the physical amplitude is singular only on a very limited real subset of the whole variety (the so-called alpha positive part) and that this subset has rather nice support properties. In fact, by increasing the order of the FRA the subset tends to move away into high energy regions.

One has then two basic options for the resulting strategy. The first is to restrict the investigation of the FRA and of the S-matrix to the real analytic domain. The second is to jump into the complex domain. The first approach has received a strong boost from the remarkable algorithm of the hyperfunctions and related microlocal calculus developed by Sato and co-workers. This method seems to fit very nicely with the general approach developed by Stapp, Iagolnitzer and others. In particular the concent of MOS seems to be a natural gluing tool in fitting together the singular behaviour of the partial FRA. The second strategy has been put into practice on the S-matrix by the Cambridge group (Olive, Eden, Polkinghorne, Landshoff et. al.) in a rather unsystematic fashion because of the tremendous difficulties involved. A related approach by Pham and Fotiadi, Lascoux and Froissart has used the machinery of algebraic topology on individual FRA. Finally, in a set of related papers Ponzano, Speer, Westwater and myself (PSWR) have developed an algebraic approach on FRA which works on selected classes of FRA (single loop graphs, 2-vertex self-energy graphs, necklace graphs). None of these last methods has been applied to the S-matrix itself.

What are the relative virtues of the two strategies? As far as I can see, the real analytic approach has the merit of working with quantities having a direct physical meaning and direct physical intuition is certainly of great help. Also the novel mathematical techniques of the Japanese school hold great promise of the future development. On the negative side I would place the still pending problem of classifying all the Landau varieties. The position of these depends on the masses of the interacting particles. By varying these the relative ordering of the singular loci changes often in an unpredictable manner. The occurrence of cusps and of accidental degeneracies follows from a set of complicated inequalities among masses which appear difficult to control in an ambitious global theory, but perhaps manageable in a local context.

In the complex domain the problem of controlling the Landau curve is less crucial. There is however an enormously larger freedom of carrying out analytic continuations. The scattering amplitude (or at this stage more modestly an individual FRA) is initially defined in a suitable neighborhood and then continued outside. The resulting function turns out to be an element of a Nillson class and the set of all possible analytic continuation is identified with the fundamental group of the compliment of the Landau varities. By restricting the theory to a thin slab around the real domain it is possible to translate part of the results and concepts of the complex approach into those of the real analytic approach. There are in fact, differential equations in the complex domain for the FRA which are strongly related to the MOS proposed by the Sato group. The drawback is, of course, the lack of direct physical interpretation of the amplitudes and the still unsolved problem of formalizing unitarity in the complex domain.

3) The non-Landau singularities. A common difficulty lying ahead in both the complex and real analytic approach is however the occurrence of non-Landau singularities. As we have stated the set of all Landau singularities is actually infinite. We cannot exclude that it may be everywhere dense in the complex domain or perhaps even in the real domain. The study of individual FRA has shown that there are other singularity loci which were not included in the original Landau treatment and do not seem to have a simple physical meaning. These singularities are perhaps best understood by looking at unitarity relations. Generally the cycle of integration in a unitarity integral is compact. It may happen however that for particular values of the physical variable the cycle becomes non-compact and that a "pinch" occurs at infinity with the appearance of a new singularity. In a FRA this is not a disaster for the resuting singularity is tractable with the same techniques as for the standard ones, the relevant exception being the non-validity of the Cutkosky rules. In a S-matrix theory the points at infinity are really the accumulation of an indefinite sequence of ordinary singularities, all of which are "pinched" with the integration cycle thus producing an infinite set of Landau singularities accumulating around the non-Landau one. In fact a non-Landau singularity is just the accumulation of infinitely many ordinary ones. The process can be repeated and iterated and the resulting structure becomes hopelessly complicated. A non-Landau locus is then, by abuse of language, just as far away as infinity. In a FRA they appear simpler because all accumulating Landau singularities are collapsed into a single locus. In a way in this case the non-Landau locus is simply

a way to describe the effect of all the singularities of the other graphs. I feel that this basic difficulty will appear in both approaches.

4) A more basic question concerns the meaning and the kind of information one hopes to extract from an S-matrix theory. In the ideal case one would like to estabilish a set of axioms which uniquely characterize the scattering amplitudes and stable states of the theory.

In all applications of the theory however, the most significant results have correlated the amplitudes for relatively simple scattering processes to the masses of a few elementary particles. Is the theory supposed to give, it its final form, also the masses of all particles and not simply to correlate them to other interesting predictions? In other words, are there infinitely many parameters left into the theoretical frame, which need to be specified before we attempt comparison with experiment? In the early views of Chew this was not the case but it may happen that a physical theory merely predicts a manifold of possible physical worlds, and that the one we live in is chosen because it is the only one allowing biochemistry.

In this view the theory would merely provide us with a set of relations between masses, phase shifts, binding energies and the like without specifying all of them. Further axioms, not germane to conventional S-matrix theory, would be needed in order to establish uniqueness of the predictions.

5) The construction of an S-matrix is enormously complicated by the multi-sheeted structure of the amplitudes and by the intricated nature of the singular set. In the PRSW approach the main hope was that the singular set and the monodromy group could be calculated directly from the FRA and that a convenient generalization of Hibert 21st problem would provide the natural frame for an axiomatic approach to the S-matrix.

The program has faced the following, still unresolved difficulties.

A) The singular locus, the fundamental group of its complement and the monodromy group can be calculated only for very restricted classes of FRA. For instance multi-loops diagrams generally are out of reach.

B) Although FRA characterize the local behaviour of the S-Matrix no attempt was made to cast this correspondence in a more systematic form. In particular, the fundamental group of the complement of the set of all Landau singularities can be defined probably only through a limiting procedure (projective limit) on the set of infinite discrete groups, which is not free of dangers. This comment applies even more on monodromy group.

C) Even granting the complete construction of the fundamental and monodromy groups the proper setting and even the existence of a solution of the 21st Hilbert problem has not been proved for unions of inifnitely many algebraic varieties.

D) Finally, one needs to specify the local Laurent expansion near the singular locus and not only the monodromy group. In a sense therefore problem is not the one proposed by Hilbert but rather the type of investigation carried out in particular cases by Lappo-Danilevsky, Schlesinger and others.

It has to be noted however, that the fundamental group is just a very neat and convenient way to classify all Riemann sheets of multivalued function and that will appear in a natural way whenever this task is attempted. In particular, even the real analytic approach will have to face the need for a systematic treatment of the multi-sheeted structure of the amplitudes and will encounter the fundamental group in a more or less disguised form. Also, quite remarkably, although the analysis in PRSW was ostensibly done in the complex domain, the results really could have been obtained by restricting it to a neighourhood of the real domain with no relevant changes.

For this reason I feel that there is no real incompatibility between the real analytic and the complex approach.

6) I shall recall in this section a few definitions and relevant ideas of the complex analysis. In dealing with a given FRA there is no restriction in considering it as a Nillson function $A(z)$ defined over $\boldsymbol{C}^n - L^{n-1}$ or possibly $\boldsymbol{CP}^n - L^{n-1}$, where L^{n-1} is the singular locus consisting of the union of a large, but finite number of irreducible algebraic varieties.

The physical amplitude is initially defined in a neighbourhood U of a base point P, not intersecting L. Suppose now that the following conditions are satisfied:

a) $A(z)$ can be extended to a holomorphic function over the universal covering space of $C^n - L^{n+1}$. This means that A can be continued along any element g of $\pi_1(C^n - L^{n-1})$ yielding an analytic continuation A_g defined in U.

b) There are growth conditions implying that the extended A is bounded by a polynomial in the neighbourhood of L.

c) All A_g are linear combinations, with constant coefficients $\lambda_i(g)$ of a basic set of amplitudes A_i within U:

$$A_g = \sum_{i=1}^{N} \lambda_i(g) A_i . \tag{1}$$

By acting with an element of π_1 on A_i we find therefore that

$$(g\, A_i)(z) = \sum_{k=1}^{N} \mu_{ki}(g) A_k(z) \quad i=1, \cdots, N. \tag{2}$$

The matrix $\mu_{ki}(g)$ is readily seen to provide a linear representation of π, which is by definition the monodromy group G.

The general results in this field imply that if G is given then A is specified only as follows:

$$A(z) = \sum_{e=1}^{N} R_e(z) A^e \tag{3}$$

where the R_e are generic rational functions and the A^e form a set of N independent (over the field of rational functions) solutions of the problem. The derivative $\frac{dA^e}{dz}$ with respect to any variable z has still the same monodromy group, therefore we must have a differential equation of the type:

$$\frac{dA^e}{dz} = \sum_k R_k{}^e A^k . \tag{4}$$

On each amplitude A and course for A itself implies a single differential equation most obviously be related maximally overdetermined systems (MOS) in Sato's approach. Conversely the existence of such a system implies the Nillson property for its solution. There is no easy way to deduce the monodromy group from the differential equation, and vice-versa that I know of besides solving the equations themselves in terms of known functions, if possible.

From the brilliant analysis of the Japanese school the MOS for a given FRA is essentially known from the topology of the graph. The MOS pseudo differential operator however, corresponds roughly to the leading order of the conventional complex differential equations and we have no equally powerful algorithm for the constriction of the complete system. This is the main reason why in PRSW the effort was directed toward the construction of the monodromy group which appears to be more tractable. Clearly the differential equation plus boundary conditions in the form of N constants given at P determines completely the solution and therefore a specific amplitude. Differential equations on one side and monodromy group plus growth conditions on the other contain the same information packed in a radically different way and the connection between the two should be further elucidated before any attempts on the S-matrix are carried out in this fashion.

7) A last difficulty arises from the presence of infinitely many Landau singularities. This difficulty is common to all approaches. Using Sato conjecture one has to consider pseudodifferential operators of infinite order. This is not a novelty in the theory. What is a novelty is perhaps the occurrence of second kind singularities with associated accumulation of conventional Landau singularities which will require an ad hoc supplementary analysis.

Perhaps the best strategy is to give up entirely the idea of defining S-matrix amplitudes globally with a single equation of a single integral. The axiomatic field theory of Haag and Araki tries to reach a definition of the complete algebra of observables by nesting together local C^* algebras defined on open sets of space-time in a manner consistent with invariance and causality. Similarly either MOS or (in the complex) monodromy groups and associated structures should be defined only on selected classes of subsets in momentum space which are then glued together with appropriate axioms of identification. In this sense one has a set of privileged subsets of the space of scattering variables for which the monodromy group, the fundamental group and when the case, MOS are defined in the usual fashion. Sets of this sort need a definition of "compactness" in order to avoid an actual infinity of singular algebraic varieties with intractable pathologies. It is not difficult to introduce this notion of compactness. A set is "compact" if the amplitude satisfies a MOS of finite order in it or (in the complex) if the amplitude continues into a Nillson function with N finite and a finite number of Landau singularities (and therefore no non-Landau singularity).

The difficult part is to create an algorithm out of this definition. The intersection of compact sets is itself compact but the union is not necessarily so and this limits greatly the usefulness of the definition. Finally, there is always the last hurdle on how to treat unitarity efficiently in terms of these linear structures. In spite of the tremendous difficulties lying ahead I fell confident that the S-matrix theory is far from dead and that much is to be learned from the new mathematical methods and in turn much new interesting mathematics will be created by attempting to formalize it.

Acknowledgements

The author is deeply indebted to Prof. Sato for the hospitality at RIMS and the invitation to join the Oji International Seminar on Algebraic Analysis. The author also thanks Dr. Kawai, Dr. Stapp, Dr. Westwater and of course, Dr. Sato for many interesting discussions.

References

[1] Regge, T., *Actes du Congres International des Methematiciens*, 1/10 Septembre 1970/Nice, France, p. 9.

[2] Ponzano, G., Regge, T., Speer, E. R. and Westwater, M. J., The monodromy rings of a class of self-energy graphs, *Comm. Math. Phys.* **15** (1969), 83–132.

[3] Regge, T., "The fundamental group of Poincaré and the Analytic properties of Feynman relativistic amplitudes", *Nobèl Symposium 8, Elementary Particle Theory*, ed., Nils Svartholm, N. Y., Interscience, 1969.

[4] Ponzano, G., Regge, T., Speer, E. R. and Westwater, M. J., The monodromy rings of one loop Feynman integrals, *Comm. Math. Phys.* **18** (1970), 1–64.

PART III

Regge Calculus and Discrete Gravity

The Ponzano–Regge Model

Annalisa Marzuoli
Dipartimento di Matematica "F. Casorati", Università degli Studi di Pavia
via Ferrata 5, 27100 Pavia-I
INFN, Sezione di Pavia, via Bassi 6, 27100 Pavia-I
E-mail: annalisa.marzuoli@unipv.it

We present an overview of the role of the (Racah–)Wigner $6j$ symbol as the basic building block underlying such different fields as state sum models for quantum geometry, topological quantum field theory, statistical lattice models and quantum computing. Focusing on the geometric side, the results found in the seminal paper *Semiclassical Limit of Racah Coefficients* were recognized only many years later as a fundamental breakthrough for 3D discretized quantum gravity and will be addressed in some details. Selected — and by no means complete — lists of further applications and interconnections are included.

1. Introduction

Recall first that the (re)coupling theory of many $SU(2)$ angular momenta — framed mathematically in the structure of the Racah–Wigner tensor algebra — is the most exhaustive formalism in dealing with interacting many-angular momenta quantum systems.[1,2] It suffices here to mention the basic work of Wigner, Racah, Fano and others (see the collection of reprints[3] and the Racah memorial volume quoted in Ref. 4). As such it has been over the years a common tool in advanced applications in atomic and molecular physics, nuclear physics as well as in mathematical physics. In the last three decades there has been a deep interest in applying (extensions of) such notions and techniques to the branch of theoretical physics known as topological quantum field theory, as well as in related discretized models for three-dimensional quantum gravity. More recently the same techniques have been employed for establishing a new framework for quantum computing, the so-called "spin network" quantum simulator.[5]

In section 2 the $6j$ is looked at as a geometric tetrahedron — emerging in the Ponzano–Regge view — the basic *magic brick* in constructing three-dimensional quantum geometries of the Regge type, while in section 3 a few remarks on some developments and connection with other fields are briefly reported. Standard definitions and results on the Wigner $6j$ symbol and its asymptotics are collected in Appendices.

This contribution is largely based on the review paper.[6]

2. The 6j symbol and 3D quantum gravity

From a historical viewpoint the Ponzano–Regge asymptotic formula for the $6j$ symbol,[4] reproduced in (14) of Appendix A.1, together with the seminal paper[7] in which Regge calculus was founded, are no doubt at the basis of all discretized approaches to general relativity, both at the classical and at the quantum level.

In Regge's approach the edge lengths of a triangulated spacetime are taken as discrete counterparts of the metric and a Regge spacetime is a piecewise linear (PL) manifold of dimension D dissected into *simplices*, namely triangles in $D = 2$, tetrahedra in $D = 3$, 4-simplices in $D = 4$ and so on. Inside each simplex either an Euclidean or a Minkowskian metric can be assigned: accordingly, PL manifolds obtained by gluing together D-dimensional simplices acquire an overall PL metric of Riemannian or Lorentzian signature.[a]
Consider a particular triangulation $\mathcal{T}^D(\ell) \to \mathcal{M}^D$, where $\mathcal{M}^D$ is a closed, locally Euclidean manifold of fixed topology and ℓ denotes collectively the (finite) set of edge lengths of the simplices in $\mathcal{T}^D$. The Regge action is given explicitly by (units are chosen such that the Newton constant G is equal to 1)

$$S(\mathcal{T}^D(\ell)) \equiv S^D(\ell) = \sum_{\sigma_i} \mathrm{Vol}^{(D-2)}(\sigma_i)\, \epsilon_i\,, \tag{1}$$

where the sum is over $(D-2)$-dimensional simplices $\sigma_i \in \mathcal{T}^D$ (called hinges or "bones"), $\mathrm{Vol}^{(D-2)}(\sigma_i)$ are their $(D-2)$-dimensional volumes expressed in terms of the edge lengths and ϵ_i represent the deficit angles at σ_i. The latter are defined, for each i, as $2\pi - \sum_k \theta_{i,k}$, where $\theta_{i,k}$ are the dihedral angles between pairs of $(D-1)$-simplices meeting at σ_i and labeled by some k. Thus a positive [negative or null] value of the deficit angle ϵ_i corresponds to a positive [negative or null] curvature to be assigned to the bone i, detected for instance by moving a D-vector along a closed path around the bone i and measuring the angle of rotation. Even such a sketchy description of Regge geometry should make it clear that a discretized spacetime is flat (zero curvature) inside each D-simplex, while the curvature is concentrated at the bones which represent singular subspaces. It can be proven that (under suitable technical conditions) the limit of the Regge action (1) when the edge lengths become smaller and smaller gives the usual Einstein–Hilbert action for a smooth spacetime. Regge equations — the discretized analog of vacuum Einstein's field equations — can be derived from the classical action by varying it with respect to the dynamical variables, *i.e.* the set $\{\ell\}$ of edge lengths of $\mathcal{T}^D(\ell)$, according to Hamilton principle of classical field theory (refer to[8] for a bibliography and brief review on Regge calculus from its beginning up to the 1990's and to the contribution by Ruth M. Williams in the present volume).

[a]Einstein's General Relativity corresponds to the physically significant case of a four-dimensional spacetime endowed with a smooth Lorentzian metric. However, models formulated in non-physical dimensions such as $D = 2, 3$ turn out to be highly non-trivial and very useful in a variety of applications, ranging from conformal field theories and associated statistical models in $D = 2$ to the study of geometric topology of 3-manifolds. Moreover, the most commonly used quantization procedure of such theories has a chance of being well-defined only when the underlying geometry is (locally) Euclidean, see further remarks below.

Recall that Regge calculus gave rise in the early 1980's to a novel approach to quantization of general relativity known as simplicial quantum gravity (see Refs. 8–10 and references therein). The quantization procedure most commonly adopted is the Euclidean path-sum approach, namely a discretized version of Feynman's path-integral describing D-dimensional Regge geometries undergoing "quantum fluctuations" (in Wheeler's words a "sum over histories",[11] formalized for gravity in the so-called Hawking–Hartle prescription[12]). Without entering into technical details, the discretized path-sum approach turns out to be very useful in addressing a number of conceptual open questions in the approach relying on the geometry of smooth spacetimes, although the most significant improvements have been achieved for the $D = 3$ case, which we are going to address in some details in the rest of this section.

Coming to the interpretation of the Ponzano–Regge asymptotic formula for the $6j$ symbol given in (14) of Appendix A.1, we realize that it represents the semiclassical functional, namely the semiclassical limit of a path-sum over all quantum fluctuations, to be associated with the simplest three-dimensional spacetime, an Euclidean tetrahedron T. In fact the (positive frequency part of) argument in the exponential reproduces the Regge action $S^3(\ell)$ for T since in the present case $(D-2)$ simplices are one-dimensional (edges) and $\mathrm{Vol}^{(D-2)}(\sigma_i)$ in (1) are looked at as the associated edge lengths, see the introductory part of Appendix A.

More in general, we denote by $\mathcal{T}^3(j) \to \mathcal{M}^3$ a particular triangulation of a closed three-dimensional Regge manifold $\mathcal{M}^3$ (of fixed topology) obtained by assigning $SU(2)$ spin variables $\{j\}$ to the edges of $\mathcal{T}^3$. The assignment must satisfy a number of conditions, better illustrated if we introduce the *state functional* associated with $\mathcal{T}^3(j)$, namely

$$\mathbf{Z}[\mathcal{T}^3(j) \to \mathcal{M}^3; L] = \Lambda(L)^{-N_0} \prod_{A=1}^{N_1} (-1)^{2j_A} \mathsf{w}_A \prod_{B=1}^{N_3} \phi_B \begin{Bmatrix} j_1 & j_2 & j_3 \\ j_4 & j_5 & j_6 \end{Bmatrix}_B \tag{2}$$

where N_0, N_1, N_3 are the number of vertices, edges and tetrahedra in $\mathcal{T}^3(j)$, $\Lambda(L) = 4L^3/3C$ (L is a fixed length and C an arbitrary constant), $\mathsf{w}_A \doteq (2j_A + 1)$ are the dimensions of irreducible representations of $SU(2)$ which weigh the edges, $\phi_B = (-1)^{\sum_{p=1}^{6} j_p}$ and $\{:::\}_B$ are $6j$ symbols to be associated with the tetrahedra of the triangulation. Finally, the Ponzano–Regge *state sum* is obtained by summing over triangulations corresponding to all assignments of spin variables $\{j\}$ bounded by the cut-off L

$$\mathbf{Z}_{PR}\,[\mathcal{M}^3] \;=\; \lim_{L\to\infty} \sum_{\{j\}\leq L} \mathbf{Z}\,[\mathcal{T}^3(j) \to \mathcal{M}^3; L]\,, \tag{3}$$

where the cut-off is formally removed by taking the limit in front of the sum.

3. Remarks and outcomes

It is impossible to review in short, or even mention, the huge number of implications, outcomes and further improvements of the Ponzano–Regge state sum functional (3), as well as its deep and somehow surprising relationships with so many different issues in modern theoretical physics and in pure mathematics. We are going to present in the rest of this section a limited number of items and a few references, whose selection is made mainly on the basis of the longstanding interest in spin networks of the author and her collaborators.

It is worth to stress however the crucial role of this model in the so-called Loop approach to quantum gravity, see[13,14] and references therein, and also the older review paper[10]).

(a) As implicitly stated in,[4] the state sum $\mathbf{Z}_{PR}\,[\mathcal{M}^3]$ is a topological invariant of the manifold $\mathcal{M}^3$, owing to the fact that its value is actually independent of the particular triangulation, namely does not change under suitable combinatorial transformations. Remarkably, these "moves" are expressed algebraically in terms of the relations given in Appendix A.2, namely the Biedenharn–Elliott identity (17) — representing the moves (2 *tetrahedra*) $\leftrightarrow$ (3 *tetrahedra*) — and of both the Biedenharn–Elliott identity and the orthogonality conditions (18) for $6j$ symbols, which represent the barycentric move together its inverse, namely (1 *tetrahedra*) $\leftrightarrow$ (4 *tetrahedra*).

(b) In[15] a "regularized" version of (3) — based on representation theory of a quantum deformation of the group $SU(2)$ — was proposed and shown to be a well-defined *quantum invariant* for closed 3-manifolds.[b]
Its expression reads

$$\mathbf{Z}_{TV}\,[\mathcal{M}^3; q] = \sum_{\{j\}} \mathbf{w}^{-N_0} \prod_{A=1}^{N_1} \mathbf{w}_A \prod_{B=1}^{N_3} \begin{vmatrix} j_1 & j_2 & j_3 \\ j_4 & j_5 & j_6 \end{vmatrix}_B, \tag{4}$$

where the summation is over all $\{j\}$ labeling highest weight irreducible representations of $SU(2)_q$ ($q = \exp\{2\pi i/r\}$, with $\{j = 0, 1/2, 1\ldots, r-1\}$), $\mathbf{w}_A \doteq (-1)^{2j_A}[2j_A+1]_q$ where $[\,]_q$ denote a quantum integer, $\mathbf{w} = 2r/(q-q^{-1})^2$ and $|\ ::: \ |_B$ represents here the q-$6j$ symbol whose entries are the angular momenta $j_i, i = 1, \ldots, 6$ associated with tetrahedron B. If the deformation parameter q is set to 1 one gets $\mathbf{Z}_{TV}\,[\mathcal{M}^3; 1] = \mathbf{Z}_{PR}\,[\mathcal{M}^3]$.
It is worth noting that the q-Racah polynomial — associated with the q-$6j$ by a procedure that matches with what can be done in the $SU(2)$ case, see (16) in Appendix A.2 — stands at the top of Askey's q-hierarchy collecting orthogonal q-polynomials of one discrete or continuous variable. On the other hand, the discovery of the Turaev–Viro invariant has provided major developments in the branch of mathematics known as geometric topology.[16] It is still a very active field of research.

(c) The Turaev–Viro or Ponzano–Regge state sums can be generalized in many directions. For instance, they can be extended to simplicial 3-manifold endowed with a two-dimensional boundary,[17] to D-manifolds[18] (giving rise to topological invariants related to suitable discretized topological quantum field theory of the Schwarz type and to not semi-simple Lie groups, just to mention a few.

(d) The fact that the Turaev–Viro state sum is a topological invariant of the underlying (closed) 3-manifold reflects a crucial physical property of gravity in dimension 3 which makes it different from the $D = 4$ case. Loosely speaking, the gravitational field does not possess local degrees of freedom in $D = 3$, and thus any quantized functional can depend only on global features of the manifold encoded into its overall

[b]The adjective quantum refers here to "deformations" of semi-simple Lie groups introduced by the Russian School of theoretical physics in the 1980's in connection with inverse scattering theory. From the mathematical viewpoint the Turaev–Viro invariant, unlike the Ponzano–Regge state sum functional, is always finite and has been evaluated explicitly for some classes of 3-manifolds.

topology. Actually the invariant (4) can be shown to be equal to the square of the modulus of the Witten–Reshetikhin–Turaev invariant, which in turn represents a quantum path-integral of an $SU(2)$ Chern–Simons topological field theory — whose classical action can be shown to be equivalent to Einstein–Hilbert action[19] — written for a closed oriented manifold $\mathcal{M}^3$.[20,21] Then there exists a correspondence

$$\mathbf{Z}_{TV}\,[\mathcal{M}^3;q] \longleftrightarrow |\,\mathbf{Z}_{WRT}\,[\mathcal{M}^3;k]\,|^2\,, \tag{5}$$

where the level k of the Chern–Simons functional is related to the deformation parameter q of the quantum group.

Despite the "topological" nature of the Turaev–Viro (Ponzano–Regge) state sum and Witten–Reshetikhin–Turaev functionals in the case of closed 3-manifolds, whenever a $2D$-dimensional boundary occurs in $\mathcal{M}^3$, giving rise to a pair $(\mathcal{M}^3, \Sigma)$, where Σ is an oriented surface (or possibly the disjoint union of a finite number of surfaces), things change radically. For instance, if we add a boundary to the manifold in Witten–Reshetikhin–Turaev quantum functional, the theory induced on Σ is a Wess–Zumino–Witten (WZW)-type conformal field theory (CFT),[19] endowed with non-trivial quantum degrees of freedom. In particular, the frameworks outlined above can be exploit to establish a direct correspondence between $2D$ Regge triangulations and punctured Riemann surfaces, thus providing a novel characterization of the WZW model on triangulated surfaces on any genus[22] at a fixed level k. We cannot enter here into many technical details on developments on this topic, and refer to the monograph[23] for an upgraded view on the geometry of polytopes and their moduli spaces.

(e) In[24] a $(2 + 1)$-dimensional decomposition of Euclidean gravity (which takes into account the correspondence (5)) is shown to be equivalent, under mild topological assumptions, to a Gaussian $2D$ fermionic system, whose partition function takes into account the underlying $3D$ topology. More precisely, the partition function for free fermions propagating along "knotted loops" inside a three-dimensional sphere corresponds to a $3D$ Ising model on so-called knot-graph lattices. On the other hand, the formal expression of the $3D$ Ising partition function for a dimer covering of the underlying graph lattice can be shown to coincide with the permanent of the generalized incidence matrix of the lattice.[25,26] Recall first that the permanent of an $n \times n$ matrix A is given by

$$\mathrm{per}[A] \;=\; \sum_{\sigma \in \mathsf{S}_n} \prod_{i=1}^{n} a_{i,\sigma(i)} \tag{6}$$

where $a_{i,\sigma(i)}$ are minors of the matrix, $\sigma(i)$ is a permutation of the index $i = 1, 2, \ldots, n$ and S_n is the symmetric group on n elements. A graph lattice $\mathfrak{G}$ associated with a fixed orientable surfaces Σ of genus g embedded in S^3 may be constructed by resorting to the so-called surgery link-presentation. Then the incidence matrix of such piecewise linear graph with, say, $\mathfrak{n}$ vertices, is defined as an $\mathfrak{n} \times \mathfrak{n}$ matrix $A = (a_{ij})$ with entries in $(1, 0)$ according to whether vertices i, j are connected by an edge or not. Finally, the Ising partition function turns out to be a weighted sum — over all possible configurations of knot-graph lattices — of suitable determinants of generalized forms of the incidence matrices which take into account the topology of the underlying manifold. We skip however other technical details and refer to[27,28] for a discussion of algorithmic questions related in turn with item (f) below.

Note finally that the deep relationship between $3D$ quantum field theories that share a topological nature and (solvable) lattice models in $2D$, sketched in the last item by resorting to a specific example, was indeed predicted in the pioneering paper by E. Witten.[29] Not so surprisingly, the basic quantum functional that realizes this connection was identified there with *the expectation value of a certain tetrahedral configuration of braided Wilson lines*, where Wilson lines are quantum observables associated with particle trajectories that in general look like sheafs of braided strands propagating from a surface Σ_1 to another Σ_2, both embedded in a $3D$ background.

(f) The model for universal quantum computation proposed in Ref. 5, the spin network simulator, is based on the (re)coupling theory of $SU(2)$ angular momenta as formulated in the basic texts[1,2] on the quantum theory of angular momentum and the Racah–Wigner algebra respectively. At the first glance the spin network simulator can be thought of as a non-Boolean generalization of the Boolean *quantum circuit model*[c,30] with finite-dimensional, binary coupled computational Hilbert spaces associated with N mutually commuting angular momentum operators and unitary gates expressed in terms of:
i) recoupling coefficients ($3nj$ symbols) between inequivalent binary coupling schemes of $N = (n+1)$ $SU(2)$-angular momentum variables (j-gates);
ii) Wigner rotations in the eigenspace of the total angular momentum $\mathbf{J}$ (M-gates).

The spin network simulator is actually the discretized counterpart of the so-called topological approach to quantum computing developed in Ref. 31 (based, by the way, on the Witten–Reshetikhin–Turaev approach quoted in item (d)).
The role of the model in quantum circuit theory is addressed in Ref. 32.

A few years ago, in collaboration with S. Garnerone and M. Rasetti, we have developed, by resorting to q-deformed quantum automata, a new approach to dealing with classes of algorithmic problems that classically admit only exponential time algorithms. The problems in question arise in the physical context of $3D$ topological quantum field theories discussed above in the light of the fundamental result relating a topological invariant of knots, the Jones polynomial,[33] with a quantum observable given by the vacuum expectation value of a Wilson loop operator[34] associated with closed knotted curves in the Witten–Reshetikhin–Turaev background model.
Without entering into technical details, efficient (polynomial time) quantum algorithms for approximating (with an error that can be made as small as desired) generalizations of Jones polynomial have been found in,[35,36] while the case of topological invariants of 3-manifolds

[c]Recall that this scheme is the quantum version of the classical Boolean circuit in which strings of the basic binary alphabet $(0,1)$ are replaced by collections of qubits, namely quantum states in $(\mathbb{C}^2)^{\otimes N}$, and the gates are unitary transformations that can be expressed, similarly to what happens in the classical case, as suitable sequences of elementary gates associated with the Boolean logic operations *and, or, not.*

has been addressed in.[37] The relevance in having solved this kind of problems stems from the fact that an approximation of the Jones polynomial is sufficient to simulate any polynomial quantum computation.[38]

Appendix A: the Wigner 6j symbol and its symmetries

Given three angular momentum operators $\mathbf{J}_1, \mathbf{J}_2, \mathbf{J}_3$ — associated with three kinematically independent quantum systems — the Wigner-coupled Hilbert space of the composite system is an eigenstate of the total angular momentum

$$\mathbf{J}_1 + \mathbf{J}_2 + \mathbf{J}_3 \doteq \mathbf{J} \tag{7}$$

and of its projection J_z along the quantization axis. The degeneracy can be completely removed by considering binary coupling schemes such as $(\mathbf{J}_1 + \mathbf{J}_2) + \mathbf{J}_3$ and $\mathbf{J}_1 + (\mathbf{J}_2 + \mathbf{J}_3)$, and by introducing intermediate angular momentum operators defined by

$$(\mathbf{J}_1 + \mathbf{J}_2) = \mathbf{J}_{12}; \; \mathbf{J}_{12} + \mathbf{J}_3 = \mathbf{J} \tag{8}$$

and

$$(\mathbf{J}_2 + \mathbf{J}_3) = \mathbf{J}_{23}; \; \mathbf{J}_1 + \mathbf{J}_{23} = \mathbf{J}, \tag{9}$$

respectively. In Dirac notation the simultaneous eigenspaces of the two complete sets of commuting operators are spanned by basis vectors

$$|j_1 j_2 j_{12} j_3; jm\rangle \text{ and } |j_1 j_2 j_3 j_{23}; jm\rangle, \tag{10}$$

where j_1, j_2, j_3 denote eigenvalues of the corresponding operators, j is the eigenvalue of $\mathbf{J}$ and m is the total magnetic quantum number with range $-j \leq m \leq j$ in integer steps. Note that j_1, j_2, j_3 run over $\{0, \frac{1}{2}, 1, \frac{3}{2}, 2, \dots\}$ (labels of $SU(2)$ irreducible representations), while $|j_1 - j_2| \leq j_{12} \leq j_1 + j_2$ and $|j_2 - j_3| \leq j_{23} \leq j_2 + j_3$ (all quantum numbers are in $\hbar$ units).

The Wigner $6j$ symbol expresses the transformation between the two schemes (8) and (9), namely

$$|j_1 j_2 j_{12} j_3; jm\rangle = \sum_{j_{23}} [(2j_{12}+1)(2j_{23}+1)]^{1/2} \begin{Bmatrix} j_1 & j_2 & j_{12} \\ j_3 & j & j_{23} \end{Bmatrix} |j_1 j_2 j_3 j_{23}; jm\rangle \tag{11}$$

apart from a phase factor.[d] It follows that the quantum mechanical probability

$$P = [(2j_{12}+1)(2j_{23}+1)] \begin{Bmatrix} j_1 & j_2 & j_{12} \\ j_3 & j & j_{23} \end{Bmatrix}^2 \tag{12}$$

represents the probability that a system prepared in a state of the coupling scheme (8), where j_1, j_2, j_3, j_{12}, j have definite magnitudes, will be measured to be in a state of the coupling scheme (9).

[d] Actually this expression should contain the Racah W-coefficient $W(j_1 j_2 j_3 j; j_{12} j_{23})$ which differs from the $6j$ by the factor $(-)^{j_1+j_2+j_3+j}$. Recall that $(2j_{12}+1)$ and $(2j_{23}+1)$ are the dimensions of the representations labeled by j_{12} and j_{23}, respectively.

The $6j$ symbol may be written as sums of products of four Clebsch–Gordan coefficients or their symmetric counterparts, the Wigner $3j$ symbols. The relations between $6j$ and $3j$ symbols are given explicitly by (see *e.g.* Ref. 39)

$$\begin{Bmatrix} a & b & c \\ d & e & f \end{Bmatrix} = \sum (-)^{\Phi} \begin{pmatrix} a & b & c \\ \alpha & \beta & -\gamma \end{pmatrix} \begin{pmatrix} a & e & f \\ \alpha & \epsilon & -\varphi \end{pmatrix} \begin{pmatrix} d & b & f \\ -\delta & \beta & \varphi \end{pmatrix} \begin{pmatrix} d & e & c \\ \delta & -\epsilon & \gamma \end{pmatrix} \tag{13}$$

where $\Phi = d + e + f + \delta + \epsilon + \varphi$. Here Latin letters stand for j-type labels (integer or half-integers non-negative numbers) while Greek letters denote the associated magnetic quantum numbers (each varying in integer steps between $-j$ and j, $j \in \{a, b, c, d, e, f\}$). The sum is over all possible values of α, β, γ, δ, ϵ, φ with only three summation indices being independent.
On the basis of the above decomposition it can be shown that the $6j$ symbol is invariant under any permutation of its columns or under interchange the upper and lower arguments in each of any two columns. These algebraic relations involve $3! \times 4 = 24$ different $6j$ with the same value and are referred to as *classical symmetries* as opposite to *Regge symmetries* to be discussed in A.2. On the geometric side the classical symmetries of the $6j$ symbol encode the *tetrahedral symmetry* since each $3j$ (or Clebsch–Gordan) coefficient vanishes unless its j-type entries satisfy the triangular conditions, namely $|b - c| \le a \le b + c$, *etc.* This suggests that each of the four $3j$'s in (13) can be be associated with either a 3-valent vertex or a triangle. Here we adopt the Wigner–Ponzano–Regge three-dimensional picture used in Ref. 4, rather than Yutsis' dual representation as a complete graph on four vertices.[40] Accordingly, the $6j$ can be thought of as a real solid tetrahedron T with edge lengths $\ell_1 = a + \frac{1}{2}, \ell_2 = b + \frac{1}{2}$, $\ldots, \ell_6 = f + \frac{1}{2}$ in $\hbar$ units[e] and triangular faces associated with the triads (abc), (aef), (dbf), (dec). This implies in particular that the quantities $q_1 = a + b + c$, $q_2 = a + e + f$, $q_3 = b + d + f$, $q_4 = c + d + e$ (sums of the edge lengths of each face), $p_1 = a + b + d + e$, $p_2 = a + c + d + f$, $p_3 = b + c + e + f$ are all integer with $p_h \ge q_k$ ($h = 1, 2, 3$, $k = 1, 2, 3, 4$). The conditions addressed so far are in general sufficient to guarantee the existence of a non-vanishing $6j$ symbol, but they are not enough to ensure the existence of a geometric tetrahedron T living in Euclidean 3-space with the given edges. More precisely, T exists in this sense if (*and only if*, see the discussion in the introduction of Ref. 4) its square volume $V(T)^2 \equiv V^2$, evaluated by means of the Cayley–Menger determinant, is positive.

A.1 The Ponzano–Regge asymptotic formula

The Ponzano–Regge asymptotic formula for the $6j$ symbol reads[4]

$$\begin{Bmatrix} a & b & d \\ c & f & e \end{Bmatrix} \sim \frac{1}{\sqrt{24\pi V}} \exp \left\{ i \left(\sum_{r=1}^{6} \ell_r \theta_r + \frac{\pi}{4} \right) \right\} \tag{14}$$

where the limit is taken for all entries $\gg 1$ (recall that $\hbar = 1$) and $\ell_r \equiv j_r + 1/2$ with $\{j_r\} = \{a, b, c, d, e, f\}$. V is the Euclidean volume of the tetrahedron T and θ_r is the angle between the outer normals to the faces which share the edge ℓ_r. From a quantum mechanical viewpoint, this above probability amplitude has the form of a semiclassical (wave) function

[e]The $\frac{1}{2}$-shift is shown to be crucial in the analysis developed in:[4] for high quantum numbers the length $[j(j+1)]^{1/2}$ of an angular momentum vector is closer to $j + \frac{1}{2}$ in the semiclassical limit.

since the factor $1/\sqrt{24\pi V}$ is slowly varying with respect to the spin variables while the exponential is a rapidly oscillating dynamical phase. Such kind of asymptotic behavior complies with Wigner's semiclassical estimate for the probability, namely $\left\{\begin{smallmatrix} a & b & d \\ c & f & e \end{smallmatrix}\right\}^2 \sim 1/12\pi V$, to be compared with the quantum probability given in (12). Moreover, according to Feynman's path sum interpretation of quantum mechanics, the argument of the exponential in (14) must represent a classical action, and indeed it can be read as $\sum \mathsf{p}\dot{\mathsf{q}}$ for pairs (p, q) of canonical variables (angular momenta and conjugate angles). Such an interpretation has been improved recently by resorting to multidimensional WKB theory for integrable systems and geometric quantization methods.[41]

A.2 The Racah hypergeometric polynomial

The generalized hypergeometric series, denoted by ${}_pF_q$, is defined on p real or complex numerator parameters $a_1, a_2, \ldots, a_p$, q real or complex denominator parameters $b_1, b_2, \ldots, b_q$ and a single variable z by

$$ {}_pF_q \begin{pmatrix} a_1 \ldots a_p & \\ & ; z \\ b_1 \ldots b_q & \end{pmatrix} = \sum_{n=0}^{\infty} \frac{(a_1)_n \cdots (a_p)_n}{(b_1)_n \cdots (b_p)_n} \frac{z^n}{n!}, \tag{15} $$

where $(a)_n = a(a+1)(a+2)\cdots(a+n-1)$ denotes a rising factorial with $(a)_0 = 1$. If one of the numerator parameter is a negative integer, as actually happens in the following formula, the series terminates and the function is a polynomial in z.
The key expression for relating the $6j$ symbol to hypergeometric functions is given by the well-known Racah sum rule (see *e.g.*, Ref. 2 topic 11 and Ref. 39, Ch. 9 also for the original references). The final form of the so-called *Racah polynomial* can be recasted in terms of the ${}_4F_3$ hypergeometric function evaluated at $z = 1$ according to

$$ \begin{Bmatrix} a & b & d \\ c & f & e \end{Bmatrix} = \Delta(abe)\,\Delta(cde)\,\Delta(acf)\,\Delta(bdf)\,(-)^{\beta_1}(\beta_1+1)! $$

$$ \times \frac{{}_4F_3 \begin{pmatrix} \alpha_1-\beta_1 & \alpha_2-\beta_1 & \alpha_3-\beta_1 & \alpha_4-\beta_1 & \\ & & & & ; 1 \\ -\beta_1-1 & \beta_2-\beta_1+1 & \beta_3-\beta_1+1 & & \end{pmatrix}}{(\beta_2-\beta_1)!(\beta_3-\beta_1)!(\beta_1-\alpha_1)!(\beta_1-\alpha_2)!(\beta_1-\beta_3)!(\beta_1-\alpha_4)!}, \tag{16} $$

where

$$ \beta_1 = \min(a+b+c+d; a+d+e+f; b+c+e+f) $$

and the parameters β_2, β_3 are identified in either way with the pair remaining in the 3-tuple $(a+b+c+d; a+d+e+f; b+c+e+f)$ after deleting β_1. The four α's may be identified with any permutation of $(a+b+e; c+d+e;\ a+c+f; b+d+f)$. Finally, the Δ-factors in front of ${}_4F_3$ are defined, for any triad (abc) as

$$ \Delta\,(abc) = \left[\frac{(a+b-c)!(a-b+c)!(-a+b+c)!}{(a+b+c+1)!}\right]^{1/2}. $$

Such a seemly complicated notation is indeed the most convenient for the purpose of listing further interesting properties of the Wigner $6j$ symbol.

- The Racah polynomial is placed at the top of the Askey hierarchy including all of hypergeometric orthogonal polynomials of one (discrete or continuous) variable.[42] Most commonly encountered families of special functions in quantum mechanics are obtained from the Racah polynomial by applying suitable limiting procedures, as reviewed in.[43] Such an unified scheme provides in a straightforward way the algebraic *defining relations* of the Wigner $6j$ symbol viewed as an orthogonal polynomial of one discrete variable, *cfr.* (16). By resorting to standard notation from the quantum theory of angular momentum, the defining relations are:
the Biedenharn–Elliott identity ($R = a+b+c+d+e+f+p+q+r$):

$$\sum_x (-)^{R+x}\,(2x+1) \begin{Bmatrix} a & b & x \\ c & d & p \end{Bmatrix} \begin{Bmatrix} c & d & x \\ e & f & q \end{Bmatrix} \begin{Bmatrix} e & f & x \\ b & a & r \end{Bmatrix} = \begin{Bmatrix} p & q & r \\ e & a & d \end{Bmatrix} \begin{Bmatrix} p & q & r \\ f & b & c \end{Bmatrix}; \tag{17}$$

the orthogonality relation (δ is the Kronecker delta)

$$\sum_x (2x+1) \begin{Bmatrix} a & b & x \\ c & d & p \end{Bmatrix} \begin{Bmatrix} c & d & x \\ a & b & q \end{Bmatrix} = \frac{\delta_{pq}}{(2p+1)}. \tag{18}$$

- Given the relation (16), the unexpected new symmetry of the $6j$ symbol discovered by Regge[44] (see also[1,39]) is recognized as a set of permutations on the parameters α, β that leaves ${}_4F_3$ invariant. Combining the Regge symmetry and the classical, tetrahedral ones, one get a total number of 144 algebraic symmetries for the $6j$. Implications of Regge symmetry on the the study of the geometry of the quantum tetrahedron and its semiclassical Hamiltonian dynamics have been extensively addressed in.[45,46]
- The Askey hierarchy of orthogonal polynomials can be extended to a q-hierarchy,[42] on the top of which the q-${}_4F_3$ polynomial stands. Recall that when dealing with quantum invariants of knots and 3-manifolds formulated in the framework of unitary quantum field theory, the deformation parameter q must be a complex root of unity, the case $q = 1$ being considered as the trivial one. We refer to[47,48] for accounts on the theory of q-special functions and q-tensor algebras.

References

1. Biedenharn, L. C.; Louck, J. D. *Angular Momentum in Quantum Physics, Theory and Applications*; Encyclopedia of Mathematics and its Applications Vol 8, Rota, G–C. (Ed.); Addison–Wesley Publ. Co.: Reading MA, 1981.
2. Biedenharn, L. C.; Louck, J. D. *The Racah–Wigner Algebra in Quantum Theory*; Encyclopedia of Mathematics and its Applications Vol 9 Rota, G–C. (Ed.); Addison–Wesley Publ. Co.: Reading MA, 1981.
3. Biedenharn, L. C.; Van Dam, H. (Eds.) *Quantum Theory of Angular Momentum*; Academic Press: New York, 1965.
4. Ponzano, G.; Regge, T. *Semiclassical Limit of Racah coefficients*, in: Bloch, F. *et al.* (Eds.), *Spectroscopic and Group Theoretical Methods in Physics*; North–Holland: Amsterdam, 1968, p. 1.

5. Marzuoli, A.; Rasetti, M. *Ann. Phys.* **2005**, *318*, 345.
6. Carfora, M.; Marzuoli, A.; Rasetti, M. *J. Phys. Chem.*; **A2009** *113*, 15376.
7. Regge, T. *Nuovo Cimento*; **1961** *19*, 558.
8. Williams, R. M.; Tuckey, P. A. *Class. Quant. Grav.* **1992**, *9*, 1409.
9. Ambjorn, J.; Carfora, M.; Marzuoli, A. *The Geometry of Dynamical Triangulations*; Lect. Notes in Phys., m 50, Springer–Verlag: Berlin, 1997.
10. Regge, T.; Williams, R. M. *J. Math. Phys.* **2000**, *41*, 3964.
11. Misner, C. W.; Thorne, K. S.; Zurek, W. H. *Physics Today* **2009**, April, p. 40.
12. Hawking, S. W. *Nucl. Phys. B* **1978**, *144*, 349; Hartle, J. B. *J. Math. Phys.* **1985**, *26*, 804.
13. Rovelli, C. *Quantum Gravity*; Cambridge University Press: Cambridge, 2004.
14. Thiemann, T. *Modern Canonical Quantum Gravity General Relativity*; Cambridge University Press: Cambridge, 2008.
15. Turaev, V. G.; Viro, O. Ya. *Topology* **1992**, *31*, 865.
16. Ohtsuki, T. (Ed.) *Problems on invariants of knots and 3-manifolds*; RIMS Geometry and Topology Monographs, Vol. 4 (*eprint* arXiv:math.GT/0406190).
17. Carbone, G.; Carfora, M.; Marzuoli, A. *Commun. Math. Phys.* **2000**, *212*, 571.
18. Carbone, G.; Carfora, M.; Marzuoli, A. *Nucl. Phys. B* **2001**, *595*, 654.
19. Carlip, S. *Quantum Gravity in 2+1 Dimensions*; Cambridge University Press: Cambridge, 1998.
20. Witten, E. *Nucl. Phys. B* **1988/1989**, *311*, 49.
21. Reshetikhin, N.; Turaev, V. G. *Invent. Math.* **1991**, *103*, 547.
22. Arcioni, G.; Carfora, M.; Dappiaggi, C.; Marzuoli, A. *J. Geom. Phys.* **2004**, *52*, 137.
23. Carfora, M.; Marzuoli, A. *Quantum Triangulations*; Lecture Notes in Physics **942** Springer Nature, Switzerland, 2017.
24. Martellini, M.; Rasetti, M. *Int. J. Mod. Phys. B* **1996**, *10*, 2217.
25. Ceresole, A.; Rasetti, M.; Zecchina, R. *Rivista Nuovo Cimento* **1998**, *21*, 1.
26. Regge, T.; Zecchina, R. *J. Phys. A: Math. Gen.* **2000**, *33*, 741.
27. Marzuoli, A.; Rasetti, M. *Int. J. Quant. Inform.* **2007**, *5*, 223.
28. Loebl, M.; Moffat, I. *Adv. Appl. Math.* **2011**, *21*, 659.
29. Witten, E. *Nucl. Phys. B* **1989**, *322*, 629.
30. Nielsen, M. A.; Chuang, I. L. *Quantum Computation and Quantum Information*; Cambridge University Press: Cambridge, 2000.
31. Freedman, M.; Kitaev, A.; Larsen, M.; Wang, Z. *Bull. Amer. Math. Soc.* **2002**, *40*, 31.
32. Marzuoli, A.; Rasetti, M. *Int. J. Circ. Theor. Appl.* **2017**.
33. Jones, V. F. R. *Bull. Amer. Math. Soc.* **1985**, *12*, 103.
34. Witten, E. *Commun. Math. Phys.* **1989**, *121*, 351.
35. Garnerone, S.; Marzuoli, A.; Rasetti, M. *J. Phys. A: Math. Theor.* **2007**, *40*, 3047.
36. Garnerone, S.; Marzuoli, A.; Rasetti, M. *Quant. Inform. Comput.* **2007**, *7*, 479.
37. Garnerone, S.; Marzuoli, A.; Rasetti, M. *Adv. Theor. Math. Phys.* **2009**, 1.
38. Bordewich, M.; Freedman, M.; Lovasz, L.; Welsh, D. *Combinatorics, Probability and Computing* **2005**, *14*, 737.
39. Varshalovich, D. A.; Moskalev, A. N.; Khersonskii, V. K. *Quantum Theory of Angular Momentum*; World Scientific: Singapore, 1988.
40. Yutsis, A. P.; Levinson, I. B.; Vanagas, V. V. *The Mathematical Apparatus of the Theory of Angular Momentum*; Israel Program for Sci. Transl. Ltd.: Jerusalem, 1962.
41. Aquilanti, V.; Haggard, H. M.; Littlejohn, R. G. Yu, L. *J. Phys. A: Math. Theor.* **2007**, *40*, 5637.
42. Askey, R. *Ortogonal Polynomials and Special Functions*; SIAM: Philadelphia PE, 1975; Koekoek, R.; Swarttouw, R. F. *The Askey-Scheme of Hypergeometric Orthogonal Polynomials and its q-Analogue*; Technische Universiteit Delft: Delft, the Netherlands, 1998.
43. Ragni, M.; Bitencourt, A. P. C.; da S. Ferreira, C.; Aquilanti, V.; Anderson, R.; Littlejohn, R.G. *Int. J. Quant. Chem.* **2010** , 731.

44. Regge, T. *Nuovo Cimento* **1959**, *11*, 116.
45. Aquilanti, V.; Marinelli, D.; Marzuoli, A. *J. Phys. A: Math. Theor.* **2013**, *46*, 175303.
46. Aquilanti, V.; Marzuoli, A. *J. Phys.: Conf. Series* **2018**, *965*, 012005.
47. Nikiforov, A. F.; Suslov, S. K.; Uvarov, V. B. *Classical Orthogonal Polyomials of a Discrete Variable*; Springer–Verlag: Berlin, 1991.
48. Biedenharn, L. C.; Lohe, M. A. *Quantum Group Symmetry and q-Tensor Algebras*; World Scientific: Singapore, 1995.

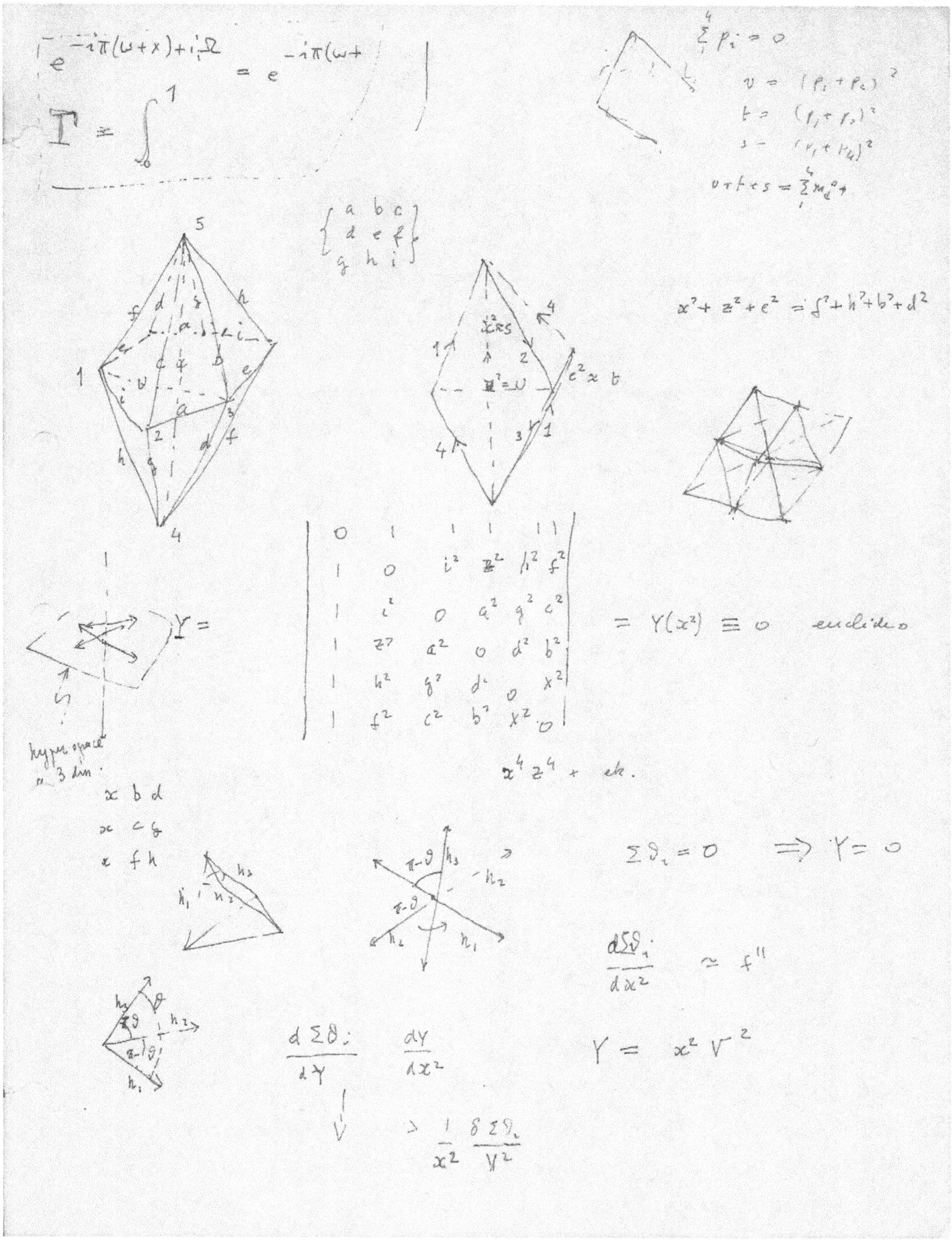

Image from the handwritten original draft of G. Ponzano and T. Regge.

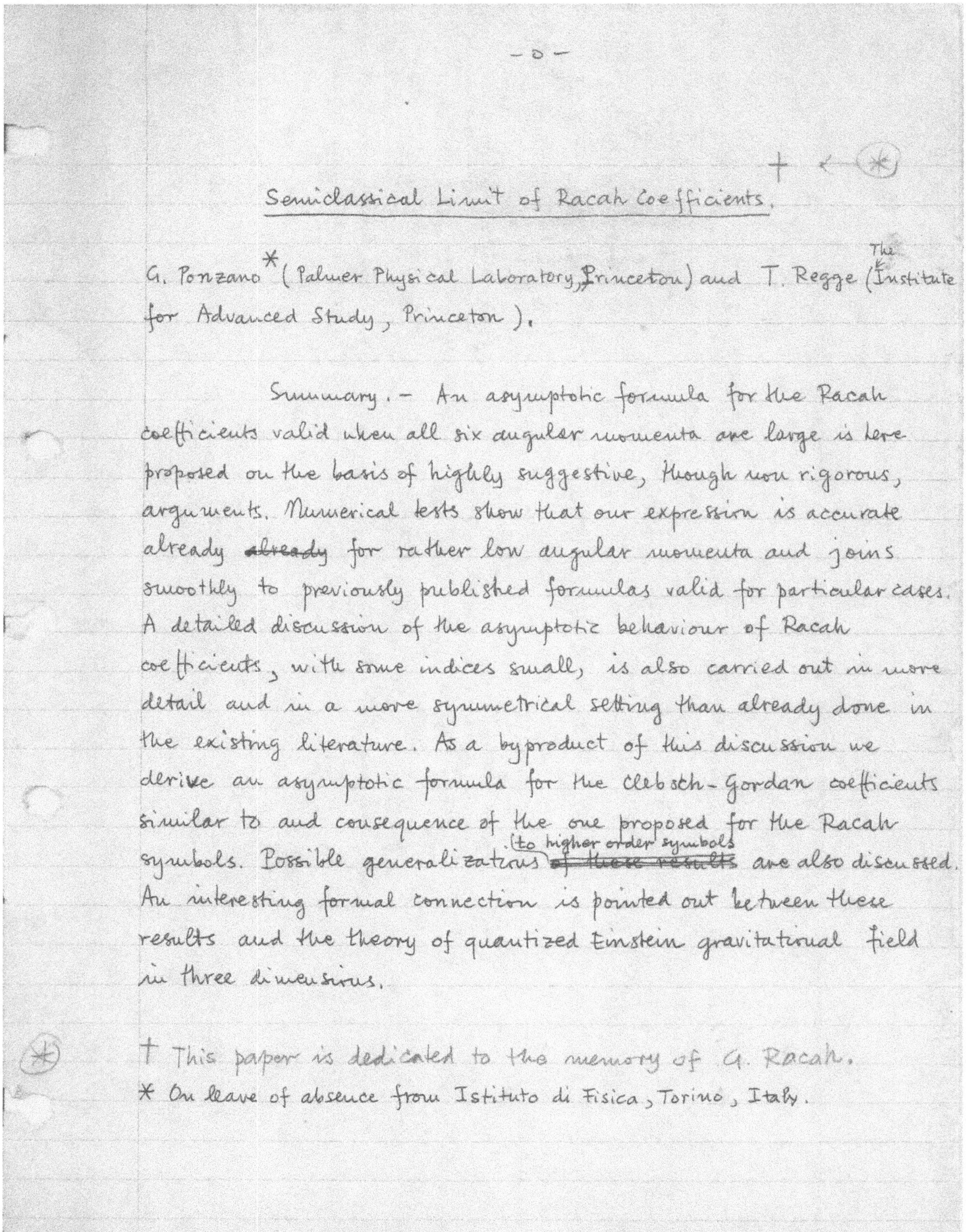

– 0 –

† ← (*)

Semiclassical Limit of Racah Coefficients.

G. Ponzano* (Palmer Physical Laboratory, Princeton) and T. Regge (The Institute for Advanced Study, Princeton).

Summary. – An asymptotic formula for the Racah coefficients valid when all six angular momenta are large is here proposed on the basis of highly suggestive, though non rigorous, arguments. Numerical tests show that our expression is accurate already ~~already~~ for rather low angular momenta and joins smoothly to previously published formulas valid for particular cases. A detailed discussion of the asymptotic behaviour of Racah coefficients, with some indices small, is also carried out in more detail and in a more symmetrical setting than already done in the existing literature. As a byproduct of this discussion we derive an asymptotic formula for the Clebsch-Gordan coefficients similar to and consequence of the one proposed for the Racah symbols. Possible generalizations (to higher order symbols ~~of these results~~ are also discussed. An interesting formal connection is pointed out between these results and the theory of quantized Einstein gravitational field in three dimensions.

(*) † This paper is dedicated to the memory of G. Racah.

* On leave of absence from Istituto di Fisica, Torino, Italy.

Image from the handwritten original draft of G. Ponzano and T. Regge.

Spectroscopic and group theoretical methods in physics

SEMICLASSICAL LIMIT OF RACAH COEFFICIENTS†

G. PONZANO*

Palmer Physical Laboratory, Princeton University, Princeton, New Jersey, U.S.A.

and

T. REGGE

Institute for Advanced Study, Princeton, New Jersey, U.S.A.

1. Introduction

Almost twenty-five years ago the W-coefficient appeared for the first time in a paper by Racah[1]) as an auxiliary tool for the computation of matrix elements in the theory of complex spectra. Today there is hardly any branch of physics involving angular momenta[2]) where the use of W-coefficients is not needed in order to carry out the simplest computation. Yet we feel that the W-coefficient is something more than an extremely successful computational tool and a beautiful toy for theoretical physicists to play with. In fact, a complete understanding of the properties of this remarkable function may very well yield to a new insight into the theory of angular momenta.

We think, and we know that our view is shared by others[3]), that a complete investigation of the semiclassical limit of W-coefficients and related functions is a prerequisite for a deeper understanding of their properties.

The present paper contains a heuristic derivation of an asymptotic formula, or better, of a set of asymptotic formulae with separate ranges of validity, for the W-coefficient. These formulae are certainly a useful complement to the existing tables of Racah coefficients, since they are remarkably accurate for surprisingly low values of the angular momenta involved.

A similar point of view could be adopted for the Clebsch–Gordan coefficients. However, since their definition depends on the particular labelling

† Work supported in part by EOAR under grant 66–29.

* On leave of absence from Istituto di Fisica, Torino, Italy. N.A.T.O. visiting fellow.

adopted for the vector basis of the representations, and since, moreover, they can be deduced as a particular limit carried on the W-coefficients, we are definitely tempted to regard them as subsidiary quantities in this paper.

Coming back to the W-coefficient or, rather, to the symmetric version of it, i.e. the 6*j*-symbol defined by Wigner[4]), it has been for years a normal practice to associate to it a diagram or graph which exhibits the symmetry properties in a most obvious way. A further advantage of these graphs is that they can be generalized to the higher order 3*nj*-symbols defined by Wigner[4]) and others[5,6]). There are at least three different versions[7]) of these graphical algorithms, all having approximately the same content, the

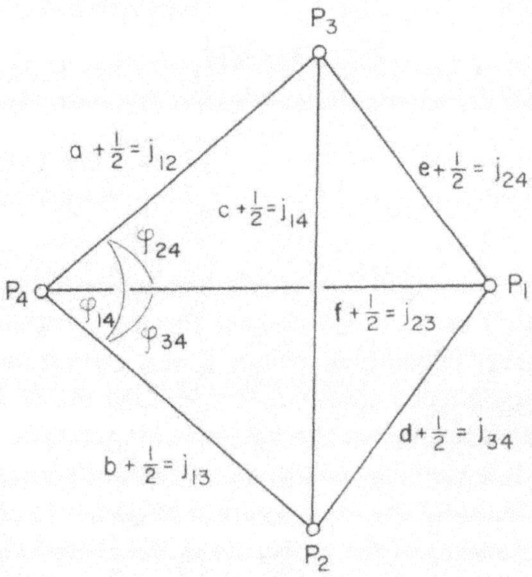

Fig. 1. Three-dimensional representation of the 6*j*-symbol $\begin{Bmatrix} a & b & c \\ d & e & f \end{Bmatrix}$.

translation of one into the others being achieved through some principle of plane or space duality. The reason for choosing any one of them is rather sentimental and largely related to individual habits.

We shall prefer here a three-dimensional representation in which angular momenta appear as vectors satisfying "bona fide" graphical composition rules. In this particular calculus the 6*j*-symbol $\begin{Bmatrix} a & b & c \\ d & e & f \end{Bmatrix}$ is associated to the tetrahedron shown in fig. 1. So far the tetrahedron is just a mnemonical device. However, we may think about a real solid T whose edges are just $a+\frac{1}{2}$, $b+\frac{1}{2}$, etc.[8]). With reference to fig. 1, we shall use also the following notation for the edges: $j_{12}=a+\frac{1}{2}$, $j_{13}=b+\frac{1}{2}$, $j_{14}=c+\frac{1}{2}$, $j_{34}=d+\frac{1}{2}$, $j_{24}=e+\frac{1}{2}$, $j_{23}=f+\frac{1}{2}$, and $j_{hh}=0$, $j_{hk}=j_{kh}(h, k=1, 2, 3, 4)$.

We shall restrict ourselves to values of the angular momenta which satisfy the triangular inequalities, i.e., $|b-c|\leqslant a\leqslant b+c$, etc. for the triads (abc), (aef), (dbf), (dec). Therefore, in each face there must be an even number of half-integer angular momenta. This entails that the sums $q_1=a+b+c$, $q_2=a+e+f$, $q_3=b+d+f$, $q_4=c+d+e$, $p_1=a+b+d+e$, $p_2=a+c+d+f$, $p_3=b+c+e+f$, are all integer. Moreover, because of the triangular inequalities, we have

$$p_h \geqslant q_k\,, \qquad h, k = 1, 2, 3, 4\,. \tag{1.1}$$

While these conditions are in general sufficient to guarantee the existence of a non-vanishing $6j$-symbol, they are not enough to ensure the existence of the tetrahedron T with the given edges. Since the two cases (A) T exists, (B) T does not exist, deserve radically different asymptotic treatments, we must give necessary and sufficient conditions for the existence of T.

It is known since Tartaglia[9]) and Jungius[10]) that the square of the volume of a tetrahedron is given by a polynomial in the square of its edges; in a more symmetrical setting, given by Cayley in his first published paper[11]), we have indeed

$$2^3(3!)^2\, V^2 = \begin{vmatrix} 0 & j_{34}^2 & j_{24}^2 & j_{23}^2 & 1 \\ j_{34}^2 & 0 & j_{14}^2 & j_{13}^2 & 1 \\ j_{24}^2 & j_{14}^2 & 0 & j_{12}^2 & 1 \\ j_{23}^2 & j_{13}^2 & j_{12}^2 & 0 & 1 \\ 1 & 1 & 1 & 1 & 0 \end{vmatrix}\,. \tag{1.2}$$

Therefore we see that $V^2\geqslant 0$ is a necessary condition. It can be proved to be also sufficient. In fact let us keep all edges fixed but, for instance, j_{12} and let $j_{12}=x$. Then V^2 is a second order polynomial in x^2 which will have two roots $(x_<)^2<(x_>)^2$. Since $\partial^2[V^2(x^2)]/[\partial x^2]^2=-(j_{34})^2/72$, we shall have $V^2(x^2)>0$ if $x_<<x<x_>$. A more elaborate discussion would show, in addition, that $x_<>x_{\mathrm{m}}-\frac{1}{2}$ where x_{m} is the largest between $|b-c|+\frac{1}{2}$ and $|e-f|+\frac{1}{2}$, and that $x_><x_{\mathrm{M}}+\frac{1}{2}$ where x_{M} is the smallest between $b+c+\frac{1}{2}$ and $e+f+\frac{1}{2}$. Therefore the condition $V^2>0$ is stronger than (1.1). We shall accordingly distinguish between the above mentioned cases: (A) $V^2>0$, and (B) $V^2<0$. The third possibility, $V^2=0$, which would correspond to a flat tetrahedron, is purely academical, for it can be proved that if p_h and q_k are all integer then $V^2\neq 0$ (ref. 12). A "tetrahedron" in (B) will be referred to as a hyperflat tetrahedron. Let us start with case:

(A) We expect T to be relevant in describing the properties of $6j$-symbols.

In fact a result due to Wigner[13]) states that for large angular momenta

$$\begin{Bmatrix} a & b & c \\ d & e & f \end{Bmatrix}^2 \sim \frac{1}{24\pi V}, \tag{1.3a}$$

where V is the volume of T. This formula, which has been a guiding principle in the present investigation, is very interesting because it relates the numerical value of the $6j$-symbol directly to a geometric property of T. However, as stressed in the same ref. 13, Wigner's asymptotic estimate cannot be accepted at face value. Inspection of numerical tables shows in fact that the symbol is a rapidly oscillating function of the indices and that the r.h.s. of (1.3a) more correctly approximates the average of the l.h.s. over several contiguous values of the indices.

A correct statement would be

$$\begin{Bmatrix} a & b & c \\ d & e & f \end{Bmatrix} \simeq \frac{1}{\sqrt{12\pi V}} \mathscr{C}, \tag{1.3b}$$

where $\mathscr{C}$ is a rapidly oscillating function so that the average $\mathscr{C}^2$ over a large enough interval is $\frac{1}{2}$. We claim that

$$\begin{Bmatrix} a & b & c \\ d & e & f \end{Bmatrix} \simeq \frac{1}{\sqrt{12\pi V}} \cos\left(\sum_{h,k=1}^{4} j_{hk}\theta_{hk} + \tfrac{1}{4}\pi \right), \tag{1.4}$$

where $\theta_{hk} = \theta_{kh} (k \neq h = 1, 2, 3, 4)$ are the angles between the outer normals of the two faces which belong to j_{hk}. Let A_h be the area of the face opposite to the vertex h (fig. 1); then we have (appendix B)

$$A_h A_k \sin \theta_{hk} = \tfrac{3}{2} V j_{hk}, \qquad h \neq k = 1, 2, 3, 4. \tag{1.5}$$

(B) Wigner's argument yields

$$\begin{Bmatrix} a & b & c \\ d & e & f \end{Bmatrix}^2 \sim 0. \tag{1.6}$$

This result could be loosely described as the impossibility of having six angular momenta forming a non-existing tetrahedral scheme. A closer scrutiny of numerical tables, however, shows that the symbols in (B), although, as a rule, smaller than in (A), are still non-vanishing.

Any attempt to use (1.4) in this region leads to a meaningless result. In fact, relations (1.1) guarantee that A_h are real; since $V^2 < 0$, V is imaginary and (1.5) implies that $\theta_{hk} = n\pi + \mathrm{i}\,\mathrm{Im}\,\theta_{hk}$. However, as it stands, (1.4) bears a strong resemblance with some formulae familiar from the WKB method[14]). Although we know of no differential equation from which in general (1.4) might be deduced, there are particular instances in which this can be done (section 5). This suggests that we may use the connection formulae of the WKB method to go across the transition points $x_<$, $x_>$. We define

$$\Phi = \sum_{h,k=1}^{4} (j_{hk} - \tfrac{1}{2}) \operatorname{Re} \theta_{hk} \tag{1.7}$$

which, for physical values of j_{hk}, is always an integer multiple of π. According to the WKB connection formulae (appendix G), we find for physical angular momenta

$$\begin{Bmatrix} a & b & c \\ d & e & f \end{Bmatrix} \simeq \frac{1}{2\sqrt{12\pi|V|}} \cos\Phi \exp\left(-\left|\sum_{h,k=1}^{4} j_{hk} \operatorname{Im}\theta_{hk}\right|\right), \tag{1.8}$$

where the sign of Im θ_{hk} must be chosen according to the rules explained in section 5. Also this formula turns out to be in remarkably satisfactory agreement with numerical tables. The exponential decrease shown by (1.8) clearly describes a quantum tunnel effect into the classically forbidden region (B).

We expect (1.4) and (1.8) to be inaccurate in the neighbourhood of $x_{\gtrless}$. In fact, the error is here considerably large, although not disastrous. Transition formulae involving Airy functions have been worked out for this region (section 5) and found to be accurate.

In spite of these numerical checks, a sound proof of our formulae is still missing. However there are other arguments in favour of (1.4) and (1.8). For instance, (1.4) has the right symmetry properties, including the extra symmetries discovered by one of us[15]) and satisfies asymptotically the recursion relations as well as the identities of the 6*j*-symbols. It is also consistent with the previously investigated particular cases of asymptotic behaviours[16]).

Relations (1.4) and (1.8), together with a transition formula, solve completely the analysis of 6*j*-symbols when all angular momenta are large. However, it is also interesting to investigate the case in which one or more edges remain constant and finite while the others increase. We may picture the limiting process as one in which one or more vertices of T go to infinity either separately or in clusters. Therefore there are as many ways to carry out the process as decompositions of 4 into sums of natural integers, i.e., $1+1+1+1$ (all edges large), $1+1+2$ (one small edge), $2+2$ (two small edges), $1+3$ (three small edges). The $1+1+2$ case has been widely studied[16]), while we have found no reference to $2+2$ and accordingly we solve this uninteresting case in the present paper. Some examples of the $1+3$ case, which yield a connection between 6*j*- and 3*j*-symbols, have been discussed by Brussaard and Tolhoek[16]). Our treatment, however, is quite general and shows how the symmetries of the 3*j*-symbol can be derived from those of the 6*j*-coefficient. As a by-product of this analysis, we obtain a new asymptotic formula for Clebsch–Gordan coefficients.

2. Asymptotic connection with the Clebsch–Gordan coefficients

Of some interest is the $1+3$ case, which occurs when we take the positive integer R large in $\begin{Bmatrix} a & b & c \\ d+R & e+R & f+R \end{Bmatrix}$. The related limits $\begin{Bmatrix} a & b+R & c+R \\ d+R & e & f \end{Bmatrix}$ etc. can be reduced to it by symmetry. The starting point of our discussion is Racah's formula (A.4), which, using $\xi = x - 2R$ as summation variable becomes

$$\begin{Bmatrix} a & b & c \\ d+R & e+R & f+R \end{Bmatrix} = [\Delta(abc)\,\Delta(d+R, e+R, c) \times \\ \times \Delta(e+R, f+R, a)\,\Delta(f+R, d+R, b)]^{\frac{1}{2}} \sum_{\xi} (-1)^{\xi+2R} \times \\ \times (\xi + 2R + 1)!\,[(a+b+d+e-\xi)!\,(b+c+e+f-\xi)! \times \\ \times (c+a+f+d-\xi)!(\xi - a - b - c + 2R)! \times \\ \times (\xi - a - e - f)!\,(\xi - d - e - c)!\,(\xi - d - b - f)!]^{-1}. \tag{2.1}$$

From Stirling's formula we find, for instance, that for large R

$$\frac{(\xi + 2R + 1)!}{(\xi + 2R - a - b - c)!} \simeq \xi^{a+b+c+1}, \tag{2.2}$$

which entails for example

$$\Delta(a, e+R, f+R) \simeq (a+e-f)!\,(a-e+f)!\,(2R)^{-2a-1}. \tag{2.3}$$

By means of (2.2), (2.3) and similar relations, (2.1) transforms into

$$\begin{Bmatrix} a & b & c \\ d+R & e+R & f+R \end{Bmatrix} \simeq \left[\frac{\Delta(abc)}{2R}\right]^{\frac{1}{2}} [(a+e-f)!\,(a-e+f)! \times \\ \times (b+f-d)!\,(b-f+d)!\,(c+d-e)!\,(c-d+e)!]^{\frac{1}{2}} \times \\ \times \sum_{\xi} (-1)^{\xi} [(a+b+d+e-\xi)!\,(b+c+e+f-\xi)! \times \\ \times (c+a+f+d-\xi)!\,(\xi - a - e - f)! \times \\ \times (\xi - b - d - f)!\,(\xi - c - d - e)!]^{-1}. \tag{2.4}$$

Looking at (A.1) we realize that the r.h.s. of (2.4) can be written in terms of a $3j$-symbol (this result is quoted by K. Alder et al.[16]):

$$\begin{Bmatrix} a & b & c \\ d+R & e+R & f+R \end{Bmatrix} \simeq (-1)^{a+b+c+2(d+e+f)} (2R)^{-\frac{1}{2}} \begin{pmatrix} a & b & c \\ e-f & f-d & d-e \end{pmatrix} \tag{2.5a}$$

or using the pattern notation of appendix A

$$\begin{vmatrix} a+b-c & b+f-d & a+f-e \\ b+d-f & b+c-a & c+d-e \\ a+e-f & e+c-d & c+a-b \\ e+d-c+2R & e+f-a+2R & d+f-b+2R \end{vmatrix} \simeq$$

$$\simeq (-1)^{a+b+c+2(d+e+f)} (2R)^{-\frac{1}{2}} \begin{vmatrix} b+c-a & c+a-b & a+b-c \\ a-e+f & b-f+d & c-d+e \\ a+e-f & b+f-d & c+d-e \end{vmatrix} . \tag{2.5b}$$

From (2.5b) it is easy to check that the symmetry

$$\begin{Bmatrix} a & b & c \\ d+R & e+R & f+R \end{Bmatrix} =$$

$$= \begin{Bmatrix} a & \frac{1}{2}(f+c+b-e) & \frac{1}{2}(b+e+c-f) \\ d+R & \frac{1}{2}(f+c+e-b)+R & \frac{1}{2}(b+e+f-c)+R \end{Bmatrix}$$

entails

$$\begin{vmatrix} b+c-a & c+a-b & a+b-c \\ a-e+f & b-f+d & c-d+e \\ a+e-f & b+f-d & c+d-e \end{vmatrix} = \begin{vmatrix} b+c-a & a+e-f & a-e+f \\ a+b-c & c+d-e & c-d+e \\ c+a-b & b+f-d & b-f+d \end{vmatrix} ,$$

which is one of the extra symmetries of the $3j$-symbol pointed out by one of us[15]). Actually, (2.5b) relates the subgroup of R_1 (see appendix A) formed by all the even 36 symmetries of the $3j$-symbol to the subgroup of R_2 corresponding to permutations of columns and/or of the upper three lines of the pattern in the l.h.s. of (2.5b). The geometrical and physical content of these symmetries is still to be understood and they remain a puzzling feature of the theory of angular momenta. Therefore it is a pleasant result to be able to reduce the problem of their interpretation to the Racah coefficient only.

From (2.5a) we obtain also an expression for the $3j$-symbol for large quantum numbers. Our derivation of this result is rather heuristic as it involves the exchange of different limiting processes. Just as for (1.4) the formula which we are going to present has a rather "a posteriori" validity, for it satisfies all possible consistency checks. We obtain it from (2.5a) by supposing a, b, c, d, e, f large and finite. Using (1.4) in the l.h.s. of (2.5a) and performing the limit $R \to \infty$, we find

$$\begin{pmatrix} a & b & c \\ m_a & m_b & m_c \end{pmatrix} \simeq (2\pi A)^{-\frac{1}{2}}(-1)^{a+b-c+1} \times$$
$$\times \cos\left[(a+\tfrac{1}{2})A + (b+\tfrac{1}{2})B + (c+\tfrac{1}{2})C - m_b D + m_a E + \tfrac{1}{4}\pi\right], \qquad (2.6)$$

where $m_a = e - f$, $m_b = f - d$, $m_c = d - e$ are the third components of a, b, c along the direction in which P_4 was sent to infinity (fig. 2). A is the area of the shaded triangle in fig. 2 which corresponds to the projection of $P_1P_2P_3$ on a plane perpendicular to the z axis. The angles $A, \ldots, D, \ldots,$ defined according to (1.4) are given in terms of m_a, m_b, m_c by

$$\cos A = \frac{2(a+\frac{1}{2})^2 m_c + m_a[(c+\frac{1}{2})^2 + (a+\frac{1}{2})^2 - (b+\frac{1}{2})^2]}{([(a+\frac{1}{2})^2 - m_a^2]\{4(c+\frac{1}{2})^2(a+\frac{1}{2})^2 - [(c+\frac{1}{2})^2 + (a+\frac{1}{2})^2 - (b+\frac{1}{2})^2]^2\})^{\frac{1}{2}}}. \qquad (2.7)$$

$$\cos D = \frac{1}{2}\frac{(a+\frac{1}{2})^2 - (b+\frac{1}{2})^2 - (c+\frac{1}{2})^2 - 2m_b m_c}{\{[(b+\frac{1}{2})^2 - m_b^2][(c+\frac{1}{2})^2 - m_c^2]\}^{\frac{1}{2}}}, \qquad (2.8)$$

$\cos B$, $\cos C$ and $\cos E$, $\cos F$ are deduced respectively from (2.7), (2,8) by circular permutations of the labels, a, b, c; note that $D+E+F=2\pi$. When

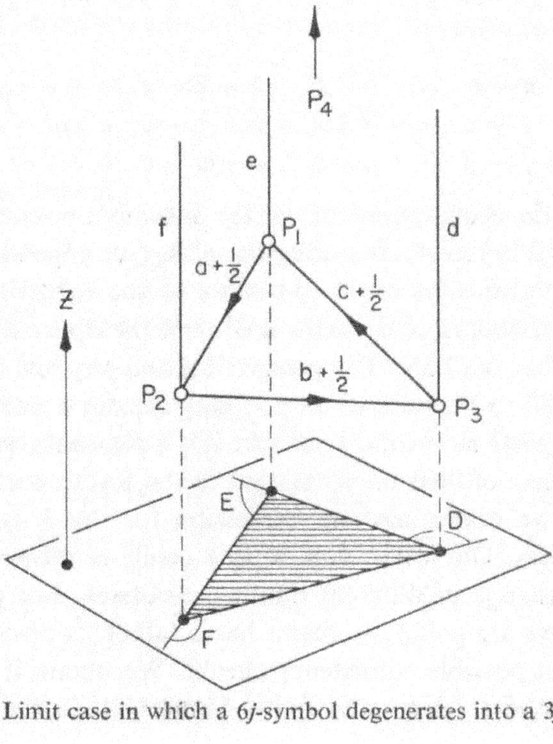

Fig. 2. Limit case in which a $6j$-symbol degenerates into a $3j$-symbol.

$m_a = m_b = m_c = 0$, the plane $P_1P_2P_3$ is perpendicular to z and $A = B = C = \frac{1}{2}\pi$; (2.6) reduces to the known[16] result

$$\begin{pmatrix} a & b & c \\ 0 & 0 & 0 \end{pmatrix} \simeq \tfrac{1}{2}[1 + (-1)^{a+b+c}](-1)^{\frac{1}{2}(a+b+c)}(2\pi A)^{-\frac{1}{2}}. \qquad (2.9)$$

The 2+2 case can be dealt with in much the same way. Using once more $\xi = x - 2R$ as summation variable, from (A.4), (2.2) we obtain

$$\begin{Bmatrix} a & b+R & c+R \\ d & e+R & f+R \end{Bmatrix} \simeq [(a+b-c)!\,(a-b+c)!\,(a+e-f)!\,\times$$
$$\times (a-e+f)!\,(d+e-c)!\,(d-e+c)!\,(d+b-f)!\,(d-b+f)!]^{\frac{1}{2}} \times$$
$$\times (2R)^{-2a-2d-b-c-e-f-1} \sum_{\xi} (-1)^{\xi} (2R)^{2\xi} [(a+d+b+e-\xi)!\,\times$$
$$\times (a+d+c+f-\xi)!\,(\xi-a-b-c)!\,\times$$
$$\times (\xi-a-e-f)!\,(\xi-c-d-e)!\,(\xi-b-d-f)!]^{-1}. \qquad (2.10)$$

For R large the main contribution to this summation comes from the largest allowed value of ξ, i.e. from the minimum between $a+d+b+c$ and $a+d+c+f$; therefore

$$\begin{Bmatrix} a & b+R & c+R \\ d & e+R & f+R \end{Bmatrix} \simeq (-1)^{a+d+\min(b+e,\,c+f)} \times$$
$$\times \left[\frac{(a-b+c)!\,(a-e+f)!\,(d-e+c)!\,(d-b+f)!}{(a+b-c)!\,(a+e-f)!\,(d+e-c)!\,(d+b-f)!}\right]^{\frac{1}{2}\mathrm{sign}(c+f-b-e)} \times$$
$$\times \frac{(2R)^{-|b+e-c-f|-1}}{|b+e-c-f|!}[1 + O(R^{-2})], \qquad (2.11)$$

where sign $(x) = +1, -1$ according to $x \geqslant 0$, < 0. It must be pointed out that unless $b+e = c+f$, the corresponding tetrahedron becomes hyperflat; in fact it turns out that $144\,V^2 = -4(b+c-e-f)^2R^4 + O(R^3)$.

The remaining particular case 2+1+1 will be discussed in some detail in appendix H.

3. Improvement of Wigner asymptotic formula

According to Wigner[13], the physical interpretation of the 6*j*-symbol is clearly related to its definition as a recoupling coefficient

$$|(j_1,(j_2,j_3)j_{23})J\rangle =$$
$$= \sum_{j_{12}} [(2j_{12}+1)(2j_{23}+1)]^{\frac{1}{2}}(-1)^{j_1+j_2+j_3+J} \begin{Bmatrix} j_1 & j_2 & j_{12} \\ j_3 & J & j_{23} \end{Bmatrix} \times$$
$$\times |((j_1,j_2)j_{12},j_3)J\rangle. \qquad (3.1)$$

It follows that

$$(2j_{12}+1)(2j_{23}+1)\begin{Bmatrix} j_1 & j_2 & j_{12} \\ j_3 & J & j_{23} \end{Bmatrix}^2 \mathrm{d}j_{23}, \tag{3.2}$$

is the probability that the sum of the angular momenta $\boldsymbol{j}_2$ and $\boldsymbol{j}_3$ has the length in the interval $j_{23}, j_{33}+\mathrm{d}j_{23}$ whenever $\boldsymbol{j}_1$ and $\boldsymbol{j}_2$ have sum of length j_{12} and $\boldsymbol{j}_{12}=\boldsymbol{j}_1+\boldsymbol{j}_2$ is coupled with $\boldsymbol{j}_3$ to a vector $\boldsymbol{J}$ of length J. As anticipated in the introduction, the mutual relationship of these vectors is best seen on a diagram (fig. 3).

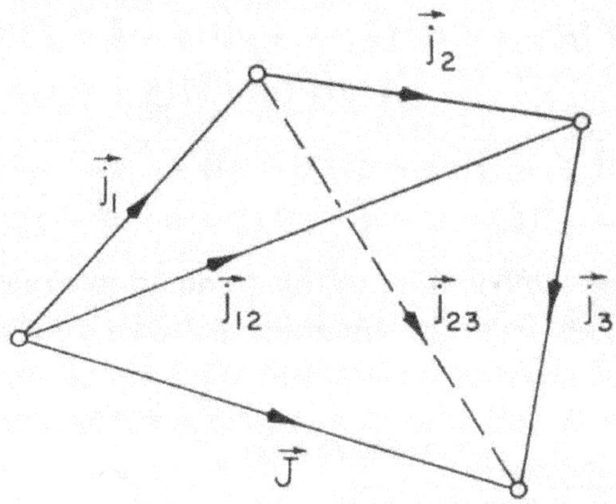

Fig. 3. Recoupling scheme corresponding to (3.1).

Let j_1, j_2, j_3, j_{12}, J (i.e. all the quantum numbers in the l.h.s. of (3.1)) be fixed; then the angle ψ between the plane of the vectors $\boldsymbol{j}_1, \boldsymbol{j}_2$ and the plane of the vectors $\boldsymbol{j}_3, \boldsymbol{J}$ is still undetermined, for both of them can rotate around the common axis $\boldsymbol{j}_{12}$. During this rotation the point P describes a circle; we assume[13]) that every point of this circle has equal probability. Then the probability that the length of $\boldsymbol{j}_{23}$ falls into the interval $j_{23}, j_{23}+\mathrm{d}j_{23}$ is just $\{2|\mathrm{d}\psi/\mathrm{d}j_{23}|/(2\pi)\}\,\mathrm{d}j_{23}$, the factor 2 being needed because there are two configurations corresponding to ψ and $2\pi-\psi$ which yield the same j_{23}. An elementary computation (appendix B) shows that

$$\frac{\mathrm{d}\psi}{\mathrm{d}j_{23}} = \frac{j_{12}j_{23}}{6V}, \tag{3.3}$$

where V is the volume of the tetrahedron in fig. 3. Therefore, for large angular momenta, we obtain Wigner's result (1.3a). If j_{23} is such that $V^2<0$, it is impossible to reach the prescribed value of j_{23} by varying ψ in the real interval $0-2\pi$.

The arguments given so far are clearly heuristic since they assume as granted a uniform probability distribution in ψ. A rigorous justification of this statement would take us too far and would destroy the simplicity of the discussion.

It is, however, interesting to notice that there is a variation to Wigner's argument which seems to have escaped detection so far. Let us suppose that in fig. 3 three of the vertices of the tetrahedron are held fixed, while the remaining one P is allowed to vary. We use J^2, j_{23}^2, j_3^2 as coordinates of P instead of the usual Euclidean coordinates (i.e., for instance, the components of $\boldsymbol{J}$: J_x, J_y, J_z). Notice that there are two points P corresponding to the same set J^2, j_{23}^2, j_3^2. We assume that the "a priori" probability for P to lie in the small volume $\mathrm{d}V = \mathrm{d}J_x\, \mathrm{d}J_y\, \mathrm{d}J_z$ does not depend on P. In this case, the probability that the "tricentrical" coordinates of P: J^2, j_{23}^2, j_3^2 lie in the interval $\mathrm{d}J^2\, \mathrm{d}j_{23}^2\, \mathrm{d}j_3^2$ is

$$2\left|\frac{\partial(J^2, j_{23}^2, j_3^2)}{\partial(J_x, J_y, J_z)}\right|^{-1} \mathrm{d}J^2\, \mathrm{d}j_{23}^2\, \mathrm{d}j_3^2 = 2\mathscr{J}\, \mathrm{d}J^2\, \mathrm{d}j_{23}^2\, \mathrm{d}j_3^2, \tag{3.4}$$

and since

$$\left|\frac{\partial(J^2, j_{23}^2, j_3^2)}{\partial(J_x, J_y, J_z)}\right| = 8\,|\boldsymbol{J} \times \boldsymbol{j}_{23} \cdot \boldsymbol{j}_3| = 48V,$$

using (1.3a) we have

$$\begin{Bmatrix} j_1 & j_2 & j_{12} \\ j_3 & J & j_{23} \end{Bmatrix}^2 \sim 2\frac{\mathscr{J}}{\pi}. \tag{3.5}$$

This second argument has the advantage that it can be formally generalized to higher $3nj$-symbols.

An interesting discussion, which leads to a generalization of Wigner's formula, can be developed by the combined use of the known results[16] for the 1+1+2 case and of the Biedenharn–Elliott identity[17]. According to Edmonds (eq. (A.2.2) of ref. 16) we have with our conventions of appendix A

$$\begin{Bmatrix} c & a & b \\ f & b+\delta & a+\delta' \end{Bmatrix} \simeq \frac{(-1)^{a+b+c+f+\delta}}{[(2a+1)(2b+1)]^{\frac{1}{2}}}\, d^{(f)}_{\delta,\delta'}(\theta), \tag{3.6}$$

where a, b, c are large in comparison with f, δ, δ' and (see fig. 4)

$$\cos\theta = \frac{a(a+1) + b(b+1) - c(c+1)}{2[a(a+1)b(b+1)]^{\frac{1}{2}}}, \qquad 0 \leqslant \theta \leqslant \pi. \tag{3.7}$$

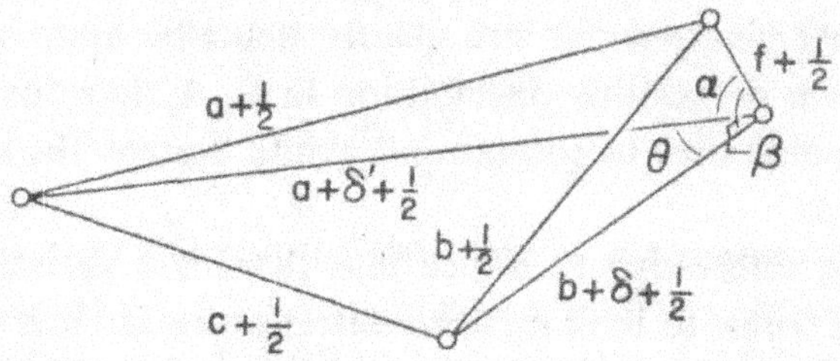

Fig. 4. Particular case in which f is small with respect to the other edges.

Let us recall here the just mentioned identity

$$Y \equiv \begin{Bmatrix} g & h & j \\ e & a & d \end{Bmatrix} \begin{Bmatrix} g & h & j \\ e' & a' & d' \end{Bmatrix} = \sum_x (-1)^{\varphi_x} (2x+1) \begin{Bmatrix} a & a' & x \\ d' & d & g \end{Bmatrix} \begin{Bmatrix} d & d' & x \\ e' & e & h \end{Bmatrix} \begin{Bmatrix} e & e' & x \\ a' & a & j \end{Bmatrix}, \tag{3.8}$$

$$\varphi_x = g + h + j + e + a + d + e' + a' + d' + x .$$

Now we intend to use (3.8) under the following conditions:

i) $g, h, j, e, a, d, e', a', d'$, are large;

ii) $e' - e = \eta$, $a' - a = \alpha$, $d' - d = \delta$ are small with respect to the parameters quoted in i). It follows that the $6j$-symbols which appear under summation in (3.8) are of a form suitable for the use of (3.6); we have:

$$\begin{Bmatrix} a & a' & x \\ d' & d & g \end{Bmatrix} \simeq \frac{(-1)^{a+d+g+\delta+x}}{[(2a+1)(2d+1)]^{\frac{1}{2}}} d^{(x)}_{\delta,\alpha}(\gamma),$$

$$\cos\gamma = \frac{a(a+1) + d(d+1) - g(g+1)}{2[a(a+1)d(d+1)]^{\frac{1}{2}}}, \quad 0 \leqslant \gamma \leqslant \pi, \tag{3.9}$$

$$\begin{Bmatrix} d & d' & x \\ e' & e & h \end{Bmatrix} \simeq \frac{(-1)^{d+e+h+\eta+x}}{[(2d+1)(2e+1)]^{\frac{1}{2}}} d^{(x)}_{\eta,\delta}(\chi),$$

$$\cos\chi = \frac{d(d+1) + e(e+1) - h(h+1)}{2[d(d+1)e(e+1)]^{\frac{1}{2}}}; \quad 0 \leqslant \chi \leqslant \pi, \tag{3.10}$$

$$\begin{Bmatrix} e & e' & x \\ a' & a & j \end{Bmatrix} \simeq \frac{(-1)^{e+a+j+\alpha+x}}{[(2e+1)(2a+1)]^{\frac{1}{2}}} d^{(x)}_{\alpha,\eta}(\iota),$$

$$\cos\iota = \frac{e(e+1) + a(a+1) - j(j+1)}{2[e(e+1)a(a+1)]^{\frac{1}{2}}}, \quad 0 \leqslant \iota \leqslant \pi . \tag{3.11}$$

If we now replace (3.9)–(3.11) into (3.8) and assume that we may perform the asymptotic limit under the infinite summation on x, we obtain an expression for the product of symbols in the l.h.s. of (3.8) where *all* angular momenta are large. Clearly this procedure is incorrect, but, nevertheless, it turns out that it is very instructive. We find

$$\begin{Bmatrix} g & h & j \\ e & a & d \end{Bmatrix} \begin{Bmatrix} g & h & j \\ e+\eta & a+\alpha & d+\delta \end{Bmatrix} \sim$$
$$\sim [(2a+1)(2e+1)(2d+1)]^{-1} \sum_{x=0}^{\infty} (2x+1)\, d^{(x)}_{\alpha,\eta}(\iota)\, d^{(x)}_{\eta,\delta}(\chi)\, d^{(x)}_{\delta,\alpha}(\gamma). \tag{3.12}$$

The sum in the r.h.s. can be performed by exploiting the group representation properties of the functions $d^{(x)}_{\alpha,\eta}(\iota)$ as shown in appendix C. The result is

$$\begin{Bmatrix} g & h & j \\ e & a & d \end{Bmatrix} \begin{Bmatrix} g & h & j \\ e+\eta & a+\alpha & d+\delta \end{Bmatrix} \sim \frac{\Theta(V^2)}{24\pi V} \cos(\eta E + \alpha A + \delta D), \tag{3.13}$$

where $\Theta(V^2)=0$ or 1 according to $V^2<0$ or $V^2>0$; V is the volume of the tetrahedron T with edges $g+\frac{1}{2}$, $h+\frac{1}{2}$ etc. The angles E, A, D are defined by

$$\cos E = \frac{\cos\chi \cos\iota - \cos\gamma}{\sin\chi \sin i},$$
$$\cos A = \frac{\cos\iota\cos\gamma - \cos\chi}{\sin i \sin\gamma}, \quad \cos D = \frac{\cos\gamma\cos\chi - \cos\iota}{\sin\gamma\sin\chi} \tag{3.14}$$

and A, for instance, can be interpreted as the angle between outer normals of the two faces of T which have a as common edge. If $\delta=\alpha=\eta=0$, we find once again Wigner's result.

This procedure is in part disappointing because it fails to yield a complete description of the rapidly oscillating term $\mathscr{C}$ in (1.3b). However the result (3.13), when δ, α, $\eta\neq0$, is very illuminating because the r.h.s. contains the interference term $\cos(\eta E+\alpha A+\delta D)$ which, according to the point of view exposed in section 1, is an average over the product of the two rapidly oscillating factors of the two 6j-symbols.

In order to reconstruct the original expression, let us introduce the function

$$\Omega'(T) \equiv \Omega'\begin{pmatrix} a & b & c \\ d & e & f \end{pmatrix} = \sum_{h,k=1}^{4} j_{hk}\theta_{hk}, \tag{3.15a}$$

with notations defined in section 1. An interesting property of $\Omega'(T)$ is that, with obvious notations[18])

$$\Omega'(T+\delta T) - \Omega'(T) = \sum_{h,k=1}^{4} \delta j_{hk}\theta_{hk} \quad \text{or} \quad \frac{\partial\Omega'}{\partial j_{hk}} = \theta_{hk}, \tag{3.16}$$

i.e. we may vary the parameters in Ω' as if θ_{hk} were constant. Therefore we deduce readily that

$$\delta\Omega' = \Omega'\begin{pmatrix} g & h & j \\ e+\eta & a+\alpha & d+\delta \end{pmatrix} - \Omega'\begin{pmatrix} g & h & j \\ e & a & d \end{pmatrix} = \eta E + \alpha A + \delta D,$$

and the r.h.s. is just the argument of the cosine in (3.13). Notice that V and $\delta\Omega'$ are slowly varying functions of the edges as compared to Ω' itself. This result suggests a formula of the following type:

$$\begin{Bmatrix} a & b & c \\ d & e & f \end{Bmatrix} \simeq \frac{1}{\sqrt{12\pi V}} \cos(\Omega' + \omega), \tag{3.17}$$

where ω is a yet unknown constant phase. ω can be determined by matching (3.17) to the particular $1+1+2$ case studied first by Racah[1]):

$$\begin{Bmatrix} a & b & c \\ b & a & f \end{Bmatrix} \simeq \frac{(-1)^{a+b+c+f}}{\sqrt{(2a+1)(2b+1)}} P_f(\cos\theta), \tag{3.18}$$

where $\cos\theta$ is given by (3.7). If f is large, but small with respect to a, b, c, we may replace the Legendre polynomial with its asymptotic behaviour[19]):

$$P_f(\cos\theta) \simeq \left[\frac{2}{\pi(f+\frac{1}{2})\sin\theta}\right]^{\frac{1}{2}} \cos\left[(f+\tfrac{1}{2})\theta - \tfrac{1}{4}\pi\right], \tag{3.19}$$

from which we deduce:

$$\begin{Bmatrix} a & b & c \\ b & a & f \end{Bmatrix} \simeq \frac{(-1)^{a+b+c+f}}{\sqrt{12\pi V}} \cos\left[(f+\tfrac{1}{2})\theta - \tfrac{1}{4}\pi\right], \tag{3.20a}$$

having noticed that $6V \simeq (a+\frac{1}{2})(b+\frac{1}{2})(f+\frac{1}{2})\sin\theta$. On the other hand, in order to work out how Ω' depends on f, we can write $\Omega'(f+\frac{1}{2})$ explicitly for this particular case. We have (fig. 4 with $\delta=\delta'=0$): $\Omega'(0)=\pi(a+b+c+\frac{3}{2})$ and $(\partial\Omega'/\partial f)_{f+\frac{1}{2}=0}=\pi-\theta$. Therefore,

$$\Omega'(f+\tfrac{1}{2}) \simeq \Omega'(0) + \left(\frac{\partial\Omega}{\partial f}\right)_{f+\frac{1}{2}=0}(f+\tfrac{1}{2}) \simeq$$
$$\simeq \pi(a+b+c+\tfrac{3}{2}) + (\pi-\theta)(f+\tfrac{1}{2}). \tag{3.21}$$

Taking into account these results, we see that (3.20a) can be rewritten as

$$\begin{Bmatrix} a & b & c \\ b & a & f \end{Bmatrix} \simeq \frac{1}{(12\pi V)^{\frac{1}{2}}} \cos(\Omega' + \tfrac{1}{4}\pi), \tag{3.20b}$$

which shows not only that (3.1) is compatible with this particular case, but also tells us that $\omega=\frac{1}{4}\pi$. We have reached, therefore, the important general formula

$$\begin{Bmatrix} a & b & c \\ d & e & f \end{Bmatrix} \simeq \frac{1}{(12\pi V)^{\frac{1}{2}}} \cos \Omega, \tag{1.4a}$$

$$\Omega = \sum_{h,k=1}^{4} j_{hk}\theta_{hk} + \tfrac{1}{4}\pi, \tag{3.15b}$$

which must be supplemented with explicit formulae for the angles θ_{hk} (appendix B):

$$\cos \theta_{hk} = -\frac{9}{A_h A_k} \frac{\partial V^2}{\partial (j_{rs}^2)}, \qquad 0 \leqslant \theta_{rs} \leqslant \pi \tag{3.22}$$

where $h \neq k \neq r \neq s = 1, 2, 3, 4$; A_h are defined in (1.5).

TABLE 1

Numerical examples of $6j$-symbols showing the degree of approximation of our asymptotic formulae. The cases denoted with I, II, III correspond to the use of (1.4), (5.6), or (5.7), (1.8) respectively. The indices $a, b, \ldots, f$ are the same as in fig. 1. The exact values of this table as well as of table 2 are taken from A. F. Nikiforov et al., *Tables of Racah coefficients* (New York, 1965); as usual, .1–01, for instance, means 10^{-2}

a	b	c	d	e	f	Exact value	Approximate value	
1	1	1	1/2	1/2	1/2	−.33333–00	−.37828–00	I
1	1	1	1	1	1	.16667–00	.16683–00	I
7/2	7	9/2	17/2	5	5/2	−.41785–01	−.41520–01	I
17/2	15/2	10	15/2	15/2	4	.16494–01	.16422–01	I
13/2	8	9/2	13/2	6	15/2	.25518–01	.25506–01	I
5	8	12	9	7	6	−.22441–01	−.22422–01	I
9	9	9	9	9	9	−.15647–01	−.15640–01	I
1/2	1	1/2	1/2	1	1/2	.16667–00	.16026–00	II
7/2	7	9/2	17/2	5	17/2	.19826–01	.18954–01	II
4	13/2	15/2	8	9/2	17/2	−.20120–01	−.19897–01	II
17/2	15/2	9	15/2	15/2	13	−.13801–01	−.13773–01	II
17/2	15/2	10	15/2	13/2	12	.99633–02	.97944–02	II
5	8	10	9	7	10	−.21100–01	−.19971–01	II
8	9	14	9	8	10	.13420–01	.13296–01	II
9/2	5	1/2	11/2	6	3/2	.13222–01	.13549–01	III
17/2	6	9/2	17/2	6	7/2	.10386–02	.10411–02	III
5	8	13	9	7	9	−.19671–02	−.19726–02	III
17/2	15/2	10	15/2	13/2	14	.49191–03	.49301–03	III
5	8	13	9	7	12	−.11052–04	−.10443–04	III
8	9	17	9	8	13	.59756–06	.56161–06	III
8	9	17	9	8	16	.31622–09	.28579–09	III

4. Arguments in favor of the proposed asymptotic formula

We list here some arguments in favour of (1.4).

i) Our formula is numerically accurate, as can be seen from tables 1–4 and figs. 5–7. On the average, the accuracy improves as the values of the angular momenta increase.

TABLE 2

Numerical results for the 6*j*-symbol with $a=7$, $b=8$, $c=9$, $d=6$, $e=9$ as $f \equiv J$ assumes all permissible values

J	Exact value	Approximate value		J	Exact value	Approximate value	
2	.76018–02	.76923–02	III	9	−.20540–01	−.20578–01	I
3	−.22627–01	−.22171–01	II	10	−.31376–02	−.36479–02	I
4	.29469–01	.30474–01	I	11	.20586–01	.20528–01	I
5	−.13704–01	−.13212–01	I	12	.20068–01	.19019–01	II
6	−.13529–01	−.13944–01	I	13	.84777–02	.78298–02	II
7	.20388–01	.20345–01	I	14	.16637–02	.16638–02	III
8	.34782–02	.37774–02	I				

TABLE 3

Case with $a=13$, $b=15$, $c=24$, $d=29/2$, $e=33/2$ and $f \equiv J$ variable. The exact values of this table as well as of table 4 were obtained by means of the recursion formula (4.3)

J	Exact value	Approximate value		J	Exact value	Approximate value	
7/2	−.17136–01	−.18812–01	I	35/2	−.66442–02	−.66033–02	II
9/2	−.94803–02	−.94381–02	I	37/2	−.38116–02	−.37155–02	II
11/2	.59559–02	.60861–02	I	39/2	−.17980–02	−.18980–02	III
13/2	.11188–01	.11222–01	I	41/2	−.70668–03	−.72809–03	III
15/2	.25847–02	.25778–02	I	43/2	−.23231–03	−.23652–03	III
17/2	−.78677–02	−.78773–02	I	45/2	−.63708–04	−.64401–04	III
19/2	−.85291–02	−.85619–02	I	47/2	−.14450–04	−.14536–04	III
21/2	−.13122–03	−.19186–03	I	49/2	−.26704–05	−.26756–05	III
23/2	.77958–02	.77680–02	I	51/2	−.39262–06	−.39197–06	III
25/2	.81257–02	.82093–02	I	53/2	−.44251–07	−.43997–07	III
27/2	.17590–02	.19590–02	I	55/2	−.35949–08	−.35537–08	III
29/2	−.56229–02	−.54326–02	I	57/2	−.18760–09	−.18332–09	III
31/2	−.95185–02	−.94911–02	II	59/2	−.47264–11	−.44115–11	III
33/2	−.92382–02	−.87903–02	II				

TABLE 4

Case with $a = 35/2$, $b = 39/2$, $c = 9$, $d = 31/2$, $e = 37/2$ and $f \equiv J$ variable

J	Exact value	Approximate value		J	Exact value	Approximate value	
4	.21149–03	.20909–03	III	20	−.66008–02	−.65600–02	I
5	−.69682–03	−.71066–03	III	21	.29687–02	.29011–02	I
6	.16116–02	.16753–02	III	22	.16649–02	.17306–02	I
7	−.30296–02	−.32386–02	III	23	−.57355–02	−.57788–02	I
8	.49034–02	.47498–02	II	24	.76269–02	.76402–02	I
9	−.70293–02	−.70467–02	II	25	−.63213–02	−.63099–02	I
10	.90494–02	.88460–02	II	26	.20313–02	.20080–02	I
11	−.10502–01	−.99536–02	II	27	.34992–02	.35214–02	I
12	.10918–01	.11351–01	I	28	−.73734–02	−.73873–02	I
13	−.99508–02	−.99826–02	I	29	.68216–02	.68250–02	I
14	.75021–02	.73395–02	I	30	−.12056–02	−.11953–02	I
15	−.38167–02	−.35839–02	I	31	−.61644–02	−.61915–02	I
16	−.49920–03	−.72271–03	I	32	−.83350–02	−.83482–02	I
17	.45603–02	.47270–02	I	33	.10743–03	.24555–03	I
18	−.73894–02	−.74789–02	I	34	−.11326–01	−.11727–01	I
19	.81842–02	.81987–02	I	35	−.58537–02	−.56155–02	II

ii) (1.4) is obviously invariant under the exchange of the vertices of T. As stated before, this is only a subgroup of the full symmetry group R_2 of $6j$-symbols (appendix A). We checked, however, in a somewhat laborious way, that both V and Ω are actually invariant under R_2. The proof is sketched in appendix D.

iii) It has been shown [20]) that the identities (A.6), (A.7), (2.8) together with the tetrahedral symmetries, are enough to derive all properties as well as the numerical values of $6j$-symbols, apart from an overall phase. In particular, the Biedenharn–Elliott identity and the recursion relations which follow from it are a distinctive feature of Racah coefficients. Therefore it is a highly significant result that our formula satisfies asymptotically not only these recursion relations, but also the above mentioned identities.

An intuitive understanding of these formulae can be reached by a correspondence principle of the form:

$$\frac{1}{\mathrm{i}} \frac{\partial}{\partial j_{hk}} \sim \theta_{hk} ,$$

where θ_{hk} are the angles appearing in Ω. By the same token we define the

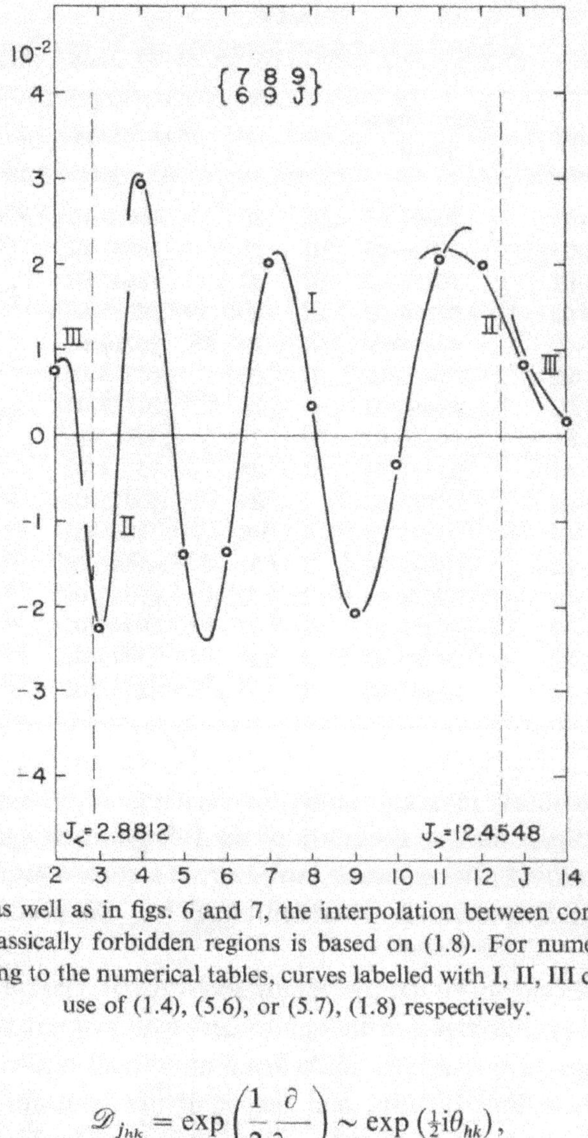

Fig. 5. Here, as well as in figs. 6 and 7, the interpolation between contiguous physical points in the classically forbidden regions is based on (1.8). For numerical values, see table 2. According to the numerical tables, curves labelled with I, II, III correspond to the use of (1.4), (5.6), or (5.7), (1.8) respectively.

operator

$$\mathscr{D}_{j_{hk}} = \exp\left(\frac{1}{2}\frac{\partial}{\partial_{j_{hk}}}\right) \sim \exp\left(\tfrac{1}{2}\mathrm{i}\theta_{hk}\right), \tag{4.1}$$

so that, for instance:

$$\mathscr{D}_a^2 \begin{Bmatrix} a & b & c \\ d & e & f \end{Bmatrix} = \begin{Bmatrix} a+1 & b & c \\ d & e & f \end{Bmatrix}$$

and from (4.1)

$$(\mathscr{D}_a^2 + \mathscr{D}_a^{-2}) \begin{Bmatrix} a & b & c \\ d & e & f \end{Bmatrix} = $$

$$= \begin{Bmatrix} a+1 & b & c \\ d & e & f \end{Bmatrix} + \begin{Bmatrix} a-1 & b & c \\ d & e & f \end{Bmatrix} \sim 2\cos\theta_a \begin{Bmatrix} a & b & c \\ d & e & f \end{Bmatrix}. \tag{4.2}$$

Let us consider[21])

$$a[(a+b+c+2)(a-b+c+1)(a+b-c+1)\times \\ \times(b+c-a)(a+e+f+2)(a-e+f+1)(a+e-f+1)\times \\ \times(e+f-a)]^{\frac{1}{2}}\begin{Bmatrix} a+1 & b & c \\ d & e & f \end{Bmatrix}+(a+1)[(a+b+c+1)\times \\ \times(a-b+c)(a+b-c)(b+c-a+1)(a+e+f+1)\times \\ \times(a-e+f)(a+e-f)(e+f-a+1)]^{\frac{1}{2}}\begin{Bmatrix} a-1 & b & c \\ d & e & f \end{Bmatrix}= \\ =(2a+1)\{2[a(a+1)d(d+1)-b(b+1)e(e+1)-c(c+1)\times \\ \times f(f+1)]+[a(a+1)-b(b+1)-c(c+1)][a(a+1)+ \\ -e(e+1)-f(f+1)]\}\begin{Bmatrix} a & b & c \\ d & e & f \end{Bmatrix}; \tag{4.3a}$$

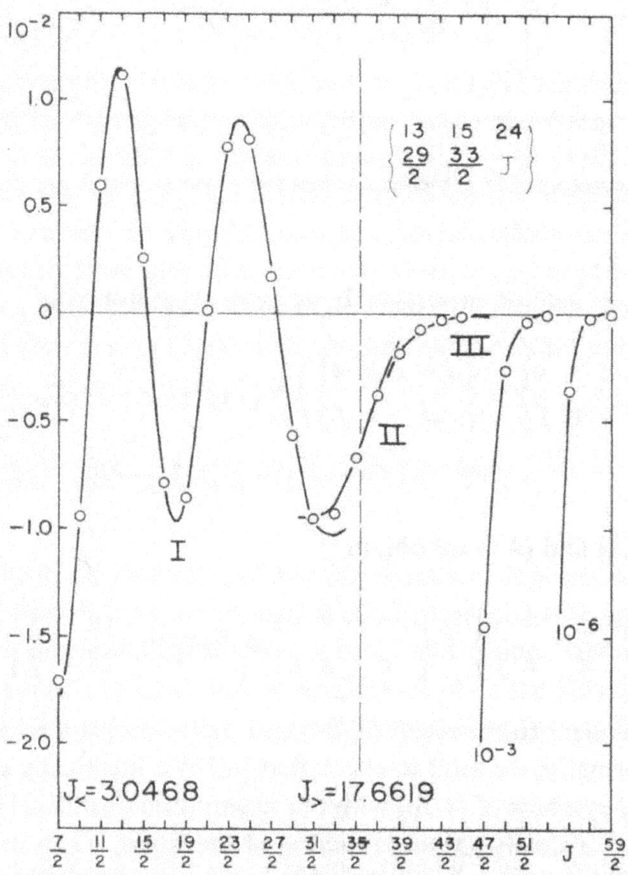

Fig. 6. In this case, which corresponds to table 3, the exponential decrease is particularly emphasized.

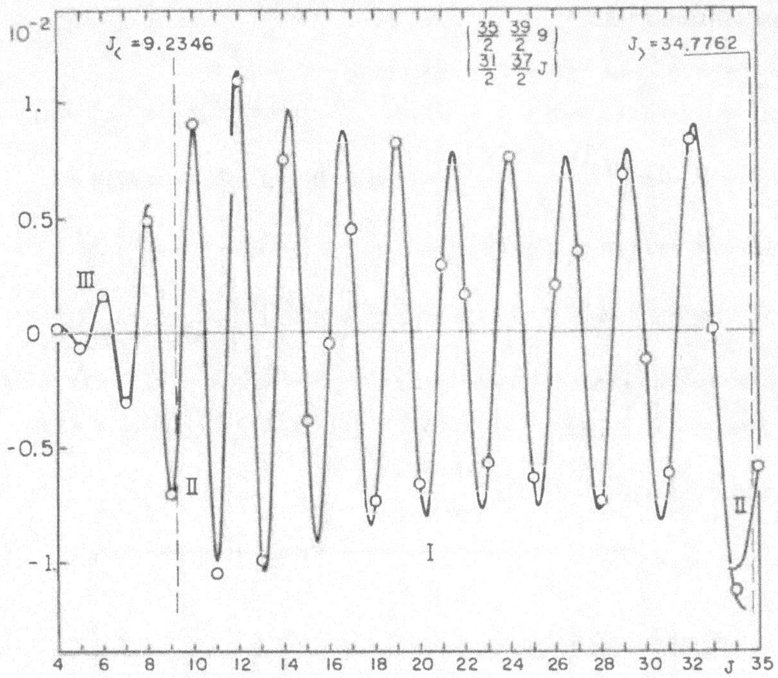

Fig. 7. The corresponding numerical values for physical points are given in table 4.

using notations defined in section 1, we have asymptotically

$$8A_1A_2\left(\begin{Bmatrix} a+1 & b & c\\ d & e & f\end{Bmatrix}+\begin{Bmatrix} a-1 & b & c\\ d & e & f\end{Bmatrix}\right)\simeq[2(j_{12}^2j_{34}^2-j_{13}^2j_{24}^2-j_{14}^2j_{23}^2)+$$
$$+(j_{12}^2-j_{13}^2-j_{14}^2)(j_{12}^2-j_{24}^2-j_{23}^2)]\begin{Bmatrix} a & b & c\\ d & e & f\end{Bmatrix}.$$

Recalling (B.2) and (4.1) we obtain

$$\begin{Bmatrix} a+1 & b & c\\ d & e & f\end{Bmatrix}+\begin{Bmatrix} a-1 & b & c\\ d & e & f\end{Bmatrix}\simeq 2\cos\theta_{12}\begin{Bmatrix} a & b & c\\ d & e & f\end{Bmatrix}, \tag{4.3b}$$

which is equivalent to the result (4.2) based on the correspondence principle (4.1). Less formally, we want to check that (4.3b) is identically satisfied if we replace the 6*j*-symbols of (4.3b) with our asymptotic formula (1.4). Since the volume of T is a slowly varying function of the edges, it can be regarded as constant in the three symbols of (4.3b); in this case (4.3b) transforms into the trivial identity $\cos(\Omega+\theta_{12})+\cos(\Omega-\theta_{12})\simeq 2\cos\theta_{12}\cos\Omega$.

In a similar way, the recursion relation

$$[(a+b+c+1)(b+c-a)(c+d+e+1)(c+d-e)]^{\frac{1}{2}} \begin{Bmatrix} a & b & c \\ d & e & f \end{Bmatrix} +$$
$$- [(a+b-c+1)(a-b+c)(e+d-c+1)(e+c-d)]^{\frac{1}{2}} \times$$
$$\times \begin{Bmatrix} a & b & c-1 \\ d & e & f \end{Bmatrix} = -2c[(b+d+f+1)(b+d-f)]^{\frac{1}{2}} \times$$
$$\times \begin{Bmatrix} a & e & f \\ d-\frac{1}{2} & b-\frac{1}{2} & c-\frac{1}{2} \end{Bmatrix}$$

by means of (1.4) becomes asymptotically

$$[(j_{13}+j_{12}+j_{14})(j_{13}+j_{14}-j_{12})(j_{14}+j_{24}+j_{34}) \times$$
$$\times (j_{14}+j_{34}-j_{24})]^{\frac{1}{2}} e^{\frac{1}{2}i(\theta_{13}+\theta_{14}+\theta_{34})} - [(j_{12}+j_{13}-j_{14}) \times$$
$$\times (j_{12}+j_{14}-j_{13})(j_{24}+j_{34}-j_{14})(j_{24}+j_{14}-j_{34})]^{\frac{1}{2}} e^{\frac{1}{2}i(\theta_{13}-\theta_{14}+\theta_{34})} \simeq$$
$$\simeq -2j_{14}[(j_{13}+j_{34}+j_{23})(j_{13}+j_{34}-j_{23})]^{\frac{1}{2}},$$

which, using Delambre's relations [22]), reduces to simple identities.

A much more involved computation is needed to show that also the full Biedenharn–Elliott identity is satisfied asymptotically by (1.4). Introducing (1.4) into the r.h.s. of (3.8), we realize that inside the summation over x there appears a rapidly varying function of x. Its behaviour can be displayed most transparently if we split all cosines into positive and negative frequency parts according to Euler's formula. Let T_j ($j=1, 2, 3$) be the tetrahedra corresponding to the r.h.s. of (3.8); with obvious notations we have

$$\prod_{j=1}^{3} \cos \Omega_j = 2^{-3} \times$$
$$\times [e^{i(\Omega_1+\Omega_2+\Omega_3)} + e^{i(\Omega_1+\Omega_2-\Omega_3)} + e^{i(\Omega_1-\Omega_2+\Omega_3)} + e^{i(\Omega_1-\Omega_2-\Omega_3)} + \text{c.c.}]. \tag{4.4}$$

Given the heuristic character of our investigation, it is reasonable to assume that the discrete summation over x can be replaced with an integration whose most important contribution arises from points where the phase $\Gamma(x)=\Omega_1(x)+\Omega_2(x)+\Omega_3(x)$ and its analogs of (4.4) are slowly varying as functions of x. We are led quite naturally to consider, for example

$$\frac{\partial \Gamma(x)}{\partial x} = 0. \tag{4.5}$$

Denoting the supplementary dihedral angles relative to x with θ_x^1, θ_x^2, θ_x^3, a heartening result is that, since in any first order variation of the edges we may consider all angles as constants, we have as necessary condition for (4.5)

$$\theta_x^1 + \theta_x^2 + \theta_x^3 - \pi = 0, \tag{4.6}$$

having taken into account the phase $(-1)^{\varphi_x}$ of (3.8). If we consider the other terms in (4.4) as well, we find that the bulk of the contribution to the integral may come only from those values of x such that

$$\pm\theta_x^1 \pm \theta_x^2 \pm \theta_x^3 = \pi. \tag{4.7}$$

The conditions (4.7) have an immediate geometrical interpretation if we look at the diagram in fig. 8a. There one sees that by leaving out in turn any one of the five points $P_1, \ldots, P_5$, the remaining points form five tetrahedra $T_1, \ldots, T_5$ which are just those appearing in (3.8). Note that there are ten edges connecting five points in all possible ways and in fact there are ten angular momenta appearing in (3.8) including x. As it has been long known, if five points are imbedded into a three dimensional-Euclidean space, their mutual distances are not independent. The explicit form of their dependence was discovered by Cayley[11]) and can be written as

$$-2^4(4!)^2 I^2 \equiv \begin{vmatrix} 0 & j_{12}^2 & j_{13}^2 & j_{14}^2 & j_{15}^2 & 1 \\ j_{12}^2 & 0 & j_{23}^2 & j_{24}^2 & j_{25}^2 & 1 \\ j_{13}^2 & j_{23}^2 & 0 & j_{34}^2 & j_{35}^2 & 1 \\ j_{14}^2 & j_{24}^2 & j_{34}^2 & 0 & j_{45}^2 & 1 \\ j_{15}^2 & j_{25}^2 & j_{35}^2 & j_{45}^2 & 0 & 1 \\ 1 & 1 & 1 & 1 & 1 & 0 \end{vmatrix} = 0, \tag{4.8}$$

where j_{hk} is the distance between P_h and P_k. It is crucial to understand that (4.8) is in fact equivalent to (4.7). Indeed (4.7) implies that the sum of the internal dihedral angles $\pi - \theta_x^j$ around the edge x is a multiple of 2π, as expected if the diagram is drawn in three dimensions. The ambiguity in the signs of the angles arises from the different possible orientations of the five involved tetrahedra, as exemplified by fig. 8b.

Therefore, the only asymptotical contribution to the integral comes from the configuration of the diagram in fig. 8a and the like, which are three-dimensional, i.e. from those values of x such that $I(x^2)=0$. First we notice that the range of summation is restricted to $x>0$ and, "a fortiori," to $x^2>0$. Secondly, $I(x^2)$ is a quadratic polynomial in x^2 and it has therefore two roots $(x_<)^2$ and $(x_>)^2$. There are rigorous arguments showing that, if T_4, T_5 are physical tetrahedra, i.e. $V_4^2, V_5^2>0$, then $x_<$ and $x_>$ are real. To us it will be enough to note that, quite obviously, the smaller root $x_<$ corresponds to the configuration in which P_4, P_5 lie on the same side of the plane

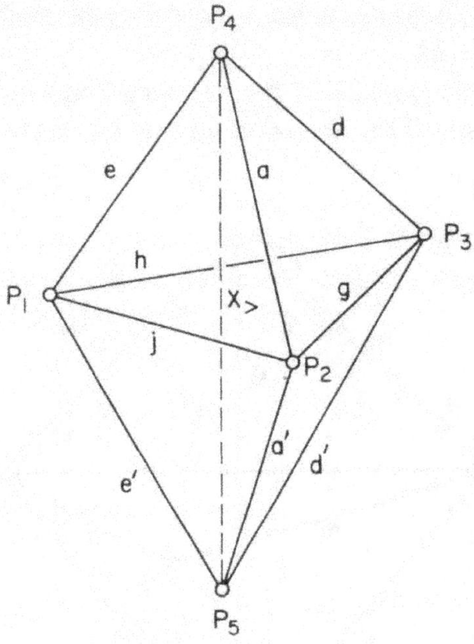

(a)

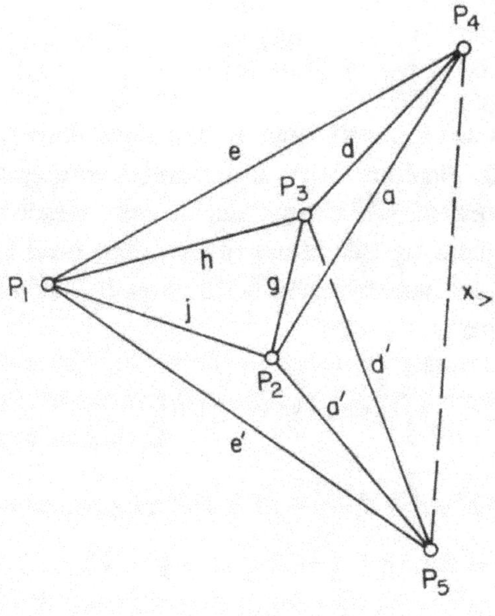

(b)

Fig. 8.

$P_1P_2P_3$ as depicted in fig. 9, while $x_>$ corresponds to P_4,P_5 being on opposite sides as in fig. 8a.

Since in general θ_x^j $(j=1, 2, 3)$ are not vanishing, or are equal to some multiple of π, only one of the choices of signs in (4.7) is valid for a given root.

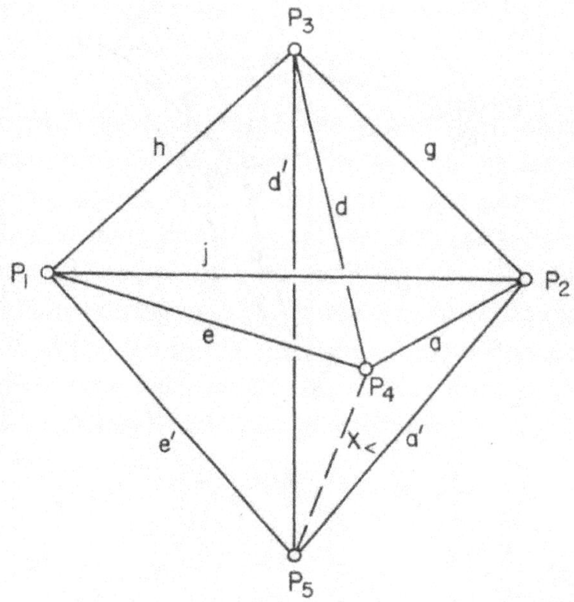

Fig. 9.

It follows that, for $x=x_<$, only one of the eight terms arising from the decomposition (4.4), together with its complex conjugate, does actually contribute to the integral. We cannot decide here which term contributes, because this will depend on the values of the other fixed angular momenta. Let us suppose they are such that for both roots $\theta_x^1+\theta_x^2+\theta_x^3=\pi$.

From (3.8) we have

$$Y \simeq \sum_x \frac{\pi^{-\frac{3}{2}}}{96\sqrt{3}} e^{i\pi(\omega+x)} x \left[V_1(x)\, V_2(x)\, V_3(x)\right]^{-\frac{1}{2}} \left[e^{i\Gamma(x)} + e^{-i\Gamma(x)}\right], \tag{4.9}$$

where $\omega=a+e+d+a'+e'+d'+g+j+h$. Let us write now

$$x+\tfrac{1}{2}=R\xi;\;\; a+\tfrac{1}{2}=R\alpha,\;\; e+\tfrac{1}{2}=R\eta,\;\; d+\tfrac{1}{2}=R\delta,\;\; \ldots,\;\; d'+\tfrac{1}{2}=R\delta' \tag{4.10}$$

with R large and ξ, α, η, δ, α', η', δ' finite; note that the angles θ_{hk} of the five tetrahedra are independent of R. The spacing in the summation on ξ is now

R^{-1} and, as anticipated, we may replace the summation with an integral on ξ

$$Y \simeq \frac{\pi^{-\frac{3}{2}}}{96\sqrt{3}} \mathrm{e}^{-\mathrm{i}\pi(\omega-\frac{3}{4})} R^2 \int_{\Xi} \mathrm{d}\xi\, \xi [V_1(\xi)\, V_2(\xi)\, V_3(\xi)]^{-\frac{1}{2}}\, \mathrm{e}^{\mathrm{i}R[\tilde{\Gamma}(\xi)-\pi\xi]} + \text{c.c.} \tag{4.11}$$

having used the fact that $\omega + x$ is integer; here $\Gamma(x) = R\tilde{\Gamma}(\xi) + \frac{1}{4}\pi$. In the limit $R \to \infty$ this integral can be computed with the steepest descent method[23]), which yields for an integral of the form

$$\mathscr{I} = \int_{\alpha}^{\beta} g(\xi)\, \mathrm{e}^{\mathrm{i}Rf(\xi)}\, \mathrm{d}\xi, \tag{4.12}$$

with f and g real, the approximate result

$$\mathscr{I}_{R\to\infty} \cong \sum_j \left(\frac{2\pi}{R\,|f''(\xi_j)|}\right)^{\frac{1}{2}} g(\xi_j)\, \mathrm{e}^{\mathrm{i}[Rf(\xi_j)\pm\frac{1}{4}\pi]}; \tag{4.13}$$

here ξ_j are such that $f'(\xi_j) = 0$ and $\alpha < \xi_j < \beta$. In (4.13) the phases $\pm\frac{1}{4}\pi$ must be chosen according to $f''(\xi_j) \gtrless 0$. If $\xi_{\gtrless}$ correspond to $x_{\gtrless}$, since we have supposed

$$\left[\frac{\partial[\tilde{\Gamma}(\xi) - \pi\xi]}{\partial\xi}\right]_{\xi=\xi_{\gtrless}} = [\theta_x^1 + \theta_x^2 + \theta_x^3 - \pi]_{x=x_{\gtrless}} = 0, \tag{4.14}$$

we find in our case

$$f(\xi) = \tilde{\Gamma}(\xi) - \pi\xi; \quad g(\xi) = \xi[V_1(\xi)\, V_2(\xi)\, V_3(\xi)]^{-\frac{1}{2}},$$
$$f''(\xi) = \frac{\partial(\theta_x^1 + \theta_x^2 + \theta_x^3 - \pi)}{\partial\xi}. \tag{4.15}$$

A naive computation of $f''(\xi)$ is out of question because of the lengthy and uninspiring algebra involved. We rather take $I^2(x^2)$, defined in (4.8) as independent variable and write

$$f''(\xi_{\gtrless}) = \lim_{\xi\to\xi_{\gtrless}} \left[\frac{\partial(\theta_x^1 + \theta_x^2 + \theta_x^3 - \pi)}{\partial I^2(x^2)} \frac{\partial I^2(x^2)}{\partial\xi}\right]; \tag{4.16}$$

some manipulation of determinants (appendix D) shows that

$$f''(\xi_{\gtrless}) = \mp\tfrac{1}{6}R^3(\xi_{\gtrless})^2 \left[\frac{V_4 V_5}{V_1 V_2 V_3}\right]_{\xi=\xi_{\gtrless}}. \tag{4.17}$$

The evaluation of $f(\xi_{\gtrless})$ is straightforward and yields

$$Rf(\xi_{\gtrless}) \mp \tfrac{1}{4}\pi = \Omega_4 \pm \Omega_5 + \pi(\omega - \tfrac{3}{4}). \tag{4.18}$$

From (4.11)–(4.18) we obtain

$$Y \simeq \frac{1}{4}\frac{1}{12\pi(V_4V_5)^{\frac{1}{2}}}\left[e^{i(\Omega_4+\Omega_5)} + e^{i(\Omega_4-\Omega_5)} + \text{c.c.}\right] \tag{4.19}$$

$$Y \simeq \frac{1}{(12\pi V_4)^{\frac{1}{2}}}\cos\Omega_4 \frac{1}{(12\pi V_5)^{\frac{1}{2}}}\cos\Omega_5\,,$$

which is in obvious agreement with the straightforward use of (1.4) in the l.h.s.

We indulged somewhat more than strictly necessary on the proof of the Biedenharn–Elliott identity in the asymptotic limit, because we felt that the mechanism involved is illuminating and more general than shown by this case. We do not discuss whether the other identities of Racah coefficients: (A.6), (A.7) are satisfied by (1.4), since these proofs follow quite easily from the stationary-phase method. The same procedure can be extended in principle to the computation of asymptotic 9*j*-symbols.

5. A formal analogy with the WKB method

In the previous section we strived to provide as many as possible independent checks and counterchecks for the validity of (1.4) in the region $A(V^2>0)$. As stated in our Introduction, a complete description of the behaviour of a 6*j*-symbol for large angular momenta must include of necessity a similar formula for the region $B(V^2<0)$ and makes it highly desirable to have one for the transitional region $V^2 \simeq 0$ as well.

The guiding idea in this section is a formal analogy with a WKB approximation for the solution of a differential equation. This analogy was prompted by the actual existence of a differential equation at least in the $1+1+2$ case, where in fact the symbol can be expressed as a Jacobi polynomial.

Unfortunately we have not been able to derive a single differential equation valid for unrestricted large parameters, with the possible exception of the transitional region. Our disappointment is somehow mitigated by noticing that, after all, we need a differential equation only in order to fix unambiguously the prosecution of (1.4) through a transition point. This we have achieved and the resulting formulae satisfy properties which are similar to the ones listed under i, ii, iii in section 4. However, it must be stated that the WKB analogy is just a formal device which cannot be accepted as a proof.

Before proceeding further along this analogy, a more detailed investigation of the region B is necessary. From (B.3) we see that if $V^2=0$ and $\prod_{h=1}^{4} A_h \neq 0$, then the angles θ_{hk} are all multiple of π. Since we always choose $0 \leqslant \theta_{hk} \leqslant \pi$, we have that either $\theta_{hk}=0$ or $\theta_{hk}=\pi$; the ambiguity can be settled by means of (3.22). Here, as in the following, we always assume that the areas A_h are represented by positive square roots of radicands like (B.1). Because of the conditions satisfied by angular momenta and taking into account our definition of j_{hk}, it is easy to see that these radicands are indeed always non-negative and may vanish only if $V^2=0$, in which case (see footnote 12), at least one angular momentum assumes a non-physical value.

If $V^2=0$, the four vertices of T lie in the same plane; Let Q be the convex plane set generated by the four vertices. Depending on their relative position, Q may have three or four edges (see figs. 10 and 1). As the reader can

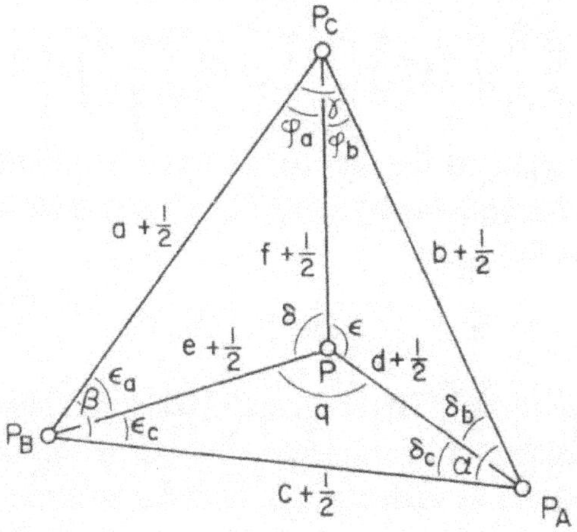

Fig. 10.

easily check, the rule is then: $\theta_{hk}=\pi$ for the edges of Q and $\theta_{hk}=0$ for the others. We may use the symbol $\begin{Bmatrix} \pi & \pi & \pi \\ 0 & 0 & 0 \end{Bmatrix}$ to denote the set $\theta_{12}=\theta_{13}=\theta_{14}=\pi$, $\theta_{34}=\theta_{24}=\theta_{23}=0$. The only possibilities are $\begin{Bmatrix} \pi & \pi & \pi \\ 0 & 0 & 0 \end{Bmatrix}$, $\begin{Bmatrix} \pi & 0 & 0 \\ 0 & \pi & \pi \end{Bmatrix}$, $\begin{Bmatrix} 0 & 0 & \pi \\ \pi & \pi & 0 \end{Bmatrix}$, $\begin{Bmatrix} 0 & \pi & 0 \\ \pi & 0 & \pi \end{Bmatrix}$, $\begin{Bmatrix} \pi & \pi & 0 \\ \pi & \pi & 0 \end{Bmatrix}$, $\begin{Bmatrix} \pi & 0 & \pi \\ \pi & 0 & \pi \end{Bmatrix}$, $\begin{Bmatrix} 0 & \pi & \pi \\ 0 & \pi & \pi \end{Bmatrix}$. The subsets with some $A_h=0$ lie on the boundary of the above sets. For instance, the case $j_{12}=j_{13}+j_{14}$ is the common boundary of $\begin{Bmatrix} \pi & 0 & 0 \\ 0 & \pi & \pi \end{Bmatrix}$, $\begin{Bmatrix} 0 & \pi & \pi \\ 0 & \pi & \pi \end{Bmatrix}$ and, in fact, here θ_{12}, θ_{13}, θ_{14} are discontinuous.

If $V^2 < 0$, then the angles θ_{hk} are all of the form $n\pi + i\xi_{hk}$. Because of continuity, it follows that $\mathrm{Re}\,\theta_{hk}$ are constants in B whenever $\prod_{h=1}^{4} A_h \neq 0$. Therefore we may subdivide B into non-overlapping regions, all points in the same region having the same values of $\mathrm{Re}\,\theta_{hk}$. If we let $V \to 0$ and $\prod_{h=1}^{4} A_h \neq 0$, we see that $\lim_{V \to 0} \mathrm{Re}\,\theta_{hk} = \theta_{hk}(V=0)$ so the value of $\mathrm{Re}\,\theta_{hk}$ is determined by the limiting process for $V \to 0$ and is the same as listed above. We may subdivide B into regions $B\begin{pmatrix}\pi & \pi & \pi \\ 0 & 0 & 0\end{pmatrix}$, $B\begin{pmatrix}\pi & \pi & 0 \\ \pi & \pi & 0\end{pmatrix}$, etc. such that we have, for instance, in $B\begin{pmatrix}\pi & \pi & \pi \\ 0 & 0 & 0\end{pmatrix}$: $\mathrm{Re}\,\theta_{12} = \mathrm{Re}\,\theta_{13} = \mathrm{Re}\,\theta_{14} = \pi$, $\mathrm{Re}\,\theta_{23} = \mathrm{Re}\,\theta_{24} = \mathrm{Re}\,\theta_{34} = 0$; it is also convenient to use the following abbreviations:

$$B_3 \equiv B\begin{pmatrix}\pi & \pi & \pi \\ 0 & 0 & 0\end{pmatrix} \cup B\begin{pmatrix}\pi & 0 & 0 \\ 0 & \pi & \pi\end{pmatrix} \cup B\begin{pmatrix}0 & 0 & \pi \\ \pi & \pi & 0\end{pmatrix} \cup B\begin{pmatrix}0 & \pi & 0 \\ \pi & 0 & \pi\end{pmatrix},$$

$$B_4 \equiv B\begin{pmatrix}\pi & \pi & 0 \\ \pi & \pi & 0\end{pmatrix} \cup B\begin{pmatrix}\pi & 0 & \pi \\ \pi & 0 & \pi\end{pmatrix} \cup B\begin{pmatrix}0 & \pi & \pi \\ 0 & \pi & \pi\end{pmatrix}.$$

These regions are not closed, for they have common boundary points in which some A_h's vanish. As anticipated in (1.7), in every such region we define a phase function Φ as follows

$$\Phi = \sum_{h,k=1}^{4} (j_{hk} - \tfrac{1}{2})\,\mathrm{Re}\,\theta_{hk}\,.$$

For physical values of the angular momenta, Φ is always an integer multiple of π.

We fix now the value of all parameters but one. Without loss of generality we may always choose a as variable and, in order to stress this point, we use x instead of a. We do not restrict ourselves to physical values of x and of the remaining parameters in order to retain some, albeit nominal, freedom in the final results. If $V^2(x_{\gtrless}) = 0$ we have

$$\begin{aligned} \Omega(x_{\gtrless}) &= \Phi - \tfrac{1}{4}\pi \quad \text{approaching } B_3\,, \\ \Omega(x_{\gtrless}) &= \Phi + \tfrac{1}{4}\pi \quad \text{approaching } B_4\,. \end{aligned} \tag{5.1}$$

In appendix F it is shown that for $x \to x_{\gtrless}$:

$$\Omega - \Omega(x_{\gtrless}) = \begin{cases} +\dfrac{9}{2}\dfrac{V^3}{\prod_{h=1}^{4} A_h} & \text{approaching } B_3\,, \\[2ex] -\dfrac{9}{2}\dfrac{V^3}{\prod_{h=1}^{4} A_h} & \text{approaching } B_4\,. \end{cases} \tag{5.2}$$

It follows that in the neighbourhood of a transition point we have

$$\begin{Bmatrix} a & b & c \\ d & e & f \end{Bmatrix} \simeq \frac{1}{\sqrt{12\pi V}} \cos\left(\frac{9}{2}\frac{V^3}{\prod_{h=1}^{4} A_h} + \Phi - \tfrac{1}{4}\pi\right); \tag{5.3}$$

obviously this formula is incorrect if one gets too close to the transition points. We notice however that, as it stands, (5.3) is the WKB asymptotic approximation to the following differential equation (see (G.6)):

$$\frac{\mathrm{d}}{\mathrm{d}(V^2)}\frac{\mathrm{d}}{\mathrm{d}(V^2)}\psi = \left(\frac{27}{4\prod_{h=1}^{4} A_h}\right)^2 V^2\psi, \tag{5.4}$$

where the independent variable is V, $\prod_{h=1}^{4} A_h$ being treated as a constant[24]. The general solution of (5.4) is (appendix G):

$$\psi \sim V\left[c_1 J_{\frac{1}{3}}\left(\frac{9}{2}\frac{V^3}{\prod_{h=1}^{4} A_h}\right) + c_2 J_{-\frac{1}{3}}\left(\frac{9}{2}\frac{V^3}{\prod_{h=1}^{4} A_h}\right)\right] \tag{5.5}$$

and the one which joins smoothly with (5.3) for large values of $V^2>0$ is:

$$\begin{Bmatrix} a & b & c \\ d & e & f \end{Bmatrix} \simeq 2^{-\frac{4}{3}}\left(\prod_{h=1}^{4} A_h\right)^{-\frac{1}{6}} \times$$
$$\times \left\{\cos\Phi\, \mathrm{Ai}\left[-\left(\frac{(3V)^2}{\left(4\prod_{h=1}^{4} A_h\right)^{\frac{2}{3}}}\right)\right] + \sin\Phi\, \mathrm{Bi}\left[-\left(\frac{(3V)^2}{\left(4\prod_{h=1}^{4} A_h\right)^{\frac{2}{3}}}\right)\right]\right\}. \tag{5.6}$$

in terms of Airy functions[25]. We assume (5.6) to be the correct asymptotic form of the Racah coefficient in the transitional region. The soundness of this assumption is of course at this stage purely aesthetical. However, our conjecture is borne out by comparison with published tables (see tables 1–4).

According to the standard procedure, we may continue (5.6) into the region B; here the resulting formula in the neighbourhood of a transition point is:

$$\begin{Bmatrix} a & b & c \\ d & e & f \end{Bmatrix} \simeq 2^{-\frac{4}{3}}\left(\prod_{h=1}^{4} A_h\right)^{-\frac{1}{6}} \times$$
$$\times \left\{\cos\Phi\, \mathrm{Ai}\left[\frac{(3|V|)^2}{\left(4\prod_{h=1}^{4} A_h\right)^{\frac{2}{3}}}\right] + \sin\Phi\, \mathrm{Bi}\left[\frac{(3|V|)^2}{\left(4\prod_{h=1}^{4} A_h\right)^{\frac{2}{3}}}\right]\right\}. \tag{5.7}$$

For large values of $|V|^2$ this function joins smoothly with

$$\begin{Bmatrix} a & b & c \\ d & e & f \end{Bmatrix} \simeq \frac{1}{2(12\pi|V|)^{\frac{1}{2}}} \{2 \sin \Phi \, \mathrm{e}^{|\mathrm{Im}\, \Omega|} + \cos \Phi \, \mathrm{e}^{-|\mathrm{Im}\, \Omega|}\}, \tag{5.8}$$

where $\mathrm{Im}\, \Omega = \sum_{h,k=1}^{4} \mathrm{Im}\, \theta_{hk}$. Choosing conventionally in region B: $V = \mathrm{i}|V|$, the imaginary part of θ_{hk} can be retrieved by means of the following formulae:

$$\cosh(\mathrm{Im}\, \theta_{hk}) = -\cos(\mathrm{Re}\, \theta_{hk}) \frac{9}{A_h A_k} \frac{\partial V^2}{\partial j_{rs}^2}, \quad h \neq k \neq r \neq s, \tag{3.22a}$$

$$\sinh(\mathrm{Im}\, \theta_{hk}) = \cos(\mathrm{Re}\, \theta_{hk}) \frac{3\, j_{hk}}{2\, A_h A_k} |V|, \quad h \neq k. \tag{1.5a}$$

According to (5.8), the $6j$-symbol would be represented in region B by a superposition of decreasing and increasing exponentials; however, the coefficient $\sin \Phi$ of the increasing exponential vanishes at the physical points, where (5.8) reduces to (1.8). It turns out that (5.8) is numerically accurate when applied to physical angular momenta.

It is worth noticing that the coefficient $\cos \Phi = (-1)^{\Phi/\pi}$ of the decreasing exponential gives instead a determination of the sign of the $6j$-symbol. This sign is in complete agreement with numerical tables, as well as with the one obtained from the limiting case of a stretched tetrahedron where one edge reaches its maximum permissible value. As an example, let $a = b + c$; from (A.4) we have:

$$\text{sign of} \begin{Bmatrix} b+c & b & c \\ d & e & f \end{Bmatrix} = (-1)^{b+c+e+f}. \tag{5.9}$$

Since $b + c > x_> - \frac{1}{2}$, it is a simple exercise to see that if a increases through $x_>$, then we enter either one of the regions $B\begin{pmatrix} 0 & \pi & \pi \\ 0 & \pi & \pi \end{pmatrix}$, $B\begin{pmatrix} \pi & 0 & 0 \\ 0 & \pi & \pi \end{pmatrix}$, $B\begin{pmatrix} \pi & \pi & \pi \\ 0 & 0 & 0 \end{pmatrix}$. The region $B\begin{pmatrix} \pi & \pi & \pi \\ 0 & 0 & 0 \end{pmatrix}$ is exluded because the point $a = b + c$ is almost on the boundary between $B\begin{pmatrix} 0 & \pi & \pi \\ 0 & \pi & \pi \end{pmatrix}$ and $B\begin{pmatrix} \pi & 0 & 0 \\ 0 & \pi & \pi \end{pmatrix}$; actually, it would be exactly on this boundary if: $a + \frac{1}{2} = b + \frac{1}{2} + c + \frac{1}{2}$, or $a - b - c = \frac{1}{2}$ which is prevented by selection rules. Therefore our phase is $\Phi = \pi(b + c + e + f) = \pi(a + e + f)$ and it agrees with (5.9).

Further evidence in favour of (5.6), (5.8) is offered by the discussion of the particular case: $2 + 1 + 1$; this analysis is carried out in appendix H where it is shown that the behaviour of (3.6) in region A as well as in B is in agreement with (5.8). We notice also that the arguments of the Airy functions in (5.6), (5.7) as well as Φ are invariant under R_2.

We have now a complete set of conjectured asymptotic formulae valid for every range of large *physical* angular momenta. Numerical examples deduced from these formulae are shown in tables 1–4 and in figs. 5–7.

6. The 3*nj*-symbols

The problem of extending our results to higher 3*nj*-symbols is certainly very difficult and we were not able to reach a solution within the frame of this paper. Yet some of the intuitive arguments presented here provide fairly interesting information about the general problem.

In dealing with 3*nj*-symbols where *n* is large the use of diagrams becomes imperative. At first the diagrams, as in the current literature[7]), are just a mnemonical device in order to keep track of the growing complexities of the symbols. They provide an information, which is purely combinatorial, on how angular momenta are coupled in a given scheme. In this sense a very natural language for diagrams is provided by combinatorial topology.

A diagram is essentially a shorthand notation for the expansion of a 3*nj*-symbol in terms of 3*j*-symbols. Let [D] be the 3*nj*-symbol corresponding to the diagram D. [D] can be expressed as:

$$[\mathrm{D}] = \sum_{\text{all } m} \begin{pmatrix} l_1 & l_2 & l_3 \\ m_1 & m_2 & m_3 \end{pmatrix} \cdots \begin{pmatrix} l_i & l_j & l_k \\ m_i & m_j & m_k \end{pmatrix} \begin{pmatrix} & l_k & \\ m_k & & m_k' \end{pmatrix} \begin{pmatrix} l_k & l_p & l_q \\ m_k' & m_p & m_q \end{pmatrix} \cdots, \qquad (6.1)$$

where[4]):

$$\begin{pmatrix} & l_k & \\ m_k & & m_k' \end{pmatrix} = (-1)^{l_k - m_k} \delta_{m_k, -m'_k} . \qquad (6.2)$$

D can be retrieved from (6.1) by means of the following rules:

a) D is a 2-dimensional combinatorial manifold.

b) There is a one-to-one correspondence between 1-simplexes (edges) of D and angular momenta in the r.h.s. of (6.1).

c) There is a one-to-one correspondence between 3*j*-symbols in (6.1) and 2-simplexes (faces) of D.

d) The boundary of a face is the sum of the three edges appearing as angular momenta in the corresponding 3*j*-symbol.

e) We work for the time being with homology modulo 2, i.e. we forget about the orientation of the simplexes.

From the general structure of [D] we see that there are 2*n* faces and 3*n* edges in D. We have no "a priori" conditions on 0-simplexes (vertices) and in fact they have no physical meaning. This brings in a certain amount of arbitrariness in the construction of D which is lacking in the customary definition of diagrams[6]) (figs. 12a, b). We regard as equivalent diagrams those which yield the same symbol. Since D is a manifold, we may define its Euler characteristic:

$$M = -f + e - v + 2 = n - v + 2,$$

where f, e, v are the number of faces, edges, vertices respectively. Since $M \geqslant 0$, we have $v \leqslant n+2$. For a $6j$-symbol, $M=0$, while in the case of a $9j$-symbol we have: $n=3$, $v=4$ and $M=1$ according to the example sketched in fig. 11. When $M=1$ the diagram is the triangulation of a one-sided surface,

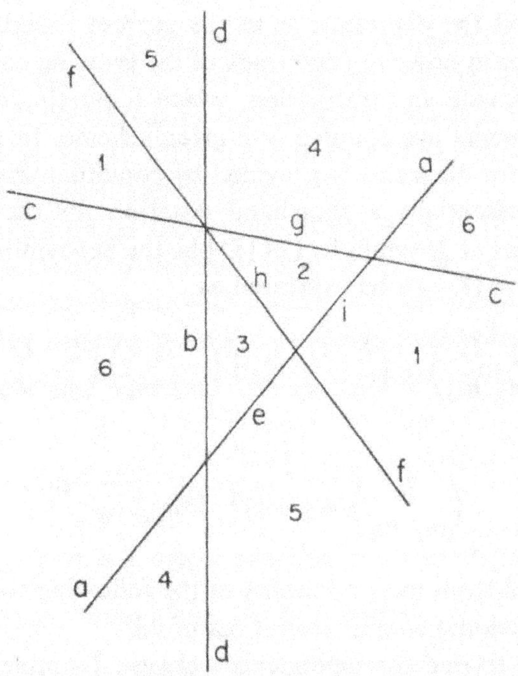

Fig. 11. Planar graphical representation of the $9j$-symbol with triads (abc), (def), (ghi). (adg), (beh), (cfi).

in our case the real projective plane. Since the sphere where opposite points are identified is a homeomorph of the projective plane, it is possible to exhibit the $9j$-symbol as a double hexagonal pyramid as shown in fig. 12a; here each edge and face are repeated twice and the whole diagram has a centre of symmetry. We shall prefer this representation to the one given in fig. 11.

In fact, in dealing with the asymptotic behaviour and with the semiclassical limit of $3nj$-symbols, we expect that it will be possible to represent angular momenta as vectors and the coupling of angular momenta as the addition of classical vectors. It follows that it will be convenient to think of a diagram as a polyhedron imbedded into a 3-dimensional Euclidean space. If one

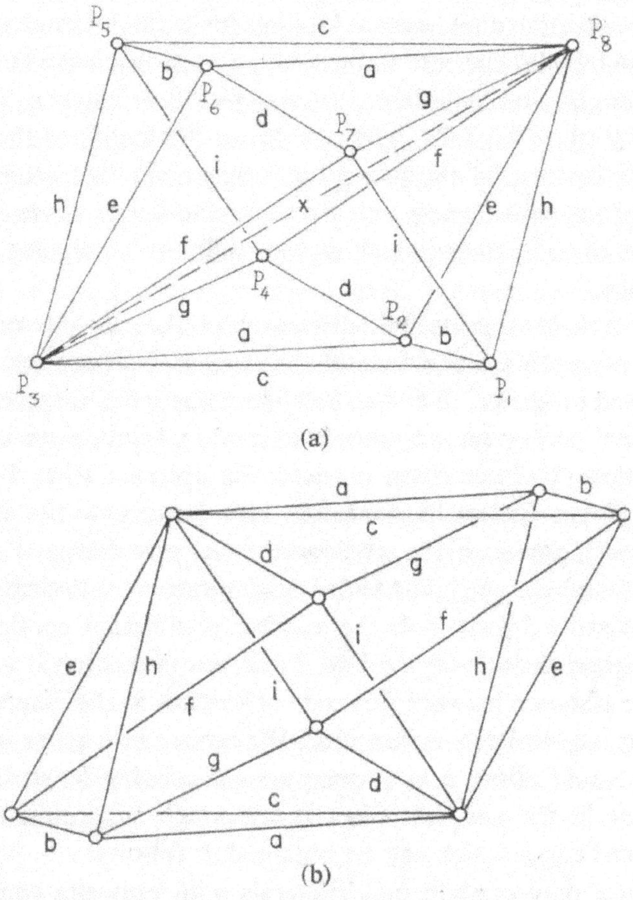

Fig. 12. Different three-dimensional diagrams for the $9j$ of fig. 11; note that the edges in case a have the same length and direction as in b.

such a construction exists with prescribed lengths of the edges, we shall speak of a configuration of the diagram.

An interesting phenomenon is that there are several different configurations with identical edges for the same diagram. This ambiguity is connected with the fact that there is uniqueness for a given polyhedron of given topology and given edges only if the polyhedron is convex. A trivial example is a pair of mirror-like right-handed and left-handed tetrahedra. A less trivial example arises already with the $9j$-symbol and we expect it to become more and more important for higher symbols. The technical reasons of this multiplicity can be best seen in the $9j$-symbols. Here the geometrical shape of the object would be completely determined if we knew all the relative angles of all the edges of the diagram. Elementary theorems tell us that this can be achieved if the angle we are looking for is the internal angle of some triangle of which all edges are known; this is true in particular for all faces in the diagram. A similar attempt to compute, for instance, the angle between b and f (fig. 12a) fails unless we know the length of the diagonal x. In the particular case of the $9j$-symbol it turns out that all angles can be computed, provided we know this only missing length x; as we shall see, there may be in principle as much as four different configurations with the same topology.

Since the classification and the discussion of these configurations is relevant to the asymptotic behaviour of the symbol, it is convenient to refine the so far used language. We shall introduce the word diagram when only the topological properties are considered and in doing so we identify equivalent diagrams. Configuration is instead a diagram with the additional information about the angles needed in order to remove the above ambiguities. We may introduce the additional word orientation if a distinction between left-handed and right-handed configurations is desired.

For complicated $3nj$-symbols the number of different configurations for the same diagram grows very rapidly. From our discussion it is clear that if one gives the distance between any pair of vertices in the diagram, then the configuration is completely determined. We cannot give here a general set of rules which would allow one to compute the missing lengths. We found, however, that in the simplest cases it is enough to exploit the relations among squared edges which can be obtained as follows:

a) the diagram may contain quadrilaterals with opposite equal edges; in this case the sum of the squared diagonals is twice the sum of the two different squared edges, a known and elementary relation;

b) since the diagram is imbedded into a 3-dimensional space, one can write for any choice of five vertices the Cayley identity (4.8).

Not all these relations are actually independent. Once a complete and consistent set of identities has been written, one finds a set of algebraic equations for the missing lengths; to each solution of these equations we associate a configuration.

We come now to the general problem of computing the $3nj$-symbols.

In the available literature[7]) explicit formulae are written for any symbol with the aid of diagrams in terms of lower order coefficients. The diagrams used in these previous works are not the same as ours; however it is not difficult to translate one language into the other.

To this purpose we introduce another diagram $\mathscr{D}$(D) which corresponds to the familiar procedure of dissecting the interior of the polyhedron into tetrahedra; more formally, $\mathscr{D}$(D) can be conveniently defined as a 3-dimensional combinatorial manifold with boundary D. We name cells the 3-simplexes of $\mathscr{D}$. The edges, faces and vertices of $\mathscr{D}$ will be named external if they belong to D, else internal. Let the set of cells be labelled by T_k, $k=1, 2, \ldots, p$.

From the definition of the symbol [D], we know that it is a function of as many variables as different edges of D. These variables take up integer or half-integer values with the selection rule that the sum of the variables along the boundary of any 2-cycle is always integer. We assume the function [D] to be given by the usual Racah formula (A.4) when D is a tetrahedron. In what follows we shall give a sketchy account of a set of rules which allow the computation of [D] for any D.

In order to evaluate [D], we first construct a given $\mathscr{D}$(D). We associate variables x_i, $i=1, 2, \ldots, q$ to all internal edges of $\mathscr{D}$(D) and variables l_j, $j=1, 2, \ldots r$ to the external ones. In this way, to each cell T_k, considered as a diagram, we may associate $[T_k]$ which is clearly a function of the internal and, possibly, of the external edges of $\mathscr{D}$(D). Then we form the product:

$$A(x_1, \ldots, x_q) = \prod_{k=1}^{p} [T_k]\cdot(-1)^{\chi} \prod_{i=1}^{q} (2x_i + 1). \tag{6.3}$$

We found so far no combinatorial rule to construct χ, which applies to any diagram. If D and $\mathscr{D}$(D) are homeomorphs of a 2-sphere and of a 3-ball, then we have in general:

$$\chi = \sum_{j=1}^{q} (n_j - 2)x_j + \chi_0,$$

where n_j is the number of tetrahedra belonging to x_j and χ_0 is a fixed phase chosen in order to make χ integer. For simplicity we shall limit ourselves to this case. Let us now consider the sum over all internal variables

$$S = \sum_{x_1} \ldots \sum_{x_q} A(x_1, \ldots, x_q). \tag{6.4}$$

If there are no internal vertices, the sum is finite and $S=[D]$. On the contrary, if there are internal vertices, the sum is infinite but it is still possible to renormalize it in such a way to obtain [D]. When the immediate goal is just the computation of [D] with the simplest possible method, then there is no need to introduce this additional complication, for one can always find a $\mathscr{D}$(D) with no internal vertices. However the more general case is relevant in suggesting a formal analogy with the Feynman summation over histories[26]) in connection with the theory of relativity; we shall discuss this point later.

Coming back to (6.3), (6.4) and supposing that there are no internal vertices, we may attempt to evaluate the summation in (6.4) using the same methods already tested for the Biedenharn–Elliott identity (section 4). In this case we shall replace each $[T_k]$ with its asymptotic behaviour according to (1.4). Moreover, we shall split each cosine according to Euler's formula. The function A will then appear as the sum of 2^p pairwise conjugate terms. It is also convenient to replace each factor $(-1)^{n_j x_j}$ with $\exp[\pm i\,\pi(n_j-2)\,x_j]$. This procedure, which is clearly correct only for integer x_j, can be easily extended also to half-integer summation indices. Therefore A will contain, among the others, a term of the form

$$\prod_{j=1}^{q}(2x_j+1)\exp\left\{i\left[\left(\sum_{k=1}^{p_j}\theta_j^k\right)-\pi p_j+2\pi\right]x_j\right\}. \tag{6.5}$$

As before, we may try to replace the summation with an integral; we expect that the most important contributions to the integral will arise from the points where the phase is stationary with respect to the q variables x_j. Imposing the stationary phase condition, we find:

$$\sum_{k=1}^{p_j}(\pi-\theta_j^k)=2\pi, \tag{6.6}$$

which means that the sum of internal angles around x_j is just 2π. A discussion similar to the one carried out in section 4 brings to the conclusion that (6.6) implies the existence of a configuration, in the sense defined above, imbedded in a 3-dimensional Euclidean space, where the internal and external lengths are well specified. Because of the lack of internal vertices, the internal edges connect external vertices of the diagram and in fact they are sufficient to specify the configuration completely. A similar discussion can be carried out for the other 2^p-1 terms. In this way we see that the final result will be a sum of contributions from each configuration. We do not know of any simple rule to compute the general form of the partial second derivatives:

$$\partial\left[\sum_{k=1}^{p_j}\theta_j^k\right]/\partial x_j \tag{6.7}$$

needed in order to carry out, up to the end, the evaluation of (6.4). This lack of knowledge stops us here short of the final result. However the above discussion shows already that the study of configurations is certainly relevant to a complete understanding of the semiclassical limit of the $3nj$-symbols.

As anticipated, we point out a curious connection between our asymptotic formulae and a simplified quantization "à la Feynman" occurring in a 3-dimensional Euclidean theory of gravitation. The classical counterpart of this theory is trivial because the Einstein field equation for empty space

$$R_{\mu\nu} - \tfrac{1}{2}g_{\mu\nu}R = 0 \tag{6.8}$$

implies that the complete Riemann tensor vanishes, i.e., the space is flat. However, the connection we point out may be relevant for further generalizations to less academical cases.

We begin by discussing the sum (6.4) when there are internal vertices. As stated, this sum is infinite, but it is rather interesting to see how this infinity actually arises. For simplicity we restrict ourselves to the case when D is a tetrahedron and there is only one internal vertex P_5 (fig. 13). In this case (6.4) reads:

$$S = \sum_{xyzt} (2x+1)(2y+1)(2z+1)(2t+1)(-1)^{x+y+z+t+a+b+c+d+e+f} \times$$
$$\times \begin{Bmatrix} a & b & c \\ x & y & z \end{Bmatrix} \begin{Bmatrix} f & e & a \\ y & z & t \end{Bmatrix} \begin{Bmatrix} d & b & f \\ z & t & x \end{Bmatrix} \begin{Bmatrix} c & e & d \\ t & x & y \end{Bmatrix} \tag{6.9}$$

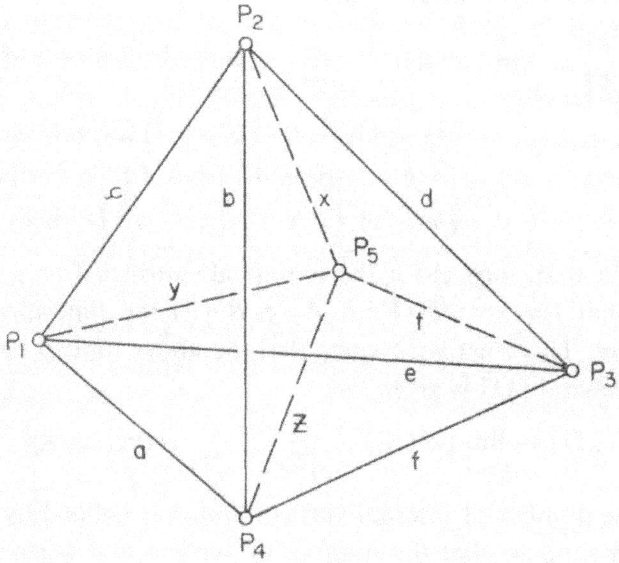

Fig. 13.

having chosen $\chi_0 = a+b+c+d+e+f$; the summation is carried out on all x, y, z, t which are compatible with the selection rules. Without loss of generality we may suppose x integer; using (3.8) the summation over t yields:

$$S = \begin{Bmatrix} a & b & c \\ d & e & f \end{Bmatrix} \sum_{xyz} (2x+1)(2y+1)(2z+1) \begin{Bmatrix} a & b & c \\ x & y & z \end{Bmatrix}^2, \tag{6.10}$$

and from (A.6):

$$S = \begin{Bmatrix} a & b & c \\ d & e & f \end{Bmatrix} \sum_{xy} \frac{\delta_{xyc}}{2c+1} (2x+1)(2y+1), \tag{6.11}$$

where δ_{xyc} is a triangular delta which is equal to unity if x, y, c satisfy triangular inequalities and zero otherwise. Since

$$\sum_{y=|x-c|}^{x+c} (2y+1) = (2x+1)(2c+1),$$

we have

$$S = \begin{Bmatrix} a & b & c \\ d & e & f \end{Bmatrix} \sum_{x=0}^{\infty} (2x+1)^2, \tag{6.12}$$

which is infinite and correspondingly meaningless. Let us limit the summation on x up to $x=R$, with R large; in this case we find

$$\mathscr{R}(R) \equiv \sum_{x=0}^{R} (2x+1)^2 \simeq \frac{1}{\pi}\frac{4\pi R^3}{3}. \tag{6.13}$$

This result hints that we may write:

$$\begin{aligned}[\mathrm{D}] \equiv \begin{Bmatrix} a & b & c \\ d & e & f \end{Bmatrix} = \lim_{R\to\infty} (\mathscr{R}(R))^{-1} \sum_{x,y,z,t<R} (-1)^{a+b+c+d+e+f+x+y+z+t} \times \\ \times (2x+1)(2y+1)(2z+1)(2t+1) \times \\ \times \begin{Bmatrix} a & b & c \\ x & y & z \end{Bmatrix} \begin{Bmatrix} f & e & a \\ y & z & t \end{Bmatrix} \begin{Bmatrix} d & b & f \\ z & t & x \end{Bmatrix} \begin{Bmatrix} c & e & d \\ t & x & y \end{Bmatrix}.\end{aligned} \tag{6.14}$$

In fact we have that, for fixed x, the summand vanishes if $z>x+b$, $y>x+c$, $t>x+d$ so that for $x<\min(R-b, R-c, R-d)$ the limitations $y, z, t<R$ have no weight. Therefore we assume that the above limit is correct and expect that in general [D] is given by:

$$[\mathrm{D}] = \lim_{R\to\infty} (\mathscr{R}(R))^{-P} \sum_{x_1<R} \dots \sum_{x_q<R} A(x_1, \dots, x_q) \tag{6.15}$$

where P is the number of internal vertices and A is defined by (6.3).

Now let us suppose that the number of vertices and edges in D and in $\mathscr{D}$(D) is very high. Let also the complex $\mathscr{D}$(D) approach a differentiable manifold $\mathscr{M}$ with boundary D. According to a discussion carried out in a previous paper by one of us[27]), the sum $\sum_{j=1}^{q}[\sum_{k=1}^{p_j}(\pi-\theta_j^k)]x_j$ approaches the integral $8\pi\mathscr{L}(\mathscr{M})=\frac{1}{2}\int_{\mathscr{M}}R\,\mathrm{d}V$, where $\mathrm{d}V$ is the volume element on $\mathscr{M}$ and R the scalar Riemann curvature of $\mathscr{M}$. We see therefore that the positive frequency part S^+ of S in some sense looks like:

$$S^+ = \frac{1}{\mathscr{R}^P}\int_{\partial\mathscr{M}=\mathrm{D}} \mathrm{e}^{\mathrm{i}\mathscr{L}(\mathscr{M})}\,\mathrm{d}\mu(\mathscr{M}), \tag{6.16}$$

where the summations over the variables x_j have been interpreted as an integration over all the manifolds $\mathscr{M}$ with fixed boundary D. The measure $\mathrm{d}\mu$ is here not defined in any precise mathematical sense since all the discussion carried out so far is clearly heuristic in character. In this form, S^+ strongly resembles a Feynman summation over histories with density of Lagrangian $\mathscr{L}$ as in a 3-dimensional Einstein theory. In a more conventional 4-dimensional theory with pseudo Euclidean metric, the corresponding summation would be[28]):

$$S(\Sigma_1, \Sigma_2) = \int \mathrm{d}\mu \exp\left(\mathrm{i}\int_{\Sigma_1}^{\Sigma_2} R\,\mathrm{d}^4x\sqrt{-g}\right), \tag{6.17}$$

the integral on the coordinates being performed in the slab between the space-like hypersurfaces Σ_1, Σ_2. The other terms, other than the positive frequency part, are related to different orientations of the tetrahedra T_j and have a similar interpretation, although their precise meaning is still unclear. It is plausible that in the transition to a smooth manifold $\mathscr{M}$, they will give no essential contribution to the final result.

Finally we report an interesting conjecture over possible extensions of Wigner's result for the 6*j*-symbol. For simplicity we limit ourselves to the 9*j*-symbol, further generalizations being obvious. In the diagram of fig. 12a we keep the vertices P_1, P_2, P_3 of one face fixed; in this way we fix also a, b, c. It can be easily realized that to determine D completely it is enough to give the points P_4 and P_5; in fact, from P_1, P_5 we deduce the symmetry centre $\frac{1}{2}(\mathrm{P}_1+\mathrm{P}_5)$ of D and from it all the remaining vertices. We may use as coordinates for P_4, P_5 either their six Euclidean coordinates $\boldsymbol{r}_4$, $\boldsymbol{r}_5$ or the six remaining lengths d, e, f, g, h, i. Our conjecture, suggested by (3.5), is that:

$$\begin{Bmatrix} a & b & c \\ d & e & f \\ g & h & i \end{Bmatrix}^2 \sim C\sum_{k=1}^{M}\left|\frac{\partial(\boldsymbol{r}_4, \boldsymbol{r}_5)}{\partial(d^2, e^2, f^2, g^2, h^2, i^2)}\right|_{\boldsymbol{r}_4=\boldsymbol{r}_4^{(k)},\, \boldsymbol{r}_5=\boldsymbol{r}_5^{(k)}}, \tag{6.18}$$

where the summation is carried on all the M different configurations $\boldsymbol{r}_4^{(k)}$,

$r_5^{(k)}$ labelled by k, which correspond to the same values of the lengths appearing in the symbol. Cayley identity (4.8) when applied, for instance, to P_3, P_5, P_6, P_7, P_8, yields a fourth order equation in x^2, which means that, in general, there are four different contributions to (6.18). The constant C can be determined by evaluating

$$\mathfrak{N} = \sum_{a,b,c,d,e,f<R} (2a+1)(2b+1)(2c+1)(2d+1)\times \times(2e+1)(2f+1)\begin{Bmatrix} a & b & c \\ d & e & f \\ g & h & i \end{Bmatrix}^2 \tag{6.19}$$

by means of the orthogonality of $9j$-symbols[6]); following a procedure similar to the one which led to (6.12), we obtain $\mathfrak{N}=(\frac{4}{3}R^3)^2$. On the other hand, we have by means of (6.18)

$$\mathfrak{N} \simeq \int_{a,b,\ldots,f<R} \mathrm{d}(a^2)\,\mathrm{d}(b^2)\ldots\mathrm{d}(f^2)\begin{Bmatrix} a & b & c \\ d & e & f \\ g & h & i \end{Bmatrix}^2 \simeq \simeq \tfrac{1}{2}C \int_{|r_4|<R} \mathrm{d}^3 r_4 \int_{|r_5|<R} \mathrm{d}^3 r_5 \simeq \tfrac{1}{2}C\,(\tfrac{4}{3}\pi R^3)^2, \tag{6.20}$$

from which is follows: $C=2(\pi)^{-2}$. The inverse of the Jacobian appearing in (6.18) can be evaluated easily in terms of volumes of tetrahedra; we hope to present elsewhere more detailed results for this as well as for higher order symbols.

Acknowledgments

The authors are indebted to Prof. E. P. Wigner, Prof. F. J. Dyson, Dr. S. Adler and Dr. J. Lascoux for many interesting discussions.

Appendix A. Racah algebra "in nuce" for SU(2)

This appendix will be devoted to a brief summary of formulae and properties of coupling and recoupling coefficients, as well as of matrix representations for the unitary unimodular group in two dimensions[29]).

The $3j$-symbol is defined by[4]):

$$\begin{pmatrix} a & b & c \\ \alpha & \beta & \gamma \end{pmatrix} \equiv \begin{vmatrix} b+c-a & c+a-b & a+b-c \\ a-\alpha & b-\beta & c-\gamma \\ a+\alpha & b+\beta & c+\gamma \end{vmatrix} = $$

$$= (-1)^{a-b-\gamma}[\Delta(abc)(a+\alpha)!\,(a-\alpha)!\,(b+\beta)!\,(b-\beta)!\,(c+\gamma)!\,\times \times (c-\gamma)!]^{\frac{1}{2}}\sum_x (-1)^x [x!\,(c-b+\alpha+x)!\,(c-a-\beta+x)!\,\times \times(a+b-c-x)!\,(a-\alpha-x)!\,(b+\beta-x)!]^{-1}, \tag{A.1}$$

with

$$\Delta(abc) = \frac{(a+b-c)!\,(b+c-a)!\,(c+a-b)!}{(a+b+c+1)!}. \tag{A.2}$$

The positive integers or half-integers a, b, c satisfy triangular inequalities: $|a-b| \leqslant c \leqslant a+b$ etc. and $\alpha+\beta+\gamma=0$; $a-|\alpha|$ etc. must be natural integers. If anyone of these conditions is not fulfilled, the value of the symbol is assumed to be zero. The 72 elements[15]) of the symmetry group R_1 of this coefficient are the permutations of lines and/or columns of the square symbol in (A.1) (which yield the phase $(P)^S$ with $S=a+b+c$ and $P=+1,-1$ according to even, odd permutations), and the exchange of lines with columns. For unitarity and orthogonality of $3j$-symbols, see Edmonds[16]).

From the definition (3.1) it turns out[4]) that the $6j$-symbol is given in terms of $3j$-coefficients by:

$$\begin{Bmatrix} a & b & c \\ d & e & f \end{Bmatrix} = \sum_{\substack{\alpha\beta\gamma \\ \delta\varepsilon\varphi}} (-1)^{d+e+f+\delta+\varepsilon+\varphi} \times$$
$$\times \begin{pmatrix} d & e & c \\ \delta & -\varepsilon & \gamma \end{pmatrix} \begin{pmatrix} e & f & a \\ \varepsilon & -\varphi & \alpha \end{pmatrix} \begin{pmatrix} f & d & b \\ \varphi & -\delta & \beta \end{pmatrix} \begin{pmatrix} a & b & c \\ \alpha & \beta & \gamma \end{pmatrix}. \tag{A.3}$$

Racah's treatment of this formula[1]) yields:

$$\begin{Bmatrix} a & b & c \\ d & e & f \end{Bmatrix} = [\Delta(abc)\,\Delta(aef)\,\Delta(cde)\,\Delta(bdf)]^{\frac{1}{2}} \times$$
$$\times \sum_x (-1)^x (x+1)!\,[(a+b+d+e-x)!\,(a+c+d+f-x)! \times$$
$$\times (b+c+e+f-x)!\,(x-a-b-c)!\,(x-a-e-f)! \times$$
$$\times (x-c-d-e)!\,(x-b-d-f)!]^{-1}. \tag{A.4}$$

The symbol is assumed to vanish if anyone of the triads (abc), (cde), (aef), (bdf) does not satisfy triangular inequalities. In addition to the well known symmetries of the associated tetrahedron (fig. 1), the $6j$-symbol has also the less evident symmetry[15]):

$$\begin{Bmatrix} a & b & c \\ d & e & f \end{Bmatrix} \equiv \begin{vmatrix} a+b-c & b+f-d & f+a-e \\ d+b-f & b+c-a & c+d-e \\ a+e-f & e+c-d & c+a-b \\ d+e-c & e+f-a & f+d-b \end{vmatrix} =$$
$$= \begin{Bmatrix} a & \frac{1}{2}(c+f+e-b) & \frac{1}{2}(b+e+f-c) \\ d & \frac{1}{2}(c+f+b-e) & \frac{1}{2}(b+e+c-f) \end{Bmatrix}, \tag{A.5}$$

which entails that this coefficient is invariant under the 144 elements of a group R_2 which correspond to the permutations of lines and/or columns of the 3×4 pattern in (A.5). In addition to the Biedenharn–Elliott identity (3.8), the $6j$-symbol satisfies also the following relations[30]:

$$\sum_x (2x+1)\begin{Bmatrix} a & b & x \\ d & e & f \end{Bmatrix}\begin{Bmatrix} a & b & x \\ d & e & f' \end{Bmatrix} = \frac{\delta_{ff'}}{2f+1}, \tag{A.6}$$

$$\sum_x (-1)^{f+g+x}(2x+1)\begin{Bmatrix} a & b & x \\ d & e & f \end{Bmatrix}\begin{Bmatrix} d & e & x \\ b & a & g \end{Bmatrix} = \begin{Bmatrix} a & e & f \\ b & d & g \end{Bmatrix}. \tag{A.7}$$

Wigner's $9j$-symbol[4]), which is proportional to the transformation matrix between different coupling schemes of four angular momenta, is given in terms of $6j$-coefficients by:

$$\begin{Bmatrix} a & b & c \\ d & e & f \\ g & h & i \end{Bmatrix} = \sum_x (-1)^{2x}(2x+1)\begin{Bmatrix} a & i & x \\ f & b & c \end{Bmatrix}\begin{Bmatrix} f & b & x \\ h & d & e \end{Bmatrix}\begin{Bmatrix} h & d & x \\ a & i & g \end{Bmatrix}. \tag{A.8}$$

Its known symmetries are formally the same as those mentioned above for the $3j$-symbol, with $S=a+b+c+d+e+f+g+h+i$. For a more comprehensive account of relations involving these coefficients, see ref. 6.

According to the conventions and notations of ref. 30 for basis vectors, angular momentum operators and Euler angles, the rotation operator

$$\hat{D}(\alpha\beta\gamma) = e^{-i\alpha\hat{Y}_Z}\, e^{-i\beta\hat{Y}_Y}\, e^{-i\gamma\hat{Y}_Z} \tag{A.9}$$

has matrix elements in the $2J+1$ dimensional representation which can be written as

$$D^{(J)}_{MM'}(\alpha\beta\gamma) \equiv \langle JM|\hat{D}(\alpha\beta\gamma)|JM'\rangle = e^{-i\alpha M}\, d^{(J)}_{MM'}(\beta)\, e^{-i\gamma M'} \tag{A.10}$$

in terms of the real matrix: $d^{(J)}_{MM'}(\beta)=D^{(J)}_{MM'}(0\beta0)$, The following properties hold

$$[D^{(J)}(\alpha\beta\gamma)]^+ = [D^{(J)}(\alpha\beta\gamma)]^{-1} = D^{(J)}(-\gamma, -\beta, -\alpha), \tag{A.11}$$

$$\sum_M D^{(J)}_{MM'}\, D^{(J)*}_{MM''} = \delta_{M'M''}; \qquad \sum_M D^{(J)}_{M'M}\, D^{(J)*}_{M''M} = \delta_{M'M''}, \tag{A.12}$$

$$\frac{1}{8\pi^2}\int_0^{2\pi} d\alpha \int_0^{2\pi} d\gamma \int_0^{\pi} d\beta \sin\beta\, D^{(J)*}_{M_1M_2}(\alpha\beta\gamma)\, D^{(J')}_{M_1'M_2'}(\alpha\beta\gamma) = \frac{\delta_{JJ'}\,\delta_{M_1M_1'}\,\delta_{M_2M_2'}}{2J+1}, \tag{A.13}$$

in addition to the symmetry

$$D^{(J)*}_{MM'} = (-1)^{M-M'}\, D^{(J)}_{-M,-M'}. \tag{A.14}$$

In general[4])

$$d^{(J)}_{-M,M'}(\beta) = \sum_x (-1)^x \frac{[(J-M)!\,(J+M)!\,(J-M')!\,(J+M')!]^{\frac{1}{2}}}{(J-M-x)!\,(J-M'-x)\,x!\,(x+M+M')!} \times$$
$$\times (\cos \tfrac{1}{2}\beta)^{2J-M-M'-2x} (\sin \tfrac{1}{2}\beta)^{2x+M+M'}, \qquad \text{(A.15)}$$

which leads in particular to

$$d^{(L)}_{M,0}(\beta) = (-1)^M \left[\frac{(L-M)!}{(L+M)!}\right]^{\frac{1}{2}} P^M_L(\cos\beta); \qquad \text{(A.16)}$$

the following symmetries are very useful

$$d^{(J)}_{M,M'}(\beta) = (-1)^{M-M'} d^{(J)}_{-M,-M'}(\beta) =$$
$$= (-1)^{M-M'} d^{(J)}_{M',M}(\beta) = (-1)^{J-M'} d^{(J)}_{-M',M}(\pi-\beta). \qquad \text{(A.17)}$$

The connection with Jacobi polynomials is given by

$$d^{(J)}_{MM'}(\beta) = (-1)^{M-M'} \left[\frac{(J+M)!\,(J-M)!}{(J+M')!\,(J-M')!}\right]^{\frac{1}{2}} \times$$
$$\times (\cos\tfrac{1}{2}\beta)^{M+M'} (\sin\tfrac{1}{2}\beta)^{M-M'} P^{(M-M',\,M+M')}_{J-M}(\cos\beta). \qquad \text{(A.18)}$$

Appendix B. Elementary geometry of tetrahedra

Heron's formula for the content A of a triangle of edges j_1, j_2, j_3 can be written as

$$A^2 = \tfrac{1}{16}(j_1+j_2+j_3)(j_1+j_2-j_3)(j_1-j_2+j_3)(-j_1+j_2+j_3) =$$
$$= -\frac{1}{16}\begin{vmatrix} 0 & j_1^2 & j_2^2 & 1 \\ j_1^2 & 0 & j_3^2 & 1 \\ j_2^2 & j_3^2 & 0 & 1 \\ 1 & 1 & 1 & 0 \end{vmatrix}. \qquad \text{(B.1)}$$

We have already given in (1.2) Cayley's form for the content V of a tetrahedron. Performing in (1.2) the derivation of V^2 with respect to $(j_{rs})^2$ and denoting the (r, s) algebraic minor of the Cayley determinant with C_{rs}, the relation[31])

$$-C_{rs} = 16 A_h A_k \cos\theta_{hk}, \qquad r \neq s \neq h \neq k \qquad \text{(B.2)}$$

yields (3.22); in (B.2) h, k, r, s are any permutation of 1, 2, 3, 4. Using (3.22) and the obvious identity

$$A_h A_k \sin\theta_{hk} = \tfrac{3}{2} V j_{hk} \qquad \text{(B.3)}$$

we obtain the differentiation formulae

$$-\frac{d\theta_{hk}}{dj_{rs}} = \frac{j_{hk}j_{rs}}{6V}, \qquad h \neq k \neq r \neq s \tag{B.4}$$

Looking at fig. 1 we see that

$$V = \tfrac{1}{6}j_{12}j_{23}j_{13} \sin\varphi_{14} \sin\varphi_{24} \sin\theta_{12} \tag{B.5}$$

or

$$V = \tfrac{1}{6}j_{12}j_{13}j_{23} \begin{vmatrix} 1 & \cos\varphi_{14} & \cos\varphi_{24} \\ \cos\varphi_{14} & 1 & \cos\varphi_{34} \\ \cos\varphi_{24} & \cos\varphi_{34} & 1 \end{vmatrix}^{\frac{1}{2}} \tag{B.6}$$

since $\cos\varphi_{34} = \cos\varphi_{14}\cos\varphi_{24} - \sin\varphi_{14}\sin\varphi_{24}\cos\theta_{12}$. Now, if we define with reference to P_4:

$$2\sigma_4 = \theta_{12} + \theta_{13} + \theta_{23}, \tag{B.7}$$

$$\Sigma_4 = [\sin\sigma_4 \sin(\sigma_4 - \theta_{12}) \sin(\sigma_4 - \theta_{13}) \sin(\sigma_4 - \theta_{23})]^{\frac{1}{2}}, \tag{B.8}$$

then it follows[22]):

$$2\Sigma_4 = \sin\theta_{12}\sin\theta_{23}\sin\varphi_{24} = \sin\theta_{13}\sin\theta_{23}\sin\varphi_{34} = \\ = \sin\theta_{12}\sin\theta_{13}\sin\varphi_{14}, \tag{B.9}$$

and

$$\begin{aligned} \Sigma_4 \operatorname{tg} \tfrac{1}{2}\varphi_{14} &= \sin\sigma_4 \sin(\sigma_4 - \theta_{23}), \\ \Sigma_4 \operatorname{tg} \tfrac{1}{2}\varphi_{24} &= \sin\sigma_4 \sin(\sigma_4 - \theta_{13}), \\ \Sigma_4 \operatorname{tg} \tfrac{1}{2}\varphi_{34} &= \sin\sigma_4 \sin(\sigma_4 - \theta_{12}). \end{aligned} \tag{B.10}$$

It turns out that $K = \Sigma_h/A_h$ is independent of h; therefore, from (B.5), (B.9):

$$V = \tfrac{2}{3}\left(\prod_{h=1}^{4} A_h\right)^{\frac{1}{2}} K^{\frac{1}{2}}. \tag{B.11}$$

Appendix C. A summation property of Jacobi polynomials

The purpose of this appendix is to prove the relation

$$\sum_{L=0}^{\infty} (2L+1)\, d^{(L)}_{M_1M_2}(\beta_3)\, d^{(L)}_{M_2M_3}(\beta_1)\, d^{(L)}_{M_3M_1}(\beta_2) = \\ = \frac{2\Theta(\mathscr{B})}{\pi[\mathscr{B}]^{\frac{1}{2}}} \cos\left(\sum_{i=1}^{3} M_i\delta_i\right), \tag{C.1}$$

where Θ is the step function used in (3.13), while the angles β_i, δ_i ($i = 1, 2, 3$)

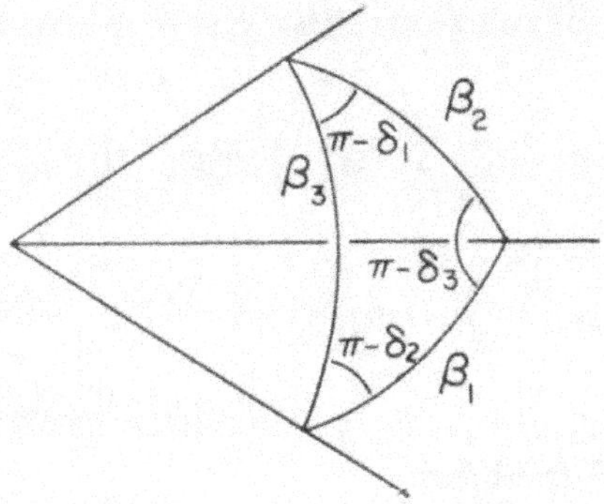

Fig. C1.

are shown in fig. C1 and

$$\mathscr{B} = \begin{vmatrix} 1 & \cos\beta_3 & \cos\beta_2 \\ \cos\beta_3 & 1 & \cos\beta_1 \\ \cos\beta_2 & \cos\beta_1 & 1 \end{vmatrix} . \tag{C.2}$$

The addition property of the rotation group yields:

$$D^{(L)}_{M_1M_2}(\alpha_3\beta_3\gamma_3) = \sum_{M_3} D^{(L)}_{M_1M_3}(\alpha_2\beta_2\gamma_2)\, D^{(L)}_{M_3M_2}(\alpha_1\beta_1\gamma_1), \tag{C.3}$$

from which, using unitarity

$$D^{(L)}_{M_3M_2}(\alpha_1\beta_1\gamma_1) = \sum_{M_1} D^{(L)}_{M_1M_2}(\alpha_3\beta_3\gamma_3)\, D^{(L)*}_{M_1M_3}(\alpha_2\beta_2\gamma_2), \tag{C.4}$$

$$D^{(L)}_{M_1M_3}(\alpha_2\beta_2\gamma_2) = \sum_{M_2} D^{(L)*}_{M_3M_2}(\alpha_1\beta_1\gamma_1)\, D^{(L)}_{M_1M_2}(\alpha_3\beta_3\gamma_3). \tag{C.5}$$

By exploiting (C.3)–(C.5) for low values of L, we obtain

$$\cos\beta_i = \cos\beta_j \cos\beta_k - \sin\beta_j \sin\beta_k \cos\delta_i, \quad i \neq j \neq k = 1, 2, 3 \tag{C.6}$$

$$\sin\beta_i/\sin\delta_i = \sin\beta_j/\sin\delta_j, \quad i \neq j \tag{C.7}$$

where

$$\delta_1 = \alpha_2 - \alpha_3 + \pi, \quad \delta_2 = \gamma_1 - \gamma_3 - \pi, \quad \delta_3 = \alpha_1 + \gamma_2. \tag{C.8}$$

By means of (A.13) and (C.3) we find

$$\int_0^{2\pi} d\alpha_2 \int_0^{2\pi} d\gamma_2 \int_0^{\pi} d\beta_2 \sin\beta_2\, D^{(L')*}_{M_1'M_3'}(\alpha_2\beta_2\gamma_2)\, D^{(L)}_{M_1M_2}(\alpha_3\beta_3\gamma_3) =$$

$$= \frac{8\pi^2}{2L+1}\, \delta_{LL'}\, \delta_{M_1M_1'}\, \delta_{M_3M_3'}\, D^{(L)}_{M_3M_2}(\alpha_1\beta_1\gamma_1); \tag{C.9a}$$

(A.10) and the reality of functions $d^{(L)}_{MM'}$ give

$$\int_0^{2\pi} d\alpha_2 \int_0^{2\pi} d\gamma_2 \int_0^{\pi} d\beta_2 \sin\beta_2 \cos\left(\sum_{j=1}^{3} M_j\delta_j\right) d^{(L)}_{M_1M_3}(\beta_2)\, d^{(L)}_{M_1M_2}(\beta_3) = \frac{8\pi^2}{2L+1}(-1)^{M_1-M_2}\, d^{(L)}_{M_3M_2}(\beta_1). \quad \text{(C.9b)}$$

We notice that, once $\alpha_1, \beta_1, \gamma_1, \beta_2, \gamma_2$ have been fixed, β_3 and δ_i $(i=1, 2, 3)$ are independent of α_2; therefore:

$$\int_0^{2\pi} d\gamma_2 \int_0^{\pi} d\beta_2 \sin\beta_2 \cos\left(\sum_{j=1}^{3} M_j\delta_j\right) d^{(L)}_{M_1M_3}(\beta_2)\, d^{(L)}_{M_1M_2}(\beta_3) = \frac{4\pi}{2L+1}(-1)^{M_1-M_2}\, d^{(L)}_{M_3M_2}(\beta_1). \quad \text{(C.10)}$$

Going from the integration variables β_2, γ_2 to $\cos\beta_2, \cos\beta_3$, we find:

$$\frac{\partial(\beta_2, \gamma_2)}{\partial(\cos\beta_2, \cos\beta_3)} = (\sin\beta_2)^{-1}\mathscr{B}^{-\frac{1}{2}} \quad \text{(C.11)}$$

and

$$\int_{-1}^{+1} d(\cos\beta_2) \int_{-1}^{+1} d(\cos\beta_3) \frac{\Theta(\mathscr{B})}{\mathscr{B}^{\frac{1}{2}}} \cos\left(\sum_{j=1}^{3} M_j\delta_j\right) d^{(L)}_{M_1M_3}(\beta_2) \times d^{(L)}_{M_1M_2}(\beta_3) = \frac{2\pi}{2L+1}(-1)^{M_1-M_2}\, d^{(L)}_{M_3M_2}(\beta_1), \quad \text{(C.12)}$$

having noticed that a given point of the integration domain in (C.12) corresponds to two different points in (C.10). Since by means of the orthogonal polynomials $d^{(L)}_{MM'}$ we can define the following expansions:

$$f_{M_1M_2M_1'M_2'}(\beta, \beta') = \sum_{L, L'} \frac{2L+1}{2}\,\frac{2L'+1}{2}\, f_{LL'}^{M_1M_2M_1'M_2'}\, d^{(L)}_{M_1M_2}(\beta)\, d^{(L')}_{M_1'M_2'}(\beta'), \quad \text{(C.13)}$$

with coefficients:

$$f_{LL'}^{M_1M_2M_1'M_2'} = \int_{-1}^{+1} d(\cos\beta) \int_{-1}^{+1} d(\cos\beta')\, f_{M_1M_2M_1'M_2'}(\beta, \beta') \times d^{(L)}_{M_1M_2}(\beta)\, d^{(L)}_{M_1'M_2'}(\beta'), \quad \text{(C.14)}$$

which are non-vanishing only if $L \geqslant \max(|M_1|, M_2|)$, $L' \geqslant \max(|M_1'|, |M_2'|)$, we have from (C.12)

$$\frac{\Theta(\mathscr{B})}{\mathscr{B}^{\frac{1}{2}}} \cos\left(\sum_{j=1}^{3} M_j \delta_j\right)(-1)^{M_1 - M_2} =$$

$$= \tfrac{1}{2}\pi \sum_{L=0}^{\infty} (2L+1)\, d^{(L)}_{M_3 M_2}(\beta_1)\, d^{(L)}_{M_1 M_3}(\beta_2)\, d^{(L)}_{M_1 M_2}(\beta_3).$$

The symmetries (A.17) lead us to (C.1).

In order to complete the proof of (3.13), we have just to recall (B.6).

Appendix D. Invariance of Ω under R_2

It is straightforward, though very tedious, to verify that the volume of T is invariant also under the symmetry (A.5). This can be worked out by expanding (1.2) and replacing j_{13} with $\frac{1}{2}(j_{14}+j_{24}+j_{23}-j_{13})$ etc. Incidentally, we note that in the same way one can check also the invariance of $\prod_{h=1}^{4} A_h$ under R_2. Here we prefer to sketch how the symmetry of Ω can be proved.

Let us multiply eqs. (B.10) among themselves. We have for instance

$$(\sin\sigma_4)^2 = \Sigma_4 \operatorname{tg}\tfrac{1}{2}\varphi_{14} \operatorname{tg}\tfrac{1}{2}\varphi_{24} \operatorname{tg}\tfrac{1}{2}\varphi_{34}\,; \tag{D.1}$$

then choosing for example the first of (B.10), we have

$$[\sin\tfrac{1}{2}(\theta_{12}+\theta_{13}-\theta_{23})]^2 = \Sigma_4 \operatorname{tg}\tfrac{1}{2}\varphi_{14} \operatorname{cotg}\tfrac{1}{2}\varphi_{24} \operatorname{cotg}\tfrac{1}{2}\varphi_{34}. \tag{D.2}$$

In order to compute the r.h.s. of (D.2) it is convenient to introduce, with reference to fig. D1, the following notations

$$\begin{aligned} q_1 &= j_{12}+j_{13}+j_{14}, & q_2 &= j_{12}+j_{23}+j_{24}, \\ q_3 &= j_{13}+j_{23}+j_{34}, & q_4 &= j_{14}+j_{24}+j_{34}; \end{aligned} \tag{D.3}$$

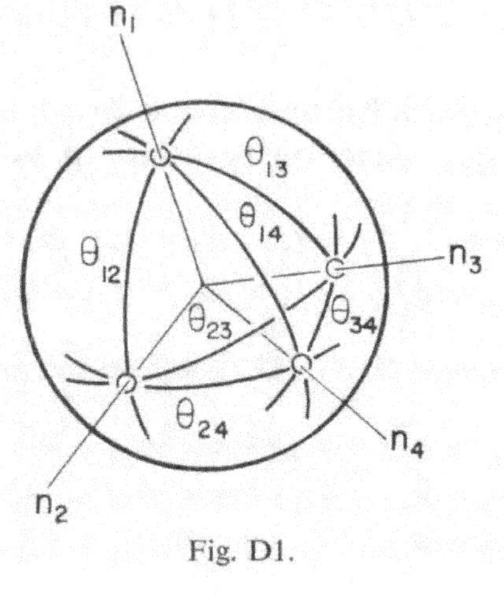

Fig. D1.

$$p_{14} = p_{23} = j_{12} + j_{13} + j_{34} + j_{24}, \quad p_{13} = p_{24} = j_{12} + j_{34} + j_{14} + j_{23},$$
$$p_{12} = p_{34} = j_{13} + j_{24} + j_{14} + j_{23}; \tag{D.4}$$

as well as the two patterns

$$r_{st} \equiv \begin{bmatrix} p_{14} - q_1 & p_{13} - q_1 & p_{12} - q_1 & q_1 \\ p_{14} - q_2 & p_{13} - q_2 & p_{12} - q_2 & q_2 \\ p_{14} - q_3 & p_{13} - q_3 & p_{12} - q_3 & q_3 \\ p_{14} - q_4 & p_{13} - q_4 & p_{12} - q_4 & q_4 \end{bmatrix};$$
$$\tfrac{1}{2}\vartheta_{st} \equiv \begin{bmatrix} \sigma_4 - \theta_{23} & \sigma_3 - \theta_{24} & \sigma_2 - \theta_{34} & \sigma_1 \\ \sigma_3 - \theta_{14} & \sigma_4 - \theta_{13} & \sigma_1 - \theta_{34} & \sigma_2 \\ \sigma_2 - \theta_{14} & \sigma_1 - \theta_{24} & \sigma_4 - \theta_{12} & \sigma_3 \\ \sigma_1 - \theta_{23} & \sigma_2 - \theta_{13} & \sigma_3 - \theta_{12} & \sigma_4 \end{bmatrix}. \tag{D.5}$$

We note that from (B.11)

$$K = \frac{\Sigma_4}{A_4} = \frac{9}{4}\,\frac{V^2}{\prod_{h=1}^{4} A_h}. \tag{D.6}$$

Since, for example

$$\operatorname{tg} \tfrac{1}{2}\varphi_{14} = \left[\frac{(p_{13} - q_3)(p_{12} - q_2)}{q_1(p_{14} - q_4)}\right]^{\frac{1}{2}},$$

(D.2) becomes

$$\cos(\theta_{12} + \theta_{13} - \theta_{23}) = 1 - 2^5 3^2 V^2 \,[(p_{13} - q_1) \times$$
$$\times (p_{12} - q_1) q_1 (p_{14} - q_2)(p_{14} - q_3)(p_{14} - q_4)]^{-1}$$

and generally for $s \neq t = 1, 2, 3, 4$

$$\cos \vartheta_{st} = 1 - 2^3 (3!)^2 V^2 \left\{ \left(\prod_{n=1}^{4} r_{sn} \right) \left(\prod_{n=1}^{4} r_{nt} \right) \right\}^{-1} (r_{st})^2, \tag{D.7}$$

which gives, correctly, for a flat tetrahedron, $\vartheta_{st} = 0$ or $\vartheta_{st} = 2\pi$.

It is easy to check that, under the symmetry (A.5)

$$q'_1 = q_2, \quad q'_2 = q_1, \quad q'_3 = q_4, \quad q'_4 = q_3;$$
$$p'_{14} = p_{13}, \quad p'_{13} = p_{14}, \quad p'_{12} = p_{12}, \tag{D.8}$$

which, taking into account (D.5), (D.7), entails for instance

$$\cos(\theta'_{12} + \theta'_{13} - \theta'_{23}) = \cos(\theta_{12} + \theta_{13} - \theta_{23}),$$
$$\cos(\theta'_{13} + \theta'_{23} - \theta'_{12}) = \cos(\theta_{14} + \theta_{24} - \theta_{12}), \tag{D.9}$$
$$\cos(\theta'_{12} + \theta'_{23} - \theta'_{13}) = \cos(\theta_{12} + \theta_{13} - \theta_{23}).$$

Let us suppose the tetrahedron to be almost regular; asymptotically, this property is not destroyed by the symmetry (A.5); in this case: $0 \leqslant \theta_{hk} \leqslant \frac{1}{2}\pi$ and similarly for the transformed θ'_{hk}. Moreover, looking at fig. D1 we see for example that $\theta_{12}, \theta_{13}, \theta_{23}$ satisfy spherical triangular inequalities; therefore $0 \leqslant \vartheta_{st} < \pi$, $0 \leqslant \vartheta'_{st} < \pi$. Then from (D.9) and similar relations, we obtain

$$\begin{aligned} \theta'_{13} &= \tfrac{1}{2}(\theta_{14}+\theta_{24}+\theta_{23}-\theta_{13}), \\ \theta'_{14} &= \tfrac{1}{2}(\theta_{13}+\theta_{23}+\theta_{24}-\theta_{14}), \qquad \theta'_{12}=\theta_{12}, \\ \theta'_{24} &= \tfrac{1}{2}(\theta_{13}+\theta_{23}+\theta_{14}-\theta_{24}), \\ \theta'_{23} &= \tfrac{1}{2}(\theta_{14}+\theta_{24}+\theta_{13}-\theta_{23}), \qquad \theta'_{34}=\theta_{34}. \end{aligned} \tag{D.10}$$

We conclude that (A.5) induces the same linear transformation on j_{hk} as well as on θ_{hk}; the unitarity of this transformation entails the invariance of $\Omega = \sum_{h,k} j_{hk}\,\theta_{hk} + \frac{1}{4}\pi$ under R_2. Finally, the constraint $\theta_{hk} < \frac{1}{2}\pi$ can be dropped by invoking analytical continuation in θ_{hk}.

Appendix E. Evaluation of $[\partial(\theta_x^1+\theta_x^2+\theta_x^3)/\partial(x^2)]_{x^2=(x_\gtrless)^2}$

According to (4.16), first let us calculate $\partial I^2(x^2)/\partial(x^2)|_{x^2=(x_\gtrless)^2}$. Write (4.8) as follows:

$$C = c_2(x^2)^2 + c_1(x^2) + c_0 = 0, \tag{E.1}$$

where $-2^4(4!)^2 I^2(x^2) = C$; then the solutions of (E.1) are obviously

$$(x_\gtrless)^2 = \frac{\pm\sqrt{c_1^2-4c_0c_2}-c_1}{2c_2}, \tag{E.2}$$

where c_0, c_1, c_2 are determinants extracted from C which, for the sake of brevity, we do not write explicitly. Since the tetrahedra T_4, T_5 are supposed to be physical, it follows $(x_\gtrless)^2 > 0$. Then

$$\left[\frac{\partial C}{\partial(x^2)}\right]_{x^2=(x_\gtrless)^2} = \pm\sqrt{c_1^2-4c_0c_2}. \tag{E.3}$$

From known properties of determinants[32]), it turns out that

$$\tfrac{1}{4}(c_1^2-4c_0c_2) = 2^6(3!)^4 V_4^2 V_5^2, \tag{E.4}$$

therefore

$$\left[\frac{\partial I^2(x^2)}{\partial(x^2)}\right]_{x^2=(x_\gtrless)^2} = \mp\frac{V_4V_5}{16}. \tag{E.5}$$

In order to evaluate $\partial^2 I^2(x^2)/\partial(\theta_x^1+\theta_x^2+\theta_x^3)|_{x^2=(x_\gtrless)^2}$, suppose to imbed the simplex P_1, P_2, P_3, P_4, P_5 of fig. 8a into a 4-dimensional Euclidean space.

The content I of this simplex is

$$(4!)^2 I^2 = \begin{vmatrix} x & 0 & 0 & 0 \\ 0 & h_{11} & h_{12} & h_{13} \\ 0 & h_{21} & h_{22} & h_{23} \\ 0 & h_{31} & h_{32} & h_{33} \end{vmatrix}^2 \tag{E.6}$$

where $h_{1j}, h_{2j}, h_{3j} (j=1, 2, 3)$ are the components of the distances from P_1, P_2, P_3 to x. The direction of x has been chosen as the first axis of refer-

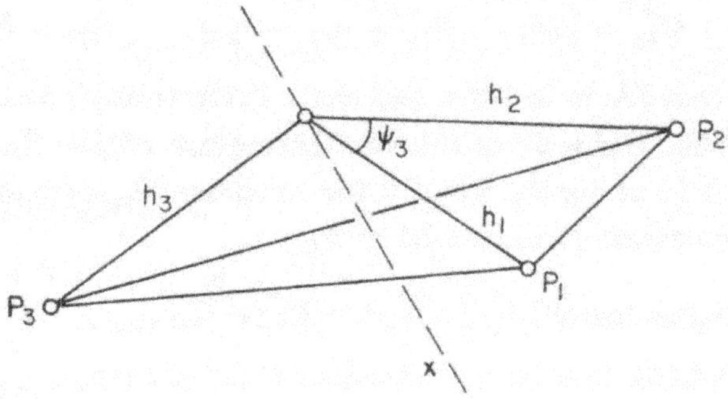

Fig. E1.

ence in this hyperspace. Let H be the volume of the tetrahedron defined by h_1, h_2, h_3 (see fig. E1). Then we are led to the simple formula

$$I = \tfrac{1}{4} x H\,. \tag{E.7}$$

Now we notice that the angle between, for instance, h_1 and h_2 is $\pi - \theta_x^3$. Therefore from (B.6) we have

$$\begin{aligned} \delta(H^2) = \tfrac{1}{18}(h_1 h_2 h_3)^2 [&\sin\theta_x^1 (\cos\theta_x^2 \cos\theta_x^3 + \cos\theta_x^1)\,\delta\theta_x^1 + \\ &+ \sin\theta_x^2 (\cos\theta_x^3 \cos\theta_x^1 + \cos\theta_x^2)\,\delta\theta_x^2 + \\ &+ \sin\theta_x^3 (\cos\theta_x^1 \cos\theta_x^2 + \cos\theta_x^3)\,\delta\theta_x^3]\,. \end{aligned} \tag{E.8}$$

Since (E.8) must be evaluated when $\theta_x^1 + \theta_x^2 + \theta_x^3 = \pi$, i.e. when h_1, h_2, h_3 are coplanar, we find

$$\left[\frac{\partial(H^2)}{\partial(\theta_x^1 + \theta_x^2 + \theta_x^3)}\right]_{x^2=(x_{\gtrless})^2} = \tfrac{1}{18}(h_1 h_2 h_3)^2 \sin\theta_x^1 \sin\theta_x^2 \sin\theta_x^3\,. \tag{E.9}$$

Moreover, from (B.5) we obtain

$$V_i = \tfrac{1}{6} h_1 h_2 h_3 x \frac{\sin\theta_x^i}{h_i}, \qquad i = 1, 2, 3 \tag{E.10}$$

and finally, by means of (E.7)

$$\left[\frac{\partial I^2(x^2)}{\partial(\theta_x^1+\theta_x^2+\theta_x^3)}\right]_{x^2=(x\gtrless)^2}=\frac{3}{4}\left[\frac{V_1V_2V_3}{x}\right]_{x^2=(x\gtrless)^2}. \tag{E.11}$$

Relations (E.5) and (E.11) yield (4.17).

Appendix F. Evaluation of $\lim_{V^2\to 0}\Omega(V^2)$

The behaviour of Ω in the neighbourhood of a transition point deserves a rather careful investigation. To begin with, let us suppose that we approach the region B_3 as $V^2\to 0$. In fig. 10 we show the notations which will be used in the sequel; moreover we indicate with A, B, C the *internal* dihedral angles between faces belonging to $a+\frac{1}{2}$, $b+\frac{1}{2}$, $c+\frac{1}{2}$ and with $\pi-D$, $\pi-E$, $\pi-F$ the corresponding ones relative to $d+\frac{1}{2}$, $e+\frac{1}{2}$, $f+\frac{1}{2}$. Also the following shorthand notation will be convenient:

$$\begin{aligned} &\mu_{bc}=\cos\alpha\,, &&\mu_{bd}=\cos\delta_b\,, &&\mu_{cd}=\cos\delta_c\,,\\ &\mu_{ca}=\cos\beta\,, &&\mu_{ce}=\cos\varepsilon_c\,, &&\mu_{ae}=\cos\varepsilon_a\,,\\ &\mu_{ab}=\cos\gamma\,, &&\mu_{af}=\cos\varphi_a\,, &&\mu_{bf}=\cos\varphi_b\,. \end{aligned} \tag{F.1}$$

When P lies very near to the plane of the triangle $P_AP_BP_C$ and within its boundary we have $\overset{0}{\mu}_{bc}=\cos(\delta_c+\delta_b)$, $\overset{0}{\mu}_{ca}=\cos(\varepsilon_a+\varepsilon_c)$, $\overset{0}{\mu}_{ab}=\cos(\varphi_a+\varphi_b)$ and $A=B=C=D=E=F=0$. Since $a+\frac{1}{2}=(e+\frac{1}{2})\,\mu_{ae}+(f+\frac{1}{2})\,\mu_{af}$ etc., we obtain in general from the definition (3.15b)

$$\Omega=\pi(a+b+c+\tfrac{7}{4})+(d+\tfrac{1}{2})\Psi_A+(e+\tfrac{1}{2})\Psi_B+(f+\tfrac{1}{2})\Psi_C, \tag{F.2}$$

where

$$\begin{aligned}\Psi_A=D-B\mu_{bd}-C\mu_{cd}\,,\quad \Psi_B=E-C\mu_{ce}-A\mu_{ae}\,,\\ \Psi_C=F-A\mu_{af}-B\mu_{bf}\,.\end{aligned} \tag{F.3}$$

In order to evaluate Ω when $V^2\sim 0$, it is useful to exploit the following integral representation for Ψ_A:

$$\Psi_A=\int_{\mu_{bc}{}^0}^{\mu_{bc}}(1-\xi^2)^{-1}(1+2\mu_{bd}\mu_{cd}\xi-\mu_{bd}^2-\mu_{cd}^2-\xi^2)^{\frac{1}{2}}\,\mathrm{d}\xi\,, \tag{F.4}$$

and similar ones for Ψ_B, Ψ_C. Noticing that $\Psi_A(\mu_{bc}=\overset{0}{\mu}_{bc})=0$ and using the relations

$$D=\arc\cos\left\{\frac{\mu_{bd}\mu_{cd}-\mu_{bc}}{[(1-\mu_{bd}^2)(1-\mu_{cd}^2)]^{\frac{1}{2}}}\right\},\quad B=\arc\cos\left\{\frac{\mu_{cd}-\mu_{bd}\mu_{bc}}{[(1-\mu_{bd}^2)(1-\mu_{bc}^2)]^{\frac{1}{2}}}\right\},$$

$$C=\arc\cos\left\{\frac{\mu_{bd}-\mu_{cd}\mu_{bc}}{[(1-\mu_{cd}^2)(1-\mu_{bc}^2)]^{\frac{1}{2}}}\right\}, \tag{F.5}$$

it is easy to verify that the derivatives with respect to μ_{bc} of both sides of (F.4) are equal. Now let us put

$$1 + 2\mu_{bd}\mu_{cd}\xi - \mu_{bd}^2 - \mu_{cd}^2 - \xi^2 = (\xi - \mu_{bc}^0)(\mu_{bc}^1 - \xi), \tag{F.6}$$

where μ_{bc}^1 is the value assumed by μ_{bc} when the tetrahedron is flat with P outside $P_AP_BP_C$. In this case $\alpha = |\delta_b - \delta_c|$ and $\mu_{bc}^1 > \mu_{bc}^0$. In the neighbourhood of $\mu_{bc} = \mu_{bc}^0$ we can approximate the integrand of (F.4) as follows:

$$\Psi_A \simeq \int_{\mu_{bc}^0}^{\mu_{bc}} (1 - \xi^2)^{-1}[(\xi - \mu_{bc}^0)(\mu_{bc}^1 - \mu_{bc}^0)]^{\frac{1}{2}}\, d\xi \simeq \frac{2}{3}\frac{(\mu_{bc} - \mu_{bc}^0)^{\frac{3}{2}}}{1 - (\mu_{bc}^0)^2}(\mu_{bc}^1 - \mu_{bc}^0)^{\frac{1}{2}}, \tag{F.7}$$

and using (F.6)

$$\Psi_A \simeq \tfrac{2}{3}(\mu_{bc} - \mu_{bc}^0)[1 - (\mu_{bc}^0)^2]^{-1}(1 + 2\mu_{bd}\mu_{cd}\mu_{bc} - \mu_{bd}^2 - \mu_{cd}^2 - \mu_{bc}^2)^{\frac{1}{2}}. \tag{F.8}$$

Since $\mu_{bc} - \mu_{bc}^0 = 2\sin\frac{1}{2}(\alpha + \delta_b + \delta_c)\sin\frac{1}{2}(-\alpha + \delta_b + \delta_c) \simeq -(\alpha - \delta_b - \delta_c)\sin\alpha$, we obtain from (F.8), (B.6), (B.5)

$$\begin{aligned}\Psi_A &\simeq \tfrac{2}{3}\sin\alpha(\delta_b + \delta_c - \alpha)\frac{6V}{(b+\frac{1}{2})(c+\frac{1}{2})(d+\frac{1}{2})}(\sin\alpha)^{-2} \simeq \\ &\simeq \tfrac{2}{3}(\delta_b + \delta_c - \alpha)\sin\delta_c \sin C.\end{aligned} \tag{F.9}$$

If h is the distance of P from the plane $P_AP_BP_C$ we have also

$$(d + \tfrac{1}{2})\Psi_A \simeq \tfrac{2}{3}h(\delta_b + \delta_c - \alpha). \tag{F.10}$$

Therefore from (F.10) and similar relations for Ψ_B, Ψ_C we find

$$\Omega \simeq \pi(a + b + c - \tfrac{1}{4}) + \tfrac{2}{3}h(2\pi - \delta - \varepsilon - \varphi). \tag{F.11}$$

Our result (E.9) yields in the neighbourhood of $V^2 = 0$

$$V^2 = \tfrac{1}{18}[(d + \tfrac{1}{2})(e + \tfrac{1}{2})(f + \tfrac{1}{2})]^2 \sin\delta \sin\varepsilon \sin\varphi\,(2\pi - \delta - \varepsilon - \varphi), \tag{F.12}$$

having taken into account the different definition of δ, ε, φ. Using the proportionality between h and V, we obtain finally from (F.11), (F. 12)

$$\Omega \simeq \pi(a + b + c - \tfrac{1}{4}) + \tfrac{9}{2}V^3\left(\prod_{k=1}^{4} A_k\right)^{-1}. \tag{F.13}$$

In a similar way, when we approach B_4 as $V^2 \to 0$, we find in the case corresponding to fig. 1

$$\Omega \simeq \pi(a + b + d + e + \tfrac{1}{4}) - \tfrac{9}{2}V^3\left(\prod_{k=1}^{4} A_k\right)^{-1}. \tag{F.14}$$

Appendix G. WKB approximation for $d^{(J)}_{\delta\delta'}(\theta)$

Let us recall here the main features of the WKB solutions of

$$\frac{d^2 g}{dz^2} + Q^2(z)\, g = 0, \tag{G.1}$$

without discussing their degree of approximation[33]); in (G.1) $Q^2(z) = \mathcal{Q}^2(z)(z-z_1)(z_2-z)$ with $z_1 < z_2$ and $\mathcal{Q}^2 > 0$. We know that the function

$$z \leqslant z_1, \quad g_1(z) = \frac{2}{\pi}\left(\frac{|\mathrm{Im}\, t_1|}{|Q|}\right)^{\frac{1}{2}} \times \\ \times \{\pi \sin \eta_1\, I_{\frac{1}{3}}(|\mathrm{Im}\, t_1|) + \cos(\tfrac{1}{3}\pi - \eta_1)\, K_{\frac{1}{3}}(|\mathrm{Im}\, t_1|)\} \tag{G.2}$$

joins smoothly with

$$z_1 < z \leqslant z_2 \begin{cases} g_2(z) = \left(\dfrac{4t_1}{3Q}\right)^{\frac{1}{2}} \{\cos(\tfrac{1}{3}\pi + \eta_1)\, J_{\frac{1}{3}}(t_1) + \cos(\tfrac{1}{3}\pi - \eta_1)\, J_{-\frac{1}{3}}(t_1)\}, & \text{(G.3)} \\ t_1(z) = \displaystyle\int_{z_1}^{z} Q(\xi)\, d\xi, \quad Q(\xi) = +[Q^2(z)]^{\frac{1}{2}}, & \text{(G.4)} \end{cases}$$

if we choose for $z \leqslant z_1$: $Q(z) = i|Q(z)|$, $|\mathrm{Im}\, t_1| = \int_z^{z_1} |Q(\xi)|\, d\xi$; far from z_1 we have

$$g_1(z) \simeq (2\pi|Q|)^{-\frac{1}{2}} \{2 \sin \eta_1\, e^{|\mathrm{Im}\, t_1|} + \cos \eta_1\, e^{-|\mathrm{Im}\, t_1|}\}, \tag{G.5}$$

$$g_2(z) \simeq \left(\frac{2}{\pi Q}\right)^{\frac{1}{2}} \cos(t_1 + \eta_1 - \tfrac{1}{4}\pi). \tag{G.6}$$

Similarly, continuity through z_2 can be achieved by means of

$$z_1 < z \leqslant z_2 \begin{cases} g_2'(z) = \left(\dfrac{4t_2}{3Q}\right)^{\frac{1}{2}} \{\cos(\tfrac{1}{3}\pi + \eta_2)\, J_{\frac{1}{3}}(t_2) + \cos(\tfrac{1}{3}\pi - \eta_2)\, J_{-\frac{1}{3}}(t_2)\}, & \text{(G.7)} \\ t_2(z) = \displaystyle\int_{z}^{z_2} Q(\xi)\, d\xi, \quad Q(\xi) = +[Q^2(\xi)]^{\frac{1}{2}}, & \text{(G.8)} \end{cases}$$

$$z_2 < z, \quad g_3(z) = \frac{2}{\pi}\left(\frac{|\mathrm{Im}\, t_2|}{|Q|}\right)^{\frac{1}{2}} \times \\ \times \{\pi \sin \eta_2\, I_{\frac{1}{3}}(|\mathrm{Im}\, t_2|) + \cos(\tfrac{1}{3}\pi - \eta_2)\, K_{\frac{1}{3}}(|\mathrm{Im}\, t_2|)\}, \tag{G.9}$$

choosing for

$$z_2 < z: \quad Q(z) = \mathrm{i}\,|Q(z)|, \quad |\mathrm{Im}\, t_2| = \int_{z_2}^{z} |Q(\xi)|\, \mathrm{d}\xi\,;$$

for large $|z - z_2|$ these solutions behave according to (G.6), (G.5) with the obvious replacements: $t_1 \to t_2$, $\eta_1 \to \eta_2$. We note incidentally that, for instance $g_2(z)$ can be written in terms of Airy functions[25]):

$$g_2(z) = \left(\frac{3t_1}{Qy}\right)^{\frac{1}{2}} \{\cos \eta_1\, \mathrm{Ai}(-y) + \sin \eta_1\, \mathrm{Bi}(-y)\} \qquad \text{(G.10)}$$

$$\mathrm{Ai}(-y) = \tfrac{1}{3} y^{\frac{1}{2}} \left[J_{-\frac{1}{3}}(t_1) + J_{\frac{1}{3}}(t_1)\right], \quad \mathrm{Bi}(-y) = \left(\tfrac{1}{3} y\right)^{\frac{1}{2}} \left[J_{-\frac{1}{3}}(t_1) - J_{\frac{1}{3}}(t_1)\right], \qquad \text{(G.11)}$$

where $y = (\frac{3}{2} t_1)^{\frac{2}{3}}$. We see from (G.6) that the consistency condition $g_2(z) = g_2'(z)$ can be fulfilled for $z_1 \ll z \ll z_2$ if

$$\int_{z_1}^{z_2} Q(\xi)\, \mathrm{d}\xi = -\eta_1 - \eta_2 + (2N + \tfrac{1}{2})\pi, \quad N \gg 1 \qquad \text{(G.12)}$$

which provides a relation between η_1 and η_2.

We know (Brussaard and Tolhoek[16])) that, for large f, $d^{(f)}_{\delta\delta'}(\theta)$ satisfies

$$\left\{\frac{\mathrm{d}}{\mathrm{d}(\cos\theta)} \sin^2\theta \frac{\mathrm{d}}{\mathrm{d}(\cos\theta)} + (f + \tfrac{1}{2})^2 \sin^2\theta \left[(1 - \mu^2) \times \right.\right.$$
$$\left.\left. \times (1 - \nu^2) - (\cos\theta - \mu\nu)^2\right]\right\} d^{(f)}_{\delta\delta'}(\theta) = 0, \qquad \text{(G.13)}$$

where $(f + \frac{1}{2})\,\mu = \delta'$, $(f + \frac{1}{2})\,\nu = \delta$; we shall consider only the domain $|\cos\theta| \leqslant 1$. If we put $\cos\theta = \mathrm{tgh}\, z$, $d^{(f)}_{\delta\delta'}(\theta) = g(z)$, then we obtain (G.1) with

$$Q^2(z) = (f + \tfrac{1}{2})^2 \left[(1 - \mu^2)(1 - \nu^2) - (\mathrm{tgh}\, z - \mu\nu)^2\right], \qquad \text{(G.14)}$$

and transition points

$$\cos\theta_{1,2} \equiv \mathrm{tgh}\, z_{1,2} = \mu\nu \mp \sqrt{(1 - \mu^2)(1 - \nu^2)}; \qquad \text{(G.15)}$$

when $\mu = \nu$, z_2 corresponds to $\cos\theta_2 = 1$; therefore we shall limit ourselves to $\delta \neq \delta'$.

In order to apply (G.2)–(G.9), we must provide t_1, t_2, η_1 and η_2 explicitly. From (G.4), (G.14) and (F.4) we have

$$\int^{z} (f + \tfrac{1}{2}) \left[1 - \mu^2 - \nu^2 + 2\mu\nu\, \mathrm{tgh}\, z - (\mathrm{tgh}\, z)^2\right]^{\frac{1}{2}} \mathrm{d}z =$$
$$= (f + \tfrac{1}{2}) \left[F(z) - \mu A(z) - \nu B(z)\right], \qquad \text{(G.16)}$$

where $\pi - F$, A, B are the *internal* dihedral angles between the planes belonging respectively to f, $a+\delta'$, $b+\delta$ (fig. 4 with $\mu=\cos\alpha$, $\nu=\cos\beta$). We notice from (G.15) that z_1 corresponds to $\theta_1=\alpha+\beta$, $A=B=F=0$. On the other hand z_2 is associated to the case $\theta_2=|\alpha-\beta|$; we have for $\alpha>\beta$: $B=F=\pi$, $A=0$ and for $\alpha<\beta$: $A=F=\pi$, $B=0$. Therefore[34]:

$$\begin{gathered} t_1(\theta) \equiv t = (f+\tfrac{1}{2})(F - A\mu - B\nu), \\ F = \arc\cos\left(\frac{\mu\nu - \cos\theta}{|\sin\alpha\sin\beta|}\right), \quad A = \arc\cos\left(\frac{\nu - \mu\cos\theta}{|\sin\alpha\sin\theta|}\right), \\ B = \arc\cos\left(\frac{\mu - \nu\cos\theta}{|\sin\beta\sin\theta|}\right), \end{gathered} \tag{G.17}$$

and similarly for t_2. Furthermore (G.12) yields now:

$$\begin{aligned} -\eta_1 - \eta_2 + (2M+\tfrac{1}{2})\pi &= \int_{z_1}^{z_2} Q(\xi)\,\mathrm{d}\xi = \\ &= \begin{cases} (f+\tfrac{1}{2})(\pi - \pi\nu), & \alpha>\beta, \quad (\delta>\delta'), \\ (f+\tfrac{1}{2})(\pi-\pi\mu), & \alpha<\beta, \quad (\delta<\delta'). \end{cases} \end{aligned} \tag{G.18}$$

From the asymptotic behaviour of Jacobi polynomials[19]) and relation (A.18) it is easy to obtain:

$$d^{(f)}_{\delta\delta'}(\theta) \simeq \left[\frac{2}{\pi(f+\frac{1}{2})\sin\theta}\right]^{\frac{1}{2}} \cos\left\{(f+\tfrac{1}{2})\theta + \tfrac{1}{2}\pi(\delta-\delta') - \tfrac{1}{4}\pi\right\}, \tag{G.19}$$

valid when f is large and $|\delta|, |\delta'| \ll f$; performing this limit in (G.17), we have: $A\simeq B\simeq\frac{1}{2}\pi$, $F\simeq\pi-\theta$ and consequently $t_1(\theta)\simeq(f+\frac{1}{2})(\pi-\theta)-\frac{1}{2}\pi(\delta+\delta')$. Using this result and identifying the arguments of the cosines in (G.6) and (G.19), we obtain

$$\eta_1 = \pi(\delta' - f). \tag{G.20}$$

Therefore from (G.18)

$$\eta_2 = \begin{cases} \pi(\delta-\delta') & \delta > \delta', \\ 0 & \delta<\delta'; \end{cases} \tag{G.21}$$

obviously η_1 and η_2 are determined modulo $2N\pi$. It must be stressed that only for "physical" values of δ, δ' the phases η_1, η_2 (except when $\delta<\delta'$) are integer multiples of π; in this case the exponentially increasing term in (G.5) is ruled out.

The overall normalization in (G.2), (G.3) and (G.9) has been chosen in agreement with (A.15).

Appendix H. The 1 + 1 + 2 case

In this last appendix we want to show that (3.6) is in agreement with our asymptotic formulae (1.4), (1.8). To this end we consider the limiting case in which f, δ, δ', though large, are still small with respect to a, b, c; therefore we may use in (3.6) our results of appendix G.

In the oscillatory region we deduce from (3.6), (G.6), (G.17), (G.20)

$$\begin{Bmatrix} c & a & b \\ f & b+\delta & a+\delta' \end{Bmatrix} \simeq \frac{(-1)^{a+b+c+f+\delta}}{[(2a+1)(2b+1)]^{\frac{1}{2}}} \left[\frac{2}{\pi(f+\frac{1}{2})}\right]^{\frac{1}{2}} \times$$
$$\times [(1-\mu^2)(1-\nu^2)-(\cos\theta-\mu\nu)^2]^{-\frac{1}{4}} \cos\{t+\pi(\delta'-f)-\tfrac{1}{4}\pi\}$$

and recalling (B.6):

$$\begin{Bmatrix} c & a & b \\ f & b+\delta & a+\delta' \end{Bmatrix} \simeq \frac{(-1)^{a+b+c+\delta+\delta'}}{[12\pi V]^{\frac{1}{2}}} \cos(t-\tfrac{1}{4}\pi). \tag{H.1}$$

On the other hand we note that as a, b, c increase in the tetrahedron of fig. 4, the two faces belonging to c become almost parallel; therefore: $\theta_c \simeq \pi$, $\theta_a \equiv A \simeq \pi - \theta_{a+\delta'}$, $\theta_b \equiv B \simeq \pi - \theta_{b+\delta}$ and from the definition (3.15b) we obtain

$$\begin{aligned} \Omega &= (a+b+c+\delta+\delta'-\tfrac{1}{4})\pi + (f+\tfrac{1}{2})F - \delta' A - \delta B = \\ &= t + (a+b+c+\delta+\delta'-\tfrac{1}{4})\pi, \end{aligned} \tag{H.2}$$

which, when introduced into (1.4a) leads to (H.1).

Let us now consider the classically forbidden regions. First we notice that the transition points $z_{1,2}$ relative to $d^{(f)}_{\delta\delta'}(\theta)$ discussed in appendix G correspond to configurations in which the tetrahedron of fig. 4 becomes flat; this can be checked by means of (G.14) and (B.6). More precisely, since in z_1: $\theta = \alpha + \beta$, the tetrahedron enters the region B_3 with

$$z_1) \quad \Phi = (a+b+c+\delta+\delta')\pi, \tag{H.3}$$

while in z_2, $\theta = |\alpha - \beta|$ and it enters B_4 in either one of the two ways

$$z_2) \quad \Phi = \begin{cases} (a+b+c+f+\delta')\pi & \delta > \delta', \\ (a+b+c+f+\delta)\pi & \delta < \delta'. \end{cases} \tag{H.4}$$

On the other hand, from (G.5) and (3.6) we have for the forbidden region relative to z_1

$$\begin{Bmatrix} c & a & b \\ f & b+\delta & a+\delta' \end{Bmatrix} \simeq \frac{(-1)^{a+b+c+f+\delta}}{2[12\pi|V|]^{\frac{1}{2}}} \times$$
$$\times \{2\sin\pi(\delta'-f)\,e^{|\mathrm{Im}\,t|} + \cos\pi(\delta'-f)\,e^{-|\mathrm{Im}\,t|}\}. \tag{H.5}$$

Noticing that $|\text{Im}\, t| = |\text{Im}\, \Omega|$, we see from (5.8), (H.2), (H.3) and (H.5) that for physical values of the angular momenta (H.5) and (5.8) become identical even in sign. The same holds also for the forbidden region relative to z_2; in fact we have from (G.21):

$$(-1)^{\eta_2/\pi + a+b+c+f+\delta} = \begin{cases} (-1)^{a+b+c+f+\delta'} & \delta > \delta' \\ (-1)^{a+b+c+f+\delta} & \delta < \delta' \end{cases}$$

in agreement with (H.4).

References and footnotes

1) G. Racah, Phys. Rev. **61**, 186 (1942); **62**, 438 (1942); **63**, 367 (1943); **76**, 1352 (1949).
2) See the exhaustive bibliography in L. C. Biedenharn and H. Van Dam, *Quantum Theory of Angular Momentum* (Academic Press, New York, 1965).
3) L. C. Biedenharn, private communication to one of us (T.R.), 1958.
4) E. P. Wigner, Am. J. Math. **63**, 57 (1941); On the Matrices which Reduce the Kronecker Products of Representations of S. R. Groups (unpublished, 1940); hectographed paper (Princeton, 1951) (see ref. 2).
5) W. T. Sharp, Some Formal Properties of the 12*j*-symbol, Chalk River Report, TPJ-81, Ontario, 1955.
6) A. P. Yutsis, I. B. Levinson and V. V. Vanagas, *The Mathematical Apparatus of the Theory of Angular Momentum* (translated from Russian) (Jerusalem, 1962); A. P. Yutsis and A. A. Bandzaitis, *Teoria Momenta Kolichestva Dvizheniya Kvantovoi Mekhanike* (Vilnius, 1965) (in Russian).
7) In addition to the three-dimensional representation which will be used in this paper, we recall the graphical techniques developed by A. Yutsis and coworkers[6]) as well as its dual, exemplified by U. Fano and G. Racah, *Irreducible Tensorial Sets* (Academic Press, New York, 1959) appendices.
8) We introduce $a+\frac{1}{2}$ rather than a... as length of the edge corresponding to a... because, for high quantum numbers, the length $[a(a+1)]^{\frac{1}{2}}$ of the angular momentum vector is closer to $a+\frac{1}{2}$ in the semiclassical limit; it turns out that any other choice is inconsistent with numerical results and with existing formulae.
9) N. F. Tartaglia, *General trattato de numeri et misure* (Venezia, 1560).
10) See the footnote 142 of ref. 22.
11) A. Cayley, Cambridge Math. J. **2**, 267 (1841).
12) S. Adler has kindly pointed out to us that:

$$2^9(3!)^2V^2 \equiv -(2\alpha+1)(2\gamma+1)(2\varepsilon+1) - 2[2q_4+1]_2[2q_2+1]_2[2q_1+1]_2 + 1, \text{ mod. } 8$$

where q_1, q_2, q_4 are defined above and $[A]_2 = 0$, 1 if A is even, odd;

$$\alpha = [2a+1]_2[2q_4+1]_2, \quad \gamma = [2c+1]_2[2q_2+1]_2, \quad \varepsilon = [2e+1]_2[2q_1+1]_2.$$

It follows that if all the q's are integer and $\alpha, \gamma, \varepsilon = 0$, or 1:

$$2^9(3!)^2V^2 \equiv -(2\alpha+1)(2\gamma+1)(2\varepsilon+1) - 1 \equiv -2, -4 \text{ mod. } 8;$$

therefore $V^2 \neq 0$.

13) E. P. Wigner, *Group Theory* (Academic Press, New York, 1959).
14) W. Pauli, in: *Handbuch der Physik*, Band 24/1 (Springer-Verlag, Berlin, 1933) p. 170.
15) T. Regge, Nuovo Cimento **10**, 296 (1958); **11**, 116 (1959).
16) G. Racah, Phys. Rev. **84**, 910 (1951); *Lectures on Group Theory* (Institute for Advanced Study, Princeton, 1951); L. C. Biedenharn, J. Math. Phys. **31**, 287 (1953); K. Alder, A. Bohr, T. Huus, B. Mottelson and A. Winther, Rev. Mod. Phys. **28**, 432 (1956); P. J. Brussaard and H. A. Tolhoek, Physica **23**, 955 (1957); A. R. Edmonds, *Angular Momentum in Quantum Mechanics* (Princeton University Press, 1957).
17) L. C. Biedenharn, ref. 16, and J. P. Elliott, Proc. Royal Soc. (London) A **218**, 370 (1953).
18) This property stems from a particular case of a theorem by Schläfli on elliptic tetrahedra; see H. S. M. Coxeter, *Non-Euclidean Geometry* (Toronto, 1957).
19) G. Szegö, *Orthogonal Polynomials* (American Mathematical Society, New York, 1959).
20) A discussion of this theorem, due originally to G. Racah, can be found in the appendix to the book by U. Fano and G. Racah, ref. 7.
21) See for instance eq. (16.8) ref. 6.
22) E. Hammer, *Trigonometrie* (Stuttgart, 1923).
23) A. Erdélyi, *Asymptotic Expansions* (Dover Publications, New York, 1956).
24) We do not expect to be true in general that the $6j$-symbol is a function of the variables V and $\Pi_{h=1}{}^4 A_h$ only.
25) J. C. P. Miller, *The Airy Integral* (Cambridge, 1946).
26) R. P. Feynman and A. R. Hibbs, *Quantum Mechanics and Path Integrals* (McGraw-Hill, New York, 1965).
27) T. Regge, Nuovo Cimento **19**, 551 (1961).
28) J. A. Wheeler, Geometrodynamics and the Issue of the Final State (1963), in: *Relativity, Groups and Topology*, Eds. C. and B. DeWitt (Gordon and Breach, New York, 1964).
29) B. L. Van der Waerden, *Die gruppentheoretische Methode in der Quantenmechanik* (Springer-Verlag, Berlin, 1932).
30) A. Messiah, *Quantum Mechanics* (North-Holland Publ. Co., Amsterdam, 1964) appendix C.
31) R. F. Scott, *The Theory of Determinants* (Cambridge, 1904) chapter XVII.
32) T. Regge and G. Barucchi, Nuovo Cimento **34**, 106 (1964).
33) R. E. Langer, Phys. Rev. **51**, 669 (1937).
34) The discussion which is carried out here as well as in appendix H corresponds actually to $\delta > 0$, $\delta' > 0$; however, it can be easily extended to the other cases yielding slight modifications in t_1, t_2, η_1, η_2. For instance, when $\delta < 0$, $\delta' < 0$ one would find $\theta_1 = 2\pi - \alpha - \beta$, $t_1(\theta) = (f + \frac{1}{2})\,[F - (A - \pi)\,\mu - (B - \pi)\,\nu]$, $\eta_1 = -\pi(f + \delta)$, while η_2 would be the same as in (G.21) for "physical" values of δ, δ'.

IL NUOVO CIMENTO Vol. XIX, N. 3 1° Febbraio 1961

General Relativity without Coordinates.

T. Regge

Palmer Physical Laboratory, Princeton University - Princeton, N. J. (*)

(ricevuto il 17 Ottobre 1960)

Summary. — In this paper we develop an approach to the theory of Riemannian manifolds which avoids the use of co-ordinates. Curved spaces are approximated by higher-dimensional analogs of polyhedra. Among the advantages of this procedure we may list the possibility of condensing into a simplified model the essential features of topologies like Wheeler's wormhole and a deeper geometrical insight.

1. - Polyhedra.

In this section we shall first describe our approach for the simple case of 2-dimensional manifold (surfaces). Following Aleksandrov [1] we develop the theory of intrinsic curvature on polyhedra. A general surface is then considered as the limit of a suitable sequence of polyhedra with an increasing number of faces. A rigorous definition of limit is not given here since it would involve a treatment of the topology on the set of all polyhedra and this would carry us too far. It is to be expected however that any surface can be arbitrarily approximated, as closely as wanted, by a suitable polyhedron. The approximation will be bad if we look at the details to the picture but an observer looking at the broad details only will find it quite satisfactory. On any surface we can define an integral Gaussian curvature by carrying out curvature experiments with geodesic triangles.

Let t be one such a triangle and let α, β, γ be its internal angles. If the geometry inside the triangle is not euclidean we have in general $\alpha+\beta+\gamma\neq\pi$.

(*) Nov at the University, Torino.

(1) P. S. Aleksandrov: *Topologia combinatoria* (Torino, 1957).

The Gaussian integral curvature ε_t of t is then defined by $\varepsilon_t = \alpha + \beta + \gamma - \pi$. On a sphere we have $\varepsilon_t = A_t/R^2$, where R is the radius of the sphere and A_t the area of t. If t shrinks to a point P and the limit $\lim_{A\to 0} \varepsilon_t/A_t = K(P)$ exists and is independent of the particular limiting procedure we take $K(P)$ as definition of local Gaussian curvature in P. On a sphere obviously $K(P) = 1/R^2$. It is well known that $\varepsilon_t = \int_t K(P)\,\mathrm{d}A$, where $\mathrm{d}A$ is the area element on the surface and the integration is carried inside t. Let it be clear however that, while the integral curvature exists under very broad conditions, the local curvature may not exist at all as an ordinary function but rather only as a measure.

This last point is essential in defining the curvature of a polyhedron M. It is obvious that if t lies entirely on a face of M we have $\varepsilon_t = 0$. The same result holds if t does not contain any vertex of M. Therefore $K(P) = 0$ if P is not a vertex.

If t contains a vertex, say V, and no other vertices, it will not depend on the explicit form of t but only on the particular choice of V. In other words $\varepsilon_t = \varepsilon_V$ is a characteristic constant of V (the deficency of V) ε_V can be found as follows. Take the sum of all internal angles of the faces of M with vertex V. This sum equals $2\pi - \varepsilon_V$. If several vertices V, X, etc., are inside t we have $\varepsilon_t = \varepsilon_V + \varepsilon_X + \varepsilon_Y + \text{etc.}$

All these results can be condensed into the original formula $\varepsilon_t = \int_t K(P)\,\mathrm{d}A$, provided one understands $K(P)$ as a Dirac type distribution, having the vertices as supports. The integral $\int_M f(P)K(P)\,\mathrm{d}A$, where $f(P)$ is a continuous function, is then to be calculated as $\sum_n f(V_n)\varepsilon_n$, where ε_n is the deficency of V_n, and the summation is carried out on all vertices of M.

If M is a compact (*i.e.* finite closed) polyhedron Gauss-Bonnet's integral curvature theorem can be written as

$$\sum_n \varepsilon_n = 2\pi(2 - N)\,,$$

N is here the genus of M, $N = 0$ for a sphere and $N = 2$ for a torus. There is no loss in generality in supposing that all faces of M are triangles.

Under this form the connection of this theorem with Euler's formula for the genus is most evident. Indeed let σ_{fn} be the internal angle of the face f with the vertex n. We have:

$$\sum_f \sigma_{fn} = 2\pi - \varepsilon_n\,,$$

where the summation is carried out on all faces f having V_n as vertex. It is also clear that $\sum_n \sigma_{fn} = \pi$, where the summation now is carried out on all vertices belonging to the face f. The sum $\sum_{fn} \sigma_{fn}$ can now be carried out in two different ways by summing first on faces and using then Gauss-Bonnet's theorem or summing on vertices first. With the first method we obtain:

$$\sum_{fn} \sigma_{fn} = 2\pi v - 2\pi(2 - N) .$$

with the second

$$\sum_{fn} \sigma_{fn} = \sum_n \pi = \pi f ,$$

where v, e, f are the number of vertices, edges, faces of M respectively. Since all faces of M are triangles we have $2e = 3f$. Therefore the equation $2v - 2(2 - N) = f$ is simply Euler's formula $v - e + f = 2 - N$.

This example shows how elementary is the treatment of curvature on polyhedra.

Since we are chiefly interested in the intrinsic geometry of manyfolds we are not particularly interested in the edges of M and we regard them as a rather immaterial convention for dividing M into triangles, any other convention being just as good.

It is interesting to notice that the intrinsic geometry of M is completely fixed by the connection matrix and the length of all edges. The connection matrix is essentially a list of all faces, edges and vertices of M and a list of their mutual relationship *i.e.* by reading it one can decide which vertices, edges belong to a given face, etc. The connection matrix supplies us with all the topological information needed in the construction of M.

Once we have constructed a simplectic net on M (in plain words a division of M into triangles) the knowledge of the lengths of all edges of M implies the knowledge of all angles σ_{fn} and therefore of the deficencies ε_n.

The notion of simplectic net replaces therefore the notion of a co-ordinate net. The metric tensor on the other hand is replaced by the lengths of the edges.

If the number of vertices, faces, etc., of M increases the local Gaussian curvature will approximate a continuous function $K(P) = \varrho\varepsilon$, where ϱ is the density of vertices and ε their deficiency, provided this product varies slowly within the test triangle t.

In dealing with higher dimensional analogs of polyhedra a geodesic triangle is no longer convenient in testing curvature. We replace it with the notion of Levi-Civita's parallel transportation (shorthly LCT). Let P and Q be points of M. Let the arc a join P to Q. LCT is then an orthogonal mapping between the vector space S_p in P and the vector space S_q in Q. If in particular $P = Q$

and a is a loop LCT maps S_p into itself. This mapping is a rotation of S around P by some angle $\varepsilon(a)$ which depends on a. By inspection it can be checked that $\varepsilon(a) = \int_a K(P)\,\mathrm{d}A$ and that therefore $\varepsilon(t) = \varepsilon_t$. ε is clearly an additive function of the loop. Take namely two loops a, b with the same end point P. Let us perform LCT around their «product» ab, ab being the loop obtained by sticking end to end the loops a and b. We get $\varepsilon(ab) = \varepsilon(a) + \varepsilon(b)$. On a polyhedron M we shall not consider the case when a vertex lies on a loop because LCT cannot be defined unambiguously.

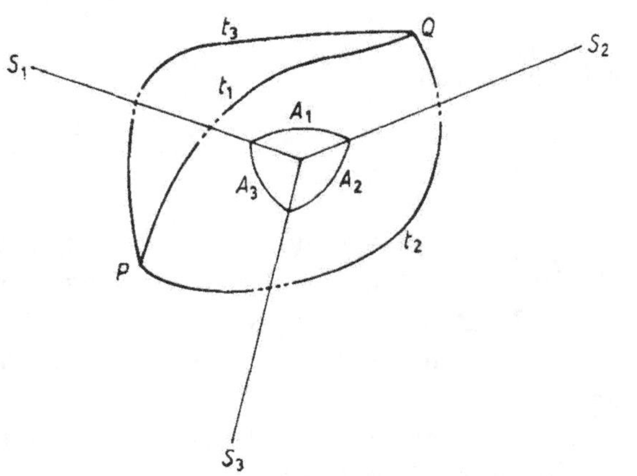

Fig. 1.

If a loop a can be deformed continuously into a loop b by keeping P fixed without encountering the above exceptional situation then $\varepsilon(a) = \varepsilon(b)$, and we write $a = b$. In the following w denotes the set of all vertices of M then $a \approx b$ is simply a homotopy in $M - w$ and is an equivalence relation. A more general and rigorous definition of homotopy will be given later. On the moment it is important to realize that the existence of a curvature set w entails the existence of equivalence classes in the set of loops with a given end point P.

2. - ε-cones.

Let us now consider a polyhedron M with one vertex only. We can visualize M as a one-sheeted cone or pyramid. There is a simple mathematical way of characterizing M. Take namely in the plane polar co-ordinates ϱ and θ so that the metric is given by $\mathrm{d}s^2 = \mathrm{d}\varrho^2 + \varrho^2 \mathrm{d}\theta^2$. In the euclidean plane it is understood that two points with the same ϱ and anomalies θ differing by a multiple of 2π are identical. We obtain a manifold with the same intrinsic geometry of a cone by replacing 2π with $2\pi - \varepsilon$.

The origin is then a vertex with deficiency ε. This manifold, hereafter named the ε-cone, is everywhere euclidean with the exception of the vertex.

These ideas can be readily extended to 3 (or higher) dimensional spaces. Let R be the real line of the variable z and let the direct product $R \times \varepsilon$-cone be metrized as follows:

$$ds^2 = dz^2 + d\varrho^2 + \varrho^2 d\theta^2 .$$

This space is euclidean with the exception of the straight line $\varrho = 0$. We name it the $\varepsilon - 3$-cone. If the product $R^{n-2} \times \varepsilon$-cone is considered with the metric:

$$ds^2 = dz_1^2 + dz_2^2 + \ldots + dz_{n-2}^2 + d\varrho^2 + \varrho^2 d\theta^2 ,$$

we get the $\varepsilon - n$-cone. This manifold is euclidean with the exception of the $n-2$ dimensional flat subset $\varrho = 0$.

The $\varepsilon - n$-cone is the building stone of our structure. The $\varepsilon - 2$-cone is the previously defined ε-cone. If the dimension n is otherwise known we shall drop it from the writing.

3. – Fundamental group.

Take now a path in a Riemannian n-dimensional manifold M_n. We define a path p as a point-valued continuous function $P(s)$ of a parameter s, $0 \leqslant s \leqslant 1$, $P(0)$, $P(1)$ being the end point of the path. If $P(0) = P(1) = P$ then the path is called a loop. We shall regard as identical the loops $P(s)$ and $P(s'(s))$ where $s'(s)$ is non decreasing and continuous and $s(0) = 0$, $s(1) = 1$. If we have two paths a, b, defined by the functions $A(s)$, $B(s)$ such that $A(1) = B(0)$ then the product $c = ab$ is defined by the function $C(s) = A(2s)$ if $s \leqslant 0.5$, $C(s) = B(2s - 1)$ if $s \geqslant 0.5$. The product $c = ab$ joins $A(0)$ to $B(1)$ through $A(1) = B(0)$. This product is associative but in general is not commutative. If a, b are loops with the same end point P then ab and ba both exist and are loops with end point P. In general, however, $ab \neq ba$. a^{-1}, the reverse of a, is defined by $A(1-s)$. Suppose now that our manifold is almost everywhere euclidean except a closed subset w which carries curvature, to be described in detail later. Take two paths a, b with the same end points, defined by $A(s)$ and $B(s)$. We always suppose $B(s) \notin w$, $A(s) \notin w$. If there is a point-valued function $P(s, t)$ of s, $0 \leqslant s \leqslant 1$, and t, $0 \leqslant t \leqslant 1$, continuous in both s and t, such that $P(s, 0) = A(s)$, $P(s, 1) = B(s)$ we say that we have deformed a into b, and the function $P(s, t)$ is called a deformation. If for given a, b it is possible to find a deformation such that $P(s, t) \notin w$ for all s, t then we write $a \sim b$. The symbol $\sim$ satisfies the formal properties of an equivalence (holonomy). Indeed $a \sim a$ because one such a deformation is given by $P(s, t) = A(s)$. If $a \sim b$ then $P(s, 1-t)$ deforms b into a and $P(s, 1-t) \notin w$ it follows $a \sim b$. Similarly if $P(s, t)$ deforms a into b and $Q(s, t)$ deforms b into c and therefore $a \sim b$, $b \sim c$, the function $R(s, t) = P(s, 2t)$ if $t < 0.5$ and $R(s, t) = Q(s, 2t-1)$ if $t > 0.5$ deforms a into c and $R(s, t) \notin w$, $a \sim c$. These definitions apply as well to loops, with the additional requirement that loops remain such during the deformation or that $P(0, t) = P(1, t)$ for every t. If moreover we have $P(0, t) = P(1, t) = P(0, 0) = P$ then we write $a \approx b$. Clearly $a \approx b$ implies $a \sim b$ but the converse is not true. The symbol $\approx$ applies to loops only.

The symbol $\approx$ also defines an equivalence (homotopy in $M_n - w$). The proof is trivial. If we only know that $a \sim b$ then generally the function $C(s) = P(0, s) = P(1, s)$ defines a loop c because $P(0, 0) = P(0, 1) = P$. Surprisingly enough we have b $b \approx c^{-1}ac$. This relation can be checked from the deformation $S(s, t)$ defined as:

$$\begin{aligned} S(s, t) &= C(1 - 3ts) && s < \tfrac{1}{3}, \\ &= P(3s - 1, 1 - t) && \tfrac{1}{3} < s < \tfrac{2}{3}, \\ &= C(1 + 3(s - 1)t) && \tfrac{2}{3} < s < 1 . \end{aligned}$$

We define next the unit loop u in the point P through the function $U(s) = P = \text{const.}$ Clearly $aa^{-1} \approx u$. Suppose now that $a \approx a'$ and $b \approx b'$. Let $P(s, t)$ deform a into a' and $Q(s, t)$ deform b into b'. The function $R(s, t) = P(2s, t)$, $(s < 0.5)$ and $R(s, t) = Q(2s - 1, t)$, $(s > 0.5)$ is then a deformation with $R(0, t) = R(1, t) = p$ of ab into $a'b'$ and $R(s, t) \notin w$. Therefore from $a \approx a'$ and $b \approx b'$ it follows $ab \approx a'b'$.

Take now a set of loops with end point P. This set can be partitioned into equivalence classes. A class is identified by one element belonging to it and from now on, unless differently stated, we use a loop to label the whole class. If a and b are classes then all products $a'b'$ where $a \approx a'$ and $b \approx b'$ are homotopic and belong to a new class which we define as the product of the classes a and b and write ab. Moreover if a is a class and u the unit class then $au = a = ua$.

To any class a we can associate the inverse class a^{-1} and $aa^{-1} = a^{-1}a = u$. The product of classes is associative. We see therefore that the set of all classes satisfies the axioms of a group, the fundamental group. Now to each loop a, we can associate an orthogonal matrix $S(a)$ as follows. Take a vector $\boldsymbol{A}$ in P and carry $\boldsymbol{A}$ around a according to LCT. There will be a linear, norm-preserving, mapping between the initial and final position of $\boldsymbol{A}$. This mapping can be therefore represented by an orthogonal matrix if we choose in P a local orthogonal frame of reference. Of course, as we shall require in the applications, if our space is locally pseudoeuclidean we shall have Lorentz matrices. Now the important point is that homotopic loops correspond to the same matrix or, otherwise said, the orthogonal matrix is a function of classes. This in turn implies that the mapping classes matrices are a representation of the fundamental group on the vector space of P. To prove this we first notice that all loops $\approx u$ correspond to unit matrices. If $v \approx u$ then there is a deformation function $T(s, t)$ which carries v into u. The set of all points

$T(s, t)$ is a simply connected surface Σ and we can choose on Σ and in a suitable neighborhood of Σ a globally euclidean frame of reference and the result of L-C-transportation along v calculated in this frame is the identity.

It is furthermore evident that if $c = ab$ and if the matrices $S(a)$, $S(b)$, $S(c)$ correspond to a, b, c we have $S(c) = S(b)S(a)$. Take namely a vector $\boldsymbol{V}$ in P and let us carry it along a. The result is by definition $S(a)\boldsymbol{V}$. We then carry this vector along b. We obtain $S(b)S(a)\boldsymbol{V}$. But in so doing we have carried $\boldsymbol{V}$ along c, therefore $S(v)\boldsymbol{V} = S(b)S(a)\boldsymbol{V}$ and the result follows. If then $a \approx a'$ we know that $a = va'$ where $v \approx u$. It follows $S(a) = S(v)S(a')$ but $S(v) = 1$ and $S(a) = S(a')$. The variable in $S(a)$ is then understood to be an equivalence class. The fundamental group has in general infinitely many elements. The example of the ε-cone is instructive. If we take a loop in the ε-cone we can give the angle θ of a point on the loop as function of s. Let $\theta(0) = 0$ without loss of generality. We know that $\theta(1) = N(2\pi - \varepsilon)$. N gives the number of times the loop has encircled the line $\varrho = 0$ or bone, which coincides with the subset w. Loops with the same N are clearly homotopic. Let $a(N)$ be the equivalence class with $\theta(1) = N(2\pi - \varepsilon)$. We have $a(N)a(M) = a(N+M)$. The fundamental group is then isomorphic to the group of integers under addition. In order to find $S(a(n))$, shortly $S(n)$ and $S(0) = S$, we notice that $S(N) = S^N$. S will be called the generator of the bone. Take then a vector $\boldsymbol{V}$ in P, and carry it along $a(1)$. V can be split into orthogonal components, one lying into the $\varepsilon - 2$-cone, the other into R^{n-2}. The latter is unaffected by the process, but the first is rotated by the angle ε. The total effect is that of an orthogonal $n \times n$ matrix with R^{n-2} as invariant subspace. In a $\varepsilon - 3$-cone, R^{n-2} is just a real line orthogonal to an $\varepsilon - 2$-cone. S is a rotation with axis $R^{n-2} = R$. This particular fundamental group is abelian but, in general, this is not true.

4. – We shall define now n-dimensional generalizations of polyhedra, shortly skeleton spaces, where curvature is carried by a $n-2$ subset w, the skeleton. One such structure is the $\varepsilon - n$-cone.

We start from a simplectic decomposition of an n-dimensional space M_n. This decomposition determines the topology of M_n but not the metric. We define a metric in M_n wtih the following axioms.

A) The metric in the interior of any n-dimensional closed simplex T_n is euclidean. This means that we can calculate the distance of any two points inside T_n, define on T_n a cartesian system of co-ordinates, and give the co-ordinates of the points of the boundary of T_n in this frame.

B) In the metric of T_n the boundary of T_n is decomposable into the sum of $n+1$ closed simplexes T_{n-1} and these simplexes are flat.

C) If a simplex T_{n-1} is common boundary of T_n and $T'_{n'}$ the distance of any two points of T_{n-1} is the same in both frames of T_n and $T'_{n'}$.

D) If $P \in T_n$, $P' \in T'_n$ and P, P' are close enough to T_{n-1} we define the distance PP' as the infimum of $PQ+QP'$ for all $Q \in T_{n-1}$.

With this definition the metric is euclidean also on T_{n-1} except perhaps on T_{n-2}'s on the boundary of T_{n-1}. Our definition works on a strictly local scale only but this is all we need to join smoothly neighbouring simplexes. If we want to introduce indefinite metrics a similar but more complicated definition can be used with the same results. Our manifold is now everywhere euclidean outside the union w of all T_{n-2} boundary simplexes (bones). In a M_3 the bones are straight segments while the T_{n-3} are points. Each bone connects two T_0's. If the length of the bones is given, the geometrical structure of the simplexes and of M_3 is determined and in particular their internal dihedral angles can be calculated from elementary formulas of spherical trigonometry. If there would be no curvature on a bone, then the sum of all dihedral angles around the bone should be 2π. But if these lengths are chosen at random, this sum for the i-th bone will be $2\pi - \varepsilon_i$. Here we recognize that the cross section of such a bone is locally an ε_i—2-cone and that therefore locally the bone is an ε_i—3-cone. In a S_n the T_{n-2} will carry along locally the same geometry of an $\varepsilon - n$-cone. Take now a T_0 in M_3. In general this point is common end of several bones $T_1^1 \dots T_1^m$. We orient the bones in such a way as to have the positive direction outgoing from T_0. We refer then to T_0 as to an oriented m-joint.

In an m-joint there is a remarkable identity among the $S_p p:1 \dots m$, S_p being the generator of the p-th bone. In order to find it we may deal with an idealized isolated joint, where all departing bones stretch up to infinity. We then select a particular cyclic order of the bones, for instance the one defined by the index p, (0 follows m) the particular choice being irrelevant. We consider then the set of plane sectors, A_p having two contiguous bones T^p, T^{p+1} as sides. We suppose that no two such sectors have common points except perhaps a common boundary bone. These sectors then cut the space in two regions M' and M''. Select the test point P in one of these regions, say M', and a point Q in M''. Take now m paths t_p, joining P to Q, and therefore intersecting the boundary between M' and M''. Let t_p intersect A_p only and only once. Now $t_p t_{p-1}^{-1}$ is a loop a_p with end point P encircling T_1^p once. a_p is a suitable candidate as generator of T_1^p. It must be pointed out however that any other $a'_p \sim a_p$ is just as good and that our procedure merely selects unambiguously one such a_p. We have than the fundamental identity:

$$\text{(1)} \qquad \begin{cases} a_1 a_2 a_3 \dots a_p \approx u \,, \\ \text{or} \\ S_1 S_2 \dots S_p = 1 \,. \end{cases}$$

The latter identity binds the deficiencies ε_p of the bones and their mutual angles. This can be clearly seen if $m=3$. In this case:

$$S_1S_2S_3 = 1\,.$$

This identity can be visualized with the help of a theorem of Euler which states that if in a spherical triangle the sides are in the order the angles of $A_1A_2A_3$, then the opposite internal angles are $\pi-\varepsilon_3$, $\pi-\varepsilon_2$, $\pi-\varepsilon_1$. Therefore if we know the deficiency we know the mutual angles of the bones and viceversa.

As we shall see later, in our geometrical approach (1) plays the role of Bianchi's identity in a differentiable Riemannian manifold.

There are corresponding generalizations of (1) in higher dimensional spaces. More precisely there is one such identity at each T_{n-3}, binding the generators of all T_{n-2} bones having this T_{n-3} as boundary. Not all these identities are independent since it is enough to write them for all T_{n-3} having a given T_{n-4} as boundary but one, this one being a consequence of the others. In a sense a T_{n-4} acts as a negative constraint by diminishing the number of Bianchi identities. We have not carried our investigation as far as to show what the actual number of Bianchi's identities is and in particular what is the role of T_{n-5} etc., in determining it since the problem does not arise in M_4.

5. – Differentiable manifolds.

The transition of a skeleton space into a differentiable manifold can be accomplished if we increase the density ϱ of bones while at the same time keeping $\varrho\varepsilon$ slowly varying. We have already shown how this procedure works on a surface. Here we shall proceed in similar fashion on a M_n, $n>2$. To begin with let's deal with a boundle of parallel bones on a M_3. We suppose that the curvature induced by the bones is small so that we may regards M_3 as approximately euclidean. Let $\boldsymbol{U}$ be a unit vector parallel to the bones. We test the curvature by carrying a vector $\boldsymbol{A}$ around a small loop of area Σ and unit normal $\boldsymbol{n}$, $\boldsymbol{\Sigma}=\Sigma\boldsymbol{n}$. At the end of the test $\boldsymbol{A}$ is found to have rotated around U by the angle $\sigma=N\varepsilon$, where N is the number of bones entangled in the loop. We have $N=\varrho(\boldsymbol{U},\boldsymbol{\Sigma})$, being the number of bones passing through the loop $\boldsymbol{\Sigma}=\boldsymbol{U}$. The final result is then:

$$\boldsymbol{A}=\varrho\varepsilon(\boldsymbol{U}\wedge\boldsymbol{A})(\boldsymbol{U},\boldsymbol{\Sigma})\,.$$

On the other hand from standard textbooks we know that:

$$\begin{aligned}
A_\mu &= R^\sigma_{\varrho\cdot\alpha\beta}\,\Sigma^{\alpha\beta}A_\sigma\,,\\
\Sigma^{\alpha\beta} &= \varepsilon^{\alpha\beta\gamma}\Sigma_\gamma\,,\\
\varepsilon^{\alpha\beta\gamma} &= \pm 1 = -\,\varepsilon^{\beta\alpha\gamma}\,, \quad \text{etc.}
\end{aligned}$$

By comparing these results we get:

$$R_{\mu\sigma\,\alpha\beta} = \varrho\varepsilon U_{\mu\sigma} U_{\alpha\beta}\,; \qquad U_{\mu\sigma} = \varepsilon_{\mu\sigma\lambda} U^{\lambda}\,. \tag{2}$$

This approximation has all the symmetry properties of the full Riemann tensor. If several bundles are present their effects at this stage are merely additive. In M_n a bone has an orientation determined by a skew symmetric tensor $U_{\mu\sigma}$, $U_{\mu\sigma}U^{\mu\sigma} = 2$, $U_{\mu\sigma}U_{\alpha\beta} + U_{\mu\alpha}U_{\beta\sigma} + U_{\mu\beta}U_{\sigma\alpha} = 0$. If we deal with a bundle of parallel bones we may chose co-ordinates x_1, x_2 perpendicular to the bone, $x_3 \dots x_n$ parallel to it. In this case $U_{12} = -U_{21} = 1$, while all other components vanish. Eq. (2) still holds provided μ, σ, α, β are allowed to take all values $1 \dots n$. Independently of n the scalar curvature R is given by $2\varrho\varepsilon$. We shall now show the connection between eq. (1) and Bianchi's identities in this approximation. If the ε's are small (1) can be written as:

$$\sum_1^m \varepsilon_p U^p_{\mu\sigma} = 0 \quad \text{or} \quad \sum_1^m \varepsilon_p \boldsymbol{U}^p = 0\,; \qquad \text{for an } m\text{-joint in } \boldsymbol{M}_3.$$

Suppose next that there is a distribution of identical m-joints of density ϱ. This means also that we have m bundles of parallel bones, the p-th bundle having defect ε_p, direction $\boldsymbol{U}^p$, density ϱ_p. Of course ϱ_p cannot be considered constant because an m-joint produces a decay or creation (depending on the direction) of a bone into or from the others. The rate of decay depends on ϱ. This rate can be estimated as follows. By definition ϱ_p is the density of flux of the outgoing bundle of bones through a surface Σ orthogonal to $\boldsymbol{U}_p$.

Let the position of Σ be determined by an abscissa s computed along the direction of $\boldsymbol{U}_p$. Let C be a cylinder of height ds and basis $\Sigma(s)$. Each joint in C increases by a unit the flux of ϱ_p bones through $\Sigma(s+\mathrm{d}s)$. There are $\Sigma\cdot\varrho\cdot\mathrm{d}s$ joints in C and therefore:

$$\Sigma\,[\varrho_p(s+\mathrm{d}s) - \varrho_p(s)] = \Sigma\,\varrho(s)\,\mathrm{d}s \qquad \text{or} \qquad \frac{\mathrm{d}\varrho_p}{\mathrm{d}s} = \varrho\,,$$

it follows

$$(\boldsymbol{U}_p\,\mathbf{grad}\,\varrho_p) = \varrho\,.$$

The Riemann tensor is then given by eq. (2) as:

$$R_{\alpha\beta\,\lambda\delta} = \Sigma_p \varepsilon_p \varrho_p U^p_{\alpha\beta} U^p_{\lambda\delta}\,. \tag{3}$$

In our approximation covariant derivatives reduce to normal derivatives. Bianchi's identities can be written as:

$$B_{\alpha\beta;\lambda\delta\varepsilon} = \frac{\partial}{\partial x_\varepsilon} R_{\alpha\beta,\lambda\delta} + \frac{\partial}{\partial x_\lambda} R_{\alpha\beta,\delta\varepsilon} + \frac{\partial}{\partial x_\delta} R_{\alpha\beta,\varepsilon\lambda} = 0\,. \tag{4}$$

We introduce (3) into (4) and treat ε_p and $\boldsymbol{U}_p$ as constants. Use in made of operator identity:

$$U_{\lambda\delta}\frac{\partial}{\partial x_\varepsilon}+U_{\delta\varepsilon}\frac{\partial}{\partial x_\gamma}+U_{\varepsilon\lambda}\frac{\partial}{\partial x_\delta}=(\boldsymbol{U},\,\mathbf{grad})\varepsilon_{\nu\lambda\delta}\,,$$

the result is then:

$$B_{\alpha\beta.\lambda\delta\varepsilon}=\sum_p(\boldsymbol{U}_p,\,\mathbf{grad}\,\varrho_p)\varepsilon_p U^p_{\alpha\beta}\varepsilon_{\lambda\delta\varepsilon}=\varepsilon_{\lambda\delta\varepsilon}\varrho\sum_p\varepsilon_p U^p_{\alpha\beta}=0\,,$$

by virtue of (1). The connection between (1) and Bianchi's identities is then evident.

6. – Lorentz manifolds.

If the metric of a skeleton space is not positive definite then we must replace orthogonal matrices with Lorentz matrices. Correspondingly the defects may become imaginary. Also the usual triangular inequalities among sides of a simplex cease to hold, for instance in a triangle where all sides are timelike one side is larger than the sum of the others. In general, however, it is evident how to modify ordinary trigonometry in order to fit the needs. In an M_4 bones are triangles and, if the metric is indefinite with one time co-ordinate, there are three kinds of bones:

1) Spacelike. Every vector on the bone is spacelike. It is imaginary.

2) Null. Every vector on the bone is a linear combination of a null vector and a spacelike vector orthogonal to it. In this case the corresponding Lorentz matrices cannot be diagonalized. We have $\varepsilon=0$.

3) Timelike. There are both timelike and spacelike vectors on the bone. ε is real.

For simplicity we shall restrict ourselves to timelike bones only or at most null bones.

A useful parameter in the following is the area L of the bone. Let A_μ, B_μ be two sides of the bone. We define L as follows:

$$4L^2=(A_\mu B^\mu)^2-(A_\mu A^\mu)(B_\mu B^\mu)\,.$$

L is therefore always a real quantity. If the bone is timelike, L is real and we choose always $L>0$. We restrict ourselves to timelike bones also in order to avoid ambiguities in the definition of L. If the bone is null then $L=0$.

7. – Variational principles.

The most direct way of approximating an Einstein space with a skeleton space is to use a variational principle. The lagrangian is evidently

$$\mathscr{L} = \frac{1}{16\pi}\int R\, \mathrm{d}^4x\, \sqrt{-g}\,,$$

calculated on the skeleton and we vary this lagrangian in the lengths of all T_1. This choice is particularly happy because we know that these lengths yield the same type of information as the metric tensor. Moreover there is a simple expression for $\mathscr{L}$ in terms of the deficiencies and areas af the bones. Indeed we know that within the T_4 and T_3 there is no contribution to the integral because R is a distribution with the support w. L is an additive function of the bones. Let n label the bones in the skeleton. Then $\mathscr{L} = \sum_n F(T_2^n)$. The function F is of course the same function for all bones. Since the bone is homogeneous, $F(T_2^n)$ is proportional to the area of the bone $F(T_2^n) = L_n f(\varepsilon_n)$, where f depends on ε_n only. T_2^n can be considered as the superposition of two bones $T_2^{n'}$ and $T_2^{n''}$ with the same shape and area and such that $\varepsilon_n = \varepsilon_n' + \varepsilon_n''$. It follows $f(\varepsilon_n) = f(\varepsilon_n') + f(\varepsilon_n'')$. The latter condition implies $f(\varepsilon) = C\varepsilon$, where C is a constant. We may write then $\mathscr{L} = C\sum_n \varepsilon_n L_n$. This result has to be compared with $\mathscr{L} = (1/8\pi)\int \varepsilon \underset{\sim}{\varrho}\, \mathrm{d}^4x$ in the limit discussed in Section **5** and for a bundle of identical bones of equal area. From the comparison we get $C = 1/8\pi$ and $\mathscr{L} = (1/8\pi)\sum_n L_n \varepsilon_n$. In M_m the corresponding result is $\mathscr{L} = C\sum_n \varepsilon_n L_n^m$, where L_n^m is the $m-2$ dimensional measure of T_{m-2}^m.

The next task is then the variation of $\mathscr{L}$ in the lengths l_p of T_1^p: While it is clear how L_n depends on l_p, the dependence of ε_n on l_p is rather involved. However quite remarkably we can carry out the variation of $\mathscr{L}$ as if the ε_n were constants:

$$\delta\mathscr{L} = \frac{1}{8\pi}\sum_n \varepsilon_n\, \delta L_n\,,$$

(See the Appendix for the proof; the result holds in any dimension.)

In an M_3 the measures L_p^3 are simply the lengths l_p of T_1^p, which are all independent, the field equations are then simply $\varepsilon_n = 0$, *i.e.* all 3 dimensional Einstein skeletons are flat, as expected.

In M_4 the field equations are instead:

$$\sum_n \varepsilon_n \frac{\partial L_n}{\partial l_p} = 0\,.$$

The summation is now extended on all n such that l_p is a side of L_n. We have. then:

$$\frac{\partial L_n}{\partial l_p} = l_p \operatorname{ctg} \theta_{pn} ,$$

θ_{pn} being the angle which is opposite to l_p in T^n_2. Finally:

$$\sum_n \varepsilon_n \operatorname{ctg} \theta_{pn} = 0 \tag{5}$$

are Einstein's equations for an empty skeleton space. They can be shown, with the methods of Section **5**, to degenerate into $R_{\mu\nu} = 0$ for a differential manifold.

If M_4 is a compact manifold, and the number of simplexes in the decomposition is finite, say $1 \leqslant p \leqslant P$, we have P equations (5) in P variables l_p. One such variable can be however fixed at will since if l_p is a solution of (5), also λl_p is a solution. In other words (5) is a set of P homogeneous equations in the l_p's and in general we should have no solution to (5) except the trivial $l_p = 0$. However we have no proof that one or more eq. (5) are not independent of the others and the question of the existence of a compact Einstein skeleton space is still open. Any answer to the question in our simplified model would bring strong evidence in favor of a similar answer for differentiable manifolds. Also (5) is formally suitable for relaxation methods or it could be programmed on a computer if the question of the existence of a solution is first met.

APPENDIX

We shall show now that for infinitesimal variations of l_j we have in M_m

$$\sum_p \delta\varepsilon_p L^m_p = 0 , \tag{6}$$

L^m_p is the $m-2$ dimensional measure of the n-th bone.

Let T_n be a simplex and let in T_n be defined a cartesian set of co-ordinates. T_n has $n+1$, T^r_{n-1} boundary simplexes, $r:1 \dots n+1$. Any two simplexes T^r_{n-1}, T^s_{n-1} have a common boundary T^{rs}_{n-2}. Let U^r_μ be the unit vector normal to T^r_{n-1}, $r:0 \dots n-1$, and $U^r_\mu U^{r,\mu} = 1$. If θ_{rs} is the angle between T^r_{n-1} and T^s_{n-1} we have:

$$\cos \theta_{rs} = U^r_\mu U^{s,\mu} .$$

Next we consider the skew symmetric tensor

$$U^{rs}_{\mu\lambda} = - U^{rs}_{\lambda\mu} : \sin \theta_{rs} U^{rs} = U^r_\mu U^s_\lambda - U^r_\lambda U^s_\mu .$$

Clearly we have $U^{rs}_{\lambda\mu} U^{rs,\lambda\mu} = 2$ so that $U^{rs}_{\lambda\mu}$ is exactly the tensor used in Section 5, eq. (2) and following, and defines the orientation of T^{rs}_{n-2}. If L_{rs} is the measure of T^{rs}_{n-2} we have the following identity:

$$\sum_s L_{rs} U^{rs} = 0 : \tag{7}$$

take namely a frame where $U^r_0 = 1$, $U^r_{\mu\neq 0} = 0$. Then $V^s_\lambda = U^{rs}_{\lambda 0}$, $\lambda \neq 0$ are the only non vanishing components of $U^{rs}_{\lambda\mu}$ and $V^s_\lambda V^{s,\lambda} = 1$. In T^r_{n-1} the quantity $\sum L_{rs} V^s_\lambda A^\lambda$ is the flux of the constant vector A^λ through the boundary of T^r_{n-1} and therefore vanishes.

Take now the formula:

$$\delta\theta_{rs} = \frac{-1}{\sin\theta_{rs}} (U^r_\mu \delta U^{s,\mu} + U^s_\mu \delta U^{r,\mu}) ,$$

and transform it using the identity:

$$U^{rs}_{\lambda\mu} U^{r,\lambda} = \frac{1}{\sin\theta_{rs}} (\cos\theta_{rs} U^r_\mu - U^s_\mu) ,$$

and:

$$U^{rs}_{\lambda\mu} U^{r,\lambda} \delta U^{r,\mu} = -\frac{1}{\sin\theta_{rs}} (U^s_\mu \delta U^{r,\mu}) ,$$

having used $U^r_\mu \delta U^{r,\mu} = 0$. We get:

$$\delta\theta_{rs} = U^{rs}_{\lambda\mu}(U^{r,\lambda}\delta U^{r,\mu} + U^{s,\lambda}\delta U^{s,\mu}) ,$$

so that by (7):

$$\sum_{r,s} L_{rs} \delta\theta_{rs} = 0 .$$

(6) then follows by taking the sum of the above identity on all simplexes of the skeleton.

* * *

The author wishes to thank Prof. J. A. WHEELER and Dr. C. MISNER for many suggestions and encouragement.

RIASSUNTO

In questo lavoro viene sviluppato il formalismo matematico della relatività generalizzata in modo da evitare l'uso di coordinate. A tal uopo si introducono decomposizioni simpliciali di varietà rimanniane che costituiscono l'analogo iperdimensionale di poliedri. Si spera che questo nuovo formalismo renda possibile la discussione di soluzioni delle equazioni di Einstein corrispondenti a topologie di elevata complessità.

Simplicial Gravity with Coordinates

A. D'Adda
INFN, Sezione di Torino
and
Arnold–Regge Center,
via P. Giuria 1, 10125 Torino, Italy

Putting a field theory on a lattice or on a discretized structure is often a powerful tool to perform computer simulations that would be otherwise impossible, but it is also a way to get a deeper insight of the theory and of its symmetries. In Einstein's theory of gravity the invariance under general coordinate transformations is needed in the continuum formulation to ensure that the physical degrees of freedom are geometrical entities, which do not depend on the choice of coordinates. Regge's simplicial gravity on the other hand does away with the notion of coordinates altogether and formulates gravity purely in terms of geometrical quantities. This is in itself an extraordinary achievement, and it is not a chance that the title of Regge's seminal paper[1] "General relativity without coordinates" emphasizes this point.

The basic ideas of Regge Calculus are well known: the smooth d-dimensional space-time manifold of the continuum formulation is replaced by a triangulated manifold made of piecewise flat d-dimensional simplices glued together by identifying in pairs their $d-1$-dimensional faces. The geometrical properties of this manifold are determined by the lengths of all its one-dimensional edges: in fact each d-dimensional simplex is completely fixed by the lengths of its $\frac{d(d+1)}{2}$ edges.

The curvature is associated to the $d-2$-dimensional subspaces, the hinges, and is given for each hinge h by the deficit angle θ_h defined as 2π minus the sum of the dihedral angles between the faces of the simplices which the hinge h belongs to. In a flat space θ_h is zero for all h, as clearly shown by the two-dimensional case, where the hinges are points (dimension zero) and θ_h the complement to 2π of the sum of the angles meeting at that point.

The discrete version of Einstein action is then given by:

$$S_R = \sum_h |V|_h \theta_h \tag{1}$$

where $|V|_h$ is the volume of the hinge h.

Several different formulations of discrete gravity followed the original paper from

Regge,[a] some maintaining a purely geometrical approach, like the so-called Area Regge Calculus[2] where in four dimensions the areas of the triangles are chosen as fundamental degrees of freedom, some reintroducing local symmetries while keeping the simplicial discretization of space-time. Among the latter I want to mention the approach of Ref. 4, which is closely related to the one presented here, where a Euclidean reference frame is introduced in each simplex, and the degrees of freedom are defined on the links of the dual lattice as the Poincaré transformations needed to transform the reference frame defined on a simplex α into the one of a contiguous simplex β. The action is the one of a gauge theory on the dual lattice with a local Poincaré invariance, but it is eventually equivalent to the Regge action of Eq. (1) although with the deficit angle θ_h replaced by $\sin(\theta_h)$. A first order formalism is possible in this framework, and the presence of a local Lorentz group makes the coupling of fermions to gravity possible.

In this article, written to celebrate the work of Tullio Regge on discrete gravity, I shall present an approach which, while keeping the original Regge's triangulation of space-time, preserves the notion of space-time coordinates and is endowed with a discrete version of the invariance under general coordinate transformations. In this formalism simplicial gravity follows step by step the classical formalism of general relativity which we have been used to from university courses, and it includes a discretized version of tensor calculus, of parallel transport, and of Riemann curvature tensor.

As the title of this paper suggests, I follow here an almost opposite attitude with respect to Regge's original paper by preserving the invariance under coordinates transformations at the cost of introducing unphysical degrees of freedom and of loosing some of the beautiful geometrical simplicity of Regge calculus.

My aim in developing this approach has not been to provide a new tool for calculations, although I am not ruling out the possibility that the formalism presented in this paper might in the future have some concrete application, but to answer a question that has come to my mind in the past, namely whether it is possible to formulate gravity on a lattice keeping the invariance under general coordinate transformations and following closely the standard formalism of Einstein gravitational theory.

1. Simplicial gravity with coordinates

Let α be a d-dimensional simplex and $i, j, \ldots$ labels for its $d+1$ vertices. In Regge calculus the simplex is completely identified by giving the $\frac{d(d+1)}{2}$ lengths l_{ij} of the edges joining the vertices i and j. The lengths l_{ij} have to satisfy triangular inequalities, but are otherwise arbitrary. They constitute the fundamental degrees of freedom of Regge's discrete gravity.

There are alternative ways to identify the simplex α. One is to associate a coordinate x_i^μ ($\mu = 1, 2, \ldots d$) to each vertex i *and* a constant, but in general not Euclidean, metric $g_{\mu\nu}(\alpha)$ to each simplex α. The lengths l_{ij} of the edges are then determined and given by:

$$l_{ij}^2 = g_{\mu\nu}(\alpha) \left(x_i^\mu - x_j^\mu\right) \left(x_i^\nu - x_j^\nu\right) . \tag{2}$$

Conversely, if the lengths l_{ij} are given and the coordinates x_i^μ of the vertices are chosen in an arbitrary way, then Eq. (2) provide a set of $d(d+1)/2$ equations in the $d(d+1)/2$ unknown components of the metric $g_{\mu\nu}(\alpha)$. These equations have a unique solution provided the determinant of their coefficients is not vanishing. The latter condition is satisfied iff the

[a]For a review and an exhaustive list of references see Ref. 3.

determinant of the differences $x_i^\mu - x_{d+1}^\mu$ is different from zero:[b]

$$\det\left(x_i^\mu - x_{d+1}^\mu\right) \neq 0 \qquad i = 1, \ldots, d. \tag{3}$$

So a simplicial manifold can be characterized by assigning the coordinates of all the vertices and the metric of each simplex, instead of the edges' lengths as in Regge calculus. This is essentially the same as in the usual formulation of Einstein's gravity in the continuum. As in the continuum the choice of the coordinates is arbitrary, provided for each simplex the determinant condition (3) is satisfied, and we expect the theory to be invariant under general coordinate transformations, that is to depend only on geometrical invariants such as l_{ij}.

Notice however that the components of the metric tensors belonging to different simplices are not all independent. Consider in fact two simplices α and β which have in common a $d-p$-dimensional sub-simplex. Their common edges have lengths l_{ij} that cannot depend upon the fact of being considered as part of α or part of β. Then from (2) we have:

$$[g_{\mu\nu}(\alpha) - g_{\mu\nu}(\beta)]\left(x_i^\mu - x_j^\mu\right)\left(x_i^\nu - x_j^\nu\right) = 0 \qquad i, j \in \alpha \cap \beta. \tag{4}$$

In particular Eq. (4) applies when α and β have a face in common, and it should be regarded as a constraint whenever the metric is varied or is integrated upon in a functional integral.

Let us consider a general coordinate transformation on the simplicial manifold:

$$x_i^\mu \Longrightarrow x_i'^\mu \qquad i = 1, \ldots, N \tag{5}$$

where N is now the total number of vertices in the manifold. The metric $g_{\mu\nu}(\alpha)$ of each simplex α should transform under (5) into a new metric $g'_{\mu\nu}(\alpha)$ in such a way to leave all the edges lengths l_{ij} invariant.

Let now x_i^μ be the coordinates on the vertices of a specific simplex α. Then the general coordinate transformation (5), restricted to the vertices in α can always be written as:

$$x_i'^\mu = \Lambda^\mu{}_\nu(\alpha) x_i^\nu + \Lambda^\mu(\alpha) \qquad i \in \alpha \tag{6}$$

which also implies:

$$x_i'^\mu - x_j'^\mu = \Lambda^\mu{}_\nu(\alpha)\left(x_i^\nu - x_j^\nu\right) \qquad i, j \in \alpha. \tag{7}$$

The matrix $\Lambda^\mu{}_\nu(\alpha)$ is the discrete analogue of $\frac{\partial x'^\mu}{\partial x^\mu}$ and Eq. (7) defines the transformation properties of a contravariant vector under general coordinate transformation on the simplicial manifold:

$$A'^\mu(\alpha) = \Lambda^\mu{}_\nu(\alpha) A^\nu(\alpha). \tag{8}$$

We shall assume that $\det \Lambda^\mu{}_\nu(\alpha) \neq 0$ for all α. In fact it is clear from (7) and (3) that this is the necessary and sufficient condition for (3) to be preserved under (5). Notice that if i and j in (7) belong to both simplices α and β then from (7) we have:

$$[\Lambda^\mu{}_\nu(\alpha) - \Lambda^\mu{}_\nu(\beta)]\left(x_i^\nu - x_j^\nu\right) = 0 \qquad i, j \in \alpha \cap \beta \tag{9}$$

which can be regarded as a constraint on Λ.

[b]The proof is omitted for brevity.

Tensor calculus can now be formulated on the lattice: tensors with covariant and contravariant indices can be defined as quantities that transform with Λ for each contravariant index and with Λ^{-1} for each covariant index:

$$A'^{\mu_1\cdots\mu_h}{}_{\nu_1\cdots\nu_k}(\alpha) = \Lambda^{\mu_1}{}_{\rho_1}(\alpha)\cdots\Lambda^{\mu_h}{}_{\rho_h}(\alpha)(\Lambda^{-1})^{\sigma_1}{}_{\nu_1}(\alpha)\cdots(\Lambda^{-1})^{\sigma_k}{}_{\nu_k}(\alpha)\; A^{\rho_1\cdots\rho_h}{}_{\sigma_1\cdots\sigma_k}(\alpha). \quad (10)$$

The metric $g_{\mu\nu}(\alpha)$ transforms as a covariant tensor of rank 2. In fact from (7) the requirement that the lengths l_{ij} given in (2) are invariant we have:

$$g'_{\mu\nu}(\alpha) = (\Lambda^{-1})^{\rho}{}_{\mu}(\alpha)\;(\Lambda^{-1})^{\sigma}{}_{\nu}(\alpha)\; g_{\rho\sigma}(\alpha). \quad (11)$$

Quantities that are invariant under general coordinate transformations, and hence have an intrinsic geometrical meaning can now be constructed. The simplest is the volume $V(\alpha)$ of the simplex, which is the discrete analogue of the invariant volume $\sqrt{g}\;d^dx$ of the continuum theory, and is given by:

$$V(\alpha) = \frac{1}{d!}\left|\epsilon_{\mu_1\mu_2\cdots\mu_d}\left(x_1^{\mu_1} - x_{d+1}^{\mu_1}\right)\cdots\left(x_d^{\mu_d} - x_{d+1}^{\mu_d}\right)\right|\sqrt{\det g_{\mu\nu}(\alpha)} \qquad i \in \alpha, \quad (12)$$

where the absolute value at the r.h.s. removes the sign ambiguity due to the arbitrary ordering of the vertices under the antisymmetric $\epsilon_{\mu_1\mu_2\cdots\mu_d}$ tensor. The invariance of $V(\alpha)$ under general coordinate transformations follows from (7) and (11).

In a similar fashion one can define a covariant vector associated to any $d-1$-dimensional subsimplex. Let α and β be two simplices with a common face that we denote $\alpha \cap \beta$. We define:

$$V_\mu^{(\alpha\cap\beta)}(\alpha) = \frac{\sqrt{\det g_{\mu\nu}(\alpha)}}{(d-1)!}\epsilon_{\mu\nu_1\ldots\nu_{d-1}}(x_1^{\nu_1} - x_d^{\nu_1})\ldots(x_{d-1}^{\nu_{d-1}} - x_d^{\nu_{d-1}}) \quad (13)$$

where $x_1^\mu \cdots x_d^\mu$ are the vertices of $\alpha \cap \beta$. The vector $V_\mu^{(\alpha\cap\beta)}(\alpha)$ is orthogonal to the face $\alpha \cap \beta$:

$$\left(x_i^\mu - x_j^\mu\right) V_\mu^{(\alpha\cap\beta)}(\alpha) = 0 \qquad i,j \in \alpha\cap\beta \quad (14)$$

and is chosen to be pointing toward the outside of α. The modulus of $V_\mu^{(\alpha\cap\beta)}(\alpha)$ is equal to the $d-1$-dimensional volume $V(\alpha\cap\beta)$ of $\alpha\cap\beta$, so that we can write:

$$V_\mu^{(\alpha\cap\beta)}(\alpha) = V(\alpha\cap\beta)\; n_\mu^{(\alpha\cap\beta)}(\alpha) \quad (15)$$

where $n_\mu^{(\alpha\cap\beta)}(\alpha)$ is a unit vector orthogonal to $\alpha\cap\beta$ and pointing to the outside of α.

If x_{d+1}^μ is the vertex of α that does not belong to $\alpha\cap\beta$, then we also have:

$$\left(x_{d+1}^\mu - x_i^\mu\right) V_\mu^{(\alpha\cap\beta)}(\alpha) = d\; V(\alpha) \qquad \forall i \in \alpha\cap\beta. \quad (16)$$

Let us consider now a $d-2$-dimensional hinge h with $h \subset \alpha$. The hinge h can be identified as the intersection of α and two simplices β_1 and β_2 that have each a face in common with α.

We can now define the covariant tensor

$$V_{\mu_1\mu_2}^{(h)}(\alpha) = \frac{\sqrt{\det g_{\mu\nu}(\alpha)}}{(d-2)!}\epsilon_{\mu_1\mu_2\nu_1\ldots\nu_{d-2}}\left(x_1^{\nu_1} - x_{d-1}^{\nu_1}\right)\cdots\left(x_{d-2}^{\nu_{d-2}} - x_{d-1}^{\nu_{d-2}}\right) \quad (17)$$

where $x_1^\mu, x_2^\mu \ldots x_{d-1}^\mu$ are the coordinates of the hinge's vertices. $V_{\mu_1\mu_2}^{(h)}(\alpha)$ is orthogonal to all the directions of the hinge h:

$$\left(x_i^\mu - x_j^\mu\right) V_{\mu\nu}^{(h)}(\alpha) = 0 \qquad i,j \in h \quad (18)$$

and it is entirely contained in the two-dimensional subspace spanned by $V_\mu^{(\alpha\cap\beta_1)}(\alpha)$ and $V_\mu^{(\alpha\cap\beta_2)}(\alpha)$. It is convenient to introduce in this two-dimensional subspace an orthonormal base by introducing two unit vectors $n_{a\mu}^{(h)}(\alpha)$ with $a = 1,2$ such that:

$$n_{a\mu}^{(h)}(\alpha) g^{\mu\nu}(\alpha) n_{b\nu}^{(h)}(\alpha) = \delta_{ab} \qquad a, b \in \{1,2\}. \tag{19}$$

The covariant tensor of Eq. (17) can then be written as:

$$V_{\mu_1\mu_2}^{(h)}(\alpha) = \frac{1}{2}\left(n_{1\mu_1}^{(h)}(\alpha) n_{2\mu_2}^{(h)}(\alpha) - n_{1\mu_2}^{(h)}(\alpha) n_{2\mu_1}^{(h)}(\alpha)\right) V(h) \tag{20}$$

where $V(h)$ is the $d-2$-dimensional volume of the hinge.

In order to proceed in parallel with the Einstein theory of gravity we have now to introduce the notion of parallel transport. Consider a contravariant vector $A^\mu(\alpha)$. If α and β are two simplices with one face in common, the derivative of $A^\mu(\alpha)$ should be proportional on the lattice to the difference $A^\mu(\alpha) - A^\mu(\beta)$, which however is not a vector since the two terms of the difference transform with different matrices $\Lambda(\alpha)$ and $\Lambda(\beta)$ under general coordinate transformations.

The parallel transport of a contravariant vector $A^\mu(\beta)$ from a simplex β onto a neighbouring simplex α is introduced by defining the transported vector $A^\mu_{(\alpha)}(\beta)$ as:

$$A^\mu(\beta) \quad \Longrightarrow \quad A^\mu_{(\alpha)}(\beta) = K^\mu{}_\nu(\alpha|\beta) A^\nu(\beta) \tag{21}$$

where $K^\mu{}_\nu(\alpha|\beta)$ is entirely determined by the following properties:[c]

- If $g_{\mu\nu}(\alpha) = g_{\mu\nu}(\beta)$ then $K^\mu{}_\nu(\alpha|\beta) = \delta^\mu{}_\nu$.
- $A^\mu_{(\alpha)}(\beta)$ transforms as a contravariant vector with $\Lambda^\mu{}_\nu(\alpha)$ under general coordinate transformations, namely

$$A'^\mu_{(\alpha)}(\beta) = \Lambda^\mu{}_\nu(\alpha) A^\nu_{(\alpha)}(\beta). \tag{22}$$

The matrix $K^\mu{}_\nu(\alpha|\beta)$ does not transform as a tensor but rather as link variable in the dual lattice. In fact from (22) and (21) one easily finds:

$$K'^\mu{}_\nu(\alpha|\beta) = \Lambda^\mu{}_\rho(\alpha) K^\rho{}_\sigma(\alpha|\beta) \Lambda^{-1\sigma}{}_\nu(\beta). \tag{23}$$

Notice that if we start in (23) from a coordinate system where $g_{\mu\nu}(\alpha) = g_{\mu\nu}(\beta)$, then $K^\rho{}_\sigma(\alpha|\beta) = \delta^\rho{}_\sigma$ and for a generic coordinate system we have:

$$K'^\mu{}_\nu(\alpha|\beta) = \Lambda^\mu{}_\rho(\alpha) \Lambda^{-1\rho}{}_\nu(\beta). \tag{24}$$

It follows from (24) that $K'^\mu{}_\nu(\beta|\alpha)$ is the inverse of $K'^\mu{}_\nu(\alpha|\beta)$:

$$K'^\mu{}_\nu(\alpha|\beta) K'^\nu{}_\rho(\beta|\alpha) = \delta^\mu{}_\rho. \tag{25}$$

Scalar quantities are obviously invariant under parallel transform. This determines, together with (21), the parallel transport of a covariant vector:

$$A_\mu(\beta) \quad \Longrightarrow \quad A_{(\alpha)\mu}(\beta) = A_\nu(\beta) K^\nu{}_\mu(\beta|\alpha). \tag{26}$$

Equations (21) and (26) can easily be generalised to tensors of arbitrary rank: in particular it follows from the definition of parallel transport that the parallel transport of $g_{\mu\nu}(\beta)$ to a neighbour simplex α coincides with $g_{\mu\nu}(\alpha)$, so that the following identity holds:

$$g_{\mu\nu}(\alpha) = g_{\rho\sigma}(\beta) K^\rho{}_\mu(\beta|\alpha) K^\sigma{}_\nu(\beta|\alpha). \tag{27}$$

[c]We only consider here the case of zero torsion.

The last equation defines implicitly $K^{\rho}{}_{\mu}(\alpha|\beta)$ as a function of the metric tensor, but its exact explicit solution does not seem to be possible (unlike its continuum analogue) unless it is linearized by neglecting higher order terms in the lattice constant defined below. In a first order formulation where $K^{\rho}{}_{\mu}(\alpha|\beta)$ and $g_{\mu\nu}(\alpha)$ are treated as independent dynamical variable Eq. (27) may arise from the eq.s of motion as in the Palatini formalism of the continuum theory, but this has not been checked in the present case.

The variation of $A^{\mu}(\beta)$ as an effect of the parallel transport from β to α is then given by:

$$\delta_{\beta\to\alpha}A^{\mu}(\beta) = A^{\mu}_{(\alpha)}(\beta) - A^{\mu}(\beta) = [K^{\mu}{}_{\nu}(\alpha|\beta) - \delta^{\mu}{}_{\nu}]\, A^{\nu}(\alpha). \tag{28}$$

Notice that in (28) only the component of $A^{\nu}(\alpha)$ orthogonal to the face $\alpha\cap\beta$ contributes to the variation. In fact from (24) and (9) we have:

$$[K^{\mu}{}_{\nu}(\alpha|\beta) - \delta^{\mu}{}_{\nu}]\,(x^{\nu}_{i} - x^{\nu}_{j}) = 0 \qquad i,j\in\alpha\cap\beta. \tag{29}$$

We are almost ready at this point to define the lattice analogue of the Christofell symbol and of the covariant derivative, but it is convenient first to discuss the ordinary derivative of a scalar field $\varphi(\alpha)$. Let us start from the standard formula of the continuum:

$$d\varphi(x) = \partial_{\mu}\varphi(x)dx^{\mu}. \tag{30}$$

The analogue of the differential $d\varphi(x)$ on the simplicial lattice is the finite difference $\varphi(\alpha) - \varphi(\beta)$, however it is not obvious how to split that into a product of a covariant vector (the derivative) and a contravariant vector (the coordinate differential).

The coordinate differential can be defined on the lattice as a contravariant vector $\Delta x^{\mu}(\alpha|\beta)$ orthogonal to the face $\alpha\cap\beta$ and pointing from α toward β:

$$\Delta x^{\mu}(\alpha|\beta) = l(\alpha|\beta)n^{(\alpha\cap\beta)\mu}(\alpha) \tag{31}$$

where $n^{(\alpha\cap\beta)\mu}(\alpha)$ is the unit vector defined in (15)[d] and $l(\alpha|\beta)$ is a lattice spacing defined by:

$$l(\alpha|\beta) = n^{(\alpha\cap\beta)}_{\mu}(\alpha)(x^{\mu}_{d+1} - x^{\mu}_{i}) = \frac{d\; V(\alpha)}{V(\alpha\cap\beta)} \tag{32}$$

where x^{μ}_{d+1} are the coordinates of the vertex of α that does not belong to $\alpha\cap\beta$. The last equality in (32) follows from (15) and (16).

The derivative can now be consistently defined as:

$$\partial_{\mu}\varphi(\alpha|\beta) = \frac{[\varphi(\alpha) - \varphi(\beta)]}{l(\alpha|\beta)}n^{(\alpha\cap\beta)}_{\mu}(\alpha). \tag{33}$$

Notice that with this choice of derivative there is no symmetry under exchange of α and β, namely since $l(\alpha|\beta) \neq l(\beta|\alpha)$ we have $\partial_{\mu}\varphi(\alpha|\beta) \neq -\partial_{\mu}\varphi(\beta|\alpha)$ even if we choose the coordinates in such a way to have $g_{\mu\nu}(\alpha) = g_{\mu\nu}(\beta)$.[e]

[d]The raising and lowering of indices is done, as in the continuum theory, by the metric tensor $g_{\mu\nu}(\alpha)$ and its inverse $g^{\mu\nu}(\alpha)$.

[e]There are other possible choices that preserve such symmetry but are not so easily expressed in terms of the coordinates of the vertices. For instance one could define the spacing $l(\alpha|\beta)$ as the distance between the center of α and the center of β, where by center I mean the center of the sphere that touches all the vertices.

Consider now a contravariant vector $A^\mu(\alpha)$. The covariant differential between two neighbouring simplices α and β is given by:

$$DA^\mu(\alpha|\beta) = A^\mu(\alpha) - A^\mu_{(\alpha)}(\beta) = A^\mu(\alpha) - A^\mu(\beta) - \delta_{\beta\to\alpha}A^\mu(\beta). \tag{34}$$

Following the previous discussion the covariant derivative can be written as:

$$D_\nu A^\mu(\alpha|\beta) = \partial_\nu A^\mu(\alpha|\beta) + \Gamma_\nu{}^\mu{}_\rho(\alpha|\beta)A^\rho(\beta) \tag{35}$$

where $\Gamma^\mu{}_{\nu\rho}$ is the lattice analogue of the Christoffel symbol and is given by:

$$\Gamma^\mu{}_{\nu\rho}(\alpha|\beta) = \frac{K^\mu{}_\nu(\alpha|\beta) - \delta^\mu{}_\nu}{l(\alpha|\beta)} n^{(\alpha\cap\beta)}_\rho(\alpha). \tag{36}$$

Notice that according to Eqs. (29) and (14) $\Gamma^\mu{}_{\nu\rho}(\alpha|\beta)$ is different from zero only when both the index ν and the index ρ are orthogonal to the face $\alpha \cap \beta$, hence the Christoffel symbol is symmetric in the two covariant indices as in the continuum case.

2. Curvature and discrete Einstein action

Let h be a $d-2$-dimensional hinge and α a simplex that contains h as a subsimplex. We shall denote with β_i with $i = 1, 2, \dots r$ the other simplices that share with α the hinge h, in particular we shall denote with β_1 and β_r the simplices that have a face in common with α.

Given a contravariant vector $A^\mu(\alpha)$ we can define according to the definitions of the previous section the parallel transport of $A^\mu(\alpha)$ around the hinge h starting and arriving in α, namely following the path $\gamma_{h|\alpha} \equiv \alpha \to \beta_1 \to \cdots \to \beta_r \to \alpha$. The variation of $A^\mu(\alpha)$ under the parallel transport along $\gamma_{h|\alpha}$ is given by:

$$\delta_{\gamma_{h|\alpha}} A^\mu(\alpha) = R^\mu{}_\rho(h|\alpha)A^\rho(\alpha) \tag{37}$$

with

$$R^\mu{}_\rho(h|\alpha) = K^\mu{}_{\nu_1}(\alpha|\beta_1)K^{\nu_1}{}_{\nu_2}(\beta_1|\beta_2)\dots K^{\nu_r}{}_\rho(\beta_r|\alpha) - \delta^\mu{}_\rho. \tag{38}$$

Repeated use of Eq. (29) shows that only the components of $A^\mu(\alpha)$ orthogonal to the hinge are varied under the parallel transport along $\gamma_{h|\alpha}$, namely that the only non vanishing elements of the curvature matrix $R^\mu{}_\rho(h|\alpha)$ are the ones where both indices μ and ρ are orthogonal to the hinge h:

$$R^\mu{}_\rho(h|\alpha)(x^\rho_i - x^\rho_j) = 0 \qquad i, j \in h. \tag{39}$$

The curvature $R^\mu_\rho(h|\alpha)$ is then entirely contained in the two dimensional subspace spanned by the orthonormal base vectors $n^{(h)}_{a\mu}(\alpha)$ introduced in (19) and (20). Since the orthonormality relations are preserved under parallel transport, its effect can only be a rotation by an angle θ_h of the orthonormal base vectors. The angle θ_h can be identified as the deficit angle of the Regge calculus.

An explicit expression for $R^\mu{}_\rho(h|\alpha)$ in terms of the deficit angle and of the vectors $n^{(h)}_{a\mu}(\alpha)$ can then be written, and reads:

$$R_{\mu\rho}(h|\alpha) = \sum_{a=1}^{2} (\cos\theta_h - 1)\, n^{(h)}_{a\mu}(\alpha)n^{(h)}_{a\rho}(\alpha) + \sum_{a,b=1}^{2} \sin\theta_h \epsilon_{ab} n^{(h)}_{a\mu}(\alpha)n^{(h)}_{b\rho}(\alpha). \tag{40}$$

In order to write the curvature as a tensor with four indices we use a straightforward generalization of the procedure used to define the lattice derivative in Eq. (33). First we introduce the "area" of the loop $\gamma_{h|\alpha}$ defined as:

$$l_2(\alpha|h) = \left(x^{\mu}_{d+1} - x^{\mu_1}_i\right)\left(x^{\nu}_d - x^{\mu_2}_i\right) n^{(h)}_{\mu_1\mu_2}(\alpha) = \frac{d(d+1)V(\alpha)}{V(h)} \tag{41}$$

where

$$n^{(h)}_{\mu_1\mu_2}(\alpha) = \frac{1}{2}\left(n^{(h)}_{1\mu_1}(\alpha)n^{(h)}_{2\mu_2}(\alpha) - n^{(h)}_{1\mu_2}(\alpha)n^{(h)}_{2\mu_1}(\alpha)\right). \tag{42}$$

The curvature tensor with four covariant indices can be defined as:

$$R_{\mu\nu,\rho\sigma}(h|\alpha) = \frac{R_{\mu\nu}(h|\alpha)}{l_2(\alpha|h)} n^{(h)}_{\rho\sigma}(\alpha). \tag{43}$$

Unlike its continuum analogue the curvature tensor (18) is not antisymmetric in the first two indices as a result of the first term in (40). However the symmetric part becomes negligible (order θ_h^2) compared to the antisymmetric part (order θ_h) in the limit of small deficit angle. Let us denote by $R_{[\mu\nu],\rho\sigma}(h|\alpha)$ the curvature tensor where only the antisymmetric term proportional to $\sin\theta_h$ has been kept. Then it is easy to check that the usual algebraic symmetries of the Riemann tensor are identically satisfied, namely the invariance under exchange of the two pairs of indices and the first Bianchi identity:

$$R_{[\mu\nu],\rho\sigma}(h|\alpha) + R_{[\mu\rho],\sigma\nu}(h|\alpha) + R_{[\mu\sigma],\nu\rho}(h|\alpha) = 0. \tag{44}$$

The second Bianchi identity involves derivatives, and its lattice generalization is less obvious and is related to the existence of closed paths on the lattice contractable to a point. We are not going to discuss it here.

By contracting pairs of indices in $R_{\mu\nu,\rho\sigma}(h|\alpha)$ with the inverse metric $g^{\mu\rho}(\alpha)$ one obtains the Ricci tensor (which has now an antisymmetric part of order θ_h^2) and the curvature scalar. This is given by:

$$R(h|\alpha) = \frac{\sin\theta_h}{l_2(h|\alpha)} = \frac{\sin\theta_h\ V(h)}{d(d+1)V(\alpha)}. \tag{45}$$

The above expression gives the contribution to the curvature scalar from a particular hinge h, choosing the simplex α as the starting and ending point of the parallel transport. A more symmetric expression for the curvature scalar, which is a function only of the simplex α, can be obtained by taking a weighted sum over all the $\frac{d(d+1)}{2}$ hinges that belong to α, namely:

$$R(\alpha) = \sum_{h\in\alpha} \frac{1}{k_h} R(h|\alpha) = \frac{1}{d(d+1)V(\alpha)} \sum_{h\in\alpha} \frac{\sin\theta_h V(h)}{k_h} \tag{46}$$

where k_h is the number of simplices that insist on the hinge h.

$R(\alpha)$ is the true lattice analogue od the curvature scalar $R(x)$ of Einstein continuum theory. Given the correspondence between the simplex volume $V(\alpha)$ (see Eq. (12)) and the integration volume of the continuum $\sqrt{g}d^dx$ the lattice equivalent of the Einstein–Hilbet action is then:

$$S_l = k\sum_{\alpha} R(\alpha)V(\alpha) = \frac{k}{d(d+1)} \sum_{h} \sin\theta_h V(h) \tag{47}$$

where k is the Newton constant. The last expression in (47) coincides with the Regge's action in the limit of small deficit angles. The appearance of $\sin\theta_h$ in place of θ_h is not a new feature, and it seems a natural one when the analogy of gravity with gauge theories is made explicit on the lattice.

3. Vierbein and local Lorentz invariance

Given a simplex α it is always possible to perform a change of coordinates of the form (6) that transforms the metric $g_{\mu\nu}(\alpha)$ into the flat diagonal metric η_{ab} (the flat indices will from now on be denoted by the letters $a, b, \ldots$ to distinguish them from the "curved" indices $\mu, \nu, \ldots$). If we name ξ_i^a the new coordinates of the vertices of α the transformation (6) now reads:

$$\xi_i^a = \Lambda_\nu^a(\alpha) x_i^\nu + \Lambda^a(\alpha) \qquad i \in \alpha \tag{48}$$

and according to Eq. (11) the metric is given in terms of $\Lambda_\nu^a(\alpha)$ by:

$$g_{\mu\nu}(\alpha) = \Lambda_\mu^a(\alpha)\Lambda_\nu^b(\alpha)\eta_{ab}. \tag{49}$$

It is clear from (49) that $\Lambda_\mu^a(\alpha)$ can be interpreted as a vierbein (or d-bein), and that given the metric $g_{\mu\nu}(\alpha)$ the vierbein $\Lambda_\mu^a(\alpha)$ is determined only up to a Lorentz transformation (or rotation in Euclidean space-time) acting on the flat index a. Similarly the coordinates ξ_i^a are determined up to a Poincaré transformation whose translational part is given by $\Lambda^a(\alpha)$ in (48).

We assume that the transformation (48) is done separately and independently in each simplex,[f] so that the coordinates ξ_i^a of a vertex i regarded as part of a simplex α are in general different from the coordinates of the same vertex regarded as part of a neighbouring simplex β. In the following we shall denote these coordinates $\xi_i^a(\alpha)$ to avoid ambiguities.

At the end of this procedure, through a transformation of the form (48) each simplex of the simplicial manifold is endowed with a Euclidean reference frame determined up to a Poincaré transformation. Geometrical entities are not affected by the choice of the local frame and the resulting theory will exhibit a local Poincaré invariance.

Consider now two neighbouring simplices α and β with a face in common. The transition from the Euclidean reference frame in α to the one in β is described by a Poincaré transformation, so that the coordinates of a generic point in the reference frame of α and β are related by:

$$\xi^a(\beta) = \Omega^a{}_b(\beta|\alpha)\xi^b(\alpha) + \Omega^a(\beta|\alpha). \tag{50}$$

The rotation matrix $\Omega^a{}_b(\beta|\alpha)$ (which is a Lorentz rotation in a Minkowski metric) is closely related to the matrix $K^\mu{}_\nu(\beta|\alpha)$ that defines the parallel transform. In fact if in Eq. (23) we replace the curved indices μ and ν with flat ones we have:

$$\Omega^a{}_b(\beta|\alpha) = \Lambda^a{}_\rho(\beta)K^\rho{}_\sigma(\beta|\alpha)\Lambda^{-1\sigma}{}_b(\alpha) \tag{51}$$

where $\Lambda^{-1\sigma}{}_b(\alpha)$ is the inverse of the vierbein.

It is clear from Eq. (51) that the curvature and all the geometrical entities described in the previous subsection can be expressed in terms of the vierbeins and of the Lorentz connection $\Omega^a{}_b(\beta|\alpha)$. For instance the curvature of Eq. (38) can be written as:

$$R^a{}_b(h|\alpha) = \Lambda^a{}_\mu(\alpha)R^\mu{}_\rho(h|\alpha)\Lambda^{-1\rho}{}_b(\alpha) = \Omega^a{}_{c_1}(\alpha|\beta_1)\Omega^{c_1}{}_{c_2}(\beta_1|\beta_2)\ldots\Omega^{c_r}{}_b(\beta_r|\alpha) - \delta^a{}_b \tag{52}$$

The resulting formalism exhibits a local Lorentz invariance and it coincides with the one introduced many years ago in the paper,[4] to which we refer the reader for further details.

[f]This means that the transformation (48) is not the restriction to the simplex α of a general coordinate transformation (5).

4. Some concluding remarks

We have shown that it is indeed possible to construct a simplicial gravity with coordinates, which is invariant under general coordinate transformations and that closely mimicks the standard formalism of continuum Einstein gravity.

Terms of higher order in the lattice length $l(\alpha|\beta)$ defined in (32) may however modify on the lattice some of the characteristic features of the continuum formulation. We already remarked for instance that some symmetries of the Riemann tensor are violated on the lattice by higher order terms in the deficit angle θ_h although these terms are uninfluential on the action. Another example is the relation between the derivative of the metric tensor and the Christoffel symbol. In fact if we write Eq. (27) in terms of the Christoffel symbol (36) we find;

$$\begin{aligned}\frac{g_{\mu\nu}(\alpha)-g_{\mu\nu}(\beta)}{l(\alpha|\beta)} &= n^{(\alpha\cap\beta)\tau}(\alpha)\left[\Gamma^{\rho}{}_{\mu\tau}(\alpha|\beta)g_{\rho\nu}(\beta)+\Gamma^{\rho}{}_{\nu\tau}(\alpha|\beta)g_{\rho\mu}(\beta)\right]\\ &\quad + l(\alpha|\beta)n^{(\alpha\cap\beta)\tau}(\alpha)n^{(\alpha\cap\beta)\tau'}(\alpha)\Gamma^{\rho}{}_{\mu\tau}(\alpha|\beta)\Gamma^{\sigma}{}_{\nu\tau'}(\alpha|\beta)g_{\rho\sigma}(\beta) \qquad (53)\end{aligned}$$

where the l.h.s. is the derivative of the metric tensor along a direction orthogonal to the face $\alpha\cap\beta$ and the r.h.s. is the usual term, linear in the Christoffel symbol, plus a quadratic term which is of order $l(\alpha|\beta)$ and hence vanishes in the continuum limit. The presence of this term makes it apparently impossible to invert Eq. (84) and express the Christoffel symbol in terms of derivatives of the metric tensor.

A more substantial difference between the present lattice formulation and the continuum formalism is given by the constraint (4), that tells us that the degrees of freedom $g_{\mu\nu}(\alpha)$, unlike $g_{\mu\nu}(x)$ in the continuum theory, are not all independent. Implementing (4) in the equations of motion and in the functional integral may be one of the main obstacles in view of practical applications of the present formalism. In spite of these limitations the formalism described here may provide a useful link between the continuum theory and Regge calculus. In particular, it could be useful in understanding the relation between the integration measure of the functional integral over the metrics in the continuum and the measure of integration over the lengths l_{ij} in Regge calculus.

References

1. T. Regge, Nuovo Cim. **19**, 558 (1961).
2. J. W. Barrett, M. Rocek and R. M. Williams, Class. Quant. Grav. **16**, 1373 (1999).
3. T. Regge and R. M. Williams, J. Math. Phys. **41**, 3964 (2000).
4. M. Caselle, A. D'Adda and L. Magnea, Phys. Lett. B **232**, 457 (1989).

Computer art by T. Regge, Johann Sebastian Beach (see `http://www.arnoldreggecenter.com/about/bio-regge/` for colour version and more drawings).

Tullio Regge's Legacy: Regge Calculus and Discrete (Quantum) Gravity

John W. Barrett

School of Mathematical Sciences, University Park, Nottingham NG7 2RD, UK, EU

Daniele Oriti

Max Planck Institute for Gravitational Physics (Albert Einstein Institute), Am Mühlenberg 1, D-14476 Golm, Germany, EU,
II Institute for Theoretical Physics, University of Hamburg, Luruper Chaussee 149, D-22761 Hamburg, Germany, EU and
Arnold-Sommerfeld-Center for Theoretical Physics, Ludwig-Maximilians-Universität, Theresienstrasse 37, D-80333 München, Germany, EU

Ruth M. Williams

Girton College, Cambridge CB3 0JG, UK, EU and
DAMTP, CMS, Wilberforce Road, Cambridge CB3 0WA, UK. EU

The review paper "Discrete Structures in Physics",[1] written in 2000, describes how Regge's discretization of Einstein's theory has been applied in classical relativity and quantum gravity. Here, developments since 2000 are reviewed briefly, with particular emphasis on progress in quantum gravity through spin foam models and group field theories.

1. Introduction

Regge's seminal 1961 paper, "General Relativity without Coordinates",[2] is a brief and rigorous description of how to approximate curved spaces and space-times by simplicial complexes, with the curvature distributed on simplicial nets of co-dimension 2. This was motivated by the desire to model complicated topologies and to obtain a deeper geometrical insight. The action for a simplicial space was written down and the analogue of the Einstein equations derived, making use of the Schläfli identity. In the last sentence, Regge mentions that the approach could be useful in numerical work. The paper shows clearly Regge's profound geometrical intuition; he admitted later that, having seen how to triangulate space-time manifolds, he was really interested in triangulating group manifolds. His approach became known as Regge calculus.

The paper opened up whole new areas of research. At first it was used for calculations of the evolution of model universes in classical general relativity, and later in efforts to formulate a theory of simplicial quantum gravity. Regge himself never worked extensively in these fields, but maintained an interest and in fact made a connection which led to one of the most important approaches to discrete quantum gravity.

In the classical theory, in an unpublished paper with Lund, "Simplicial Approximation to some Homogeneous Cosmologies",[3] Regge contributed to the continuous time three-plus-one formulation of Regge calculus by writing down the form of the action for homogeneous and isotropic spaces, giving an explicit form for the Hamiltonian constraint for such spaces. (The momentum constraints are identically satisfied in this case, but finding a simplicial form for them is more challenging.)

The paper which proved so influential in simplicial quantum gravity was written with Ponzano in 1968, and entitled "Semiclassical limit of Racah coefficients",[4] these coefficients being a very useful tool in the calculation of matrix elements in atomic physics. Almost in passing, the authors point out the relationship between a sum involving the asymptotic values of 6j-symbols associated with a triangulated three-manifold, and the path integral for three-dimensional simplicial gravity with the Regge action. This work was largely neglected until the 1990s when mathematicians[5] were writing down invariants of three-manifolds, hoping that they would help in the classification, and it was realised that there was a very close connection with the work of Ponzano and Regge. In fact the the Turaev–Viro model appeared to be a regularised version of the Ponzano–Regge model.

There were many attempts to generalise this work to four dimensions, in the general language of spin foam models, the first being the Barrett–Crane model, which will be discussed in a later section. This, in its turn, renewed interest in group field theories, first introduced for the Ponzano–Regge–Turaev–Viro models, which are described in the penultimate section.

Regge was clearly still thinking about discretization, and sometime in the 1970s produced a draft paper entitled "Discrete Yang–Mills theories". (He gave his address as 'The New Jersey Mental Hospital for Retired Physicists and Compulsory Psychopaths' — in mirror writing on one copy; he was at the Institute for Advanced Study in Princeton at the time!) A final version was published in the Festschrift for Yuval Ne'eman.[6]

For the millenium edition of the Journal of Mathematical Physics, Regge co-authored a review paper[1] on discrete structures in physics. This describes in detail forty years of development of Regge calculus and formulations of quantum gravity which have grown out of it. (When the review was being written, Regge was fascinated by computer-generated art work, and produced a picture he called "Johann Sebastian Beach", with piano keys forming the sea wall!) This article will not attempt to cover the material in that review, but rather to describe some of the developments since it was written, under the headings Classical Formalism, Numerical Relativity, Simplicial Quantum Gravity, the Barrett–Crane Model and Group Field Theories.

2. Classical formalism

Many of the developments in classical Regge calculus since 2000 have come from Warner Miller and his collaborators. Gentle and Miller[7] developed an algorithm which produced spacelike simplicial hypersurfaces with constant mean curvature. The equations resulting from a variational approach to this problem were solved together with the Regge equations, and a formulation was proposed which was compatible with Sorkin evolutions.[8] McDonald and Miller[9] have emphasised the important role that the dual lattice plays in Regge calculus. They derived a vertex-based scalar curvature using a new lattice obtained both from the simplicial lattice and its dual; this was a vertex-based weighted average of deficit angles per weighted average of dual areas. Mean curvature on a simplicial lattice was also discussed by Conbove, Miller and Ray.[10]

Turning their attention to non-vacuum space-times, Gentle, Kheyfets, McDonald and Miller[11] derived a conservation law in Regge calculus by equating the discrete Bianchi identity to a sum of components of the stress-energy tensor projected along the edges of the simplicial lattice. In principle, this extends Regge calculus in a natural way to non-vacuum space-times, provided an appropriate form of the stress-energy tensor can be written down. McDonald and Miller[12] considered the coupling of non-gravitational fields to simplicial space-times, and constructed the lattice action for scalar fields, the Maxwell field tensor and Dirac particles, using discrete differential forms.

An exact form for the Bianchi identity on a simplicial lattice has been given by Hamber and Kegel.[13] It is valid for arbitrarily curved manifolds, but is not linear in the curvatures in general.

Arjwahjoedi and Zen[14] have focussed on (2+1)-Regge calculus and obtained expressions for discrete curvatures, the Bianchi identity and the Gauss–Codazzi equation, in an attempt to give geometrical clarification to earlier work. It was shown that the standard formulae for these can be obtained in the continuum limit. The main result is that the Gauss–Codazzi equation is very closely related to the dihedral angle formula, which relates an n-dimensional angle between two $(n-1)$-dimensional simplices, to $(n-1)$-dimensional angles between two $(n-2)$-dimensional simplices. This work should now be extended to 3+1 dimensions.

In a contrasting approach, Höhn[15] has studied canonical Regge calculus expanded to linear order about a flat background. He showed how to use the Pachner moves in the evolution of a spacelike hypersurface and identified the gauge and 'graviton' degrees of freedom. The constraints generating the vertex displacement symmetry are consistent with the dynamics and are preserved by the Pachner moves. However it seems that the gauge symmetries will be broken in higher order approximations.

'Area Regge calculus' is an alternative approach to Regge's scheme, first suggested by Rovelli[16] and motivated by loop quantum gravity and spinfoam models. In this, the triangle areas rather than the edge lengths are the fundamental variables. The variational principle leads to vanishing deficit angles, and metric discontinuities. Wainwright and Williams[17] showed how a simple class of geometries with a discontinuity across a hyperplane leads to refractive wave geometries, which are generalised solutions of general relativity.[18]

Neiman[19] posed the question of whether 'area Regge calculus' was a valid discretisation of general relativity. It was found that in cases of interest (Euclidean, or Lorentzian with spacelike tetrahedra) the distributional scalar curvature is non-zero and has the same sign round all triangles, so cannot average to zero, which would be required in general relativity. A non-zero cosmological constant does not solve the problem. If all tetrahedra were null, the argument would not hold, but it is combinatorically impossible to triangulate space-time with null tetrahedra. A carefully-constructed space-time with both timelike and spacelike tetrahedra might solve the problem but that seems unlikely. Neiman's result has implications for the Barrett–Crane model.

Dittrich and Speziale[20] introduced a modified version of Regge calculus in four dimensions where the fundamental variables are areas and a certain class of angles. The constraints are local on the triangulation, so this formulation solves the long-standing problem of how to implement the area constraints.

In a series of papers, Bahr and Dittrich[21] constructed a further interesting variation on conventional Regge calculus. Motivated by the fact that diffeomorphism symmetry is broken if the simplicial space-time is not flat, which leads to pseudo-constraints in the canonical theory, they considered the replacement of flat simplices by ones with constant sectional curvature (consistent with the presence of a cosmological constant). The lengths were replaced by dihedral angles as the basic variables and there were no constraints in the first order formalism.

Christiansen[22] studied the convergence of linearised three-dimensional Regge calculus, and showed that the first non-trivial terms in the Regge action (the second variations) agree with the Einstein–Hilbert action.

Regge calculus as originally formulated is a torsion-free theory. Schmidt and Kohler[23] have generalised it to include dislocations in the simplicial lattice, corresponding to torsion singularities. Similarly to curvature, torsion is distributed on the simplices of co-dimension two, and in four dimensions the contribution to the action (a discrete version of the Einstein–Cartan action) involves the square of $b_{(i)}$, the part of the Burgers vector parallel to the triangle. The variables of the theory are taken to be not only the edge lengths but also the $b_{(i)}$, and the field equations show that the Burgers vector couples algebraically to the matter term. Thus the torsion vanishes in the absence of matter. See also work by Drummond[24] and Barrett[25] on torsion in Regge calculus and work by Xue[26] on the Einstein–Cartan theory in the quantum Regge calculus context.

Teleparallel gravity is a version of gravity where the curvature vanishes but the torsion is non-zero and acts as a force. Pereira and Vargas[27] formulated the teleparallel version of Regge calculus, with the action again proportional to the square of the Burgers vector parallel to the triangular hinge. The variables are taken to be just the edge lengths and, in the variation, the Burgers vector is treated as constant. The result is that in this theory, the Burgers vector does not necessarily vanish in the absence of matter and torsion is a propagating field.

3. Numerical relativity

Many of the applications of Regge calculus in numerical relativity have been relatively small-scale calculations, where the space-time has considerable symmetry and where, in many cases, the continuum solution is known. Examples of such work will be described first, then there will be brief more general comments about Regge calculus as a tool in numerical relativity.

A suitable cosmology for testing Regge calculus is the Kasner model, which was first examined in this way by Lewis,[28] who was unable to obtain the full set of Kasner–Einstein equations in the continuum limit. Gentle[29] re-examined the problem and discovered that Lewis had neglected one type of curvature. With this included, the full set of equations was derived in the continuum limit and accurate numerical solutions obtained. Brewin and Gentle,[30] again in the context of the Kasner cosmology, investigated how the simplicial solutions were second-order accurate approximations to the continuum solution, but the residual of the Regge equations evaluated on the continuum solution did not converge, because of a wave-like disturbance with high frequency but low amplitude in the simplicial solution. The amplitude of this wave converged to zero as the discretisation was refined, and it did not affect the overall second-order accuracy of the simplicial solution. Brewin[31] also looked at evolution of the Kasner cosmology using both Regge calculus and his smooth lattice method; both produced convergent approximations to the exact solution, but Regge calculus was two orders of magnitude slower.

Closed FLRW universes with positive cosmological constant were considered in three dimensions by Tsuda and Fujiwara.[32] The Cauchy surfaces were taken to be the surfaces of the regular polyhedral solids in three dimensions, with prisms connecting the surfaces at subsequent times. The numerical solution was found to deviate from the continuum solution at large times. The triangles on the Cauchy surface were then subdivided and the new vertices projected onto a sphere to give a geodesic dome. It was found that the numerical solution agreed with the continuum one in the limit of infinite subdivision. This work should now be extended to four dimensions, and matter included.

In a series of papers, Liu and Williams discussed various aspects of the time evolution of simple model universes. In the context of a closed empty universe with positive cosmological constant, they investigated whether to apply the variational principle to the action then impose symmetries (local variation) or to impose symmetries then vary (global variation). It emerged that local variation does not generally lead to a viable set of Regge equations.[33] They then considered lattice universes with point masses distributed on a regular lattice on the Cauchy surface. Constraints were obtained on the distribution of the masses for the model to be stable. The evolution resembled that of a closed FLRW dust-filled universe. When one mass was perturbed, the model's evolution was well-behaved, with the expansion increasing in magnitude as the perturbation increased.[34]

Progress with the classical evolution of a spacelike hypersurface was made in the early 1990s when it was realised that, in general, the Regge equations decouple into a collection of much smaller groups, leading to the so-called 'Sorkin evolution',[8] already mentioned. De Felice and Fabri[35] showed that the vertices in the 600-tetrahedron tessellation of the three-sphere fall into *five* distinct classes, not four as in the original paper on the Parallelisable

Implicit Evolution Scheme.[8] They also investigated the so-called 'stopping point' which has plagued many numerical calculations of the evolution of model universes using Regge calculus; this point is where the evolution stops well before the spatial volume becomes zero. Brewin[36] had suggested that this occurs when the 'vertical' edges become spacelike as the evolution is so fast. De Felice and Fabri showed that it is a causality-breaking singularity in the effective metric in Regge calculus in such calculations. The issue of causality in this approach was also investigated by Khavari,[37] who generalised the triangle inequalities in Euclidean space to inequalities on edges of triangles in Minkowski space. When causality is included in this way in the Parallelisable Implicit Evolution Scheme, the 'stopping point' problem is resolved, which is significant progress. The revised algorithm was applied successfully to the FLRW universe.[38] Khavari also set up a Regge calculus version of the Raychaudhuri equation.[37] In both 2+1 and 3+1 dimensions, analogues for the average expansion and the shear scalar were found.

More generally, in a brief review, Gentle and Miller[39] set out a programme for large-scale numerical work using Regge calculus and progress was discussed by Gentle.[40] In the spherically symmetric case, the static Schwarzschild solution was modelled to high accuracy. Axisymmetric initial data was constructed for Brill waves, and for a black hole with Brill wave perturbations. Regge calculus methods produced a very accurate approximation to Misner's analytic solution for initial data for the head-on collision of two equal mass non-rotating black holes. The generic axisymmetric code developed by Gentle, Miller and collaborators was used to obtain the time evolution of the Brill wave initial data, the first successful time evolution of gravitational radiation on a lattice. These successes are very encouraging but there is still much work to be done, particularly on the inclusion of matter and the relation of lattice approaches to standard finite element methods for solving differential equations. Because of the investment of time and computer power needed to develop Regge lattice methods further, it seems that current work in numerical relativity relies more on the very successful standard finite difference techniques to perform calculations in astrophysics, including the recent focus on gravitational waves.

4. Simplicial quantum gravity

This section reports on some of the recent work in quantum Regge calculus, but excludes the Barrett–Crane model and group field theories, which are discussed separately.

In a new approach to quantum Regge calculus, Xue[26] translated the Einstein–Cartan theory to a Euclidean lattice, with the tetrad field, the gauge field and the spin connection assigned to the edges. The partition function and effective action are constructed, and the vacuum expectation values of diffeomorphism and local gauge-invariant quantities can in principle be calculated. Some calculations in two dimensions are presented.

In a paper dedicated to Rafael Sorkin, one of the pioneers of Regge calculus, Gambini and Pullin[41] apply their 'consistent discretization' approach to Regge calculus. They set up a canonical formalism which is free of constraints, with the time evolution equivalent to a canonical transformation. Some of the edge lengths are effectively Lagrange multipliers and can be eliminated using their equations of motion. This seems to avoid the common problem of 'spikes' (thin, elongated simplices) in the Lorentzian case. Quantization is achieved by writing down a path integral with the 'Lagrange multipliers' eliminated; the measure is determined uniquely by the unitary transformation that implements the dynamics. It is straightforward to include topology change in this framework.

Work by Dittrich and Höhn[42] generalizes the 'consistent discretization' approach to allow for arbitrary triangulations and for a change of triangulation in time. Quantization of this framework is discussed further by Höhn.[43]

Bonzom and Dittrich[44] and then Asante, Dittrich and Haggard[45] used Regge calculus to calculate one-loop partition functions for gravity (for regions with boundary) and to define holographic duals for three-dimensional gravity and for the flat sector of four-dimensional gravity. It turns out that Regge calculus is well-suited to defining the partition function for regions with boundary and provides a regularization for computing the one-loop correction. A fully non-perturbative set-up and calculations using the Ponzano–Regge model are described by Dittrich, Göller, Livine and Riello.[46]

Although it is perhaps more relevant to the section on the Barrett–Crane model, it is worth mentioning the relation between Regge calculus and BF theory obtained by Kisielowski.[47] A smooth manifold is obtained by removing the hinges of a simplicial complex. The Regge geometry is then encoded in a BF theory on the boundary of this manifold. The process amounts to replacing the degrees of freedom of Regge calculus with discrete degrees of freedom of topological BF theory. Bonzom[48] has also explored the relation between BF theory and Regge calculus, showing that together, the gluing relations and the simplicity constraints which turn BF theory into simplicial gravity, contain the constraints of area-angle Regge calculus.[21] The action includes the contribution from the Immirzi parameter.

The possibility that the strength of gravitational interactions might increase with distance has been explored in a series of papers by Hamber, Toriumi and Williams. A set of effective field equations is formulated incorporating the gravitational, vacuum-polarization induced running of Newton's constant, G. This results in an accelerated power law expansion for the Robertson–Walker universe, at times of the order of the inverse of the Hubble constant.[49] The implications for cosmological density perturbations are also considered.[50] The requirement of general covariance for the effective field equations restricts the value of the gravitational scaling exponent to be an integer greater than one.[51] The running of the cosmological constant, on the other hand, is shown in a number of approaches to be inconsistent with general covariance.[52] To make the connection with experiment, Hamber[53] argues that the lattice results suggest that the growth of Newton's constant, G, with distance, should become observable only on very large distance scales, comparable to the observed scaled cosmological constant. The hope is that future high precision satellite experiments will be able to detect this small quantum correction.

Quantum gravity in the limit of a large number of space-time dimensions has been considered by Hamber and Williams.[54] For a simplicial lattice dual to a hypercube, a critical point was found, separating a weak coupling from a strong coupling phase, where dominant contributions to the curvature correlation functions were described by large closed random polygonal surfaces.

Hamber and Williams[55] have used the gravitational Wilson loop to obtain information about the large-scale curvature properties of the geometry. By comparing the resulting quantum averages to expected semi-classical forms valid for macroscopic observers, it is possible to identify the gravitational correlation length in the Wilson loop with the observed large-scale curvature. The results imply a positive effective cosmological constant at large distances.

Most of the quantum gravity work of Hamber and collaborators has used the Euclidean lattice path integral approach, with numerical simulations using Monte Carlo methods. To complement this, Hamber and Williams[56] have obtained a discrete form of the Wheeler–De Witt equation for the quantum wave functional of the lattice. In the strong coupling limit, the wave functional depends only on geometric quantities such as areas and volumes. Explicit solutions in 2+1 dimensions are found;[57] a finite correlation length emerges, that cuts off any infra-red divergences, and there seems to be no weak-coupling perturbative phase. By contrast, in 3+1 dimensions,[58] the critical point in G has a non-zero value, but the weak-coupling perturbative ground state appears to be non-perturbatively unstable. The results obtained seem to suggest that the Lorentzian and Euclidean formulations of lattice quantum gravity belong to the same field-theoretic universality class.

In work related to that of Khavari,[37] using the causal structure in Lorentzian space-time and the fact that certain simplices are not constrained by the triangle inequalities, Tate and Visser[59] set up Lorentzian signature models of quantum Regge calculus, showing that these are not related to Euclidean models by a simple Wick rotation. The lack of the triangle inequality constraints means that it is easier to do analytical calculations and that numerical simulations are more computationally efficient. They set up the path integral for a model in 1+1 dimensions, obtaining scaling relations for the path integral and showing that spikes are absent, then discuss the model in higher dimensions.

Causal dynamical triangulations are based on a Lorentzian simplicial lattice and use the Regge action, but rather than integrating over the edge lengths in the path integral, sums are performed over triangulations using the Pachner moves. A sophisticated and detailed analysis of the phase structure in 3+1 dimensions has been performed and we refer the reader to reviews by Ambjorn, Loll and collaborators.[60]

5. The Barrett–Crane model

The original development of the Barrett–Crane model is explained in detail in the earlier review.[1] The two versions of the model are based on representations of the group SO(4) in the Euclidean case or SO(3, 1) in the Lorentzian case. After the initial flurry of excitement of the model as a potential four-dimensional quantum gravity theory, the hard work of establishing its properties continued at a slower pace. Different proposals for the normalisation factors on the lower-dimensional faces were examined, arriving at a reasonable scheme after numerical simulations of several proposals.[61] The large-spin asymptotics of the amplitude of a single 4-simplex was established in a precise way by analytical means[62,63] and confirmed by numerical means.[64] This showed that although the desired four-dimensional 'Regge calculus' geometries are present and contribute the Regge action to the amplitude, some degenerate configurations typically dominate. This presents a physical picture that is not so clear overall.

The issue of understanding the geometry of several 4-simplices glued together proves to be a hard problem. The main issue is that the data that is matched on a pair of 4-simplices is the areas of the four common triangles, but this information is not enough to guarantee a continuous metric spanning the two. A geometrical interpretation of this gluing is lacking, except perhaps across a null surface in the Lorentzian case.[18]

The Pachner moves for the gluing have been analysed recently[65] and it is argued there that the action of 'area Regge calculus' (see section 2) provides a plausible semiclassical model for the phase of the Barrett–Crane amplitude. There is evidence that when more than two simplices are considered, the possible metric discontinuity is rather more restricted; for example along planes the discontinuity vanishes except in the null case.[17] There is much about the geometry of the Barrett–Crane model that is still to be understood.

Discontent with the gluing conditions led to some substantial extensions of the Barrett–Crane model. These new models were formulated by Freidel–Krasnov (FK)[66] and by Engle–Levine–Pereira–Rovelli (EPRL),[67] and in the Lorentzian case by FK and Pereira.[66,68] The idea is to allow a vector space of intertwiners on each tetrahedron, so that the gluing between two 4-simplices involves an inner product in this intertwiner space. Another closely related spin foam model for 4d quantum gravity based on the imposition of simplicity constraints via non-commutative flux variables was proposed, to deal with these and other issues, by Baratin–Oriti[69] (so far only in the Riemannian case).

Representations of SO(4) are determined by a pair of half-integers (j, j') which label the total spin of each factor in the spin covering $\mathrm{Spin}(4) \cong \mathrm{SU}(2) \times \mathrm{SU}(2)$. Choosing to sum over all representations (j, j') on each triangle and the full intertwiner space on each tetrahedron leads back to the Ooguri model (or Crane–Yetter model in the q-deformed case), whereas limiting to the simple representations (j, j) and the one-dimensional intertwiner space determined by the 'canonical vertex' gives the Barrett–Crane model again. The idea of the new models is to take an intermediate case where $j' = cj$ for a global constant c, and on each tetrahedron a vector space of intertwiners for the group SU(2) (not SO(4)). These intertwiners are supposed to give a more complete description of the quantum geometry of a tetrahedron. This means that these degrees of freedom can propagate into the neighbouring 4-simplex in a manifold, giving a more geometric gluing.

The constant c is determined by an Immirzi parameter, as introduced in loop quantum gravity, giving an asymmetry in the action between the self-dual and anti-self-dual parts of the curvature. For some values of c, the EPRL and FK models are identical, and for other values they are different but similar in construction.

As before, the first semiclassical analysis of the amplitudes for the new models was for the single 4-simplex, in the Euclidean case[70] and in the Lorentzian case.[71] These papers also provide a more precise definition of the models themselves. The boundary data is now more complicated because, beside the areas of the triangles, there is also data for the geometry of each tetrahedron. The amplitude for most boundary data is exponentially damped because the integral representation for the amplitude contains no stationary points. The interesting cases are where there are stationary points so that the amplitudes are not damped. The analysis shows that there is a class of such boundary data called 'Regge-like' where all of the data is consistent with a flat metric in the 4-simplex. In terms of this metric, the phase of the 4-simplex amplitude is again the Regge action for a 4-simplex. There are however some other important configurations, in the same way as in the original Barrett–Crane model, but with different details. These other configurations are three-dimensional 'vector geometries' or even lower-dimensional geometries, and do not have a Regge calculus interpretation.

Probably the simplest explanation of this asymptotics is given by considering the 4-simplex amplitudes as squares of 15j symbols and analysing the rather simpler asymptotics of the 15j symbols.[72] Amazingly, the 15j symbols can also have Regge-like boundary data with asymptotics determined by the Regge calculus action, even though these symbols belong to the Ooguri model, which is not normally regarded as a gravity model at all.

Numerical methods to compute the 4-simplex amplitudes were developed for the 15j symbols,[73] and for the Lorentzian EPRL model,[74] giving confirmations of the asymptotic formulae. These papers also derive a number of generalisations of the asymptotic formulae together with their geometric interpretations.

The EPRL and FK models still have difficulties with gluing. Considering a spatial slice of spacetime (i.e., a 3-manifold), each tetrahedron now has a quantum state space that can distinguish between the flat geometries with the same area for each triangle. But unfortunately, this data does not always glue together continuously from one tetrahedron to another across a common triangle. In essence, it is like the problem with the original Barrett–Crane model but in one lower dimension. It is also similar to the gluing problem for the geometries associated to states in loop quantum gravity.[75]

Several works have analysed the EPRL/FK models when a number of 4-simplices are glued together to form a manifold. Replacing each 4-simplex amplitude with its asymptotic expression for large spins results in an amplitude for a triangulated manifold whose phase part is the Regge calculus action, but with additional independent variables.[76,77] This procedure is only a heuristic one as the spin variables on interior triangles are summed over all values, so that it is not clear whether the asymptotic formula is a useful approximation. Nevertheless, continuing with this formula and looking for the stationary points when the spin is varied leads to the naive conclusion that the geometry is flat, at least in an asymptotic limit.[78] More careful analyses are underway[79,80] but it is too soon for a definitive conclusion.

6. Group field theories

Group field theories (GFTs)[81] bring Regge's intuition about describing space-time, geometry and gravity in terms of piecewise-flat, simplicial structures to a whole new level. They extend some of the simplicial quantum gravity approaches discussed above, merging them into a single framework. They were first introduced[82] as an enrichment of tensor models, themselves a generalization of matrix models for two-dimensional gravity, to higher dimensions, and a generating functional for Euclidean dynamical triangulations. This enrichment defined instead a generating functional for sums over triangulations weighted by the state sum models for topological BF theory in three dimensions (Ponzano–Regge model) and four dimensions, based on the representation theory of $SU(2)$. Then, group field theories were found[83] to provide a generating functional for the Barrett–Crane model and a tentative definition of the dynamics of four-dimensional quantum gravity, in the language of loop quantum gravity and spin foam models. Many developments over the last twenty years form now a massive body of work making group field theories a promising formalism for a quantum gravity theory based on discrete structures. They have also led to a new perspective, in which spacetime and geometry are emergent notions, to be extracted from the collective behaviour of the discrete GFT structures, seen as fundamental entities.[84]

GFTs can be understood as quantum field theories for elementary objects (the 'quanta' of the GFT field) corresponding to quantum tetrahedra. They rely on a description of simplicial geometry in terms of group- and representation-theoretic data,[85] also important in the context of spin foam models and loop quantum gravity. The relevant phase spaces are cotangent bundles of group manifolds (usually $SU(2)$, $Spin(4)$ or $SL(2,\mathbb{C})$) and the quantization leads to Hilbert spaces which admit a complete orthonormal basis of spin network states. In fact, the fundamental tetrahedra (in four-dimensional models) can be seen as dual to spin network vertices, with the outgoing links corresponding to the faces of the tetrahedra. The variables encoding the geometry of the simplicial structures are then group elements (corresponding to a discretized connection) and group representations, corresponding to eigenvalues of quantized face areas. A non-commutative description of the states can also be given,[86] in terms of Lie algebra elements corresponding to metric variables associated to the same faces. Generic quantum states are then many-body states formed by several quantum tetrahedra, including those in which the tetrahedra are glued to form extended three-dimensional triangulations. A Fock space description of the GFT Hilbert space has been given by Oriti,[87] clarifying also the nature of these models as second quantized descriptions of loop quantum gravity-type theories.

On these kinematical states, different dynamical prescriptions can be imposed. They are encoded in a choice of GFT action for a field whose arguments are the same data characterizing the geometry of a tetrahedron. The defining feature is a non-local pairing of field arguments in the interaction terms, reflecting the gluing of the tetrahedra to form the boundary of four-dimensional cells, taken as the fundamental building block of their interaction processes. The perturbative expansion of the GFT partition function defines a sum over Feynman diagrams dual to four-dimensional cellular complexes, with the interaction vertices defining the constituting cells and the propagator enforcing their gluing. One route towards model building[88] is based on the fact that GFT fields can be seen as generalised tensors with indices on the group manifold, transforming under unitary groups acting on the same indices. Thus, a theory space can be defined including all possible unitary invariants as interaction terms. Such tensorial GFTs can then take advantage directly of the many results obtained in the context of tensor models,[88] concerning the control over the topology of the GFT interaction diagrams, and their statistical analysis and critical behaviour. In fact, when the Lie group defining the domain of GFT fields is replaced by a finite group or simply by some finite set, then GFT models become tensor models, and their Feynman amplitudes depend only on the combinatorics of the underlying diagram; they actually become akin to those of the dynamical triangulations approach. Beside this tensorial setting, there have been few analyses of symmetries in GFT models,[89] and thus there is little control over possible theory spaces. Model building has been based, then, on the choice of having Feynman diagrams corresponding to triangulations (which suggests interactions with the combinatorics of 4-simplices) and on the desire to have the corresponding Feynman amplitudes coinciding with the most promising spin foam models, proposed in the context of simplicial quantum gravity and loop quantum gravity,[69,81,83] and inspired by the formulation of (continuum and discrete) GR as a constrained BF theory. The latter requirement translates into specific choices of kinetic and interaction terms in the GFT action, for each desired model. In fact, one can show that the correspondence between GFTs and spin foam models is generic:[90] for any given spin foam model defined on a given cellular complex, there is a GFT model whose Feynman amplitudes coincide with it on the Feynman diagram dual to the same complex (and vice versa, any given GFT model defines a corresponding spin foam model, in its perturbative expansion). Similarly generic is the fact that, when one formulates the same GFT models in terms of Lie algebra variables, the corresponding Feynman amplitudes take the form of (non-commutative) simplicial gravity path integrals in first order variables, on the same complex.[69]

From the point of view of simplicial quantum gravity, therefore, group field theories define in their perturbative expansion a quantum dynamics combining a sum over discrete geometric data corresponding to a first order version of quantum Regge calculus, associated to a given lattice, with a sum over lattices, including very refined ones, in the spirit of dynamical triangulations. The problem becomes, of course, to show the consistency of such quantum theory, i.e. to control analytically the combined sum, and to 'resum it', i.e. to define the full partition function, thus the continuum limit. This is where the quantum field theory nature of GFTs becomes crucial. Both problems, in fact, become problems in the renormalization of GFT models: to prove their perturbative renormalizability, and to define their full RG flow and continuum phase diagram. This has become a very active research direction over the last ten years,[91] in parallel with the developments in tensor models. Among the many results, it has led to: a better understanding of the divergences of four-dimensional quantum gravity models for constrained BF theories,[92] and a complete characterization of them for topological BF models;[93] the rigorous proof of renormalizability of several tensorial GFT models,[94] both abelian and non-abelian, in several dimensions; constructive analyses of the same tensorial GFT models;[95] the application of functional renormalization group techniques with a first characterization of their flows and phase diagrams (in simple truncations), confirming their asymptotic freedom or safety, and providing hints of new phases.[96] A first analysis of inequivalent (coherent) representations of GFT observables algebras, possibly corresponding to different phases of GFT systems, has been done by Kegeles, Oriti and Tomlin.[97]

The extraction of effective physics from GFT models can be pursued using methods from simplicial (quantum) gravity and spin foam models, at the level of GFT perturbation theory. The generating functional for superpositions of discrete structures defined by GFT models, as well as their second quantized formalism, are useful tools to do so. They also offer, however, new avenues to study emergent continuum physics bypassing to some extent the discrete gravity picture, and aiming at coarse-grained information encoding many discrete gravity degrees of freedom. In one such line of research,[98] effective cosmological physics is extracted from the hydrodynamics of GFT models and more precisely from the mean field description corresponding to simple GFT condensate states. Among the many results, we cite the evidence that (within this hydrodynamic approximation): a homogeneous and isotropic universe satisfies a modified Friedmann equation with the correct classical limit but with a bouncing regime replacing the classical big bang singularity,[99] similar to what is found in loop quantum cosmology; GFT interactions can modify this dynamics to give an extended period of acceleration following the bounce (a sort of quantum gravity-driven inflation), without the need to introduce additional (inflaton-like) degrees of freedom, and a later recollapse and cyclic evolution;[100] the extension of the formalism to include anisotropies[101] and inhomogeneities,[102] with approximate scale invariance found in the spectrum of cosmological perturbations.

7. Conclusions

Applying the geometric concept of piecewise linear spaces to general relativity, Regge provided the basis for very fruitful research in a number of areas, including calculations of the classical evolution of model universes and formulations of quantum gravity. Since the review paper of 2000, in particular there has been outstanding progress in the development of quantum gravity models, which arise directly from Regge's work. These have the potential to provide a consistent theory of quantum gravity with wide-ranging implications in physics.

Acknowledgments

The work of RMW has been partially supported by STFC consolidated grant ST/P000681/1.

References

1. T. Regge and R. M. Williams, J. Math. Phys. **41**, 3964 (2000), arXiv:gr-qc/0012035.
2. T. Regge, Il Nuovo Cimento **19**, 558 (1961).
3. F. Lund and T. Regge, *Simplicial Approximation to some Homogeneous Cosmologies*, unpublished (1974).
4. G. Ponzano and T. Regge, in *Spectroscopic and Group Theoretical Methods in Physics*, (North Holland, 1968).
5. V. G. Turaev and O. Y. Viro, Topology **31**, 865 (1992).
6. T. Regge, in *From SU(3) to Gravity: Festschrift in Honor of Yuval Ne'eman*, ed. E. Gotsman and G. Tauber, (Cambridge University Press, 2016).
7. A. P. Gentle and W. A. Miller, *Generating constant mean curvature hypersurfaces in Regge calculus*, unpublished (2000).
8. J. W. Barrett, M. Galassi, W. A. Miller, R. Sorkin, P. A. Tucker and R. M. Williams, Int. J. Theor. Phys. **36**, 809 (1997), arXiv:gr-qc/9411008.
9. J. R. McDonald and W. A. Miller, Class. Quant. Grav. **25**, 195017 (2008), arXiv:0804.0279 [gr-qc].
10. R. Conboye, W. A. Miller and S. Ray, Class. Quant. Grav. **32**, 185009 (2015), arXiv:1502.07782 [gr-qc].
11. A. P. Gentle, A. Kheyfets, J. R. McDonald and W. A. Miller, Class. Quant. Grav. **26**, 015005 (2009), arXiv:0807.3041 [gr-qc].
12. J. R. McDonald and W. A. Miller, Class. Quant. Grav. **27**, 095011 (2010), arXiv:1002.5001 [gr-qc].
13. H. W. Hamber and G. Kagel, Class. Quant. Grav. **21** 5915 (2004), arXiv:gr-qc/0107031.
14. S. Ariwahjoedi and F. P. Zen, arXiv:1709.08373 [gr-qc] (2017).
15. P. A. Höhn, Phys. Rev. **D91**, 124034 (2015), arXiv:1411.5672 [gr-qc].
16. C. Rovelli, Phys. Rev. **D48**, 2702 (1993), arXiv:hep-th/9304164.
17. C. Wainwright and R. M. Williams, Class. Quant. Grav. **21**, 4865 (2004), arXiv:gr-qc/0405031.
18. J. W. Barrett, arXiv:gr-qc/0011051 (2001).

19. Y. Neiman, arXiv:1308.1012 [gr-qc] (2012).
20. B. Dittrich and S. Speziale, New J. Phys. **10**, 083006 (2008), arXiv:0802.0864 [gr-qc].
21. B. Bahr and B. Dittrich, Class. Quant. Grav. **26**, 225011 (2009), arXiv:0905.1670 [gr-qc]; New J. Phys. **12**, 033010 (2010), arXiv:0907.4325 [gr-qc]; AIP Conf. Proc. **1196**, 10 (2009), arXiv:0909.5688 [gr-qc].
22. S. H. Christiansen, Num. Math. **119**, 613 (2011), arXiv:1106.4266 [math.NA].
23. J. Schmidt and C. Kohler, Gen. Rel. Grav. **33**, 1799 (2001), arXiv:gr-qc/0103111.
24. I. T. Drummond, Nucl. Phys. **B273**, 125 (1986).
25. J. W. Barrett, Class. Quant. Grav. **4**, 1565 (1987).
26. S.-S. Xue, Phys. Lett. **B682**, 300 (2009), arXiv:0902.3407 [hep-th].
27. J. G. Pereira and T. Vargas, Class. Quant. Grav. **19**, 4807 (2002), arXiv:gr-qc/0208036.
28. S. M. Lewis, Phys. Rev. **D25**, 306 (1982).
29. A. P. Gentle, Class. Quant. Grav. **30**, 085004 (2013), arXiv:1208.1502 [gr-qc].
30. L. C. Brewin and A. P. Gentle, Class. Quant. Grav. **18**, 517 (2001), arXiv:gr-qc/0006017.
31. L. C. Brewin, Class. Quant. Grav. **32**, 195008 (2015), arXiv:1505.00067 [gr-qc].
32. R. Tsuda and T. Fujiwara, Prog. in Theor. and Exp. Phys. **2017**, 073E01 (2017), arXiv:1612.06536 [gr-qc].
33. R. G. Liu and R. M. Williams, Phys. Rev. **D93**, 024032 (2016), arXiv:1501.07614 [gr-qc].
34. R. G. Liu and R. M. Williams, Phys. Rev. **D93**, 023502 (2016), arXiv:1502.03000 [gr-qc]; R. G. Liu and R. M. Williams, arXiv:1510.05771 [gr-qc] (2016).
35. A. De Felice and E. Fabri, arXiv:gr-qc/0009093 (2000); arXiv:gr-qc/0106077 (2001).
36. L. C. Brewin, Class. Quant. Grav. **4**, 899 (1987).
37. P. Khavari, University of Toronto thesis (2009).
38. P. Khavari and C. C. Dyer, arXiv:0809.1815 [gr-qc] (2008).
39. A. P. Gentle and W. A. Miller, in *On Recent Developments in Theoretical and Experimental General Relativity, Gravitation and Relativistic Field Theories*, Proceedings of the Ninth Marcel Grossmann Meeting (World Scientific, 2002).
40. A. P. Gentle, Gen. Rel. Grav. **34**, 1701 (2002), arXiv:gr-qc/0408006.
41. R. Gambini and J. Pullin, Int. J. Mod. Phys. **D15**, 1699 (2006), arXiv:gr-qc/0511096.
42. B. Dittrich and P. A. Höhn, Class. Quant. Grav. **27**, 155001 (2010), arXiv:0912.1817 [gr-qc]; Class. Quant. Grav. **29**, 115009 (2012), arXiv:1108.1974 [gr-qc]; J. Math. Phys. **54**, 093505 (2013), arXiv:1303.4294 [math-ph].
43. P. A. Höhn, J. Math. Phys. **55**, 083508 (2014), arXiv:1401.6062 [gr-qc]; J. Math. Phys. **55**, 103507 (2014), arXiv:1401.7731 [gr-qc].
44. V. Bonzom and B. Dittrich, arXiv:1511.05441 [hep-th].
45. S. A. Asante, B. Dittrich and H. M. Haggard, arXiv:1811.11744 [hep-th].
46. B. Dittrich, C. Göller, E. R. Livine and A. Riello, arXiv:1710.04202 [hep-th]; arXiv:1710.04237 [hep-th]; Class. Quant. Grav. **35**, 13LT01 (2018), arXiv:1803.02759 [hep-th].
47. M. Kisielowski, arXiv:1704.00998 [gr-qc] (2017).
48. V. Bonzom, Class. Quant. Grav. **26**, 155020 (2009), arXiv:0903.0267 [gr-qc].
49. H. W. Hamber and R. M. Williams, Mod. Phys. Lett. **A21**, 735 (2006), arXiv:gr-qc/0506137; Phys. Rev. **D72**, 044026 (2005), arXiv:hep-th/0507017; Phys. Rev. **D75**, 084014 (2007), hep-th/0607228.
50. H. W. Hamber and R. Toriumi, Phys. Rev. **D82**, 043518 (2010), arXiv:1006.5214 [gr-qc]; Phys. Rev. **D84**, 103507 (2010), arXiv:1109.1437 [gr-qc].
51. H. W. Hamber and R. M. Williams, Phys. Lett. **B643**, 228 (2006), arXiv:gr-qc/0607131.
52. H. W. Hamber and R. Toriumi, Int. J. Mod. Phys. **D22**, 1330023 (2013), arXiv:1301.6259 [hep-th].
53. H. W. Hamber, Phys. Rev. **D92**, 064017 (2015), arXiv:1506.07795 [hep-th].
54. H. W. Hamber and R. M. Williams, Phys. Rev. **D73**, 044031 (2006), arXiv:hep-th/0512003.
55. H. W. Hamber and R. M. Williams, Phys. Rev. **D76**, 084008 (2007), arXiv:0706.2342 [hep-th]; Phys. Rev. **D81**, 084048 (2010), arXiv:0907.2652 [hep-th].

56. H. W. Hamber and R. M. Williams, Phys. Rev. **D84**, 104033 (2011), arXiv:1109.2530 [hep-th].
57. H. W. Hamber, R. Toriumi and R. M. Williams, Phys. Rev. **D86**, 084010 (2012), arXiv:1207.3759 [hep-th].
58. H. W. Hamber, R. Toriumi and R. M. Williams, Phys. Rev. **D88**, 084012 (2012), arXiv:1212.3492 [hep-th].
59. K. Tate and M. Visser, J.H.E.P. **1111**, 072 (2011), arXiv:1108.4965 [gr-qc].
60. J. Ambjorn, D. Coumbe, J. Gizbert-Studnicki and J. Jurkiewicz, in *On Recent Developments in Theoretical and Experimental General Relativity, Astrophysics and Relativistic Field Theories*, Proceedings of the Fourteenth Marcel Grossmann Meeting (World Scientific, 2017), arXiv:1509.08788 [gr-qc]; J. Ambjorn, J. Gizbert-Studnicki, A. Görlich, J. Jurkiewicz, N. Klitgaard and R. Loll, Eur. J. Phys. **C77**, 1 (2017), arXiv:1610.05245 [hep-th].
61. J. C. Baez, J. D. Christensen, T. R. Halford and D. C. Tsang, Class. Quant. Grav. **19**, 4627 (2002), arXiv:gr-qc/0202017.
62. J. W. Barrett and C. M. Steele, Class. Quant. Grav. **20**, 1341 (2003), arXiv:gr-qc/0209023.
63. L. Freidel and D. Louapre, Class. Quant. Grav. **20**, 1267 (2003), arXiv:hep-th/0209134.
64. J. C. Baez, J. D. Christensen and G. Egan, Class. Quant. Grav. **19**, 6489 (2002), arXiv:gr-qc/0208010.
65. S. K. Asante, B. Dittrich and H. M. Haggard, Class. Quant. Grav. **35**, 135009 (2018), arXiv:1802.09551 [gr-qc].
66. L. Freidel and K. Krasnov, Class. Quant. Grav. **25**, 125018 (2008), arXiv:0708.1595 [gr-qc].
67. J. Engle, E. Livine, R. Pereira and C. Rovelli, Nucl. Phys. B **799**, 136 (2008), arXiv:0711.0146 [gr-qc].
68. R. Pereira, Class. Quant. Grav. **25**, 085013 (2008), arXiv:0710.5043 [gr-qc].
69. A. Baratin and D. Oriti, Phys. Rev. **D85**, 044003 (2012), arXiv:1111.5842 [hep-th].
70. J. W. Barrett, R. J. Dowdall, W. J. Fairbairn, H. Gomes and F. Hellmann, J. Math. Phys. **50**, 112504 (2009), arXiv:0902.1170 [gr-qc].
71. J. W. Barrett, R. J. Dowdall, W. J. Fairbairn, F. Hellmann and R. Pereira, Class. Quant. Grav. **27**, 165009 (2010), arXiv:0907.2440 [gr-qc].
72. J. W. Barrett, W. J. Fairbairn and F. Hellmann, Int. J. Mod. Phys. A **25**, 2897 (2010), arXiv:0912.4907 [gr-qc].
73. P. Donà, M. Fanizza, G. Sarno and S. Speziale, Class. Quant. Grav. **35** (2018) no. 4, 045011 [arXiv:1708.01727 [gr-qc]].
74. P. Donà, M. Fanizza, G. Sarno and S. Speziale, arXiv:1903.12624 [gr-qc].
75. L. Freidel and S. Speziale, Phys. Rev. D**82**, 084040 (2010), arXiv:1001.2748 [gr-qc].
76. M. X. Han and M. Zhang, Class. Quant. Grav. **29**, 165004 (2012), arXiv:1109.0500 [gr-qc].
77. M. Han and M. Zhang, Class. Quant. Grav. **30**, 165012 (2013), arXiv:1109.0499 [gr-qc].
78. V. Bonzom, Phys. Rev. D **80**, 064028 (2009). arXiv:0905.1501 [gr-qc].
79. F. Hellmann and W. Kaminski, JHEP **1310**, 165 (2013), arXiv:1307.1679 [gr-qc].
80. J. R. Oliveira, Class. Quant. Grav. **35**, 095003 (2018), arXiv:1704.04817 [gr-qc].
81. D. Oriti, in *Foundations of space and time*, G. Ellis, J. Murugan, A. Weltman (eds.), (Cambridge University Press, 2012), arXiv:1110.5606 [hep-th]; T. Krajewski, PoS QGQGS2011 (2011) 005, arXiv:1210.6257 [gr-qc]; D. Oriti, in *Loop Quantum Gravity*, A. Ashtekar, J. Pullin (eds.), (World Scientific 2017), arXiv:1408.7112 [gr-qc].
82. D. V. Boulatov, Mod. Phys. Lett. **A7**, 1629 (1992), arXiv:hep-th/9202074; H. Ooguri, Mod. Phys. Lett. **A7**, 2799 (1992), arXiv:hep-th/9205090.
83. R. De Pietri, L. Freidel, K. Krasnov, C. Rovelli, Nucl. Phys. **B574**, 785 (2000), arXiv:hep-th/9907154; A. Perez, C. Rovelli, Nucl. Phys. **B599**, 255 (2001), arXiv:gr-qc/0006107.
84. D. Oriti, Stud. Hist. Philos. Mod. Phys. **46**, 186 (2014), arXiv:1302.2849 [physics.hist-ph]; K. Crowther, arXiv:1410.0345 [physics.hist-ph].
85. J. C. Baez and J. W. Barrett, Adv. Theor. Math. Phys. **3**, 815 (1999), arXiv:gr-qc/9903060; F. Conrady and L. Freidel, J. Math. Phys. **50**, 123510 (2009), arXiv:0902.0351 [gr-qc].

86. A. Baratin and D. Oriti, Phys. Rev. Lett. **105**, 221302 (2010), arXiv:1002.4723 [hep-th]; A. Baratin, B. Dittrich, D. Oriti and J. Tambornino, Class. Quant. Grav. **28**, 175011 (2011), arXiv:1004.3450 [hep-th].
87. D. Oriti, Class. Quant. Grav. **33**, 085005 (2016), arXiv:1310.7786 [gr-qc].
88. R. Gurau and J. Ryan, SIGMA **8**, 20 (2012), arXiv:1109.4812 [hep-th]; V. Rivasseau, arXiv:1407.0284 [hep-th].
89. J. Ben Geloun, J. Math. Phys. **53**, 022901 (2012), arXiv:1107.3122 [hep-th]; A. Baratin, F. Girelli and D. Oriti, Phys. Rev. **D83**, 104051 (2011), arXiv:1101.0590 [hep-th]; A. Kegeles and D. Oriti, J. Phys. **A50**, 125402 (2017), arXiv:1608.00296 [gr-qc].
90. M. P. Reisenberger and C. Rovelli, Class. Quant. Grav. **18**, 121 (2001), arXiv:gr-qc/0002095.
91. S. Carrozza, SIGMA **12**, 070 (2016), arXiv:1603.01902 [gr-qc].
92. V. Bonzom and B. Dittrich, Phys. Rev. **D88**, 124021 (2013), arXiv:1304.6632 [gr-qc]; A. Riello, Phys. Rev. **D88**, 024011 (2013), arXiv:1302.1781 [gr-qc]; P. Dona, Class. Quant. Grav. 35 (2018) 17, 175019, arXiv:1803.00835 [gr-qc].
93. V. Bonzom and M. Smerlak, Annales Henri Poincare **13**, 185 (2012), arXiv:1103.3961 [gr-qc].
94. J. Ben Geloun and V. Rivasseau, Commun. Math. Phys. **318**, 69 (2013), arXiv:1111.4997 [hep-th]; S. Carrozza, D. Oriti and V. Rivasseau, Commun. Math. Phys. **330**, 581 (2014), arXiv: 1303.6772 [hep-th]; S. Carrozza, V. Lahoche and D. Oriti, Phys. Rev. **D96**, 066007 (2017), arXiv:1703.06729 [gr-qc].
95. V. Rivasseau, SIGMA **12**, 69 (2016), arXiv:1603.07278 [math-ph].
96. D. Benedetti, J. Ben Geloun, and D. Oriti, JHEP **03**, 084 (2015), arXiv:1411.3180 [hep-th]; J. Ben Geloun and T. Koslowski, arXiv:1606.04044 [gr-qc]; V. Lahoche and D. Ousmane Samary, arXiv:1803.09902 [hep-th].
97. A. Kegeles, D. Oriti and C. Tomlin, Class. Quant. Grav. 35 (2018) 12, 125011, arXiv:1709.00161 [gr-qc].
98. S. Gielen and L. Sindoni, SIGMA **12** 082 (2016), arXiv:1602.08104 [gr-qc]; D. Oriti, Comptes Rendus Physique **18**, 235 (2016?), arXiv:1612.09521 [gr-qc].
99. D. Oriti, L. Sindoni and E. Wilson-Ewing, Class. Quant. Grav. **33**, 224001 (2016), arXiv:1602.05881 [gr-qc].
100. M. De Cesare, A. Pithis and M. Sakellariadou, Phys. Rev. **D94**, 064051 (2016), arXiv:1606.00352 [gr-qc].
101. M. De Cesare, D. Oriti, A. Pithis, M. Sakellariadou, Class. Quant. Grav. 35 (2018) 1, 015014, arXiv:1709.00994 [gr-qc].
102. D. Oriti and S. Gielen, Phys. Rev. D98 (2018)10, 106019, arXiv:1709.01095 [gr-qc].

JOURNAL OF MATHEMATICAL PHYSICS VOLUME 41, NUMBER 6 JUNE 2000

Discrete structures in gravity

Tullio Regge[a)]
Dipartimento di Fisica, Politecnico di Torino, Corso Duca degli Abruzzi, 10129 Torino, Italy

Ruth M. Williams[b)]
Girton College, Cambridge CB3 0JG, United Kingdom and DAMTP, CMS, Wilberforce Road, Cambridge CB3 0WA, United Kingdom

(Received 8 October 1999; accepted for publication 31 January 2000)

Discrete approaches to gravity, both classical and quantum, are reviewed briefly, with emphasis on the method using piecewise-linear spaces. Models of three-dimensional quantum gravity involving $6j$-symbols are then described, and progress in generalizing these models to four dimensions is discussed, as is the relationship of these models in both three and four dimensions to topological theories. Finally, the repercussions of the generalizations are explored for the original formulation of discrete gravity using edge-length variables. © *2000 American Institute of Physics.* [S0022-2488(00)00206-1]

I. INTRODUCTION TO DISCRETE GRAVITY

A. Basic formalism

The original motivation for the development of a discrete formalism for gravity[1] arose from a number of problems with the continuum formulation of general relativity. These included the difficulty of solving Einstein's equations for general systems without a large degree of symmetry, the problems of representing complicated topologies, and the need for considerable geometric insight and capacity for visualization. It turned out, as we shall see, that the discretization scheme to be described not only helped with these problems but also found a vital role in numerical relativity and in attempts at a formulation of quantum gravity.

The related branches of mathematics which found their application to physics in this formulation of gravity are those of piecewise-linear spaces and topology and the geometric notion of intrinsic curvature on polyhedra. The immediate aim was to develop an approach to general relativity which avoided the use of coordinates, since the physical predictions of the theory are coordinate independent. The basic idea of the approach, which has subsequently become known as *Regge calculus*, is as follows. Rather than considering spaces (or space–times) with continuously varying curvature, we deal with spaces where the curvature is restricted to subspaces of codimension two. This is achieved by considering collections of n-dimensional blocks, which are glued together by identification of their flat $(n-1)$-dimensional faces. The curvature lies on the $(n-2)$-dimensional subspaces, known as *hinges* or *bones*. For technical reasons, it is convenient to use blocks which are *simplices* (triangles, tetrahedra, and their higher dimensional analogs).

Consider first the realization of these ideas in two dimensions. Here we have examples in everyday life, geodesic domes; these consist of networks of flat triangles which are fitted together to approximate curved surfaces, usually parts of a sphere. Since two triangles with a common edge can be flattened out without distortion, there is no curvature on the edges. However, when a collection of triangles meeting at a vertex is flattened, there will be a gap, indicating the presence of curvature at the vertex. The amount of curvature there depends simply on the size of the gap or *deficit angle*.

[a)]Electronic mail: regge@polito.it
[b)]Electronic mail: r.m.williams@damtp.cam.ac.uk

It is relatively simple to visualize the generalization of a triangulated surface to three dimensions, where a collection of flat tetrahedra are glued together on their flat triangular faces. In general, the tetrahedra at an edge will not fit together exactly in flat space, so there will be a deficit angle at that edge giving a measure of the curvature there. In four dimensions, the curvature is restricted to the triangles between the tetrahedra where the four-simplices meet. And so on in higher dimensions. Thus we have a set of flat simplices glued together to approximate a curved space.

There is another way of viewing the scheme that has just been described. Piecewise-flat spaces are interesting in their own right, so in addition to using them as an approximation scheme for some curved "reality," we may also study such spaces for their own sake. It has been argued (e.g., by Friedberg and Lee[2]) that space–time is actually discrete at the smallest scales, so one could also regard curved spaces as approximations to a discrete reality. À chacun ses goûts!

In order for the piecewise-flat spaces to be of any practical use in relativity, beyond ease of visualization, it must be possible to calculate geometric quantities like curvature and volume, and in particular to evaluate the Einstein action of such a space. In Ref. 1 it was shown heuristically that the analog of the Einstein action

$$I=\frac{1}{2}\int R\sqrt{g}d^nx, \tag{1}$$

is given by

$$I_R=\sum_{\text{hinges } i}|\sigma^i|\epsilon_i, \tag{2}$$

where $|\sigma^i|$ is the measure of a hinge σ^i and ϵ_i is the deficit angle there, equal to 2π minus the sum of the dihedral angles between the faces of the simplices meeting at that hinge. Rigorous justification for this formula followed in Ref. 3, where it was shown that it converges to the continuum form of the action, in the sense of measures, provided that certain conditions on the fatness of the simplices are satisfied. Friedberg and Lee[4] approached the problem from the opposite direction, deriving the Regge action from a sequence of continuum spaces approaching a discrete one.

The reason for choosing the building blocks to be simplices is that the geometry of a flat simplex is completely determined by the specification of its edge lengths, so a simplicial space may be described exactly by these lengths without the need for any further variables like angles. This means that the simplest choice of variables for the discrete theory is the edge lengths; clearly the action may be calculated once they are specified and they are also the obvious analogs of the metric tensor, which serves as variable in the continuum theory. There, an elegant way of deriving Einstein's equations is from the principle of stationary action, varying I with respect to the metric. The analog in Regge calculus is to vary I_R with respect to the edge lengths, giving the simplicial equivalent of Einstein's equations:

$$\sum_i \frac{\partial|\sigma^i|}{\partial l_j}\epsilon_i=0, \tag{3}$$

where we have used the result in Ref. 1, that the variation of the angular terms gives zero when summed over each simplex (Schläfli's differential identity).

At first sight, it appears that there is one equation for each variable, promising the possibility of a complete solution for the edge lengths. However the situation is not as simple as that; there are analogs of the Bianchi identities in Regge calculus,[1,5–8] which in the case of flat space provide exact relations between sets of equations, and approximate relations in the nearly flat case, so the equations may not provide sufficient information for a complete solution. In that case there is freedom to specify certain variables, in analogy with the freedom to specify lapse and shift in the $3+1$ version of continuum general relativity.

B. Classical applications

In the ten years after its formulation, Regge calculus was applied almost exclusively to problems in classical relativity, in particular to the time development of simple model universes. (Rather than give a complete list of references here, we refer the reader to the bibliography of Ref. 9, which contains a comprehensive list for the first 20 years.) The basic idea was really $3+1$ in nature: Take a triangulation of a three-dimensional (3D) surface (usually closed but not necessarily so) to represent a hypersurface at a particular moment of time and join its vertices to the corresponding vertices of a second three-dimensional triangulation, representing the same hypersurface at a later time. The edges used to join these vertices are taken to be timelike and the slice of four-dimensional space–time between the two triangulations is then divided into four-simplices by inserting appropriate diagonals. Given the edge lengths on the first 3D triangulation, and specifying the timelike edge lengths, the Regge equations may in principle be solved for the edge lengths on the second 3D triangulation. By repetition of this process, the classical evolution of the initial spacelike surface may be calculated. This sounds simple enough, but unless quite strong assumptions of symmetry are made, the numerical calculation, involving large sets of simultaneous equations for the edge lengths, can be very time consuming and complicated.

Significant progress with this approach was made in the early 1990s when, based on an idea of Sorkin,[10] it was realized that in general, the Regge equations decouple into a collection of much smaller groups. These groups of equations can then be solved in parallel, which means that the computer time required for an equivalent calculation is much less. This parallelizable implicit evolution scheme is described in detail in Ref. 11 and the basic mechanism is as follows. Consider a single vertex in a triangulated three-dimensional spacelike hypersurface and introduce a new vertex "above" this. Connect the new vertex by a "vertical" edge to the chosen vertex, and by "diagonal" edges to all the vertices in the original hypersurface to which the chosen vertex was joined. Each tetrahedron in the original surface now has based on it a four-simplex, with apex at the new vertex. Note that there is one diagonal corresponding to each edge in the original vertex radiating from the chosen vertex. We now use the Regge equations for these edges in the original surface and for the vertical edge; the only unknown edges which these equations involve are the new vertical edge and the diagonal edges, and there is precisely the same number of equations as unknowns. Thus, in principle, we can solve exactly for the unknown edge lengths. (In practice, because of the approximate relationship between the equations from the Bianchi identities, it is often more convenient to ignore some of the equations and instead specify conditions equivalent to the lapse and shift.)

We have described how to evolve vertices one by one in the Sorkin evolution scheme, and the entire hypersurface can be evolved in this way. The method is very general and can be used for a hypersurface with arbitrary topology. However, advancing the vertices one by one will not ordinarily be the most efficient way of evolving a hypersurface. If any two vertices in a hypersurface are not connected by an edge, then they can be evolved to the next surface at the same time without interfering with each other, which is why the method is obviously parallelizable.

C. Some quantum applications

The earliest application of Regge calculus to quantum gravity was in three dimensions[12] and involved $6j$-symbols. This work, and subsequent developments along those lines, will be the subject of Secs. II and III and we shall not discuss it further here.

From the early 1980s onwards, there have been many attempts to formulate a theory of quantum gravity based on Regge calculus, and we shall summarize the salient features of some of those approaches, both analytic and numerical.

The first work on quantum Regge calculus in four dimensions involved using a study of small perturbations about a flat background to relate the discrete variables with their continuum counterparts.[13] The discrete propagator was derived in the Euclidean case and shown to agree with the continuum propagator in the weak field limit. (More details of this calculation will be given in

Sec. IV.) The technique of weak field approximation has proved to be very useful not only for comparisons with the continuum theory but also as a guide in numerical calculations.

The difficulties of analytic calculations in quantum Regge calculus, coupled with the need for a nonperturbative approach and also the availability of sophisticated techniques developed in lattice gauge theories, have combined to stimulate numerical work in quantum gravity, based on Regge calculus. One approach is to start with a Regge lattice for, say, flat space, and allow it to evolve using a Monte Carlo algorithm (see e.g., Refs. 14–16). Random fluctuations are made in the edge lengths and the new configuration is rejected if it increases the action, and accepted with a certain probability if it decreases the action. The system evolves to some equilibrium configuration, about which it makes quantum fluctuations, and expectation values of various operators can be calculated. It is also possible to study the phase diagram and search for phase transitions, the nature of which will determine the vital question of whether or not the theory has a continuum limit. Many of the simulations have involved an action with an extra term, quadratic in the curvature, to avoid problems of convergence of the functional integral; some have included scalar fields coupled to gravity.[17] Recent work by Riedler and collaborators in four dimensions describes evidence for a new continuous phase transition, essential for a continuum limit, at negative gravitational coupling.[18]

The choice of measure in the functional integral is still a matter for controversy, depending both on attitude to simplicial diffeomorphisms and also on the stage at which translation from the continuum to the discrete takes place. The numerical simulations just described mainly use a simple scale invariant measure.[19] Menotti and Peirano have derived an expression for the functional measure in two-dimensional Regge gravity, starting from the DeWitt supermetric, and giving exact expressions for the Fadeev–Popov determinant for both S^2 and $S^1 \times S^1$ topologies (see Ref. 20, and references therein).

A rather different approach to numerical simulations of quantum gravity is that of *dynamical triangulations*. (For a review containing an extensive set of references, see Ref. 21). This also uses Regge lattices and the Regge action, but there are important differences. In the traditional approach, we are effectively integrating over the edge lengths in the functional integral, but in dynamical triangulations, the lattice is taken to be equilateral, with a certain length scale, and the summation is over different triangulations, which are generated by a set of *(k,l) moves*.[22] In two dimensions, there are just two possible moves (and their inverses): the reconnection of vertices in two triangles with a common edge, and the insertion of a vertex and edges in a triangle to divide it into three triangles (2-2 and 1-3 moves). There are straightforward generalizations of these moves to higher dimensions. The moves are ergodic in the sense that any combinatorially equivalent triangulation can be generated by a finite succession of these moves. It is argued that the restriction to equilateral triangulations is a way of avoiding over-counting gauge-related configurations. The approach has been very successful in two dimensions, where there are analytic results with which to compare the calculations. In three and four dimensions, there has been progress in, for example, deriving the crucially important exponential bound on the number of triangulations for a given number of vertices,[23] but there are still open questions on the continuum limit, since the phase transition appears to be first order (see the review by Loll[24]). Recently a Lorentzian version of dynamical triangulations has been formulated in $(1+1)$ dimensions.[25] Numerical simulations have revealed a new universality class for pure gravity, with Haussdorf dimension two.

Discrete gravity has also proved very useful in calculations of the wave function of the universe.[26] According to the Hartle–Hawking prescription, the wave function for a given three-geometry is obtained by a path integral over all four-geometries which have the given three-geometry as a boundary. To calculate such an object in all its glorious generality is impossible, but one can hope to capture the essential features by integrating over those four-geometries which might, for whatever reason, dominate the sum over histories. This has led to the concept of *minisuperspace* models, involving the use of a single four-geometry (or perhaps several). In the continuum theory, the calculation then becomes feasible if the chosen geometry depends only on a small number of parameters, but anything more complicated soon becomes extremely difficult. For this reason, Hartle[27] introduced the idea of summing over *simplicial* four-geometries as an approximation tool in quantum cosmology. Although this is an obvious way of reducing the number of integration variables, there are still technical difficulties: The unboundedness of the Einstein action (which persists in the discrete Regge form) leads to convergence problems for the functional integral, and it is necessary to rotate the integration contour in the complex plane to give a convergent result.[28,29]

In principle, the sum over four-geometries should include not only a sum over metrics but also a sum over manifolds with different topologies. One then runs into the problem of classifying manifolds in four and higher dimensions, which led Hartle[30] to suggest a sum over more general objects than manifolds, *unruly topologies*. Schleich and Witt[31] have explored the possibility of using conifolds, which differ from manifolds at only a finite number of points, and this has been investigated in some simple cases.[32,33] However, a sum over topologies is still very far from implementation.

Yet another area of application of Regge calculus in quantum gravity involves the study of the simplicial supermetric, the metric on the space of three-geometries. Its signature is crucial for determining spacelike surfaces in superspace, which are important in Dirac quantization and in quantum cosmology. In the continuum, there are limited results on the signature and this led to the possibility of investigating it in the discrete case,[34] where the analog is the Lund–Regge supermetric.[35] This supermetric was constructed for some simple manifolds (S^3 and T^3) and its signature calculated. The results agreed with the continuum predictions and also showed that the supermetric can become degenerate. We still do not have a complete understanding of the division of the modes into "vertical" (corresponding to metrics related by diffeomorphisms) and "horizontal" ones.

D. Other approaches to discrete gravity

Of course Regge calculus is not the only way of setting up a theory of discretized general relativity. In this section, we shall describe some alternatives.

One important class of schemes involves treating gravity as a gauge theory. For example, Mannion and Taylor[36] defined a theory of gravity on a fixed hypercubic lattice, and Kaku[37] used a fixed random lattice. However a dynamical lattice seems more appropriate in a theory aiming to describe the quantum fluctuations of space–time, and this was used in much earlier work by Weingarten.[38]

In an approach closely related to Regge calculus, Caselle, D'Adda, and Magnea[39] defined a theory of gravity on the dual lattice, giving both first- and second-order formulations. The action they obtained was a compactified form of the Regge action, involving the sine of the deficit angle. D'Adda and Gionti showed[40] that Regge calculus is a solution of the first-order formulation in the limit of small deficit angles. The action of Caselle, D'Adda, and Magnea was also used by Kawamoto and Nielsen[41] in their version of lattice gravity with fermions.

Immirzi investigated the links between canonical general relativity in the continuum, loop quantum gravity, and spin networks, in an attempt to formulate a quantized version of discrete gravity in the spirit of Regge calculus but ran into problems over hermiticity.[42]

A totally different approach to discrete gravity is 't Hooft's polygon model in $(2+1)$ dimensions.[43] This was introduced as a way to refute Gott's claim of acausality in $(2+1)$ gravity coupled to point particles.[44] 't Hooft's method is to split space–time into the direct product of cosmological time and a Cauchy surface tessellated by flat polygons. The local flatness of space–time in the pure gravity regime and the conelike structure introduced by particles, as in Regge calculus,[45] are expressed in terms of conditions on the edges and vertices of the polygons. A local Lorentz frame is attached to each polygon and two constraints imposed; these are, first, that time runs at the same rate in each polygon (which corresponds to a partial gauge fixing) and, second, that all vertices are trivalent (which is acceptable because higher order vertices can always be split into trivalent vertices connected by edges of zero length). The consequences of these conditions are that the length and velocity of an edge are the same in both polygons to which it belongs, and that the velocity of each edge is orthogonal to it in both frames. These facts result in transition rules for the vertices in the tessellation.

The method for evolving such a space–time is as follows. Initial data (lengths and velocities), subject to consistency conditions, are assigned to the edges on a polygonally tessellated hypersurface. The configuration evolves linearly until an edge collapses to zero length or a vertex crosses another edge. A transition, governed by the vertex conditions, then takes place to another configuration which will, in general, have different numbers of vertices, edges, and polygons. The new data will still satisfy the consistency conditions and the process is then repeated. When there are particles present at the vertices, there are deficit angles proportional to their masses, and the transition rules are modified accordingly.

It is not an easy task to follow the time evolution of a $(2+1)$-dimensional model with particles, even though the system has a finite number of degrees of freedom. 't Hooft did numerical simulations on a small computer, with some unexpected predictions. His big-bang and big-crunch hypotheses were based on the evolution of a Cauchy surface with S^2 or $S^1 \times S^1$ topology, tessellated by a single polygon.[46] It would be interesting to test these predictions for more complex initial configurations, and as a means to this, there has been recent work[47] in which the constraint equations have been interpreted in terms of hyperbolic geometry (see also Ref. 48), and various consistent sets of initial data set up, but the evolution calculations have not yet been completed. Part of the motivation for this work is to compare 't Hooft's method with other approaches to $(2+1)$ gravity, in particular Regge calculus. A $(2+1)$-dimensional code has been set up for Regge space–times and a number of calculations performed,[49] with a view to making detailed comparisons with the 't Hooft method. The ultimate aim is to understand the exact relationship between the two approaches, which seem rather different but have many concepts in common.

Based on the polygon approach, various toy models of $(2+1)$-dimensional gravity have been constructed,[46,50] issues of topology been addressed,[50,51] and particle decay and space–time kinematics investigated.[52] 't Hooft himself has proposed quantized models of $(2+1)$-dimensional space–time,[53] showing that gravitating particles live on a space–time lattice. For an $S^2 \times S^1$ topology, first quantization of Dirac particles is possible. Waelbroeck has suggested a similar approach, using canonical quantization in $(2+1)$ dimensions.[54]

Back at the classical level, Brewin has formulated[55] a discretization of gravity which he feels is closer to the original theory of general relativity. Preliminary calculations are encouraging. For other important work on lattice gravity by Bander, Jevicki and Ninomiya, Khatsymovsky and Lehto, Nielsen and Ninomiya, we refer the reader to the Regge calculus review and bibliography.[9] We emphasize again that this paper is not meant to be an exhaustive review of the subject.

After this rather rapid survey of applications of Regge calculus, and some other approaches to discrete gravity, we shall now concentrate on one particular approach and show how it has led to exciting new developments in the search for a quantum theory of gravity.

II. 6*j*-SYMBOLS IN THREE-DIMENSIONAL QUANTUM GRAVITY

As promised, we now look in detail at the earliest link forged between Regge calculus and quantum gravity, now known as the Ponzano–Regge model.[12] This emerged from a paper on $6j$-symbols and we will first give the background to these.

A. 6*j*-symbols

$6j$-symbols, which are generalizations of the more well-known Clebsch–Gordan coefficients, first arose as tools for the computation of matrix elements in the theory of complex spectra,[56] and are now used routinely by atomic physicists and theoretical chemists in quantum mechanical calculations involving angular momentum. In particular, they relate the possible basis wave functions when three angular momentum are added:

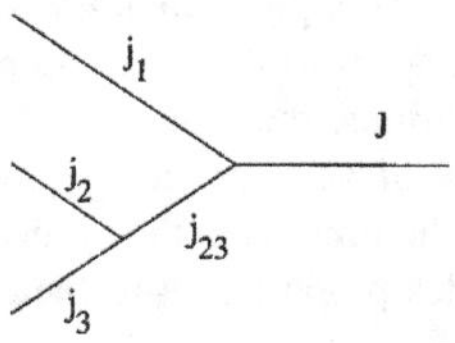

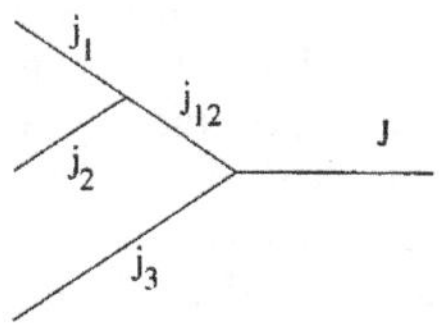

$$|j_1,(j_2,j_3)j_{23},J\rangle=\sum_{j_{12}}\sqrt{(2j_{12}+1)(2j_{23}+1)}(-1)^{j_1+j_2+j_3+J}\begin{Bmatrix} j_1 & j_2 & j_3 \\ j_{12} & j_{23} & J \end{Bmatrix}|(j_1,j_2)j_{12},j_3,J\rangle. \tag{4}$$

An alternative and useful definition involves the *recoupling* diagram:

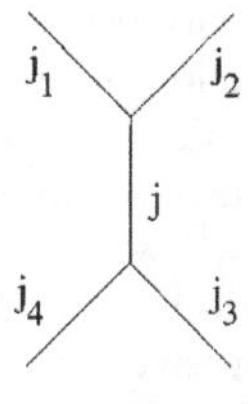

$$=\sum_J \begin{Bmatrix} j_1 & j_2 & j \\ j_3 & j_4 & J \end{Bmatrix}$$

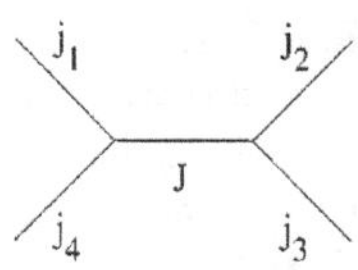

A graphical representation is obtained by associating a $6j$-symbol with a tetrahedron:

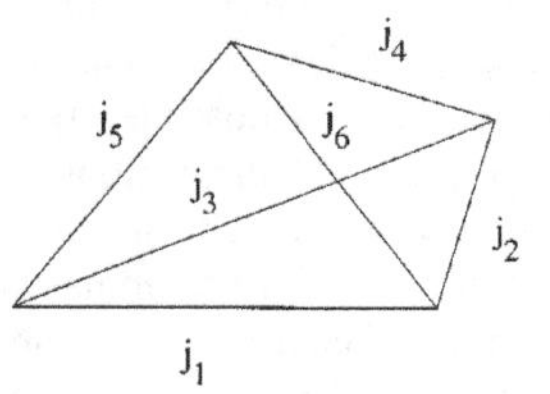

(5)

with the arguments of the $6j$-symbol corresponding to the edge lengths of the tetrahedron. (For technical reasons, it turns out to be more accurate to associate a symbol with arguments *a, b*, ... to a tetrahedron with edge lengths $a+1/2$, $b+1/2$,... .) For the $6j$-symbol to be nonzero, the arguments have to satisfy the analog of the triangle inequalities for each face of the tetrahedron:

$$j_3 \leqslant j_1+j_2, \quad \text{etc.} \tag{6}$$

They can be evaluated from the formula

$$\begin{Bmatrix} a & b & c \\ d & e & f \end{Bmatrix} = \sqrt{\Delta(a,b,c)\Delta(a,e,f)\Delta(c,d,e)\Delta(b,d,f)}$$
$$\times\sum_x (-1)^x(x+1)![(a+b+d+e-x)!(a+c+d+f-x)!(b+c+e+f-x)!$$
$$\times(x-a-b-c)!(x-a-e-f)!(x-c-d-e)!(x-b-d-f)!]^{-1}, \tag{7}$$

where

$$\Delta(a,b,c)=(a+b-c)!(b+c-a)!(c+a-b)![(a+b+c+1)!]^{-1}. \tag{8}$$

These $6j$-symbols are based on the group SU(2), but as we shall see, it is also possible to have q-deformed $6j$-symbols based on quantum groups. For example, define[57]

$$q = \exp(2\pi i/r) \tag{9}$$

and

$$[n] = \frac{(q^{n/2} - q^{-n/2})}{(q^{1/2} - q^{-1/2})}. \tag{10}$$

Then the q-deformed $6j$-symbol for $SU_q(2)$ is defined in the same way as the undeformed one, with n replaced everywhere by $[n]$. Note that $[n] \to n$ as $q \to 1$ and $r \to \infty$.

B. The Ponzano–Regge model

The main purpose of the paper by Ponzano and Regge[12] was to derive asymptotic formulas for classical (i.e., undeformed) $6j$-symbols in the limit when certain arguments became large. The case most relevant to the exposition here is when all six parameters become large. The edge lengths of the corresponding tetrahedron are really related to $j_i\hbar$ and these quantities are kept finite as $j_i \to \infty$ while $\hbar \to 0$ so this process corresponds to the semiclassical limit. This asymptotic behavior is given by

$$\begin{Bmatrix} j_1 & j_2 & j_3 \\ j_4 & j_5 & j_6 \end{Bmatrix} \sim \frac{1}{12\pi V} \cos\left(\sum_i j_i \theta_i + \pi/4 \right), \tag{11}$$

where V is the volume of the terahedron and θ_i is the exterior dihedral angle at edge i (i.e., the angle between the outward normals to the faces meeting there). This was recently proved rigorously by Roberts.[58]

To see the connection between this formula and quantum gravity, consider the following *state sum* defined in Ref. 12. Take a closed two-dimensional surface, triangulate it and divide its interior into tetrahedra, possibly inserting internal vertices. Label the internal edges by x_i and the external ones by l_i. Define

$$S(\{l_i\}) = \sum_{x_i} \prod_{\text{tetrahedra}} \{6j\}(-1)^X \prod_i (2x_i + 1), \tag{12}$$

where the X in the phase factor is a function of the edge lengths.

Although this expression is infinite in many cases, it has some extremely interesting properties. In particular, noting that in the sum over the internal edges, the large values dominate, we can replace the sum by an integral with respect to those edge lengths and use the above-stated asymptotic formula. Then the dominant contribution to the integral comes from the points of stationary phase, which are given by

$$\sum_{\text{tetrahedra } k \text{ meeting on edge } i} (\pi - \theta_i^k) = 2\pi. \tag{13}$$

This means that the sum of the dihedral angles at each edge is 2π, which is precisely the condition for local flatness in a three-dimensional simplicial space. What is more, the state sum is given approximately by

$$S \approx \frac{1}{\sqrt{12\pi}} \int \prod_i dx_i (2x_i + 1)(-1)^X \prod_{\text{tet } k} \frac{1}{\sqrt{V_k}} \cos\left(\sum_{l \in \text{tet } k} j_l \theta_l^k + \frac{\pi}{4} \right). \tag{14}$$

Now this contains a term of the form

$$\int \prod_i dx_i(2x_i+1)\exp\left(i\sum_{\text{edges } l} j_l\left(2\pi-\sum_{\text{tet } k\ni i}(\pi-\theta_l^k)\right)\right)=\int \prod_i dx_i(2x_i+1)\exp\left(i\sum j_l\epsilon_l\right), \tag{15}$$

which looks precisely like a Feynman sum over histories with the Regge calculus action in three dimensions:

$$\int \prod_i d\mu(x_i)\exp(iI_R) \tag{16}$$

with

$$I_R=\sum_l j_l\epsilon_l, \tag{17}$$

where ϵ_l is the deficit angle at edge l and $d\mu(x_i)$ is the measure on the space of edge lengths.

This result was rather puzzling and, although Hasslacher and Perry[59] emphasized the connection between spin networks and simplicial gravity, its significance was not fully appreciated until much later, when a very similar expression was written down in a different context.

C. The Turaev–Viro model

In the late 1980s and early 1990s, mathematicians put a lot of effort into searching for invariants of manifolds, the hope being, at least in part, that such quantities would help with the classification of manifolds. Without being aware of the Ponzano–Regge work, Turaev and Viro[60] defined a state sum for triangulated three-manifolds, which in many aspects was identical to that of Ponzano and Regge. The main differences were that they gave formulas for closed manifolds as well as those with boundary, they showed explicitly that the quantity obtained was independent of triangulation, and finally, they used $6j$-symbols for the quantum group $\mathrm{SL}_q(2)$. Only some of the irreducible representations of this group, the ones with j taking finite values, have suitable algebraic properties, which means that the edge lengths are not summed up to infinite values; j_i can take only integer and half-integer values from the set $(0,\ 1/2,\ 1,\ldots,(r-2)/2)$, with $r\geqslant 3$. A very important consequence of this is that the answer obtained is finite, and so the model appears to be a regularized version of the Ponzano–Regge model.

The obvious question to ask is how the Turaev–Viro state sum is connected to quantum gravity. Witten[61] conjectured that it was equivalent to a Feynman path integral with the Chern–Simons action for $\mathrm{SU}_k(2)\otimes\mathrm{SU}_{-k}(2)$, and this and equivalent results were proved by a number of people.[62–64] To see how this works,[65,66] consider the Chern–Simons Lagrangian for this group product:

$$\mathcal{L}=\frac{k}{4\pi}\int_M \mathrm{Tr}\left(A_+\wedge dA_+ +\frac{2}{3}A_+\wedge A_+\wedge A_+\right)-\frac{k}{4\pi}\int_M \mathrm{Tr}\left(A_-\wedge dA_- +\frac{2}{3}A_-\wedge A_-\wedge A_-\right), \tag{18}$$

where

$$A_\pm=A^a_{i(\pm)}T_a\,dx_i \tag{19}$$

with T_a a basis of the SU(2) Lie algebra. Making the change of variables

$$A^a_{i(\pm)}=\omega_i^a\pm\frac{1}{k}e_i^a, \tag{20}$$

where e_i^a is the dreibein and

$$\omega_i^a = \tfrac{1}{2}\epsilon^{abc}\omega_{ibc}, \tag{21}$$

with ω_{ibc} being the connection two-form, we obtain

$$\int \left(e\wedge R + \frac{\lambda_k}{3} e\wedge e\wedge e \right), \tag{22}$$

which is the Einstein–Hilbert action for gravity with cosmological constant given by

$$\lambda_k = \left(\frac{4\pi}{k} \right)^2. \tag{23}$$

(Note that the k here is equal to $r-2$, where r appears in the definition of q.) By taking the limit as $k\to\infty$, we obtain three-dimensional gravity with zero cosmological constant, i.e., the theory represented by the Ponzano–Regge model. This result is consistent with the fact that the $q\to 1$ limit of the Tuaev–Viro model is the Ponzano–Regge model.

The properties of the Turaev–Viro state sum show that the formalism is an example of a *topological quantum field theory* (see e.g., Ref. 67). This is perfectly appropriate for a theory of gravity in three dimensions where there are no local degrees of freedom. As for the Ponzano–Regge theory, the dominant classical configurations are locally flat (recall that in Chern–Simons gravity, the solutions involve the space of flat connections).

The relationship between the Turaev–Viro invariant and three-dimensional quantum gravity is an extremely important one. It means that in three dimensions, we have in principle a way of calculating the partition function for triangulated manifolds. This has been done for many of the simpler three-manifolds (see Refs. 68 and 69, for example). The Turaev–Viro expression can also be used for calculating topology-changing amplitudes in three-dimensional gravity; the method here is to construct a cobordism between two two-dimensional triangulated surfaces and then use the Turaev–Viro expression for a manifold with boundaries to evaluate the transition probability.[70]

D. Spin networks

The Turaev–Viro expression is not the only method of calculating this particular invariant of three-manifolds. Various other prescriptions have been written down, and one that is worth describing at this stage is that using spin networks. These were invented by Penrose,[71] who wanted to formulate a purely combinatorial approach to space–time. His networks had trivalent vertices and the edges of the graphs were labeled by spins. He developed a method of calculating the value of an arbitrary spin network and was able to show that this led to the usual angles of three-dimensional space.

Penrose's spin networks were later generalized in a number of ways. The edges were labeled by representations of quantum groups and it was necessary to introduce intertwining operators or *intertwiners* at the vertices.[72] In some cases a framing was introduced and the graphs became "ribbon graphs."[73] Kauffman[74] showed how to calculate the Turaev–Viro invariant by taking the graph dual to a triangulation to be a spin network; the edges of the graph inherit the labels of the triangulation edges which they cross. Spin networks have also been introduced into loop quantum gravity,[75] where they are an important calculational tool, for instance in the derivation of the spectrum of the area and volume operators.[75] [Note that Freidel and Krasnov also obtained a discrete spectrum for the volume operator in BF theory (see Sec. III D) by differentiating the Turaev–Viro amplitude with respect to the cosmological constant.[76]] As we shall see, spin networks also play a role in recent attempts at formulations of four-dimensional quantum gravity.

3974 J. Math. Phys., Vol. 41, No. 6, June 2000 T. Regge and R. M. Williams

III. EXTENSIONS TO FOUR DIMENSIONS

After it was realized that the Turaev–Viro state sum provides a finite theory of three-dimensional gravity, the search began for a generalization to four dimensions. Before this is described, we shall stop to ask what we hope to achieve by this. In classical general relativity, there are enormous qualitative differences between gravity in three and four dimensions. In particular, there are gravitons in four dimensions, but not in three, so although it seems reasonable to describe three-dimensional gravity by a topological invariant of a manifold, it seems *likely* that an invariant of a four-manifold might describe only some topological sector of gravity. We shall return to this point later.

The obvious way of setting about extending the three-dimensional model, based on $6j$-symbols, to four dimensions is by using some $3nj$-symbol for a value of n larger than 2. The $3nj$-symbols in the state sum would then be expanded in terms of $6j$-symbols, and the Ponzano–Regge formula for their asymptotic values inserted, the hope being that this would give an expression looking like a path integral with the four-dimensional Regge action. The problem with this is that the asymptotic formula involves the three-dimensional dihedral angles and it is very difficult to relate these to four-dimensional angles. This indicates that a more radical generalization may be needed.

We shall now describe some of the attempts at generalization, leading up to some recent work which seems very promising.

A. The Ooguri model

A source of inspiration for some generalizations of the Ponzano–Regge and Turaev–Viro models was Boulatov's generalized matrix model,[77] which involved a scheme for generating three-dimensional simplicial complexes as terms in a perturbative expansion. The contribution from each simplicial complex was weighted by its Ponzano–Regge or Turaev–Viro invariant, depending on the value of q. Boulatov's model was formulated in a way that it could be extended to higher dimensions, and the four-dimensional case for $q=1$ was worked out by Ooguri.[78]

The essential ingredients in Ooguri's model are the assigning of group variables to the tetrahedra and spin j labels to the triangles in the triangulated four-manifold. The terms in Ooguri's action are of two types: The first is a product of two functions of the group variables, and this represents two glued tetrahedra; the second is a product of five functions and represents the tetrahedra in a four-simplex. A Fourier decomposition is performed in terms of rotation matrices and the group variables are then integrated out, using the standard relationship between rotation matrices and $3j$-symbols, and the invariant Haar measure normalized to unity. The resulting expression has four $3j$-symbols associated with each tetrahedron; these may then be divided between the four-simplices meeting on that tetrahedron, and then each four-simplices ends up with ten $3j$-symbols which can be combined to give a $15j$-symbol. At first sight, it seems odd to associate a $15j$-symbol with a four-simplex, which has only ten triangles labeled by spin values. The way to interpret the symbol is to consider the dual graph, which has ten edges and five four-valent vertices (corresponding to each tetrahedon in the original triangulation). Each of these four-valent vertices can be split into two trivalent ones, and an extra spin label can be assigned to the edge joining them. This splitting sounds rather arbitrary but different splittings are related by $6j$-symbols (see the second diagram in the section on $6j$-symbols) and when all summations are performed, the result is independent of splitting.

The partition function is calculated by integrating the exponential of minus the action over the Fourier coefficients, and the resulting expression is

$$Z=\sum_{C}\frac{1}{N_{\mathrm{symm}}(C)}\lambda^{N_4(C)}Z(C), \tag{24}$$

with $Z(C)$ given by

$$Z(C)=\sum_{j}\prod_{\text{triangles}}(2j_t+1)\prod_{\text{tetrahedra}}\{6j\}\prod_{\text{four-simplices}}\{15j\}. \tag{25}$$

The summation in Z is over simplicial complexes C, with N_{symm} being the rank of the symmetry group of C, and N_4 the number of four-simplices in C. By writing the contributions from all the tetrahedra meeting on a particular triangle in terms of rotation matrices, one can show that the holonomy around any triangle is trivial. This ties up with the proposed link between Ooguri's model and BF theory, as we shall see later.

B. The Archer, Crane–Yetter, and Roberts models

The extension of the Ooguri model to general values of q was worked out by various people. Archer[79] showed how to construct a q-deformed topological quantum field theory in general dimension, giving realizations in three and four dimensions based on the quantum group $U_q(SLN)$, and suggesting that his theory corresponded to BF theory with a cosmological constant.

Crane and Yetter[80] outlined the construction of a q-deformed version of Ooguri's model and recognized its relationship to the work of Roberts,[64] who had defined a four-dimensional generalization of his own "chain-mail" formulation of the Turaev–Viro invariant. Roberts showed that his invariant for a four-manifold M depended on two simple functions of r, one raised to the power of $\sigma(M)$, the signature, and the other to the power of $\chi(M)$, the Euler character.

The result of Roberts was disappointing but instructive for those trying to construct a theory of four-dimensional quantum gravity by this method. Since the models do not give any new information about four-manifolds, it showed that a more radical generalization was needed.

C. The Barrett–Crane model

An important step forward in these generalization attempts has been taken recently with the formulation of the Barrett–Crane model. (Although the details of some aspects of the model, and other related models, have yet to be worked out, we consider the ideas sufficiently important to include in this review.) First came the realization that it made sense to generalize spin networks to *relativistic spin networks* appropriate to four dimensions.[81] The symmetry group SO(3) in three dimensions is replaced by SO(4) in four dimensions, which has spin covering SU(2)⊗SU(2). Barrett and Crane therefore label the triangles by two spin labels rather than one. Thus in a relativistic spin network, the edges (dual to the triangles in the four-complex) carry labels (j_1, j_2) and the vertices (dual to tetrahedra) carry the appropriate intertwiners. Barrett and Crane suggested that the two labels j_1 and j_2 should be equal to satisfy the constraints at the vertices, and Reisenberger[82] showed that this solution is unique. Thus the Barrett–Crane model is a constrained doubling of the earlier attempts described in Secs. III A and III B which can thus be regarded as just describing the self-dual section of gravity.

We now describe the Barrett–Crane model in a little more detail. Consider a single four-simplex, draw its dual graph, and then split the vertices as described for the Ooguri model. The first expression written down by Barrett and Crane for the amplitude of a four-simplex was of the form

$$I_1=\sum_{\text{extra edges}} c_j\{15j\}^2, \tag{26}$$

where c_j is a weight factor and the $15j$-symbol is squared because of the (j, j) labeling on each edge of the dual graph. It turned out to be very difficult to evaluate the asymptotic value of this expression, so Barrett and Crane tried a second approach.

Label the five tetrahedra in a four-simplex by k; the spin label on the triangle where tetrahedra k and l meet is then denoted by j_{kl}. The matrix representing the element $g\in \mathrm{SU}(2)$ in the irreducible representation of spin j_{kl} is denoted by $\rho_{kl}(g)$. Variables $h_k \in \mathrm{SU}(2)$ are assigned to the tetrahedra and the invariant I_2 (the second Barrett–Crane model) is obtained by integrating a function of these variables over each copy of SU(2):

$$I_2=(-1)^{\Sigma_{k<l}2j_{kl}}\int_{h\in \mathrm{SU}(2)^5}\prod_{k<l}\mathrm{Tr}\,\rho_{kl}(h_k h_l^{-1}). \tag{27}$$

The measure used is the Haar measure normalized to unity.

The next step is to relate this expression to the geometry of the four-simplex.[83] Using the fact that SU(2) is isomorhic to S^3, and embedding S^3 in R^4, we can regard the element $h_k\in \mathrm{SU}(2)$ as a unit vector in R^4, normal to the three-dimensional hyperplane in which tetrahedron k lies. Then according to a well-known formula in representation theory,

$$\mathrm{Tr}\ \rho(h_k h_l^{-1})=\frac{\sin(2j+1)\phi}{\sin\phi}, \tag{28}$$

where $\cos\phi=h_k\cdot h_l$, i.e., ϕ is the angle between the normals and thus the exterior angle between the two hyperplanes.

Note that the five hyperplanes define a four-simplex up to translation and an overall scale. Thus integration over the elements h_k may be interpreted as integration over all possible four-simplices.

Recalling the equivalence of the asymptotic value of the Ponzano–Regge model to a path integral with the three-dimensional Regge calculus action, we now look for a similar result here.[84] We write $\sin(2j+1)\phi$ in terms of exponentials and, for large j, use the method of stationary phase to find the asymptotic value of the integral. Setting $\epsilon_{kl}=\pm 1$, we write I_2 as

$$I_2=\frac{(-1)\Sigma_{k<l}2j_{kl}}{(2i)^{10}}\sum_{\epsilon_{kl}=\pm 1}\prod_{h\in \mathrm{SU}(2)^5}\frac{\epsilon_{kl}}{\sin\phi_{kl}}\exp\left(i\sum_{s\leqslant l}\epsilon_{kl}(2j_{kl}+1)\phi_{kl}\right), \tag{29}$$

which makes it clear that we need the stationary points of

$$I=\sum_{k\leqslant l}\epsilon_{kl}(2j_{kl}+1)\phi_{kl}. \tag{30}$$

Now the ϕ_{kl}'s for a four-simplex are not independent variables; as is shown in the original formulation of Regge calculus,[1] their variations are related by

$$\sum_{k<l}A_{kl}d\phi_{kl}=0. \tag{31}$$

Adding this constraint to I with a Lagrange multiplier μ, we find that for each triangle,

$$\epsilon_{kl}(2j_{kl}+1)=\mu A_{kl}. \tag{32}$$

The overall scale can then be fixed by taking $\mu=\pm 1$.

What has been established is that for a stationary phase point, first, the angles ϕ_{kl} are those of a geometric four-simplex with triangle areas

$$A_{kl}=2j_{kl}+1, \tag{33}$$

and, second, the integrand is $\exp(i\mu I_R)$, with

$$I_R=\sum_{\text{triangles } kl}A_{kl}\phi_{kl}, \tag{34}$$

the Regge calculus version of the Einstein action for a four-simplex, with $\mu = \pm 1$.

The formulation of this model is by no means complete. The next step is to sum over four-simplices, which is likely to be more difficult than for the first model, where the extra labels on tetrahedra could provide links between neighboring four-simplices. The resulting expression will need to be regularized by passing to representations of the quantum group $U_q(SL2)$, as in the transition from the Ponzano–Regge state sum to that of Turaev and Viro. This analogy is not precise because the Barrett–Crane amplitude is *not* independent of triangulation. The covariance lost here may perhaps be restored by summing over triangulations using a generalized matrix model approach, as suggested by De Pietri *et al.*[85] (Note that these authors refer to what we have called the "second Barrett–Crane model" as their "first version.")

We shall return to the interpretation of this model in Sec. III D, but first note that the formulation described so far is Euclidean. There have been Lorentzian models proposed recently: in $(2+1)$ dimensions, Freidel[86] has set up a version in which SU(2) is replaced by SL(2,R), for which both discrete and continuous representations are used. This results in a model in which time is discrete and space continuous. The partition function requires summation over causal structures, which obviously has no analog in the Euclidean case. The $6j$-symbols for the discrete series representation of SL(2,R) were defined first by Davids,[87] who also obtained the analogous Ponzano–Regge formula, which here involves $\exp(iI_L)$, where I_L is the Lorentzian Regge action. In $(3+1)$ dimensions, Barrett and Crane[88] have proposed versions based on the classical Lorentz group and on the quantum Lorentz algebra, but the second of these is still at a preliminary stage.

D. Relation to BF theory

This is not the place for a review of BF theory, but let us briefly mention its relevant properties. It is a gauge theory which can be defined in any dimension and is "background free" in the sense that no pre-existing metric or other geometrical structure on space–time is needed. It is a theory with no local degrees of freedom.

The action for BF theory in four dimensions is

$$I_{\mathrm{BF}} = \int_M \mathrm{Tr}(B \wedge F), \tag{35}$$

where B is a Lie algebra-valued two-form, and $F = dA + A \wedge A$, with A the connection one-form. It gives rise to the constraint $F=0$, which means that the connection A is flat. This ties up with the trivial holonomy around triangles in the Ooguri model. The other constraint, $d_A B = 0$, is the statement of a particular type of gauge symmetry in BF theory.

To understand the relationship between general relativity and BF theory in four dimensions,[89] consider the Palatini formulation of general relativity, which has action

$$I_P = \int_M \mathrm{Tr}(e \wedge e \wedge F), \tag{36}$$

with e a one-form on the manifold M, and F defined in terms of the connection as for BF theories. It is immediately apparent that there is a relationship between this Palatini formulation and BF theory with B constrained to be of the form $e \wedge e$. There is a subtle difference between the equations of motion derived from the two actions: For general relativity, we have

$$e \wedge F = 0, \quad d_A B = 0 \tag{37}$$

as compared with the BF equations

$$F = 0, \quad d_A B = 0. \tag{38}$$

Thus the equations of general relativity are weaker here than those for BF theory, which, heuristically, is why general relativity in four dimensions is more general than a topological theory. We see that general relativity in four dimensions is equivalent to BF theory with an extra constraint ($B = e\wedge e$) (giving rise to the paradoxical statement that adding a constraint produces a less restricted theory!)

We see now a further justification of why, in the Barrett–Crane model, the two spin labels on each triangle should be equal [i.e., we see the parallel between (j, j) and $e\wedge e$]. Thus the constraint which Reisenberger[82] derived may be interpreted as equivalent to the constraint which relates BF theory to general relativity in four dimensions.

Reisenberger[90] has explored further the relationship between the Barrett–Crane model and continuum theories, showing that the model corresponds to an SO(4) BF theory in which the right- and left-handed areas, defined by the self-dual and anti-self-dual components of B, are constrained to be equal.

Before considering an extension of BF theory in four dimensions, let us return to the case of three dimensions. It can be shown that three-dimensional general relativity without matter is a special case of BF theory, where the equations of motion give simply that the connection is torsion free and flat. Adding an extra term to the BF Lagrangian has a very interesting effect. Starting from the modified action

$$I'_{\mathrm{BF}} = \int_M \mathrm{Tr}\left(B\wedge F + \frac{\lambda}{6} B\wedge B\wedge B \right) \tag{39}$$

and making the transformation

$$A_{\pm} = A \pm \sqrt{\lambda} B, \tag{40}$$

we can show that I'_{BF} is equal to the difference of the two Chern–Simons actions as in Sec. II C. It was shown there that this was equivalent to three-dimensional general relativity with a cosmological constant λ related to the deformation parameter q, which gives a finite theory of quantum gravity in that dimension.[91] Thus a role of the cosmological constant is to regularize the theory.

In four dimensions, the extra term that we need seems to be slightly different. The proposed modified action is

$$I''_{\mathrm{BF}} = \int_M \mathrm{Tr}\left(B\wedge F + \frac{\lambda}{12} B\wedge B \right). \tag{41}$$

The form of this extra term was first suggested by Archer,[79] whose contribution is described earlier. It has been discussed more recently by Baez,[89,92] who gives a very comprehensive discussion of BF theory and the discrete models of quantum gravity in three and four dimensions. (Reference 89 is recommended strongly for fuller details of these issues.) Imposing the constraint $B = e\wedge e$ as before, the action becomes that for the Palatini formulation of general relativity with cosmological constant,

$$I'_P = \int_M \mathrm{Tr}\left(e\wedge e\wedge F + \frac{\lambda}{12} e\wedge e\wedge e\wedge e \right). \tag{42}$$

This suggests the possibility of finding a regularized version of four-dimensional quantum gravity by constructing a q-deformed version of the Barrett–Crane model, satisfying the relationship

$$\lambda \to 0 \quad as \quad q \to 1. \tag{43}$$

Another possible (and related) way forward is through spin foam models, as described briefly in Sec. III E.

E. Spin foam

As mentioned in the section on three-dimensional gravity, spin networks have played an important role in calculations of invariants of three-manifolds, and in loop quantum gravity, where they provide a gauge-invariant basis of states.[75,93] If we wish to describe *space–time* by this type of method, we need, as we have already remarked, an extension of the concept of spin networks. An alternative to the idea of relativistic spin networks is provided by what has been called *spin foam*,[94] because one can think of a spin foam as a soap film connecting two spin networks at different times. ''Sums over surfaces'' formulations of loop quantum gravity have been given by Reisenberger and Rovelli,[95] and Iwaski[96] has formulated the Ponzano–Regge model in terms of surfaces. Turaev and Viro[60] formulated their theory not only in terms of a triangulation of the three-manifold but also in terms of simple 2-polyhedra forming a two-complex embedded in the manifold, and we can interpret this second method as the first example of a spin foam model! The relationship between the evolution of spin networks and the approach using triangulated manifolds has been explored and illuminated by Markopoulou.[97]

The theory of spin foam is a way of formalizing the calculation of the partition function in BF theory by triangulating manifolds. Recall that a spin network is a graph with edges labeled by irreducible representations and vertices by interwiners. Imagine moving such an object through space, or rather *space–time*, so that it traces out a two-dimensional surface, a generic slice through which would be a spin network; this, heuristically, is what we mean by a spin foam. It is a two-complex, the faces of which are labeled by irreducible representations and the edges by interwiners. The dual triangulation of a manifold is an example of such an object.

Baez[89] has outlined how to calculate transition amplitudes in BF theory using sums over spin foams, and the derivation of the spin foam model from the classical action principle based on BF theory has been discussed by Freidel and Krasnov.[98] It has already been shown[99] that a particular type of spin network may be evaluated as a Feynman graph, and the idea in the evaluation of spin foam sums is to use Feynman's sum over histories approach, with BF theory playing the role of the free theory and spin foams as two-dimensional analogos of Feynman diagrams. These techniques have produced agreement with the lowest order terms in the known state sum models.[98] Markopoulou and Smolin[100] have defined a model of the time evolution of spin networks based on local causality rules, which are equivalent to those for spin foams.

Recently Smolin[101] has suggested a connection between evolving spin networks, spin foam, and such approaches related to loop quantum gravity, and string theory, where there are clearly intuitive similarities in the evolution of strings and membranes. Any precise equivalence still needs to be worked out, but Smolin's suggestion is typical of recent ideas in which a number of apparently unrelated approaches to quantum gravity seem at last to be coming together.

IV. AREA REGGE CALCULUS

It seems that those attempts at formulating a theory of quantum gravity in four dimensions described in Sec. III all need one ingredient to be at all successful; this is the assignment of labeling to the triangles instead of (or possibly as well as) the edges. (This fits in with work by Birmingham and Rakowski,[102] who constructed state sum models based on Z_p for four-dimensional triangulated manifolds. When the colorings from Z_p were assigned only to the edges, the invariant depended only on the three-dimensional boundary manifold, but when colorings were assigned also to the triangles, the invariant depended on the four-dimensional structure.) Even the spin foam description fits into this pattern when one considers the triangulation to which it is dual. By considering the asymptotic value of the amplitude of a four-simplex, we have seen that in this case, it appears to be related to the path integral with the Regge calculus action but with the triangle areas playing the most important role, rather than the edge lengths.

A. Problems with the basic idea

The idea that, in four dimensions, the triangle areas could be regarded as the basic variables in a modified form of Regge calculus was first suggested by Rovelli[103] and the possibility was discussed in some detail in Ref. 104. In this section, we shall consider the advantages and disadvantages of the approach, and report on some progress in understanding the relationship between the two types of variable.

A four-simplex not only has ten edges, it also has ten triangles. Thus at first sight, the change from edge lengths to triangles areas as basic variables looks very straightforward, but there are actually a number of problems.[104]

Consider first a single four-simplex. It is simple to express the triangle areas in terms of the edge lengths. However, to express the Regge action in terms of the new variables, we need to invert the relationship between areas and edge lengths to be able to calculate the deficit angles. Unfortunately the Jacobian is singular in cases where a number of triangles are right-angled and there is not necessarily a unique set of edge lengths corresponding to a given set of areas.[105,106] This means that, right from the start, certain regions in the space of edge lengths must be avoided.

Second, for a collection of four-simplices joined together, there will not in general be equal numbers of edges and triangles so there may be ambiguity about which is the correct number of variables.

Third, by considering two four-simplices meeting on a tetrahedron with all triangle areas assigned, we can envisage the following bizarre situation. Solve for the edge lengths of one of the four-simplices in terms of its triangle areas. Repeat this for the other four-simplex. It is possible that the edge lengths of the common tetrahedron will differ according to the four-simplex where the calculation was done (see Ref. 104 for an example). Clearly there are difficulties in interpreting the edge lengths as real physical quantities in the usual sense.

In this section, we shall now discuss possible theories in terms of equations of motion and then investigate the dynamical content of area Regge calculus by studying the weak-field expansion about a flat background in terms of variations in the areas.

B. Equations of motion

The counting of degrees of freedom in a discrete theory is never completely straightforward. In a simplicial theory, the usual argument is that in n dimensions, an n-simplex has $n(n+1)/2$ edges, which corresponds to the number of independent degrees of freedom of the metric tensor in n dimensions. If one thinks of these variables as being at some chosen point in each simplex, the counting becomes somewhat less clear when one realizes that each of the edges is shared by a number of other simplices, so the number of variables per point is quite obscure.

Given this ambiguity, we can take two attitudes to the counting problem in area Regge calculus. Either we can take the areas as the fundamental variables, worrying about the different numbers of edge lengths only inasfar as we need them to calculate deficit angles or volumes, or we can regard some of the areas as redundant variables and aim to reduce their number to the number of edge lengths in the simplicial complex.

In the theory where the areas are taken seriously as variables (which is our principal interest here since we aim thereby to understand the models described as four-dimensional generalizations of the Turaev–Viro theory), we concentrate on the restricted class of metrics where the Jacobian is nonsingular. Then the hyperdihedral angles are well defined and the Regge action may be written as

$$I_R(A_s)=\sum_t A_t\epsilon_t(A_s), \tag{44}$$

where the sum is over triangles t and ϵ_t is the deficit angle at triangle t. Variation of the action with respect to the area A_u, use of the chain rule, an interchange of the orders of summation, and use of the Regge identity[1] leads to

$$\epsilon_u=0 \quad \text{for all } u. \tag{45}$$

For details, see Ref. 104. Since all deficit angles vanish, the space is locally flat; the holonomy around any triangle is trivial. This agrees with Ooguri's state sum model for BF theory.[78] The interpretation of this result is not obvious and the investigation of such spaces using parallel transport is under way.

The other possibility, that of regarding some of the areas as redundant variables, has been investigated by Mäkelä.[107] Clearly in order to recover the conventional view of simplicial gravity where the edge lengths are real physical quantities, it is necessary to impose the condition that a given edge has the same length in whichever four-simplex that length is calculated. This leads to a large number of constraints: For each edge, there is a constraint for each pair of four-simplices meeting there. For a simplicial complex with N_1 edges and N_2 triangles, a total of $N_2 - N_1$ of these constraints will be independent, but it is not easy to give any general rule for picking out which these are. (An *ad hoc* rule has been formulated for a particular model and it is likely that there is some group-theoretic basis for the rule).[108] Mäkelä has shown that if the variations of the constraints are added in with Lagrange multipliers to the variation of the Regge action expressed in area variables, then the usual Regge calculus equations of motion are recovered.

C. Dynamics

Restricting our attention now to the area variable theory without constraints, we investigate its dynamical content by performing a weak field expansion about a flat background.[109] This is in analogy with the weak field expansion for edge length variables,[13] which we now describe briefly.

In the original calculation, a four-dimensional hypercubic lattice is divided into simplices by drawing in various diagonals, giving 15 edges per vertex. Small variations of the edge lengths about their flat space values are made by setting

$$l_i = l_i^{(0)}(1+\delta_i) \tag{46}$$

with $\delta_i \ll 1$. The second variation of the Regge action (the first nonvanishing term) is evaluated as a quadratic expression in the δ's, written as

$$\delta^2 S = \delta_i M_{ij} \delta_j , \tag{47}$$

with M_{ij} a sparse infinite dimensional matrix. A Fourier transform is then performed by relating δ in the n direction and based at the lattice point (i, j, k, l) steps in the (1, 2, 4, 8) directions from the origin (see Ref. 13 for details of the binary notation) to the corresponding δ at the origin by

$$\delta_n^{(i,j,k,l)} = \omega_1^i \omega_2^j \omega_4^k \omega_8^l \delta_n^{(0)} \tag{48}$$

with $\omega_\mu = \exp(2\pi i/n_\mu)$, where n_μ is the period in the μ direction. Acting on periodic modes, M reduces to a block diagonal matrix with 15×15 dimensional blocks, M_ω. This matrix M_ω has four zero modes, corresponding to periodic translations of points of the lattice, and a fifth zero mode corresponding to periodic fluctuations of the hyperbody diagonal. Block diagonalizing M_ω decouples four further modes; they enter without ω's and so do not contribute to the dynamics at all. Their equations of motion constrain them to vanish. We see from this that an apparent mismatch in the number of components (15 per vertex) is corrected by the dynamics of the theory, leaving ten degrees of freedom per vertex, as would be expected from the continuum theory. (The zero modes correspond of course to gauge fluctuations.)

We now perform the analogous calculation with area variables. In this case, it is necessary to use a "distorted" hypercubic lattice because the original one contains many right angles which lead to vanishing of the Jacobian when transforming between areas and edge lengths. This is obtained by squeezing each unit hypercube along its hyperbody diagonal until it has length 1 in lattice units, like the edges originally along the coordinate axes. The face and body diagonals then all have length $\sqrt{(3/2)}$. Small variations of these edge lengths about their flat space values are then made and the second variation of the action within each four-simplex calculated. These variations in edge lengths induce changes in the triangle areas represented by

$$A_i = A_i^{(0)}(1+\Delta_i) \tag{49}$$

with $\Delta_i \ll 1$. Within each four-simplex, the expressions for the Δ_i's in terms of the δ_i's are inverted (uniquely) and the second variation of the action written in terms of the Δ_i's. Adding together the contributions from all four-simplices gives

$$\delta^2 S = \Delta_i N_{ij} \Delta_j , \tag{50}$$

with N_{ij} again a sparse infinite dimensional matrix. A Fourier transform is then performed as in the edge-length variable case, and N reduces to a block diagonal matrix with 50×50 dimensional blocks N_ω (note that there are 50 triangles based at each vertex). The size of N_ω makes it necessary to investigate the modes numerically, and, somewhat contrary to our original expectations, it turns out that the number of dynamical modes is exactly the same as in the edge length case. There are again four zero modes, corresponding to periodic fluctuations of the lattice, and six further modes scaling with k^2, where k is the momentum in the Fourier transform. The remaining 40 modes enter nondynamically (they are massive and do not scale with momentum) and are constrained to vanish by their equations of motion.

Thus the theory with area variables is equivalent to the edge length variable theory from the point of view of dynamical content. This is very encouraging and gives impetus to the search for the exact correspondence between the variables in models like that of Barrett and Crane, the variables of Regge calculus and ultimately the variables of conventional general relativity. That search continues.

ACKNOWLEDGMENTS

The authors thank John Barrett and Radu Ionicioiu for help with the preparation of this paper. The work was supported in part by the UK Particle Physics and Astronomy Research Council.

[1] T. Regge, Nuovo Cimento **19**, 558 (1961).
[2] R. Friedberg and T. D. Lee, Nucl. Phys. B **225**, 1 (1983).
[3] J. Cheeger, W. Müller, and R. Schrader, Commun. Math. Phys. **92**, 405 (1984).
[4] R. Friedberg and T. D. Lee, Nucl. Phys. B **242**, 145 (1984).
[5] M. Roček and R. M. Williams, in *Quantum Structure of Space and Time*, edited by M. J. Duff and C. J. Isham (Cambridge University Press, Cambridge, 1982).
[6] W. A. Miller, in *Dynamical Spacetimes and Numerical Relativity*, edited by J. Centrella (Cambridge University Press, Cambridge, 1986).
[7] L. Brewin, Class. Quantum Grav. **5**, 839 (1988).
[8] P. A. Tuckey, Ph.D. thesis, University of Cambridge, 1988.
[9] R. M. Williams and P. A. Tuckey, Class. Quantum Grav. **9**, 1409 (1992).
[10] R. D. Sorkin, Phys. Rev. D **12**, 385 (1975).
[11] J. W. Barrett, M. Galassi, W. A. Miller, R. D. Sorkin, P. A. Tuckey, and R. M. Williams, Int. J. Theor. Phys. **36**, 809 (1997).
[12] G. Ponzano and T. Regge, in *Spectroscopic and Group Theoretical Methods in Physics*, edited by F. Block, S. G. Cohen, A. DeShalit, S. Sambursky, and I. Talmi (North-Holland, Amsterdam, 1968).
[13] M. Roček and R. M. Williams, Phys. Lett. **104B**, 31 (1981); Z. Phys. C **21**, 371 (1984).
[14] H. W. Hamber, in *Critical Phenomena, Random Systems, Gauge Theories*, edited by K. Osterwalder and R. Stora (North-Holland, Amsterdam, 1986); Nucl. Phys. B (Proc. Suppl.) **25**, 150 (1991).
[15] H. W. Hamber and R. M. Williams, Nucl. Phys. B **248**, 392 (1984); Phys. Lett. **157B**, 368 (1985); Nucl. Phys. B **267**, 482 (1986); *ibid.* **269**, 712 (1986).
[16] B. Berg, Phys. Rev. Lett. **55**, 904 (1985); Phys. Lett. B **176**, 39 (1986).
[17] H. W. Hamber and R. M. Williams, Nucl. Phys. B **415**, 463 (1994).
[18] J. Riedler, W. Beirl, E. Bittner, A. Hauke, P. Homolka, and H. Markum, Class. Quantum Grav. **16**, 1163 (1999).
[19] H. W. Hamber and R. M. Williams, Phys. Rev. D **59**, 06014 (1999).
[20] P. Menotti and R. P. Peirano, Nucl. Phys. B (Proc. Suppl.) **57**, 82 (1997).
[21] J. Ambjørn, M. Carfora, and A. Marzuoli, *The Geometry of Dynamical Triangulations*, Lecture Notes in Physics (Springer, Berlin, 1997).
[22] U. Pachner, Eur. J. Comb. **12**, 129 (1991).
[23] M. Carfora and A. Marzuoli, J. Math. Phys. **36**, 6353 (1995).
[24] R. Loll, gr-qc/9805049.
[25] J. Ambjørn and R. Loll, Nucl. Phys. B **536**, 407 (1999).
[26] J. B. Hartle and S. W. Hawking, Phys. Rev. D **28**, 2960 (1983).

[27] J. B. Hartle, J. Math. Phys. **26**, 804 (1985).
[28] J. B. Hartle, J. Math. Phys. **30**, 452 (1989).
[29] J. Louko and P. A. Tuckey, Class. Quantum Grav. **9**, 41 (1991).
[30] J. B. Hartle, Class. Quantum Grav. **2**, 707 (1985).
[31] K. Schleich and D. M. Witt, Nucl. Phys. B **402**, 411 (1983); **402**, 469 (1983).
[32] D. Birmingham, Phys. Rev. D **52**, 5760 (1995).
[33] C. L. B. Correia da Silva and R. M. Williams, Class. Quantum Grav. **16**, 2197 (1999); **16**, 2681 (1999).
[34] J. B. Hartle, W. A. Miller, and R. M. Williams, Class. Quantum Grav. **14**, 2137 (1997).
[35] F. Lund and T. Regge (unpublished).
[36] C. L. Mannion and J. G. Taylor, Phys. Lett. **100B**, 261 (1981).
[37] M. Kaku, in *Quantum Field Theory and Quantum Statistics*, edited by I. A. Batalin, C. J. Isham, and G. A. Vilkovisky (Adam Hilger, Bristol, 1987).
[38] D. Weingarten, J. Math. Phys. **18**, 165 (1977); Nucl. Phys. B **210**, 229 (1982).
[39] M. Caselle, A. D'Adda, and L. Magnea, Phys. Lett. B **232**, 457 (1989).
[40] G. Gionti, Ph.D. thesis, SISSA, Trieste, 1989.
[41] N. Kawamoto and H. B. Nielsen (unpublished).
[42] G. Immirzi, Nucl. Phys. B (Proc. Suppl.) **57**, 65 (1997).
[43] G. 't Hooft, Class. Quantum Grav. **9**, 1335 (1992).
[44] J. R. Gott, Phys. Rev. Lett. **66**, 1126 (1991).
[45] M. Roček and R. M. Williams, Class. Quantum Grav. **2**, 701 (1985).
[46] G. 't Hooft, Class. Quantum Grav. **10**, 1023 (1993).
[47] H. R. Hollmann and R. M. Williams, Class. Quantum Grav. **16**, 1503 (1999).
[48] G. 't Hooft, Class. Quantum Grav. **10**, S79 (1993).
[49] A. P. Gentle and R. M. Williams (unpublished).
[50] M. Welling, Class. Quantum Grav. **14**, 929 (1997).
[51] R. Franzosi and E. Guadagnini, Class. Quantum Grav. **13**, 433 (1996).
[52] R. Franzosi and E. Guadagnini, Nucl. Phys. B **450**, 327 (1995).
[53] G. 't Hooft, Class. Quantum Grav. **10**, 1653 (1993); **13**, 1023 (1996).
[54] H. Waelbroeck, Phys. Rev. D **50**, 4982 (1994).
[55] L. Brewin, Class. Quantum Grav. **15**, 2427 (1998).
[56] G. Racah, Phys. Rev. **61**, 186 (1942).
[57] A. N. Kirillov and N. Yu. Reshetikhin, in *Infinite Dimensional Lie Algebras and Groups*, edited by V. G. Kac (World Scientific, Singapore, 1989).
[58] J. D. Roberts, Geometry Topology **3**, 21 (1999).
[59] B. Hasslacher and M. J. Perry, Phys. Lett. **103B**, 21 (1981).
[60] V. G. Turaev and O. Y. Viro, Topology **31**, 865 (1992).
[61] E. Witten (private communication).
[62] V. G. Turaev, Lect. Notes Math. **1510**, 363 (1992); (unpublished).
[63] K. Walker (unpublished).
[64] J. D. Roberts, Topology **34**, 771 (1995).
[65] H. Ooguri and N. Sasakura, Mod. Phys. Lett. A **6**, 3591 (1991).
[66] R. M. Williams, Int. J. Mod. Phys. B **6**, 2097 (1992).
[67] D. Birmingham, M. Blau, M. Rakowski, and G. Thompson, Phys. Rep. **209**, 129 (1991).
[68] L. H. Kauffman and S. Lins, Manuscr. Math. **72**, 81 (1991).
[69] R. Ionicioiu and R. M. Williams, Class. Quantum Grav. **15**, 3469 (1998).
[70] R. Ionicioiu, Class. Quantum Grav. **15**, 1885 (1998).
[71] R. Penrose, in *Quantum Theory and Beyond*, edited by T. Bastin (Cambridge University Press, Cambridge, 1971).
[72] J. W. Barrett and B. W. Westbury, Trans. Am. Math. Soc. **348**, 3997 (1996).
[73] N. Yu. Reshetikhin and V. G. Turaev, Commun. Math. Phys. **127**, 1 (1990).
[74] L. H. Kauffman, *Knots and Physics* (World Scientific, Singapore, 1991).
[75] C. Rovelli and L. Smolin, Phys. Rev. D **52**, 5743 (1995): Nucl. Phys. B **442**, 593 (1995).
[76] L. Freidel and K. Krasnov, Class. Quantum Grav. **16**, 351 (1999).
[77] D. V. Boulatov, Mod. Phys. Lett. A **7**, 1629 (1992).
[78] H. Ooguri, Mod. Phys. Lett. A **7**, 2799 (1992).
[79] F. J. Archer, Ph.D. thesis, University of Cambridge, 1993; J. Geom. Phys. **16**, 39 (1995).
[80] L. Crane and D. Yetter, in *Quantum Topology*, edited by L. H. Kauffman and R. Baadhio (World Scientific, Singapore, 1993).
[81] J. W. Barrett and L. Crane, J. Math. Phys. **39**, 3296 (1998).
[82] M. Reisenberger, J. Math. Phys. **40**, 2046 (1999).
[83] J. W. Barrett, Adv. Theor. Math. Phys. **2**, 593 (1998).
[84] J. W. Barrett and R. M. Williams, Adv. Theor. Math. Phys. **3**, 1 (1999).
[85] R. De Pietri, L. Freidel, K. Krasnov, and C. Rovelli, hep-th/9907154.
[86] L. Freidel, Lecture at The Third International Conference on Constrained Dynamics and Quantum Gravity, Villasimius, Italy, September 1999.
[87] S. Davids, gr-qc/9807061.
[88] J. W. Barrett and L. Crane, gr-qc/9904025.

[89] J. Baez, gr-qc/9905087; in *Geometry and Quantum Physics*, edited by H. Gausterer and H. Grosse, Lecture Notes in Physics (Springer, Berlin, in press).
[90] M. Reisenberger, Class. Quantum Grav. **16**, 1357 (1999).
[91] J. W. Barrett, J. Math. Phys. **36**, 6161 (1995).
[92] J. Baez, Lett. Math. Phys. **38**, 129 (1996).
[93] J. Baez, Adv. Math. **117**, 253 (1996).
[94] J. Baez, Class. Quantum Grav. **15**, 1827 (1998).
[95] M. Reisenberger and C. Rovelli, Phys. Rev. D **56**, 3490 (1997).
[96] J. Iwasaki, J. Math. Phys. **36**, 6288 (1995).
[97] F. Markopoulou, gr-qc/9704013.
[98] L. Freidel and K. Krasnov, Adv. Theor. Math. Phys. **2**, 1221 (1998).
[99] L. Freidel and K. Krasnov, hep-th/9903192.
[100] F. Markopoulou and L. Smolin, Nucl. Phys. B **508**, 409 (1997); Phys. Rev. D **58**, 084032 (1998).
[101] L. Smolin, hep-th/9801022; hep-th/9808192.
[102] D. Birmingham and M. Rakowski, Int. J. Mod. Phys. A **10**, 1329 (1995); hep-th/9405108.
[103] C. Rovelli, Phys. Rev. D **48**, 2702 (1993).
[104] J. W. Barrett, M. Roček, and R. M. Williams, Class. Quantum Grav. **16**, 1373 (1999).
[105] J. W. Barrett, Class. Quantum Grav. **11**, 2723 (1994).
[106] P. A. Tuckey (private communication).
[107] J. Mäkelä, gr-qc/9801022.
[108] J. Mäkelä and R. M. Williams (unpublished).
[109] M. Roček and R. M. Williams (unpublished).

PART IV

Hamiltonian Formulation of General Relativity

ANNALS OF PHYSICS 88, 286–318 (1974)

Role of Surface Integrals in the Hamiltonian Formulation of General Relativity*

TULLIO REGGE

Institute for Advanced Study, Princeton, New Jersey 08540

AND

CLAUDIO TEITELBOIM

Joseph Henry Laboratories, Princeton University, Princeton, New Jersey 08540

Received April 8, 1974

It is shown that if the phase space of general relativity is defined so as to contain the trajectories representing solutions of the equations of motion then, for asymptotically flat spaces, the Hamiltonian does not vanish but its value is given rather by a nonzero surface integral. If the deformations of the surface on which the state is defined are restricted so that the surface moves asymptotically parallel to itself in the time direction, then the surface integral gives directly the energy of the system, prior to fixing the coordinates or solving the constraints. Under more general conditions (when asymptotic Poincaré transformations are allowed) the surface integrals giving the total momentum and angular momentum also contribute to the Hamiltonian. These quantities are also identified without reference to a particular fixation of the coordinates. When coordinate conditions are imposed the associated reduced Hamiltonian is unambiguously obtained by introducing the solutions of the constraints into the surface integral giving the numerical value of the unreduced Hamiltonian. In the present treatment there are therefore no divergences that cease to be divergences after coordinate conditions are imposed. The procedure of reduction of the Hamiltonian is explicitly carried out for two cases: (a) Maximal slicing, (b) ADM coordinate conditions.

A Hamiltonian formalism which is manifestly covariant under Poincaré transformations at infinity is presented. In such a formalism the ten independent variables describing the asymptotic location of the surface are introduced, together with corresponding conjugate momenta, as new canonical variables in the same footing with the g_{ij}, π^{ij}. In this context one may fix the coordinates in the "interior" but still leave open the possibility of making asymptotic Poincaré transformations. In that case all ten generators of the Poincaré group are obtained by inserting the solution of the constraints into corresponding surface integrals.

* Work partially supported by N.S.F. Grant No. GP30799X to Princeton University.

1. Introduction

It has been recognized long ago that there are drastic differences in the Hamiltonian formulation of general relativity depending upon whether three-space is open or closed. DeWitt [1] noticed that in the case of an asymptotically flat spacetime the usual Hamiltonian

$$H_0 = \int d^3x \{N(x) \mathscr{H}(x) + N^i(x) \mathscr{H}_i(x)\} \tag{1.1}$$

for general relativity must be suplemented by the addition of a surface integral at infinity [3]:

$$E[g_{ij}] = \oint d^2s_k (g_{ik,i} - g_{ii,k}) \tag{1.2}$$

in order for the (modified) Hamiltonian

$$H = H_0 + E[g_{ij}] \tag{1.3}$$

to coincide asymptotically with the usual expression of the linearized theory of gravity. However he (DeWitt) stated at the same time that: "... although such a partial integration leaves the dynamical equations unaffected it does change the definition of energy...." This point of view seems to be commonly accepted; Dirac [2], for example, also states that: "The removal of this surface integral does not disburb the validity of H_{main} [our H of Eq. (1.3) with $N = 1$ and $N_i = 0$] for giving equations of motion, but it results in H_{main} not vanishing weakly...."

We would like to point out in this paper that with a proper definition of the functional space of the gravitational fields the inclusion of the surface integral (1.2) in the Hamiltonian is not a matter of choice and that it is not necessary either to justify it by making appeal to the linearized theory. Rather, in our case, the usual Hamiltonian H_0 not only does not give the correct equations of motion (Einstein's equation) but, worse than that, it (H_0) gives no well-defined set of equations of motion at all. It will be in fact shown in Section 2 that if the phase space of the system is to include the trajectories representing the solutions of the equations of motion, which is a necessary and basic consistency requirement for any Hamiltonian theory, then the Hamiltonian for general relativity must be expression (1.3) which includes the surface integral (1.2). If one would use H_0 given by (1.1) instead of H he would find that Hamilton's principle

$$\delta \int dt\, (p_i \dot{q}^i - H) = 0 \tag{1.4}$$

has no solutions.

Once one sees that the surface integral (1.2) must be included in the Hamiltonian to start with by a fundamental reason and not by some ad hoc considerations, the logical procedure leading to the identification of the energy in general relativity becomes considerably simpler: the energy is just the (nonzero!) numerical value of the Hamiltonian which, on account of the initial-value equations

$$\mathscr{H} = 0 \tag{1.5a}$$

and

$$\mathscr{H}_i = 0 \tag{1.5b}$$

is precisely the surface integral (1.2).

In much the same way as the energy is identified as the conserved quantity (the Hamiltonian) associated to invariance of the action under time displacements at infinity one can identify the momentum

$$P^i = -2 \oint d^2s_j \, \pi^{ij} \tag{1.6}$$

and the angular momentum

$$L_i = 2 \oint d^2s_l \epsilon_{ijk} \pi^{lj} x^k \tag{1.7}$$

as the conserved quantities associated to translations and rotations at infinity, respectively (Section 3).

What we are trying to emphasize at this point is that the identification of energy, momentum and angular momentum can be achieved without solving (even in principle) the constraint equations and, consequently, before fixing the spacetime coordinates and finding the corresponding reduced Hamiltonian of the theory ("true Hamiltonian" in the terminology of ADM [3]). On the contrary, as it will be discussed in Section 4, the reduced Hamiltonian corresponding to a particular fixation of the coordinates is obtained by inserting the solutions of the constraint equations (1.5) into the surface integral (1.2). It becomes then trivial to see that any fixation of the coordinates (for limitations on the meaning of "any" see Section 4) leads to the same value for the energy. Similar remarks apply to the other conserved quantities. ADM worked in the opposite direction, reducing first the Hamiltonian and then identifying the value of the reduced Hamiltonian with the energy. It is then not obvious at all that different reductions will lead to the same value for the energy [4].

The understanding of the necessity of including the surface integral $E[g_{ij}]$ in the Hamiltonian to start with clarifies also a difficulty found by ADM, who discarded the surface integral at the beginning but found that if they would do so also at the end (i.e., after coordinate conditions are imposed) they would be left with no Hamiltonian. To bypass the difficulty they stated [3, 5] that "It should be emphasized that, while the *energy* and *momentum* are indeed divergences, the integrands in the *Hamiltonian* and in the space translation generators are *not* divergences when expressed as functions of the canonical variables." What ADM seemed to have in mind is that before solving the constraints one can assume that δg_{ij} and $\delta\pi^{ij}$ vanish sufficiently fast at infinity so that $\delta E[g_{ij}] = 0$, but that, on the other hand, one cannot make such an assumption after solving the constraints, because at that moment the behavior of some of the canonical variables is determined by the behavior of the others. As we will see, in our approach the asymptotic behavior of δg_{ij} and $\delta\pi^{ij}$ is determined from general considerations having little to do with imposing coordinate conditions. Such an asymptotic behavior stays the same before and after solving the constraints. Consequently a surface integral that can be eliminated at the beginning (as the surface terms other than $E[g_{ij}]$ in (2.6)) can also be eliminated at the end. On the other hand, a surface integral having a nonzero variation at the beginning ($E[g_{ij}]$) will also have a nonzero variation at the end and consequently cannot ever be discarded.

Another bonus of the present treatment is a clear cut proof of the correctness of Dirac's results [2] in the reduction of the Hamiltonian for maximal slicing of spacetime. (It should be emphasized that although we find the *result* to be correct we are not satisfied with the original reasoning of Dirac. In this aspect we agree with ADM [5] who found the original Dirac's procedure "logically incomplete.") For the sake of completeness we also show at the end of Section 4 how the reduction of the Hamiltonian by ADM coordinate conditions is achieved in our approach.

Section 5 is devoted to the problem of invariance of the theory under Poincaré transformations at infinity. To begin with, the results of previous sections are generalized so as to allow for asymptotic Poincaré transformations among the permissible deformations of the hypersurface. In that case nine surface integrals besides (1.3) contribute to the Hamiltonian. These integrals are the six momentum and angular momentum expression (1.6), (1.7) and three other quantities related to the "center-of-mass" motion. If one fixes the coordinates in the "interior" but still leaves open the possibility of making Poincaré transformations at infinity then the ten generators of the Poincaré group are obtained by inserting the solutions of the constraints into the surface integrals in question, in total analogy with the procedure for reducing the (less general) Hamiltonian (1.3). The steps described above are not however the end in achieving a Hamiltonian formalism which is manifestly covariant under Poincaré transformations at infinity. To reach this goal one must introduce, according to the general method of Dirac [6], the ten variables describing the asymptotic location of the surface (together with their conjugate momenta) as extra canonical variables besides, and in the same footing with, the (g_{ij}, π^{ij}). In doing so one acquires ten new constraints which enter into the Hamiltonian in the same footing with Eqs. (1.5). The new Hamiltonian obtained in this way vanishes then weakly and is this quantity which is the analog of H_0 given by (1.1) for compact spaces.

The true phase space of the gravitational field for asymptotically flat spaces is therefore the space of the (g_{ij}, π^{ij}) "*completed by the introduction of boundary conditions as canonical variables.*" We might mention, incidentally, that in this enlarged phase space the canonical transformations involving inverse Laplacians used by ADM [3] and Kuchař [7] become one to one and time independent, two properties which they did not simultaneously possess in the original space of the g_{ij}, π^{ij}.

2. Hamilton's Principle and Phase Space

When matter fields are present the normal generator $\mathscr{H}$ and the tangential generators $\mathscr{H}_i$ are constructed besides the gravitational variables g_{ik} and π^{ik} from the matter fields and their conjugate momenta. In all cases in which the theory has been put in Hamiltonian form [6, 8] the matter part of the generators is local in the metric ($\mathscr{H}^{\text{matter}}(x)$ contains g_{ik} at the same point x and no derivatives, $\mathscr{H}_i^{\text{matter}}$ is independent of g_{ik}) and independent of the π^{ik}. Therefore, varying the matter part of the Hamiltonian with respect to the gravitational variables is a well-defined operation which brings in no difficulties at all. Besides, the matter fields decrease at infinity faster than the metric (in electrodynamics, for example, **A** and **E** go as r^{-2} at infinity), which makes surface integrals at infinity containing these fields vanish. For these reasons we will restrict ourselves in the rest of this paper to the pure gravitational field without matter. This is done only for the purpose of conciseness and represents no loss of generality whatsoever. The conclusions reached (and, in particular, the expressions for energy, momentum and angular momentum) hold also in the presence of matter fields.

The essential requirement that must be met by an acceptable definition of the phase space of a dynamical system is that all physically reasonable solutions of the equations of motion must lie inside the phase space. If this is not true, the variational problem

$$\delta \int dt\,(p_i \dot{q}^i - H) = 0.$$

has no solutions because the extremal trajectories are not admitted among the original "competing curves" of the variational problem. Once the requirement of containing all extremal trajectories is met one can enlarge at will the phase space but one *cannot* mutilate it arbitrarily.

In vacuum general relativity a point in phase space is represented by twelve function variables (g_{ij}, π^{kl}). Any solution of Einstein's equations representing a physically reasonable, asymptotically flat spacetime behaves at spatial infinity in the Schwarzschild form

$$ds^2 \underset{r\to\infty}{\sim} -\left(1 - \frac{M}{8\pi r}\right) dt^2 + \left(\delta_{ij} + \frac{M}{8\pi}\frac{x^i x^j}{r^3}\right) dx^i\, dx^j. \tag{2.1}$$

The phrase "physically reasonable" here means essentially that the total mass energy of the system must be finite. This assumption will be satisfied if the system has been radiating (gravitationally or otherwise) during a finite time only. It is quite plausible however that one can admit a more general situation in this context, namely one in which the system is allowed to have been radiating during an infinite time, but doing so in such a way that the total amount of radiation remains finite. [This will be the case if the radiation rate goes as $\tau^{-(1+\epsilon)}$ when $\tau \to -\infty$, where the parameter τ is some kind of a retarded proper time associated with the source. When going to infinity along a spacelike surface the $\tau^{-(1+\epsilon)}$ behavior provides an extra damping factor $r^{-(1+\epsilon)}$ in the metric so that condition (2.2b) below will be effectively satisfied even though the metric will look oscillatory.] The coordinate system in (2.1) is such that the x^i are asymptotically cartesian and is also such that no "coordinate waves" are present (see in this context [3] and references therein). The precise form of the line element (2.1) can be altered by a change of coordinates, but no coordinate system exists such that, when $M \neq 0$, *all* components of the metric and its first spatial derivatives can be made to decrease at infinity faster than r^{-1} and r^{-2}, respectively. It follows therefore that any definition of phase space *has to contain* metric functions such that

$$g_{ij} - \delta_{ij} \sim r^{-1} \tag{2.2a}$$

and

$$g_{ij,k} \sim r^{-2}. \tag{2.2b}$$

We will need later a more precise definition of phase space, but the essential point is that one cannot do any better than (2.2).

Let us return now to Hamilton's principle keeping the asymptotic behavior (2.2) in mind. When one deals with a continuous system like the present one Hamilton's equations read

$$\dot{g}_{ij}(x) = \delta(\text{Hamiltonian})/\delta\pi^{ij}(x) \tag{2.3a}$$

and

$$\dot{\pi}^{ij}(x) = -\delta(\text{Hamiltonian})/\delta g_{ij}(x). \tag{2.3b}$$

The functional derivatives appearing on the right-hand side of (2.3) are, by definition, the coefficients of δg_{ij} and $\delta\pi^{ij}$ in a generic variation of the Hamiltonian, i.e., if

$$\delta(\text{Hamiltonian}) = \int d^3x \, \{A^{ij}(x)\, \delta g_{ij}(x) + B_{ij}(x)\, \delta\pi^{ij}(x)\} \tag{2.4a}$$

then

$$\delta(\text{Hamiltonian})/\delta g_{ij} = A^{ij} \tag{2.4b}$$

and

$$\delta(\text{Hamiltonian})/\delta\pi^{ij} = B_{ij}\,. \tag{2.4c}$$

Thus, in order for Hamilton's equations to be defined at all it is necessary that the variation of the Hamiltonian can be put in the form (2.4) for an arbitrary change in the phase space point (g_{ij}, π^{kl}). We shall see now that δH_0 (with H_0 given by (1.1)) cannot be put in the form (2.4a) and that the Hamiltonian has to be ammended by the addition of the surface integral (1.2) in order that Hamilton's equations (2.3) coincide with Einstein's equations.

The explicit form of the generators $\mathscr{H}$ and $\mathscr{H}_i$ appearing in (1.1) is

$$\mathscr{H} = g^{-1/2}(\pi_{ij}\pi^{ij} - \tfrac{1}{2}\pi^2) - g^{1/2}R \tag{2.5a}$$

and

$$\mathscr{H}_i = -2\pi_i{}^j{}_{/j}\,. \tag{2.5b}$$

Introducing expressions (2.5) into (1.1) one gets for the change in H_0, keeping all terms,

$$\begin{aligned}\delta H_0 = &\int d^3x\,\{A^{ij}(x)\,\delta g_{ij}(x) + B_{ij}(x)\,\delta\pi^{ij}(x)\}\\ &- \oint d^2s_l G^{ijkl}(N\delta g_{ij/k} - N_{,k}\,\delta g_{ij})\\ &- \oint d^2s_l\,\{2N_k\,\delta\pi^{kl} + (2N^k\pi^{jl} - N^l\pi^{jk})\,\delta g_{jk}\},\end{aligned} \tag{2.6}$$

where

$$G^{ijkl} \equiv \tfrac{1}{2}g^{1/2}(g^{ik}g^{jl} + g^{il}g^{jk} - 2g^{ij}g^{kl}). \tag{2.7}$$

The coefficients A^{ij} and B_{ij} need not be explicitly written here (they may be found, for example, from the right-hand sides of Eqs. (7-3.15a, b of [3]). We need only to observe that in order for Hamilton's equations to reproduce Einstein's equations one must identify A^{ij} and B_{ij} in Eq. (2.6) with the variational derivatives appearing in (2.3). This means that, *if the surface integrals in* (2.6) *would vanish* H_0 *would be the correct Hamiltonian.* For a closed space this is the case but as we shall see now for the open asymptotically flat situation the first surface integral in (2.6) is different from zero.

To deal with the surface integrals in (2.6) we will need a more complete specification of the asymptotic behavior than the one given by (2.2). We shall assume also the following conditions at infinity:

$$\pi^{ij} \sim r^{-2}, \tag{2.8}$$

$$N - 1 \sim r^{-1}, \tag{2.9a}$$

$$N_{,k} \sim r^{-2}, \tag{2.9b}$$

$$N^i \sim r^{-1}, \tag{2.10a}$$

$$N^i{}_{,k} \sim r^{-2}. \tag{2.10b}$$

Equation (2.8) follows from requiring that the asymptotic behavior (2.2) of the spatial metric be preserved under a deformation of the surface which involves a Lorentz transformation at infinity (see the appendix for details). Most of the proofs of this section would work however with the milder asymptotic decrease $\sim r^{-(1+\epsilon)}$ for π^{ij}. Equations (2.9) and (2.10) fix the behavior of the lapse and shift functions (which are not canonical coordinates) according to the asymptotic Schwarzschild form (2.1) and ensure that the spacetime coordinates (t, x^i) become Minkowskian at infinity. Taking into account Eqs. (2.2) and (2.8)–(2.10) we see that the only surviving surface term in (2.6) is

$$-\oint d^2s_l \, G^{ijkl} \, \delta g_{ij/k}, \tag{2.11}$$

which, by using (2.2) and (2.7), may be rewritten as

$$-\oint d^2s_l \, G^{ijkl} \, \delta g_{ij/k} = -\delta \oint d^2s_k \, (g_{ik,i} - g_{ii,k}) \equiv -E[g_{ij}]. \tag{2.12}$$

Taking into account (2.12) one may in turn rewrite (2.6) as

$$\delta(H_0 + E[g_{ij}]) = \int d^3x \, (A^{ij}(x) \, \delta g_{ij}(x) + B_{ij}(x) \, \delta\pi^{ij}(x)). \tag{2.13}$$

It should be remarked here that $E[g_{ij}]$ does not vanish and neither does its variation (even though E is a surface integral). For example, for the line element (2.1) one has, as it is well known [1]

$$E = M \tag{2.14}$$

and a generic variation of g_{ij} will have $\delta M \neq 0$.

From Eq. (2.13) we see, thus, that for asymptotically flat spacetimes the correct Hamiltonian for general relativity is

$$H = H_0 + E[g_{ij}]$$

with H_0 and given by (1.1) and (1.2), respectively. Equation (1.3) is therefore proved.

3. Conserved Quantities

A. *Energy*

It was just shown at the end of Section 2 above that the correct Hamiltonian for general relativity in an asymptotically flat spacetime is

$$H = H_0 + E[g_{ij}] \tag{1.3}$$

The Hamiltonian (1.3) does not vanish on account of the constraint equations (1.5), it takes on, rather, the value E. Since H does not depend on the time t the quantity E is a constant of the motion and is to be identified with the energy.

It is interesting to notice here that although the action is invariant under arbitrary timelike deformations of the hypersurface one obtains only *one* conserved quantity associated to this invariance, the energy E. One could say that what happens here is that there are infinitely many conserved quantities but they are all equal and consequently provide only one piece of information. Such a "degeneration" of the conserved quantities arises because, in this context, only the asymptotic part of the deformation is what counts—what happens inside is irrelevant. More precisely speaking, one verifies that the action

$$S = \int_{t_1}^{t_2} dt \left\{ \int d^3x \, (\pi^{ik}\dot{g}_{ik} - N\mathscr{H} - N^i\mathscr{H}_i) + E[g_{ij}] \right\} \tag{3.1}$$

is invariant under the transformation

$$\pi^{ik}(x, t) \to \pi^{ik}(x, t - \epsilon), \tag{3.2a}$$

$$g_{ik}(x, t) \to g_{ik}(x, t - \epsilon), \tag{3.2b}$$

$$N(x, t) \to N(x, t) + \delta N(x, t), \tag{3.2c}$$

$$t_1, t_2 \to t_1 + \epsilon, t_2 + \epsilon, \tag{3.2d}$$

with

$$\delta N(\infty) = 0, \tag{3.2e}$$

provided that the original history (g^{ik}, π^{ik}, N, N^i) is a solution of the equations of motion. The arbitrary variation (3.2b) of N means that one is allowing for an arbitrary deformation of the hypersurfaces in the interior. Asymptotically, however, the hypersurfaces are rigidly shifted in time [Eq. (3.2e)]. One notices that $\delta N(x)$ provides no information because it is multiplied by $\mathscr{H}$ which vanishes if the original trajectory satisfies the equations of motion. What happens inside is thus immaterial. Only one parameter is left in the transformation, the quantity ϵ which corresponds to the amount of proper time shift at spatial infinity. This is why one speaks of the energy in general relativity as being the conserved quantity associated to "time displacements at infinity."

B. *Momentum and Angular Momentum*

In order to stress that the conserved quantities can be unambiguously identified without solving the constraints or reducing the Hamiltonian (which makes it unnecessary to prove alfterwards that the result is independent of the particular coordinate conditions chosen to carry out the reduction) we will obtain below the expressions for the total momentum and angular momentum.

The momentum is normally defined as the conserved quantity associated to invariance of the action under translations. A translation is an operation which is defined only for a flat space—it amounts to transplant the fields from the point x^k to the point $x^k + \epsilon^k$, with ϵ^k fixed (independent of x) and where the x^k are cartesian coordinates. The corresponding (more general) operation for a curved space is to transplant in a coordinate invariant way the fields from the point x^k to the point $x^k + \xi^k(x)$. This is achieved by substracting from a field quantity ("active transformation;" move fields, leave coordinate system fixed) its Lie derivative along the vector field $\xi^k(x)$. The action (3.1) is invariant under such a transformation even if ξ^k behaves as r^{+1} at infinity (which is the case for an asymptotic rotation—see below). To check this invariance it is helpful to use the following facts: (a) The E-term is invariant separately, (b) the integrand in the spatial integral is a scalar density— its Lie derivative is then

$$\{(\pi^{ij}\dot{g}_{ij} - N\mathscr{H} - N^i\mathscr{H}_i)\, \xi^k\}_{,k}\,,$$

which transforms the variation into a surface integral $\oint d^2s_k \pi^{ij}\dot{g}_{ij}\xi^k$ (the $\mathscr{H}$ and $\mathscr{H}_i$ terms vanish because the original trajectory is assumed to satisfy the equations of motion), (c) the spatial integral vanishes on account of the asymptotic conditions (2.2), (2.8), and (2.10) [using $\dot{g}_{ij} = 2Ng^{1/2}(\pi_{ij} {-}^1/_2\pi g_{ij}) + N_{(i/j)}$].

Knowing that the action is invariant, the next step is, following the lines of Noether's theorem, to rearrange the variation of the action is the form

$$\delta S = \int_{t_1}^{t_2} dt \frac{d}{dt} \int d^3x\, (\pi^{ik}\, \delta g_{ik} + \text{terms vanishing by the equations of motion}). \tag{3.3}$$

Inserting then

$$\delta g_{ik} = -\mathscr{L}_\xi\, g_{ik} = -(\xi_{i/k} + \xi_{k/i}) \tag{3.4}$$

into (3.3) one obtains

$$\delta S = \int_{t_1}^{t_2} dt \frac{d}{dt} \int d^3x\, \{-(2\pi^{ik}\xi_k)_{/i} + 2\pi^{ik}{}_{/k}\xi_i\}. \tag{3.5}$$

The second term in (3.5) vanishes by the constraint (1.5b) and what is left may be transformed into a surface integral so that δS reads

$$\delta S = \oint d^2s_i\, (-2\pi^{ik}\xi_k)\Big|_{t_1}^{t_2}. \tag{3.6}$$

If we set $\xi_k \underset{r\to\infty}{\longrightarrow} \epsilon_k$, a constant, we are dealing with an asymptotic translation.

We can write then

$$\delta S = \epsilon_k(P^k(t_2) - P^k(t_1)) \tag{3.7}$$

with

$$P^k = -2 \oint d^2s_i \, \pi^{ik}. \tag{3.8}$$

From (3.7) we identify P^k given by (3.8) as the total linear momentum of the system. It is a constant of the motion because δS in (3.6) is zero since the action is invariant under the transformation in consideration. Again we see that only $\epsilon_k = \xi_k(\infty)$ provides nontrivial information. The invariance of the action with respect to arbitrary $\xi^k(x)$ in the interior is immaterial in this context. One speaks accordingly of the momentum (3.8) as being the conserved quantity associated to invariance under "spatial translations at infinity."

To find the linear momentum we made ξ^k to be asymptotically a translation ($\xi^k(\infty) = \text{const}$). If we are concerned with angular momentum then what we want is an asymptotic rotation by an angle $\delta\varphi$:

$$\xi_i \xrightarrow[r\to\infty]{} \epsilon_{ijk} \, \delta\varphi^j \, x^k. \tag{3.9}$$

One arrives then, instead of (3.7), to an expression

$$\delta S = \delta\varphi^j(L_j(t_2) - L_j(t_1)), \tag{3.10}$$

where

$$L_i = 2 \oint d^2s_l \, \epsilon_{ijk}\pi^{lj}x^k \tag{3.11}$$

is the conserved quantity associated with "invariance under rotations at infinity": the total angular momentum.

4. Reduced Hamiltonian

A. *General Method*

The equations of motion associated with the Hamiltonian (1.3) describe the evolution of the system under an arbitrary deformation of the hypersurface on which the field state is defined. Often one wants to answer a more limited set of questions, namely one is satisfied with being able to determine how the system evolves along a one-parameter family of surfaces. To deal with such a more restricted question one does not need the whole power of the theory based on the Hamiltonian (1.3) which contains four arbitrary functions N and N^i, but, rather, one deals with a reduced Hamiltonian which is able to give only the particular information one is asking for. The reduced Hamiltonian theory has the advantage of containing no constraints and of dealing with a lesser number of degrees of freedom than the full theory (the degrees of freedom corresponding to an arbitrary deformation of the surface become frozen in the reduction process).

The general method for reducing the Hamiltonian was invented by Dirac [6]. However some of the main proofs are not valid, unfortunately, for a system having a continuous number of degrees of freedom. It becomes therefore necessary to restate the reduction procedure and prove its consistency separately for the continuous case. We shall not attempt here to do so in general but we will restrict ourselves to the part of the problem that has immediate bearing on, and connection with, the subject of this paper. Thus, we shall prove below that *for general relativity in asymptotically flat spacetime the reduced Hamiltonian is obtained by inserting the solution of the constraints in the surface integral* $E[g_{ij}]$, which shows again that it is quite important to include E in the Hamiltonian. We shall show that such a procedure gives the correct equations of motion for a class of coordinate conditions that seem to be the one encountered in practice, namely when essentially one sets some of the momenta equal either to zero, or to prescribed functions of position which are independent of time. We will not be concerned here with a more general class of coordinate donditions but it seems quite likely that the basic statement (in italics above) is true in the more general case.

The scheme runs as follows: One assumes that the variables g_{ik} , π^{ik} can be separated by a one to one, time independent, functionally differentiable canonical transformation in two sets $(\varphi^{\alpha}(x); \pi_{\alpha}(x))$ and $(\psi^{A}(x); \pi_{A}(x))$ in-such a way that:

(a) The surface integral $E[g_{ij}]$ depends only on the φ^{α}, and the π_{α} .

(b) When the π_{α} are prescribed as functions p_{α} of x which satisfy

$$\dot{p}_{\alpha} = 0, \tag{4.1a}$$

then the constraints $\mathscr{H} = 0$, $\mathscr{H}_i = 0$ can be solved to express the φ^{α} as functionals

$$\varphi^{\alpha} = f^{\alpha}[\psi^{A}; \pi_{A}] \tag{4.1b}$$

of the remaining canonical variables. The functional derivaties of f^{α} with respect to ψ^{A}, π_{A} are assumed to exist.

If the above conditions are true then Hamilton's equations for the Hamiltonian

$$H^{\text{reduced}}[\psi^{A}; \pi_{A}] = E[g_{ij}]|_{\substack{\varphi^{\alpha}=f^{\alpha} \\ \pi_{\alpha}=p_{\alpha}}} \tag{4.2}$$

together with Eqs. (4.1) are equivalent to Einstein's equations in the particular frame defined by $\pi_{\alpha} = p_{\alpha}$.

Proof. Recalling that Poisson brackets are invariant under canonical transformations and since, moreover, the Hamiltonian is unchanged in value if the canonical transformation is independent of time, we have

$$\dot{\psi}^A(x) = \left.\frac{\delta H}{\delta \pi_A(x)}\right|_{\substack{\varphi^\alpha = f^\alpha \\ \pi_\alpha = p_\alpha}}. \tag{4.3}$$

On the other hand

$$H[\varphi^\alpha; \pi_\alpha, \psi^A; \pi_A]\Big|_{\substack{\varphi^\alpha = f^\alpha \\ \pi_\alpha = p_\alpha}} = E[\varphi^\alpha; \pi_\alpha]\Big|_{\substack{\varphi^\alpha = f^\alpha \\ \pi_\alpha = p_\alpha}} = H^{\text{reduced}}[\psi^A; \pi_A]. \tag{4.4}$$

Next, differentiating (4.4) with respect to π_A we get

$$\int d^3y \left.\frac{\delta H}{\delta \varphi^\alpha(y)}\right|_{\substack{\varphi^\alpha = f^\alpha \\ \varphi_\alpha = p_\alpha}} \frac{\delta f^\alpha(y)}{\delta \pi_A(x)} + \left.\frac{\delta H}{\delta \pi_A(x)}\right|_{\substack{\varphi^\alpha = f^\alpha \\ \pi_\alpha = p_\alpha}} = \frac{\delta H^{\text{reduced}}}{\delta \pi_A(x)}. \tag{4.5}$$

However, by Eq. (4.1a),

$$\dot{\pi}_\alpha(x) = \left.\frac{\delta H}{\delta \varphi^\alpha(y)}\right|_{\substack{\varphi^\alpha = f^\alpha \\ \pi_\alpha = p_\alpha}} = 0$$

whence

$$\left.\frac{\delta H}{\delta \pi_A(x)}\right|_{\substack{\varphi^\alpha = f^\alpha \\ \pi_\alpha = p_\alpha}} = \frac{\delta H^{\text{reduced}}}{\delta \pi_A(x)}, \tag{4.6}$$

a result that, when inserted back into (4.3), shows that H^{reduced} generates the correct equation of motion for ψ^A. In a completely analogous way one shows that the good equation of motion is also obtained for π_A. The evolution of φ^α as calculated from (4.1b) will agree with the one given by the unreduced formalism because the constraints are preserved in time.

Note that we have not allowed for an explicit dependence of E on ψ^A, π_A (i.e., a dependence other than the one induced by $\varphi^\alpha = f^\alpha$) because such a dependence would result in general on the variational derivatives $\delta H^{\text{reduced}}/\delta\psi^A$, $\delta H^{\text{reduced}}/\delta\pi_A$ not being well defined. This follows from the point that we have been emphasizing all over, namely that H_0 and E do not separately have well defined functional derivatives, only the sum $H_0 + E$ does. Fro an explicit example of the issue discussed here, see the discussion of (4.27) below.

In practice, the new phase space coordinates $(\varphi^\alpha; \pi_\alpha)$, $(\psi_A; \pi^A)$ are not generally canonical but they almost are. One can easily verify, along the same lines of the above proof, that the theorem is still correct (i.e., that H^{reduced} is obtained by inserting the solution of the constraints into $E[g_{ij}]$) provided that the following restrictions hold:

(a) $$[\varphi^\alpha(x), \psi^A(x')] = [\varphi^\alpha(x), \pi_A(x')] = [\pi_\alpha(x), \psi^A(x')] = [\pi_\alpha(x), \pi_A(x')] = [\pi_\alpha(x), \pi_\beta(x')] = 0. \tag{4.7}$$

(b) The "matrix" $[\pi_\alpha(x), \varphi^\beta(x')]$ is invertible, i.e.,

$$\int d^3x' \, [\pi_\alpha(x), \varphi^\beta(x')] f_\beta(x') = 0 \Rightarrow f_\beta = 0. \tag{4.8}$$

If Eqs. (4.7) and (4.8) hold, then H^{reduced} given by (4.2) will generate the correct equations of motion

$$\dot{\psi}^A = [\psi^A, H^{\text{reduced}}], \qquad \dot{\pi}_A = [\pi_A, H^{\text{reduced}}] \tag{4.9}$$

provided that the Poisson brackets are computed correctly by the general formula

$$[F, G] = [Q^a, Q^b] \frac{\partial F}{\partial Q^a} \frac{\partial G}{\partial Q^b} + [P_a, P_b] \frac{\partial F}{\partial P_a} \frac{\partial G}{\partial P_b} + [Q^a, P_b] \left(\frac{\partial F}{\partial Q^a} \frac{\partial G}{\partial P_b} - \frac{\partial G}{\partial Q^a} \frac{\partial F}{\partial P_b} \right), \tag{4.10}$$

which holds for a general (not necessarily canonical) set of phase space coordinates (Q^a, P_a).

B. *Maximal Slicing*

The one parameter family of surfaces on which the evolution is being observed may be fixed by the maximal slicing condition

$$\pi \equiv \pi_i{}^i = 0. \tag{4.11}$$

Condition (4.11), being invariant under changes of coordinates in the surface does not restrict the spatial coordinate system. The Hamiltonian will therefore be only *partially* reduced by condition (4.11) i.e., it will still contain a term $\int N^i \mathscr{H}_i$ and also three constraints $\mathscr{H}_i = 0$ will be still present, corresponding to the freedom of making arbitrary tangential deformations (coordinate changes) in the surface. To achieve total reduction one can still impose three coordinate conditions, but we shall not worry about that here.

The slicing condition (4.11) has been previously treated by Dirac [2] but his procedure was obscured by an unclear and apparently arbitrary handling of the surface integral $E[g_{ij}]$ and his results were subsequently questioned by ADM [5]. In Dirac's procedure it seemed that the inclusion or deletion of the surface term

$E[g_{ij}]$ (or of any other surface term!) was just a matter of taste (see quotation in the introduction). However the final result turned out to depend critically on the inclusion of the surface term. In fact, if one follows Dirac's line of reasoning without including the surface term one arrives at a reduced Hamiltonian which is identically zero! He also set $N = 1$ everywhere arbitrarily (now we can see that it worked because H^{reduced} is independent of the value of N inside). In spite of all the unjustified steps, Dirac arrived at the correct reduced Hamiltonian—a situation not unfamiliar in theoretical physics. We shall now recover Dirac's results by application of the conclusions reached earlier in this paper and giving special consideration to the question of boundary conditions. Proceeding in this way no confusion arises and the result becomes unquestionable

As Dirac realized one may rearrange the canonical variables in the following way: (a) One pair $(\varphi; \pi)$ and (b) five other pairs $(\tilde{g}_{ij}; \tilde{\pi}^{ij})$. The quantities in consideration are defined as follows

$$\varphi \equiv \tfrac{1}{3} \log g, \tag{4.12}$$

$$\pi \equiv g_{ij}\pi^{ij}, \tag{4.13}$$

$$\tilde{g}_{ij} \equiv g^{-1/3} g_{ij}, \tag{4.14}$$

$$\tilde{\pi}^{ij} = g^{1/3}(\pi^{ij} - \tfrac{1}{3}\pi g^{ij}), \tag{4.15}$$

and they have the following Poisson brackets:

$$[\varphi(x), \pi(x')] = \delta(x, x'), \tag{4.16}$$

$$[\tilde{g}_{ij}(x), \tilde{\pi}^{kl}(x')] = \tilde{\delta}_{ij}^{kl}\delta(x, x'), \tag{4.17}$$

with

$$\tilde{\delta}_{ij}^{kl} = \tfrac{1}{2}(\delta_i{}^k\delta_j{}^l + \delta_i{}^l\delta_j{}^k) - \tfrac{1}{3}\tilde{g}_{ij}\tilde{g}^{kl}, \tag{4.18}$$

and

$$[\tilde{\pi}^{ij}(x), \tilde{\pi}^{kl}(x')] = \tfrac{1}{3}(\tilde{\pi}^{ij}\tilde{g}^{kl} - \tilde{\pi}^{kl}\tilde{g}^{ij})\, \delta(x, x'). \tag{4.19}$$

All other Poisson brackets are zero.

The quantity $\tilde{g}^{ij}$ appearing in (4.14) and (4.15) is the inverse of the conformal metric $\tilde{g}_{ij}$, i.e., $\tilde{g}^{lm}\tilde{g}_{ms} = \delta_s{}^l$, it is related to the full metric by $\tilde{g}^{ij} = g^{1/3}g^{ij}$. The "conformal Kronecker delta" defined by (4.18) has the following good properties:

$$\tilde{\delta}_{ij}^{ij} = 5, \tag{4.20}$$

$$\tilde{\delta}_{kl}^{ij}\tilde{\delta}_{mn}^{kl} = \tilde{\delta}_{mn}^{ij}, \tag{4.21}$$

$$\tilde{\delta}_{ij}^{kl}\tilde{g}_{kl} = \tilde{\delta}_{ij}^{kl}\tilde{g}^{ij} = 0, \tag{4.22a}$$

$$\tilde{\delta}_{ij}^{kl}\tilde{\pi}^{ij} = \tilde{\pi}^{kl}. \tag{4.22b}$$

Equation (4.20) says that there are only five independent canonical pairs $(\tilde{g}_{ij}, \tilde{\pi}^{ij})$ per space point, in accordance with the conditions

$$\tilde{g} \equiv \det \| \tilde{g}_{ij} \| = 1, \tag{4.23}$$

$$\tilde{\pi} \equiv \tilde{g}_{ij}\tilde{\pi}^{ij} = 0. \tag{4.24}$$

From Eqs. (4.16)–(4.19) we see immediately that the new phase space coordinates satisfy the conditions (4.7)–(4.8) of the reduction theorem. According to Eqs. (4.10) and (4.16)–(4.19) the Poisson bracket is computed in the new variables by the equation

$$[F, G] = \int d^3x \left\{ \frac{\delta F}{\delta \varphi} \frac{\delta G}{\delta \pi} - \frac{\delta G}{\delta \varphi} \frac{\delta F}{\delta \pi} + \delta_{ij}^{kl} \left(\frac{\delta F}{\delta \tilde{g}_{ij}} \frac{\delta G}{\delta \tilde{\pi}^{kl}} - \frac{\delta G}{\delta \tilde{g}_{ij}} \frac{\delta F}{\delta \tilde{\pi}^{kl}} \right) \right.$$
$$\left. + \frac{1}{3} (\tilde{\pi}^{ij}\tilde{g}^{kl} - \tilde{\pi}^{kl}\tilde{g}^{ij}) \frac{\delta F}{\delta \tilde{\pi}^{ij}} \frac{\delta G}{\delta \tilde{\pi}^{kl}} \right\}. \tag{4.25}$$

One can check from (4.25) that $\tilde{g}$ and $\tilde{\pi}$ have zero Poisson bracket with everything and, consequently, they can be set equal to unity and zero respectively either before or after calculating a Poisson bracket. This circumstance permits one to preserve the symmetry in all indices even though one is dealing with more variables than needed [remember Eqs. (4.23) and (4.24)].

In order to reduce the Hamiltonian we have to express the energy E as a functional of the new variables, which gives

$$E = \oint d^2s_k \, (g_{ik,i} - g_{ii,k}) = \oint d^2s_k \, (\tilde{g}_{ik,i} - 2g^{1/3}{}_{,k}). \tag{4.26}$$

From (4.26) we see that to have well-defined variational derivatives of H^{reduced} with respect to $\tilde{g}_{ik}$ we have to get rid of the $\tilde{g}_{ik,i}$ term in (4.26). In other words, the reduction of the Hamiltonian by means of the maximal slicing condition $\pi = 0$ is not possible unless one fixes the asymptotic spatial coordinate system a bit more tightly than in (2.2b), namely we shall require

$$\tilde{g}_{ik,j} \sim r^{-(2+\epsilon)}. \tag{4.27}$$

The asymptotic form (2.1) is thus not allowed in this context, but the line element

$$ds^2 \xrightarrow[r\to\infty]{} -\left(1 - \frac{M}{8\pi r}\right) dt^2 + \left(1 + \frac{M}{8\pi r}\right) \delta_{ij} \, dx^i \, dx^j, \tag{4.28}$$

which is obtained from (2.1) by a change of coordinates

$$\bar{x}^i \underset{r\to\infty}{\sim} x^i \left(1 - \frac{M}{16\pi r}\right) \tag{4.29}$$

satisfies (4.27).

Taking into account (4.27) the energy becomes

$$E = -2 \oint d^2s_k\, g^{1/3}{}_{,k}\,. \tag{4.30}$$

One may check that, when inserted into (4.30), the metric (4.28) gives correctly $E = M$. The final step in the reduction procedure is to solve the Hamiltonian constraint $\mathscr{H} = 0$ to express $g^{1/3}$ in (4.30) as a functional of the $\tilde{g}_{ij}$ and $\tilde{\pi}^{ij}$. This leads to an equation, first examined by Lichnerowicz [9] which has recently been extensively studied by Choquet-Bruhat, York, and O'Murchadha [10], namely,

$$\tilde{\Delta}\phi + \tfrac{1}{8}\tilde{\pi}_{ij}\tilde{\pi}^{ij}\phi^{-7} - \tfrac{1}{8}\tilde{R}\phi = 0, \tag{4.31}$$

with $\phi \equiv g^{1/12}$. (Here $\tilde{\pi}_{ij} = \tilde{g}_{il}\, g_{jk}\tilde{\pi}^{lk}$ and $\tilde{R}$, $\tilde{\Delta}$ are, respectively, the curvature and the Laplacian in the metric $\tilde{g}_{ij}$.) At this stage the reduced Hamiltonian is given therefore by

$$H^{\text{reduced}}[\tilde{g}_{ij}\,;\,\tilde{\pi}^{ij}] = -2 \oint d^2s_k\, g^{1/3}{}_{,k}[\tilde{g}_{ij}\,;\,\tilde{\pi}^{ij}] + \int d^3x\, N^i(x)\, \tilde{\mathscr{H}}_i(x), \tag{4.32}$$

subject to the constraints

$$\tilde{\mathscr{H}}_i = 0. \tag{4.33}$$

The tangential generator appearing in (4.32) and (4.33) is just (2.5b) reduced also by the condition $\pi = 0$. It may be written in the form

$$\tilde{\mathscr{H}}_i = -2\tilde{\pi}_{i/j}^{j} \tag{4.34}$$

with the covariant derivative being taken in the metric $\tilde{g}_{ij}$.

The first term on the right-hand side of (4.32) [which may be rewritten in a variety of ways by converting the surface integral into a volume integral and using (4.31)] is identical with Dirac's H^*_{main}. Equation (4.32) is thus Eq. (33) of [2] which is what we set to prove.

Before leaving the case of maximal slicing it is interesting to emphasize that, in spite of its strange appearance, the right-hand side of (4.32) *is* invariant under changes of coordinates on the surface (provided that the boundary conditions (4.27) are preserved), as it should be the case since the surface coordinates have not been fixed. Thus, "the time part" of the reduced Hamiltonian

$$-2 \oint d^2s_k\, g^{1/3}{}_{,k}[\tilde{g}_{ij}\,;\,\tilde{\pi}^{ij}] \equiv -2 \int d^3x\, g^{1/3}{}_{,kk}[\tilde{g}_{ij}\,;\,\tilde{\pi}^{ij}],$$

has zero Poisson-bracket with the (reduced) tangential generator $\tilde{\mathscr{H}}_i$ given by (4.34) (the Poisson bracket being understood to be evaluated according to (4.25)) and, consequently, the constraints (4.33) are preserved during the evolution of the system.

To finish this paragraph we write more explicitly the equations of motion for $\tilde{g}_{ij}$ and $\tilde{\pi}^{ij}$. From (4.25) these equations are seen to read

$$\dot{\tilde{g}}_{ij} = \tilde{\delta}_{ij}^{kl} \frac{\delta H^{\text{reduced}}}{\delta \tilde{\pi}^{kl}}, \tag{4.35a}$$

$$\dot{\tilde{\pi}}^{kl} = -\tilde{\delta}_{ij}^{kl} \frac{\delta H^{\text{reduced}}}{\delta \tilde{g}_{ij}} + \frac{1}{3}(\tilde{\pi}^{kl}\tilde{g}^{ij} - \tilde{\pi}^{ij}\tilde{g}^{kl}) \frac{\delta H^{\text{reduced}}}{\delta \tilde{\pi}^{ij}}, \tag{4.35b}$$

with $\tilde{\delta}_{ij}^{kl}$ and H^{reduced} given by (4.18) and (4.32), respectively. Note that the presence of the projection operator $\tilde{\delta}_{kl}^{ij}$ in (4.35) ensures that in order to determine the motion of the system one needs to know H^{reduced} only for those values of the function arguments $\tilde{g}_{ij}, \tilde{\pi}^{ij}$ which satisfy (4.23) and (4.24). The reduced Hamiltonian may therefore be considered undefined outside the domain $\| \tilde{g}_{ij} \| = 1$, $\tilde{g}_{ij}\tilde{\pi}^{ij} = 0$ as it should be the case.

C. *ADM Coordinate Conditions*

We shall show here how the well-known results of ADM [3] are recovered in our procedure. ADM separated the canonical variables in two sets as follows: (a) Four pairs $(g^T; \pi^i; \pi^T; g_i)$ and (b) two (independent) pairs $(g_{ij}^{TT}; \pi^{ijTT})$. The new variables are defined by applying to both, $g_{ij} - \delta_{ij}$ and π^{ij}, the splitting

$$f_{ij} = f_{ij}^{TT} + f_{ij}^{T} + f_{i,j} + f_{j,i}, \tag{4.36}$$

where

$$f_i = \frac{1}{\nabla^2}\left(f_{ij,j} - \frac{1}{2\nabla^2} f_{kj,kji}\right), \tag{4.37}$$

$$f_{ij}^{T} = \frac{1}{2}\left[\delta_{ij} f^T - \frac{1}{\nabla^2} f^T{}_{,ij}\right], \tag{4.38}$$

and

$$f^T = f_{ii} - \frac{1}{\nabla^2} f_{ij,ij}. \tag{4.39}$$

Here $1/\nabla^2$ is the inverse of the flat space Laplacian which vanishes at infinity. The fixation of coordinates is achieved by imposing

$$\pi^T = 0, \tag{4.40}$$

$$g_i = x^i. \tag{4.41}$$

Thus g^T and π^i correspond to the φ^α of Eq. (4.7) and π^T and g_i to the π_α. The role of $(\psi^A; \pi_A)$ is then played by g_{ij}^{TT} and π^{ijTT}.

The Poisson brackets of the new variables are

$$[g^T(x), \pi^T(x')] = 2\delta(x, x'), \tag{4.42}$$

$$[\pi^i(x), g_k(x')] = \frac{1}{2\nabla^2}\left(\delta_{ik} - \frac{1}{2\nabla^2}\partial_i\partial_k\right)\delta(x, x') \tag{4.43}$$

$$[g_{ij}^{TT}(x), \pi^{klTT}(x')] = \delta_{ij}^{TTkl}\delta(x, x'), \tag{4.44}$$

all others being zero.

The symbol δ_{ij}^{TTkl} in (4.44) represents the differential operator which projects a symmetric tensor onto its "T–T" part, namely,

$$\delta_{ij}^{TTkl} \equiv \frac{1}{2}\left\{\left(\delta_{il} - \frac{\partial_i\partial_l}{\nabla^2}\right)\left(\delta_{jk} - \frac{\partial_j\partial_k}{\nabla^2}\right) + \left(\delta_{ik} - \frac{\partial_i\partial_k}{\nabla^2}\right)\left(\delta_{jl} - \frac{\partial_j\partial_l}{\nabla^2}\right) - \left(\delta_{kl} - \frac{\partial_k\partial_l}{\nabla^2}\right)\left(\delta_{ij} - \frac{\partial_i\partial_j}{\nabla^2}\right)\right\}. \tag{4.45}$$

The operator (4.45) satisfies the following relations, some of which are the analogs of (4.20)–(4.22):

$$\delta_{ij}^{TTij} = 2, \tag{4.45a}$$

$$\delta_{ij}^{TTkl}\delta_{kl}^{TTmn} = \delta_{ij}^{TTmn}, \tag{4.45b}$$

$$\delta_{ij}^{TTkl}f^{TTij} = f^{TTkl}, \tag{4.45c}$$

$$\delta_{ij}^{TTkl} = \delta_{kl}^{TTij}, \tag{4.45d}$$

$$\delta_{ii}^{TTkl} = \partial_i\delta_{im}^{TTkl} = 0. \tag{4.45e}$$

Equation (4.45a) expresses the fact that there are only two independent canonical pairs $(g_{ij}^{TT}; \pi^{klTT})$ per space point. From Eq. (4.42)–(4.44) one easily checks that conditions (4.7) and (4.8) are satisfied. [The only not immediately obvious result is that

$$\frac{1}{2\nabla^2}\left(\delta_{ik} - \frac{1}{2\nabla^2}\partial_i\partial_k\right)f^k = 0 \Rightarrow f^k = 0,$$

but this is also easily checked by making appropriate contractions and recalling that, with zero boundary conditions at infinity, $(1/\nabla^2)(0) = 0$.]

In the new variables the surface integral $E[g_{ij}]$ reads simply

$$E[g_{ij}] \equiv \oint d^2s_k\,(g_{ik,i} - g_{ii,k}) = -\oint d^2s_k\,g^T{}_{,k}\,. \tag{4.46}$$

If, following ADM we assume that, when conditions (4.40) and (4.41) hold, the constraints can be solved to express

$$g^T = f[g_{ij}^{TT}, \pi^{klTT}], \tag{4.47}$$

then the reduced Hamiltonian is

$$H^{\text{reduced}}[g_{ij}^{TT}, \pi^{klTT}] = -\oint d^2s_k\, g^T{}_{,k}[g_{ij}^{TT}, \pi^{klTT}] \tag{4.48}$$

which is ADM's result (usually written by transforming the surface integral in (4.48) into a volume integral).

According to (4.10) and (4.42)–(4.44) the equations of motion for g_{ij}^{TT} and π^{klTT} will then read:

$$\dot{g}_{ij}^{TT} = \delta_{ij}^{TTkl} \frac{\delta H^{\text{reduced}}}{\delta \pi^{klTT}}, \tag{4.49}$$

$$\dot{\pi}^{klTT} = \delta_{ij}^{TTkl} \frac{\delta H^{\text{reduced}}}{\delta g_{ij}^{TT}}, \tag{4.50}$$

with δ_{ij}^{TTkl} and H^{reduced} given by (4.45) and (4.48), respectively. Here again—in total analogy with the situation in (4.35)—the presence of the projection operator δ_{ij}^{TTkl} in (4.49) and (4.50) ensures that in order to find the motion of the system one needs to know H^{reduced} only for those values of the function arguments g_{ij}^{TT}, π^{TTkl} which are transverse and traceless ("TT"). The reduced Hamiltonian may therefore be considered undefined outside the domain $g_{ij,j}^{TT} = g_{ii}^{TT} = \pi^{TTkl}_{,l} = \pi^{TTkk} = 0$—as it should be the case.

5. Asymptotic Poincaré Invariance

We obtained in Section 3 the expressions for the total energy, momentum and angular momentum. A natural question to ask is whether

$$P^{\mu} = (E, P^i) \tag{5.1}$$

behaves as a four vector under asymptotic Lorentz transformations. An infinitesimal Lorentz transformation at infinity is characterized by lapse and shift functions of the form

$$N \xrightarrow[r\to\infty]{} \beta^{\perp}{}_r x^r, \tag{5.2a}$$

$$N_s \xrightarrow[r\to\infty]{} \beta_{sr} x^r, \tag{5.2b}$$

with

$$\beta_{sr} = -\beta_{rs} . \tag{5.2c}$$

[Actually (5.2a) describes a boost and (5.2b) a spatial rotation.] By introducing (as in 5.1) Greek lower case indices running over the range $\perp$, 1, 2, 3 (where $\perp$ denotes the unit normal to the surface with an arbitrarily prescribed future orientation) one may rewrite (5.2) in the compact way

$$N^\mu \xrightarrow[r\to\infty]{} \beta_r{}^\mu x^r \tag{5.3}$$

with

$$\beta_{\mu\nu} = -\beta_{\nu\mu} . \tag{5.4}$$

The indices μ, ν are raised and lowered with the Minkowskian metric $\eta_{\alpha\beta} =$ diag $(-1, +1, +1, +1)$.

The straightforward way of answering whether P^μ defined by (5.1) transforms as a four vector would be to introduce the lapse and shift (5.3) into the equations of motion (2.3) for g_{ij} and π^{ij} and check whether

$$\dot{P}^\mu = -\beta^\mu_\nu P^\nu \tag{5.5}$$

holds or not. However, such a procedure besides from being cumbersome would be logicaly unsatisfactory, because one would be using the equations of motion to an extent larger than what they follow from Hamilton's principle. The problem comes simply from the fact that the N^μ given by (5.3) do not fulfill the boundary conditions (2.9), (2.10) which were essential in showing that (1.3) is the correct Hamiltonian. To meet this difficulty we shall develop in this section a more general formulation than the one considered in the previous sections. The improved formulation will be manifestly covariant under Poincaré transformations at infinity and the way in which the Poincaré group acts on the various asymptotically defined quantities will be obvious. Furthermore, it s possible in this context to reduce the Hamiltonian by eliminating the degrees of freedom corresponding to arbitrary deformations in the interior, but still leaving open the possibility of making arbitrary Poincaré transformations at infinity. *In that case all ten generators of asymptotic Poincaré transformations are obtained by introducing the solutions of the constraints in appropriate surface integrals*, which are unambiguously identified below. The method is thus a direct generalization of the one used in Section 4 under more restrictive conditions. As an example we may say now that the reduced generator of spatial translations is obtained by introducing the solution of the constraints into the surface integral which gives the total momentum, in much the same way as the reduced Hamiltonian of Section 4 was obtained from the energy expression $E[g_{ij}]$.

In order to deal at once with the full (proper ortochronous) Poincaré group we will add on top of the Lorentz transformation (5.3) an asymptotic translation. The lapse and shift functions will then behave asymptotically as

$$N^\mu - (\alpha^\mu + \beta_r{}^\mu x^r) \underset{r\to\infty}{\sim} r^{-1}. \tag{5.6}$$

[See part C of the Appendix for a more thorough discussion of (5.6).]

If we go back now to the variation (2.6) of the usual Hamiltonian H_0 and take into account (5.6) we arrive at the equation

$$\begin{aligned}\delta(H_0 - \alpha^\mu P_\mu + \tfrac{1}{2}\beta_{rs}M^{rs}) = &\int d^3x\,(A^{ij}(x)\,\delta g_{ij}(x) + B_{ij}(x)\,\delta\pi^{ij}(x)) \\ &- \beta^{\perp}r \oint d^2s_l\, G^{ijkl}(x^r\,\delta g_{ij/k} - \delta^r_k\,\delta g_{ij}) \\ &- \beta_{sr} \oint d^2s_l\, x^r(2\delta^k_s\pi^{jl} - \delta^l_s\pi^{jk})\,\delta g_{jk}\,, \end{aligned} \tag{5.7}$$

with

$$P^{\perp} \equiv \oint d^2s_l\,(g_{rl,r} - g_{rr,l}), \tag{5.8}$$

$$P^r \equiv -2\oint d^2s_l\,\pi^{rl}, \tag{5.9}$$

$$M^{rs} \equiv -2\oint d^2s_l\,(x^r\pi^{ls} - x^s\pi^{lr}). \tag{5.10}$$

Ultimately we want to eliminate the surface integrals still left on the right-hand side of (5.7) by converting them to variations of some quantity which is to be added to H_0, in the same way as it has been already done with P_μ and M_{rs}. Consider thus first the coefficient of $-\beta^{\perp}r$ in (5.7). Its integrand may be written as

$$I^{rl} \equiv \{{}^{(0)}G^{ijkl} + {}^{(1)}G^{ijkl}\}\{x^r[(\delta^{(1)}g_{ij})_{,k} + (\delta^{(2)}g_{ij})_{,k}] - \delta^r_k[\delta^{(1)}g_{ij} + \delta^{(2)}g_{ij}]\}, \tag{5.11}$$

where the left superscripts indicate order in powers of r^{-1} (so that $\delta^{(1)}g_{ij}$ goes as r^{-1} and $(\delta^{(1)}g_{ij})_{,k}$ goes as r^{-2}). Now, it is shown in the appendix that the (apparently divergent) contribution to the surface integral coming from the r^{-1} part of I^{rl} vanishes, if the asymptotic form of the spatial coordinate system is restricted. What is left is then only:

$$\begin{aligned}&\oint d^2s_l^{(0)}\, G^{ijkl}[x^r(\delta^{(2)}g_{ij})_{,k} - \delta_k{}^r\delta^{(2)}g_{ij}] \\ &= \delta\oint d^2s_l^{(0)}\, G^{ijkl}(x^{r(2)}g_{ij,k} - \delta^{r(2)}_k g_{ij}) \equiv \delta M_{\perp r}\,, \end{aligned} \tag{5.12}$$

where $M_{\perp r}$ is defined as

$$M_{\perp r} \equiv \oint d^2s_l\,[x^r(g_{sl,s} - g_{ss,l}) - g_{rl} + g_{ss}\delta_{rl}]. \tag{5.13}$$

Note that we have eliminated the superscripts$^{(2)}$ of g_{ij} in (5.13) because, by the reasons given in the appendix, $^{(0)}g_{ij}$ and $^{(1)}g_{ij}$ do not contribute to the right-hand side of (5.13). Finally, the integral multiplying β_{sr} in (5.7) may also be shown to vanish (see the appendix). Taking into account the preceding discussion and setting by definition

$$M_{r\perp} \equiv -M_{\perp r}, \tag{5.14}$$

we can write (5.7) as

$$\delta H_E = \int d^3x\, (A^{ij}(x)\, \delta g_{ij}(x) + B_{ij}(x)\, \delta\pi^{ij}(x)), \tag{5.15}$$

where the extended Hamiltonian H_E is given by

$$H_E = \int d^3x\, N^\mu(x)\, \mathscr{H}_\mu(x) - \alpha^\mu P_\mu + \tfrac{1}{2}\beta^{\mu\nu} M_{\mu\nu}. \tag{5.16}$$

Here P_μ and $M_{\mu\nu}$ are the ten surface integrals (5.8)–(5.10), (5.13). Equation (5.15) shows that H_E is the correct Hamiltonian to use when asymptotic Poincaré transformations are allowed among the permissible deformations of the hypersurface.

If we take H_E as our Hamiltonian it would be legitimate to use the equations of motion (2.3) with the asymptotic behavior (5.6) in order to find how the quantities P_μ, $M_{\rho\sigma}$ behave under Poincaré transformations, but such a procedure would still be a bit awkward. We will follow instead a different route which, besides from making clear the Poincaré behavior of P_μ and $M_{\rho\sigma}$ will brin to light other aspects of the problem.

Since the spacetime is asymptotically Minkowskian we can introduce a system of rectangular spacetime coordinates at infinity. These coordinates will be denoted, following Dirac [6], by y^A ($A = 0, 1, 2, 3$). The surface on which the state is defined will have then, asymptotically, the equation

$$y^A = a^A + b^A_r x^r, \tag{5.17}$$

with

$$b^A_r b_{As} = \delta_{rs}. \tag{5.18}$$

Equation (5.18) ensures that the coordinates x^r on the surface are rectangular. Note however that, on account of (5.17) the x^r do not have to match the spatial coordinates y^l—a useful feature to deal with translations and rotations at infinity.

According to the basic idea of Dirac's generalized Hamiltonian dynamics one should consider the variables describing the location of the surface in the same footing as the "truly dynamical" variables themselves. For the "interior" of the surface such inclusion has already been achieved implicitly [7, 11] through the six canonical pairs (g_{ij} ; π^{ij}) (there is an excess of four over the required number for a theory with two degrees of freedom per point), and correspondingly one has the constraints $\mathscr{H}_\mu = 0$. However, the asymptotic location of the surface, governed by the ten independent quantities among the sixteen a^A, b^A_r (remember condition (5.18)) cannot be determined from a knowledge of g_{ij} and π^{ij}. *One must therefore introduce a^A and b^A_r together with corresponding conjugate momenta π_A, $\pi_A{}^r$ as additional canonical variable in the same footing with the g_{ij}, π^{ij}*. After this is done one will have a Hamiltonian formalism which is manifestly invariant under Poincaré transformations at infinity.

The introduction of the b^A_r and their conjugate momenta $\pi_\Gamma{}^s$ must be done taking due care of the additional constraint (5.18) which requires a modification in the Poisson brackets of these quantities from their canonical value $\delta^A_\Gamma\delta^s_r$. However, for the purposes of this article, there is no need to do this modification explicitly and consequently we shall not worry about the matter any further. (We refer the reader to [12] for a discussion related to this point.)

At the moment of introducing ten new pairs of canonical variables we must gain, unavoidably, ten new constraints. Each of these constraints will enter the Hamiltonian with an arbitrary Lagrange multiplier. The multipliers in question will describe the amount of hypersurface deformation at infinity and they will be given therefore precisely by the α^μ and $\beta^{\mu\nu}$ appearing in (5.6). The new, total Hamiltonian, H_T will then be of the form

$$H_T = \int d^3x\, N^\mu(x)\, \mathscr{H}_\mu(x) + \alpha^\mu(p_\mu - P_\mu) + \tfrac{1}{2}\beta^{\mu\nu}(m_{\mu\nu} + M_{\mu\nu})$$

$$= H_E + \alpha^\mu p_\mu + \tfrac{1}{2}\beta^{\mu\nu} m_{\mu\nu}\,, \tag{5.19a}$$

with

$$N^\mu \xrightarrow[r\to\infty]{} \alpha^\mu + \beta^\mu_r x^r. \tag{5.19b}$$

The Hamiltonian H_T is supposed to give the change in the (g_{ij} ; π^{ij}; a^A; $b^A_r\pi_A$; $\pi_A{}^r$) under an arbitrary deformation of the hypersurface which preserves its asymptotic flatness. The change in (g_{ij} ; π^{ij}) will be governed by H_E and the one of the asymptotic surface variables by the extra terms $\alpha^\mu p_\mu + \frac{1}{2}\beta^{\mu\nu}m_{\mu\nu}$ in H_T. The $\int N^\mu\mathscr{H}_\mu$ term in (5.19) governs the change in (g_{ij} ; π^{ij}) in the interior, in the sense that if the deformation vanishes asymptotically ($\alpha^\mu = \beta^{\mu\nu} = 0$) only this term is left in the Hamiltonian. Thus, speaking somewhat lossely, the arbitrary functions (Lagrange multipliers) associated to the $\mathscr{H}_\mu$ are the N^μ "without their asymptotic part" (If we vary N^μ in (5.19) keeping the asymptotic part fixed we get $\mathscr{H}_\mu$ as the variational coefficient.) On the other hand the ($p_\mu - P_\mu$), ($m_{\mu\nu} + M_{\mu\nu}$) quantities may be thought of as governing the change induced by the asymptotic part of the deformation, and associated to them are the arbitrary coefficients α^μ and $\beta^{\mu\nu}$. In much the same way as variation of N^μ keeping the asymptotic part fixed yields $\mathscr{H}_\mu = 0$ via Hamilton's principle, one gets the constraints

$$p_\mu - P_\mu = 0 \tag{5.20a}$$

and

$$m_{\mu\nu} + M_{\mu\nu} = 0 \tag{5.20b}$$

by varying the asymptotic part of the deformation. There are therefore as many constraints as arbitrary parameters in the theory, as it should be the case.

The above remarks, which speak of the "asymptotic part" of the deformation as a separate entity from its "interior part" provide a good feeling of what is going on but they should not be taken too literally. It is true that one can vary the interior part of the deformation keeping the asymptotic part fixed but, for continuity reasons, one cannot do the inverse, i.e., vary only the asympttotic part without altering what happens in the finite. When translated into the canonical formalism these remarks are reflected into the fact that, for N^μ going to zero at infinity, the $\int N^\mu \mathscr{H}_\mu$ term has well-defined Poisson brackets whereas, on the other hand, the quantities P_μ and $M_{\rho\sigma}$ do not posses well-defined functional derivatives with respect to g_{ij} and π^{ij} in any case.

As we remarked above one can go quite far without knowing the precise way in which the p_μ, $m_{\rho\sigma}$ are constructed from canonical variables. We shall now proceed to determine their Poisson brackets which is what will determine the Poincaré invariance of the theory. We just mentioned that the constraints (5.20) do not have well-defined Poisson brackets, which means that *to deal correctly with the Poisson brackets of the constraints one must treat all of them at once*, by writing

$$\begin{aligned} H_T(\mathrm{N}) &\equiv \int N^\mu \mathscr{H}_\mu + \alpha^\mu_{(\mathrm{N})}(p_\mu - P_\mu) + \tfrac{1}{2}\beta^{\mu\nu}_{(\mathrm{N})}(m_{\mu\nu} + M_{\mu\nu}) \\ &\equiv H_E(\mathrm{N}) + \alpha^\mu_{(\mathrm{N})} p_\mu + \tfrac{1}{2}\beta^{\mu\nu}_{(\mathrm{N})} m_{\mu\nu} = 0. \end{aligned} \tag{5.21}$$

The subscript (N) in α^μ and $\beta^{\mu\nu}$ is intended to remind us that these quantities are related to N^μ through the asymptotic formula (5.19b).

It has been shown in general by one of us [8, 13] that a theory of the kind being discussed here will predict a dynamical evolution consistent with the Riemannian structure of spacetime if and only if

$$[H_T(\xi), H_T(\eta)] = H_T(\zeta), \tag{5.22}$$

where the "compensating deformation" ζ, resulting from commuting the elementary deformations ξ and η, has the following components:

$$\zeta^\perp = -(\eta^r \xi^\perp{}_{,r} - \xi^r \eta^\perp{}_{,r}), \tag{5.23a}$$

$$\zeta^r = -g^{rs}(\eta^\perp \xi^\perp{}_{,s} - \xi^\perp \eta^\perp{}_{,s}) + (\xi^s \eta^r{}_{,s} - \eta^s \xi^r{}_{,s}). \tag{5.23b}$$

Condition (5.22) ensures, in particular, that the constraints (5.21) are preserved under deformations at the hypersurface. Now, expanding the P.B. on the left-hand side of (5.22) and noticing that

$$[H_E, p_\mu] = [H_E, m_{\rho\sigma}] = 0, \tag{5.24}$$

which holds because H_E and p_μ, $m_{\rho\sigma}$ are constructed from different sets of canonical variables, one sees that (5.22) implies

$$\begin{aligned}&[H_E(\xi), H_E(\eta)] - H_E(\zeta)\\ &\quad = \alpha^\mu_{(\zeta)} p_\mu + \tfrac{1}{2}\beta^{\mu\nu}_{(\zeta)} m_{\mu\nu} - [\alpha^\mu_{(\xi)} p_\mu + \tfrac{1}{2}\beta^{\mu\nu}_{(\xi)} m_{\mu\nu}, \alpha^\rho_{(\eta)} p_\rho + \tfrac{1}{2}\beta^{\rho\sigma}_{(\eta)} m_{\rho\sigma}].\end{aligned} \tag{5.25}$$

Now, by arguing again that both sides of (5.25) are constructed from disjoint sets of canonical variables one sees that (5.25) can hold only if both members of the equation vanish separately. One may check by direct calculation that indeed

$$[H_E(\xi), H_E(\eta)] = H_E(\zeta). \tag{5.26}$$

[Moreover, one can actually use (5.26) to derive the form of the $\mathscr{H}_\mu$, (Ref. [14]) but that need not concern us here.] Going back to the present objects of interest, one finds upon inserting (5.23) in the (vanishing) right-hand side of (5.25) and comparing coefficients that

$$[p_\mu, p_\nu] = 0, \tag{5.27a}$$

$$[p_\mu, m_{\rho\sigma}] = \eta_{\mu\sigma} p_\rho - \eta_{\mu\rho} p_\sigma, \tag{5.27b}$$

$$[m_{\mu\nu}, m_{\rho\sigma}] = \eta_{\mu\rho} m_{\nu\sigma} - \eta_{\nu\rho} m_{\mu\sigma} - \eta_{\mu\sigma} m_{\nu\rho} + \eta_{\nu\sigma} m_{\mu\rho}. \tag{5.27c}$$

Thus the p_μ, $m_{\rho\sigma}$ satisfy the commutation relations for the generators of the Poincaré group—not an unexpected result!

Once we have Eqs (5.27) we can see right away that P_μ and $M_{\rho\sigma}$ have the correct Poincaré behavior. In fact it is enough to recall that the constraints (5.20) are preserved under deformations of the hypersurface to see that P_μ and $M_{\rho\sigma}$ have the same behavior as p_μ and $m_{\rho\sigma}$, respectively: the transformation properties of P_μ and $M_{\rho\sigma}$ have become obvious once we deal with a manifestly Poincaré-covariant formalism.

Let us say now a few words about the reduced formalism. One may eliminate in this context the degrees of freedom corresponding to an arbitrary deformation of the surface in exactly the same way as it was done in Section 4 working with the more restrictive Hamiltonian H_0. Doing so one obtains now not only the generator of asymptotic time translations (the H^{reduced} of Eq. (5.2)) but all ten generators of the Poincaré group. In fact, by the same reasoning followed in Section 4 one concludes now that the reduced Hamiltonian becomes

$$H_E^{\text{reduced}}[\psi^A;\ \pi_A\ ;\ a^A;\ b_r^A\ ;\ \pi_A\ ;\ \pi_A{}^r]$$

$$= \alpha^\mu(p_\mu - P_\mu^{\text{reduced}}) + \tfrac{1}{2}\beta^{\mu\nu}(m_{\mu\nu} + M_{\mu\nu}^{\text{reduced}}), \tag{5.28}$$

with

$$P_\mu^{\text{reduced}} = P_\mu[g_{ij}\ ;\ \pi^{ij}]\Big|_{\substack{\varphi^\alpha=f^\alpha \\ \pi_\alpha=p_\alpha}}, \tag{5.29a}$$

and

$$M_{\mu\nu}^{\text{reduced}} = M_{\mu\nu}[g_{ij}\ ;\ \pi^{ij}]\Big|_{\substack{\varphi^\alpha=f^\alpha \\ \pi_\alpha=p_\alpha}}. \tag{5.29b}$$

The obtention of the reduced generators (5.29) for the examples of coordinate conditions treated in Section 4 is thus a straightforward repetition of the techniques used there to find H^{reduced} (which is $P^{\perp\text{reduced}}$) and will be left to the reader.

It is important to emphasize here that the reduced generators (5.29) will have well-defined P.B.'s in the variables ψ^A, π_A (or in any other equivalent set), even though the original unreduced quantities P_μ, $M_{\rho\sigma}$ do not have well-defined variational derivatives with respect to g_{ij} and π^{ij}—a phenomenon already encountered in Section 4 when dealing with H^{reduced} as derived from $E[g_{ij}]$. Geometrically speaking the nonexistence of the functional derivatives of P^μ and $M_{\rho\sigma}$ follows, as we already said above, from the impossibility of deforming a surface at infinity without altering its shape in the finite, which implies that the generators of the changes in g_{ij} and π^{ij} induced by a "deformation which vanishes in the finite but differs from zero asymptotically" (precisely what P_μ and $M_{\rho\sigma}$ are candidates for) cannot be defined. On the other hand, if one freezes the degrees of freedom which allow for an arbitrary deformation in the finite, then one can reconstruct the deformation everywhere from a knowledge of its asymptotic value: The parameters α^μ, $\beta^{\mu\nu}$ fix this time the deformation not only asymptotically but globally, and it becomes meaningful to ask for the generators of such a deformation, which will be well-defined objects. These generators, which are precisely the quantities (5.29), will then have well-defined P.B.'s with ψ^A, π_A (otherwise they would be unable to act as generators!). We may also point out that the generators (5.29) will satisfy the same algebra (5.27) as the p_μ, $m_{\rho\sigma}$ (otherwise the constraints (5.20) would not remain first class in the reduced formalism).

These comments complete our discussion of the manifestly Poincaré-covariant Hamiltonian formalism. In the next and final section we summarize the results and conclusion of the paper.

6. Concluding Remarks

We have seen that with a correct definition of the functional phase space of general relativity there are certain surface integrals that must be included in the Hamiltonian, which then does not vanish (weakly) anymore. The criterion for correctness in the definition of phase space is that it must contain every reasonable solution of the equations of motion. The qualification "reasonable"here means that a finite amount of radiation must have been emitted from the distant past up to the present moment. The possibility that the system might have been radiating for an infinitely long time is therefore not excluded provided that the radiation rate decreases sufficiently fast with time in the remote past [as $\tau^{-(1+\epsilon)}$]. The surface terms in the Hamiltonian provide definitions for the total energy, momentum and angular momentum which are independent of any particular fixation of the space-time coordinates. Also, the procedure for fixing the coordinates becomes particularly simple in this context: one needs only to introduce the solution of the constraints into the surface integrals in question in order to obtain the reduced Hamiltonian of the theory. This procedure permits one to clarify some aspects of previous work by Dirac [2] and Arnowitt, Deser and Misner [3–5], mainly in what concerns the handling of surface terms in the reduction of the Hamiltonian.

If one wants to examine the behavior of the theory under asymptotic Poincaré transformations within the Hamiltonian formalism, one must enlarge the phase space of the system by introducing, besides the original variables g_{ij}, π^{ij} (and the matter variables) a new set of ten independent canonical pairs which describe the asymptotic location of the spacelike surface on which the state is defined. One gains in this process ten new constraints which must be included in the Hamiltonian following the general method of Dirac [6]. The new Hamiltonian obtained in this way vanishes then weakly and it is this quantity which is the analog of the expression

$$\int d^3x \, (N(x) \, \mathscr{H}(x) + N^i(x) \, \mathscr{H}_i(x)),$$

for compact spaces. Thus the configuration space of general relativity for open spaces is not Wheeler's "Superspace" but rather the product of Superspace with the space spanned by the ten additional variables describing the location of the surface. That the asymptotic coordinates also play a role for open spaces is not new (it was already noticed by Higgs [15]). However what does not seem to have been previously done is the introduction of the "boundary conditions as *canonical variables*," a procedure which has the advantage of permitting one to build a Hamiltonian formalism which is manifestly Poincaré-covariant at infinity. Finally we might mention again here that the inclusion of the asymptotic coordinates as canonical variables makes also the use of canonical transformations involving inverse Laplacians "cleaner" in the sense that in the enlarged phase space these transformations become at the same time one to one and time independent, two properties that they did not simultaneously possess in the original space of the g_{ij}, π^{ij}.

APPENDIX: Asymptotic Form of Canonical Variables

The purpose of this appendix is to discuss more thoroughly than in the main text the conditions we demand on the asymptotic form of the metric g_{ij} and its conjugate momentum π^{ij}. These asymptotic conditions will then be used to show that certain surface integrals which were neglected in Section 5 are indeed zero. The coordinate invariance of the four-momentum P^μ and the angular momentum tensor $M_{\mu\nu}$ will also be established.

A. *Asymptotic Conditions and their Poincaré Invariance*

The basis requirement we impose on the asymptotic behavior of g_{ij} and π^{ij} is that we only allow for coordinate systems such that

$$g_{ij} \xrightarrow[r\to\infty]{} \delta_{ij} + \frac{{}^{(1)}h_{ij}(\mathbf{n})}{r} + \frac{{}^{(2)}h_{ij}(\mathbf{n})}{r^2} + O(r^{-(2+\epsilon)}), \tag{A.1a}$$

$$\pi^{ij} \xrightarrow[r\to\infty]{} \frac{{}^{(2)}p^{ij}(\mathbf{n})}{r^2} + \frac{{}^{(3)}p^{ij}(\mathbf{n})}{r^3} + O(r^{-(3+\epsilon)}), \tag{A.1b}$$

with

$${}^{(1)}h_{ij}(-\mathbf{n}) = {}^{(1)}h_{ij}(\mathbf{n}), \tag{A.2a}$$

and

$${}^{(2)}p^{ij}(-\mathbf{n}) = -{}^{(2)}p^{ij}(\mathbf{n}). \tag{A.2b}$$

[Here and in what follows $\mathbf{a}\cdot\mathbf{b} = a^1b^1 + a^2b^2 + a^3b^3$, $r = (\mathbf{x}\cdot\mathbf{x})^{1/2}$ and $\mathbf{n} = r^{-1}(x^1, x^2, x^3)$.]

Since we want to be able to discuss Poincaré invariance within the Hamiltonian formalism we must check that conditions (A.1-2) are invariant under Poincaré transformations at infinity (if this would not be the case the phase space would not contain all states we are interested in). Let us discuss first the invariance of (A.1-2) under boosts and time translations. Taking the shift vector to be zero and the lapse to behave asymptotically as $N^\perp \to \alpha^\perp + \boldsymbol{\beta}\cdot\mathbf{x} + O(r^{-1})$ the only not immediately obvious property is whether the parity conditions (A.2) are preserved or not. However, this is indeed seen to be the case from the expressions

$$\dot{g}_{ij} \underset{r\to\infty}{\sim} 2(\boldsymbol{\beta}^{\perp} \cdot \mathbf{x})(\pi_{ij} - \tfrac{1}{2}\pi g_{ij}), \tag{A.3a}$$

$$\dot{\pi}^{ij} \underset{r\to\infty}{\sim} -(\boldsymbol{\beta}^{\perp} \cdot \mathbf{x})\, R^{ij} + (\Gamma^{s}_{ij} - \delta_{ij}\Gamma^{s}_{kk})\, \beta^{\perp}{}_{s}\,, \tag{A.3b}$$

which give (from the equations of motion) the change in g_{ij} and π^{ij} to the order thought for. In fact since ${}^{(2)}\pi_{ij}$ is odd it follows immediately from (A.3a) that ${}^{(1)}\dot{g}_{ij}$ is even, so the parity of ${}^{(1)}g_{ij}$ is preserved. On the other hand (A.1a) implies that the r^{-2} part of Γ^{k}_{ij} is odd and the r^{-3} part of g_{ij} is even which makes ${}^{(2)}\dot{\pi}^{ij}$ in (A.36) to be odd, showing that the parity of ${}^{(2)}\pi^{ij}$ is also preserved.

An analysis similar to the precedent one but applied to the equations

$$\dot{g}_{ij} = N_{i/j} + N_{j/i}\,, \tag{A.4a}$$

$$\dot{\pi}^{ij} = (\pi^{ij}N^{m})_{,m} - \pi^{mj}N^{i}{}_{,m} - \pi^{mi}N^{j}{}_{,m}\,, \tag{A.4b}$$

which give the change (Lie derivative) of g_{ij} and π^{ij} under an infinitesimal tangential deformation **N**, shows that the asymptotic conditions (A.1-2) are also invariant under asymptotic rotations and translations, i.e., when $N^{i} \to \alpha^{i} + \beta_{j}{}^{i}x^{j}$. [See in this context the remarks in paragraph (c) of this appendix.] Conditions (A.1-2) are therefore Poincaré invariant.

B. *Vanishing of Surface Integrals*

Let us show now that the surface integrals neglected in Section 5 indeed vanish. Consider first the quantity

$$r \int d^{2}\Omega\; x^{l}x^{r}(2\delta_{s}{}^{k}\pi^{jl} - \delta_{s}{}^{l}\pi^{jk})\, \delta g_{jk} \cdot \mathbf{x}^{2} = r^{2}, \tag{A.5}$$

encountered in (5.7). Only ${}^{(2)}\pi^{jl}$ and ${}^{(1)}\delta g_{jk}$ could, a priori, contribute to (A.5) in the limit $r \to \infty$. However, since only even ${}^{(2)}g_{ij}$ are admitted, δg_{jk} in (A.5) must be also even; but, on the other hand ${}^{(2)}\pi^{jl}$ is odd, which means that the expression being integrated over solid angles in (A.4) is odd under the inversion $\mathbf{x} \to -\mathbf{x}$, and consequently its integral over $d^{2}\Omega$ vanishes.

The same kind of argument used above can be applied to (5.11). In fact it follows from (A.1-2) that the quantity

$$\{{}^{(0)}G^{ijkl} + {}^{(1)}G^{ijkl}\}\, \{x^{r}(\delta^{(1)}g_{ij,k}) - \delta_{k}{}^{r}\delta^{(1)}g_{ij}\} \tag{A.6}$$

is even under inversion, which implies that its integral over $d^{2}s_{l} \equiv r d^{2}\Omega x^{l}$ vanishes.

Equation (5.12) is therefore established. The transition from (5.12) to (5.13) is again accomplished by noticing that ${}^{(1)}g_{ij}$, being even under inversion, does not contribute to the integration in (5.13). We have thus that the surface integrals disregarded in Section 5 indeed vanish.

C. *Invariance of P^μ and $M_{\mu\nu}$ Under Changes of the Spatial Coordinates*

In order for P^μ and $M_{\mu\nu}$ to have an unambiguous physical meaning, the expression (5.8-10) and (5.13) must be independent of the particular system of spatial coordinates used to actually carry out the integration, as long as the asymptotic conditons (A.1-2) hold and as long as neither rotations nor translations at infinity are present. If the most general infinitesimal transformation of coordinates which leaves (A.1-2) invariant would be of the form $N^i \rightarrow \alpha^i + \beta_j{}^i x^j + 0(r^{-1})$ there would be nothing to prove, as we already know from the main text that for that class of shift vectors P^μ and $M_{\mu\nu}$ are conserved when $\alpha^i = \beta_j{}^i = N^\perp = 0$. [See discussion following (5.27c).] There is however a more general class of shift vectors which leaves (A.1-2) unchanged, namely those which behave asymptotically as

$$N^i \xrightarrow[r\to\infty]{} \alpha^i + \beta^i{}_j x^j + \xi^i(\mathbf{n}) + O(1/r) \tag{A.7a}$$

with

$$\xi^i(-\mathbf{n}) = -\xi^i(\mathbf{n}), \tag{A.7b}$$

and one may easily convince oneself that (A.7) is the most general possibility.

The point now is that all the derivations of the main text will still go through if we relax condition (5.6) to allow for the possibility of the extra term $\xi^i(\mathbf{n})$ in (A.7), which implies in turn that the coordinate invariance of P^μ and $M_{\mu\nu}$ will also follow from their conserved character under tangential deformations not involving rotations or translations at infinity, in the more general case (A.7). To show that the conclusions of the main text are not changed by the presence of $\xi^i(\mathbf{n})$ in (A.7) we must prove that ξ^i does not contribute to any of the surface integrals neglected in the variation of the Hamiltonian. This follows by a reasoning similar to the one employed in Section B of this appendix: The relevant integral is [from (2.6)]

$$-2 \oint d^2s_l\, \xi_k\, \delta\pi^{kl} = -2r^2 \oint d^2\Omega\, n_l\xi_k\, \delta\pi^{kl}, \tag{A.8}$$

which vanishes because $n_l\xi_k\delta\pi^{kl}$ is odd under inversion on account of (A.2b) and (A.7b).

The proof that (A.8) vanishes is not however the end of the story. One is not free to relax the restrictions on the asymptotic behavior of N^i, so as to allow for the presence of ξ^i in (4.5), without allowing for a corresponding amount of extra freedom in the lapse function $N^{\perp}$. In fact, it follows from (5.23a) that by commuting a tangential deformation of the type (4.5a) with a normal deformation which goes as $N^{\perp} \sim \alpha^{\perp} + \beta^{\perp}{}_r x^r$ at infinity, we obtain a normal deformation which behaves asymptotically as

$$N^{\perp} \underset{r\to\infty}{\sim} \alpha^{\perp} + \beta^{\perp}{}_r x^r + \xi^{\perp}(\mathbf{n}) \qquad \text{(A.9a)}$$

with

$$\xi^{\perp}(-\mathbf{n}) = -\xi^{\perp}(\mathbf{n}). \qquad \text{(A.9b)}$$

Thus, collecting (A.7) and (A.5), we see that (5.6) gets replaced by the less restrictive statement

$$N^{\mu} \underset{r\to\infty}{\sim} \alpha^{\mu} + \beta^{\mu}{}_r x^r + \xi^{\mu}(\mathbf{n}) + O(1/r) \qquad \text{(A.10a)}$$

with

$$\xi^{\mu}(-\mathbf{n}) = -\xi^{\mu}(\mathbf{n}). \qquad \text{(A.10b)}$$

A glance at (5.23) shows now that by commuting deformations of the type (A.10) one gets always a new element of the same class.

Having enlarged the class of allowed laps functions we must check that no nonvanishing surface integral is picked up in this way. The relevant integral this time is [from (2.6)]

$$-r^2 \int d^2\Omega\, n_l^{(0)} G^{ijkl}(\xi^{\perp}\, \delta^{(1)} g_{ij,k} - \xi^{\perp}{}_{,k}\, \delta^{(1)} g_{ij}), \qquad \text{(A.11)}$$

but we again find from (A.2a) and (A.10b) that the quantity being integrated over solid angles in (A.11) is odd under inversion and consequently (A.11) vanishes, which is what we wanted to prove.

To finish we should emphasize that all the conditions related to oddness or eveness of certain quantities are needed only to deal with the angular momentum $M_{\mu\nu}$. To start with conditions (A.2) ensure that $M_{\mu\nu}$ is finite! No parity conditions of the type (A.2) and (A.10b) would be necessary if we wanted to deal only with P^{μ} (i.e., if we would set $\beta_{\mu\nu} = 0$ in all equations).

Acknowledgments

We aie indebted to Stanley Deser, Robert Geroch, Karel Kuchař and James York for their comments. A particularly interesting point raised by Professsors Deser and Kuchař in stimulating discussions with one of us (C.T.) w as the desirability of relaxing the asymptotic conditions (A-1) and (A-2) in order to allow a more general class of spacetimes to fit in our scheme. We plan to investigate this question in the future. Finally, C.T. would like to thank John A. Wheeler for much encouragement.

References

1. B. S. DEWITT, *Phys. Rev.* **160** (1967), 1113.
2. P. A. M. DIRAC, *Phys. Rev.* **114** (1959), 924.
3. R. ARNOWITT, S. DESER, AND C. W. MISNER, *in* "Gravitation: An Introduction to Current Research" (L. Witten, Ed.), John Wiley and Sons, New York, 1962.
4. R. ARNOWITT, S. DESER, AND C. W. MISNER, *Nuovo Cimento* **19** (1961), 668.
5. R. ARNOWITT, S. DESER, AND C. W. MISNER, *J. Math. Phys.* **1** (1960), 434.
6. P. A. M. DIRAC, *Canad. J. Math.* **3** (1951), 1; "Lectures on Quantum Mechanics," Belfer Graduate School of Science, Yeshiva University, New York, 1964.
7. K. KUCHAŘ, *J. Math. Phys.* **11** (1970), 3322; *Phys. Rev. D* **4** (1971), 955.
8. C. TEITELBOIM, "The Hamiltonian Structure of Spacetime," Ph.D. Thesis, Princeton, 1973, unpublished.
9. A. LICHNEROWICZ, *J. Math. Pure Appl.* **23** (1944), 37.
10. I. CHOQUET-BRUHAT, *Symposia Mathematica* **12** (1973), Instituto Nazionale di Alto Matematica, Bologna; *C. R. Acad. Sci. Paris* **274** (1972), 682. J. W. YORK, *Phys. Rev. Lett.* **28** (1972), 1082; N. O'MURCHADHA, "Existence and Uniqueness of Solutions to the Hamiltonian Constraint of General Relativity," Ph.D. Thesis, Princeton, 1972, unpublished; N. O'MURCHADHA AND J. W. YORK, *J. Math. Phys.* **14** (1973), 1551.
11. R. F. BAIERLEIN, D. H. SHARP AND J. A. WHEELER, *Phys. Rev.* **126** (1962), 1864; J. A. WHEELER, *in* "Relativity, Groups and Topology," 1963 Les Houches Lectures, University of Grenoble (C. DeWitt and B. DeWitt, Eds.), Gordon and Breach, New York, 1964; Superspace, *in* "Analytic Methods in Mathematical Physics" (R. D. Gilbert and R. Newton, Eds.), Gordon and Breach, New York, 1970.
12. A. J. HANSON AND T. REGGE, *Ann. Phys.* (*N.Y.*) in press.
13. C. TEITELBOIM, *Ann. Phys.* (*N.Y.*) **79** (1973), 542.
14. S. HOJMAN, K. KUCHAŘ, AND C. TEITELBOIM, *Nature Phys. Sci.* **245** (1973), 97; *Ann. Phys.* (*N.Y.*), to be published. K. KUCHAŘ, *J. Math. Phys.*, in press.
15. P. W. HIGGS, *Phys. Rev. Lett.* **3** (1959), 66.

Regge–Teitelboim Analysis of the Symmetries of Electromagnetic and Gravitational Fields on Asymptotically Null Spacelike Surfaces[1]

To Tullio, for old times' sake

Claudio Bunster[†], Andrés Gomberoff [*,†] and Alfredo Pérez[†]

[†] *Centro de Estudios Científicos (CECs), Avenida Arturo Prat 514, Valdivia, Chile*

[*] *Facultad de Ingeniería y Ciencias, Universidad Adolfo Ibáñez, Avda. Diagonal las Torres 2640, Peñalolén, Santiago, Chile*

We present a new application of the Regge–Teitelboim method for treating symmetries which are defined asymptotically. It may be regarded as complementary to the one in their original 1974 paper. The formulation is based on replacing an asymptotic plane by the two-sheeted "hourglass" shaped surface obtained by joining smoothly an incoming hyperboloid with an outgoing one. The hyperboloids have a fixed radius, and as one moves the center of the hourglass along the time axis one covers the whole of spacetime. The motivation is to study radiation, and the hourglass is well suited to the task because it is asymptotically null, and thus is able to register the details of the process. A simple parity condition for the fields on the hyperboloid is given. It specifies that as much radiation is coming in as it is going out. With it, a Hamiltonian formulation of the symmetry of Bondi, van der Burg, Metzner and Sachs is developed for both electromagnetism and gravitation. It is indispensable for the construction to have electric-magnetic duality asymptotically. For gravitation, an electric-magnetic duality invariant formulation is not available at present; but enough rudiments of it are at hand, so that the main results can be conjectured. A definition for angular momentum which is conserved (for which the "magnetic sector" is essential) is given. It incorporates an interrelationship between spin and charge. For the gravitational field, Taub–NUT space appears as the analog of a magnetic pole.

1. Introduction

The 1974 paper by Regge and Teitelboim[1] contained two main results: (i) A completion of Dirac's analysis[2] of the role of constraints in field dynamics that was necessary in order to account for the different character of the gauge transformations which do not change the physical state ("proper gauge transformations"), from those which do ("improper gauge transformations"). It was found that, in the latter case, Dirac's "weakly vanishing" generators have to be improved by the addition of a surface integral and do not vanish weakly. The surface integral gives the value of the charge associated to the improper transformations. (ii) An application of (i) to obtain a Poincaré invariant formulation of the theory of gravitation on spacelike surfaces which are asymptotically planes.

We develop below a new application of the method which is based on replacing an asymptotic plane by the two-sheeted "hourglass" shaped surface obtained by joining smoothly an incoming hyperboloid with an outgoing one. The hyperboloids have a fixed radius, and as one moves the center of the hourglass along the time axis one covers the whole of spacetime. The motivation is to study radiation, and the hourglass is well suited to the task because it is asymptotically null, and thus is able to register the details of the process. In contradistinction, if one goes far enough on an asymptotic plane, the radiation emitted by a confined source has not enough time to reach there.

A simple parity boundary condition, with a direct physical meaning, for the fields defined on the hourglass is given. With it, a Hamiltonian formulation of the symmetry of Bondi, van der Burg, Metzner and Sachs[3–5] is developed. The formulation yields a Hamiltonian definition of the Bondi "news". For electromagnetism, an interrelationship between spin and the BMS charge is brought out, and manifest electric-magnetic duality is achieved by introducing a second vector potential. For the gravitational field, Taub–NUT space appears as a gravitational magnetic pole and the magnetic BMS charges are exhibited. However, an expression for their conjugates, the magnetic supertranslations, is lacking.

The "hourglass" construction, the parity conditions on it and the intimate connection between BMS invariance and electric-magnetic duality were reported in Ref. 6.

If one gives initial data on one hyperbolic hourglass, the Hamiltonian equations of motion determine the canonical variables on any other hyperbolic hourglass. However, the initial data should only be specified on one half of the hourglass. This is because a point, which is not at infinity, lying, say, on the outgoing half of one hourglass at a given time also lies on the incoming half of another hourglass at a later time. This double counting does not happen at infinity, and hence one needs to specify the radiation on the other half. That is, if one is giving the data on the outgoing half one must specify in addition the incoming Bondi news. But the parity condition states that they are equal to the outgoing ones, and therefore it is sufficient to specify just the data on the outgoing half of the hourglass (or, vice versa, on the incoming one). In this sense the hyperbolic hourglass is a Cauchy surface.

If compared with an hourglass formed by incoming and outgoing light cones, the one formed by hyperboloids has the advantage of being spacelike and therefore permitting direct step by step use of the Dirac's procedure with the Regge–Teitelboim complement, which has been battle-tested, and in which all the structures that appear (action, Hamiltonian, Poisson and Dirac brackets, surface deformations, most general permissible motion) are well, and tightly, defined from the start.

Furthermore, the slicing by hyperbolic hourglasses of fixed radius and varying center has the essential property of covering the whole of spacetime, in contradistinction with the foliations by hyperboloids of fixed center and varying radius, used previously by many authors, which only cover part of it.

The structure of the paper is the following. In order to make the treatment self-contained and set the notation and terminology, section 2 begins by reviewing the general procedure. Section 3 discusses the foliation by hyperbolic hourglasses of the same radius and different center. Next, section 4 contains an analysis of the asymptotic properties of the free electromagnetic field on the hyperbolic hourglass foliation. The discussion is given in detail, because practically all the results derived for electromagnetism can be translated literally to the gravitational case, whose treatment becomes then considerably lighter. Section 5 is then devoted to the gravitational case.

Three appendices are included: appendix A gives explicit expressions for the Poincaré generators on the hyperbolic hourglass foliation. Appendix B gives the details of the preservation in time of the parity boundary conditions. Appendix C provides a "dictionary" for translating in the gravitational case the variables which appear in the present Hamiltonian treatment with those employed in the original BMS light cone analysis.

2. Hamiltonian field dynamics, surface deformations, gauge transformations, surface integrals, and conservation laws

2.1. *Quick review of formulation*

In the formulation of field dynamics in which the state is defined on a general spacelike surface developed by Dirac,[2] and completed by Regge and Teitelboim[1] to incorporate symmetries which are defined asymptotically, the generator — through Poisson brackets — of the "most general permissible motion" has the form

$$H[\xi;\lambda] = H_0[\xi;\lambda] + Q[\xi;\lambda]\,, \tag{1}$$

where $H_0[\xi;\lambda]$ is an integral over the spacelike surface on which the state is defined, of the form

$$H_0[\xi;\lambda] = \int d^3x\,(\xi^\mu \mathcal{H}_\mu + \lambda^a G_a)\;, \tag{2}$$

and $Q[\xi,\lambda]$ is a surface integral over the asymptotic boundary of that spacelike surface.

The surface integral $Q[\xi;\lambda]$ is included to make well-defined the functional derivatives of $H[\xi;\lambda]$, so that one has,

$$\delta H[\xi;\lambda] = \int d^3x \left(\frac{\delta H}{\delta\phi}\delta\phi + \frac{\delta H}{\delta\pi}\delta\pi\right)\,, \tag{3}$$

without any surface terms. This means that the contribution of the asymptotic part of the field is already included in (3). Here we have abbreviated as (ϕ,π) all the canonical field variables of the theory.

In (2) the $\mathcal{H}_\mu$ are the generators of deformations of the spacelike surface in which the state is defined, while the G_a are the generators of internal gauge transformations. They are both constrained to vanish, that is, they are weakly equal to zero:

$$\mathcal{H}_\mu \approx 0, \quad G_a \approx 0\,. \tag{4}$$

2.1.1. *Proper and improper gauge transformations*

If the parameters (ξ,λ) are such that the surface integral $Q[\xi;\lambda]$ vanishes, the motion generated by $H_0[\xi;\lambda]$ is called a "proper" gauge transformation,[7] and it it is not a symmetry, but rather an expression of the fact that the system is described by variables which are redundant. This normally happens when they vanish at infinity, but there are important cases, in both electromagnetism and gravity[1] where the surface integrals vanish even though the parameters do not vanish at infinity but obey parity conditions there. In that case the transformation is still proper. For an internal symmetry one feels on safe grounds stating that a proper gauge transformation does not change the physical state on a given spacelike surface. For the case of a surface deformation this point of view may be kept for purely tangential deformations (changes of spatial coordinates), but if the deformation has a normal component the hypersurface is geometrically deformed, and therefore, if one can "perform local observations inside", one would expect the physical state to change. On the other hand if one only performs observations at infinity, then one may safely take the point of view that proper normal deformations are also gauge transformations, and do not change the physical state either. If (ξ, λ) are such that the surface integral $Q[\xi; \lambda]$ does not vanish the motion is called an "improper" gauge transformation, and one expects it to change the physical state. As a consequence of (3), the functional form of the transformation generated by (1) is the same if the transformation is proper or improper, the difference is only introduced by the asymptotic behavior of the transformation parameters.

The difference between proper and improper transformations manifests itself at the level of the action principle in which (1) is the Hamiltonian, in the fact that the constraints (4) are obtained from it by extremizing the action with respect to ξ and λ *keeping fixed the part that contributes to the surface integral* $Q[\xi, \lambda]$. Thus, (4) states that the generator of proper gauge transformations vanishes weakly. On the other hand, for improper gauge transformations this does not happen:

$$H[\xi; \lambda] \approx Q[\xi; \lambda] \neq 0 \,.$$

If one is interested in the action of an asymptotic motion, one gives the asymptotic part of the transformation and continues it inside in an arbitrary manner. The way in which one chooses to continue inside is irrelevant because any two continuations differ by a proper gauge transformation. It is however necessary to continue, because it is only the sum of the volume part of the generator and the surface term which has well-defined functional derivatives and is therefore capable of acting through a Poisson bracket. Neither the volume part alone, nor the surface integral alone has that capability. Alternatively, one may choose a particular continuation inside by fixing the gauge. If this is done, the combined set of the original gauge constraints and the gauge conditions become second class, and one can pass from the original Poisson bracket to the associated Dirac bracket, in terms of which the second class constrains vanish strongly and have zero bracket with everything. Then the surface term stands alone and is capable to act as itself as a generator through the Dirac bracket.

2.1.2. *Commutation of deformations*

The constraint-generators $\mathcal{H}_\mu$, G_a are first class. If we denote collectively by $\Lambda_1 = (\xi_1, \lambda_1)$, $\Lambda_2 = (\xi_2, \lambda_2)$, the parameters of any two motions one has, for proper gauge transformations,

$$\Big[H[\Lambda_1], H[\Lambda_2]\Big] = H[\Lambda_{12}] \,, \tag{5}$$

where the commutator $\Lambda_{12} = -\Lambda_{21}$ of the two original infinitesimal transformations is a bilinear expression of Λ_1, Λ_2 and their derivatives (in practice, first derivatives, with coefficients which in general depend on the fields). For improper transformations it may happen that Eq. (5) is relaxed by the appearance of a central extension on its right-hand side, that is, by the addition to $H[\Lambda_{12}]$ of a term that has zero Poisson brackets with all the dynamical variables. This is not allowed for proper transformations because it would spoil the first class character of the constraints; i.e. the vanishing of the constraints would not be preserved by the transformation. This obstruction to the presence of a central extension does not happen for improper transformations because the charge Q is not constrained to be zero.

Given any theory one may calculate Λ_{12} by working out directly the Poisson bracket on the left side of (5). However, if one has geometrical insight on the nature of the motions at hand one may write down the result without doing that calculation. For example, for a Yang–Mills theory, with structure constants $C^a{}_{bc}$, one has

$$\lambda_{12}^c = C^c{}_{ab}\lambda_1^a\lambda_2^b \,,$$

and for two surface deformations within an arbitrary Riemannian spacetime with Lorentzian signature one has,[8,9]

$$\xi_{12}^{\perp} = \xi_1^i\xi_{2,i}^{\perp} - \xi_2^i\xi_{1,i}^{\perp} \,, \tag{6}$$

$$\xi_{12}^i = g^{ij}\left(\xi_1^{\perp}\xi_{2,j} - \xi_2^{\perp}\xi_{1,j}\right) + \xi_1^j\xi_{2,j}^i - \xi_2^j\xi_{1,j}^i \,, \tag{7}$$

where g_{ij} is the metric of the spacelike surface.

2.1.3. *No royal road*

Lastly, we elaborate on the sentence "...The surface integral $Q[\xi;\lambda]$ is included in (1) to make well-defined the functional derivatives of $H[\xi;\lambda]$..." written above. To achieve this it is necessary to find an appropriate set of boundary conditions. There is no foolproof, inductive method for that. One rather works by trial and error and there is no guarantee of success. The procedure in practice is as follows: (i) A tentative set of asymptotic conditions is obtained by applying the asymptotic transformations that one wants to have present, to a simple field configuration that one also wants to have present. For example, in the case of gravitation, one would boost a Schwarzschild field; or, in electromagnetism, a Coulomb field. (ii) One extracts properties of the result obtained that can be formulated independently of the specific original configuration, and uses them as a starting ansatz. For example, one may retain a decay rate in inverse powers of the radial distance and a parity condition for the coefficients. (iii) One finds the most general parameters (ξ, λ) which preserve the ansatz. It may happen then that that set of parameters does not contain all the symmetries that one was interested in (for example, the complete Poincaré group). Then one relaxes the ansatz to make room. If success is achieved — meaning, in the example just given, that one has the Poincaré group or more — one checks whether the surface integral that appears in the variation of the H_0 is the variation of a finite surface integral, or, as one says colloquially, "if the δ can be taken out". If this happens one is done. If it does not, one modifies the boundary conditions in the light of the nature of the failure. With luck and dedication the process converges and one finally succeeds.

3. Foliation of Minkowski space by hyperboloids of the same radius and different centers

3.1. *Incoming and outgoing hyperboloids*

In Minkowski space it is natural to define the state on a three-dimensional spacelike hyperboloid, because that surface is mapped onto itself by a Lorentz transformation. In this sense, hyperboloids are more adequate to the special principle of relativity than the planes corresponding to inertial frames, for which the boosts are interchanged with spatial translations. This possibility was considered by Dirac in 1940[10] and he called it the "point form of field dynamics" with the term "point" referring to the center of the hyperboloid.

A spacelike hyperboloid with center at $x^{\mu}_{(0)}$ and radius τ_0 obeys the equation

$$(x^{\mu} - x^{\mu}_{(0)})(x_{\mu} - x_{(0)\mu}) = -\tau_0^2 \,. \tag{8}$$

Actually, Eq. (8) describes two disjoint hyperboloids, one with $x^0 > x^0_{(0)}$ ("outgoing hyperboloid") and another with $x^0 < x^0_{(0)}$ ("incoming hyperboloid"). Although Dirac did not discuss foliations of spacetime by means of a family of hyperboloids, this has been done by many authors but to our knowledge in all cases treated so far the foliation has been defined by keeping $x^{\mu}_{(0)}$ fixed and letting τ_0 vary as one passes from one hyperboloid to the next.

In other words, the foliations used previously have consisted of a sequence of hyperboloids with fixed center and varying radius.[a] These foliations have the advantage that the four x^{μ} are treated on the same footing so Lorentz invariance is manifest; but the price payed is extremely high, because only a small part of Minkowski space is covered, and moreover, a spurious explicit dependence on the varying τ_0, which is taken as the time τ, is introduced. Here we take the other natural option, we keep the radius τ_0 fixed and we allow the position of the center to vary. This is a direct extension of what is done with null foliations, which may be regarded as being the limit $\tau_0 \to 0$. The actual value of τ_0 will turn out to be irrelevant, since all the quantities of physical interest will incorporate naturally τ_0 in their units.

3.2. *The hyperbolic hourglass*

The hyperbolic hourglass consists of an outgoing hyperboloid with center $x^{\mu}_{(0)} = (-\tau_0, 0)$ joint to an incoming one with center $x^{\mu}_{(0)} = (\tau_0, 0)$. At time t it has spatial coordinates (r, ϑ, φ), whose range is:

$$\begin{aligned} -\infty < r < +\infty, \\ 0 \le \vartheta \le \pi, \\ 0 \le \varphi < 2\pi. \end{aligned}$$

[a]See, for example, Refs. 11–13 and also Ref. 14 and references therein. In some of these discussions timelike hyperboloids are employed (in which case $-\tau_0^2$ in (8) is replaced by λ_0^2).

It is defined parametrically from Minkowskian coordinates $(x^0, \boldsymbol{x})$ through

$$x^0 = t + r\sqrt{1 + \frac{\tau_0^2}{r^2}} + \tau_0 \quad r \leq 0 \tag{9}$$

$$x^0 = t + r\sqrt{1 + \frac{\tau_0^2}{r^2}} - \tau_0 \quad r \geq 0, \tag{10}$$

and

$$\boldsymbol{x} = r\hat{r}\ , \quad -\infty < r < +\infty, \tag{11}$$

where the unit vector $\hat{r}$ is given by

$$\hat{r} = (\cos\vartheta\cos\varphi, \cos\vartheta\sin\varphi, \sin\vartheta). \tag{12}$$

The radius τ_0 is taken to be positive. The embedding defined by the above equations is continuously differentiable. The tangent vectors are continuous at $r = 0$ and the surface has a well-defined global orientation.

The hyperbolic hourglass may be regarded as a spacelike deformation of the full (past and future) lightcone, with an orientation inherited from the propagation of a light front that comes in, goes through itself, and then comes out. Since this wave propagation process is physically smooth, fields defined on the global coordinate system just described should be smooth. (The parametric equations (9)–(11) automatically incorporate the antipodal map,[15,16] which amounts to rewriting them by using a positive r for both sheets of the hyperboloid and inverting the orientation of the two-sphere at a given r. That is, keeping (9)–(11) for $r \geq 0$ and setting, $r' = -r$, $\hat{r}' = -\hat{r}$, for $r \leq 0$.)

If one considers an incoming wave which is not spherically symmetric, then the spacetime point at which the wavefront goes through itself will be different for different $\hat{r}$'s. But in the present paper we are only interested in the analysis of the asymptotic region and therefore the details of what happens inside are irrelevant. The key aspects are the asymptotic hyperbolic shape and its orientation inherited from that of an incoming wave that goes through itself and becomes outgoing.

Figure 1 shows the embedding in Minkowski space of a single hyperbolic hourglass, Figure 2 exhibits the slicing of Minkowski space by a one parameter family of hyperbolic hourglasses and Figure 3 shows a sequence of Penrose diagrams with hyperbolic slicings of different radius τ_0.

4. Electromagnetic field in Minkowski space

We will analyze in this section the case of the electromagnetic field on a fixed Minkowskian background. Practically all the features that will be encountered in the gravitational case already appear in this technically simpler context.

The main difference which does not hinder the analogy is that, since the background is fixed, its Poincaré symmetry appears as a global symmetry rather than an asymptotic gauge symmetry. There are no constraints associated with the surface deformation ξ, which are

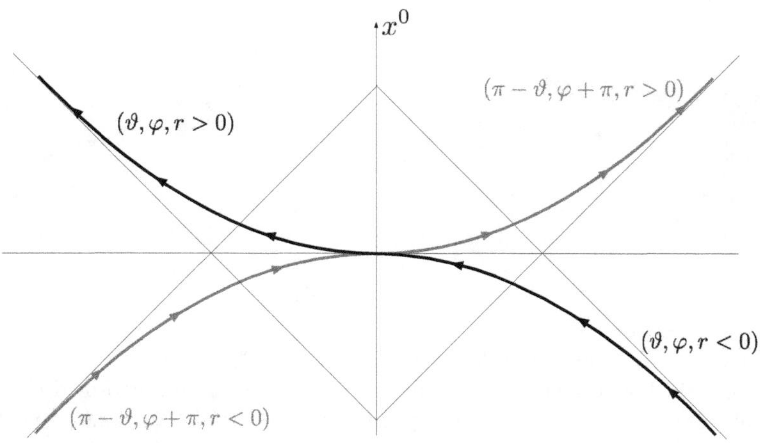

Figure 1. *The hyperbolic hourglass.* The figure shows a two-dimensional cut of an incoming hyperboloid and an outgoing one which are joint smoothly at at $r = 0$. The arrows show the direction of increasing r, which coincide asymptotically with the direction of propagation of a wave that comes in, goes through itself, and then comes out. If the incoming wave is not spherically symmetric, the spacetime point at which the wavefront goes through itself will be different for different (ϑ, φ). For the analysis of the asymptotic region the details of what happens inside are irrelevant. The key aspects are the asymptotic hyperbolic shape and its orientation inherited from that of an incoming wave that goes through itself and becomes outgoing.

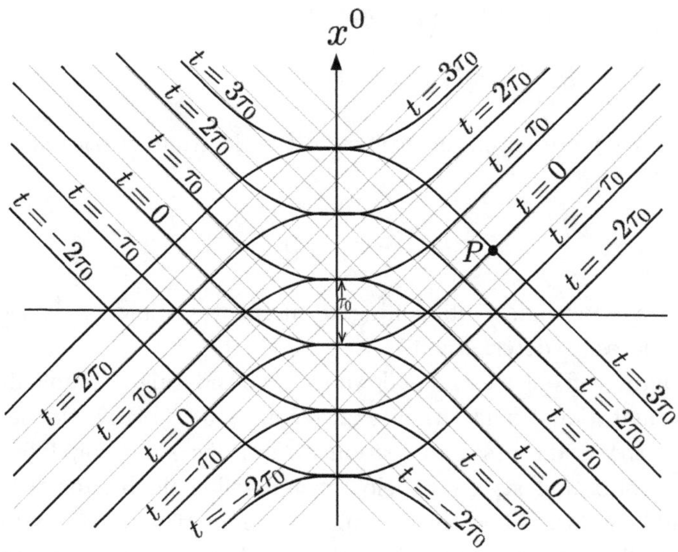

Figure 2. *Slicing by hyperbolic hourglasses.* A given spacetime point is labeled by two set of coordinates. In the case of the point P shown in the figure, these are $(t = 0, r, \vartheta, \varphi)$ and $(t = 3\tau_0, -r, \pi - \vartheta, \varphi + \pi)$.

not varied in the action principle. The $\mathcal{H}_\mu$ in (2) are replaced by the energy and momentum

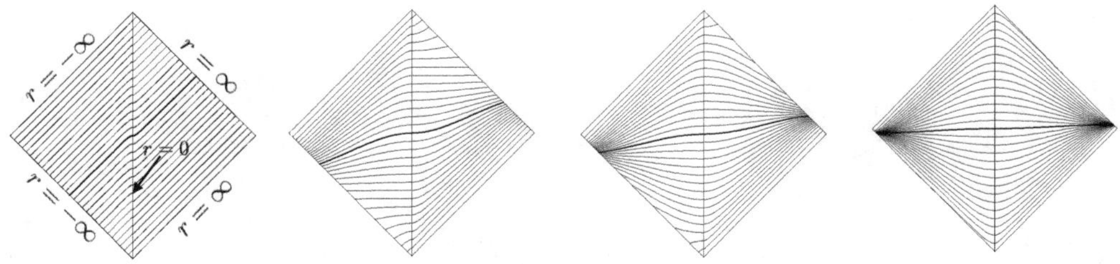

Figure 3. *Limits $\tau_0 \to 0$ and $\tau_0 \to \infty$ for Minkowski space.* The succession of conformal diagrams shows from left to right how the surfaces of the hourglass foliation are deformed from nearly light cones to nearly planes as τ_0 increases from a very small value to a very large one. To better illustrate the effect, different members of the foliation are shown in the different figures of the sequence; but, to keep track of the deformation, the surface at $t = 0$ (shown with a heavy line) in all cases. The Penrose diagram has been doubled to admit negative values of r in the left triangular area. This doubling shows how the curves of constant t, ϑ, φ are smooth spacelike curves that connect asymptotically past and future null infinities. The scale of the length τ_0 is irrelevant for the effect described in the figure, which only depends on the ratios between the different τ_0's shown.

densities of the electromagnetic field

$$\mathcal{H}_{\perp}^{(elm)} = \frac{1}{2}\left(g^{-\frac{1}{2}}\pi_i\pi^i + \frac{1}{2}g^{\frac{1}{2}}F^{ij}F_{ij}\right), \tag{13}$$

$$\mathcal{H}_i^{(elm)} = F_{ij}\pi^j . \tag{14}$$

The only gauge symmetry present in the problem is the electromagnetic one, whose generator is

$$\mathcal{G} = -\pi^i_{\ ,i} \approx 0 . \tag{15}$$

Here A_i is the vector potential, π^i its conjugate momentum, and g_{ij} is the metric on the hourglass, and g denotes its determinant.

If instead of having a fixed background we were considering dynamically coupled electromagnetic and gravitational fields, then expressions (13), (14) would be added to their gravitational counterparts discussed in section 5, and the sum would be constrained to vanish. The asymptotic analysis given below would still hold because at large distances the spacetime would be flat. Then the asymptotic symmetry transformations of the coupled Einstein–Maxwell system would be those discussed here (internal electromagnetic, and Poincaré transformations) and the additional gravitational supertranslations.

We will now discuss the Poincaré and proper and improper gauge transformations for the electromagnetic field on the hourglass slicing. In this case the time equal constant surface is left invariant under the Lorentz group, whereas it is mapped onto a different hyperboloid by spacetime translations. Thus if one compares the situation with $t =$ constant planes, one sees that the roles of spatial translations and boosts are interchanged.

4.1. *Asymptotic boundary conditions*

4.1.1. *Power expansion near* $r = \pm\infty$

By applying the procedure described at the end of Sec. 2, starting from the Coulomb field written in hyperbolic coordinates, one is led to the boundary conditions,

$$A_a = a_a^{(0)} + a_a^{(1)} r^{-1} + \mathcal{O}(r^{-2}), \tag{16}$$
$$A_r = a_r^{(2)} r^{-2} + \mathcal{O}(r^{-3}), \tag{17}$$
$$\pi^a = \pi^a_{(2)} r^{-2} + \mathcal{O}(r^{-3}), \tag{18}$$
$$\pi^r = \pi^r_{(0)} + \mathcal{O}(r^{-1}), \tag{19}$$
$$\lambda = \lambda_{(0)} + \mathcal{O}(r^{-1}), \tag{20}$$

Here λ is the Lagrange multiplier that accompanies the gauge generator (15). In addition to the power law decays (16)–(20) it is necessary to introduce parity conditions. This is achieved by splitting some of the variables in longitudinal and transverse parts as follows

$$a_a^{(0)} = \nabla_a F + \star\nabla_a \bar{G}, \tag{21}$$
$$h_a = \gamma^{\frac{1}{2}} \left(\nabla_a N + \star\nabla_a \bar{N}\right). \tag{22}$$

Here,

$$\star\, \nabla_a = \gamma^{\frac{1}{2}} \epsilon_{ab} \nabla^b, \tag{23}$$

where γ is the determinant of the metric γ_{ab} on the unit two-sphere. The "news" vector h^a in (22), which will play a central role in what follows, is defined by

$$h^a = \frac{1}{\tau_0^2}\left(\pi^a_{(2)} + \gamma^{\frac{1}{2}} \gamma^{ab} f^{(2)}_{br}\right), \tag{24}$$

for $r \to \pm\infty$, where the $f^{(2)}_{br}$ is the leading order coefficient of F_{br}. In Minkowski coordinates the news correspond to an electromagnetic field that decays as r^{-1}, that is to a wave emerging from a confined source ($r \to +\infty$), or converging towards an absorber($r \to -\infty$). For an accelerating electric charge e one has from the Lienard–Wiechert field,

$$\gamma^{\frac{1}{2}} f^{ar}_{(2)} \boldsymbol{\partial}_a = \pi^a_{(2)} \boldsymbol{\partial}_a = -2e\gamma^{\frac{1}{2}} \hat{r} \times (\hat{r} \times \boldsymbol{a}),$$

where $\boldsymbol{a}$ is the acceleration in rest frame of the emitter (outgoing wave) or absorber (incoming wave). See for example.[17,18]

On the entire hourglass this field stems from

$$F_{rad} = F_{ret} - F_{adv}$$

introduced by Dirac,[19] which is derived from the function,

$$\frac{1}{r}\delta(t-r), \qquad -\infty < r < \infty\,.$$

This function is not a Green function, but a solution of the homogeneous equation: the difference $F_{ret} - F_{adv}$. Therefore it has as much radiation coming in as going out, and obeys the parity conditions that will be discussed next.

4.1.2. *Parity conditions*

The parity conditions will be then the following,

$$(F, \lambda_{(0)}, N, \bar{N})\big|_{+\infty} = (F, \lambda_{(0)}, N, \bar{N})\big|_{-\infty} \quad \text{for each } (\vartheta, \varphi). \tag{25}$$

Parity conditions play a fundamental role in the Regge–Teitelboim discussion of Poincaré invariance on asymptotic planes. We see that when dealing with Bondi, Metzner, and Sachs invariance on hyperboloids, they again come in.[b] The reason for their appearance is discussed next.

4.1.3. *Surface deformation algebra*

The surface deformations involved in the present analysis are Poincaré deformations of the hypersurface, which are motions generated by the Killing vectors given in appendix A, and besides them, gauge transformations. If one evaluates the commutator of any two such deformations one finds the following results for the asymptotic part of the commutator:

(i) Two Poincaré deformations close according to the Poincaré group.

(ii) Spacetime translations commute with improper gauge transformations.

(iii) The commutator of a Lorentz transformation with Killing vector ξ_{Lorentz} and an improper gauge transformation with parameter $\lambda(\vartheta, \varphi)$ at infinity, is a gauge transformation with parameter,

$$\zeta = \xi^a_{\text{Lorentz}} \partial_a \lambda. \tag{26}$$

The results (i)–(iii) are the expressions, in terms of deformations, of the electromagnetic BMS algebra.

Equation (26), which technically stand from the fact that the Lorentz group acts asymptotically as a Lie derivative on the two sphere (appendix A) has a profound consequence: it mixes improper gauge transformations with spacetime motions. This provides a tantalizing confirmation of the fact that improper gauge transformations are physically relevant motions that cannot be "factored away".

4.2. *The hyperbolic hourglass as an unconventional Cauchy surface*

The physical motivation for the parity conditions is very simple. They essentially state that for a closed system (the free electromagnetic field in this case) everything that comes in must come out. That is, one allows for non-vanishing incoming and outgoing fluxes of energy, momentum, and other (BMS) charges; but requires that the net flux should be equal to zero.

This requirement, which physically is a condition connecting the remote past with the remote future, can be formulated as a fixed time statement, because the spacelike hyperbolic hourglass is asymptotically tangent to the past and future lightcones. This is the reason for bringing in the hourglass in the first place.

---

[b]The BMS symmetry has been tamed to fit a foliation by surfaces that are asymptotically planes.[20–23] This has required dexterity, since the symmetry is intimately related to radiation and its natural habitat is an asymptotically null surface, rather than a plane.

When regarded as an initial value surface, the hourglass has the unconventional feature, that a point, which is not at infinity, lying, say, on the outgoing half of one hourglass at a given time, also lies on the incoming half of another hourglass at a later time. This implies that one cannot give freely initial value data on the complete hourglass but only on half of it, the outgoing half for example. However, the double counting of points does not happen at infinity, so if one gives data on the outgoing half one should specify additionally the incoming radiation, that is one should give the news at $r = -\infty$. But this is precisely what the parity condition does, stating that the incoming news are equal to the outgoing ones. Thus it is sufficient to specify just the data on the outgoing half of the hourglass (or, vice versa, on the incoming one) if the parity condition is imposed.

Therefore one must bring in the complete hourglass in order to deal in Hamiltonian terms with the interrelationship between past and future, but one only gives initial value data on one half of it, together with asymptotic information on the other half. In this sense the hourglass is a Cauchy surface.

4.3. *Poincaré invariance of the boundary conditions. Fiber memory*

One must demand, by consistency, that the boundary conditions (16)–(20) and the parity conditions (25) should be preserved under the Poincaré group. This can be efficiently analyzed by writing the equations of motion in Hamiltonian form,

$$\dot{A}_i = -F_{ij}\xi^j + g^{-1/2}\xi^{\perp}\pi_i + \partial_i\lambda\,, \tag{27}$$

$$\dot{\pi}^i = (\pi^i\xi^j)_{,j} - \pi^j\xi^i_{,j} + (\xi^{\perp}F^{ji}g^{1/2})_{,j}\,, \tag{28}$$

and taking as the deformation parameters that multiply the generators (13), (14) to be the components, $\xi^{\perp}$ and ξ^i of the Poincaré Killing vectors given in appendix A.

The preservation of the boundary conditions (16)–(20) is straightforward. One first verifies Lorentz invariance which is simpler because the hyperbolic hourglass is Lorentz invariant. With this established, it is sufficient to check invariance under time translations, which is more laborious but straightforward.

Next, one turns to the conditions (25). The preservation of the parity conditions under Lorentz transformations is evident because they map each asymptotic region into itself. One needs then only be concerned with time translations. Under them, the equation of motion for the leading order term of A_a is,

$$\dot{a}^0_a = \gamma^{-\frac{1}{2}}h_a, \tag{29}$$

for $r \to \pm\infty$. Its longitudinal component is

$$\dot{F} = N, \tag{30}$$

which shows that the preservation in time of the parity of F is automatically satisfied because of the parity conditions on N. The parity of F is also preserved under improper gauge transformations because of the parity condition on $\lambda_{(0)}$. For $\lambda_{(0)}$ itself there is nothing to check because it has no equation of motion.

The preservation of the parity conditions for N and $\bar{N}$ pose no problem either, as it is shown in appendix B.

Equation (30) has a highly non trivial content. It shows that, even when the generator of improper gauge transformations does not act, i.e., when $\lambda_{(0)} = 0$, and one is only moving in the time t, there is still a displacement,

$$\delta F = N\delta t \tag{31}$$

along the U(1) fiber of amount $N\delta t$ when a time δt elapses. That is: (i) if there are no news (and one does not change the gauge frame) F is conserved, (ii) if there are news during a time interval the value of F changes from F_{before} to F_{after} according to the integral of (30) over the time interval. That is, F "remembers" the news, and for that reason is called the "fiber memory". Another kind of memory, "charge memory" will be encountered below in section 4.9.

It is important to realize that (31) is not just a "redefinition of $\lambda_{(0)}$ by the amount N". This is because $\lambda_{(0)}$ is not in the phase space, and can be held fixed in the variation of the Hamiltonian, whereas N is a dynamical variable, which obeys a (gauge invariant) equation of motion and hence cannot be held fixed.

4.4. *BMS charges*

4.4.1. *Electric BMS charge*

Taking into account the parity condition on $\lambda_{(0)}$ one finds that the surface integral that must be added to the electromagnetic gauge generator to include improper transformations is given by

$$\oint \lambda_{(0)} Q \tag{32}$$

where the gauge charge Q is given by

$$Q\left(\vartheta,\varphi\right) = \pi^r_{(0)}\Big|_{+\infty} - \pi^r_{(0)}\Big|_{-\infty} \equiv Q_+ + Q_-. \tag{33}$$

It is important to interpret this expression appropriately. The hourglass is a construct that enables one to keep track, within the Hamiltonian formalism, of the incoming and outgoing radiation in an economic manner, that is without introducing separate overlapping incoming and outgoing hyperbolic patches. This brings in a redundancy: one way or another space its counted twice. We just saw one instance of this above in connection with the initial value data. The redundancy strikes again in expression (33) for the charge. If one considers the Coulomb field of a particle of charge e at rest at $x^i = 0$ one finds,

$$Q_+\left(\vartheta,\varphi\right) = \frac{e}{4\pi}\sin\vartheta$$

and

$$Q_-\left(\vartheta,\varphi\right) = \frac{e}{4\pi}\sin\vartheta$$

and hence

$$Q\left(\vartheta,\varphi\right) = 2\frac{e}{4\pi}\sin\vartheta. \tag{34}$$

The factor two arises because one is counting twice: Q_+ is the charge as seen in the outgoing description of space, while Q_- is the *same* charge as seen from its incoming replica. This point will reappear below in connection with radiation rates.

4.4.2. *Magnetic BMS charge*

There is a magnetic analog of (33) given by

$$\bar{Q} = \epsilon^{ab} \nabla_a a_b^{(0)} \Big|_{-\infty}^{+\infty} = \gamma^{1/2} \nabla^2 \bar{G} \Big|_{-\infty}^{+\infty}, \tag{35}$$

which is conserved as a consequence of (29) and the parity condition for $\bar{N}$,

$$\dot{\bar{Q}} = 0. \tag{36}$$

In the electric representation this conservation law appears as an "accident", because it does not follow from a symmetry of the action. The formalism becomes complete if one introduces a second potential, so that the electric and magnetic charges are treated on the same footing. This completion of the formalism may be regarded as a matter of elegance and economy, but not of necessity, for questions that can be asked within the electric representation. But, as we will see further below, it becomes essential when one discusses Lorentz transformations. Therefore we recall it right away.

4.5. *Asymptotic two potential formulation*

One brings in a new, "magnetic" vector potential $\bar{A}$. For the present purposes it is sufficient to do so only asymptotically. The potential $\bar{A}$ satisfies,

$$\pi^r = -\epsilon^{ab} \partial_a \bar{A}_b \,, \qquad \pi^a = -\epsilon^{ab} \left(\partial_b \bar{A}_r - \partial_r \bar{A}_b \right) . \tag{37}$$

Then, equations (18), (19) are replaced by

$$\bar{A}_a = \bar{F}_{,a} - \gamma^{-\frac{1}{2}} \gamma_{ac} \epsilon^{cb} G_{,b} + \bar{a}_a^{(1)} r^{-1} + \mathcal{O}(r^{-2}) \,, \tag{38}$$

$$\bar{A}_r = \bar{a}_r^{(2)} r^{-2} + \mathcal{O}(r^{-3}) \,. \tag{39}$$

It is important to realize that the new potential $\bar{A}$ incorporates with it the *additional* variable $\bar{F}$, which was not present in the electric representation and drops out from Eqs. (37).

There are now also magnetic gauge transformations with an associated parameter $\bar{\lambda}_{(0)}$, which is independent of the "electric" $\lambda_{(0)}$. Under a magnetic BMS transformation $\bar{F}$ and G transform according to

$$\bar{F} \to \bar{F} + \bar{\lambda}_{(0)} \,, \tag{40}$$

$$G \to G \,. \tag{41}$$

Here the electric and magnetic radial momenta π^r, $\bar{\pi}^r$, are related G and $\bar{G}$ through,

$$\pi^r = \gamma^{\frac{1}{2}} \nabla^2 G, \qquad \bar{\pi}^r = \gamma^{\frac{1}{2}} \nabla^2 \bar{G}. \tag{42}$$

If one demands that G and $\bar{G}$ be regular on the sphere, there is no room for a zero mode in the electric and magnetic BMS charges. The zero modes must be introduced through Dirac string singularities.

For a magnetic pole of strength g at the origin, on has

$$A_\varphi = g(1 - \cos\vartheta) \,, \tag{43}$$

$$\bar{G} = g \log(1 + \cos\vartheta) \,. \tag{44}$$

For an electric pole of strength e, which in the electric representation has

$$\pi^r_{(0)} = \gamma^{\frac{1}{2}} e \,, \tag{45}$$

one now writes

$$\bar{A}_\varphi = e(1 - \cos\vartheta) \,, \tag{46}$$

$$G = e \log(1 + \cos\vartheta) \,. \tag{47}$$

If one admits Dirac string singularities in G and $\bar{G}$ one must also do so for F and $\bar{F}$ in order, for example, to be able to implement rotations. This is so because under a rotation the monopole potentials change by a singular gauge transformation.

4.5.1. *Electric-magnetic duality invariant notation*

It is useful to introduce a compact notation that makes electric-magnetic duality invariance of the theory manifest. This is achieved by writing

$$A_a^M = \partial_a F^M + \epsilon^M{}_N \star \nabla_a G^N, \tag{48}$$

$$A^M = \begin{pmatrix} A \\ \bar{A} \end{pmatrix}, \quad N^M = \begin{pmatrix} N \\ \bar{N} \end{pmatrix}, \quad F^M = \begin{pmatrix} F \\ \bar{F} \end{pmatrix}, \quad G^M = \begin{pmatrix} G \\ \bar{G} \end{pmatrix}. \tag{49}$$

where

$$Q^M = \gamma^{\frac{1}{2}} \nabla^2 G^M \Big|_{-\infty}^{+\infty}, \tag{50}$$

are the electric and magnetic charges.

4.6. *Spacetime translations: Improved generator*

4.6.1. *Analysis starting from the electric representation*

Rather than employing the electric-magnetic invariant formalism ab initio, we prefer to start from the "electric" representation and then use elements of duality to "patch it" in order to cast final results in a duality invariant form. This we do for expediency, but — more importantly — because in the case of gravitation, where the full duality invariant formalism has not yet been developed, one can still perform the same steps, starting from the available electric representation.

If one considers the Hamiltonian for a motion corresponding to a time translation, the surface term in the variation of the Hamiltonian

$$H_0\,[\xi] = \int d^3x \left(\xi^\perp \mathcal{H}_\perp^{(elm)} + \xi^i \mathcal{H}_i^{(elm)} \right), \tag{51}$$

is given, in the electric representation, by

$$\delta H_0 = \oint \left(\alpha^\mu k_\mu\right) h^a \delta a_a^{(0)} \Big|_{-\infty}^{\infty} \tag{52}$$

where α^μ is the amount of spacetime translation and

$$k_\mu = (-1, \hat{r}) \,. \tag{53}$$

Equation (52) may be rewritten separating the electric memory and magnetic charge variations as,

$$\delta H_0 = -\oint \partial_a \left[(\alpha_\mu k^\mu)\left(\gamma^{\frac{1}{2}}\gamma^{ab}N_{,b} + \epsilon^{ab}\bar{N}_{,b}\right)\right]\delta F\Big|_{-\infty}^{+\infty}$$
$$-\oint \partial_a \left[(\alpha_\mu k^\mu)\left(\gamma^{\frac{1}{2}}\gamma^{ab}\bar{N}_{,b} - \epsilon^{ab}N_{,b}\right)\right]\delta\bar{G}\Big|_{-\infty}^{+\infty}. \tag{54}$$

If the parity conditions are used, the first term on the right-hand side on (54) vanishes, and the second term may be written as

$$\delta H_0 = \oint \bar{\lambda}_{\text{trans}}\delta\bar{Q}, \tag{55}$$

where $\bar{\lambda}$ is given by

$$\nabla^2\bar{\lambda}_{\text{trans}} = -\partial_a\left[(\alpha_\mu k^\mu)\left(\gamma^{\frac{1}{2}}\gamma^{ab}\bar{N}_{,b} - \epsilon^{ab}N_{,b}\right)\right]. \tag{56}$$

Equation (55) shows that in order to improve H_0 one must add to it a term proportional to the *magnetic* gauge constraint,[c]

$$\int -\bar{\pi}^i_{,i}\bar{\lambda}_{\text{trans}}d^3x. \tag{57}$$

This shows that it is essential to bring in the magnetic sector in order to properly define the spacetime translation generators. *The improvement cannot be made solely within the electric sector.* In other words, a deformation consisting only of a spacetime translation by itself does not have a well-defined generator. Only when one adds to it a movement along the fiber whose magnitude is $\bar{\lambda}$ given by (56), thus the generator exists. It is this improved generator which deserves to be called $\alpha^\mu P_\mu$. The numerical value of P_μ is the same as the original H_0 because the other term (57) vanishes weakly.[d]

4.6.2. *Simplification for time translations. Magnetic fiber memory brought in*

For the case of the time translations, for which

$$-\alpha_\mu k^\mu = \alpha^0, \tag{58}$$

two simplifications occur that are worth noting and will be useful later on:

(i) The term proportional to $\epsilon^{ab}\delta F$ in (54) vanishes,

(ii) The compensating magnetic gauge transformation $\bar{\lambda}$ reads just

$$\bar{\lambda} = \bar{N}. \qquad \text{(time translation)} \tag{59}$$

Equation (59) permits to understand the need for the addition of the magnetic gauge transformation. *It simply brings in the magnetic fiber memory,* that — unlike the magnetic charge — is not present in the purely electric formulation, because only the gauge invariant curl of the magnetic potential appears in it.

[c]The magnetic gauge generator, $-\bar{\pi}^i_{,i}$, can be treated properly by keeping in Dirac's "total Hamiltonian" the full constraint $\boldsymbol{\pi}_{mag} = 0$ and $\boldsymbol{\pi}_{el} + \nabla \times \bar{\boldsymbol{A}} = 0$, whose curl is second class, while their divergence $\nabla \cdot \boldsymbol{\pi}_{mag}$ is first class. The details of that treatment will not be needed herein.

[d]One could have tried to stay within the electric sector by demanding that the magnetic charge $\bar{Q}$ should be a passive spectator given as an "external field", and not varied in the action principle. For consistency it should be given so that $\dot{\bar{Q}} = 0$ (Eq. (36)) up to a Lorentz transformation. But the boundary term in (54) would not vanish if $\delta\bar{G} = \bar{G}_{,a}\xi^a$, so this possibility is not tenable if one wants to have Lorentz invariance. Thus it is Lorentz invariance which forces one to bring in the magnetic sector with its own independent life.

4.6.3. *Full implementation of duality*

The above discussion cannot be yet complete because after just including (57) the variation of the Hamiltonian would read

$$\delta H_0 = - \oint \partial_a \left[(\alpha_\mu k^\mu) \left(\gamma^{\frac{1}{2}} \gamma^{ab} N_{,b} + \epsilon^{ab} \bar{N}_{,b} \right) \right] \delta F \Big|_{-\infty}^{+\infty} , \tag{60}$$

and this expression is not duality invariant. Had we started from the magnetic representation instead, we would have obtained

$$\delta H_0 = - \oint \partial_a \left[(\alpha_\mu k^\mu) \left(\gamma^{\frac{1}{2}} \gamma^{ab} \bar{N}_{,b} - \epsilon^{ab} N_{,b} \right) \right] \delta \bar{F} \Big|_{-\infty}^{+\infty} , \tag{61}$$

and λ_{trans} would have been given by

$$\nabla^2 \lambda_{\text{trans}} = -\partial_a \left[(\alpha_\mu k^\mu) \left(\gamma^{\frac{1}{2}} \gamma^{ab} N_{,b} + \epsilon^{ab} \bar{N}_{,b} \right) \right] , \tag{62}$$

in order to bring in the electric memory.

In (60) and (61) above, the equality means that terms proportional to the constraints, $-\pi^i_i$ and $-\bar{\pi}^i_i$ have been dropped.

It is evident, just from demanding duality invariance, that the correct result, which incorporates (60) and (61) should read,

$$\delta H_0 = - \oint \partial_a \left[(\alpha_\mu k^\mu) \left(\gamma^{\frac{1}{2}} \gamma^{ab} N^M_{,b} + \epsilon^{MN} \epsilon^{ab} N_{N,b} \right) \right] \delta F_M \Big|_{-\infty}^{+\infty} , \tag{63}$$

which for an infinitesimal time translation of magnitude α^0 simplifies to

$$\delta H_0 = - \alpha^0 \oint \gamma^{\frac{1}{2}} \nabla^a N^M \delta(\nabla_a F_M) \Big|_{-\infty}^{+\infty} . \tag{64}$$

4.7. ***Lorentz generators. Spin from charge***

We again start from the electric representation and at the end cast the results in a manifestly duality invariant form.

We have

$$H_0^{\text{Lorentz}} = - \int d^3x \xi^i F_{ij} \pi^j , \tag{65}$$

where ξ are the Lorentz Killing vectors. The surface term in its variation reads

$$\delta H_0^{\text{Lorentz}} = - \oint \xi^a \delta A_a \pi^r |_{-\infty}^{+\infty} . \tag{66}$$

To improve the generator $\mathcal{H}_0$ we add an electric gauge generator, but this time *with the surface term included*, namely,

$$\mathcal{G}^{\text{Lorentz}} = \int -\pi^i_{,i} \lambda_{\text{Lorentz}} \, d^3x + \oint \lambda_{\text{Lorentz}} (\infty) \, \pi^r \Big|_{-\infty}^{+\infty} , \tag{67}$$

with

$$\lambda_{\text{Lorentz}}(\infty) = \xi^a_{\text{Lorentz}} A_a . \tag{68}$$

One then finds that the variation of

$$H^{\text{Lorentz}}_{\text{improved}} = H_0^{\text{Lorentz}} + \mathcal{G}^{\text{Lorentz}} , \tag{69}$$

does not have a surface integral.

4.7.1. *Lie derivative restored*

The improvement of the Lorentz generator $\mathcal{H}_0^{\text{Lorentz}}$ has an important geometrical consequence, in that it restores the Lie derivative at infinity. Indeed, the change in A_i^M given by the generator $\mathcal{H}_0^{\text{Lorentz}}$ is given by

$$\delta_0 A_i = \xi^j F_{ji} = \mathcal{L}_\xi A_i - \partial_i\left(\xi^j A_j\right),$$

so that

$$\delta_{\text{improved}} A_a(\infty) = \mathcal{L}_\xi A_a(\infty).$$

Therefore, $\mathcal{H}_{\text{improved}}^{\text{Lorentz}}$ is the generator that will correctly implement the deformation algebra in section 4.1.3.

4.7.2. *Spin from charge*

The numerical value of the generator (69) which realizes the improvement of the Lorentz generator is not zero, but it is equal to the surface integral that appears in it. Therefore the numerical value of the angular momentum is not just the volume integral (65), but it includes a contribution

$$\oint \xi^a A_a \pi^r \Big|_{-\infty}^{+\infty} = \oint \pi^r \xi^a \partial_a F\Big|_{-\infty}^{+\infty} + S, \tag{70}$$

where

$$S = \oint \xi^a \star \nabla_a \bar{G} \gamma^{\frac{1}{2}} \nabla^2 G\Big|_{-\infty}^{+\infty} = -\frac{1}{2}\oint \xi^a \star \nabla_a G^M \epsilon_{MN} \gamma^{\frac{1}{2}} \nabla^2 G^N\Big|_{-\infty}^{+\infty}, \tag{71}$$

which is proportional to the electric BMS charge $\pi^r = \gamma^{1/2}\nabla^2 G$. This phenomenon is similar to the modification of the angular momentum which appears in the presence of a magnetic pole in abelian and non-abelian gauge theories. The novelty here is that it occurs already without a magnetic pole.

The spin from charge phenomenon does not happen for energy and momentum because no surface term analogous to the one appearing in (70) is included in the translation charge.

4.7.3. *Duality invariant Lorentz generator*

The improved electric Lorentz generator,

$$H_{el}[\xi] = H_0[\xi] + S + \oint \pi^r \xi^a \partial_a F\Big|_{-\infty}^{+\infty}, \tag{72}$$

Is not electric-magnetic duality invariant because, whereas H_0 and S have that property, the term proportional to π^r does not. Just as it was discussed for translations, it is evident that the appropriate expression is

$$\begin{aligned} H[\xi] &= H_{el}[\xi] + \oint \bar{\pi}^r \xi^a \partial_a \bar{F}\Big|_{-\infty}^{+\infty} = H_0[\xi] + S + \oint \pi^r_M \xi^a \partial_a F^M\Big|_{-\infty}^{+\infty} \\ &= H_0[\xi] + S + \oint Q_M \xi^a \partial_a F^M. \end{aligned} \tag{73}$$

One may think of $\oint \xi^a Q^M F_{M,a}$ as the generator of Lorentz transformations at infinity, and $H_0 + S$ as the "bulk part" (although S is a surface integral).

It will be shown below that the duality invariant angular momentum is conserved (the electric part (72) is not!). Since this has been an issue in the literature (in the case of gravitation, which will follow the same lines) it is worth some comment.

First of all one realizes that under improper electric and magnetic gauge transformation, with parameter $\lambda^M_{(0)} = \epsilon^M$, the Lorentz generator changes as,

$$H[\xi] \longrightarrow H[\xi] - \oint \epsilon^M \partial_a(\xi^a Q_M), \tag{74}$$

and the new angular momentum is also conserved.

This is just as it happens if one changes the origin for orbital angular momentum, and in our view it is not to be regarded as a difficulty, since the present formalism improper gauge transformations are on the same footing with spacetime translations. All the more so, since a "pure time translation" carries along with it a rotation along the fiber, due to the fiber memory.

4.8. *Symmetry algebra*

Applying the general formula (5) to the present case one obtains that the Poincaré generators obey the Poincaré algebra, the electromagnetic charges are abelian. Furthermore, the electric and magnetic charges $Q_M(\vartheta, \varphi)$ transform as

$$[Q_M, \mathcal{H}_{\text{Lorentz}}]^* = \partial_a \left(Q_M \xi^a_{\text{Lorentz}}\right), \tag{75}$$

under Lorentz transformations and they obey

$$[Q_M(\vartheta, \varphi), P_\mu]^* = 0, \tag{76}$$

so that they are invariant under spacetime translations.

Equation (75) may be verified by inspection from (73) because under a gauge transformation $F_M \to F_M + \epsilon_M$, $\delta J = \oint Q^M \epsilon_{,a}\xi^a = -\partial_a(Q^M \xi^a)$, which is what (75) yields, read as the action of Q_M on $\boldsymbol{J}$. Similarly for the boosts.

Equations (75) and (76) have been written in term of Dirac brackets because the terms proportional to gauge constraints that accompany the surface integrals to make a well-defined generator have been dropped.

4.9. *Emission and absorption rates. Charge memory*

4.9.1. *General formula for emission rates*

Our boundary conditions are appropriate for a closed system, whose Hamiltonian is invariant under Poincaré and improper gauge transformations, and the corresponding conservation laws hold as a consequence of the fact that as much radiation is coming in as going out.

However, the formalism provides expressions for the emission and absorption rates separately. For that purpose one realizes that

$$\dot{Q}_\alpha = -\oint h^a \delta_\alpha a^{(0)}_a \Big|_{-\infty}^{+\infty} = -\oint \gamma^{\frac{1}{2}} \nabla^a N^M \nabla_a(\delta_\alpha F_M)\Big|_{-\infty}^{+\infty}. \tag{77}$$

Here $\delta_\alpha F^M$ is the change of F^M generated by the charge Q_α. Thus $\delta F^M = \epsilon^M$ for gauge transformations, $\delta F^M = N^M$ for time translations, and $\delta F^M = F^M{}_{,a}\xi^a$ for rotations. The purely electric form does not exist for the Lorentz charges.

Then, the emission rates are read from the upper endpoint in (77) and the absorption rates from the lower one. In this way, one obtains the following results.

4.9.2. *BMS charge*

$$\dot{Q}^M = \partial_a h^a_M \Big|_{-\infty}^{+\infty} = \gamma^{\frac{1}{2}} \nabla^2 N^M \Big|_{-\infty}^{+\infty}, \tag{78}$$

where,

$$h^a_N = \begin{pmatrix} h^a \\ - \star h^a \end{pmatrix}.$$

This equation is to be interpreted as giving either $[+\dot{\pi}^r(\infty)]$, or $[-\dot{\pi}^r(-\infty)]$. These are *not* to be thought of as the rate of change of two different charges, but rather as the rates of change of one and the same charge, due to outgoing and incoming radiation respectively; which must be calculated using the two replicas of space that form the hourglass. When the parity conditions hold the Q^M are conserved.

On sees from (78), in analogy with (30), that the BMS charge also "remembers" the news and that, in this sense, the Laplacian of N is the "charge memory". We will see in Ref. 24 that when a cosmological constant is introduced the fiber and charge memories are different and that the fiber memory appears to be more fundamental.

4.9.3. *Energy*

Similarly, one finds for the energy

$$\begin{aligned} \frac{dP^0}{dt} &= - \oint \gamma^{-\frac{1}{2}} \gamma_{ab} h^a h^b \Big|_{-\infty}^{\infty} \\ &= - \oint \gamma^{-\frac{1}{2}} \nabla^a N^M \nabla_a N_M \Big|_{-\infty}^{\infty}. \end{aligned} \tag{79}$$

4.9.4. *Angular momentum*

The equations for the rate of change of the BMS charges and the energy given above can be expressed solely in terms of quantities defined in the electric sector. This is not the case for the angular momentum which as argued before, needs the magnetic sector for its very definition. Therefore, the rate can be read only from the second term on the right-hand side of Eq. (77), which yields,

$$\frac{d\boldsymbol{J}}{dt} = - \oint \gamma^{\frac{1}{2}} \nabla^a N^M \nabla_a \left(\mathcal{L}_{\boldsymbol{\xi}} F_M \right) \Big|_{-\infty}^{\infty}, \tag{80}$$

an expression that can be rewritten as,

$$\frac{d\boldsymbol{J}}{dt} = \oint \dot{Q}_M F^M_{,a} \boldsymbol{\xi}^a. \tag{81}$$

The last expression shows that when the parity conditions hold, so that $\dot{Q}^M = 0$, the angular momentum, Eq. (73), is conserved, as it was announced and discussed in Sec. 4.7.3.

Note that Eq. (80) involves the variable $\bar{F}$ which does not appear in the electric sector. This is a consequence, in turn, of the fact that the angular momentum changes under the action of the magnetic BMS charge.

The interpretation of these equations is that the left-hand sides are the rate of change of one and the same energy and angular momentum due to outgoing and incoming radiation. Therefore, the volume integrals appearing in the definition of P_0 and J (see Eq. (51)), are to be thought of as evaluated on the upper half of the hourglass in the calculation of outgoing radiation and on the lower half in the calculation of incoming radiation. One does not integrate over the whole hourglass because this would lead to the same overcounting encountered for the electromagnetic charges.

Just as it was the case with the angular momentum itself, the physical cogency of Eq. (80) giving its rate of change, deserves a brief comment. The time rate of change of F^M is invariant under (improper) gauge transformations. If one agrees to keep the gauge frame fixed, that is, if one only moves in the course of time on the fiber as dictated by the fiber memory, then $F_M(t)$ is determined by the equations of motion — in a gauge invariant manner — once $F_M(t = 0)$ is given. This means that if one were absorbing angular momentum at infinity so as to, say, make a top start spinning, then one would in principle be able to determine $F_M(t = 0)$ and thus learn how the BMS origin in (80) is shifted from the one arbitrarily chosen on the fiber.

5. Gravitational field

5.1. *Correspondence with electromagnetism*

In this section we analyze the gravitational field along the same lines that we analyzed above the electromagnetic field. The parallel between both cases is so close that it permits to make the following discussion succinct. The correspondence is as follows: The $\ell = 0$ mode of the improper gauge symmetry generated by the total electric charge Q is the analog of the $\ell = 0$, $\ell = 1$ modes of the Bondi–van der Burg–Metzner–Sachs supertranslation, which are the ordinary translations generated by P_μ. The modes with $\ell \geq 1$ of the improper gauge symmetry correspond to the modes $\ell \geq 2$ of the supertranslations. Therefore, altogether, one has the correspondence:

$$\underbrace{\left(Q\left(\vartheta,\varphi\right), P_\mu\right)}_{\text{electromagnetism}} \longleftrightarrow \underbrace{\mathcal{P}\left(\vartheta,\varphi\right)}_{\text{gravitation}} .$$

On the other hand, the Lorentz transformations play along side:

$$\underbrace{J_{\mu\nu}}_{\text{electromagnetism}} \quad \longleftrightarrow \quad \underbrace{J_{\mu\nu}}_{\text{gravitation}} .$$

There is, as emphasized before, the difference that in the gravitational case all the generators are given by surface integrals, whereas in the electromagnetic one since the background was fixed, the spacetime translations and the Lorentz transformations were not. But this is just a technical point which is easily accounted for and does not hinder at all the close correspondence between both cases.

The important concept of "news" is also present here, of course since it is the context in which it was originally introduced by Bondi.[25] The only difference is that now it is a symmetric traceless tensor h^{ab}, appropriate to describe a gravitational wave, rather than the vector h^a appropriate for an electromagnetic one. Thus, one has the correspondence:

$$\underbrace{h_a}_{\text{electromagnetism}} \quad \longleftrightarrow \quad \underbrace{h_{ab}}_{\text{gravitation}} \, .$$

Keeping this in mind, we will essentially write the corresponding equation without much discussion, because one may translate to gravitation word by word in each case the corresponding comments from electromagnetism.

5.2. *Asymptotic boundary conditions*

For the gravitational field the canonical variables are the spatial metric g_{ij} and their conjugate π^{ij}. The generators of surface deformation are given by,

$$\begin{aligned} \mathcal{H}_\perp &= \frac{2}{\sqrt{g}}\left(\pi^{ij}\pi_{ij} - \frac{1}{2}\pi^2\right) - \frac{1}{2}\sqrt{g}^{(3)}R \approx 0\,, \\ \mathcal{H}_i &= -2\pi_i{}^j{}_{|j} \approx 0. \end{aligned}$$

Here we have set the cosmological constant equal to zero, and have chosen units such that $8\pi G = 1$. The deformation parameters that multiply $\mathcal{H}_\perp$ and $\mathcal{H}_i$ in the Hamiltonian are the lapse $N^\perp$ and the shift N^i.

5.2.1. *Power expansion near $\rho = \pm\infty$*

Schwarzschild as a starting point

Since our spacelike surfaces are asymptotically null, we must take as a starting point a coordinate system for the Schwarzschild metric which incorporates this property. This is provided by the Eddington–Finkelstein coordinates in terms of which the line element reads,

$$ds^2 = -(dx^0)^2 + dr^2 + r^2\left(d\theta^2 + \sin^2\theta d\phi^2\right) + \frac{M}{4\pi r}\left(dx^0 - dr\right)^2 . \tag{82}$$

The next step is to pass to hyperbolic coordinates, through the change of variables (9)–(10), extract the asymptotic form of the resulting expression, and proceed by trial and error as explained in section 2.1.

In terms of the the dimensionless radial variable

$$\rho = \frac{r}{\tau_0}, \tag{83}$$

the resulting boundary conditions are,

$$g_{\rho\rho} = \tau_0^2 \left(\frac{1}{\rho^2} + \frac{f_{\rho\rho}^{(-4)}}{\rho^4} + \frac{f_{\rho\rho}^{(-5)}}{\rho^5} + O\left(\rho^{-6}\right) \right) , \tag{84}$$

$$g_{\rho a} = \tau_0^2 \left(\frac{f_{\rho a}^{(-2)}}{\rho^2} + \frac{f_{\rho a}^{(-3)}}{\rho^3} + O\left(\rho^{-4}\right) \right) , \tag{85}$$

$$g_{ab} = \tau_0^2 \left(\gamma_{ab}\rho^2 + f_{ab}^{(1)}\rho + f_{ab}^{(0)} + \frac{f_{ab}^{(-1)}}{\rho} + O\left(\rho^{-2}\right) \right) , \tag{86}$$

$$\pi^{\rho\rho} = -\sqrt{\gamma}\rho^3 + p_{(1)}^{\rho\rho}\rho + p_{(0)}^{\rho\rho} + O\left(\rho^{-1}\right) , \tag{87}$$

$$\pi^{\rho a} = \frac{p_{(-1)}^{\rho a}}{\rho} + \frac{p_{(-2)}^{\rho a}}{\rho^2} + O\left(\rho^{-3}\right) , \tag{88}$$

$$\pi^{ab} = -\sqrt{\gamma}\gamma^{ab}\frac{1}{\rho} + \frac{3\sqrt{\gamma}f^{ab(1)}}{4\tau_0}\frac{1}{\rho^2} + \frac{p_{(-3)}^{ab}}{\rho^3} + \frac{p_{(-4)}^{ab}}{\rho^4} + O\left(\rho^{-5}\right) . \tag{89}$$

The coefficients which are explicitly shown above are those that will appear in the surface integrals later on. They are not all independent, but must obey relations among them, in order for the action principle that will be discussed next be well-defined. They are the demand that some terms $\mathcal{H}_\mu^{(n)}$ in the asymptotic expansion of $\mathcal{H}_\mu$, should vanish strongly, namely: (i) $\mathcal{H}_\perp^{(-1)} = 0$. This ensures that the symplectic term in the action is finite.[e] (ii) $\mathcal{H}_a^{(0)} = \mathcal{H}_\rho^{(-2)} = \mathcal{H}_\rho^{(-3)} = 0$. Together with (i) conditions (ii) make the surface term in the variation of the Hamiltonian finite, permit to take the δ outside on the left-hand side of (112) below, and give the simple forms (113)–(125) for Q. These extra algebraic conditions are harmless. One could have solved them to express the boundary conditions in terms of a lesser number of coefficients which would then be all independent, but the resulting expressions are complicated, and it is more convenient to carry them along.

5.2.2. *Parity conditions*

In addition to the power law decays (84)–(89) it is necessary to introduce parity conditions. This is achieved by splitting some of the variables in longitudinal and transverse parts as follows

$$\tau_0 \tilde{f}_{ab}^{(1)} = \nabla_{ab}F + \star\nabla_{ab}\bar{G} , \tag{90}$$

$$h_{ab} = \frac{1}{2}\left(\nabla_{ab}N + \star\nabla_{ab}\bar{N}\right) . \tag{91}$$

Here the news tensor h_{ab}, given by

$$h^{ab} = 2\tilde{p}_{(-3)}^{ab} - \sqrt{\gamma}\tilde{f}^{ab(0)} , \tag{92}$$

is the gravitational analog of the electromagnetic h^a.

[e]The symplectic term will be taken to be $-\int \dot{\pi}^{ij} g_{ij}\, d^3x$. It is interesting to note that in this momentum representation the Hamiltonian action is equal to the Hilbert action up to surface terms at spatial infinity; this phenomenon only happens in four spacetime dimensions. We have not found a simple way to make finite the conjugate term $\int \pi^{ij}\dot{g}_{ij} d^3x$.

If one has a symmetric tensor s_{ab}, we denote its trace by s, i.e., we use the same letter but without indices. The traceless part will be denoted with a tilde $\tilde{s}^{ab} = s^{ab} - \frac{1}{2}\gamma^{ab}s$.

In the above equations the operators ∇_{ab} and $\star\nabla_{ab}$ given by,

$$\nabla_{ab} = 2(\nabla_a\nabla_b + \nabla_b\nabla_a - \gamma_{ab}\nabla^2), \quad \star\nabla_{ab} = 2\sqrt{\gamma}\gamma^{cd}\left(\epsilon_{ac}\nabla_b\nabla_d + \epsilon_{bc}\nabla_a\nabla_d\right), \tag{93}$$

are the tensor analogs of the vector gradient, ∇_a, and curl $\star\nabla_a$ appearing in (21). These operators were used by Regge and Wheeler in their analysis of the stability of a Schwarzschild singularity,[34] and obey the key properties

$$\nabla^{ab}\left(\star\nabla_{ab}\right) = \star\nabla_{ab}\left(\nabla^{ab}\right) = 0, \tag{94}$$

when they act on scalar functions, just as their vector counterparts. Their kernel is spanned by the $\ell = 0$ and $\ell = 1$ modes of the corresponding scalar functions on which they act.

The parity conditions will be then the following,

$$(F, N, \bar{N})\big|_{+\infty} = (F, N, \bar{N})\big|_{-\infty} \quad \text{for each } (\vartheta, \varphi), \tag{95}$$

in close analogy with Eq. (25) for electromagnetism.

5.2.3. *Preservation of the boundary conditions*

One may verify using Einstein's equations in Hamiltonian form

$$\dot{g}_{ij} = \frac{4}{\sqrt{g}}N^{\perp}\left(\pi_{ij} - \frac{\pi}{2}g_{ij}\right) + N_{i/j} + N_{j/i}\,, \tag{96}$$

$$\begin{aligned}\dot{\pi}^{ij} = &-\frac{1}{2}N^{\perp}\sqrt{g}\left({}^{(3)}R^{ij} - \frac{1}{2}g^{ij\,(3)}R\right) + \frac{N^{\perp}}{\sqrt{g}}g^{ij}\left(\pi^{kl}\pi_{kl} - \frac{1}{2}\pi^2\right) - \frac{4N^{\perp}}{\sqrt{g}}\left(\pi^{i}{}_{l}\pi^{jl} - \frac{1}{2}\pi^{ij}\pi\right)\\ &+\frac{1}{2}\sqrt{g}\left(N^{\perp/i/j} - g^{ij}N^{\perp}_{/k/k}\right) + \left(N^k\pi^{ij}\right)_{/k} - N^{i}_{/k}\pi^{kj} - N^{j}_{/k}\pi^{ki}\,,\end{aligned} \tag{97}$$

$$\mathcal{H}_\mu = 0\,, \tag{98}$$

that the most general surface deformation that preserves (84)–(89) takes the form:

$$N^{\perp} = \epsilon^{\perp}_{(1)}\rho + \frac{\epsilon^{\perp}_{(-1)}}{\rho} + O\left(\rho^{-2}\right), \tag{99}$$

$$N^{\rho} = -\frac{\epsilon^{\perp}_{(1)}}{\tau_0}\rho^2 - \frac{1}{2}\left(\nabla_a\epsilon^{a}_{(0)}\right)\rho + \epsilon^{\rho}_{(0)} + O\left(\rho^{-1}\right), \tag{100}$$

$$N^{a} = \epsilon^{a}_{(0)} - \frac{1}{\tau_0}\left(\nabla^a\epsilon^{\perp}_{(1)}\right)\frac{1}{\rho} + \frac{\epsilon^{a}_{(-2)}}{\rho^2} + O\left(\rho^{-3}\right), \tag{101}$$

where,

$$\epsilon^{\perp}_{(-1)} = \epsilon^{\perp}_{(1)}\left[\frac{1}{2}\frac{p_{(-3)}}{\sqrt{\gamma}} - \frac{1}{4}f^{(0)} + \frac{1}{4}f^{ab(1)}f^{(1)}_{ab}\right] + \frac{1}{4}\nabla^2\epsilon^{\perp}_{(1)}\,,$$

$$\epsilon^{\rho}_{(0)} = \frac{\epsilon^{\perp}_{(1)}}{\tau_0}\left[\frac{1}{4}\left(f^{(0)} - f^{ab(1)}f^{(1)}_{ab} - 2f^{(-4)}_{\rho\rho}\right) + \frac{1}{\sqrt{\gamma}}\left(p^{\rho\rho}_{(1)} - \frac{1}{2}p_{(-3)}\right)\right] + \frac{1}{4\tau_0}\nabla^2\epsilon^{\perp}_{(1)}\,,$$

$$\epsilon^{a}_{(-2)} = -\frac{1}{4}\partial^a\left(\nabla_b\epsilon^{b}_{(0)}\right) - \frac{\epsilon^{\perp}_{(1)}}{\tau_0}\gamma^{ab}f^{(-2)}_{\rho b} + \epsilon^{\perp}_{(1)}\frac{2}{\tau_0}\frac{p^{\rho a}_{(-1)}}{\sqrt{\gamma}} + \frac{1}{2\tau_0}f^{ab(1)}\partial_b\epsilon^{\perp}_{(1)}\,.$$

The coefficients in the asymptotic expansion above are written explicitly up to the order in which they appear in the surface integrals. One sees that they are determined by the functions $\epsilon^{\perp}_{(1)}(\vartheta,\varphi)$ and $\epsilon^{a}_{(0)}(\vartheta,\varphi)$ which will correspond to supertranslations and Lorentz transformations respectively. The latter appear in $\epsilon^{a}_{(0)}$ through the decomposition

$$\epsilon^{a}_{(0)} = \partial^a \Lambda + \frac{\epsilon^{ab}}{\sqrt{\gamma}} \partial_b \Omega \ , \tag{102}$$

with

$$\Lambda = \boldsymbol{\beta} \cdot \hat{r}, \quad \Omega = \boldsymbol{\omega} \cdot \hat{r}, \tag{103}$$

where $\boldsymbol{\beta}$ is the parameter of an infinitesimal boost, and $\boldsymbol{\omega}$ is the vector angle of an infinitesimal spatial rotation.[f]

In order for the parity conditions (25) to be preserved in time one must demand

$$\epsilon^{\perp}_{(1)}(+\infty) = \epsilon^{\perp}_{(1)}(-\infty). \tag{104}$$

In addition, just as in electromagnetism an infinite sequence of additional conditions appears. They present no problem here either.

5.2.4. *Surface deformation algebra*

Specializing the general surface deformation algebra (7) to two deformations $\xi_1 = \xi$ and $\xi_2 = \eta$ of the form (6), (7) one finds for the commutator $\xi_{12} = \zeta$

$$\zeta^{\perp}_{(1)} = \xi^{a}_{(0)} \partial_a \eta^{\perp}_{(1)} - \eta^{a}_{(0)} \partial_a \xi^{\perp}_{(1)} + \frac{1}{2} \left(\xi^{\perp}_{(1)} \nabla_a \eta^{a}_{(0)} - \eta^{\perp}_{(1)} \nabla_a \xi^{a}_{(0)} \right), \tag{105}$$

$$\zeta^{a}_{(0)} = \xi^{b}_{(0)} \partial_b \eta^{a}_{(0)} - \eta^{b}_{(0)} \partial_b \xi^{a}_{(0)} \, , \tag{106}$$

and that ζ^a closes in terms of ξ^a and η^a according to the Lorentz algebra.

These equations are the BMS algebra in the standard form (see for example Ref. 4).

5.2.5. *Supertranslation memory*

Consider a time translation:

$$\epsilon^{\perp}_{(1)} = 1. \tag{107}$$

The equation of motion (96) yields then

$$\dot{\bar{f}}^{(1)}_{ab} = 2h_{ab}, \tag{108}$$

which implies

$$\dot{F} = N. \tag{109}$$

Therefore, when time δt elapses a supertranslation of magnitude

$$\delta F = N \delta t, \tag{110}$$

takes place. This is the supertranslation memory effect, analogous to the fiber memory of electromagnetism discussed in section 4.3.

[f] "Superrotations" have been introduced in Refs. 26–28. They correspond to $\epsilon^{a}_{(0)}$ singular on the sphere.

5.3. *Electric and magnetic BMS charges*

We saw in the electromagnetic case that it was necessary to employ, asymptotically on the hourglass an electric-magnetic duality invariant formalism, in order to be able to improve the generators. The same will occur in gravitation. In that case we do not possess at the moment an electric-magnetic duality invariant formulation on the hourglass. However, it is reasonable to assume that such a description exists. Evidence in favor of the cogency of this assumption is provided by the duality invariant formulation for asymptotic planes constructed in references [29, 30] for the linearized theory.

Fortunately, it turns out that assuming the existence of the asymptotic electric-magnetic duality invariant description, one can conjecture by analogy some of the elements that are needed. The coherence of the results thus obtained reinforces the hypothesized existence of the electric-magnetic representation. We now pass to discuss those elements.

5.3.1. *Electric BMS charge*

If one varies the Hamiltonian

$$H_0 = \int d^3x \left(N^{\perp}\mathcal{H}_{\perp} + N^i\mathcal{H}_i\right) , \tag{111}$$

in the electric representation, with the Lorentz parameters $\boldsymbol{\omega}$, $\boldsymbol{\beta}$ set equal to zero, one finds

$$\delta H_0 = -\oint d\vartheta d\varphi \epsilon^{\perp}_{(1)}(\vartheta,\varphi)\left[\delta\mathcal{P} + \frac{1}{2}h^{ab}\delta\left(\tau_0 f^{(1)}_{ab}\right)\right]\Bigg|^{+\infty}_{-\infty} , \tag{112}$$

with

$$\mathcal{P}(\vartheta,\varphi) = \tau_0\sqrt{\gamma}\left(\nabla^a f^{(-2)}_{\rho a} + 3f^{(-5)}_{\rho\rho} + \frac{3}{2}f^{(-1)} + \frac{1}{\sqrt{\gamma}}\left(\frac{1}{2}h^{ab}f^{(1)}_{ab} + 2p_{(-4)}\right)\right) . \tag{113}$$

This identifies $Q(\vartheta,\varphi)$

$$Q = \mathcal{P}|^{+\infty}_{-\infty} , \tag{114}$$

as the (electric) supertranslation charge.[g,h] In the analogy with electromagnetism, the $l = 0$ and $l = 1$ of the charge (114) correspond to spacetime translations whereas those with $l \geq 2$ correspond to the electromagnetic charges with spherical modes $l \geq 1$. The first term on

[g]We have grouped τ_0 together with $f^{(1)}_{ab}$ because the product $\tau_0 f^{(1)}_{ab}$ is the analog of the electromagnetic $a^{(0)}_a$. They both have the property of remaining finite in the limit $\tau_0 \to 0$ which turns the hyperboloids into light cones. Conversely, the news h^{ab}, whose formula (92) does not have an explicit τ_0 in it is the analog of the electromagnetic news h^a whose definition (24) does have a τ_0 in it. It also remains finite in the limit $\tau_0 \to 0$. These differences in the units between the electromagnetic and the gravitational case are produced by the choice $8\pi G = 1$. Again in the above formulas, one should include τ_0 into each coefficient to have the natural variables.

[h]It is interesting to note the presence of the quadratic term $h^{ab}f^{(1)}_{ab}$ in $\mathcal{P}$. In a situation in which one has gravitational radiation being emitted by a confined source it would represent interference between the radiation field and the field that remains bound to the source ("near field"). This interference contribution is not emitted, but it remains bound to the source because it is part of $\mathcal{P}$. A similar phenomenon happens in electromagnetism and it occurs there in the volume integral of $\mathcal{H}^{(elm)}_{\mu}$. This has been discussed in Ref. 18.

the right-hand side of (112) may be compensated in the standard manner by defining a partially improved Hamiltonian $\tilde{H}_0^{\text{elec}}$ through

$$\tilde{H}_0^{\text{elec}} = H_0 + \oint \epsilon_{(1)}^{\perp} (\vartheta, \varphi) \, Q (\vartheta, \varphi) \,. \tag{115}$$

The Hamiltonian (115) is the analog of the Maxwell electric Hamiltonian for spacetime translations *and* improper gauge transformations, and just as that one it will need to be improved to eliminate the surface term

$$\oint \frac{1}{2} \epsilon_{(1)}^{\perp} h^{ab} \delta \left(\tau_0 . \tilde{f}_{ab}^{(1)} \right) \Big|_{-\infty}^{+\infty} = \oint \frac{1}{4} \gamma^{\frac{1}{2}} \epsilon_{(1)}^{\perp} \left(\nabla_{ab} N + \star \nabla_{ab} \bar{N} \right) \left(\nabla_{ab} \delta F + \star \nabla_{ab} \delta \bar{G} \right) \Big|_{-\infty}^{+\infty} . \tag{116}$$

The term proportional to δF on the right-hand side of Eq. (116) vanishes when the parity conditions hold but the one proportional to $\delta \bar{G}$, which reads

$$\delta H_0 = - \oint \bar{\eta} \delta \bar{G}, \tag{117}$$

with

$$\bar{\eta} = \frac{1}{4} \nabla^{ab} \left[\epsilon_{(1)}^{\perp} \left(\nabla_{ab} \bar{N} - \star \nabla_{ab} N \right) \right] \tag{118}$$

does not.

5.3.2. *Magnetic BMS charges*

In order to eliminate (117) one should supplement the Hamiltonian acting with the generator of magnetic BMS transformations, whose form we do not know, but which should be such that the surface term in its variation should read

$$- \oint \bar{\epsilon}_{(1)}^{\perp} \delta \bar{\mathcal{P}}, \tag{119}$$

where, by definition, $\bar{\mathcal{P}}$ is the magnetic BMS charge and $\bar{\epsilon}_{(1)}^{\perp}$ is the magnetic deformation parameter, so we must have,

$$\oint \bar{\epsilon}_{(1)}^{\perp} \delta \bar{\mathcal{P}} = \oint \bar{\eta} \delta \bar{G},$$

and the question is: what is the relationship between $\bar{\mathcal{P}}$ and $\bar{G}$?

This can be established by recalling from electromagnetism that one would like the parameter $\bar{\eta}$ to bring the magnetic memory. So we set

$$\epsilon_{(1)}^{\perp} = 1, \tag{120}$$

in which case the boundary term reads

$$\delta H_0 = - \oint \nabla^4 \bar{N} \delta (\gamma^{\frac{1}{2}} \bar{G}) \Big|_{-\infty}^{+\infty} = - \oint \bar{N} \delta \left(\nabla^4 (\gamma^{\frac{1}{2}} \bar{G}) \right) \Big|_{-\infty}^{+\infty}, \tag{121}$$

where

$$\nabla^4 = \nabla_{ab} \nabla^{ab} = \star \nabla_{ab} \star \nabla^{ab} = 8 \left[\left(\nabla^2 \right)^2 + 2 \nabla^2 \right] . \tag{122}$$

Comparison with the magnetic analog of Eq. (109) then gives

$$\bar{\mathcal{P}} = \gamma^{\frac{1}{2}} \nabla^4 \bar{G}. \tag{123}$$

The identification (123) will have a significative consistency check when we discuss angular momentum below.

In order to account for the $l = 0$ and $l = 1$ modes of $\bar{\mathcal{P}}$ (electric and magnetic translations generators) one needs to bring in Dirac strings into $\bar{G}$ because those modes are in the kernel of ∇^4.

5.4. *Lorentz generators*

If one works solely in the electric representation one finds that if one considers the motion corresponding to a Lorentz transformation, with infinitesimal rotation and boost parameters $\boldsymbol{\omega}$ and $\boldsymbol{\beta}$, one must improve the Hamiltonian by adding to it the surface term,

$$\boldsymbol{\omega}\cdot\boldsymbol{J}_{\text{el}}+\boldsymbol{\beta}\cdot\boldsymbol{K}_{\text{el}},$$

with,

$$\boldsymbol{J}_{\text{el}}=\oint 2\tau_0^2\epsilon^{ab}\left(\partial_b\hat{r}\right)\left(\frac{p^{\rho}{}_{a(-2)}}{\sqrt{\gamma}}-f^{(-3)}_{\rho a}\right), \tag{124}$$

$$\boldsymbol{K}_{\text{el}}=\oint \tau_0^2\sqrt{\gamma}\left\{\hat{r}\left(1+f^{(0)}+2f^{(-4)}_{\rho\rho}+\frac{p_{(-3)}}{\sqrt{\gamma}}\right)+2\left(\partial^a\hat{r}\right)\left(\frac{p^{\rho}{}_{a(-2)}}{\sqrt{\gamma}}-f^{(-3)}_{\rho a}\right)\right\}. \tag{125}$$

When this is done the generators are well-defined. No additional surface integral containing the news, analogous to (116) appears. This is quite reasonable because the Lorentz motion lies within the hourglass. The first term $\tau_0^2\sqrt{\gamma}\hat{r}$ in (125), which has no variation, has been incorporated so that for Minkowski space the numerical value of the boost generator is zero.

The electric generators thus obtained are the analog of the electromagnetic angular momentum (72).

If one takes $\epsilon^{\perp}_{(1)}=1$, and evaluates the rate of change of $\boldsymbol{J}_{el}$, one finds, either by direct calculation from Einstein's equations or, better, by using Eq. (132) below,

$$\frac{d\boldsymbol{J}_{el}}{dt}=-\oint(\gamma^{\frac{1}{2}}\nabla^4\bar{G})\bar{N}_{,a}\boldsymbol{\xi}^a\Big|_{-\infty}^{+\infty}. \tag{126}$$

Equation (126) is the analog of (80) for electromagnetism. It provides a consistency check of the definition (123) because if we bring in the magnetic analog $\bar{F}$ of F, and postulate the magnetic memory equation,

$$\dot{\bar{F}}=\bar{N}, \tag{127}$$

then,

$$\boldsymbol{J}=\boldsymbol{J}_{el}+\oint\bar{\mathcal{P}}\bar{F}_{,a}\boldsymbol{\xi}^a\Big|_{-\infty}^{+\infty} \tag{128}$$

is conserved

$$\frac{d\boldsymbol{J}}{dt}=0, \tag{129}$$

when the parity conditions hold.

So, by appealing to electric-magnetic duality one can find a conserved angular momentum in general relativity, even in the presence of radiation, but provided the net radiation flux is zero.

For boosts one must include an extra term (see comment at the end of the next subsection). Thus one has in general,

$$H_{\text{Lorentz}}=H^{\text{el}}_{\text{Lorentz}}+\oint\bar{Q}\left(\xi^a\partial_aF-\frac{3}{2}\nabla_a\xi^aF\right). \tag{130}$$

(The second term $\nabla_a\xi^a$ vanishes for rotations.)

5.5. *Symmetry algebra*

The analog of (75) and (76) for electromagnetism is

$$\begin{aligned}\left[Q^M(\vartheta,\varphi),Q^N(\vartheta',\varphi')\right]^\star &= 0\\ \left[Q^M(\vartheta,\varphi),\boldsymbol{J}\right]^\star &= \partial_a\left(\boldsymbol{\xi}^a_{\mathrm{R}}Q^M(\vartheta,\varphi)\right)\\ \left[Q^M(\vartheta,\varphi),\boldsymbol{K}\right]^\star &= \partial_a\left(\boldsymbol{\xi}^a_B Q^M(\vartheta,\varphi)\right)\\ &\quad+\frac{3}{2}Q^M(\vartheta,\varphi)\left(\nabla_a\boldsymbol{\xi}^a_B\right),\end{aligned} \tag{131}$$

while the Lorentz generator $\boldsymbol{K}$ and $\boldsymbol{J}$ close among themselves in the Lorentz algebra.

We have used Dirac brackets $[\ ,\ ]^\star$ here because, as explained in the introduction it is only through them that the surface term alone can act as a generator. If one wanted to use Poisson brackets one would have to add to the surface term the weakly vanishing volume part of the generator.

For the electric generators Q and the Lorentz generators, equations (131) can be obtained directly from the surface deformations algebra (105) and (106). It is then extended to the magnetic generators by duality.

5.6. *Emission and absorption rates. Charge memory*

5.6.1. *General formula for emission rates*

In this case we only possess the formula stemming from electric sector, that is,

$$\dot{Q}_\alpha = \oint \frac{1}{2}h^{ab}\delta_\alpha\left(\tau_0\tilde{f}^{(1)}_{ab}\right)\Big|_{-\infty}^{+\infty}. \tag{132}$$

The analog of the second expression on the right-hand side of (77) is not obvious to guess because, this time, under a duality transformation, one must turn the electric time into magnetic time. That is, one would have to compare motions that have $\epsilon^\perp_{(1)}=1$, $\bar{\epsilon}^\perp_{(1)}=0$ with those with $\epsilon^\perp_{(1)}=0$, $\bar{\epsilon}^\perp_{(1)}=1$.

By applying (132) one obtains the following results.

5.6.2. *Electric BMS charges*

$$\begin{aligned}\frac{\partial Q}{\partial t} &= \left(-\frac{1}{\sqrt{\gamma}}h^{ab}h_{ab}+\nabla_a\nabla_b h^{ab}\right)\Big|_{-\infty}^{+\infty},\\ &= -\frac{1}{4}\gamma^{\frac{1}{2}}\Big((\nabla^{ab}N)(\nabla_{ab}N)+(\nabla^{ab}\bar{N})(\nabla_{ab}\bar{N})+\\ &\quad+2(\nabla^{ab}N)(\star\nabla_{ab}\bar{N})++\gamma^{\frac{1}{2}}\nabla^4 N\Big)\Big|_{-\infty}^{+\infty}.\end{aligned} \tag{133}$$

(This equation, as well as its relationship with the emission rate, were found previously in Refs. 26 and 31; but they interpreted it as meaning that the Hamiltonian cannot be improved if $h_{ab}\neq 0$.)

5.6.3. *Magnetic BMS charge*

The (electric) time derivative of the magnetic BMS charge cannot be obtained from the purely electric sector formula (132), although $\bar{\mathcal{P}}$ does appear in the electric sector. One must resort to its definition (123) and to the equation of motion (108). This yields,

$$\frac{\partial \bar{Q}}{\partial t} = \gamma^{\frac{1}{2}} \nabla^4 \bar{N} \Big|_{-\infty}^{+\infty}. \tag{134}$$

Note that there is no symmetry between the rates of change of Q and $\bar{Q}$. That is quite alright because one should not expect any: the duality counterpart of (133) should be the rate of change of $\bar{Q}$ with respect to a *magnetic* time displacement with $\bar{\epsilon}^{\perp}_{(1)} = 1$, and $\epsilon^{\perp}_{(1)} = 0$.

In the same vein, one could define a variable G through

$$\gamma^{\frac{1}{2}} \nabla^4 G = \mathcal{P}. \tag{135}$$

Then the derivative of G with respect to electric time would not be equal to N as one can see from (133). However, one would expect its derivative with respect to magnetic time to be given by $\bar{N}$, in analogy with (134).

5.6.4. *Angular momentum*

One may write, in analogy with (81) in electromagnetism

$$\frac{d\boldsymbol{J}}{dt} = \oint \dot{\mathcal{P}}_M F^M_{\ ,a} \xi^a \tag{136}$$

where $\dot{\mathcal{P}}$ and $\dot{\bar{\mathcal{P}}}$ are given by (133) and (134).

It should be stressed that formula (136) has not been proven, but just conjectured by analogy and "informed guess". Only the conservation of $\boldsymbol{J}$ when $\dot{\mathcal{P}}$ and $\dot{\bar{\mathcal{P}}}$ vanish has been proven (once (127) has been postulated!). This is because, in the lack of a complete asymptotic two potential theory, we do not posses an analog of the second expression on the right-hand side of (77). But the presumption is that (136) will survive the complete development of the asymptotically duality invariant description.

All the comments made for the electromagnetic case in connection with the angular momentum and with its rate of change apply here as well.

5.7. *Special solutions*

There are two fundamental solutions of Einstein's equations for which is important to verify that they fit into the present treatment. They are Taub–Nut space and the Kerr solution. We pass to discuss them now.

5.7.1. *Taub–Nut space as a magnetic pole*

The gravitational analog of a magnetic pole in electromagnetism is Taub–Nut space. We will now show that it satisfies our boundary conditions, and its magnetic pole nature will be distinctly brought out.

The Taub–NUT metric in Schwarzschild coordinates,

$$ds^2 = -V(r)\left[dt + 2N(1-\cos\vartheta)\,d\varphi\right]^2 + V(r)^{-1}dr^2 + \left(r^2+N^2\right)\left(d\vartheta^2 + \sin^2\vartheta d\varphi^2\right), \tag{137}$$

with

$$V(r) = 1 - \frac{2\left(N^2 + \frac{M}{8\pi} r\right)}{r^2 + N^2},$$

may be brought by a change of coordinates to obey our boundary conditions with

$$F = 0, \tag{138}$$

and

$$\bar{G} = -N \log(1 + \cos\vartheta). \tag{139}$$

These are exactly the expressions (44) of electromagnetism for a magnetic pole of charge $g = -N$, with the Dirac string going through the south pole.[i]

To arrive at (138) and (139) one takes the following steps:

(i) Pass to the analog of the Eddington–Finkelstein coordinates

$$t = x^0 + r^* - r + \frac{2N^2 \cot^2\theta}{r}, \quad \vartheta = \theta + \frac{2N^2 \cot\theta \csc^2\theta}{r^2}, \quad \varphi = \phi + \frac{2N \cot\theta \csc\theta}{r},$$

where r^* is the "tortoise" radial coordinate,

$$r^* = \int^r \frac{dr}{V(r)}. \tag{140}$$

(ii) Pass to hyperbolic coordinates according to (9) and (10). The resulting expression is complicated in closed form but we only need the fact that its asymptotic form fits the boundary conditions (84)–(89). One finds that this is indeed the case, and, in particular,

$$\tau_0 f^{(1)}_{\theta\phi} = -2N\left[1 - 2\cos\theta + \cos^2\theta\right]. \tag{141}$$

(iii) Identify from (141) F and G through the decomposition (90), obtaining (138) and (139).

5.7.2. *Kerr solution*

To show that the Kerr solution satisfies our boundary conditions (84)–(89), one performs the following steps:

(i) Write the solution in Kerr–Schild coordinates:

$$\begin{aligned} ds^2 = &- (dx^0)^2 + dx^2 + dy^2 + dz^2 \\ &+ \frac{MR^3}{4\pi(R^4 + a^2 z^2)}\left[-dx^0 + \frac{R(xdx + ydy) - a(xdy - ydx)}{R^2 + a^2} + \frac{zdz}{R}\right]^2, \end{aligned}$$

where R is given by

$$R = \frac{1}{\sqrt{2}}\left(\sqrt{x^2 + y^2 + z^2 - a^2 + \sqrt{4a^2 z^2 + (x^2 + y^2 + z^2 - a^2)^2}}\right).$$

---

[i] When one discusses Taub–NUT on surfaces which are asymptotically planes, as it was done in Ref. 30, one finds, that in order to satisfy the Regge–Teitelboim boundary conditions which include a parity requirement, one must take half of the string to come out of the south pole and the other half to come out from the north pole. No such requirement is present here, where one can take just one string going out through any point on the sphere.

(ii) Perform the standard change of basis from Cartesian to spherical coordinates r, θ, ϕ, and then pass to hyperbolic coordinates by using the change of coordinates (9), (10) to obtain the asymptotic form (84)–(89).

One can then evaluate the charges. One find that the only non-vanishing ones are

$$P^0 = \oint \mathcal{P} = M \,,$$

$$J_z = aM \,.$$

Here, the value of the angular momentum has been calculated using the electric flux integral (124). This is alright because the magnetic contribution vanishes for it.

Acknowledgments

We express our gratitude to Professor Anna Ceresole for her continued encouragement throughout this work. The Centro de Estudios Científicos (CECs) is funded by the Chilean Government through the Centers of Excellence Base Financing Program of Conicyt. C.B. wishes to thank the Alexander von Humboldt Foundation for a Humboldt Research Award. The work of A.P. is partially funded by Fondecyt Grants N° 1171162 and N° 1181496.

Appendices

A. Poincaré generators

In this appendix we give the expressions for the Killing vectors of the Poincaré group in the foliation by hyperboloids with varying center and fixed radius. The transformations from Minkowskian coordinates (x^0, x^1, x^2, x^3) to the coordinates $(t, r, \vartheta, \varphi)$ adapted to the hyperbolic hourglass slicing is given in (9), (10). The metric in these coordinates reads

$$ds^2 = -\left(1 + \frac{r^2}{\tau_0^2}\right) dt^2 + r^2 d\Omega^2 + \\ + \frac{1}{\left(1+\frac{r^2}{\tau_0^2}\right)} \left(dr - \sqrt{\frac{r^2}{\tau_0^2}\left(1 + \frac{r^2}{\tau_0^2}\right)} \, dt\right)^2 \,. \tag{142}$$

We use the following notation: the components of a four-vector v^μ referred to the Minkowski coordinate system are grouped as $v^\mu = (v^0, \boldsymbol{v})$, while the boldface refers to vector fields defined in terms of their components with respect to the hyperbolic hourglass foliation. In this notation the rotation generators around the three spatial axes, that leave the origin $x^i = 0$ invariant for any fixed t,

$$\boldsymbol{\xi}_R^1 = \mathbf{J}_{23} = -\sin\varphi \boldsymbol{\partial}_\vartheta - \cot\vartheta \cos\varphi \boldsymbol{\partial}_\varphi \,, \tag{143}$$

$$\boldsymbol{\xi}_R^2 = \mathbf{J}_{31} = \cos\varphi \boldsymbol{\partial}_\vartheta - \cot\vartheta \sin\varphi \boldsymbol{\partial}_\varphi \,, \tag{144}$$

$$\boldsymbol{\xi}_R^3 = \mathbf{J}_{12} = \boldsymbol{\partial}_\varphi \,, \tag{145}$$

are written as

$$\boldsymbol{\xi}_R = \hat{r} \times \boldsymbol{\nabla} = \gamma^{-\frac{1}{2}} \epsilon^{ab} \nabla_b \hat{r} \; \boldsymbol{\partial}_a \,. \tag{146}$$

The boosts generators $\boldsymbol{\xi}^i_B = \mathbf{J}_{i0}$ along the three spatial directions which leave the center of the hyperboloid $t =$ constant fixed are given by

$$\boldsymbol{\xi}_B = \sqrt{1 + \frac{\tau_0^2}{r^2}} \left(r\hat{r}\boldsymbol{\partial}_r + \nabla^a \hat{r}\ \boldsymbol{\partial}_a \right) . \tag{147}$$

These spatial rotations and boosts map a given hyperboloid $t =$ constant onto itself. For $r \to \infty$ the tangential part of the boost generators (147) together with the rotations (146) close among themselves and form a realization of the Lorentz group on the two-sphere.

The translations do not map the hyperboloid onto itself. Their generators are

$$\boldsymbol{\xi}_{T\mu} = \mathbf{P}_\mu = (\boldsymbol{\partial}_t, \boldsymbol{\nabla}) , \tag{148}$$

or, expressed in terms of normal and tangential components,

$$\boldsymbol{\xi}_{T^0} = \sqrt{1 + \frac{r^2}{\tau_0^2}} \left(\mathbf{n} - \left| \frac{r}{\tau_0} \right| \boldsymbol{\partial}_r \right), \tag{149}$$

$$\boldsymbol{\xi}_T = -\hat{r} \left| \frac{r}{\tau_0} \right| \mathbf{n} + \left(1 + \frac{r^2}{\tau_0^2} \right) \hat{r}\boldsymbol{\partial}_r + \frac{1}{r} (\nabla^a \hat{r})\ \boldsymbol{\partial}_a . \tag{150}$$

Here, $\mathbf{n}$ is the future oriented unit normal to the hyperboloids $t =$ constant. The (future directed) unitary vector ortogonal to them is, which is given by

$$\mathbf{n} = \left(1 + \frac{r^2}{\tau_0^2} \right)^{-\frac{1}{2}} \boldsymbol{\partial}_t + \left| \frac{r}{\tau_0} \right| \boldsymbol{\partial}_r . \tag{151}$$

B. Initial data for the hyperbolic hourglass

The parity conditions for N and $\bar{N}$ say that the value of the electromagnetic news vector h^a at $r \to \infty$ is equal to its value at $r \to -\infty$,

$$\Psi^a_{(2)} = h^a(+\infty) - h^a(-\infty) = 0, \tag{152}$$

where the subindex (2) is used here to indicate that the condition contains coefficients of order $\mathcal{O}(r^{-2})$ in the expansion of the fields. Once this condition is imposed, one must demand that it be preserved under Poincaré transformations and gauge transformations, proper and improper. Being a vector defined on the sphere at infinity out of gauge invariant quantities, it is evident that $\Psi^a_1 = 0$ is invariant under Lorentz and gauge transformations. One only needs to be concerned with spacetime translations. In view of the Lorentz invariance, it is sufficient to consider only time translations, and demand that

$$\Psi^a_{(3)} \equiv \dot{\Psi}^a_{(2)} = 0. \tag{153}$$

In what follows we will set, for simplicity, $\tau_0 = 1$. Using the equations of motion (27), (28) with $(\xi^\perp, \xi^i)$ being the components of the time translation Killing vector (149), one finds that the expansion coefficients of the different fields satisfy,

$$\dot{a}_a^{(n)} = \gamma^{-\frac{1}{2}}\gamma_{ab}\left(\pi^b_{(n)} + \pi^b_{(n+2)}\right) + \sum_{0\le 2m\le n} b_m f_{ar}^{(n-2m+2)}, \tag{154}$$

$$\dot{a}_r^{(n)} = \gamma^{-\frac{1}{2}}\pi^r_{(n-2)}, \tag{155}$$

$$\dot{\pi}^r_{(n)} = \partial_a\left[\gamma^{\frac{1}{2}}\gamma^{ab}\left(f_{br}^{(n)} + f_{br}^{(n+2)}\right) + \sum_{0\le 2m\le n} b_m \pi^a_{(n-2m+2)}\right], \tag{156}$$

$$\begin{aligned}\dot{\pi}^a_{(n)} = (n-1)&\left[\gamma^{\frac{1}{2}}\gamma^{ab}\left(f_{br}^{(n-1)} + f_{br}^{(n+1)}\right) + \sum_{0\le 2m\le n-1} b_m \pi^a_{(n-2m+1)}\right]\\ &+ \partial_b\left(\gamma^{\frac{1}{2}}\gamma^{cb}\gamma^{da} f_{cd}^{(n-2)}\right),\end{aligned} \tag{157}$$

where b_n is the nth coefficient in the expansion of $\sqrt{1+x}$.

For F_{ar}, which will be of importance below, one has

$$\begin{aligned}\dot{f}_{ar}^{(n)} = (n-1)&\left[\gamma^{-\frac{1}{2}}\gamma_{ab}\left(\pi^b_{(n-1)} + \pi^b_{(n+1)}\right) + \sum_{0\le 2m\le n-1} b_m f_{ar}^{(n-2m+1)}\right]\\ &+\partial_a\left(\gamma^{-\frac{1}{2}}\pi^r_{(n-2)}\right).\end{aligned} \tag{158}$$

From this expressions one finds that (153) reads

$$\Psi^a_{(3)} = \left[2\left(\pi^a_{(3)} + \gamma^{\frac{1}{2}}\gamma^{ab}f_{br}^{(3)}\right) + \gamma^{\frac{1}{2}}\gamma^{ab}\partial_b\left(\gamma^{-\frac{1}{2}}\pi^r_{(0)}\right) + \partial_b\left(\gamma^{\frac{1}{2}}\gamma^{cb}\gamma^{da}f_{cd}^{(0)}\right)\right]\Big|_{-\infty}^{+\infty} = 0. \tag{159}$$

Note that the highest order coefficients entering $\Psi^a_{(3)}$ are the ones of $\mathcal{O}(r^3)$ in the transverse fields. This shows that one is free to give $\pi^r_{(0)}$ and $f_{\vartheta\varphi}^{(0)}$, together with $\pi^a_{(3)}$ and $f_{ar}^{(3)}$ on one half of the hourglass, say $r \to +\infty$, and Eq. (159) will relate them with the corresponding ones on its incoming image, $r \to -\infty$.

This phenomenon continues indefinitely, without imposing any restriction on the initial data given on one half of the hourglass with the exception, of course, of the Gauss law constraint. This is because every differentiation of (159) in time brings coefficients of the transverse fields of one additional order. That is, $\Psi^a_{(n)}$ includes $\pi^a_{(n)}$, $f_{ar}^{(n)}$ and terms of smaller order. The longitudinal coefficients $\pi^r_{(n)}$ and $f_{\vartheta\varphi}^{(n)}$, for $n \neq 0$ can be obtained from the Gauss law and the Bianchi identity. Therefore, the appropriate initial data are the transverse fields $\pi^r_{(0)}$ and $f_{ar}^{(0)}$, together with the electric and magnetic BMS charges, on one half of an hourglass.

C. Dictionary for translation to usual light cone variables

In this appendix, a dictionary between the asymptotic conditions in null coordinates in the original work of BMS,[3,4] and the asymptotic conditions in the hyperbolic foliation here introduced, is established.

The asymptotic form of the metric in a null foliation takes the form

$$ds^2 = e^{2\beta}\frac{V}{r}du^2 - 2e^{2\beta}dudr + G_{ab}\left(dx^a - U^a du\right)\left(dx^b - U^b du\right) ,$$

where

$$\beta = -\frac{c\bar{c}}{4r^2} + O\left(r^{-4}\right) ,$$

$$\frac{V}{r} = -1 + \frac{2m_B}{r} + O\left(r^{-2}\right) ,$$

$$U^z = \frac{(1+z\bar{z})\left[(1+z\bar{z})\,\partial_z\bar{c} - 2\bar{z}\bar{c}\right]}{2r^2} + \frac{2}{3r^3}\left\{(1+z\bar{z})\,\bar{c}\left[(1+z\bar{z})\,\partial_{\bar{z}}c - 2zc\right] - N^z\right\} + O\left(r^{-4}\right) ,$$

$$G_{zz} = \frac{4c}{(1+z\bar{z})^2}r + \frac{g_{zz}^{(-1)}}{r} + O\left(r^{-2}\right) ,$$

$$G_{z\bar{z}} = \frac{2r^2}{(1+z\bar{z})^2} + \frac{4c\bar{c}}{(1+z\bar{z})^2} + \frac{g_{z\bar{z}}^{(-1)}}{r} + O\left(r^{-2}\right) .$$

Here m_B, c, $\bar{c}$, N^z and $N^{\bar{z}}$ are functions of the retarded time u and the stereographic coordinates on the sphere $z, \bar{z}$, with $z = e^{i}\cot\left(\frac{\theta}{2}\right)$.

One passes to the hyperbolic coordinate by performing the changes of coordinates $u = x^0 - r$, and also (9), (10). Denoting with dots over the coefficients c, $\bar{c}$ partial derivatives with respect to their argument t_{ret}, one obtains,

$$g_{\rho\rho} = \frac{\tau_0^2}{\rho^2} - \left(\tau_0^2 + \frac{1}{2}c\bar{c}\right)\frac{1}{\rho^4} - \tau_0\left[(c\bar{c})^{\cdot} - 2m_B\right]\frac{1}{\rho^5} + O\left(\rho^{-6}\right) ,$$

$$g_{\rho z} = -\frac{\tau_0\left[(1+z\bar{z})\,\partial_{\bar{z}}c - 2zc\right]}{2\rho^2(1+z\bar{z})} - \frac{8N^{\bar{z}} + (1+z\bar{z})\left[3\tau_0^2\left((1+z\bar{z})\,\partial_{\bar{z}}\dot{c} - 2z\dot{c}\right) + 4c\left((1+z\bar{z})\,\partial_z\bar{c} - 2\bar{z}\bar{c}\right)\right]}{12(1+z\bar{z})^2\rho^3} + O\left(\rho^{-4}\right) ,$$

$$g_{zz} = \frac{4\tau_0 c}{(1+z\bar{z})^2}\rho + \frac{2\tau_0^2\dot{c}}{(1+z\bar{z})^2} + \left(\frac{\tau_0^3\ddot{c}}{2(1+z\bar{z})^2} + \frac{g_{zz}^{(-1)}}{\tau_0}\right)\frac{1}{\rho} + O\left(\rho^{-2}\right) ,$$

$$g_{z\bar{z}} = \frac{2\tau_0^2\rho^2}{(1+z\bar{z})^2} + \left[\frac{2\tau_0(c\bar{c})^{\cdot}}{(1+z\bar{z})^2} + \frac{g_{z\bar{z}}^{(-1)}}{\tau_0}\right]\frac{1}{\rho} + O\left(\rho^{-2}\right) ,$$

$$
\begin{aligned}
\pi^{\rho\rho} =& -\sqrt{\gamma}\rho^3 - \frac{1}{2}\sqrt{\gamma}\left(1+\frac{c\bar{c}}{\tau_0^2}\right)\rho \\
&+ \frac{\sqrt{\gamma}}{8\tau_0^3}\left\{\tau_0^2\left[(1+z\bar{z})\left(-2z\partial_{\bar{z}}c + (1+z\bar{z})\partial_{\bar{z}}^2 c - 2\bar{z}\partial_z\bar{c} + z\bar{z}\partial_z^2\bar{c} + \partial_z^2\bar{c}\right) + 2\bar{c}\left(\bar{z}^2 - \dot{c}\right) - 4m_B\right]\right. \\
&\quad \left. + 2\tau_0^2 c\left(z^2 - \dot{\bar{c}}\right) + 2(1+z\bar{z})^2 g_{z\bar{z}}^{(-1)}\right\} + O\left(\rho^{-1}\right), \\
\pi^{\rho z} =& -\frac{\sqrt{\gamma}}{6\tau_0^2}\left[(1+z\bar{z})\,\bar{c}\left((1+z\bar{z})\,\partial_{\bar{z}}c - 2zc\right) - N^z\right]\frac{1}{\rho^2} + O\left(\rho^{-3}\right), \\
\pi^{zz} =& \frac{3\sqrt{\gamma}(1+z\bar{z})^2\,\bar{c}}{4\tau_0\rho^2} + \frac{\sqrt{\gamma}(1+z\bar{z})^2\,\dot{\bar{c}}}{2\rho^3} + O\left(\rho^{-4}\right), \\
\pi^{z\bar{z}} =& -\frac{\sqrt{\gamma}(1+z\bar{z})^2}{2\rho} + \frac{\sqrt{\gamma}(1+z\bar{z})^2\left(2\tau_0^2 - 5c\bar{c}\right)}{8\tau_0^2\rho^3} \\
&+ \frac{\sqrt{\gamma}(1+z\bar{z})^2}{32\tau_0^3\rho^4}\left\{\tau_0^2\left[(1+z\bar{z})\left(\partial_z^2\bar{c}(1+z\bar{z}) - 2\bar{z}\partial_z\bar{c} + (1+z\bar{z})\partial_{\bar{z}}^2 c - 2z\partial_{\bar{z}}c\right)\right.\right. \\
&+ \left.\left.\left(2\bar{z}^2 - 13\dot{c}\right)\bar{c} + \left(2z^2 - 13\dot{\bar{c}}\right)c + 2m_B\right] + 6(1+z\bar{z})^2 g_{z\bar{z}}^{(-1)}\right\} + O\left(\rho^{-4}\right).
\end{aligned}
$$

As particular cases we have,

$$
f_{zz}^{(1)} = \frac{4c}{\tau_0(1+z\bar{z})^2}, \quad f_{\bar{z}\bar{z}}^{(1)} = \frac{4\bar{c}}{\tau_0(1+z\bar{z})^2},
$$

$$
h_{(1)}^{zz} = \frac{1}{2}\sqrt{\gamma}(1+z\bar{z})^2\,\dot{\bar{c}}, \quad h_{(1)}^{\bar{z}\bar{z}} = \frac{1}{2}\sqrt{\gamma}(1+z\bar{z})^2\,\dot{c}.
$$

The density associated to supertranslations $\mathcal{P}$ given in the main text is related to m_B according to

$$
\mathcal{P} = 2\sqrt{\gamma}m_B.
$$

(The factor 2 appearing on the right-hand side of the above equation is a consequence of the choice $8\pi G = 1$ used here, in contradistinction with $16\pi G = 1$ used in Refs. 3 and 4.)

References

1. T. Regge and C. Teitelboim. Role of surface integrals in the Hamiltonian formulation of General Relativity. *Annals Phys.*, 88:286, 1974.
2. P. A. M. Dirac. The theory of gravitation in Hamiltonian form. *Proc. Roy. Soc. Lond.*, A246:333–343, 1958.
3. H. Bondi, M.G.J. van der Burg, and A.W.K. Metzner. Gravitational waves in General Relativity. 7. Waves from axisymmetric isolated systems. *Proc.Roy.Soc.Lond.*, A269:21–52, 1962.
4. R. Sachs. Asymptotic symmetries in gravitational theory. *Phys.Rev.*, 128:2851–2864, 1962.
5. R. K. Sachs. Gravitational waves in General Relativity. 8. Waves in asymptotically flat space-times. *Proc. Roy. Soc. Lond.*, A270:103–126, 1962.
6. C. Bunster, A. Gomberoff and A. Pérez, "Bondi-Metzner-Sachs invariance and electric-magnetic duality," *in preparation.*
7. R. Benguria, P. Cordero, and C. Teitelboim. Aspects of the Hamiltonian dynamics of interacting gravitational gauge and Higgs fields with applications to spherical symmetry. *Nucl. Phys.*, B122:61–99, 1977.

8. C. Teitelboim. The Hamiltonian structure of space-time. *Ph.D. Thesis, Princeton, unpublished.*, 1973.
9. C. Teitelboim. How commutators of constraints reflect the space-time structure. *Annals Phys.*, 79:542–557, 1973.
10. P. A. M. Dirac. Forms of relativistic dynamics. *Rev. Mod. Phys.*, 21:392–399, Jul 1949.
11. A. Ashtekar and R. O. Hansen. A unified treatment of null and spatial infinity in general relativity. I - Universal structure, asymptotic symmetries, and conserved quantities at spatial infinity. *J. Math. Phys.*, 19:1542–1566, 1978.
12. A. Ashtekar and J.D. Romano. Spatial infinity as a boundary of spacetime. *Classical and Quantum Gravity*, 9(4):1069, 1992.
13. M. Campiglia and A. Laddha. Asymptotic symmetries of QED and Weinberg's soft photon theorem. *JHEP*, 07:115, 2015, 1505.05346.
14. C. Troessaert. The BMS4 algebra at spatial infinity. 2017, 1704.06223.
15. A. Strominger. On BMS Invariance of gravitational scattering. *JHEP*, 07:152, 2014, 1312.2229.
16. A. Strominger. Lectures on the infrared structure of gravity and gauge theory. 2017, 1703.05448.
17. F. Rohrlich. *Classical charged particles. Addison-Wesley, Boston*, 1965.
18. C. Teitelboim. Splitting of the Maxwell tensor - radiation reaction without advanced fields. *Phys. Rev.*, D1:1572–1582, 1970. [Erratum: Phys. Rev.D2,1763(1970)].
19. P. A. M. Dirac, Proc. Roy. Soc. Lond. A **167**, 148 (1938). doi:10.1098/rspa.1938.0124
20. M. Henneaux and C. Troessaert. BMS group at spatial infinity: the Hamiltonian (ADM) approach. *JHEP*, 03:147, 2018, 1801.03718.
21. M. Henneaux and C. Troessaert. Asymptotic symmetries of electromagnetism at spatial infinity. 2018, 1803.10194.
22. M. Henneaux and C. Troessaert, Hamiltonian structure and asymptotic symmetries of the Einstein-Maxwell system at spatial infinity. arXiv:1805.11288 [gr-qc].
23. M. Henneaux and C. Troessaert, The asymptotic structure of gravity at spatial infinity in four spacetime dimensions. arXiv:1904.04495 [hep-th].
24. C. Bunster, A. Gomberoff and A. Pérez, "Hamiltonian analysis of the electromagnetic field on asymptotically null spacelike surfaces in the presence of a cosmological constant," *in preparation.*
25. H. Bondi. Gravitational waves in General Relativity. *Nature*, 186(4724):535–535, 1960.
26. G. Barnich and C. Troessaert. Symmetries of asymptotically flat 4-dimensional spacetimes at null infinity revisited. *Phys.Rev.Lett.*, 105:111103, 2010, 0909.2617.
27. T. Banks. A critique of pure string theory: heterodox opinions of diverse dimensions. 2003, hep-th/0306074.
28. G. Barnich and C. Troessaert. Aspects of the BMS/CFT correspondence. *JHEP*, 05:062, 2010, 1001.1541.
29. M. Henneaux and C. Teitelboim. Duality in linearized gravity. *Phys. Rev.*, D71:024018, 2005, gr-qc/0408101.
30. C. Bunster, S. Cnockaert, M. Henneaux, and R. Portugues. Monopoles for gravitation and for higher spin fields. *Phys. Rev.*, D73:105014, 2006, hep-th/0601222.
31. G. Barnich and C. Troessaert. BMS charge algebra. *JHEP*, 12:105, 2011, 1106.0213.
32. G. Barnich and F. Brandt. Covariant theory of asymptotic symmetries, conservation laws and central charges. *Nucl. Phys.*, B633:3–82, 2002, hep-th/0111246.
33. C. W. Misner. Taub-Nut space as a counterexample to almost anything, in *Relativity Theory and Astrophysics. Vol. 1: Relativity and Cosmology*, J. Ehlers, ed., *American Mathematical Society, Rhode Island* 1967; page 160.
34. T. Regge and J. A. Wheeler. Stability of a Schwarzschild singularity. *Phys. Rev.*, 108:1063–1069, 1957.

PART V

Quantum Vortices

Quantum Vortices, Quantum Knots: Intertwining Physics, Topology and Algebra

Mario Rasetti
ISI Foundation,
Via Chisola 5, 10122 Torino, Italy

When in 1989 the seminal paper of Edward Witten[1] appeared, in which the author succeeded in putting link invariants in a field theoretical setting, showing that Jones polynomials arise as vacuum expectation values of Wilson loop operators in a three-dimensional SU(2) Chern–Simons (topological) quantum field theory — where the fundamental representation of the gauge group SU(2) lives on each component of the link — Tullio Regge's comment, that I cannot report literally because he said it in Piedmontese dialect (our favorite 'secret' language), was, more or less: "Imagine, we had it in our quiver more than fourteen years ago, and we didn't see it!". It was a pun, that masked just an imperceptible shadow of regret; in those days we were thinking about 'quiver' representations of ∞-dimensional quantum affine Lie algebras!

He was referring to an old paper of ours,[2] written in 1975, titled "Vortices in He II, Current Algebras and Quantum Knots", in which we had succeeded — resorting to Dirac's method for quantizing constrained hamiltonian systems — in developing a canonical quantization scheme for vortices, as they exist and can be seen in sophisticated experiments, in superfluid helium II, showing how this introduced in natural way in the theory the structure of the infinite Lie algebra of incomplessible flows (in fact, a current algebra related to *diff* $(\mathbb{R}^3)$).

This is the paper on which I'd like to focus this fond contribution of mine to the memory of Tullio: because it is most likely the most demanding and probably the deepest and most difficult among the many papers I wrote with Tullio, and one that paved the road for many progresses that we both did, together or separately, also in different areas of physics. For him, for example, with his PhD student Fernando Lund,[3] it was the background platform for the unification in a single theory, of a classical relativistic theory of one-dimensional extended objects (interacting through a massless scalar field, of which such objects are in turn the source), ended in a field equation that in the zero coupling limit is identical with the string theory model, whereas in the in the strong coupling limit, reproduces just our nonrelativistic vortex motion in a superfluid. In an external field that theory leads to a system of coupled nonlinear equations, which generalize the sine-Gordon system and have solitary wave solutions. For us, together, that paper was the beginning of a series of attempts to map the configuration space of our quantum tangle of vortices into that of the Ising model in 3-D, a road that never came to an end but was quite fruitful for our subsequent work in statistical mechanics. For me, finally, the somewhat visionary attempt of exporting methods and techniques from these topological approaches to the realm of data science, more precisely to the hard problem of mining huge data sets.[4]

That piece of work, which cost over a year of hard work, was to me exemplary in many different aspects. Indeed, what we accomplished was traversing the entire pathway leading from the classical description of a (incompressible) fluid in vortical motion, through the quantization of the (frictioness) vortices it could support, to the complete description of the dynamics and the configuration space of this system in terms of a current algebra whose invariants led to the capacity of deriving the full set of commuting invariant observables that rigorously describe the vortex as a (quantum!) knot. End to end.

Indeed, on the one hand we had a well defined physical problem to attack: describe the dynamics and configuration space of those quantum vortices that had been predicted by Lars Onsager in 1949[5] and by Richard Feynman six years later,[6] simply as the superfluid circulation around the vortex core, K, being quantized as integer multiple of $\hbar/m$, with $\hbar = h/2\pi$ (h the Planck's constant) and m the single Helium atom mass. They had eventually also been seen and measured experimentally by William Frank ″Joe″ Vinen in 1961.

All we had, to begin with, was the precious inheritance of the giants of fluid dynamics of the past that we could — and should — not ignore: Sir Joseph John Thomson, who in 1883 was awarded the Adam Prize for his essay on the motion of vortex rings;[7] William Thomson (Lord Kelvin), who had debated with extraordinary mathematical rigor the ′Vis-Viva of a Liquid in Motion′, including what we call now vortices;[8] Sir Horace Lamb, who in 1879 had written his omni-comprehensive ponderous treaty (706 pages!!) on ′Hydrodynamics′,[9] where for the first time vortices were indicated as the basic ingredient of turbulence. For us this first step was a taxing, though exciting, experience: for us, used to talk and discuss geometry in terms of differential forms, reading, studying and trying to metabolize that huge amount of deep physical discoveries organized in an authentic coherent patrimony of scientific knowledge, but written in a formalism where even notations like *div* or *grad* or *curl* were still far to come, was like learning a new, hard language, with the somewhat shocking realization: the whole of hydrodynamics, in its most complex forms, like vortical motion or turbulence, had never been formulated in a rigorous canonical form. We did it: this is indeed what, to begin with, that paper was about. We did it because we wanted to quantize it in rigorous way, and we wrote a consistent Lagrangian description of an incompressible fluid bearing vortices.

The further step was, just for these complex features of the underlying physics, even more challenging. In those days quantization of extended objects was a subject that just started to appear on stage, and quite timidly; the merit was mostly of the Japanese school: Tetsuo Gotō, Osamu Hara and, of course, Yoichiro Nambu. We decided to face the problem on our own terms, resorting to the elegant (and difficult!!) Dirac's method to generate a novel ′constrained Poisson bracket′ for the Hamiltonian dynamics of those massless vortex filaments, moving with no friction in an incompressible potential flow of the fluid.

The last stretch was surprising and unexpected in its outcomes, and gratifying simple and elegant — though not for this less demanding or not complex — in its formulation. The infinite dimensional current algebra of the observables of the theory, $\mathfrak{A}$, generated by current operators $J(\mathbf{b})$, where $\mathbf{b} \in \mathbb{R}^3$ are vector labels that we constructed, turned out to have a deep topological meaning: the center of its universal envelope $\mathcal{U}_{\mathfrak{A}}$ is characterized by a set of 'Casimir-like' operators that have a geometric meaning. Roughly speaking, what happens is that $\mathfrak{A}$ has an infinite ascending sequence sequence of subalgebras $\mathfrak{A}^{(p)}$, $p \in \mathbb{N}$, of degree p, all mutually commuting. The Casimir operators cannot be expressed in general as polynomials of finite degree in $\mathcal{U}_{\mathfrak{A}}$, but the subalgebras $\mathfrak{A}^{(p)}$ play an important role — in particular $\mathfrak{A}^{(2)}$ allows us to compute a most interesting invariant: even though at the quantum level the vortex is not sharply defined, we can retrieve sharply the projection of the vortex line on an arbitrarily chosen 2-D plane, as well as a sharp information on the sign of each over- and under-passing at the self-intersections of the vortex projection itself. This is nothing but the procedure whereby one extracts the Dehn's diagram of a knot in $\mathbb{R}^3$, namely the features of the vortex considered as a knot are physical observables which are conserved in the corresponding quantum representation. This property fascinated not only us, but topologists as well, such as Louis Kauffman and Samuel Lomonaco, who extensively elaborated on it. It was the first strong indication that the quantum vortex state conserved a strong memory of the classical shape of the vortex core.

Let me now trace out backward some of the steps of my narrative discussing, very concisely, a few (certainly not all!) of the developments the set of ideas emerged from that piece of work led to or at least inspired. In some cases Tullio himself had the time to be an active actor, others emerged too late for him to see them.

First is the fact that 'vorticity' and some of its effects can be easily represented in elementary quantum mechanical terms through the notion of phase and how it bears on the quantity that classical mechanics refers to as velocity. In some deep way this touched a nerve of quantum mechanics, getting to the very roots of the phase/number problem, and the role of topology in it. Let's start — indeed just to set the notation — with Schrödinger's equation. For a particle of mass m in some potential field $V(\mathbf{x},t)$ in 3-D it reads

$$i\,\partial_t\,\psi(\mathbf{x},t) = \Big(-\frac{\hbar^2}{2\,m}\,\boldsymbol{\Delta} + V(\mathbf{x},t)\Big)\,\psi(\mathbf{x},t)\;,$$

and allows us to compute the particle probability density $P(\mathbf{x},t) \equiv |\psi(\mathbf{x},t)|^2$. The latter enters the continuity equation $\partial_t\,P(\mathbf{x},t) + \nabla\cdot\mathbf{J}_\psi(\mathbf{x},t) = 0$, through the probability density current

$$\mathbf{J}_\psi(\mathbf{x},t) = -i\,\frac{\hbar}{2\,m}\Big(\psi^*\,\nabla\,\psi - \psi\,\nabla\,\psi^*\Big)\;.$$

Writing $\psi(\mathbf{x},t)$ in 'polar' form,

$$\psi(\mathbf{x},t) = \varrho(\mathbf{x},t)\,\mathrm{e}^{\,i\,\phi(\mathbf{x},t)} \quad , \quad \text{where} \quad \varrho(\mathbf{x},t) = \big[P(\mathbf{x},t)\big]^{\frac{1}{2}}\;,$$

and setting, for future convenience, $\phi(\mathbf{x},t) \doteq \hbar^{-1}\,S(\mathbf{x},t)$ one finds

$$\mathbf{v}(\mathbf{x},t) \doteq \frac{\mathbf{J}(\mathbf{x},t)}{P(\mathbf{x},t)} \equiv m^{-1}\,\nabla\,S(\mathbf{x},t) = \frac{\hbar}{m}\,\nabla\,\phi(\mathbf{x},t)\;.$$

It is quite natural — for the very way it is defined, as it was done in the so-called hydrodynamical interpretation of quantum mechanics — to identify $\mathbf{v}(\mathbf{x},t)$ with a 'velocity field'. What interests us here is the feature that $\mathbf{v}(\mathbf{x},t)$ becomes singular at those points where $P(\mathbf{x},t)=0$, unless at the same points $\mathbf{J}(\mathbf{x},t)=0$ as well and both vanish at the same rate. On the other hand, once such a field $\mathbf{v}$ is defined it is natural to associate with it a 'vorticity field',

$$\kappa(\mathbf{x},t) \doteq \nabla \wedge \mathbf{v}(\mathbf{x},t) \; ,$$

and, hence, with any bounded surface Σ associate a 'circulation'

$$K = \oint_{\partial\Sigma} \mathbf{v}\cdot \mathrm{d}\mathbf{s} = \iint_{\Sigma} \kappa \cdot \mathrm{d}\sigma \; .$$

$\kappa = m^{-1}\,\nabla\wedge\nabla S$ vanishes everywhere (the fluid is 'irrotational') except at the points where the velocity field becomes singular, i.e., the probability density vanishes; while the circulation is zero except along those closed curves $\partial\,\Sigma$ that encircle such a singularity. Thus our current algebra actually describes all possible measure preserving quantum diffeomorphisms of the ambient space — the subpace $\mathcal{V}\subset\mathbb{R}^{(3)}$ where the vortices live — coherent with the shape (the topology) of the vortices.

The second point I'd like to touch on has to do with the infinite dimensional current algebra $\mathfrak{A}$. As the shape of the vortices is left untouched by this algebra — which is essentially the algebra of the mapping class group, namely the group of diffeomorphisms modulo isotopy, of $\sqsubseteq\setminus\mathcal{K}$, $\mathcal{K}$ denoting the vortex core line — there ensues the possibility of associating with it with an infinite set of 'generalized harmonic oscillators'. Let me summarize concisely what we mean with this. $\mathfrak{A}$, as mentioned, is generated by infinitely many current operators $J(\mathbf{b})$, pairwise linked by the relations

$$\{J(\mathbf{b}),J(\mathbf{c})\}^{\ddagger} = -\frac{1}{K\,\varrho}\,J(curl\,(\mathbf{b}\wedge\mathbf{c})) \; ,$$

where $\varrho = m/|\mathcal{V}|$, $|\mathcal{V}|$ being the volume occupied by the fluid, so that ϱ^{-1} is nothing but the average volume at disposal of a single Helium atom; K is the circulation of the vortex, which are such that[2] the vortex Lagrangian $\mathcal{L}$ can be written as

$$\mathcal{L} = \tfrac{1}{3}\,K\,\varrho \int_{\mathcal{K}} \varepsilon_{kj\ell}\,x^k(\sigma)\,\frac{\mathrm{d}x^j}{\mathrm{d}\sigma}\,\frac{\mathrm{d}x^\ell}{\mathrm{d}t}\,\mathrm{d}\sigma - \tfrac{1}{2}\,\varrho \int_{\mathcal{V}} |\mathbf{v}|^2\,\mathrm{d}\mathbf{x} \; ,$$

$\sigma\in\mathbb{R}$ being a variable parametrizing the points of $\mathcal{K}$, while $\mathbf{v}$ is the velocity field, such that $curl\,\mathbf{v}=0$ in $\mathcal{V}\setminus\mathcal{K}$. What is relevant here is that $\{J,J'\}^{\ddagger}$ does not denote Poisson backets (that we denote as $\{J,J'\}_{(PB)}$) but what in the recent literature is referred to as RRDB[11] (Rasetti–Regge Dirac Brackets), namely

$$\{A,B\}^{\ddagger} = \{A,B\}_{(PB)} - \{A,\varphi_\ell\}_{(PB)}\,\left(C^{-1}\right)_{\ell j}\,\{\varphi_j,B\}_{(PB)} \; ,$$

where $C_{\ell j} = \{\varphi_\ell,\varphi_j\}_{(PB)}$. Here $\{\varphi_1,\ldots,\varphi_{2n}\}$ is the set of second-class constraints representing the (global) shape of the vortex. The explicit brackets for the case of vortices were explicitly evaluated for all necessary observable in Ref. 2. In a nutshell, quantization is thus finally obtained by imposing

$$\left[\,\hat{J}(\mathbf{b}),\hat{J}(\mathbf{c})\,\right]^{\ddagger} = i\,\frac{\hbar}{K\,\varrho}\,\hat{J}(curl\,(\mathbf{b}\wedge\mathbf{c})) = i\,\frac{\mathcal{V}}{2\,\pi}\,\hat{J}\big(\mathbf{c}\cdot\nabla\mathbf{b}-\mathbf{b}\cdot\nabla\mathbf{c}\big) \; ,$$

where the l.h.s. is obtained replacing all $\{\lambda,\mu\}_{(PB)}$ in the expression of $\{J(\mathbf{b}),J(\mathbf{c})\}^{\ddagger}$ by commutators $i\hbar\left[\hat{\lambda},\hat{\mu}\right]$ of the corresponding operators. At this point one can write the quantum Hamiltonian operator,

$$\hat{H}=\frac{K^2\,\varrho}{16\,\pi^3}\int_{\mathbb{R}^3}\frac{\mathrm{d}\mathbf{q}}{|\mathbf{q}|^2}\sum_{\mu}\hat{J}\big(b^{(\mu)}(\mathbf{q})\big)\,\hat{J}^{\dagger}\big(b^{(\mu)}(\mathbf{q})\big)\,,$$

where μ is a polarization label, $\mathbf{q}$ is a label for the modes of the field, $\hat{J}^{\dagger}\big(b^{(\mu)}(\mathbf{q})\big)=\hat{J}\big(b^{(\mu)}(-\mathbf{q})\big)$ and, in a suitable choice of the gauge that guarantees reparametrization invariance of the theory, $\mathbf{B}(\mathbf{x})$ — the classical vector field such that $\mathbf{b}(\mathbf{x})=curl\,\mathbf{B}(\mathbf{x})$ — has a form that consistently generates the desired representation of $\mathfrak{A}$.

This working scheme, unbelievably complex as it is, is indeed quite far reaching. Also in the presence of vortices, the motion of the fluid can be described in terms of a scalar velocity potential $\varphi(\mathbf{x})$, $\mathbf{v}=-\nabla\varphi$. However $\varphi(\mathbf{x})$ cannot be globally extended to a single valued function over $\mathcal{V}\setminus\mathcal{K}$, because continuing it along a closed curve Λ, $\varphi(\mathbf{x})$ must change by $\ell\,K$, $\ell\in\mathbb{Z}$ being the 'linking number' between Λ and $\mathcal{K}$. A way out is to resort to the potentials introduced by Alfred Clebsch,[12] $\eta(\mathbf{x})$, $\zeta(\mathbf{x})$, whereby

$$\mathbf{v}=\eta\,\nabla\zeta-\nabla\varphi\,,$$

and φ — in order to have an incompressible flow ($div\mathbf{v}=0$) — must satisfy the condition $\boldsymbol{\Delta}\varphi=div(\eta\,\nabla\zeta)$, which allows us to write

$$\mathbf{v}=curl\mathbf{A}\quad,\quad\text{with }\mathbf{A}(\mathbf{x})=\frac{1}{4\,\pi}\int_{\mathcal{V}}\frac{\mathrm{y}}{|\mathbf{x}-\mathbf{y}|}\,\nabla\eta(\mathbf{y})\wedge\nabla\zeta(\mathbf{y})\,.$$

Vector $\omega(\mathbf{x})=\nabla\eta(\mathbf{x})\wedge\nabla\zeta(\mathbf{x})$ describes therefore the distribution of vortical lines. In the vortex filament limit, when $\omega(\mathbf{x})$ is a distribution with support on the curve $\mathcal{K}$ and tangent to it, one retrieves the familiar gauge fixing: $\mathbf{v}(\mathbf{x})=curl\mathbf{A}(\mathbf{x})$, with $div\mathbf{A}=0$, $\mathbf{A}(\mathbf{x})=\dfrac{K}{4\,\pi}\displaystyle\int_{\mathcal{K}}\frac{\mathrm{d}\mathbf{y}}{|\mathbf{x}-\mathbf{y}|}$. An elegant way to achieve this is by selecting η and ζ as

$$\eta(\mathbf{x})=\Theta\big(f(\mathbf{x})\big)\quad,\quad\zeta(\mathbf{x})=\Theta\big(g(\mathbf{x})\big)\,,$$

where $\Theta(\bullet)$ is the Heaviside step function and $f(\mathbf{x})$, $g(\mathbf{x})$ are such that $\mathcal{K}$ is in $\mathbb{R}^{(3)}$ (or, better in $\mathcal{V}$) the intersection of the loci $\eta=0$, $\zeta=0$. With this choice the kinetic part of the Lagrangian $\mathcal{L}$ writes[2]

$$\mathcal{L}_{kin}=\tfrac{1}{2}\,K\,\varrho\int_{\mathcal{V}}\left(\frac{\mathrm{d}\zeta}{\mathrm{d}t}\,\eta(\mathbf{x})-\frac{\mathrm{d}\eta}{\mathrm{d}t}\,\zeta(\mathbf{x})\right)\,\mathrm{d}\mathbf{x}\,.$$

From this one can infer that η and ζ are mutually canonically conjugate but that the ensuing action $\mathcal{S}_{kin}=\displaystyle\int_{t_0}^{t}\mathcal{L}_{kin}(t')\,\mathrm{d}t'$, is degenerate, as it is linear in the time derivatives. Therefore constraints are implied; for π_η, π_ζ the canonical momenta conjugated, respectively, to η, ζ,

$$\pi_\eta+\tfrac{1}{2}\,K\,\varrho\,\zeta=0\quad,\quad\pi_\zeta+\tfrac{1}{2}\,K\,\varrho\,\eta=0\,.$$

It is (almost!) straightforward to compute the corresponding Dirac brackets and the associated quantum commutator

$$\{\eta(\mathbf{x}),\zeta(\mathbf{y})\}^{\ddagger}=\frac{1}{K\,\varrho}\,\delta^{(3)}(\mathbf{x},\mathbf{y})\quad,\quad[\eta(\mathbf{x}),\zeta(\mathbf{y})]^{\ddagger}=-\frac{i\,\mathcal{V}}{2\,\pi}\,\delta^{(3)}(\mathbf{x},\mathbf{y})\,.$$

All of this has two important messages: i) we face here a sort of generalized Weyl–Heisenberg algebra — generalized in the sense that $[\xi, \pi_\xi]$ is replaced by $[\xi, \pi_\xi]^\ddagger$ as algebra composition law — whose structure promises to be very rich, and which indeed in all these years has never been studied in depth; ii) the commutation relations, even though generalized, still imply that it is not possible to define sharply the position of the vortex in $\mathbb{R}^{(3)}$ (or in$\mathcal{V}$).

It is the latter point ii) that is especially intriguing: a description of the vortex based on the fields η and ζ amounts to states of the system that do not fit into the representation, borrowed from classical intuition, that one might depict in her mind, of the vortex as some sort of smoke ring in shape-preserving evolution. Even if the initial configuration of the vortex had η and ζ well represented by step functions, its evolution will eventually take it into a state where the vortex fills up the whole volume $\mathcal{V}$ (like a space filling Peano curve).

Also the corresponding quantum representation has not yet been explored in depth and could be exciting in terms of geometry and topology. Kauffman and Lomonaco conjecture[10] that this might be approximated as a quantum superposition of infinitely many knots belonging to the same equivalence class.

A few more comments: it is resorting to the subalgebra $\mathfrak{A}^{(2)} \subset \mathfrak{A}$ that one finds the result, already reported above, that the collection of all ingredients necessary to reconstruct the entire Dehn's diagram of the vortex core, considered as a knot, is fully represented by a subset of the Cartan subalgebra of $\mathfrak{A}$. It is straightforward to conjecture that working in the frame of the whole $\mathcal{U}_{\mathfrak{A}}$ one can indeed construct the whole Alexander polynomial for $\mathcal{K}$ .... but not the Jones polynomial[13] $\mathcal{P}_J$! And here the circle of my tale closes: one needs Witten's quantum topological field theory to get the universal invariant P_J. Where is the connection? We didn't fully perceive it, but a good deal of it was already there. The classical $SU(2)$ Chern–Simons action $\mathcal{S}_{CS}$ for the sphere $S^{(3)}$ (which is the simplest compact, oriented 3-manifold without boundary) is what succeeds in breaking the code of knot classification. $\mathcal{S}_{CS}$ is given by

$$k\,\mathcal{S}_{CS}(A) = \frac{k}{4\pi}\int_{S^{(3)}} \mathrm{tr}\left(A\,\mathrm{d}A + \tfrac{2}{3}\,A\wedge A\wedge A\right).$$

Here A is the connection 1-form with value in the Lie algebra $su(2)$ of the gauge group, k the coupling constant, d the exterior differential, $\wedge$ the wedge product of differential forms and the trace is taken over Lie algebra indices. The partition function of the quantum theory is obtained as a 'path integral', namely by integrating the exponential of i times the classical action $\mathcal{S}_{CS}(A)$ over the space of gauge-invariant flat $SU(2)$ connections (the field variables) through the formal expression

$$\mathbf{Z}_{CS}\left[S^{(3)};k\right] = \int [\mathcal{D}A]\,\exp\left\{i\,\frac{k}{4\pi}\,\mathcal{S}_{CS}(A)\right\},$$

where the coupling constant $k \in \mathbb{N}$ is constrained to be a positive integer by the gauge-invariant quantization procedure and is related to the deformation parameter necessary to make the theory consistent. The generating functional $\mathbf{Z}_{CS}\left[S^{(3)};k\right]$. Written for a generic compact oriented 3-manifold $\mathcal{M}^{(3)}$ with $\partial\mathcal{M}^{(3)} = \emptyset$, is a global invariant, namely depends only on the topological type.[1]

Yet another dangling string was closed here, with Annalisa Marzuoli, another beloved collaborator of Tullio and our former student Silvano Garnerone.[14] In the frame of quantum computation, we constructed a quantum algorithm to approximate efficiently — i.e., in polynomial time to any desired precision — the colored Jones polynomial of the plat presentation of any oriented link $\mathfrak{L}$. Our construction was based just on Witten's $SU(2)$ Chern–Simons topological quantum field theory and exploits the q-deformed spin network as computational background. We proved that a colored Jones polynomial can be evaluated in a number of elementary steps, bounded from above by a linear function of the number of crossings of the link, and polynomially bounded with respect to the number of link strands. This was a major achievement, as it is known that computing Jones polynomials is a computationally #**P** problem. The conceptual framework is the following. It is by now generally assumed that each physical theory supports computation models whose power is limited only by the physical features of the theory itself. Classical physics and quantum mechanics support a multitude of different implementations of the Turing machine model of computation. Michael Freedman, Alexei Kitaev and others suggested the possibility that non-Abelian topological quantum field theories might support a model of computation capable of solving hard problems of the computational class #**P** in polynomial time, and that systems characterized by a Chern–Simons action might serve as an analog computer (simulator) capable of solving such problems in a universal setting.

Ever since Gödel stated his incompleteness theorems, we know that universality requires a price to be paid. This is because, even though it is true that in mathematics axiomatic set theory has proven to be universal, in the sense that any presently known field of mathematics can — in principle — be boiled down to constructions in this theory, the price of Gödel's incompleteness theorems is in a never ending richness of models (in the sense of model theory of formal axiomatic systems) which axiomatic set theory has (and which — just by Gödel's results — no formal axiomatic theory whatever can come to terms with). In analogy with Gödel's statements, it is reasonable to expect that the any universal theory comes about with an inexhaustible plentitude of models (i.e., not simply infinitely many, but such that they cannot be described in a single close form). However, resorting to the definition of a 'quantum simulator' based on quantum geometries of the simplicial Regge type, namely making the base-space of the theory piece-wise linear, and extending to the quantum framework the notion of Regge simplicial complex, we did it.

Tullio, as usual eventually once more you were right: we (or better, you) really had it all there, in the wonderful quiver of your mind!! I miss you.

References

1. E. Witten, *Quantum field theory and the Jones polynomial*, Commun. Math. Phys. **121**, (1989) 351-399.
2. M. Rasetti and T. Regge, *Vortices in He II, Current Algebras and Quantum Knots*, Physica **A80**, (1975) 217-233;
   see also: M. Rasetti and T. Regge, *Quantum Vortices and diff* ($\mathbb{R}^3$), in *Group Theoretical Methods in Physics*, G. Denardo, G. Ghirardi and T. Weber, Eds., Springer Lecture Notes in

Physics, vol. **201**, Springer-Verlag, New York, (1984), 311-320;
and: M. Rasetti and T. Regge, *Quantization of extended vortices and* $sDiff(\mathbb{R}^{(3)})$ in *Algebraic Analysis - Papers Dedicated to Professor Mikio Sato on the Occasion of His Sixtieth Birthday*, M. Kashiwara and T. Kawai, Eds., Academic Press, Cambridge, Mass. (1989), 727-734.

3. F. Lund and T. Regge, *Unified approach to strings and vortices with soliton solutions*, Phys. Rev. **D14**, (1976), 1524-1535.
4. M. Rasetti and E. Merelli, *Topological Field Theory of Data: Mining Data Beyond Complex Networks*, in *Advances in Disordered Systems, Random Processes and Some Applications*, P.L. Contucci and C. Giardinà, Eds., Cambridge University Press, Cambridge, (2016), 1-42.
5. L. Onsager, *Statistical Hydrodynamics*, Nuovo Cimento Suppl. **6**, N. 2, (1949) 279-287.
6. R.P. Feynman, *Application of Quantum Mechanics to Liquid Helium*, in *Progress in Low Temperature Physics*, C.J. Gorter, Ed., Vol. 1, Ch. 2, North Holland Publ. Co., Amsterdam (1955), 17-53.
7. J.J. Thomson, *A treatise on the motion of vortex rings*, Macmillan & Co., London, (1883), 124 pages.
8. W. Thomson, *Notes on Hydrodynamics. On the Vis-Viva of a Liquid in Motion*, in *Mathematical and Physical Papers* [from the *Cambridge and Dublin Mathematical Journal*, February 1849], Vol. 1, Cambridge University Press, Cambridge, (1882), 107-112.
9. H. Lamb, *Hydrodynamics*, Cambridge University Press, Cambridge, (1879), 706 pages.
10. L.H. Kauffman and S.J. Lomonaco, *Quantum Knots*, in *SPIE Proc. on Quantum Information & Computation II*, E. Donkor, A.R. Pirich and H.E. Brandt, Eds., (2004), 5436-30, 268-284. see also: S.J. Lomonaco and L.H. Kauffman, *Quantum Knots and Mosaics*, Journal of Quantum Information Processing, **7**, Nos. 2-3, (2008), 85-115.
11. D.D. Holm, *Rasetti–Regge Dirac Bracket formulation of Lagrangian fluid dynamics of vortex filaments*, Mathematics and Computers in Simulation, **62**, (2003), 53-63. see also: A.D. Speliotopoulos, *A topological string: the Rasetti-Regge Lagrangian, topological quantum field theory and vortices in quantum fluids*, J. Phys. A: Math. Gen. **35**, (2002), 8859.
12. A. Clebsch, *Über eine allgemeine Transformation der Hydrodynamischen Gleichungen*, Crelle's Journal, (1859), 293-313; Reine Angnew. Math. ("Crelle"), **56**, (1859), 1-20.
13. V.F.R. Jones, *A polynomial invariant for knots via von Neumann algebra*, Bull. Amer. Math Soc, **12**, (1985), 103-111.
14. S. Garnerone, A. Marzuoli and M. Rasetti, *An efficient quantum algorithm for colored Jones polynomials*, Int. J. Quantum Information, **6**, issue 1, (2008), 773-782.

PHYSICAL REVIEW D VOLUME 14, NUMBER 6 15 SEPTEMBER 1976

Unified approach to strings and vortices with soliton solutions

Fernando Lund and Tullio Regge*
Institute for Advanced Study, Princeton, New Jersey 08540
(Received 19 May 1976)

A classical relativistic theory of one-dimensional extended objects interacting through a massless scalar field, of which they are in turn the source, is constructed. In the no-coupling limit, the string model is recovered. In another limit, the system that describes nonrelativistic vortex motion in a superfluid is obtained. The diverging self-interaction of these objects is shown to be regularizable through a renormalization of the slope of the Regge trajectories. Motion in an external field is studied in some detail and leads to a system of coupled nonlinear equations that generalizes the sine-Gordon system. Solitary wave solutions to these equations are obtained and a natural geometric interpretation to the associated linear equations of the inverse scattering method is given.

I. INTRODUCTION

In a previous paper by one of us in collaboration with Rasetti,[1] a canonical formalism was constructed that described the motion of vortices in an incompressible, inviscid fluid, of which superfluid He II is the best example. Among the problems that were opened by this work was the question of whether there exists a Lorentz-invariant theory that would generalize this nonrelativistic formalism in some sense. Also, in light of the work of Nielsen and Olesen[2] relating the string model with the vortex lines in a type-II superconductor, one would like to know what the relation is, if any, between the Nambu string[3] and vortices in a superfluid, apart from the mere fact that they are both one-dimensional extended objects.

We present here a classical Lagrangian that implements a unified theory of strings and (superfluid) vortices in the following sense: One-dimensional objects—that may indifferently be called strings or vortices—interact through a massless scalar field of which they are in turn the source. If the coupling constant vanishes, the action integral is just that of the Nambu string. On the other hand, if the slope of the Regge trajectory is infinite, that is, if the coefficient in front of the Nambu action vanishes, taking the special case of the one-dimensional object moving in a certain prescribed external field reduces the Lagrangian to the one of Ref. 1.

It turns out that our Lagrangian is the same as the one introduced some time ago by Kalb and Ramond[4] in their construction of a theory of interacting strings by analogy with the action-at-a-distance electrodynamics of Wheeler and Feynman.[5] Thus, another way of expressing our result is to say that the theory of Kalb and Ramond admits, as a particular case, the motion of a vortex in a superfluid.

The self-energy of an infinitely thin (classical) superfluid vortex is divergent. This is also true of our strings. However, the divergence is logarithmic and it can be taken care of by a renormalization of the slope of the Regge trajectories.

The equations of motion will be studied in some detail for the case of a uniform, static external field, and some pleasant surprises will emerge. In fact, the problem of the motion of the string is just the problem of finding the surface it describes in a Minkowski space-time. Taking the projection of this surface on a three-dimensional (Euclidean) hypersurface of constant time one recovers the ancient problem of embedding a surface in space, and our equation of motion is just a restriction on the admissible geometries for this surface. Under certain special initial conditions, the equations of Gauss and Codazzi, which say when a surface with a given geometry may be embedded in a three-dimensional Euclidean space, reduce to the sine-Gordon equation. This has been noticed before.[6] What apparently has not been realized is that the equations of Gauss and Weingarten—which actually build the surface—are first-order linear differential equations for the normal and tangent vectors that can be written in terms of a rotation operator, and that expressing this operator in the spin-$\frac{1}{2}$ representation, one recovers exactly the equations used by Ablowitz *et al.*[7] to solve the sine-Gordon equation. Thus we get a natural answer to the question of why a particular linear problem is of help in solving a certain nonlinear equation. For example, we might mention here that different eigenvalues of the linear problem correspond to projections of our surface along different three-dimensional spaces which are related to one another by a Lorentz transformation.

For general initial conditions, we are led to a set of two coupled nonlinear equations with an associated linear problem. We have not yet solved this problem in general. Static solutions can easily be found, however. These equations may be regarded as the dynamical equations of a theory of two coupled scalar fields in one space and one time dimensions. In this case a Lagrangian may be written, and some of the static solutions represent solitons in their rest frame.

Starting from a definite physical picture, namely, that of strings interacting through a scalar field in four-dimensional Minkowski space-time, we have been led by the mathematics of the problem to a quite different physical situation, namely, that of two coupled scalar fields in a two-dimensional Minkowski space-time. It would be very gratifying indeed if one could translate these formal analogies into a physical statement. We can only speculate at this stage, however. A serious obstacle to such a unified framework of quite different objects is that although the equations are the same, and thus locally a solution to one problem provides a solution to the other one, this is not so globally, as the boundary conditions are quite different. For example, closed strings are described by dynamical variables periodic in the space coordinate. This is not so in a field theory, the spatial variable being allowed to vary over the whole range of real numbers. Perhaps the way to look at this is to take infinite strings, but we have not yet done actual computations along this line. Another possibility is to interchange the role of the spatial and temporal coordinates, thus relating, one hopes, motions of a closed string with the so-called "breather" modes, which are periodic in time.

This work is organized as follows: In Sec. II we write the Lagrangian, the equations of motion, and show how the self-energy of a closed string may be regularized by a renormalization procedure. The canonical formalism for the theory is worked out in the Appendix. In Sec. III we study the motion of a closed string in a uniform external field and discuss the implications of the geometrical way of attacking the problem. Section IV contains some conclusions and summarizes the loose ends of this work. Quantum problems will not be discussed.

II. THE MODEL

In Ref. 1, a vortex was described by a closed curve Γ in R^3, parametrized by three functions $x^i(\sigma)$. This vortex moves in a fluid described by a divergenceless velocity field $\vec{v}(\vec{y},t)$. The dynamical ingredient is the statement that, at each point, the vortex velocity is given by the velocity of the fluid[8]

$$\frac{dx^i}{dt} = v^i\,|_{\vec{y}=\vec{x}}\,. \tag{2.1}$$

The dynamics of the system remains unchanged, however, if the parameter σ is changed in an arbitrary way, as it is just a coordinate along the vortex. In general, then, the parametrization is an arbitrary function of time, $\sigma=\sigma(t)$, and Eq. (2.1) acquires the more general form

$$\frac{\partial x^i}{\partial t} = v^i - \frac{d\sigma}{dt}\frac{\partial x^i}{\partial\sigma}\,, \tag{2.2}$$

where v^i is understood to be evaluated at $\vec{y}=\vec{x}$.

Since $d\sigma/dt$ is arbitrary, this equation says that the vectors $\partial\vec{x}/\partial\sigma$ and $\partial\vec{x}/\partial t-\vec{v}$ are parallel. In other words,

$$\epsilon^{ijk}\frac{\partial x^j}{\partial\sigma}\left(\frac{\partial x^k}{\partial t}-v^k\right)=0\,. \tag{2.3}$$

To consider a vortex line in a fluid means that the vorticity $\vec{w}$ vanishes everywhere except along that line, to which it is tangent. That is,

$$\begin{aligned}\vec{w} &\equiv \operatorname{curl}\vec{v}\\ &= k\frac{\partial\vec{x}}{\partial\sigma}\,\delta(\vec{y}-\vec{x}(\sigma))\,,\end{aligned} \tag{2.4}$$

where k is a constant. Moreover, since $\operatorname{div}\vec{v}=0$, there exists a vector potential $\vec{A}$ such that

$$\vec{v}=\operatorname{curl}\vec{A}\,. \tag{2.5}$$

Substituting into (2.4) and using the fact that $\vec{A}$ is not completely determined so that $\operatorname{div}\vec{A}=0$ may be imposed, we get Poisson's equation for $\vec{A}$, which has the well-known solution

$$\vec{A}(\vec{y})=\frac{k}{4\pi}\int_\Gamma\frac{1}{|\vec{y}-\vec{x}(\sigma)|}\frac{\partial\vec{x}}{\partial\sigma}\,d\sigma\,. \tag{2.6}$$

This expression, together with (2.5), gives the velocity $\vec{v}$ as a functional of the dynamical variables $\vec{x}(\sigma)$. The Lagrangian that, upon variation of the $\vec{x}(\sigma)$, gives the equation of motion (2.3) is[1,9]

$$\mathcal{L}=\frac{k\rho}{3}\int_\Gamma\epsilon^{ijk}x^i\frac{\partial x^j}{\partial\sigma}\frac{\partial x^k}{\partial t}\,d\sigma-\frac{\rho}{2}\int v^2 d^3y\,, \tag{2.7}$$

where ρ is the density of the fluid and $\vec{v}$ is understood to be expressed in terms of $\vec{x}(\sigma)$.

We now want to generalize the Lagrangian (2.7) in such a way that we will have a Lorentz-invariant theory. The only requirements are that the action integral be a Lorentz scalar and that it must have (2.7) as a special case. This special case is a theory in which the dynamical variables are the $x^i(\sigma)$. It is an obvious thing to do, then, to consider a theory with variables $x^\mu(\sigma)$ ($\mu=0,1,2,3$). The addition of one more variable reflects the fact that we shall have one more invariance, that which is under time reparametrizations.

In the nonrelativistic theory described by the Lagrangian (2.7) there is a field $\vec{v}(\vec{y})$ whose source is the vortex $\vec{x}(\sigma)$ that completely determines it through (2.5) and (2.6). The field $\vec{v}(\vec{y})$ does not have any life of its own, so to speak, just as the Newtonian gravitational field is completely determined by its source. This is, of course, just another way of saying that in these nonrelativistic theories interactions propagate instantaneously. We introduce, then, to replace v^i, a vector field F^μ with a Lagrangian density proportional to $F_\mu F^\mu$. How is this field coupled to $x^\mu(\sigma)$? The clue to answering this question lies in the fact that the first term in the right-hand side of the Lagrangian (2.7) is not translation invariant and must be replaced by a term that does have this invariance, i.e.,

$$A_{\mu\nu}\frac{\partial x^\mu}{\partial\sigma}\frac{\partial x^\nu}{\partial\tau}, \tag{2.8}$$

where $A_{\mu\nu}$ is an antisymmetric tensor. We have changed the notation for the time variable from t to τ to emphasize the fact that it is no longer the Newtonian time but an arbitrary parameter. It is natural now to take this term as the coupling between F^μ and x^μ, and we must find how F^μ can be related to $A_{\mu\nu}$. This is done as follows: In the nonrelativistic case, $x^0=\tau=t$ and the field $A_{\mu\nu}$ evaluated at the vortex must satisfy

$$A_{ij}|_{\vec{y}=\vec{x}}=\epsilon_{ijk}x^k. \tag{2.9}$$

This is most easily satisfied just by taking

$$A_{ij}(\vec{y})=\epsilon_{ijk}y^k. \tag{2.10}$$

On the other hand, still in this nonrelativistic case,

$$F^i=v^i=\epsilon^{ijk}\partial_j A_k. \tag{2.11}$$

The easiest way of generalizing this to four dimensions is to define

$$F^\mu\equiv\tfrac{1}{2}\epsilon^{\mu\nu\lambda\rho}\partial_\rho A_{\nu\lambda}. \tag{2.12}$$

We have been led then to the action

$$S'=f\int A_{\mu\nu}\frac{\partial x^\mu}{\partial\sigma}\frac{\partial x^\nu}{\partial\tau}d\sigma\,d\tau - \tfrac{1}{4}\int F^2 d^4y, \tag{2.13}$$

with f a coupling constant and with F^μ given in terms of $A_{\mu\nu}=-A_{\nu\mu}$ by (2.12). The first integral is taken over the surface described by the vortex in space-time and the field $A_{\mu\nu}$ in the integrand is evaluated at this world surface. This action is a Lorentz scalar and is also invariant under coordinate transformations on the surface as well as under the gauge transformation

$$A_{\mu\nu}\to A_{\mu\nu}+\partial_\mu\Lambda_\nu-\partial_\nu\Lambda_\mu. \tag{2.14}$$

This is true owing to the fact that vortices are closed or, eventually, infinitely long.

We have then a relativistic system with variables $x^\mu(\sigma,\tau)$ and $A_{\mu\nu}(y)$. If the coupling constant vanishes, this is a free massless scalar field. The apparent tensor nature of the $A_{\mu\nu}$ does not mean that it carries spin. This is most directly seen using the canonical formalism. We do this in the Appendix. What we do not know is how the vortex moves when it is decoupled from the field. We need an extra term in the action. The obvious candidate is the Nambu action[3] for a relativistic string. In fact, whether we call our one-dimensional objects strings or vortices makes no difference, as they are abstractions having little resemblance to what we intuitively understand by either string or vortex. We shall use both names indistinctively. Also, as we shall see below, it is the Nambu action that must be added if we want to be able to regularize the self-interaction of a vortex. We have then

$$S=-N\int\sqrt{-g}\,d\sigma\,d\tau + f\int A_{\mu\nu}\frac{\partial x^\mu}{\partial\sigma}\frac{\partial x^\nu}{\partial\tau}d\sigma\,d\tau - \tfrac{1}{4}\int F^2 d^4y, \tag{2.15}$$

with

$$\sqrt{-g}=\left[-\left(\frac{\partial x^\mu}{\partial\sigma}\frac{\partial x_\mu}{\partial\sigma}\right)\left(\frac{\partial x^\nu}{\partial\tau}\frac{\partial x_\nu}{\partial\tau}\right)+\left(\frac{\partial x^\mu}{\partial\sigma}\frac{\partial x_\mu}{\partial\tau}\right)^2\right]^{1/2} \tag{2.16}$$

It is clear that the action (2.15) has the same invariances as the preliminary action (2.13). The same expression has been obtained by Kalb and Ramond[4] through a different line of reasoning.

To summarize, then, we have an action that, for $f=0$, describes a system of noninteracting strings and fields. For $N=0$, that is, for Regge trajectories of infinite slope, we get the action (2.13). In the particular Lorentz frame $x^0=\tau$ and for

$$A_{i0}=A_i,\quad A_{ij}=\epsilon_{ijk}y^k \tag{2.17}$$

we recover the action (2.7) for nonrelativistic vortex motion in a superfluid, apart from an unimportant additive constant.

We now turn to the equations of motion. The variables $x^\mu(\sigma,\tau)$ describe a two-dimensional surface. Every such surface is conformally flat, and it is possible to choose coordinates in which this is apparent ("orthonormal gauge"). Since the metric of the surface is

$$ds^2 = \frac{\partial x^\mu}{\partial \sigma}\frac{\partial x_\mu}{\partial \sigma} d\sigma^2 + 2\frac{\partial x^\mu}{\partial \sigma}\frac{\partial x_\mu}{\partial \tau} d\sigma\, d\tau + \frac{\partial x^\mu}{\partial \tau}\frac{\partial x_\mu}{\partial \tau} d\tau^2 , \qquad (2.18)$$

the conditions that must be satisfied are

$$\frac{\partial x^\mu}{\partial \sigma}\frac{\partial x_\mu}{\partial \sigma} + \frac{\partial x^\mu}{\partial \tau}\frac{\partial x_\mu}{\partial \tau} = 0 , \qquad \frac{\partial x^\mu}{\partial \sigma}\frac{\partial x_\mu}{\partial \tau} = 0 . \qquad (2.19)$$

Also, considering the gauge freedom (2.14) for the field $A_{\mu\nu}$, it is possible to impose the "Lorentz condition"

$$\partial_\mu A^{\mu\nu} = 0 . \qquad (2.20)$$

These remarks are made more precise in the context of the canonical formalism in the Appendix. With the gauge choices (2.19) and (2.20), the equations of motion obtained by minimizing the action (2.15) are

$$\partial^\mu \partial_\mu A^{\alpha\beta} = -2f \int \left(\frac{\partial x^\alpha}{\partial \sigma}\frac{\partial x^\beta}{\partial \tau} - \frac{\partial x^\alpha}{\partial \tau}\frac{\partial x^\beta}{\partial \sigma} \right) \times \delta^{(4)}(x-y)\, d\sigma\, d\tau , \qquad (2.21)$$

$$N\left(\frac{\partial^2 x^\mu}{\partial \tau^2} - \frac{\partial^2 x^\mu}{\partial \sigma^2} \right) = f \epsilon^{\mu\nu\lambda\rho} F_\nu \frac{\partial x_\rho}{\partial \tau}\frac{\partial x_\lambda}{\partial \sigma} . \qquad (2.22)$$

We consider first Eq. (2.19), that gives the field produced by a moving string. It is just a wave equation with an external source and has the solution

$$A^{\mu\nu}(y) = f \int G_{\text{ret}}(y-x) \times \left(\frac{\partial x^\mu}{\partial \sigma}\frac{\partial x^\nu}{\partial \tau} - \frac{\partial x^\nu}{\partial \sigma}\frac{\partial x^\mu}{\partial \tau} \right) d\sigma\, d\tau , \qquad (2.23)$$

where

$$G_{\text{ret}}(x) = \frac{1}{2\pi}\, \delta(x^\mu x_\mu)\, \theta(x_0) \qquad (2.24)$$

is the retarded Green's function. From here one has

$$F^\mu(y) = -\frac{f}{\pi} \int \theta(y_0 - x_0)\, \delta'((y-x)^2) \times \epsilon^{\mu\nu\lambda\rho} (x-y)_\nu\, dx_\lambda \wedge dx_\rho , \qquad (2.25)$$

with

$$dx_\mu \wedge dx_\nu = \left(\frac{\partial x_\mu}{\partial \sigma}\frac{\partial x_\nu}{\partial \tau} - \frac{\partial x_\mu}{\partial \tau}\frac{\partial x_\nu}{\partial \sigma} \right) d\sigma\, d\tau . \qquad (2.26)$$

It is an illuminating exercise to show that F^μ is (locally) the gradient of a scalar field. As we said before, although we are working with an antisymmetric tensor $A_{\mu\nu}$, there is only one dynamical degree of freedom per space point. The rest is gauge. Explicitly,

$$F_\mu = \partial_\mu \phi , \qquad (2.27)$$

with

$$\phi(y) = -\frac{f}{3\pi} \int \theta(y_0 - x_0)\, \delta'((y-x)^2) \times \epsilon^{\mu\nu\lambda\rho} (x-y)_\mu\, dx_\nu \wedge dx_\lambda \wedge dx_\rho . \qquad (2.28)$$

If the vortex is static, ϕ is the velocity potential for the fluid or, if instead of a vortex we consider a static current loop, the scalar potential for the magnetic field. A short calculation shows that we recover in any case the classical result that ϕ is the solid angle subtended by the loop at the point of observation.

Consider now the interaction term of the action. Substitution into it of (2.23) gives the self-interaction of a vortex,

$$Y \equiv \frac{f^2}{4\pi} \int d\sigma\, d\sigma'\, d\tau\, d\tau'\, \theta(x_0(\sigma,\tau) - x_0(\sigma',\tau'))\, \delta([x(\sigma,\tau) - x(\sigma',\tau')]^2) \times \left(\frac{\partial x^\mu}{\partial \sigma}\frac{\partial x^\nu}{\partial \tau} - \frac{\partial x^\mu}{\partial \tau}\frac{\partial x^\nu}{\partial \sigma} \right)\left(\frac{\partial x_\mu}{\partial \sigma'}\frac{\partial x_\nu}{\partial \tau'} - \frac{\partial x_\mu}{\partial \tau'}\frac{\partial x_\nu}{\partial \sigma'} \right). \qquad (2.29)$$

The integrand in this expression diverges when $\sigma \sim \sigma'$, $\tau \sim \tau'$. To study this divergence, consider the finite expression Y_ϵ obtained from Y by replacing $\delta([x(\sigma,\tau) - x(\sigma',\tau')]^2)$ by $\delta([x(\sigma,\tau) - x(\sigma',\tau')]^2 + \epsilon^2)$. Clearly, $\lim_{\epsilon \to 0} Y_\epsilon = Y$.

Introducing the new variables

$$\Delta\sigma \equiv \sigma - \sigma' , \quad \Delta\tau \equiv \tau - \tau' , \qquad (2.30)$$

and developing the integrand of (2.29) in a power series in $\Delta\sigma$ and $\Delta\tau$ we obtain

$$Y_\epsilon = \frac{f^2}{2\pi}\int d(\Delta\sigma)d(\Delta\tau)d\sigma\, d\tau\ \delta\left(\frac{\partial x^\mu}{\partial\sigma}\frac{\partial x_\mu}{\partial\sigma}(\Delta\sigma)^2 + 2\frac{\partial x^\mu}{\partial\sigma}\frac{\partial x_\mu}{\partial\tau}\Delta\sigma\Delta\tau + \frac{\partial x^\mu}{\partial\tau}\frac{\partial x_\mu}{\partial\tau}(\Delta\tau)^2 + \epsilon^2\right)$$
$$\times\left[\left(\frac{\partial x^\mu}{\partial\sigma}\frac{\partial x_\mu}{\partial\sigma}\right)\left(\frac{\partial x^\nu}{\partial\tau}\frac{\partial x_\nu}{\partial\tau}\right) - \left(\frac{\partial x^\mu}{\partial\sigma}\frac{\partial x_\mu}{\partial\tau}\right)^2\right]$$
$$+\text{higher-order terms}. \tag{2.31}$$

The integration over $\Delta\tau$ is readily carried out with the aid of the δ function, giving

$$Y_\epsilon = \frac{f^2}{4\pi}\int d(\Delta\sigma)d\sigma\, d\tau(g)\left[g(\Delta\sigma)^2 + \epsilon^2\frac{\partial x^\mu}{\partial\tau}\frac{\partial x_\mu}{\partial\tau}\right]^{-1/2} + \text{higher-order terms}, \tag{2.32}$$

where g is given by (2.16). For small ϵ we have

$$Y_\epsilon \underset{\epsilon\to 0}{\sim} -\frac{f^2}{2\pi}\ln\left(\frac{\epsilon_0}{\epsilon}\right)\int d\sigma\, d\tau\sqrt{-g}$$
$$+\text{finite terms}, \tag{2.33}$$

with ϵ_0 being a cutoff in $\Delta\sigma$.

We see that the divergent term has exactly the same form as the Nambu action, and we may absorb it with a renormalization of the constant N appearing in (2.15),

$$N_R = N + \frac{f^2}{2\pi}\ln\left(\frac{\epsilon_0}{\epsilon}\right). \tag{2.34}$$

III. MOTION OF A STRING IN A UNIFORM, STATIC, EXTERNAL FIELD

In this section we consider the equation of motion (2.22) for the $x^\mu(\sigma,\tau)$, subject to the constraints (2.19), when the field F^μ is a given static and uniform external quantity. We study the motion in a Lorentz frame in which

$$x^0 = \tau,\quad F^0 = \bar{F},\quad F^i = 0, \tag{3.1}$$

with $\bar{F}$ a constant. In this case, the $\mu = 0$ of Eqs. (2.22) is identically satisfied. The others are

$$\frac{\partial^2\vec{x}}{\partial\tau^2} - \frac{\partial^2\vec{x}}{\partial\sigma^2} = -2c\left(\frac{\partial\vec{x}}{\partial\tau}\times\frac{\partial\vec{x}}{\partial\sigma}\right) \tag{3.2}$$

where

$$c = \frac{f\bar{F}}{2N}, \tag{3.3}$$

and the constraints (2.19) become

$$\frac{\partial\vec{x}}{\partial\sigma}\cdot\frac{\partial\vec{x}}{\partial\tau} = 0, \tag{3.4}$$

$$\left(\frac{\partial\vec{x}}{\partial\tau}\right)^2 + \left(\frac{\partial\vec{x}}{\partial\sigma}\right)^2 = 1. \tag{3.5}$$

The metric of the world surface is given by (2.18). Its projection on three-dimensional space has the metric

$$ds^2 = \left(\frac{\partial\vec{x}}{\partial\sigma}\right)^2 d\sigma^2 + 2\left(\frac{\partial\vec{x}}{\partial\sigma}\cdot\frac{\partial\vec{x}}{\partial\tau}\right)d\sigma\, d\tau + \left(\frac{\partial\vec{x}}{\partial\tau}\right)^2 d\tau^2, \tag{3.6}$$

which may be written, according to (3.4) and (3.5), as

$$ds^2 = \cos^2\theta\, d\sigma^2 + \sin^2\theta\, d\tau^2. \tag{3.7}$$

Notice that as x^0 is fixed by the choice (3.1), all we need to know is $\vec{x}(\sigma,\tau)$, and is enough to consider the surface S obtained by projecting the world surface of the string on the $x^0 = \tau$ hyperplane.

In what follows, we shall use the notation[10]

$$\left(\frac{\partial\vec{x}}{\partial\sigma}\right)^2 = E,\quad \frac{\partial\vec{x}}{\partial\sigma}\cdot\frac{\partial\vec{x}}{\partial\tau} = F,\quad \left(\frac{\partial\vec{x}}{\partial\tau}\right)^2 = G \tag{3.8}$$

and we shall call D, D', and D'' the coefficients of the second fundamental form of the surface (also called the extrinsic curvature).

At a given point on S, $\partial\vec{x}/\partial\sigma$ and $\partial\vec{x}/\partial\tau$ are vectors tangent to the τ- and σ-coordinate lines, respectively. They span the plane tangent to S at that point and their vector product is the vector normal to S there. The angle θ in (3.7) is $\frac{1}{2}$ the angle between the two tangent vectors.[11] The length of the normal vector is $(EG - F^2)^{1/2}$.

One possible definition of the coefficients of the second fundamental form is

$$D = \frac{\partial^2\vec{x}}{\partial\sigma^2}\cdot\hat{X}_3, \tag{3.9}$$

$$D' = \frac{\partial^2\vec{x}}{\partial\sigma\partial\tau}\cdot\hat{X}_3, \tag{3.10}$$

$$D'' = \frac{\partial^2\vec{x}}{\partial\tau^2}\cdot\hat{X}_3, \tag{3.11}$$

where we have introduced the unit normal

$$\hat{X}_3 = (EG - F^2)^{-1/2}\frac{\partial\vec{x}}{\partial\sigma}\times\frac{\partial\vec{x}}{\partial\tau}. \tag{3.12}$$

Analogously, we shall use the unit tangents

$$\hat{X}_1 = \frac{1}{\sqrt{E}}\frac{\partial\vec{x}}{\partial\sigma}, \tag{3.13}$$

$$\hat{X}_2 = \frac{1}{\sqrt{G}}\frac{\partial\vec{x}}{\partial\tau}. \tag{3.14}$$

Taking the dot product of (3.2) with $\hat{X}_3$ we obtain

$$D'' - D = 2c(EG - F^2)^{1/2}. \quad (3.15)$$

It is a classical result of surface theory (see Ref. 10) that, given coordinates u_1, u_2, a metric tensor

$$g_{ij} = \begin{pmatrix} E & F \\ F & G \end{pmatrix}, \quad (3.16)$$

and a second fundamental form

$$L_{ij} = \begin{pmatrix} D & D' \\ D' & D'' \end{pmatrix}, \quad (3.17)$$

the equations (of Gauss and Weingarten)

$$\frac{\partial \hat{X}_i}{\partial u_k} = \Gamma^l{}_{ik}\hat{X}_l + L_{ik}\hat{X}_3, \quad (3.18a)$$

$$\frac{\partial \hat{X}_3}{\partial u_k} = -g^{ij}L_{kj}\hat{X}_i \quad (3.18b)$$

have a unique solution for $\hat{X}_1$, $\hat{X}_2$, and $\hat{X}_3$, provided the following compatibility conditions hold:

$$\frac{\partial L_{12}}{\partial u_1} - \frac{\partial L_{11}}{\partial u_2} + \Gamma^l{}_{12}L_{l1} - \Gamma^l{}_{11}L_{l2} = 0, \quad (3.19a)$$

$$\frac{\partial L_{22}}{\partial u_1} - \frac{\partial L_{21}}{\partial u_2} + \Gamma^l{}_{22}L_{l1} - \Gamma^l{}_{21}L_{l2} = 0, \quad (3.19b)$$

$$K = \frac{DD'' - (D')^2}{EG - F^2}. \quad (3.20)$$

Here, K is the (intrinsic) scalar curvature of the surface and $\Gamma^l{}_{ik}$ are the Christoffel symbols. Equations (3.19) are called the Codazzi-Mainardi equations and (3.20) is the Gauss equation.

Moreover, these $\hat{X}_1, \hat{X}_2, \hat{X}_3$ so determined uniquely fix a surface $\vec{x}(\sigma,\tau)$, whose metric tensor is g_{ij}, whose second fundamental form is L_{ij}, and has $\hat{X}_1$ and $\hat{X}_2$ as unit tangents and $\hat{X}_3$ as unit normal.

We now go back to the equations of motion (3.2). It is easy to see that, as a consequence of Eq. (3.18a), and of the form (3.7) of the metric tensor, the three equations (3.2) are equivalent to the one equation (3.15). Our problem is then to find those surfaces with metric tensor of the form (3.7) and second fundamental form that satisfies (3.15).

Using (3.16) and (3.17), Eqs. (3.18) may be written as follows ($F=0$):

$$\frac{\partial}{\partial\sigma}\begin{pmatrix} \hat{X}_1 \\ \hat{X}_2 \\ \hat{X}_3 \end{pmatrix} = \begin{pmatrix} 0 & -\frac{1}{\sqrt{G}}\frac{\partial\sqrt{E}}{\partial\tau} & \frac{D}{\sqrt{E}} \\ \frac{1}{\sqrt{G}}\frac{\partial\sqrt{E}}{\partial\tau} & 0 & \frac{D'}{\sqrt{G}} \\ -\frac{D}{\sqrt{E}} & -\frac{D'}{\sqrt{G}} & 0 \end{pmatrix}\begin{pmatrix} \hat{X}_1 \\ \hat{X}_2 \\ \hat{X}_3 \end{pmatrix}, \quad (3.21)$$

$$\frac{\partial}{\partial\tau}\begin{pmatrix} \hat{X}_1 \\ \hat{X}_2 \\ \hat{X}_3 \end{pmatrix} = \begin{pmatrix} 0 & \frac{1}{\sqrt{E}}\frac{\partial\sqrt{G}}{\partial\sigma} & \frac{D'}{\sqrt{E}} \\ -\frac{1}{\sqrt{E}}\frac{\partial\sqrt{G}}{\partial\sigma} & 0 & \frac{D''}{\sqrt{G}} \\ -\frac{D'}{\sqrt{E}} & -\frac{D''}{\sqrt{G}} & 0 \end{pmatrix}\begin{pmatrix} \hat{X}_1 \\ \hat{X}_2 \\ \hat{X}_3 \end{pmatrix}, \quad (3.22)$$

and the integrability conditions (3.19) and (3.20), which are nothing more than the statement that crossed derivatives of (3.21) and (3.22) must coincide, become

$$\frac{\partial D}{\partial\tau} - \frac{\partial D'}{\partial\sigma} - \frac{D}{2E}\frac{\partial E}{\partial\tau} + \left(\frac{1}{2E}\frac{\partial E}{\partial\sigma} - \frac{1}{2G}\frac{\partial G}{\partial\sigma}\right)D' - \frac{1}{2G}\frac{\partial E}{\partial\tau}D'' = 0, \quad (3.23)$$

$$\frac{\partial D''}{\partial\sigma} - \frac{\partial D'}{\partial\tau} - \frac{D}{2E}\frac{\partial E}{\partial\sigma} + \left(\frac{1}{2G}\frac{\partial G}{\partial\tau} - \frac{1}{2E}\frac{\partial E}{\partial\tau}\right)D' - \frac{1}{2G}\frac{\partial G}{\partial\sigma}D'' = 0, \quad (3.24)$$

$$\frac{DD'' - (D')^2}{EG} = -\frac{1}{\sqrt{EG}}\left[\frac{\partial}{\partial\sigma}\left(\frac{1}{\sqrt{E}}\frac{\partial\sqrt{G}}{\partial\sigma}\right) + \frac{\partial}{\partial\tau}\left(\frac{1}{\sqrt{G}}\frac{\partial\sqrt{E}}{\partial\tau}\right)\right]. \quad (3.25)$$

Substitution of (3.7) into (3.15) gives

$$D'' - D = 2c\sin\theta\cos\theta, \quad (3.26)$$

and defining

$$B \equiv \tfrac{1}{2}(D'' + D), \quad (3.27)$$

the integrability conditions (3.23)–(3.25) become, using (3.7) and (3.26),

$$\frac{\partial B}{\partial\tau} - \frac{\partial D'}{\partial\sigma} + \frac{\partial\theta}{\partial\tau}\frac{B}{\sin\theta\cos\theta} - \frac{\partial\theta}{\partial\sigma}\frac{D'}{\sin\theta\cos\theta} = 0, \quad (3.28)$$

$$\frac{\partial B}{\partial\sigma} - \frac{\partial D'}{\partial\tau} + \frac{\partial\theta}{\partial\tau}\frac{D'}{\sin\theta\cos\theta} - \frac{\partial\theta}{\partial\sigma}\frac{B}{\sin\theta\cos\theta} = 0, \quad (3.29)$$

$$B^2 - (D')^2 = c^2\sin^2\theta\cos^2\theta + \sin\theta\cos\theta\left(\frac{\partial^2\theta}{\partial\tau^2} - \frac{\partial^2\theta}{\partial\sigma^2}\right). \quad (3.30)$$

The key remark now is, very simply, that (3.21) and (3.22) are three-dimensional rotations of $(\hat{X}_1, \hat{X}_2, \hat{X}_3)$ with rotation vectors

$$\vec{\omega}_\sigma = \left(\frac{D'}{\sqrt{G}}, -\frac{D}{\sqrt{E}}, -\frac{1}{\sqrt{G}}\frac{\partial\sqrt{E}}{\partial\tau}\right)$$

and

$$\vec{\omega}_\tau = \left(\frac{D''}{\sqrt{G}}, \frac{D'}{\sqrt{E}}, \frac{1}{\sqrt{E}}\frac{\partial\sqrt{G}}{\partial\sigma}\right),$$

respectively. We may then write these operations

in the spin-$\frac{1}{2}$ representation:

$$\frac{\partial\psi}{\partial\sigma}=\left(\frac{i}{2}\vec{\omega}_\sigma\cdot\vec{\sigma}\right)\psi, \tag{3.31}$$

$$\frac{\partial\psi}{\partial\tau}=\left(\frac{i}{2}\vec{\omega}_\tau\cdot\vec{\sigma}\right)\psi, \tag{3.32}$$

where ψ is a two-component spinor and $\vec{\sigma}=(\sigma_1,\sigma_2,\sigma_3)$ are the Pauli matrices.

Let us consider first the special case $B=D'=0$. In this case, the Gaussian curvature is constant and Eq. (3.30) simplifies to

$$-\frac{\partial^2\theta}{\partial\tau^2}+\frac{\partial^2\theta}{\partial\sigma^2}=c^2\sin\theta\cos\theta, \tag{3.33}$$

the familiar sine-Gordon equation. Conditions (3.28) and (3.29) are, of course, trivially satisfied. Equations (3.31) and (3.32) take the form

$$2i\frac{\partial\psi}{\partial\sigma}=-\begin{pmatrix}\frac{\partial\theta}{\partial\tau} & -ic\sin\theta\\ ic\sin\theta & -\frac{\partial\theta}{\partial\tau}\end{pmatrix}\psi, \tag{3.34}$$

$$2i\frac{\partial\psi}{\partial\tau}=-\begin{pmatrix}\frac{\partial\theta}{\partial\sigma} & c\cos\theta\\ c\cos\theta & -\frac{\partial\theta}{\partial\sigma}\end{pmatrix}\psi. \tag{3.35}$$

In light-cone coordinates

$$\alpha\equiv\tfrac{1}{2}c(\tau+\sigma),\qquad \beta\equiv\tfrac{1}{2}c(-\tau+\sigma), \tag{3.36}$$

they are

$$2i\frac{\partial\psi}{\partial\alpha}=-\begin{pmatrix}\frac{\partial\theta}{\partial\alpha} & e^{-i\theta}\\ e^{i\theta} & -\frac{\partial\theta}{\partial\alpha}\end{pmatrix}\psi, \tag{3.37}$$

$$2i\frac{\partial\psi}{\partial\beta}=\begin{pmatrix}\frac{\partial\theta}{\partial\beta} & e^{i\theta}\\ e^{-i\theta} & -\frac{\partial\theta}{\partial\beta}\end{pmatrix}\psi. \tag{3.38}$$

The terms in $\partial\theta/\partial\beta$ may be eliminated from (3.37) by a rotation in spinor space. Defining

$$\overline{\psi}\equiv e^{i\sigma_3\theta/2}\psi, \tag{3.39}$$

one gets

$$2i\frac{\partial\overline{\psi}}{\partial\beta}=\begin{pmatrix}0 & e^{i\phi}\\ e^{-i\phi} & 0\end{pmatrix}\overline{\psi}, \tag{3.40}$$

$$2i\frac{\partial\overline{\psi}}{\partial\alpha}=-\begin{pmatrix}\frac{\partial\phi}{\partial\alpha} & 1\\ 1 & -\frac{\partial\phi}{\partial\alpha}\end{pmatrix}\overline{\psi} \tag{3.41}$$

where $\phi=2\theta$. The sine-Gordon equation in these coordinates is

$$\frac{\partial^2\phi}{\partial\alpha\partial\beta}=\sin\phi. \tag{3.42}$$

Leave now this line of reasoning and go back for a moment to the metric (2.18), which may be written, using the constraints (2.19), as

$$ds^2=\left(\frac{\partial x^\mu}{\partial\sigma}\frac{\partial x_\mu}{\partial\sigma}\right)(d\sigma^2-d\tau^2). \tag{3.43}$$

This expression is manifestly conformally flat, and it remains so under Lorentz transformations on the two-dimensional σ-τ parameter space; say

$$\sigma'=\cosh a\,\sigma+\sinh a\,\tau,\qquad \tau'=\sinh a\,\sigma+\cosh a\,\tau. \tag{3.44}$$

On the other hand, the equation of motion (2.22) is also invariant under these transformations. If, instead of taking the projection of the world surface on the hypersurface $x^0=\tau$, we take it on the hypersurface $x^0=\tau'$, it is still possible to choose $F^0=F$ and $F^i=0$. The $\mu=0$ of Eqs. (2.22) will still be identically satisfied, and Eq. (3.2) will keep its form. Writing out the constraints (2.19) with this new coordinate condition for x^0, we get

$$\frac{\partial\vec{x}}{\partial\tau}\cdot\frac{\partial\vec{x}}{\partial\sigma}=\sinh a\cosh a,\qquad \left(\frac{\partial\vec{x}}{\partial\tau}\right)^2+\left(\frac{\partial\vec{x}}{\partial\sigma}\right)^2=\cosh^2 a+\sinh^2 a, \tag{3.45}$$

and, in the primed coordinates,

$$\frac{\partial\vec{x}}{\partial\tau'}\cdot\frac{\partial\vec{x}}{\partial\sigma'}=0,\qquad \left(\frac{\partial\vec{x}}{\partial\tau'}\right)^2+\left(\frac{\partial\vec{x}}{\partial\sigma'}\right)^2=1. \tag{3.46}$$

In terms of these new coordinates then, the geometrical problem is unchanged. We are still looking for the geometry of a two-dimensional surface embedded in three-dimensional space, which is the projection of a world surface that is embedded in four-dimensional space-time. What *has* changed is the three-space where we are doing the projection.

The transformation (3.44) induces the following transformation on the light-cone coordinates:

$$\alpha'=e^a\alpha,\quad \beta'=e^{-a}\beta. \tag{3.47}$$

This leaves Eq. (3.42) unchanged, and (3.40) and

(3.41) take the form

$$2i\frac{\partial\bar{\psi}}{\partial\alpha'}=-\begin{pmatrix}\frac{\partial\phi}{\partial\alpha'} & k\\ k & -\frac{\partial\phi}{\partial\alpha'}\end{pmatrix}\bar{\psi}, \tag{3.48}$$

$$2i\frac{\partial\bar{\psi}}{\partial\beta'}=-\begin{pmatrix}0 & k^{-1}e^{i\phi}\\ k^{-1}e^{-i\phi} & 0\end{pmatrix}\bar{\psi}, \tag{3.49}$$

where $k=e^{-a}$. These are the equations employed by Ablowitz *et al.*[7] to solve the sine-Gordon equation (3.42) by the inverse scattering method. (The correspondence with that paper's notation is $k=2\zeta$, $\bar{\psi}_1=v_1+iv_2$, and $\bar{\psi}_2=v_1-iv_2$.) They have also been used by Orfanidis[12] to study the relation between the Thirring and sine-Gordon models at the classical level.

It is interesting to see that there is a natural geometric interpretation for the linear eigenvalue problem which was used in Ref. 7 to solve the sine-Gordon equation: This equation expresses the "embeddability condition" that a surface with metric given by $E=\cos^2\theta$, $F=0$, $G=\sin^2\theta$, and extrinsic curvature $D''=-D=c\sin\theta\cos\theta$, $D'=0$, must satisfy in order to be embeddable in a three-dimensional Euclidean space. The linear equations (3.48) and (3.49) give the actual construction of the tangent and normal vectors to that surface, thus actually constructing it. Different values of the parameter k give different surfaces. These surfaces are related, however, all of them being projections of one and the same surface embedded in a four-dimensional space-time along different three-dimensional spaces. These spaces are, in turn, transformed into each other by Lorentz transformations.

The sine-Gordon equation has the interesting feature of admitting solitary waves among its solutions. It can also be considered as the evolution equation for a scalar field in one space and one time dimension with a highly nonlinear self-interaction. Our geometrical framework provides for an immediate generalization of this theory. Namely, one can ask what results if one considers surfaces whose extrinsic curvature in the three-dimensional space is not restricted by $B=D'=0$. In this case, the (nonlinear) embeddability conditions are (3.28)–(3.30) and the (linear) equations for the tangent and normal vectors are (3.31) and (3.32). Equations (3.28) and (3.29) may be written as follows:

$$\frac{\partial}{\partial\tau}(\tan\theta\, B)=\frac{\partial}{\partial\sigma}(\tan\theta\, D'), \tag{3.50}$$

$$\frac{\partial}{\partial\sigma}(\cot\theta\, B)=\frac{\partial}{\partial\tau}(\cot\theta\, D'). \tag{3.51}$$

From (3.50) it follows that there exist λ such that

$$B=\cot\theta\frac{\partial\lambda}{\partial\sigma},\quad D'=\cot\theta\frac{\partial\lambda}{\partial\tau}. \tag{3.52}$$

Substitution into (3.51) yields an equation for λ:

$$\frac{\partial}{\partial\sigma}\left(\cot^2\theta\frac{\partial\lambda}{\partial\sigma}\right)=\frac{\partial}{\partial\tau}\left(\cot^2\theta\frac{\partial\lambda}{\partial\tau}\right). \tag{3.53}$$

The other embeddability condition, Eq. (3.30) is, according to (3.52),

$$\frac{\partial^2\theta}{\partial\tau^2}-\frac{\partial^2\theta}{\partial\sigma^2}+c^2\sin\theta\cos\theta+\frac{\cos\theta}{\sin^3\theta}\left[\left(\frac{\partial\lambda}{\partial\tau}\right)^2-\left(\frac{\partial\lambda}{\partial\sigma}\right)^2\right]=0. \tag{3.54}$$

The last two equations can now be interpreted as the dynamical equations for two coupled scalar fields. They can be derived from the Lagrangian

$$\mathcal{L}=\frac{1}{2}\left[\left(\frac{\partial\theta}{\partial\sigma}\right)^2-\left(\frac{\partial\theta}{\partial\tau}\right)^2\right]+\frac{c^2}{2}\sin^2\theta+\tfrac{1}{2}\cot^2\theta\left[\left(\frac{\partial\lambda}{\partial\sigma}\right)^2-\left(\frac{\partial\lambda}{\partial\tau}\right)^2\right], \tag{3.55}$$

which is a Lorentz scalar in σ-τ space-time. The linear equations (3.31) and (3.32) are now

$$2i\frac{\partial\psi}{\partial\sigma}=-\begin{pmatrix}\frac{\partial\theta}{\partial\tau} & \frac{\cos\theta}{\sin^2\theta}\frac{\partial\lambda}{\partial\tau}+\frac{i}{\sin\theta}\frac{\partial\lambda}{\partial\sigma}-ic\sin\theta\\ \frac{\cos\theta}{\sin^2\theta}\frac{\partial\lambda}{\partial\tau}-\frac{i}{\sin\theta}\frac{\partial\lambda}{\partial\sigma}+ic\sin\theta & -\frac{\partial\theta}{\partial\tau}\end{pmatrix}\psi, \tag{3.56}$$

$$2i\frac{\partial\psi}{\partial\tau}=-\begin{pmatrix}\frac{\partial\theta}{\partial\sigma} & \frac{\cos\theta}{\sin^2\theta}\frac{\partial\lambda}{\partial\sigma}+c\cos\theta-\frac{i}{\sin\theta}\frac{\partial\lambda}{\partial\tau}\\ \frac{\cos\theta}{\sin^2\theta}\frac{\partial\lambda}{\partial\sigma}+c\cos\theta+\frac{i}{\sin\theta}\frac{\partial\lambda}{\partial\tau} & -\frac{\partial\theta}{\partial\sigma}\end{pmatrix}\psi. \tag{3.57}$$

It is easy to check directly by cross-differentiation that (3.53) and (3.54) are indeed the integrability conditions for (3.56) and (3.57).

We now look for static (τ-independent) solutions to the field equations. They are, in this case,

$$\frac{\partial}{\partial\sigma}\left(\cot^2\theta\frac{\partial\lambda}{\partial\sigma}\right)=0\,, \tag{3.58}$$

$$\frac{\partial^2\theta}{\partial\sigma^2}-c^2\sin\theta\cos\theta+\frac{\cos\theta}{\sin^3\theta}\left(\frac{\partial\lambda}{\partial\sigma}\right)^2=0\,. \tag{3.59}$$

It is easy to solve these equations using elliptic functions to obtain periodic waves. A particular solution that vanishes at infinity is given by

$$\theta=\arcsin\left((1-A^2)^{1/2}\sin\{2\tan^{-1}\exp[\sigma(1-A^2)^{1/2}]\}\right)\,, \tag{3.60}$$

$$\frac{\partial\lambda}{\partial\sigma}=A\tan^2\theta\,, \tag{3.61}$$

where A is a constant. The form of θ is sketched in Fig. 1 and that of λ in Fig. 2. Performing a Lorentz transformation we get a localized wave traveling at a constant velocity and remaining localized for all time. When $A\to 0$ the well-known sine-Gordon soliton is recovered.

IV. CONCLUDING REMARKS

We have constructed a classical theory of one-dimensional extended objects interacting through a scalar field. Special cases are the Nambu string and vortices in a superfluid. The divergence of the self-energy of a vortex may be regularized by a renormalization of the slope of the Regge trajectories. The action integral describing our system is the same as the one found previously by Kalb and Ramond[4] in a somewhat different context. At this stage, our contributions are then the relation of this action with the one describing nonrelativistic vortex motion in a superfluid and the regularization of the self-energy.

The study of the motion of a closed string in an external field led us to a geometrical problem throwing some light on the sine-Gordon equation: We found a natural geometric interpretation to the linear eigenvalue problem used to solve this equation by the inverse scattering method. This system was obtained under certain simplifying assumptions on the motion of the string. In the general case we were led to a set of two coupled nonlinear equations. This relativistic system, as does the sine-Gordon equations, admits solitary wave solutions, whose explicit form we have determined. The Lagrangian for this system was also found.

At this point, several problems are left open. Firstly, the system of nonlinear coupled equations (3.53) and (3.54) can be studied as a purely mathematical problem. Can the inverse scattering problem for the associated linear system (3.56) and (3.57) be solved? It is unfortunate that, at least at first sight, this set of equations does not seem to fit among the general class considered by Ablowitz *et al.*[13] in their systematic search for solutions to nonlinear equations. One should be able at least to determine whether a Bäcklund transformation exists and whether there is an infinite set of conservation laws[14] or not. The fact that we have a linear problem naturally associated with the nonlinear equations should be of help in resolving these points.

Secondly, one should find out whether the Lagrangian (3.55) describes a system with any relation to the physics of elementary particles. The first question to be settled in this context is whether the theory described by this Lagrangian

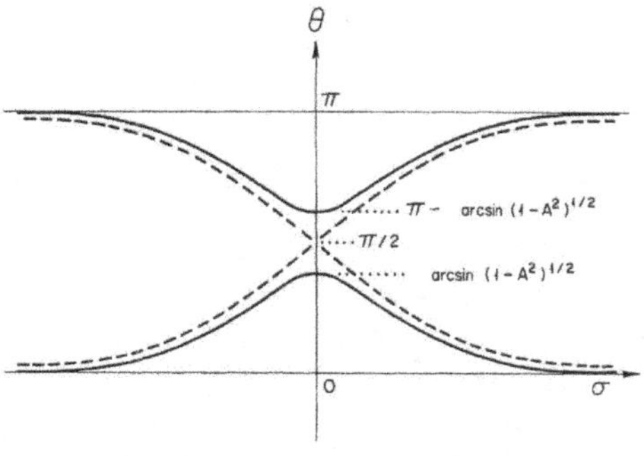

FIG. 1. The solid line is the solitary wave given by (3.60). In the limit $A\to 0$ it goes over to the sine-Gordon soliton (dashed line). Notice that if θ is a solution, so is $\pi-\theta$. Both cases are drawn here.

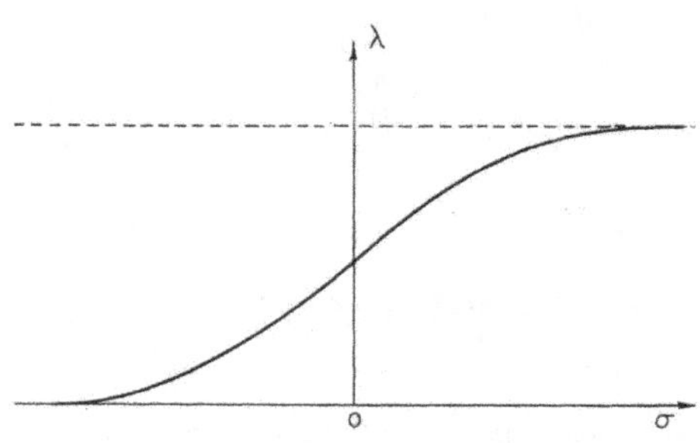

FIG. 2. Shape of field λ given by (3.61) when θ is a solitary wave.

is renormalizable.

Thirdly, as we pointed out in the Introduction, we have started from one physical system, namely strings in four dimensions, and through the mathematics of the problem we have arrived at quite a different one, coupled fields in two dimensions. It would be very interesting to find whether there is more than just formalism in this relation.

Finally, we have not tried to quantize either the system described by the Lagrangian (2.15) or the one described by the Lagrangian (3.55).

Note added in proof. We have been informed by B. Julia that the Lagrangian (3.55) has also been obtained by K. Pohlmeyer, Commun. Math. Phys. 46, 207 (1976).

ACKNOWLEDGMENTS

Stimulating discussions with B. Julia are gratefully acknowledged. Thanks are also due to T. Damour, R. Dashen, M. Kruskal, A. Mueller, H. Szu, and M. Virasoro for informative conversations.

APPENDIX

Here we shall consider the canonical formalism for the system described by the action (in units such that $N=1$)

$$S=-\int \sqrt{-g}\,d\sigma\,d\tau+f\int A_{\mu\nu}\frac{\partial x^{\mu}}{\partial\sigma}\frac{\partial x^{\nu}}{\partial\tau}d\sigma\,d\tau -\tfrac{1}{4}\int F_{\mu}F^{\mu}d^{4}y\,. \tag{A1}$$

There are two gauge freedoms: one is the invariance under reparametrizations of the surface $x^{\mu}(\sigma,\tau)$ and the other is the invariance under

$$A_{\mu\nu}\rightarrow A_{\mu\nu}+\partial_{\mu}\Lambda_{\nu}-\partial_{\nu}\Lambda_{\mu}\,. \tag{A2}$$

This is reflected by the presence of constraints among the canonical variables. We shall follow the general treatment of Dirac[15] for such systems. First we consider the variables $x^{\mu}(\sigma)$. The canonical momenta are (the overdot means τ derivative)

$$\mathcal{P}_{\mu}(\sigma)\equiv\frac{\delta S}{\delta\dot{x}^{\mu}(\sigma)}, \tag{A3}$$

and it is simple to see that they satisfy the (primary) first-class constraints

$$\psi_1\equiv\left(\mathcal{P}_{\mu}-fA_{\nu\mu}\frac{\partial x^{\nu}}{\partial\sigma}\right)\left(\mathcal{P}^{\mu}-fA^{\lambda\mu}\frac{\partial x_{\lambda}}{\partial\sigma}\right)+\frac{\partial x^{\mu}}{\partial\sigma}\frac{\partial x_{\mu}}{\partial\sigma}=0\,, \tag{A4}$$

$$\psi_2\equiv\left(\mathcal{P}_{\mu}-fA_{\nu\mu}\frac{\partial x^{\nu}}{\partial\sigma}\right)\frac{\partial x^{\mu}}{\partial\sigma}=0\,. \tag{A5}$$

This shows that this system is obtained from the free string by the "minimal coupling" $\mathcal{P}_{\mu}\rightarrow\mathcal{P}_{\mu}-fA_{\nu\mu}\partial x^{\nu}/\partial\sigma$. In fact, with this subsitution the constraints (A4) and (A5) are those of the free string. It is also simple to see that there are no more constraints: ψ_1 and ψ_2 form a closed algebra under Poisson-bracket multiplication. Since the Lagrangian is homogeneous of the first degree in the velocity $\dot{x}^{\mu}$, the quantity $(\mathcal{L}-\mathcal{P}_{\mu}\dot{x}^{\mu})$ vanishes and the Hamiltonian is given by

$$H=\int d\sigma(\lambda_1\psi_1+\lambda_2\psi_2)\,, \tag{A6}$$

where λ_1 and λ_2 are arbitrary functions, reflecting the reparametrization invariance of the system. The equations of motion are then

$$\dot{x}^{\mu}=\frac{\delta H}{\delta\mathcal{P}_{\mu}}=2\lambda_1\left(\mathcal{P}^{\mu}-fA^{\lambda\mu}\frac{\partial x_{\lambda}}{\partial\sigma}\right)+\lambda_2\frac{\partial x^{\mu}}{\partial\sigma}, \tag{A7}$$

$$\dot{\mathcal{P}}_{\mu}=-\frac{\delta H}{\delta x^{\mu}}=2\frac{\partial}{\partial\sigma}\left(\lambda_1\frac{\partial x_{\mu}}{\partial\sigma}\right)-2f\frac{\partial}{\partial\sigma}\left[\lambda_1A_{\mu\nu}\left(\mathcal{P}^{\nu}-fA^{\lambda\nu}\frac{\partial x_{\lambda}}{\partial\sigma}\right)\right]+\frac{\partial}{\partial\sigma}\left[\lambda_2\left(\mathcal{P}_{\mu}-fA_{\nu\mu}\frac{\partial x^{\nu}}{\partial\sigma}\right)\right]-f\frac{\partial}{\partial\sigma}\left(\lambda_2A_{\mu\nu}\frac{\partial\lambda^{\nu}}{\partial\sigma}\right). \tag{A8}$$

The orthonormal gauge is obtained with $\lambda_1=\frac{1}{2}$, $\lambda_2=0$. In fact, in this case

$$\dot{x}^{\mu}=\mathcal{P}^{\mu}-fA^{\nu\mu}\frac{\partial x_{\nu}}{\partial\sigma} \tag{A9}$$

and (A8) reduces to (2.22) and (A4) and (A5) to (2.19).

Now we take the field $A_{\mu\nu}=-A_{\nu\mu}$. The canonical momenta are

$$
\begin{aligned}
\Pi^{\alpha\beta} &\equiv \frac{\delta S}{\delta A_{\alpha\beta,0}} \\
&= -\tfrac{1}{8}(\partial^0 A^{\alpha\beta} + \partial^\beta A^{0\alpha} + \partial^\alpha A^{\beta 0}) \\
&= -\Pi^{\beta\alpha}. \qquad \text{(A10)}
\end{aligned}
$$

We immediately have three primary constraints,

$$\Pi^{0\beta} = 0. \qquad \text{(A11)}$$

The Hamiltonian

$$H = \int (\tfrac{1}{2}\Pi^{\alpha\beta} A_{\alpha\beta,0} - \mathcal{L}) d^3y$$

is given by (Latin indices take the values 1, 2, 3)

$$
\begin{aligned}
H = \int d^3y [\Pi_{ij}\Pi^{ij} + \tfrac{1}{4}(F_0)^2 \\
+ \Pi^{0i} A_{0i,0} - A_{0i}\Pi^{ij}{}_{,j}]. \qquad \text{(A12)}
\end{aligned}
$$

This Hamiltonian is not uniquely defined, as an arbitrary linear combination of the constraints may be added to it to obtain

$$H' = H + \int u_j \Pi^{0j} d^3y. \qquad \text{(A13)}$$

The constraints (A11) must be preserved in time. Thus, their time derivative, given by the Poisson bracket with H', must vanish. This leads to the additional (secondary) constraints

$$\Pi^{ij}{}_{,j} = 0. \qquad \text{(A14)}$$

Since $\Pi^{ij}{}_{,ij} \equiv 0$, these are two independent constraints only. The preservation in time of these constraints leads to no further ones. It is also easily seen that both (A11) and (A14) are first class. That is, they generate gauge transformations. It is easy to see now that, as asserted in the text, $A_{\mu\nu}$ corresponds to a scalar field. In fact, there are twelve canonical variables: six coordinates and six conjugate momenta. These variables must satisfy five constraints and five gauge-fixing conditions, leaving two independent variables, one coordinate and its conjugate momentum. That is, one degree of freedom per space point.

The extended Hamiltonian H_E, in the notation of Dirac,[15] is

$$
\begin{aligned}
H_E = \int d^3y (\Pi_{ij}\Pi^{ij} + \tfrac{1}{4}F_0{}^2) \\
+ \int d^3y (v_j \Pi^{0j} + w_i \Pi^{ij}{}_{,j}), \qquad \text{(A15)}
\end{aligned}
$$

where v_j are three arbitrary functions and w_i are three functions subject to one condition, for example $w_{i,i} = 0$. The gauge-independent part of (A15) is positive-definite, as it should be.

---

*Research sponsored in part by the National Science Foundation under Grant No. GP-40768X.

[1]M. Rasetti and T. Regge, Physica 80A, 217 (1975).

[2]H. Nielsen and P. Olesen, Nucl. Phys. B57, 367 (1973).

[3]Y. Nambu, Lectures at the Copenhagen Symposium, 1970 (unpublished); P. Goddard, J. Goldstone, C. Rebbi, and C. B. Thorn, Nucl. Phys. B56, 109 (1973). There are a number of very good reviews on the subject of the string model. See, for example, C. Rebbi, Phys. Rep. 12C, 1 (1974); J. Scherk, Rev. Mod. Phys. 47, 123 (1975); S. Mandelstam, Phys. Rep. 13C, 259 (1974).

[4]M. Kalb and P. Ramond, Phys. Rev. D 9, 2273 (1974).

[5]J. A. Wheeler and R. P. Feynman, Rev. Mod. Phys. 21, 425 (1949).

[6]What we mean is that it has been noted by physicists working in fields quite removed from differential geometry. See, for example, S. Coleman, in Proceedings of Lectures given at the 1975 International Summer School "Ettore Majorana," Erice, Italy (unpublished).

[7]M. Ablowitz, D. J. Kaup, A. C. Newell, and H. Segur, Phys. Rev. Lett. 30, 1262 (1973).

[8]This statement is in turn a consequence of Kelvin's circulation theorem. See S. Putterman, *Superfluid Hydrodynamics* (North-Holland, Amsterdam, 1974); H. Lamb, *Hydrodynamics* (Dover, New York, 1945).

[9]It is perhaps not out of place to sketch how the variation of the second term in the right-hand side of Eq. (2.7) leads to the second term in the equation of motion (2.3):

$$
\begin{aligned}
\frac{1}{2}\frac{\delta}{\delta x^i(\sigma)}\int v^2 d^3y &= \int d^3y\, v^k(y) \frac{\delta v^k}{\delta x^i(\sigma)} \\
&= \frac{k}{4\pi}\int d^3y\, v^k(y)\epsilon^{klr}\frac{\partial}{\partial y^l} \\
&\quad \times \left[\frac{\partial}{\partial x^i}\left(\frac{1}{|\vec{y}-\vec{x}|}\right)\frac{\partial x^r}{\partial\sigma} - \delta_{ir}\frac{\partial}{\partial\sigma}\left(\frac{1}{|\vec{y}-\vec{x}|}\right)\right].
\end{aligned}
$$

Since

$$\frac{\partial}{\partial x^i}\left(\frac{1}{|\vec{y}-\vec{x}|}\right) = -\frac{\partial}{\partial y^i}\left(\frac{1}{|\vec{y}-\vec{x}|}\right),$$

one gets

$$
\begin{aligned}
\frac{1}{2}\frac{\delta}{\delta x^i(\sigma)}\int v^2 d^3y = \frac{k}{4\pi}\frac{\partial x^p}{\partial\sigma}\left(\epsilon^{kli}\frac{\partial}{\partial x^p} - \epsilon^{klp}\frac{\partial}{\partial x^i}\right) \\
\times \frac{\partial}{\partial x^l}\int d^3y \frac{v^k(\vec{y})}{|\vec{y}-\vec{x}(\sigma)|}.
\end{aligned}
$$

Using now the identity

$$\epsilon^{kli}\frac{\partial}{\partial x^p} - \epsilon^{klp}\frac{\partial}{\partial x^i} = \epsilon^{pik}\frac{\partial}{\partial x^l} - \epsilon^{pil}\frac{\partial}{\partial x^k}$$

and integrating by parts, one obtains finally

$$\frac{1}{2}\frac{\delta}{\delta x^i(\sigma)}\int v^2 d^3y = \epsilon^{ijk}\frac{\partial x^j}{\partial\sigma}v^k .$$

[10]This notation was first introduced by Gauss in his *Disquisitiones Circa Superficies Curvas*, and is used by geometers to this day.

[11]For this and other facts of differential geometry of surfaces the most complete reference is L. Bianchi, *Lezione di Geometria Differenziale* (Pisa, 1922). Two modern introductory texts are J. J. Stoker, *Differential Geometry* (Wiley, New York, 1969) and M. Spivak, *A Comprehensive Introduction to Differential Geometry* (Publish or Perish, Boston, 1970–1975), Vols. 1–5.

[12]S. J. Orfanidis, Phys. Rev. D 14, 472 (1976).

[13]M. Ablowitz, D. J. Kaup, A. C. Newell, and H. Segur, Stud. Appl. Math. 53, 249 (1974).

[14]For a review of the subject of Bäcklund transformations and conservation laws associated with solutions of nonlinear equations, see A. C. Scott, F. Y. F. Chu, and D. W. McLaughlin, Proc. IEEE 61, 1443 (1973).

[15]P. A. M. Dirac, *Lectures on Quantum Mechanics* (Yeshiva Univ., New York, 1964). See also A. J. Hanson, T. Regge, and C. Teitelboim, *Constrained Hamiltonian Systems* (Accademia dei Lincei, Rome, 1976).

PART VI

The Group Manifold Approach to (Super-) Gravity Theories

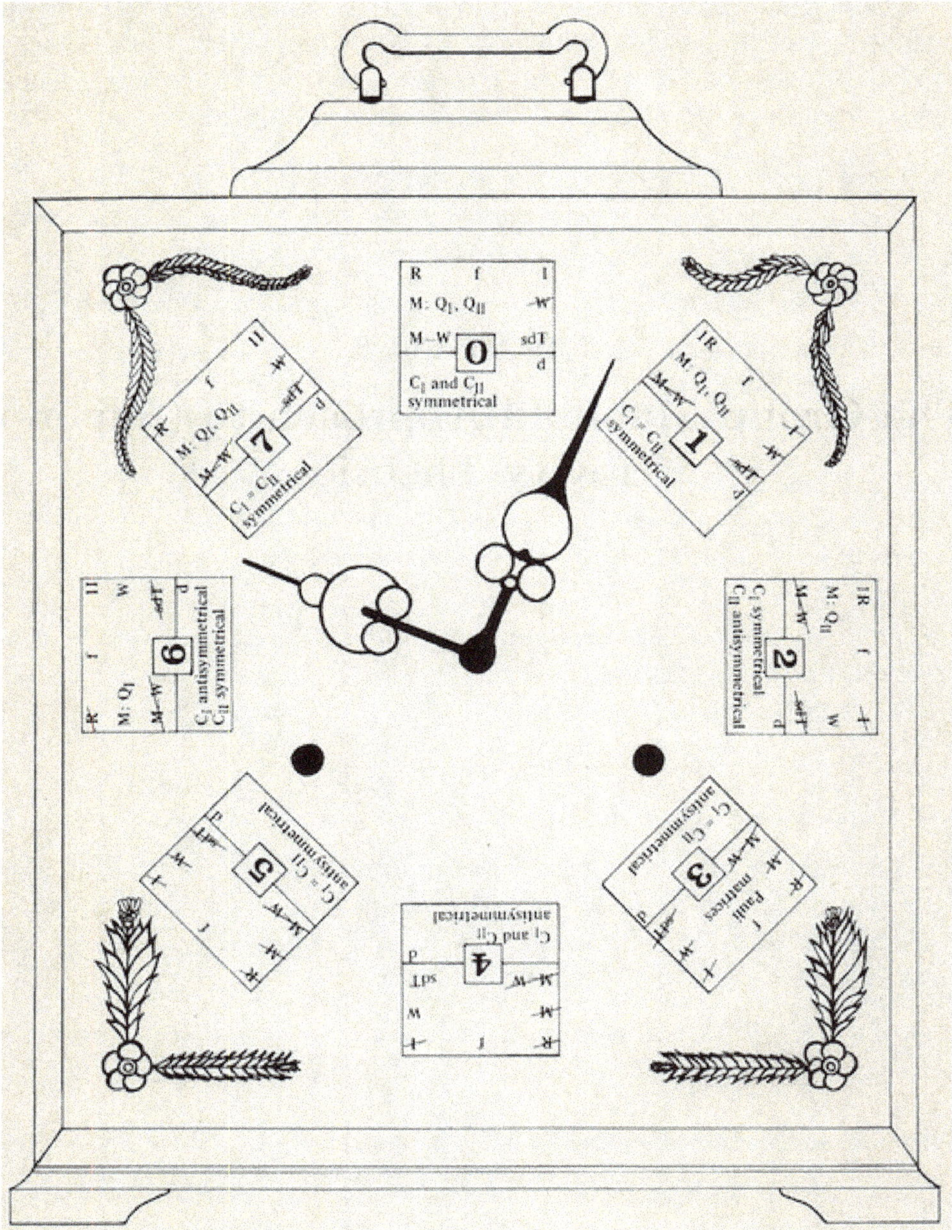

Regge's spinor clock, with the short hand pointing to the number $d = n + m$ of spacetime dimensions, with n times and m spaces, and the long hand pointing to $f = n - m$. Abbreviations: R, real irrep(s) (only one if the symbol is preceded by the number 1); I, imaginary irrep(s) (one if preceded by 1); M, Majorana spinors, further specified Q_I and Q_{II}; W, Weyl spinors; M-W, Weyl spinors which are Majorana as well; sdT, selfdual tensors. If a species does not exist at the value of f indicated, its symbol is crossed out.

From the Proceedings of Les Houches Session XL, 1983, North Holland (1984).

Geometric Supergravity

R. D'Auria
Politecnico di Torino, DISAT Department,
Corso Duca degli Abruzzi 24, 10129 Torino, Italy.

1. Introduction

In the year 1978 Y. Ne'eman and T. Regge[1] proposed a new approach to the mathematical description of gauge theories, specifically gravity and supergravity, based on the formalism introduced by E. Cartan for Riemannian geometry in a completely geometrical setting. Cartan's approach implies a new, more geometrical and group-theoretical way of formulating general relativity. Indeed, as the adopted formalism relies basically and consistently on the use of differential forms, the Cartan's beautiful setting is independent of different coordinate frames, that is of the group of general coordinate transformations (GCTG). At the same time, it gives a prominent role to the gauge invariance of the theory under the Lorentz group which emerges quite naturally from the formalism. As a matter of fact, in Cartan's view, Riemannian geometry must be seen as pertaining to finite Lie groups rather than to infinite group of general coordinate transformations. In the latter case it could be difficult to see how gravitation could be unified with gauge theories of other interactions, what instead seems quite natural in the geometrical formalism developed by him.
Following this line of approach, Y. Ne'eman and T. Regge further developed the Cartan's formalism, proposing that in principle any diffeomorphic and gauge invariant theory should be constructed directly on *the group manifold* G defining the Lie algebra valued gauge fields in the coadjoint representation of the group. This is consequence of the following considerations.

Referring to the pure gravity theory, the spin connection ω^{ab} and the vierbein V^a 1-form gauge fields are just fragments of the adjoint multiplet μ^A, $(A = 1, \dots, 10)$, of the Poincaré Lie algebra with indices decomposed with respect to the Lorentz subgroup SO(1,3), and spanning a basis of the cotangent plane of the Poincaré group, namely $\mu^A = (\omega^{ab}, V^a)$, $(a, b = 0, \dots, 3)$. However in gravity theory, there is an essential difference between the two fields: While the vierbein $V^a = V^a_\mu \, dx^\mu$ propagate, this is not true for the spin connection.

This is a consequence of the well-known fact that Riemannian geometry requires that the "curvature" of V^a, namely the torsion 2-form, must vanish. This makes the spin connection a functional of the vierbein and its derivatives. This disparity is essentially due to the *factorization hypothesis* which breaks from the very beginning the symmetry of the group, since it factorizes in a trivial way the dependence of the gauge fields from the Lorentz coordinates. In fact in the Cartan formulation of gravity the fields (ω^{ab}, V^a) live on the principal fiber bundle $[\mathcal{M}_4/\,SO(1,3)]$, where the base space $\mathcal{M}_4$ is the physical space-time which, in the vacuum configuration, reduces to the coset base space $[\mathrm{G}/\mathrm{SO}(1,3)]$ with $\mathrm{G} = \mathrm{ISO}(1,3)$.

It follows that if we now do not assume factorization, the dependence of the fields $\mu^A = (\omega^{ab}, V^a)$ on the group coordinates must be dictated by the field equations (and boundary conditions). It is then natural to try to construct gravity theory directly on the group manifold G, (ISO(1,3) in the gravity case), where the fields are represented by the Cartan–Maurer 1-forms in the coadjoint representation of the group. This implies that the gauge fields will depend not only on the coordinates x^μ $(\mu = 0, \dots, 3)$ related to the translation group, but also on the coordinates of the Lorentz group $y^{\mu\nu}$. Out of the vacuum the left-invariant 1-forms, spanning a basis on the cotangent space of G, do not obey anymore the Cartan–Maurer equations, but become dynamical, that is they acquire *curvatures*, namely the field-strengths of the dynamical fields.

Quite generally, as can be shown in general and we will show in the simplest cases, when fields are defined directly on a (graded-)group manifold, this new approach makes it possible to give a more geometrical formulation of the theory. Actually, both for gravity theory as for supergravity in any dimensions D, *the Lorentz invariance can be retrieved from the equations of motion* as a result of the action principle, even if the D-form Lagrangian has to be integrated on a group manifold whose dimensionality, $\dim \mathrm{G}$, is greater than the form degree D of the Lagrangian. In fact we will see that the integration of a D-form as a submanifold can be consistently performed as a result of its invariance under the GCTG, and the resulting equations of motion give *horizontality* of the curvatures in the Lorentz directions, leading to factorization of the Lorentz parameters. While this way to obtain Lorentz invariance starting from the whole group manifold G seems to be only of academic interest, its extension to graded groups (or supergroup, SG in the following) leads to a *geometric interpretation of supersymmetry.* Indeed, in this case, referring for simplicity to the $\mathcal{N} = 1$, $D = 4$ case, the coadjoint supermultiplet now contains an extra fermionic vielbein, namely $\mu^A = (\omega^{ab}, V^a, \psi^\alpha)$, $a, b = 0, \dots, 3, \alpha = 1, \dots, 4$, where ψ^α is a Majorana spinor 1-form, so that the coadjoint multiplet will now depend, besides translation and Lorentz coordinates, also on the odd fermionic Grassman parameters θ^α. In this case the Lorentz factorization, obtained as a result of the field equations, is not sufficient alone to leave us on space-time, but rather on *superspace* defined in the vacuum configuration as $\mathrm{R}^{4|4} = \overline{\mathrm{OSp}(1,4)/SO(1,3)}$, while out of vacuum superspace is a bundle whose base space is physical space-time. The superspace equations of motion, besides horizontality of the supercurvatures in the Lorentz directions, also give constraints on the super-curvatures in the "fermionic" directions, which allow to restrict the theory to space-time only. Indeed one finds that these components are linearly expressible in terms of the components of the super-curvatures on the (cotangent plane of the) space-time manifold. It is this property, dubbed *rheonomy*, which allows a complete geometrical interpretation of supersymmetry[a] and moreover allows the interpretation of the *superspace diffeomorphisms* as supersymmetry transformations on space-time.

In the following we shall try to give a short, albeit almost complete, account of these properties in the simplest case of $\mathcal{N} = 1$, $D = 4$ pure supergravity. This is however sufficient, since, as we will explain, they work exactly in the same way for any other supergravity independently of the supergroup G the number of supersymmetry generators, space-time dimensionality and/or matter couplings, even if they often exhibit a much more intricate structure. Notwithstanding this the geometrical interpretation of supersymmetry can be shown to remain the same.

The development of this approach was proposed by T. Regge as soon as he came back to Torino from Princeton IAS. The developments of his ideas were initially a result of his collaboration with R. D'Auria and P. Fré and partly with A. D'Adda. The Torino group then developed the Regge initial ideas during many years[b] and his geometric method has become one of the principal tools for investigation not only of supergravity, but also of any other topic where the geometrical and group-theoretical approach can be useful or even essential. Indeed a long series of achievements using the basic idea of Tullio Regge has been realized from the very beginning till our days and will probably continue also in the future.

Coming back to supergravity, let me just mention which other advantages of a purely geometric approach have emerged during the developments of the method.

- An important step has been the construction of the Lagrangian. Indeed, using the building principles of *geometricity* to be discussed below together with other obvious requirements like the presence of the Einstein term, the construction turns out to be essentially algorithmic and unique. Moreover, as outlined before, using the Ne'eman Regge's geometrical action principle the superspace equations of motion give in one stroke, besides the space-time field equations, also linear relations between the components of the supercurvatures leading to the supersymmetry of the space-time Lagrangian.
- Besides, the steady use of the geometric approach also for theories having anti-symmetric tensors in the gravitational multiplet has led the authors of Ref. 2 to develop in a geometrical way a new structure for their treatment, by generalizing the Maurer–Cartan equations to integrable structures containing higher degree p-forms.[c]
- As is well known, it is possible to develop supergravity theories in superspace using

[a]The same mechanisms of factorization and rheonomy also work in rigid theories. However, for lack of space we shall not consider theories where rigid supersymmetry is present.

[b]Torino group was essentially composed by D'Auria and Fré in the beginning, and later enlarged to L. Castellani and many other collaborators among whom an important role has been played by A. Ceresole. During the development of the approach many other collaborators joined our group in view of solving specific problems using, at least in part, our techniques, the most assiduous and important being S. Ferrara, L. Andrianopoli and lately M. Trigiante.

[c]The new structures generalizing Maurer–Cartan equations have led to a formulation of a mathematical structure which is nowadays recognized as a first example of the mathematical theory of L_∞ algebras (see e.g. Refs. 3, 4).

Bianchi identities. Since the geometric approach is essentially a superspace approach it is possible to derive directly (that is without writing an action principle) the equations of motion of any supergravity using only the Bianchi identities of the super-curvatures and using a priori *rheonomy* as a principle of the construction, as we shall explain in the following.

2. Pure gravity in Cartan formalism

In this section we shall first remind some of the most important properties of the Cartan formulation of the Einstein gravity in order to establish the notations and thus setting the stage for the formulation of its extension to the Poincaré group manifold. This is a preparatory discussion in view of discussing the geometrical interpretation of supersymmetry (also called *rheonomy*) in supergravity theories.

In the Cartan geometrical framework the gauge fields are to be identified with the left-invariant differential forms $\{\sigma^A\}$ dual to the generators T_A of the Lie algebra of the Poincaré group G= ISO(1,3), namely we have $\sigma^A(T_B) = \delta^A_B$. The indices A, B run on the (co-)adjoint representation of a Lie algebra. The left-invariant 1-forms satisfy the celebrated Maurer–Cartan equations which are a dual formulation of the Lie algebra:[d]

$$d\sigma^A + \frac{1}{2} C^A_{BC} \sigma^B \sigma^C = 0. \tag{1}$$

Here C^A_{BC} are the structure constants of the Poincaré Lie algebra satisfying the Jacobi identities as a consequence the integrability condition $d^2 = 0$ of Eq. (1).
The left-invariant 1-forms σ^A describe the configuration of the physical vacuum, that is, vanishing field-strengths. In order to have non-vanishing field-strengths one needs a non-vanishing right-hand side in Eq. (1), that is we must endow G with a set of *non left-invariant forms* μ^A, so that they can develop non-vanishing curvatures R^A:

$$R^A = d\mu^A + \frac{1}{2} C^A_{BC} \mu^B \mu^C, \tag{2}$$

R^A being defined as the coadjoint multiplet of curvatures. In particular, the 1-forms μ^A will be now dual to the non left-invariant vector fields $\tilde{T}_A$ closing a Lie algebra of vector fields with structure functions instead of structure constants, namely $C^A_{BC} \longrightarrow C^A_{BC} + R^A_{BC}$, where R^A_{BC} are the components of the curvature 2-forms along $\mu^B \mu^C$. Following Cartan's geometrical setting, the 1-forms μ^A, indexed in the coadjoint representation of the Poincaré group, can be decomposed into their Lorentz content, that is the index A is decomposed with respect to indices of the Lorentz subgroup $\mathrm{SO}(1,3)$. In this way one defines the spin connection and the vierbein 1-forms: $\mu^A \equiv \{\omega^a_{\ b}, V^a\}$, $(a, b = 0, 1, 2, 3)$. Correspondingly, the curvatures or field-strengths of $\omega^a_{\ b}$, V^a take the following form:

$$R^a_{\ b} = d\omega^a_{\ b} - \omega^a_{\ c} \wedge \omega^c_{\ b}, \tag{3}$$

$$T^a = dV^a - \omega^a_{\ b} \wedge V^b . \tag{4}$$

The R^{ab} 2-form is named the Lorentz curvature and when expanded along a vierbein basis $R^{ab} = \frac{1}{2} R^{ab}_{\ \ cd} V^c V^d$ its components coincide with minus the Riemann tensor in Lorentz

[d]Here and in the following we shall omit the wedge symbol for p-forms product, unless where it can be useful for clarity and avoid confusion.

indices.[e] The T^a 2-form is named the torsion. Moreover, the curvatures satisfy the Bianchi identities

$$\mathcal{D}R^{ab} = 0, \tag{5}$$

$$\mathcal{D}T^a - R^{ab}\,V_b = 0, \tag{6}$$

being $\mathcal{D} = d - \omega$ the Lorentz covariant derivative. Using these definitions, the Cartan gravitational Lagrangian can be written as follows:

$$\mathcal{L} \simeq R^{ab} \wedge V^c \wedge V^d \epsilon_{abcd}\,. \tag{7}$$

Note that the integrand of the action being a 4-form is automatically invariant under diffeomorphisms and, as it is shown below, it *formally* coincides with the usual Einstein–Hilbert Lagrangian written in terms of the metric and Levi–Civita connection.

In the original Cartan construction of the Lagrangian (7) the gauge fields $\{\omega^a_{\ b},\, V^a\}$ depend only on the space-time coordinates while the dependence on the Lorentz parameters is factorized. In other words, the total space is taken to be a principal fiber bundle $[\mathcal{M}_4, \mathrm{H}]$, where $\mathrm{H} = \mathrm{SO}(1,3)$.

As the fiber bundle structure implies a factorization of the Lorentz parameters of the fiber, $\mathcal{M}_4$ can be identified with a, generally non-flat, four manifold namely *space-time.* Therefore the gravitational action is obtained by integrating on M_4 the Lagrangian (7):

$$\mathcal{A} = \frac{1}{4\kappa^2}\int_{\mathrm{M}_4} R^{ab} \wedge V^c \wedge V^d \epsilon_{abcd}\,, \tag{8}$$

where $\kappa = \sqrt{8\pi\, G}$, and G is the gravitational constant.

Let us remind some of the properties of the Cartan Lagrangian (7).
First we show that it is *formally* equivalent to the traditional Einstein–Hilbert Lagrangian. Indeed[f]

$$R^{ab} \wedge V^c \wedge V^d \epsilon_{abcd} = \frac{1}{2} R^{ab}{}_{ij} V^i V^j V^c V^d \epsilon_{abcd} = \frac{1}{2} R^{ab}{}_{ij} V^i_{\ \mu} V^j_{\ \nu} V^c_{\ \rho} V^d_{\ \sigma} \epsilon_{abcd} dx^\mu dx^\nu dx^\rho dx^\sigma =$$
$$\frac{1}{2} R^{ab}{}_{ij} V^i_{\ \mu} V^j_{\ \nu} V^c_{\ \rho} V^d_{\ \sigma} \epsilon^{\mu\nu\rho\sigma} \epsilon_{abcd} d^4x = \frac{1}{2} R^{ab}{}_{ij} \epsilon^{ijcd} \epsilon_{abcd} \mathrm{det} V d^4x = -2 R^{ij}{}_{ij} \mathrm{det} V d^4x\,. \tag{9}$$

If we denote world-indices by Greek letters, we have

$$R^{ij}{}_{ij} \equiv R^{\mu\nu}{}_{\mu\nu} = R, \tag{10}$$

where R is the scalar curvature and $\mathrm{det}V = \sqrt{-g}$ is the square root of the metric determinant ($g = \mathrm{det} g_{\mu\nu}$). Hence we get:

$$\int_{M_4} R^{ab} \wedge V^c \wedge V^d \epsilon_{abcd} = -2\int_{M_4} R\sqrt{-g}\, dx^4\,. \tag{11}$$

Let us now observe that the formal equivalence between the Cartan and Einstein–Hilbert formulations just shown does not mean that they are completely equivalent.
First of all Cartan vierbein formalism, showing explicitly the gauge invariance of the theory

[e] We note that our definition of curvature and torsion differ by the sign of the spin connection ω^{ab} compared to the more common definition in the literature. Since $\omega^{ab} \to -\omega^{ab}$ implies $R^{ab} \to -R^{ab}$, the relation between R^{ab} and the Riemann tensor is $R^{ab}{}_{cd} = -V^a_\mu\, V^{b\nu}\, V^\rho_c\, V^\sigma_d\; R^\mu_{\nu\rho\sigma}$.

[f] Here and in the following we are using a mostly minus Minkowski metric.

under Lorentz transformations, allows to introduce spinors in the general relativity framework, contrary to what happens in the usual formalism. Indeed in the world index setting, tensors transform under GL(4, R) while spinors are in a SO(1, 3) representation and therefore they can be naturally coupled in a formalism where Lorentz SO(1, 3) covariance is present.
Furthermore, it is a *first order Lagrangian*, that is the gauge fields ω^{ab}, V^a, being member of the same adjoint multiplet, are off-shell independent as it is natural in a geometric Lagrangian like (7). By *geometric* we mean that, besides containing the Einstein term, it is built only in terms of forms and wedge products.
Moreover it can be easily ascertained that requiring the Lagrangian to be *geometric* makes it *uniquely determined.* Indeed any other Lorentz invariant 4-forms to be added to Eq. (7) would be a wedge product of curvature 2-forms and would give rise to Chern characteristic classes[g] (note that the term $T^a \wedge T_a$ is easily seen to reduce to the Cartan Lagrangian by partial integration).
Alternatively, uniqueness of the gravity Lagrangian is also obtained by requiring that no dimensional constant should enter the Lagrangian. In this case the (wedge) product of any two curvatures $R^A = (R^{ab}, T^a)$ would have different scaling with respect to the Lagrangian (7) and should therefore be omitted. Of course dimensional constants should appear in theories where coupling to matter, gauging and scalar potential appear.[h]

Let us now write down the equations of motion derived from the action (8). Varying the action with respect to ω^{ab} and V^d we find, respectively:

$$T^c \wedge V^d \epsilon_{abcd} = 0, \tag{12}$$

$$R^{ab} \wedge V^c \epsilon_{abcd} = 0, \tag{13}$$

where $T^a = \mathcal{D}V^a$ is the torsion 2-form, $\mathcal{D}V^a \equiv dV^a - \omega^a{}_b V^b$ denoting the Lorentz covariant derivative.

Before proceeding with the solutions to the above equations, it is important to stress that, besides the obvious invariance under GCTG, even if all the fields are valued in the Poincaré group, *the action is invariant only with respect to the subgroup of the (local) Lorentz transformations.*
This can be easily checked writing down the infinitesimal action of the Poincaré group on the gauge fields ω^{ab}, V^a. Defining $\epsilon^A = \epsilon^{ab}, \epsilon^a$, being ϵ^{ab} and ϵ^a the parameters of the infinitesimal Lorentz and translation gauge transformations, respectively, we have:

$$\delta^{(gauge)} \mu^A = (\nabla \epsilon)^A , \tag{14}$$

where ∇ denotes the Poincaré gauge covariant differential. Decomposing the (co)-adjoint

[g]By curvature we mean both the R^{ab} and T^a 2-forms.

[h]For pure theories, however, like those described in terms of massless fields only, the pure gravity case being the simplest case, one-dimensional constant of dimensions mass squared is allowed, adding the term $1/3\Lambda\epsilon_{abcd}V^a V^b V^c V^d$ with Λ having the dimension of a mass squared. This gives rise to a Einstein Lagrangian with a cosmological term. This kind of extensions, however, can be easily shown to be equivalent to starting with the group manifold of a (anti) de Sitter group instead of the Poincaré group and will not change anything in the mechanisms we are going to discuss both for gravity as for supergravity. Indeed we may note that the Poincaré group ISO(1, 3) is an Inönü–Wigner contraction of the SO(2, 3) group.

index A in indices of the Lorentz subgroup, from (14) follows

$$\delta^{(gauge)}\omega^{ab} = \mathcal{D}\epsilon^{ab},$$
$$\delta^{(gauge)}V^{a} = \mathcal{D}\epsilon^{a} + \epsilon^{ab}V_b,$$

where $\mathcal{D} = d - \omega$ denotes the Lorentz covariant differential. It is then easy to see that the Lagrangian (7) and the equations of motion are invariant under a local Lorentz transformation, but are not invariant under a local translation. Indeed performing an infinitesimal translation on (19), using the Bianchi identities (5) and (6) and integrating by parts, we have

$$\delta\int R^{ab}V^cV^d\epsilon_{abcd} = 2\int R^{ab}\mathcal{D}\epsilon^cV^d\epsilon_{abcd} = -2\int \epsilon^c R^{ab}T^d\epsilon_{abcd} \neq 0\,. \tag{15}$$

The non-invariance under translations of the equations of motion (12) and (13) can be checked in an analogous way. We will see in the next section that the fact that a torsionless vierbein can acquire torsion under the action of a translation can be best understood using the notion of Lie derivative.

Let us now solve the equations of motion (12) and (13). From (12), expanding the torsion 2-form along the vierbein basis, $T^a = T^a{}_{bc}V^b\,V^c$, it is easy to find that the components $T^a{}_{bc}$ have a vanishing trace, $T^a{}_{ab} = 0$. This implies in turn $T^a{}_{bc} = 0$ as the unique solution. Therefore

$$T^a = 0 \longrightarrow dV^a - \omega^a{}_bV^b = 0. \tag{16}$$

Expanding along the differentials, we write

$$\partial_{[\mu}V^a_{\nu]} = \omega^{ab}_{[\mu}V_{\nu]b}. \tag{17}$$

Equation (17) can be solved in an analogous way as the Levi–Civita connection is solved in terms of the metric and its derivatives, obtaining ω^{ab}_{μ} in terms of the vierbein components and its derivatives. Therefore, implementing the purely algebraic equation of motion of the spin connection allows us to express $\omega^a{}_{b|\mu}$ as a functional of the vierbein and its first derivatives. This is strictly analogous to what happens in the first order Palatini formalism in the usual metric approach.

From the second equation, expanding the 2-form $R_{ab} = \frac{1}{2}R^{ab}{}_{cd}V^cV^d$, we also find, after some algebraic manipulation, the Einstein equation[i]

$$R_{ab} - \frac{1}{2}\eta_{ab}R = 0, \tag{18}$$

where now R_{ab} is the Ricci tensor $R_{ab} = R^{am}{}_{bm}$ and $R \equiv R^a{}_a$ is the scalar curvature.

2.1. *Extending the theory from* G/H *to* G

We have seen in the previous section that the Einstein–Cartan Lagrangian is invariant under gauge Lorentz gauge transformations. It is therefore convenient to use the language of fiber bundles.

[i]Note that we write Einstein equation $R_{\mu\nu} - 1/2g_{\mu\nu}R = 0$ using rigid vierbein indices. Indeed, unless necessary, we are using rigid tangent indices throughout instead of coordinate indices. Of course the rigid indices can be traded with the world indices using the vierbein matrix in the usual way.

We will refer in the following to the Poincaré group, but most of the considerations are exactly the same for other gravity or supergravity theories, replacing the Poincaré group with a general (super-)group G and the Lorentz group with a general gauge group H.
We know that in the vacuum configuration, the fiber bundle is $[\mathrm{ISO}(1,3)/\mathrm{SO}(1,3)]$, while for a general configuration the gauge fields $\mu^A = \left(\omega^{ab},\, V^a\right)$ of the Cartan–Einstein Lagrangian (7) live on a fiber bundle $[\mathrm{M}_4, \mathrm{H}]$, where M_4 is the base space and $\mathrm{H} = \mathrm{SO}(1,3)$ is the fiber. It is then convenient and natural to consider the base space $\mathcal{M}_4$ as a deformation of the vacuum base space $\mathrm{ISO}(1,3)$. Indeed the base space out the vacuum can be obtained when the left-invariant 1-forms are deformed into a *non-left-invariant* dynamical forms μ^A enjoying *curvatures.* In this case the base space can be thought of as a space G/H on which a dynamical metric has been defined, constructed out of the on left-invariant 1-forms μ^A, which no longer has G as isometry group. Equivalently we may also say that the group G has been deformed into a $\tilde{\mathrm{G}}$ so has to have dynamical fields. $\tilde{\mathrm{G}}$ is often referred to as a *soft group manifold.* With this nomenclature we can write the general structure of the fiber bundle as $\tilde{\mathrm{G}}/\mathrm{H}$.

Let us now take the point of view of Ref. 1 and assume that the set of 1-forms $\mu^A = (\omega^{ab},\, V^a)$ is defined, right from the beginning, on the whole Poincaré group G=ISO(1,3). In this case factorization of the Lorentz coordinates is absent since the gauge fields $\mu^A = \left(\omega^{ab},\, V^a\right)$ will now depend on the full set of ten-dimensional group coordinates, namely the x^μ parameters of translations and the coordinates $y^{\mu\nu}$ associated with the Lorentz transformations of $\mathrm{SO}(1,3)$. The non left-invariant μ^A 1-forms, generalizing the σ^A, are naturally viewed as a set of vielbein spanning a local reference frame on the ten-dimensional cotangent plane at each point of $\tilde{\mathrm{G}}/\mathrm{H}$.

We consider the Lagrangian $\mathcal{L} = R^{ab} \wedge V^c \wedge V^d \epsilon_{abcd}$ where now both the vielbein and curvatures are not factorized, but depend of the full set of the G-coordinates $(x^\mu,\, y^{\mu\nu})$.

Even if the *Lagrangian* (7) is formally the same when the fields 1-form and curvatures 2-forms are defined on the full group manifold of G, there is, however, a problem to write down a suitable *action,* since we have to integrate the Lagrangian 4-form in the d-dimensional space of the group $\tilde{\mathrm{G}}$, ($d = 10$ in the Poincaré case).
A simple way out would be to embed the four-dimensional space-time as a four-dimensional hypersurface $\mathcal{M}_4$ (with boundary value M_4) in $\tilde{\mathrm{G}}$. However, the very presence of $\mathcal{M}_4$ in the variational principle makes it a dynamical variable, thus subjected to variation. Indeed new fields enter in the action corresponding to the embedding functions of $\mathcal{M}_4$ in $\tilde{\mathrm{G}}$. This implies the added complication that the equations describing the embedding of $\mathcal{M}_4$ in $\tilde{\mathrm{G}}$ should include arbitrary functions which must be considered as fields, what would of course spoil the geometric nature of the Cartan geometric approach.
The crucial observation given in[1] is that one can safely ignore the variation of $\mathcal{M}_4$, since *any variation of* $\mathcal{M}_4$ *can be compensated by a change of coordinates* $x^M = (x^\mu,\, \eta^{\mu\nu})$ *in* $\tilde{\mathrm{G}}$ under which the Lagrangian, built only in terms of differential forms (and wedge products among them), is invariant. When considering theories more general than pure gravity, this of course requires that the Lagrangian be *geometrical* in a larger sense as was previously explained. Indeed, besides the requirements of being built using only differential forms, wedge products and the differential operator $d : d^2 = 0$, we must also add the requirement of *excluding the presence of the Hodge duality operator.* Indeed as the hypersurface $\mathcal{M}_4$

on which the integration is performed can be chosen arbitrarily, due to invariance under diffeomorphisms, the equations of motion will hold on the whole $\tilde{\mathrm{G}}$. The presence of the Hodge duality operator, instead, would give a dependence on the hypersurface $\mathcal{M}_4$ and its metric.

In the case under study, namely simple gravity in the Cartan formalism, see Eq. (7), the requirement of geometricity in this larger sense is obviously satisfied and the action will be now written as follows:

$$\mathcal{A} = \frac{1}{4\kappa^2} \int_{\mathcal{M}_4 \subset \tilde{\mathrm{G}}} R^{ab} \wedge V^c \wedge V^d \epsilon_{abcd}, \tag{19}$$

where $\mathcal{M}_4$ is an arbitrary surface immersed in $\tilde{\mathrm{G}} = \widetilde{\mathrm{ISO}}(1,3)$.
The equations of motion derived from the action (19) are *formally* the same as Eqs. (12) and (13), but are now valid on the whole group manifold, since, as already observed, the hypersurface $\mathcal{M}_4$ is arbitrary. Because of that, in analyzing their content, the Lorentz curvature and the torsion 2-forms must be expanded not only in terms of the vierbein $V^a \wedge V^b$, but also in terms of the 2-forms $\omega^{ab} \wedge V^c$ and $\omega^{ab} \wedge \omega^{cd}$, which are part of the $\mu^A \wedge \mu^B$ basis of the cotangent plane to $\tilde{\mathrm{G}}$ at each point of $\mathcal{M}_4$. The projection along $V^a \wedge V^b$ leads of course to the same equations as in the Cartan original setting. On the other hand, expanding the curvature and the torsion equations also along the other two 2-forms of the basis containing at least one ω^{ab} 1-form, it is almost immediate to recognize that we obtain for their corresponding components the solution:

$$R^{ab}_{c|(lm)} = R^{ab}_{(lm)|(pq)} = T^a_{c|(lm)} = T^a_{(lm)|(pq)} = 0. \tag{20}$$

These equations assert that the curvatures are *horizontal* along the Lorentz subgroup 1-forms containing at least one Lorentz gauge field ω^{ab}, implying that *the group manifold* $\tilde{\mathrm{G}}$ *has acquired dynamically the structure of a fiber bundle, namely* $[\tilde{\mathrm{G}}/\mathrm{H}, \mathrm{H}]$, where in the present case $\tilde{\mathrm{G}} = \widetilde{\mathrm{ISO}}(1,3)$, $\mathrm{H} = \mathrm{SO}(1,3)$. Indeed we recall that an equivalent way to say that a manifold $\tilde{\mathrm{G}}$ has the structure of a fiber bundle $[\tilde{\mathrm{G}}/\mathrm{H}, \mathrm{H}]$ is to require that the curvatures R^A are *horizontal*, that is that their components along the H directions spanned by the 1-forms dual to the generators of H vanish. This is in fact the content of Eq. (105) since the curvature 2-forms $\left(R^{ab}, T^a\right)$ vanish along components $\omega^{ab} \wedge \omega^{cd}$ or $\omega^{ab} \wedge V^c$. It follows that only the components along two vielbein $R^A = R^A_{ab} V^a \wedge V^b$ will survive, that is, identifying $\mathcal{M}_4$ with the space-time and projecting them along the differentials, namely the $R^A_{\mu\nu}$ components.
Thus we have reobtained from the same Lagrangian, but an enlarged action principle, the same equations of motion as in the classical treatment of the Cartan Lagrangian given before. The new procedure of defining fields on the group manifold and of considering space-time as a hypersurface immersed in G, gives a conceptual advantage with respect to the usual formulation since the factorization of the Lorentz coordinates and the Lorentz invariance of the Lagrangian are a result of the field equations. When this is done the connection forms and the vierbein appear as a single g-bein on $\tilde{\mathrm{G}}$ (g being the dimensions of G) which also play the role of connections in computing covariant derivatives and curvatures.
On the other hand, implementing the equations of motion derived from the action principle of (19) one finds in a dynamical way that the theory lives effectively on a fiber bundle $[\tilde{\mathrm{G}}/\mathrm{H}, \mathrm{H}]$.

Note that even if in pure gravity the Hodge duality operator does not appear, in more general theories, like matter coupled gravity, it is precisely the absence of the Hodge duality operator which implies that the choice of $\mathcal{M}_4$ turns out to be irrelevant. Actually any other $\mathcal{M}'_4$ could work equally well and the physics would be the same on any of them. Indeed a diffeomorphism in the direction orthogonal to space-time, considered from a *passive* point of view, can be considered as a lifting of the hypersurfaces $\mathcal{M}_4$ to another hypersurfaces $\mathcal{M}'_4$ and corresponds to the same theory in a different Lorentz frame. In other words the lifting from $\mathcal{M}_4$ to $\mathcal{M}'_4$ corresponds to a Lorentz transformation. From an *active* point of view, however, restricting the Lorentz factorized theory to a fixed $\mathcal{M}_4$, identified as the physical space-time, the same diffeomorphism corresponds to consider the theory together with all possible Lorentz transformations. Indeed, as we will explain in the next subsection, a diffeomorphism on the group manifold reduces to a Lorentz gauge transformation when the curvatures along the Lorentz subgroup are horizontal.

Let us note that the requirements of geometricity given before for treating the simple case of pure Poincaré gravity work equally well in more complicated theories like matter coupled gravity theories in four or even higher dimensions.

In all these more realistic cases the requirement of the absence of the Hodge duality operator seems, at first sight, to be too strong. Indeed when any extended gravity theory is coupled to scalars or vector fields their kinetic terms require the presence of the Hodge duality operator. For example, a kinetic term for a vector field should be written as proportional to

$$-\int F_{\mu\nu}\, F^{\mu\nu}\sqrt{-g}d^4x = \frac{1}{2}\int F\wedge *F, \tag{21}$$

where $F_{\mu\nu} = \partial_{[\mu}A_{\nu]}$, $F = F_{ab}\, V^a\, V^b$ and $*F = \frac{1}{2}F^{ab}\epsilon_{abcd}V^c\, V^d$. This apparent obstacle can be easily overcome introducing a *first order formalism* for the vector field. Namely, we introduce a 0-form antisymmetric Lorentz tensor $\hat{F}_{ab}$ ($\hat{F}_{ab} = -\hat{F}_{ba}$) and write for the kinetic term of the 2-form F the following action:

$$\mathcal{A}_{\text{vec}} = -\int \hat{F}^{ab}\,\hat{F}_{ab}\Omega + \alpha\int \hat{F}^{ab}\, F\, V^c\, V^d\epsilon_{abcd}, \tag{22}$$

where Ω is the four-dimensional volume element $(-\frac{1}{4!}\epsilon_{pqrs}V^p\, V^q\, V^r\, V^s)$. Varying the 0-form $\hat{F}_{ab}$ we find that choosing $\alpha = -\frac{1}{2}$ we obtain

$$\hat{F}_{ab} = F_{ab}, \tag{23}$$

where F_{ab} are the components along two vielbein of the 2-form F. Varying next the gauge field A, from the second term of the action (22) we find the usual equation of motion:

$$\mathcal{D}_a F^{ab} = 0 \rightarrow \mathcal{D}_\mu F^{\mu\nu} = 0. \tag{24}$$

In this way we see that using a first order formalism for the vector fields Lagrangian, the kinetic term can be obtained from the equation of motion starting from a geometric Lagrangian where the Hodge operator is absent. The same trick can be obviously used for the kinetic term of scalar fields.

Quite generally the same requirement of geometricity should be made for any extension of the theory with additional fields (matter coupled gravity) and in particular in supergravity where, besides the mechanism of Lorentz coordinates factorization, the requirement of a *geometric Lagrangian* turns out to be necessary in order to implement in a geometrical way local supersymmetry. It is precisely this fact that makes the rather academic result of obtaining the factorization of Lorentz coordinates as a result of the field equations a simple standard example to understand in a similar way the supersymmetry transformations in supergravity.

2.2. *Gauge transformations and diffeomorphisms*

It is interesting at this point to show in an explicit way how a diffeomorphism reduces to a gauge transformation when the curvatures are horizontal, while it differs by curvature terms in the general case. We perform the derivation in a general group-theoretical setting so that it may apply to any (softened) group or supergroup $\tilde{\mathrm{G}}$.
An infinitesimal element of the GCTG on $\tilde{\mathrm{G}}$ is given by a tangent vector on $\tilde{\mathrm{G}}$, $\boldsymbol{t} = \epsilon^M T_M$, with $\epsilon^M = \delta x^M$, where the middle alphabet Latin capital indices are coordinate indices on $\tilde{\mathrm{G}}$. Using the vielbein μ^A of the whole (soft) group $\tilde{\mathrm{G}}$ we can rewrite a tangent vector $\boldsymbol{t}$ as follows:

$$\epsilon = \epsilon^A \tilde{T}_A, \tag{25}$$

where $\epsilon^A = \epsilon^M \mu^A_M$, and $\tilde{T}_A = T_M\, \mu^M_A$. Here $\tilde{T}_A$ is the vector field generator dual to the non left-invariant 1-form μ^A, $\mu^A(\tilde{T}_B) = \delta^A_B$, and $\epsilon^A = \delta x^A$ is the infinitesimal parameter associated to the shift. An infinitesimal generator of diffeomorphisms generated by ϵ^A is given by the Lie derivative

$$\ell_\epsilon \mu^A = (\iota_\epsilon d + d\iota_\epsilon)\, \mu^A, \tag{26}$$

where ι_ϵ is the contraction operator along ϵ.
On the other hand the Lie derivative (26) can be also rewritten as follows:

$$\begin{aligned} \ell_\epsilon \mu^A &= (\iota_\epsilon d + d\iota_\epsilon)\, \mu^A \\ &= \iota_\epsilon d\mu^A + d\left(\iota_{(\epsilon^B \tilde{T}_B)} \mu^A\right) \\ &= \iota_\epsilon d\mu^A + d\epsilon^A\,. \end{aligned} \tag{27}$$

Adding and subtracting $C^A{}_{BC}\mu^B \wedge \mu^C$ to $d\mu^A$ and using the definition of the covariant derivative

$$\nabla \epsilon^A = d\epsilon^A + C^A{}_{BC}\mu^B\, \epsilon^C, \tag{28}$$

we find:

$$\ell_\epsilon \mu^A = \iota_\epsilon \left(d\mu^A + \frac{1}{2} C^A{}_{BC}\mu^B \wedge \mu^C\right) - \epsilon^B C^A{}_{BC}\mu^C + d\epsilon^A, \tag{29}$$

where we have used the antisymmetry of $C^A{}_{BC}$ in the lower indices. The terms in brackets define the curvature R^A while the other two terms, using the antisymmetry of the structure constants in (B, C) define the gauge covariant differential of ϵ^A. Therefore, using the *anholonomized* parameter[j] ϵ^A, the Lie derivative can be written as follows:

$$\ell_\epsilon \mu^A = (\nabla \epsilon)^A + \iota_\epsilon R^A\,. \tag{30}$$

[j] By anholonomized parameter we mean that we are using the rigid group index of the vielbein μ^A.

Hence *an infinitesimal diffeomorphism on the manifold* $\tilde{G}$ *is a* G*-gauge transformation plus curvature correction terms.*

In particular, if the curvature R^A has vanishing projection along a vector $\epsilon^{\tilde{B}}T_{\tilde{B}}$, where $\tilde{B}$ is an adjoint index of the subgroup $H \subset \tilde{G}$ so that

$$\iota_\epsilon R^A \equiv \epsilon^{\tilde{B}} R^A{}_{\tilde{B}C}\mu^C = 0, \tag{31}$$

then *the action of the Lie derivative* ℓ_ϵ *coincides with a gauge transformation.* In this case we recover the result that the curvatures are *horizontal* along the H directions and the group manifold acquires the structure of a principal fiber bundle whose base manifold is $\tilde{G}/H$, and H the gauge group.
In conclusion, if the theory on $\tilde{G}$ can predict horizontal curvatures, it lives on $[\tilde{G}/H, H]$ and it is equivalent to the original Cartan approach. This is in fact what we have found in the case of the Poincaré group.
We stress once again that the derivation of the formula AGCT, Eq. (30), makes no explicit reference to the specific group $\tilde{G}$. It holds for any group, including supergroups, as we shall see in the supergravity case.

3. Geometric supergravity

The fact that the fiber bundle structure in H of a gravity theory can be obtained dynamically from a suitable action principle is certainly an interesting feature of these theories, since it sheds light on the geometrical origin of the theory and on the power of the action principle. However, from a purely physical point of view it does not seem to add anything important to our understanding of the theory. After all to write a theory possessing *ab initio* a fiber bundle structure does not change anything in the development of the theory and in its physical results.
The value of the previous detailed description of factorization of Lorentz parameters lies in the possibility to give a geometrical interpretation of supersymmetry analogous to the geometrical mechanism of factorization of the Lorentz coordinates shown in the pure gravity case.
Indeed we will show, using the simple example of $\mathcal{N} = 1$, $D = 4$ supergravity, that the invariance of the supergravity Lagrangian is due to the behavior of the supergroup curvatures along the fermionic components of the supervielbein in superspace. Indeed, while the curvatures in the direction of the "Lorentzian" vielbein ω^{ab} of the supergroup have to be zero in order to obtain a fiber bundle structure for the H subgroup, exactly as it happens in gravity, this will not happen for the supercurvatures in the direction of the fermionic vielbein ψ^α.[k]
What actually happens is the following: The dependence of the fields on the supergroup or superspace odd coordinates θ^α, does not imply their complete factorization, rather one finds that such components of the curvature 2-forms can be expressed algebraically, actually *linearly*, in terms of the curvatures restricted to the bosonic cotangent plane of the embedded space-time hypersurface, namely in terms of $V^a\, V^b$, the basis on the cotangent space of

[k]Here and in the following α is a spinor index in the relevant representation of SO(1,D-1).

ordinary space-time. Inserting this result into the Lie derivative formula (30), one obtains a geometrical interpretation of the local supersymmetry transformations.
For the sake of brevity and simplicity we will show how this happens in the simple example of pure $\mathcal{N}=1$, $D=4$ supergravity. However, the relevant results hold exactly in the same way for any supergravity theory, pure or matter coupled, in any dimension $4 \leq D \leq 11$ and for any number $1 \leq \mathcal{N} \leq 8$ of supersymmetry generators in the Lie superalgebra.

3.1. $\mathcal{N}=1$, $D=4$ *supergravity in the geometric approach*

We will now give the explicit description of the group manifold formalism for the $\mathcal{N}=1$, $D=4$ pure supergravity theory.

The graded group of $\mathcal{N}=1$, $D=4$ supergravity is the super-Poincaré group $\tilde{\mathrm{G}} = \overline{\mathrm{OSp}(1|4)}$, where the bar over OSp(1|4) means Inönü–Wigner contraction of the super Anti-de Sitter group (recall that $\mathrm{SO}(2,3) \simeq \mathrm{Sp}(4)$). The coadjoint multiplet of left-invariant 1-forms is

$$\sigma^A = (\mathring{\omega}^{ab}, \mathring{V}^a, \mathring{\psi}^\alpha), \tag{32}$$

where $\mathring{\psi}^\alpha$ is a Majorana spinor 1-form in the fermionic direction of the supergroup. The upper little ring means that, being left-invariant, they satisfy the Cartan–Maurer equations (vanishing curvatures):

$$d\sigma^A + \frac{1}{2} C^A{}_{BC}\, \sigma^B \wedge \sigma^C = 0. \tag{33}$$

These 1-form fields will now depend on the coordinates $(x^\mu, \eta^{\mu\nu}, \theta^\alpha)$ of the supergroup. Deforming the σ^A into the non left-invariant 1-form $\mu^A = (\omega^{ab},\ V^a\ , \psi^\alpha)$ allows us to define the following multiplet of super-curvatures:[1]

$$R^a{}_b \equiv d\omega^a{}_b - \omega^a{}_c \wedge \omega^c{}_b, \tag{34}$$

$$\hat{T}^a \equiv dV^a - \omega^a{}_b \wedge V^b - \frac{\mathrm{i}}{2}\bar{\psi}\gamma^a\psi, \tag{35}$$

$$\rho \equiv \mathcal{D}\psi = d\psi - \frac{1}{4}\gamma_{ab}\psi\omega^{ab}, \tag{36}$$

where we adopt a matrix notation for the spinor current in Eq. (35) and we have denoted by ρ the curvature 2-form of ψ; γ_a and γ_{ab} are Dirac gamma matrices in four dimensions. Moreover we have set a hat on the supertorsion, $\hat{T}^a$, to avoid confusion with the purely bosonic torsion T^a. The curvatures satisfy the following Bianchi identities:

$$\mathcal{D}R^{ab} = 0, \tag{37}$$

$$\mathcal{D}\hat{T}^a + R^a_b\, V^b - \mathrm{i}\bar{\psi}\gamma^a\, \rho = 0, \tag{38}$$

$$\mathcal{D}\rho - \frac{1}{4}\gamma_{ab}\psi\, R^{ab} = 0. \tag{39}$$

Note that all terms in the definition of the curvatures and in the Bianchi identities scale homogenously since ω^{ab}, V^a, ψ and their curvatures have length scaling $[L^0]$, $[L^1]$ and $[L^{1/2}]$, respectively.

In order to write down the Lagrangian we require that it is *geometric*. For the sake of clarity let us repeat here what it amounts to, adding some more obvious requirements:

[1]Here and in the following we will mostly omit the spinor index α on the spinors. Moreover, if there is no risk of confusion, we shall often refer to the super-curvatures simply as curvatures.

- It must be constructed using only differential forms, wedge products among them, and the d exterior differential;
- It must not contain the Hodge duality operator;
- As the Einstein term, which must be always present, scales as $[L^2]$, ($[L^{D-2}]$ in D dimensions), all the terms must scale in the same way;

To these requirements one usually adds the following one:

- If all the curvatures R^A are zero (left-invariant σ^A or vacuum configuration), the Lagrangian and the equations of motion must vanish identically.

This last requirement is useful in constructing Lagrangians more complicated compared to the present $\mathcal{N}=1$, $D=4$ theory.
It is easily seen that the only term we can add satisfying the first three requirements, for the $G=\overline{\mathrm{OSp}(1|4)}$ group, is the Rarita–Schwinger kinetic term (written in terms of differential forms). Thus we obtain

$$\mathcal{A}^{\mathcal{N}=1}_{D=4}=\frac{1}{4\kappa^2}\int_{\mathcal{M}_4\subset\overline{\mathbf{OSp}(1|4)}}\left[R^{ab}\,V^c\,V^d\epsilon_{abcd}+\alpha\overline{\psi}\gamma^5\gamma_a\mathcal{D}\psi\,V^a\right], \tag{40}$$

where instead of a tilde we have used a boldface character to denote the soft supergroup manifold $\overline{\mathrm{OSp}(1|4)}$. This supermanifold has ten bosonic and four fermionic coordinates, namely $(x^\mu,\eta^{\mu\nu},\theta^\alpha)$. The coefficient α between the Einstein and Rarita–Schwinger is related to the normalization of the gravitino 1-form ψ and will be fixed in a moment. The equations of motion obtained by varying ω^{ab}, V^a, and ψ are respectively:

$$\epsilon_{abcd}\left(\mathcal{D}V^a+\frac{\mathrm{i}\alpha}{8}\,\bar{\psi}\,\gamma^a\,\psi\right)\wedge V^d=0, \tag{41}$$

$$2R^{ab}\wedge V^c\epsilon_{abcd}-\alpha\overline{\psi}\wedge\gamma^5\gamma_d\mathcal{D}\psi=0, \tag{42}$$

$$2\gamma^5\gamma_a\mathcal{D}\psi\wedge V^a-\gamma^5\gamma_a\psi\wedge\hat{T}^a=0\,. \tag{43}$$

As the equations of motion have to vanish identically when all the (super-)curvatures are zero (fourth requirement), we see that we must set in the left-hand side of Eq. (41) $\alpha=4$ in order to have the super-torsion 2-form $\hat{T}^a$ as defined in (35). With this value of α Eq. (41) takes the form

$$\hat{T}^c\wedge V^d\epsilon_{abcd}=0 \tag{44}$$

and we see that when all the supercurvatures are zero the equations of motion vanish identically.

To analyze the content of Eqs. (42), (43) and (44) we expand the curvatures 2-forms along the basis $\mu^A\wedge\mu^B$. Let us first consider their components along the basis 2-form containing at least on ω^{ab}, namely (omitting the wedge symbol):

$$\omega^{ab}\,\omega^{cd};\quad\omega^{ab}\,V^c;\quad\omega^{ab}\,\psi\,.$$

It is then immediate to see that the equations of motion give *horizontality* of all curvatures R^{ab}, $\hat{T}^a$, ρ in the Lorentz directions, namely

$$R^A_{(ab)|(cd)}=R^A_{(ab)|c}=R^A_{(ab)|\alpha}=0, \tag{45}$$

exactly as in the pure gravity case. In this case, however, the supermanifold where the theory lives has a base space the $\overline{\mathbf{OSp}(1|4)/\mathrm{SO}(1,3)}$ that is the *softened super-coset of the vacuum* $\overline{\mathrm{OSp}(1|4)/SO(1,3)}$. We identify $\overline{\mathbf{OSp}(1|4)/\mathrm{SO}(1,3)}$ as the *superspace* ad will be denoted $\mathrm{R}^{(4|4)}$. Superspace has now as coordinates x^μ, θ^α, since the Lorentz coordinates $\eta^{\mu\nu}$ have been factorized.
However as the physical space-time $\mathcal{M}_4$ is four-dimensional while superspace has four extra fermionic dimensions, we are left with the same problem as in the bosonic case of pure gravity, namely to construct a suitable action for a Lagrangian 4-form in the eight-dimensional superspace. It is then natural to resort to the same procedure used for the pure gravity theory. Namely we can identify the space-time with *any* four-dimensional bosonic hypersurface $\mathcal{M}_4$ embedded in superspace. Indeed, as the Lagrangian is completely geometrical and it does not contain the Hodge duality operator, the invariance under diffeomorphisms of the Lagrangian allows arbitrary deformations of $\mathcal{M}_4$ in superspace. Therefore the equations of motion, being independent of the particular hypersurface chosen, will hold on the full superspace. This is completely analogous to the pure gravity case, the only difference being that the diffeomorphisms on the bosonic group manifold have been replaced by diffeomorphisms in superspace.[m]

Since we reduced ourselves to the study of a Lorentz invariant theory on superspace, $\mathrm{R}^{(4|4)}$, let us shortly describe the geometric structure of the theory in superspace.

Since the Lorentz coordinates have already been factorized, on $\mathrm{R}^{(4|4)}$ the general base of 1-forms is given by the set of super-vielbein E^A, namely $E^A = (V^a, \psi^\alpha)$, where ψ^α, $\alpha = 1, \ldots 4$, is the fermionic vielbein, that is a Majorana spinor 1-form named gravitino. The action (40) is now reduced to the following form:

$$\mathcal{A} = \frac{1}{4\kappa^2} \int_{\mathcal{M}_4 \subset \mathrm{R}^{(4|4)}} \left[R^{ab}\, V^c\, V^d \epsilon_{abcd} + 4\bar{\psi}\gamma^5\gamma_a \mathcal{D}\psi\, V^a \right] . \tag{46}$$

Let us introduce for the sake of brevity the following notation. We denote by $R^A_{(p,q)}$, $A = (ab, a, \alpha)$, the components of the curvature along p bosonic vielbein V^a and q fermionic vielbein ψ. Moreover we call *outer* all the components were $q \neq 0$, that is those components having at last one index along the ψ direction, while when $q = 0$, that is when the only non vanishing components are along the bosonic vielbein, they will be called *inner*.

To analyze Eqs. (42), (43), and (44), now restricted to superspace, we must expand the curvatures along a complete basis of 2-forms in superspace. In this case we have

$$E^A \wedge E^B = \{(V^a\, V^b); \quad (V^a\, \psi^\alpha); \quad (\psi^\alpha, \psi^\beta)\}. \tag{47}$$

Let us first work out Eq. (44). Expanding $\hat{T}^a$ we have

$$\hat{T}^a = \hat{\hat{T}}{}^a_{bc} V^b\, V^c + \bar{H}^a_{\ c}\psi\, V^c + \bar{\psi} K^a\, \psi, \tag{48}$$

where $\hat{T}^a_{(1,1)} = H^a_{\ c}$ is a spinor and K^a, defining the $\hat{T}^a_{(0,2)}$ component, is proportional to a gamma matrix in four dimensions. Considering Eq. (44), one easily conclude that the components of $\hat{T}^a_{(1,1)}$ must be zero, while $\hat{T}^a_{(0,2)}$ would only change the normalization of the gravitino 1-form in the definition (35), so we can put them also to zero. Then, Eq. (44) restricted to the $V\,V$ ($\hat{T}^a_{(2,0)}$) components has exactly the same form as in the pure gravity

[m]If the factorization of the Lorentz coordinates have not yet been implemented then we deform the hypersurface $\mathcal{M}_4$ on the full graded group manifold as it was done in the pure gravity case.

case, that is (16), provided we replace T^a with $\hat{T}^a$. Note, however, that in this case solving for the spin connection $\omega^{ab}{}_\mu$ with the usual procedure gives a spin connection which depends not only from the vielbein and their derivatives, but also from gravitino bilinears.[n] With exactly the same computations as those made for pure gravity, one easily obtains the vanishing of the $\tilde{\hat{T}}^a_{bc}$ components and therefore the whole super-torsion 2-form is zero.
In the same way, in order to solve Eqs. (42) and (43) we expand the curvatures ρ and R^{ab} along a complete basis of 2-forms in superspace. As $\hat{T}^a = 0$ Eq. (43) takes the form

$$\gamma_a \mathcal{D}\psi \, V^a = 0 \tag{49}$$

and expanding $\rho \equiv \mathcal{D}\psi$ as

$$\rho^\alpha = \tilde{\rho}^\alpha_{ab} V^a \, V^b + H_a \psi^\alpha \, V^a + \Omega_{\alpha\beta} \psi^\alpha \, \psi^\beta, \tag{50}$$

where, for the sake of clarity, we have made the spinor index explicit. From Eq. (49) one easily realizes that $H_a = \Omega_{\alpha\beta} = 0$, so that the 2-form ρ has only components $\rho_{(2,0)}$ on the cotangent space of $\mathcal{M}_4$, namely

$$\rho = \tilde{\rho}_{ab} V^a \, V^b \,. \tag{51}$$

We warn the reader that since we are now in superspace the rigid indices cannot be traded with coordinate indices using the bosonic vierbein V^a_μ. Indeed, the full set of supervielbein is now given by $E^A = (V^a, \psi^\alpha)$ and we should invert the matrix E^A_μ to find the space time components. This is in fact the reason why we have denoted with a tilde the components of the supercurvatures along two bosonic vierbein. A simpler way to find the space components is to project the equations on the space-time basis $dx^\mu \wedge dx^\nu$. For example from Eq. (50), projecting on the space-time basis we obtain

$$\rho^\alpha_{\mu\nu} = \tilde{\rho}^\alpha_{ab} V^a_\mu \, V^b_\nu + H_a \psi^\alpha_\mu \, V^a_\nu + \Omega_{\alpha\beta} \psi^\alpha_\mu \, \psi^\beta_\nu, \tag{52}$$

where the indices $\mu\nu$ are understood to be antisymmetric. We see that the tilded components of $\tilde{\rho}_{\mu\nu}$ differ from the the real space time components $\rho_{\mu\nu}$ by terms in the gravitino fields, namely *outer terms*. They are commonly named in the literature as *supercovariant field strengths.*[o] However, in our case, as far the $\hat{T}^a$ and ρ curvatures are concerned, we can easily convert rigid Lorentz indices in world indices as usual, since in the present case they do not have *outer* components $(V \wedge \psi)$ and $(\psi \wedge \psi)$. Then, for the components of the aforementioned curvature 2-forms we can neglect the tilde symbol. Instead, as we will see in a while and further discuss in the sequel, the space-time components of the Lorentz curvature do not coincide with the components along $(V^c \wedge V^d)$ expanded along the differentials of the coordinates.
Indeed, expanding R^{ab},

$$R^{ab} = \tilde{R}^{ab}_{cd} V^c \, V^d + \overline{\Theta}^{ab}_c \, \psi \, V^c + \bar{\psi} K^{ab} \psi, \tag{53}$$

from Eq. (42) we find

$$\overline{\Theta}^{ab}_c = -\epsilon^{abrs} \bar{\rho}_{rs} \gamma^5 \gamma_c - \delta^{[a}_c \epsilon^{b]mst} \bar{\rho}_{st} \gamma^5 \gamma_m \tag{54}$$

and $K^{ab} = 0$.

[n] See e.g. Ref. 5.

[o] The name *supercovariant* means that their supersymmetry transformation law does not contain derivatives of the supersymmetry parameter ϵ^α.

Note that the value for the outer component of $R^{ab}_{(1,1)} = \overline{\Theta}^{ab}_c \psi V^c$ given in (53) is written in terms of the components of the gravitino curvature along $V^a V^b$, namely ρ_{ab}. Thus an *outer* component of the Lorentz curvature 2-form is linearly expressed, on-shell, in terms of the *inner* components of the fermionic curvature ρ. This property is called *rheonomy* and will be discussed more generally in the following. Physically it just means that no new degree of freedom is introduced in the theory other than those already present on space-time. Actually, if rheonomy is assumed *a priori*, and we take advantage of the fact that all the coefficients in the expansion along the supervielbein must give rise to terms with same scale as the corresponding curvatures, then one easily recognizes that the previous results for the outer components of the curvature multiplet are easily recovered (as for the possibility to have dimensional constant, see footnote 8). In conclusion, the solution of the equations of motion (41), (42), (43) for the *outer* and *inner* projections of the curvature multiplet gives:

$$R^{ab} = \tilde{R}^{ab}_{cd} V^c V^d + \overline{\Theta}^{ab}_c \psi V^c, \tag{55}$$
$$\hat{T}^a = 0, \tag{56}$$
$$\rho = \rho_{ab} V^a V^b. \tag{57}$$

Finally, inserting the parameterizations (55), (56), (57) in the equations of motion (41), (42), (43), we find that the components of the equations of motion along $V^a V^b V^c$, that is $R^A_{(3,0)}$, give the space-time equations of motion:

$$\tilde{R}^{am}_{bm} - \frac{1}{2}\delta^a_b \tilde{R}^{mn}_{mn} = 0, \tag{58}$$
$$\hat{T}^a_{mn} = 0, \tag{59}$$
$$\epsilon^{mnrs}\gamma^5\gamma_s\rho_{mn} = 0. \tag{60}$$

To find the relation between the components on space-time of these equations, as observed before instead of inverting the supermatrix E^A_α we project Eqs. (55), (56), (57) on the space-time 2-form differentials $dx^\mu \wedge dx^\nu$. We obtain

$$R^{ab}_{\mu\nu} = \tilde{R}^{ab}_{cd} V^c_\mu V^d_\nu + \overline{\Theta}^{ab}_c \psi_{[\mu} V^c_{\nu]}, \tag{61}$$
$$\hat{T}^a_{\mu\nu} = \hat{T}^a_{bc} V^b_\mu V^c_\nu, \tag{62}$$
$$\rho_{\mu\nu} = \rho_{ab} V^a_\mu V^b_\nu. \tag{63}$$

We have already seen that the space-time components of $\hat{T}^a$ and ρ curvatures are obtained as usual converting rigid indices in curve ones using the bosonic vierbein V^a_μ since their parametrization does not contain ψ fields. Instead the space-time components of the Lorentz curvature expanded along the differentials of the coordinates, namely $R^{ab}_{\mu\nu}$ do not coincide with the components along $(V^c \wedge V^d)$. In fact writing Eq. (61) as

$$\tilde{R}^{ab}_{\mu\nu} = R^{ab}_{\mu\nu} - \overline{\Theta}^{ab}_c \psi_{[\mu} V^c_{\nu]}, \tag{64}$$

where $\tilde{R}^{ab}_{\mu\nu} \equiv R^{ab}_{cd} V^c_\mu V^d_\nu$, we see that the Einstein equation of motion, written in terms of the $\tilde{R}^{ab}_{\mu\nu}$, contains extra terms linear in the inner components $\rho_{rs} \equiv \rho_{\mu\nu} V^\mu_r V^\nu_s$. These terms give rise to the energy-momentum tensor of the gravitino field ψ_μ.

3.2. *Invariance of the Lagrangian*

Let us now check the supersymmetry invariance of the Lagrangian. In the geometric approach, the supersymmetry invariance of the Lagrangian is expressed by the vanishing of the Lie derivative of the Lagrangian for infinitesimal diffeomprphisms in the fermionic directions of superspace. Using the Lie derivative with a tangent vector $\boldsymbol{\epsilon} = \epsilon^\alpha \boldsymbol{D}_\alpha$, where $\boldsymbol{D}_\alpha$ is the tangent vector dual to ψ^β, we must have

$$\delta_\epsilon \mathcal{L} \equiv \ell_\epsilon \mathcal{L} = \iota_\epsilon d\mathcal{L} = 0, \tag{65}$$

where we have discarded the total derivative term $d(\iota_\epsilon \mathcal{L})$ and other possible exact 4-forms on the right-hand side as we are assuming that the fields vanish at infinite so that any exact form does not contribute to the action.[p] Taking into account the definition (35), a simple computation gives

$$\begin{aligned} d\mathcal{L} &= 2R^{ab}\hat{T}^c V^d \epsilon_{abcd} + \mathrm{i}\, R^{ab}\bar{\psi}\gamma^c\psi V^d \epsilon_{abcd} \\ &+4\bar{\rho}\gamma^5\gamma_a\rho V^a + \bar{\psi}\gamma^5\gamma_c\gamma_{ab}\psi R^{ab} V^c - 4\bar{\psi}\gamma^5\gamma_a\rho\hat{T}^a - 2\,\mathrm{i}\bar{\psi}\gamma^5\gamma_a\rho\bar{\psi}\gamma^a\psi, \end{aligned} \tag{66}$$

where we have used the Bianchi identities and $\iota_\epsilon\psi = \epsilon$, $\iota_\epsilon V^a = 0$. Using the Fierz identity for 1-form spinors $\gamma^a\psi\bar{\psi}\gamma_a\psi = 0$, (see Ref. 5) and performing the gamma matrix algebra, one finds:

$$d\mathcal{L} = R^{ab}\hat{T}^c V^d \epsilon_{abcd} + \bar{\rho}\gamma^5\gamma_a\rho V^a - 4\bar{\psi}\gamma^5\gamma_a\rho\hat{T}^a\,. \tag{67}$$

Finally, contracting with the tangent vector $\epsilon^\alpha \boldsymbol{D}_\alpha$, we obtain

$$\begin{aligned} \iota_{\boldsymbol{\epsilon}}\, d\mathcal{L} &= 2(\iota_\epsilon R^{ab})\hat{T}^c V^d \epsilon_{abcd} + 2R^{ab}(\iota_\epsilon \hat{T}^c) V^d \epsilon_{abcd} + 8(\iota_\epsilon\bar{\rho})\gamma^5\gamma_a\rho V^a \\ &-4\bar{\epsilon}\gamma^5\gamma_a\rho\hat{T}^a - 4\bar{\psi}\gamma^5\gamma_a(\iota_\epsilon\rho)\hat{T}^a - 4\bar{\psi}\gamma^5\gamma_a\rho(\iota_\epsilon\hat{T}^a) = d(3-\text{form}). \end{aligned} \tag{68}$$

From (68) we see that we can have an invariant action if we require constraints on the components of the curvatures. Indeed, if we set

$$\iota_\epsilon T^a = 0; \quad \iota_\epsilon \rho = 0 \tag{69}$$

and furthermore

$$2\left(\iota_\epsilon R^{ab}\right) V^d \epsilon_{abcd} + 4\bar{\epsilon}\gamma^5\gamma_c\rho = 0\,, \tag{70}$$

we find $\delta_\epsilon \mathcal{L} = 0$, that is *invariance of the Lagrangian under supersymmetry.*

We note that the requirements (69) and (70) are the same of the on-shell constraints (56) and (57) found from the equations of motion. In particular (70) gives the solution

$$\iota_\epsilon R^{ab} = \bar{\epsilon}\Theta^{ab}_c V^c, \tag{71}$$

where Θ^{ab}_c has been defined in Eq. (54).
In other words we retrieve exactly the same constraints on the curvatures as those found from the equations of motion.

We conclude that the *supergravity Lagrangian is invariant under (local) supersymmetry transformations when the superspace curvatures are defined by the Eqs. (55)–(57).* However, this restricted form of the curvatures in superspace imply that the supersymmetry transformations given below leaving the Lagrangian invariant do not form a closed algebra, unless

[p] Note that the left-hand side of Eq. (65) is not zero since we are now in the (4+4)-dimensional superspace.

one uses the equations of motion.[q] This is best understood looking at the supersymmetry transformation laws.

Indeed, since we have found the *on-shell* value of the curvatures, we may apply the Lie derivative formula (30) to write down the superspace diffeomorphisms of the gauge fields ω^{ab}, V^a, ψ using a generic tangent vector $\boldsymbol{\epsilon} = \epsilon^{ab} D_{ab} + \epsilon^a D_a + \epsilon^\alpha D_\alpha$, where the tangent vectors $D_{ab}\, D_a, D_\alpha$ are dual to the gauge field 1-forms ω^{ab}, V^a, ψ^α. We find:

$$\delta_\epsilon \omega^{ab} = (\nabla \epsilon)^{ab} + \epsilon^r V^s R^{ab}_{rs} + \overline{\Theta}^{ab}_r \psi \epsilon^r + \overline{\Theta}^{ab}_r \epsilon V^r, \tag{72}$$

$$\delta_\epsilon V^a = (\nabla \epsilon)^a, \tag{73}$$

$$\delta_\epsilon \psi^\alpha = (\nabla \epsilon)^\alpha + \epsilon^r \rho^\alpha_{rs} V^s. \tag{74}$$

Restricting ourselves to the Lie derivative along the fermionic supersymmetry parameter ϵ only, that is setting $\epsilon^{ab} = \epsilon^a = 0$, we have

$$\delta_\epsilon \omega^{ab} = (\nabla \epsilon)^{ab} + \overline{\Theta}^{ab}_r \epsilon V^r, \tag{75}$$

$$\delta_\epsilon V^a = (\nabla \epsilon)^a, \tag{76}$$

$$\delta_\epsilon \psi^\alpha = (\nabla \epsilon)^\alpha. \tag{77}$$

Here the symbol ∇ denotes the *gauge covariant derivative* of the coadjoint multiplet of OSp(1|4). The Lorentz content of the gauge covariant derivative when acting on the OSp(1|4) adjoint multiplet can be read off directly from the Bianchi identities (39). Indeed both the parameters ϵ^A and the curvatures are in the coadjoint multiplet of the supergroup. Therefore:

$$\delta^{(gauge)}_\epsilon \omega^{ab} = (\nabla \epsilon)^{ab} = \mathcal{D}\epsilon^{ab}, \tag{78}$$

$$\delta^{(gauge)}_\epsilon V^a = (\nabla \epsilon)^a = \mathcal{D}\epsilon^{ab} + \epsilon^{ab} V_b - \mathrm{i}\bar{\psi}\gamma^a \epsilon, \tag{79}$$

$$\delta^{(gauge)}_\epsilon \psi = (\nabla \epsilon) = \mathcal{D}\epsilon - \frac{1}{4}\epsilon^{ab}\gamma_{ab}\psi, \tag{80}$$

where $\mathcal{D}$ denotes the Lorentz covariant derivative. Setting again $\epsilon^{ab} = \epsilon^a = 0$ and substituting in (75), (76), (77) we find the final form of the supersymmetry transformations:

$$\delta_\epsilon \omega^{ab} = \overline{\Theta}^{ab}_r \epsilon V^r \tag{81}$$

$$\delta_\epsilon V^a = -\mathrm{i}\bar{\psi}\gamma^a \epsilon, \tag{82}$$

$$\delta_\epsilon \psi = \mathcal{D}\epsilon\,. \tag{83}$$

Now we recall that the Lie derivative along tangent vectors $\tilde{T}_A$ satisfy an algebra isomorphic to the Lie algebra of the vector fields $[\tilde{T}_A, \tilde{T}_B] = \left(C^A_{BC} + R^A_{BC}\right) \tilde{T}_C$, namely

$$[\ell_{\tilde{T}_A}, \ell_{\tilde{T}_B}] = \ell_{[\tilde{T}_A, \tilde{T}_B]}, \tag{84}$$

if the supercurvatures R^A_{BC} are completely general, that is if they do not satisfy any constraint. In our case they satisfy the constraints (55)–(57) and in general the Lie derivative

[q] As it is well known there exist theories in $D = 4$ and $D = 5$ which admit *auxiliary fields*, that is fields that added to the coadjoint supermultiplet make the supersymmetry transformations, besides leaving the Lagrangian invariant, to close the supersymmetry algebra *off-shell*. This is related to the fact that their introduction pairs the number of off-shell degrees of freedom between boson and fermions. Moreover they are not dynamical as their equations of motion make them to vanish. However, it does not seem possible to extend their introduction to higher-dimensional theories nor to matter coupled supergravities. Therefore we do not treat them in this short review.

algebra, namely the algebra of supersymmetry transformations, *cannot close off-shell.*[r] Actually, as will be discussed in subsection 3.3, requiring that the Bianchi identities on the constrained curvatures be satisfied, one finds that their components on the bosonic cotangent plane R^A_{rs} satisfy the equations of motion of the theory. It follows that the supersymmetry algebra of the transformations leaving the Lagrangian invariant, associated to the two tangent vectors $\epsilon^\alpha D_\alpha$, will in general only *close on-shell*, that is, only if the equations of motion are satisfied.

3.3. *Supersymmetry as diffeomorphisms in superspace and rheonomy*

Let us now discuss the results obtained so far.

- Even if the supercurvatures $\hat{T}^a$ and ρ, Eqs. (56) and (57), respectively, have no components along the fermionic vielbein ψ, a non-vanishing component along $\psi \wedge V^a$ does appear in the on-shell value of the Lorentz supercurvature, that is (54). *This is sufficient to exclude factorization of the odd fermionic coordinates.* Indeed its presence makes the supersymmetry transformation a *diffeomorphism* in superspace and *not a gauge transformation.*
  It must also be noted that the absence of such fermionic components in the (on-shell) gravitino curvature ρ implies that the supersymmetry variation of ψ, given in Eq. (83), makes the transformation of the gravitino gauge field the same as if the Lagrangian were invariant under supersymmetry gauge transformations. However, as the supersymmetry transformations of the Lagrangian do not close an algebra, the gravitino transformation law is actually a diffeomorphism, and the Lagrangian cannot be a true gauge symmetry because of the absence of factorization, as we have previously shown.[s]
  The point is that such behavior of the gravitino transformation law is due to the very simple form of the minimal $\mathcal{N} = 1$, $D = 4$ pure supergravity. Any other supergravity with $\mathcal{N} > 1$ or $D > 4$ or even the same theory $\mathcal{N} = 1$, $D = 4$ coupled to matter multiplets exhibits a gravitino curvature with components $\rho_{(1,1)} \neq 0$ so that the $\delta_\epsilon \psi$ will have, besides the Lorentz covariant derivative of the supersymmetry parameter, also terms along $\psi \wedge V^a$.
  As an example, let us consider $\mathcal{N} = 2$, $D = 4$ pure supergravity. Here the supergroup is $\overline{\mathrm{OSp}(2|4)}$. The coadjoint gauge supermultiplet is now given by $\mu^A = (\omega^{ab}, V^a, \psi_A, \mathcal{A})$, where $\mathcal{A}$ is a U(1) gauge field 1-form and the index A enumerates the gravitinos in the two-dimensional representation of SO(2). The definition of the associated supercurvatures are obtained as always starting from the Maurer–Cartan equations dual to the algebra of the (anti-)commutation generators and deforming the left-invariant 1-forms into non left-invariant ones. Without giving the derivation, we write, besides the definitions of the supercurvatures on the left-hand side, also their on-shell parametrization as found from the analysis of the

[r]Unless the constraints coincide with horizontality of the full set of curvatures as it happens for the Lorentz gauge invariance.

[s]Note that if the Lagrangian were invariant under supersymmetry gauge transformations the superfields would only depend on the x^μ coordinates.

equations of motion:

$$
\begin{aligned}
R^{ab} &\equiv d\omega^{a}{}_{b} + \omega^{a}{}_{c}\omega^{c}{}_{b} \\
&= \tilde{R}^{ab}_{cd}V^{c}V^{d} + \overline{\Theta}^{ab}_{A|c}\psi^{A}V^{c} - \bar{\psi}_{A}\left(F^{ab} + \mathrm{i}^{*}F^{ab}\gamma^{5}\right)\psi_{B}\epsilon^{AB}, && (85)\\
\hat{T}^{a} &\equiv \mathcal{D}V^{a} - \frac{\mathrm{i}}{2}\bar{\psi}_{A}\gamma\psi^{A} = 0, && (86)\\
F &\equiv \mathcal{F} + \epsilon^{AB}\bar{\psi}_{A}\psi_{B} = F_{ab}V^{a}V^{b}, && (87)\\
\rho_{A} &\equiv \mathcal{D}\psi_{A} = \tilde{\rho}_{A|ab}V^{a}V^{b} + \left(\gamma^{a}F_{ab} + \mathrm{i}\gamma^{5}\gamma^{a*}F_{ab}\right)\epsilon_{AB}\bar{\psi}^{B}V^{b}, && (88)
\end{aligned}
$$

where $\mathcal{F} = dA$ and F is the supercurvature, and ${}^{*}F_{ab}$ is the Hodge dual of F_{ab}. The important thing to note is that the parametrization of the curvature 2-forms given by the equations of motion are all given in terms of their inner components, namely $\tilde{R}^{ab}_{cd}$, $\tilde{\rho}_{A|ab}$, and F_{ab} ($\hat{T}^{a}_{bc}$ is zero).[t]

Since the on-shell values of the supercurvatures is known, the supersymmetry transformation laws of the coadjoint supermultiplet, now containing also $\mathcal{A}$, can be obtained at once from the general formula (29). Looking at the Lie derivative formula, we see that the transformation laws of the multiplet of fields can be simply obtained performing the contraction of the on-shell curvatures with respect to the tangent vector $\bar{\epsilon}\, D$ and adding to the gravitino transformation the Lorentz covariant derivative of the supersymmetry parameter as it happens in the $\overline{\mathrm{OSp}(1|4)}$ case. We find:

$$
\begin{aligned}
\delta_{\epsilon}\omega^{ab} &= \overline{\Theta}^{ab}_{A|r}\epsilon^{A}V^{r}, && (89)\\
\delta_{\epsilon}V^{a} &= -\mathrm{i}\bar{\psi}_{A}\gamma^{a}\epsilon^{A}, && (90)\\
\delta_{\epsilon}\psi_{A} &= \mathcal{D}\epsilon_{A} + \mathrm{i}\,\epsilon_{AB}F^{ab}V^{b}\gamma^{a}\epsilon_{B} + \mathrm{i}\,\frac{1}{2}\epsilon_{AB}\epsilon_{abcd}F^{cd}V^{b}\gamma^{5}\gamma^{a}\epsilon^{B}, && (91)\\
\delta_{\epsilon}\mathcal{A} &= 2\epsilon^{AB}\bar{\psi}_{A}\,\epsilon_{B}. && (92)
\end{aligned}
$$

From this example we see that in general not only the Lorentz curvature R^{ab}, but also the other supercurvatures have non-vanishing components along the ψ-directions.

- We can now resume our analysis in the following way:
  Supersymmetry can be interpreted geometrically as the requirement that the superspace equations of motion imply that *the outer components of the super-curvatures are expressible algebraically (actually linearly) in terms of the components along two inner vielbein.* As already mentioned this property has been called *rheonomy.* Note that *rheonomy is just a geometrical interpretation of supersymmetry originally introduced on space-time.* Explicitly, the occurrence of *rhenomy* can be written as follows:

$$R^{A}_{\alpha\, C} = C^{A|mn}_{\alpha\, C|B}\, R^{B}_{mn}, \tag{93}$$

where $C^{A|mn}_{\alpha\, C|B}$ are suitable invariant tensors of the supergroup $\tilde{\mathrm{SG}}$ defining the basic superalgebra on which the theory is constructed, $\tilde{\mathrm{SG}} = \overline{\mathrm{OSp}(1|4)}$ in our case. The geometric meaning of this property can be better understood if we use the Lie

[t]Note that F_{ab} has no tilde since F has components only along $V^{a}\, V^{b}$.

derivative formula (30) in superspace. Inserting (93) in the Lie derivative formula (30) for a supergroup $S\tilde{G}$ we obtain:

$$\delta\mu^A = (\nabla\epsilon)^A + 2\bar{\epsilon}\, C^{A|mn}_{\alpha\, C|B}\, R^B_{mn}. \tag{94}$$

On the other hand, the Lie derivative can be interpreted either from the *passive* or from the *active* point of view. From the passive point of view the supersymmetry transformation along the $\epsilon^\alpha = \delta\theta^\alpha$ parameter is interpreted as the lift from a given $\mathcal{M}_4$ to an infinitesimally close $\mathcal{M}'_4$ which does not change the physical content of the theory, since it is described by the same Lagrangian after a supersymmetry transformation (and a Lorentz gauge transformation) has been made.[u] From the active point of view, however, it transforms a given configuration on $\mathcal{M}_4$, which we can take as space-time, setting $\theta^\alpha = \delta\theta^\alpha = 0$, to another physically equivalent configuration on the same space-time hypersurface. This property allows us to restrict the theory, the Lagrangian, and the equations of motion to any such arbitrarily chosen hypersurface $\mathcal{M}_4\,(\theta^\alpha = d\theta^\alpha = 0)$ embedded in superspace and identified with space-time.

One can now appreciate why we have illustrated in detail the mechanism of the Lorentz coordinate factorization in the gravity case defined on the Poincaré manifold.
Actually the interpretation of the rheonomy mechanism just illustrated is quite analogous to the interpretation of Lorentz transformations for gravity constructed directly on a group manifold. Indeed, in the case of pure gravity, we have seen that a transfer of information from any $\mathcal{M}_4 \subset \tilde{G}$ to any other $\mathcal{M}'_4 \subset \tilde{G}$ implies a $SO(1,3)$ transformation or, equivalently, a change of Lorentz configuration on the fixed space-time hypersurface.
On the other hand, in our example of $\mathcal{N} = 1$, $D = 4$ supergravity, besides deducing the factorization of the Lorentz coordinates exactly as in the pure gravity case, we have further illustrated that the equations of motion allow us to deduce that the transfer of information concerns not only Lorentz gauge transformations but, what is our main goal, also *supersymmetry.*
However the difference between $SO(1,3)$ transformations and supersymmetry is that in the first case, due to the horizontality of the curvatures in the Lorentz directions, ω^{ab} the supergroup $\tilde{G}$ acquires the structure of the fiber bundle $[\tilde{G}/H, H]$, and the Lie derivative reduces to a Lorentz gauge transformation. On the other hand, in the case of supersymmetry, curvatures are not horizontal along the ψ gauge fields and the Lie derivative gives the geometric interpretation of supersymmetry. Actually, it is the rheonomic mechanism one is interested in, and in fact, quite generally, in the construction of any supergravity theory the fiber bundle structure with a Lorentz fiber is assumed a priori as it can be considered of academic interest to obtain it from the variational principle. In a way, restricting a supergravity theory to a factorized superspace $\tilde{G}/SO(1,3)$ includes on a $\mathcal{M}_4$ slice, identified as space-time, all possible supersymmetry related Lagrangians.

In conclusion, the entire physics is contained in any single $\mathcal{M}_4$ or, equivalently, the supersymmetry transformations relate the fields on $\mathcal{M}_4$ to the fields on any other

[u]The passive interpretation of the Lie derivative explains the world *rheonomy* given to this geometrical interpretation of supersymmetry. Indeed, referring to the lift $\mathcal{M}_4 \to \mathcal{M}'_4$, in ancient Greek "rhein" means flow and "nomos" means law.

submanifold $\mathcal{M}'_4$. It must be kept in mind that, since supersymmetry is a Lie derivative (super-diffeomorphism) in superspace, it *is not a gauge symmetry.* Indeed, as we have seen, its algebra does not even close off-shell.

3.4. *Building rules of a general Lagrangian*

Our previous detailed examples give us simple rules for the construction of a general (geometric) Lagrangian in D dimensions:

- Starting from the Cartan–Maurer equations of a supergroup one defines the supercurvatures of the supergroup on which the theory is based. More precisely one writes the dual form of its super-Lie algebra and deforms the left-invariant 1-forms so that we can define the super-curvatures.
- The Lagrangian must be a D-form built in terms of the 1-form gauge fields μ^A and their curvatures R^A. It must contain the Einstein term $R^{ab}V^{c_1}\dots V^{c_{D-2}}\epsilon_{abc_1\dots c_{D-2}}$ whose scale is $[L^{D-2}]$. All the other terms should also scale as the Einstein term. At this point the Lagrangian contains a set of undetermined coefficients which will be fixed from the superspace equations of motion.
- The Hodge operator must be absent; the kinetic terms of scalar and vector fields must be written in first order formalism, as shown in the example of Section 2.1.
- All the possible terms satisfying the previous requirements must be present.
- If gauge invariance under an H subgroup of $\tilde{\mathrm{G}}$ is imposed a priori, where H is the gauge group of the theory containing the Lorentz group as a factor $\mathrm{H} = \mathrm{SO}(1, \mathrm{D}-1)\otimes\mathrm{H}'$, the action is obtained by integrating the Lagrangian on a (bosonic) hypersurface embedded in superspace, defined as the Deformed coset $\tilde{\mathrm{G}}/\mathrm{H}$.[v] Alternatively, we could start integrating on the whole $\tilde{\mathrm{G}}$ and obtaining the factorization of the coordinates of H as field equations. Since factorization of the gauge group H is actually always true if the Lagrangian is H-invariant, the customary way to proceed is to start with a H invariant Lagrangian on superspace $\tilde{\mathrm{G}}/\mathrm{H}$.
- The field equations must reduce to identities if all the curvatures are zero, that is if we are in the vacuum configuration with left-invariant 1-forms $\mu^A = \sigma^A$.

The field equations derived from such Lagrangian give equations of two types:

(i) Field equations relating *outer* components of the curvatures linearly in terms of the *inner* ones. These are the rheonomic conditions equivalent to saying that the Lagrangian is supersymmetric. Indeed as we have seen supersymmetry is an invariance of the Lagrangian under diffeomorphisms in superspace and the the *rhenomic* relations simply express the fact that the theory is independent of the space-time identification of the hypersurface $\mathcal{M}_D$ embedded in superspace. Note that in solving these equations all undetermined coefficients of the various terms become fixed.

[v]Indeed in most supergravity theories we may have a larger group of gauge invariance other than the Lorentz one. When this happens, and we want to start from the full supergroup manifold, the factorization of the extra coordinates belonging to H′ can be obtained from the action principle exactly as for the Lorentz group coordinates.

(ii) Field equations which are differential equations in superspace. Restricted to the bosonic hypersurface $\mathcal{M}_D$ they are the space-time equations of motion.

3.5. *The role of the Bianchi identities*

Till now we have extensively explained how to give a geometrical interpretation to the supersymmetry transformations of supergravity either constructed on superspace, when factorization of the Lorentz coordinates is assumed a priori, or directly starting from the group manifold locally identified by the underlying supersymmetry algebra.
There is however an equivalent and powerful approach to the construction of the equations of motion and transformation laws which is based on a systematic use of the Bianchi identities *assuming rheonomy* from the very beginning.

To understand this point we note that the Bianchi identities are true identities only if no constraint is assumed among the supervielbein components of the curvatures. However what rheonomy does is exactly to give relations among the outer components $R^A_{B,\alpha}$ (those along at least one fermionic vielbein ψ) and inner components R^A_{ab} (namely the components on the cotangent space to space-time). Moreover, in almost any case, one also assumes a further constraint called kinematical constraint, namely vanishing super-torsion $\hat{T}^a = 0$. The Bianchi identities then assume the form of differential constraints among the space-time components. These differential constraints, on the other hand, can be nothing else than the equations of motion, since Bianchi identities cannot conflict with the differential equations obtained from the Lagrangian. Once the field equations are obtained, the Lagrangian, if desired, can be easily reconstructed. In the actual computations one usually couples the two methods, namely the Lagrangian approach and the Bianchi identities equations, to arrive in the simplest way to the final determination of the parametrization of the curvatures in superspace (and thus to the supersymmetry transformation laws) and to the determination of all the coefficients in the Lagrangian.

4. Results of the geometric approach and some remarks

Most of the previous considerations have been dedicated to the geometrical interpretation of supersymmetry and to the explicit geometric construction of a supergravity theory. A natural question is now what has been the impact of this kind of approach from the physical point of view. We cannot of course enter in a detailed exposition of the results obtained from the very beginning till nowadays. We limit ourselves to give some remarks and observations concerning its power in treating higher-dimensional supergravities and matter coupled theories containing antisymmetric tensor fields, together with some unexpected properties of superspace. As far as the most relevant results obtained in the geometric approach is concerned, we limit ourself to give a short list of them in the Appendix. Here is a couple of short and hopefully interesting comments:

- Among the most interesting results there is certainly the reduction of theories containing antisymmetric tensors to equivalent theories formulated in terms of super-Lie algebras. This result was obtained in Ref. 2 from the analysis of the geometrical formulation of $D = 11$ supergravity formulated on space-time in Ref. 6. As $D = 11$ supergravity is thought to be the low energy limit of the so-called M-theory, this result can have a special relevance for a better understanding of the group theoretical structure of M-theory. Because of its importance, we shall devote the next section to a short review of this approach in the case of $D = 11$ supergravity.

- It is undeniable that the systematic use of the geometric and group-theoretical approach has been an essential tool to obtain many other interesting results. For example, we can say that, in general, the introduction of matter coupling to pure supergravities allows to put in light the global and local symmetries inherent nonlinear interaction structure of the coupling to matter multiplets. Indeed very often the use of the geometrical approach has allowed to arrive to a complete answer to problems where other approaches often had given only limited answers.[w]
A typical example was the full construction of the $\mathcal{N} = 2$, $D = 4$ matter coupled supergravity,[5,9,14] which was previously formulated using the superconformal approach in a coordinate dependent way.[10] The geometrical approach provides a complete Lagrangian and transformation laws quite independently of the coordinate used for the scalar fields description of the σ model and it makes the introduction of the notion of Special Geometry, the geometry of the Special and Quaternionic manifolds, the momentum maps and the related gaugings, together with a complete description of the scalar potential, very natural. These results also give insight into the related superconformal two-dimensional theories and Calabi–Yau compactifications in string theory. Further developments are also to be found in Ref. 7.
- Another interesting observation is the following: The geometric approach discussed in this pedagogical review is naturally formulated in superspace. One could ask whether this approach is exactly equivalent to the purely space-time approach. This seems not be the case in some theories, like $\mathcal{N} = 1$, $D = 6$,[11] and $D = 10$, IIB.[12]
In the $D = 6$ supergravity the field multiplet contains the sechsbein, a Weyl gravitino, and a 2-form (that is an antisymmetric two-index tensor), it was shown that a consistent theory on the group manifold might have no counterpart in the usual Noether approach. In our geometric $D = 6$ superspace model, the self-duality of the 2-form field-strength, necessary to match the Bose–Fermi on-shell degrees of freedom, follows from group manifold/superspace variational equations, but not from their x-space restriction. As a consequence, the theory is consistent, although the x-space Lagrangian is not supersymmetry invariant. Exactly in the same way can be treated the $D = 10$, IIB theory, so that also in this case the self-duality of the 5-form can be retrieved from the superspace equations of motion. These results hint to extra properties of superspace which can be traded on the embedded hypersurface $\mathcal{M}_4$ only after the superspace equations of motion have been implemented, that is they are not visible using a purely space-time approach.

[w]Most of these developments and results can be found in the excellent review.[7]

5. Higher p-forms supergravities and their hidden supergroups

We have often stressed that the mechanism of rheonomy actually holds in all supergravities, independently of the number of supersymmetries, the dimensionality of space-time, and their matter couplings, if any. However, apart from few exceptions, most of the higher dimensionality theories have a gravitational multiplet containing antisymmetric tensors of higher rank, mostly of rank two. Similarly, matter supermultiplets also can have higher rank tensors. In these cases the group manifold interpretation presented before as a possible starting point for supergravities whose fields are defined on a group manifold cannot be maintained. Indeed the coadjoint multiplet of a (super-) group consists of 1-forms dual to the group generators, with no room for higher p-forms.
In the present section we will show, referring mainly to the case of $D = 11$ supergravity where this development was first presented,[2] that:

- The Maurer–Cartan equations can be generalized to more general structures, called Free Differential Algebras (FDAs), admitting in their multiplet also forms of degree higher than one. They represent a natural extension of Lie algebras in their dual formulation and can accommodate supermultiplets containing higher p-forms satisfying the integrability requirement $d^2 = 0$.
- Each higher p-form $A^{(p)}$ can be decomposed in terms of a set of trilinear (wedge) products of p 1-forms, where besides the supervielbein basis (V^a, ψ) there appear new 1-forms valued in tensor or spinor representations of the Lorentz group. The new 1-forms obey extra Maurer–Cartan equations, besides those of the super-Poincaré group. The decomposition can be done in such a way that the coefficients of the polynomial written as a sum of products of p 1-forms assure the integrability of the original FDA equation for the p-form $A^{(p)}$.
- Together with the super-Poincaré dual generators, the new Maurer–Cartan equations describe the dual form of a new super-Lie algebra which, at least locally, describes a group manifold called the *hidden supergroup* of the FDA which in a sense can be considered as the group-theoretical starting point for a construction of the supergravity theories possessing higher p-forms in their gravitational multiplet.
- Among the new 1-forms needed to assure that the given decomposition reproduces the integrable equation of the FDA, there appear extra spinor 1-forms (one in the case of $D = 11$ supergravity) whose dual generators Q'_α in the Lie superalgebra are nilpotent. Their role, as recently clarified,[13,14] is to assure that the new 1-form fields thus introduced are gauge fields living on the fiber bundle whose base space is ordinary superspace, so that their dependence on the new coordinates are completely factorized. This means that their curvatures are horizontal and do not add new degrees of freedom other than those already present in the original FDA. This property works exactly in the same way for all higher-dimensional theories with $D < 11$ whose gravitational multiplet contain antisymmetric tensor fields, for example in $\mathcal{N} = 2$, $D = 7$ supergravity where two such nilpotent spinor generators are present.

These results were obtained in[2] by R. D'Auria and P. Fré in the case of the maximal

$D = 11$ supergravity trying to give a fully geometrical interpretation of the space-time formulation of the theory.[6] Indeed, in this theory there appears an antisymmetric tensor of rank three in the gravitational multiplet. In their approach, R. D'Auria and P. Fré introduced for the first time the generalization of the Maurer–Cartan equations for integrable systems containing higher p-forms which they called Cartan Integrable Systems (CIS). Only later it was realized that structures of this kind were already introduced in mathematics and called *Free Differential Algebras*,[15] which is the name now universally accepted.
Even if we shall not give any account of the underlying mathematics, it must be said that the relation between the FDA and groups or supergroups relies on the *Chevalley–Eilenberg (super)-Lie algebra cohomology groups.* The procedure we previously alluded to of decomposing a higher p-form in a polynomial of Lorentz valued 1-forms is the inverse of the construction of a FDA starting from a (super-)Lie algebra, which, as far as I know, was not treated in the Lie algebra cohomology theory.

In the following, we give a short account of the FDA of $D = 11$ supergravity and its resolution as a *hidden* ordinary Lie supergroup.

6. Free differential algebra and hidden supergroup of $D = 11$ supergravity

Eleven-dimensional supergravity can be founded on the following FDA:

$$R^{ab} \equiv d\omega^{ab} - \omega^{ac} \wedge \omega_c^{\ b} = 0\,, \tag{95}$$

$$T^a \equiv DV^a - \frac{\mathrm{i}}{2}\overline{\Psi} \wedge \Gamma^a \Psi = 0\,, \tag{96}$$

$$\rho \equiv D\Psi = 0\,, \tag{97}$$

$$F^{(4)} \equiv dA^{(3)} - \frac{1}{2}\overline{\Psi} \wedge \Gamma_{ab}\Psi \wedge V^a \wedge V^b = 0\,, \tag{98}$$

$$F^{(7)} \equiv dB^{(6)} - 15A^{(3)} \wedge dA^{(3)} - \frac{\mathrm{i}}{2}\overline{\Psi} \wedge \Gamma_{a_1\dots a_5}\Psi \wedge V^{a_1} \wedge \dots \wedge V^{a_5} = 0\,. \tag{99}$$

Note that this differential system is an extension of the usual Maurer–Cartan equations and as such it only describes the structure of the physical vacuum of the theory.
In our case the super-Poincaré Maurer–Cartan equations in $D = 11$ with the addition of the higher order differential of the 3-form $A^{(3)}$ and 6-form $B^{(6)}$.[x]
The consistency of the FDA requires the integrability of the last two equations, $d^2A^{(3)} = 0$, $d^2B^{(6)} = 0$. It can be shown that this is in fact satisfied as a consequence of 3-fermion Fierz identities obeyed by the gravitino 1-form field in eleven dimensions (see e.g. Ref. 5).[y]

[x]In the original paper the last Eq. (99) was not present. Actually, it was almost immediately realized (see e.g. Ref. 5, Vol. 2) that, besides the simplest FDA including as exterior form only $A^{(3)}$, one can extend the FDA to include also a (*magnetic*) 6-form potential $B^{(6)}$, related to $A^{(3)}$ by Hodge-duality of the corresponding field-strengths on space-time. Indeed there is a constructive procedure based on the Chevalley–Eilenberg Lie algebra cohomology to arrive to the maximal extension of the super-Poincaré algebra given by the FDA which implies as a maximal extension the presence of the $B^{(6)}$.[5]

[y]Note that here and in the following we do not elaborate on the theory out of vacuum, namely the interacting theory, since the topological (and cohomological) structure of the theory, which will be the object of the present investigation, is fully caught by the ground state of the FDA.

The authors of[2] asked themselves whether one could trade the FDA structure on which the theory is based with an ordinary Lie superalgebra, written in its dual Cartan form, that is in terms of 1-form gauge fields which turn out to be valued in non-trivial tensor and spinor representations of Lorentz group SO(1, 10). This would allow to disclose the fully extended Lie superalgebra hidden in the supersymmetric FDA. This was proven to be true, and the hidden superalgebra underlying the FDA describing $D = 11$ supergravity was presented for the first time.

It was shown that this is indeed possible by associating, to the forms $A^{(3)}$ and $B^{(6)}$, the bosonic 1-forms B_{ab} and $B_{a_1 \cdots a_5}$, in the antisymmetric representations of SO(1, 10), and furthermore an extra *spinor* 1-form η. The Maurer–Cartan equations satisfied by these new 1-forms are:

$$\mathcal{D}B_{a_1a_2} = \frac{1}{2}\overline{\Psi} \wedge \Gamma_{a_1a_2}\Psi, \tag{100}$$

$$\mathcal{D}B_{a_1\ldots a_5} = \frac{i}{2}\overline{\Psi} \wedge \Gamma_{a_1\ldots a5}\Psi \tag{101}$$

$$\mathcal{D}\eta = iE_1\Gamma_a\Psi \wedge V^a + E_2\Gamma^{ab}\Psi \wedge B_{ab} + iE_3\Gamma^{a_1\cdots a_5}\Psi \wedge B_{a_1\ldots a_5}\,, \tag{102}$$

$\mathcal{D}$ being the Lorentz-covariant derivatives. Of course the whole consistence of this approach also requires the d^2 closure of the Maurer–Cartan newly introduced fields B_{ab}, $B_{a_1\cdots a_5}$ and η. For the two bosonic 1-form fields the d^2 closure is obvious in the ground state, because of the vanishing of the curvatures R^{ab} and $\rho = \mathcal{D}\psi$, while $\mathcal{D}^2\eta = 0$ requires the further condition:

$$E_1 + 10E_2 - 720E_3 = 0\,, \tag{103}$$

which can be derived using the Fierz identities of the wedge product of three gravitino 1-forms in superspace.[2]
In Ref. 2 the most general decomposition of the 3-form $A^{(3)}$ in terms of the 1-forms B_{ab}, $B_{a_1\ldots a_5}$ and η was presented. It has the following form:

$$\begin{aligned} A^{(3)} &= T_0B_{ab} \wedge V^a \wedge V^b + T_1B_{ab} \wedge B^b_{\ c} \wedge B^{ca} \\ &+ T_2B_{b_1a_1\ldots a_4} \wedge B^{b_1}_{\ b_2} \wedge B^{b_2a_1\ldots a_4} + T_3\epsilon_{a_1\ldots a_5b_1\ldots b_5m}B^{a_1\ldots a_5} \wedge B^{b_1\ldots b_5} \wedge V^m \\ &+ T_4\epsilon_{m_1\ldots m_6n_1\ldots n_5}B^{m_1m_2m_3p_1p_2} \wedge B^{m_4m_5m_6p_1p_2} \wedge B^{n_1\ldots n_5} \\ &+ iS_1\overline{\Psi}\Gamma_a\eta \wedge V^a + S_2\overline{\Psi}\Gamma^{ab}\eta \wedge B_{ab} + iS_3\overline{\Psi}\Gamma^{a_1\ldots a_5}\eta \wedge B_{a_1\ldots a_5}\,, \end{aligned} \tag{104}$$

where $\mathrm{T_i}$ and $\mathrm{S_j}$ are numerical coefficients. To show the equivalence of the FDA with a ordinary super-Lie algebra (in dual form) it is required that the integrability condition in superspace of the 3-form, $dA^{(3)}$, computed in terms of differentials of the new 1-forms gives the same results as in the case of Eq. (98), namely $dA^{(3)} - \frac{1}{2}\overline{\Psi} \wedge \$\tilde{\mathrm{G}}\mathrm{amma_{ab}}\Psi \wedge \mathrm{V^a} \wedge \mathrm{V^b} = 0$. To obtain such integrability the extra terms containing the currents involving the *extra spinor 1-form η turn out to be necessary.* The Ansatz (104) restricted to the bosonic 1-forms does not work. In other words the inclusion of the spinor 1-form field η enters in the decomposition of the 3-form $A^{(3)}$ in such a way to properly reproduce the vacuum FDA on ordinary superspace.

When the integrability is implemented all the coefficients in the decomposition become fixed in terms of the ratio E_3/E_2.[z]

[z]In[2] the first coefficient T_0 was arbitrarily fixed to $T_0 = 1$ giving only 2 possible solutions for the

In this way, one arrives at the following full set of Maurer–Cartan equations for the left-invariant 1-forms $(\omega^{ab}\,, V^a\,, \psi\,, B_{ab}\,, B_{a_1\dots a_5}\,, \eta)$:

$$\begin{aligned} d\omega^{ab} &= \omega^{ac} \wedge \omega_c^{\ b}, && (105)\\ DV^a &= \frac{\mathrm{i}}{2}\overline{\Psi} \wedge \Gamma^a \Psi, && (106)\\ D\Psi &= 0, && (107)\\ DB_{a_1 a_2} &= \frac{1}{2}\overline{\Psi} \wedge \Gamma_{a_1 a_2}\Psi, && (108)\\ DB_{a_1\dots a_5} &= \frac{\mathrm{i}}{2}\overline{\Psi} \wedge \Gamma_{a_1\dots a_5}\Psi, && (109)\\ D\eta &= \mathrm{i}E_1 \Gamma_a \Psi \wedge V^a + E_2 \Gamma^{ab}\Psi \wedge B_{ab} + \mathrm{i}E_3 \Gamma^{a_1\dots a_5}\Psi \wedge B_{a_1\dots a_5}\,. && (110) \end{aligned}$$

This set of Maurer–Cartan equations identifies the supergroup which is (locally) described, as anticipated, by the hidden super-Lie algebra underlying the theory (this same superalgebra written in terms of commutators is given below).
It must be noted the if we neglect the presence of the η spinor 1-form in the decomposition (104) we can obtain a closed algebra which, however, is not equivalent to the FDA we started from since it fails to give the closure $d^2 A^{(3)} = 0$. Only when the extra currents containing η are present in (104) we obtain such integrability.
Let us finally write down the hidden superalgebra in terms of generators closing a set of (anti)commutation relations.

To recover the superalgebra in terms of (anti)commutators of the dual Lie superalgebra generators

$$T_A \equiv \{P_a, Q, J_{ab}, Z^{ab}, Z^{a_1\dots a_5}, Q'\}\,, \tag{111}$$

which are dual to the 1-forms $\left(V^a,\ \psi,\ \omega^{ab}\,, B_{ab},\ B_{a_1\dots a_5},\ \eta\right)$ respectively, one uses the duality between 1-forms and generators and finds that the $D = 11$ FDA corresponds to the following hidden superalgebra (besides the Poincaré algebra):

$$\begin{aligned} \{Q,\bar{Q}\} &= -\left(\mathrm{i}\Gamma^a P_a + \frac{1}{2}\Gamma^{ab} Z_{ab} + \frac{\mathrm{i}}{5!}\Gamma^{a_1\dots a_5} Z_{a_1\dots a_5}\right), && (112)\\ \left[Q',\bar{Q}'\right] &= 0\,,\\ [Q,P_a] &= -2\mathrm{i}E_1 \Gamma_a Q'\,,\\ \left[Q,Z^{ab}\right] &= -4E_2 \Gamma^{ab} Q'\,,\\ \left[Q,Z^{a_1\dots a_5}\right] &= -2\,(5!)\mathrm{i}E_3 \Gamma^{a_1\dots a_5} Q'\,,\\ \left[J_{ab},Z^{cd}\right] &= -8\delta^{[c}_{[a} Z_{b]}^{\ d]}\,,\\ \left[J_{ab},Z^{c_1\dots c_5}\right] &= -20\delta^{[c_1}_{[a} Z_{b]}^{\ c_2\dots c_5]}\,,\\ [J_{ab},Q] &= -\Gamma_{ab} Q\,,\\ [J_{ab},Q'] &= -\Gamma_{ab} Q'\,. \end{aligned}$$

set of parameters $\{T_i, S_j, E_k\}$. It was pointed out later in Refs. 16 and 17 that this restriction can be relaxed thus giving a more general solution in terms of one parameter. Indeed, as observed in the quoted reference, one of the E_i can be reabsorbed in the normalization of η, so that, owing to the relation (102), we are left with one free parameter, say E_3/E_2.

All the other commutators (beyond the Poincaré part) vanish. We shall identify the super-Lie algebra in either of the two dual forms given as Maurer–Cartan equations or (anti-commutators) as D'Auria–Fré algebra (DF-algebra in the following). In the Lie algebra version of the dual Maurer–Cartan equations the 1-forms $B_{a_1 \dots a_5}$ and B_{ab} are the 1-forms dual to the central generators $Z^{a_1 \cdots a_5}$ and Z^{ab}, respectively, of a central extension of the supersymmetry algebra given by the usual $D = 11$ super-Poincaré algebra and including the extra nilpotent generator Q' (dual to the spinor 1-form η).[aa] Actually, as shown in Ref. 14, from a cohomological point of view, to reproduce the integrability of $dA^{(3)}$, the presence of the 1-form $B_{a_1 \dots a_5}$ in the decomposition (104) is not necessary, since all the terms where it appears sum up to give an exact 3-form. However, as we have seen, even if cohomologically trivial, its addition allows to extend the Lie superalgebra in a non-trivial way.[bb]

6.1. $D = 11$ *supergravity and M-theory*

After several years, on the basis of different considerations, the same algebra, but *without the inclusion of the extra nilpotent generator* was rediscovered. This superalgebra, actually a subalgebra of the hidden superalgebra (112), was named M-algebra.[18–22]

The presence in the relations (105)–(110) of the bosonic hidden 1-forms $B_{ab}, B_{a_1 \dots a_5}$ can be considered as a generalization of the centrally extended supersymmetry algebra of[23] (where the central generators were associated with electric and magnetic charges), and, as such, have in fact a topological meaning. This was recognized in[18] and,[24] where it was shown they are to be associated with extended objects (2-brane and 5-brane charges, respectively) in space-time. The M-algebra is commonly considered as the super-Lie algebra underlying M-theory[25–27] in its low energy limit, corresponding to $D = 11$ supergravity in the presence of non-trivial M-brane sources.[24,28–32]

A field theory based on the M-algebra, however, is naturally described on the *enlarged superspace* whose cotangent space is spanned, besides the gravitino 1-form, also by bosonic fields $\{V^a, B_{ab}, B_{a_1 \dots a_5}\}$. If we hold on the idea that the low energy limit of M-theory should be based on the same ordinary superspace, spanned by the supervielbein (V^a, ψ), as in the original formulation of $D = 11$ supergravity,[6] then the M-algebra cannot be the final answer since it does not contain the extra 1-form η dual to the nilpotent generator fermionic generator Q'. Indeed we have shown that in order to reproduce the FDA on which $D = 11$ supergravity is based the presence of η among the 1-form generators is necessary. Actually the DF-algebra, (105)–(110) or (112), contains the M-algebra as a subalgebra since it also

[aa]Here and in the following the term "central" for the charges Z^{ab}, $Z^{a_1 \cdots a_5}$, and for the spinorial charge Q' refers to their commutators with all the generators apart from the Lorentz generator J_{ab}. The commutation relations with J_{ab} are obviously dictated by their Lorentz index structure.

[bb]More precisely in Ref. 14 it was shown that once formulated in terms of its hidden superalgebra of 1-forms, $A^{(3)}$ can be actually decomposed into the sum of two parts having different group-theoretical meaning: One of them does not depend on $B_{a_1 \dots a_5}$ and allows to reproduce the FDA describing $D = 11$ supergravity, while the second one does not contribute to the 4-form cohomology, being a closed 3-form in the vacuum; however, the second part defines a one parameter family of trilinear forms invariant under a symmetry algebra that is related to $\mathfrak{osp}(1|32)$ by redefining the spin connection and adding a new Maurer–Cartan equation. Correspondingly, also the spinor 1-form η can be analogously split into two different spinors appearing each one in just one of the two parts in which $A^{(3)}$ is decomposed.

includes a nilpotent ($Q'^2 = 0$) *fermionic generator* Q' dual to the spinor 1-form η whose contribution to the Maurer–Cartan equations of the DF-algebra is given by Eq. (110).
In other words the DF-algebra underlying the formulation of the eleven-dimensional FDA on superspace reproduces the eleven-dimensional theory on space-time introduced in Ref. 6, if and only if the decomposition of the 3-form $A^{(3)}$ *also includes the 1-form* η.
As it was shown in Ref. 13, this in turn implies that the group manifold generated by the DF-algebra has a fiber bundle structure whose base space is ordinary superspace, while the fiber is spanned, besides the Lorentz spin connection ω^{ab}, also by the bosonic 1-form generators $B_{ab}, B_{a_1,\dots a_5}$. It follows that if we would start from the group manifold generated by the DF-algebra, the coadjoint multiplet would now be $\mu^A = (\omega^{ab}, B_{ab}, B_{a_1\dots a_5}, V^a, \psi, \eta)$ and the action principle would factorize the coordinates associated to the first three 1-forms so that their degrees of freedom would not enter into the equations of motion. Indeed the presence of Q', dual to the 1-form η, allows to consider the extra 1-forms B_{ab} and $B_{a_1\dots a_5}$ as gauge fields in ordinary superspace instead of additional vielbeins of an enlarged superspace, that is, their curvatures on the fiber are *horizontal*. This is due to the dynamical cancellation of their unphysical contributions to the supersymmetry and gauge transformations with the supersymmetry and gauge transformations of η.[cc]

7. Conclusions

We have shown how the original idea formulated by Y. Ne'eman and T. Regge of defining gravity and supergravity theories directly on a (super-)group manifold $\tilde{\text{G}}$ actually allows a completely geometrical and group-theoretical formulations of any gravity or supergravity extended theory.

We have explained how the generalized action principle of integrating the Lagrangian on a submanifold of $\tilde{\text{G}}$ is capable of obtaining the following results:

Factorization of coordinates belonging to gauge subgroups of $\tilde{\text{G}}$ can be obtained by using a generalized action principle where space-time is represented by a bosonic submanifold immersed in the (super-)group, so that $\tilde{\text{G}}$ actually becomes endowed by a fiber bundle structure.

For supergroups, where the notion of superspace as base space of the fiber bundle appears, the same procedure based on the extended action principle allows to add, to the notion of factorization, the notion of supersymmetry whose geometrical meaning is that the field-strengths (curvatures) in superspace are not horizontal, but can be expressed linearly in terms of the space time field-strengths. The corresponding geometrical interpretation has been named *rheonomy*.

It is possible to give very simple *building principles* that allow an almost algorithmic procedure for the construction of any supergravity Lagrangian, the most important principle being *geometricity*.

[cc]As observed in Ref. 13, all the above procedure of enlarging the field space to recover a well defined description of the physical degrees of freedom is strongly reminiscent of the BRST-procedure, and the behavior of η is such that it can be actually thought of as a ghost for the 3-form gauge symmetry, when the 3-form is parametrized in terms of 1-forms.

Besides the original results obtained not only in supergravity, but also in related questions like duality, Calabi–Yau compactifications, mononodromies etc, an important step has been the discovery of the Free Differential Algebras as a geometrical way to formulate supergravities containing higher p-forms fields by a suitable generalization of the Maurer–Cartan equations to higher p-forms. In particular using the notion of Lie algebra cohomology it has been possible to show that any such supergravity containing higher p-forms can be reduced to an ordinary supergravity based on an ordinary Lie superalgebra.

Acknowledgments

I thank my friends and colleagues L. Andrianopoli, L. Ravera and M. Trigiante for their critical reading of the manuscript and interesting suggestions.

Appendix

We give a short list of the most relevant and interesting results obtained in the geometric approach.

Besides giving the geometrical structure of the already existing supergravity theories, some supergravity theories were first obtained in the geometric approach, see e.g. $\mathcal{N} = 3$, $D = 4$ and $\mathcal{N} = 2$, $D = 6$, F_4 matter coupled supergravities.[33,34]
Construction of central and matter charges and symplectic structure of all $D = 4$, $\mathcal{N}$ extended theories.[35]
Symplectic invariant coupling of scalar-tensors[7,36,37] and vector-tensor multiplets[38] in $\mathcal{N} = 2$ supergravity and the role of magnetic charges.
Derivation of the $\mathcal{N} = 1$ and $\mathcal{N} = 2$, $D = 4$ supergravity Lagrangians in the presence of a boundary.[39]
Unexpected interesting relation between $\mathcal{N} = 2$ supergravity in $D = 4$ and a three-dimensional theory describing the graphene electronic properties.[40]
Other relevant results are the following: Anomaly Free supergravity in $D = 10$;[41] Duality transformations in supersymmetric Yang–Mills theories coupled to supergravity;[19] A detailed analysis of the role played by the Picard–Fuchs equations in supergravity;[6,7,9] Symplectic structure of $\mathcal{N} = 2$ supergravity and its central extension;[16] Symplectic structure and monodromy group for the Calabi–Yau two moduli space[1,48] (as far as rigid supersymmetry is concerned, the necessity of introducing a Chern–Simons term in $D = 10$ super Yang–Mills $\mathcal{N} = 1$ Lagrangian was first realized using the geometric approach, see Ref. 49).
Finally, even if not concerning supersymmetry, an important result has been obtained in the theory of gravitation using a completely geometric Lagrangian coupled to a pseudo-scalar field in a non-canonical way. Using this Lagrangian it has been possible to show the existence of *symptotically flat gravitational instantons in gravity.*[50] Further results in this approach were also obtained in Ref. 51.

References

1. Y. Ne'eman and T. Regge, "Gauge Theory of Gravity and Supergravity on a Group Manifold," Riv. Nuovo Cim. **1N5** (1978) 1, and Phys. Lett. 74B (1978) 54-56;
2. R. D'Auria and P. Fré, "Geometric Supergravity in d = 11 and Its Hidden Supergroup," Nucl. Phys. B **201** (1982) 101 Erratum: [Nucl. Phys. B **206** (1982) 496].
3. H. Sati and U. Schreiber, "Lie n-algebras of BPS charges," JHEP **1703** (2017) 087 doi:10.1007/JHEP03(2017)087 [arXiv:1507.08692 [math-ph]].
4. https://ncatlab.org/nlab/show/L-infinity+algebras+in+physics
5. L. Castellani, R. D'Auria and P. Fré, "Supergravity and superstrings: A Geometric perspective," Vol. 1 and 2. Singapore: World Scientific (1991).
6. E. Cremmer, B. Julia and J. Scherk, "Supergravity Theory in Eleven-Dimensions," Phys. Lett. B **76** (1978) 409 [Phys. Lett. **76B** (1978) 409].
7. M. Trigiante, "Gauged Supergravities," Phys. Rept. **680** (2017) 1 doi:10.1016/j.physrep.2017.03.001 [arXiv:1609.09745 [hep-th]].
8. L. Andrianopoli, M. Bertolini, A. Ceresole, R. D'Auria, S. Ferrara, P. Fré and T. Magri, "N=2 supergravity and N=2 superYang-Mills theory on general scalar manifolds: Symplectic covariance, gaugings and the momentum map," J. Geom. Phys. **23** (1997) 111 doi:10.1016/S0393-0440(97)00002-8 [hep-th/9605032].
9. R. D'Auria, S. Ferrara and P. Fré, "Special and quaternionic isometries: General couplings in N=2 supergravity and the scalar potential," Nucl. Phys. B **359** (1991) 705. doi:10.1016/0550-3213(91)90077-B.
10. B. de Wit, P. G. Lauwers and A. Van Proeyen, "Lagrangians of N=2 Supergravity - Matter Systems," Nucl. Phys. B **255** (1985) 569. doi:10.1016/0550-3213(85)90154-3.
11. R. D'Auria, P. Fré and T. Regge, "Consistent Supergravity in Six-dimensions Without Action Invariance," Phys. Lett. **128B** (1983) 44. doi:10.1016/0370-2693(83)90070-9.
12. L. Castellani and I. Pesando, "The Complete superspace action of chiral D = 10, N=2 supergravity," Int. J. Mod. Phys. A **8** (1993) 1125.
13. L. Andrianopoli, R. D'Auria and L. Ravera, "Hidden Gauge Structure of Supersymmetric Free Differential Algebras," JHEP **1608** (2016) 095. doi:10.1007/JHEP08(2016)095 [arXiv:1606.07328 [hep-th]].
14. L. Andrianopoli, R. D'Auria and L. Ravera, "More on the Hidden Symmetries of 11D Supergravity," Phys. Lett. B **772** (2017) 578 doi:10.1016/j.physletb.2017.07.016 [arXiv:1705.06251 [hep-th]].
15. D. Sullivan, "Infinitesimal computations in topology". Publications Mathématiques de l'IHES, 47 (1977), p. 269-331.
16. I. A. Bandos, J. A. de Azcarraga, J. M. Izquierdo, M. Picon and O. Varela, "On the underlying gauge group structure of D=11 supergravity," Phys. Lett. B **596** (2004) 145 doi:10.1016/j.physletb.2004.06.079 [hep-th/0406020].
17. I. A. Bandos, J. A. de Azcarraga, M. Picon and O. Varela, "On the formulation of D = 11 supergravity and the composite nature of its three-form gauge field," Annals Phys. **317** (2005) 238 doi:10.1016/j.aop.2004.11.016 [hep-th/0409100].
18. J. A. de Azcarraga, J. P. Gauntlett, J. M. Izquierdo and P. K. Townsend, "Topological Extensions of the Supersymmetry Algebra for Extended Objects," Phys. Rev. Lett. **63** (1989) 2443. doi:10.1103/PhysRevLett.63.2443.
19. E. Sezgin, "The M algebra," Phys. Lett. B **392** (1997) 323 doi:10.1016/S0370-2693(96)01576-6 [hep-th/9609086].
20. P. K. Townsend, "M theory from its superalgebra," In *Cargese 1997, Strings, branes and dualities* 141-177 [hep-th/9712004].
21. M. Hassaine, R. Troncoso and J. Zanelli, "Poincare invariant gravity with local supersymmetry as a gauge theory for the M-algebra," Phys. Lett. B **596** (2004) 132 doi:10.1016/j.physletb.2004.06.067 [hep-th/0306258].

22. M. Hassaine, R. Troncoso and J. Zanelli, "11D supergravity as a gauge theory for the M-algebra," PoS WC **2004** (2005) 006 [hep-th/0503220].
23. E. Witten and D. I. Olive, "Supersymmetry Algebras That Include Topological Charges," Phys. Lett. B **78** (1978) 97. doi:10.1016/0370-2693(78)90357-X.
24. A. Achucarro, J. M. Evans, P. K. Townsend and D. L. Wiltshire, "Super p-Branes," Phys. Lett. B **198** (1987) 441. doi:10.1016/0370-2693(87)90896-3.
25. J. H. Schwarz, "The power of M theory," Phys. Lett. B **367** (1996) 97 doi:10.1016/0370-2693(95)01429-2 [hep-th/9510086].
26. M. J. Duff, "M theory (The Theory formerly known as strings)," Int. J. Mod. Phys. A **11** (1996) 5623 [Subnucl. Ser. **34** (1997) 324] [Nucl. Phys. Proc. Suppl. **52** (1997) no.1-2, 314] doi:10.1142/S0217751X96002583 [hep-th/9608117].
27. P. K. Townsend, "Four lectures on M theory," In *Trieste 1996, High energy physics and cosmology* 385-438 [hep-th/9612121].
28. E. Bergshoeff, E. Sezgin and P. K. Townsend, "Supermembranes and Eleven-Dimensional Supergravity," Phys. Lett. B **189** (1987) 75. doi:10.1016/0370-2693(87)91272-X.
29. M. J. Duff, P. S. Howe, T. Inami and K. S. Stelle, "Superstrings in D=10 from Supermembranes in D=11," Phys. Lett. B **191** (1987) 70. doi:10.1016/0370-2693(87)91323-2.
30. E. Bergshoeff, E. Sezgin and P. K. Townsend, "Properties of the Eleven-Dimensional Super Membrane Theory," Annals Phys. **185** (1988) 330. doi:10.1016/0003-4916(88)90050-4.
31. P. K. Townsend, "The eleven-dimensional supermembrane revisited," Phys. Lett. B **350** (1995) 184 doi:10.1016/0370-2693(95)00397-4 [hep-th/9501068].
32. P. K. Townsend, "P-brane democracy," In *Duff, M.J. (ed.): The world in eleven dimensions* 375-389 [hep-th/9507048].
33. L. Castellani, A. Ceresole, S. Ferrara, R. D'Auria, P. Fre and E. Maina, "The Complete $N = 3$ Matter Coupled Supergravity," Nucl. Phys. B **268** (1986) 317.
34. R. D'Auria, S. Ferrara and S. Vaula, "Matter coupled F(4) supergravity and the AdS(6) / CFT(5) correspondence," JHEP **0010** (2000) 013 doi:10.1088/1126-6708/2000/10/013.
35. L. Andrianopoli, R. D'Auria and S. Ferrara, "U duality and central charges in various dimensions revisited," Int. J. Mod. Phys. A **13** (1998) 431 doi:10.1142/S0217751X98000196.
36. G. Dall'Agata, R. D'Auria, L. Sommovigo and S. Vaula, "D = 4, N=2 gauged supergravity in the presence of tensor multiplets," Nucl. Phys. B **682** (2004) 243 doi:10.1016/j.nuclphysb.2004.01.014 [hep-th/0312210].
37. R. D'Auria, L. Sommovigo and S. Vaula, "N = 2 supergravity Lagrangian coupled to tensor multiplets with electric and magnetic fluxes," JHEP **0411** (2004) 028 doi:10.1088/1126-6708/2004/11/028 [hep-th/0409097].
38. L. Andrianopoli, R. D'Auria, L. Sommovigo and M. Trigiante, "D=4, N=2 Gauged Supergravity coupled to Vector-Tensor Multiplets," Nucl. Phys. B **851** (2011) 1 doi:10.1016/j.nuclphysb.2011.05.007 [arXiv:1103.4813 [hep-th]].
39. L. Andrianopoli and R. D'Auria, "N=1 and N=2 pure supergravities on a manifold with boundary," JHEP **1408** (2014) 012 doi:10.1007/JHEP08(2014)012 [arXiv:1405.2010 [hep-th]].
40. L. Andrianopoli, B. L. Cerchiai, R. D.'Auria and M. Trigiante, "Unconventional supersymmetry at the boundary of AdS_4 supergravity," JHEP **1804** (2018) 007 doi:10.1007/JHEP04(2018)007 [arXiv:1801.08081 [hep-th]].
41. R. D'Auria, P. Fré, M. Raciti and F. Riva, "Anomaly Free Supergravity in $D = 10$. 1. The Bianchi Identities and the Bosonic Lagrangian," Int. J. Mod. Phys. A **3** (1988) 953. doi:10.1142/S0217751X88000436.
42. A. Ceresole, R. D'Auria, S. Ferrara and A. Van Proeyen, "Duality transformations in supersymmetric Yang-Mills theories coupled to supergravity," Nucl. Phys. B **444** (1995) 92 doi:10.1016/0550-3213(95)00175-R [hep-th/9502072].
43. A. Ceresole, R. D'Auria, S. Ferrara, W. Lerche and J. Louis, "Picard-Fuchs equations and special geometry," Int. J. Mod. Phys. A **8** (1993) 79 doi:10.1142/S0217751X93000047 [hep-th/9204035].

44. A. Ceresole, R. D'Auria, S. Ferrara, W. Lerche, J. Louis and T. Regge, "Picard-Fuchs equations, special geometry and target space duality," AMS/IP Stud. Adv. Math. **1** (1996) 281.
45. M. Billò, A. Ceresole, R. D'Auria, S. Ferrara, P. Fré, T. Regge, P. Soriani and A. Van Proeyen, "A Search for nonperturbative dualities of local N=2 Yang-Mills theories from Calabi-Yau threefolds," Class. Quant. Grav. **13** (1996) 831 doi:10.1088/0264-9381/13/5/007 [hep-th/9506075].
46. A. Ceresole, R. D'Auria and S. Ferrara, "The Symplectic structure of N=2 supergravity and its central extension," Nucl. Phys. Proc. Suppl. **46** (1996) 67 doi:10.1016/0920-5632(96)00008-4 [hep-th/9509160].
47. A. Ceresole, R. D'Auria and T. Regge, "Duality group for Calabi-Yau 2 moduli space," Nucl. Phys. B **414** (1994) 517 doi:10.1016/0550-3213(94)90439-1 [hep-th/9307151].
48. A. C. Cadavid, A. Ceresole, R. D'Auria and S. Ferrara, "Eleven-dimensional supergravity compactified on Calabi-Yau threefolds," Phys. Lett. B **357** (1995) 76 doi:10.1016/0370-2693(95)00891-N [hep-th/9506144].
49. R. D'Auria, P. Fré and A. J. da Silva, "Geometric Structure of $N = 1D = 10$ and $N = 4D = 4$ Superyang-mills Theory," Nucl. Phys. B **196** (1982) 205. doi:10.1016/0550-3213(82)90036-0.
50. R. D'Auria and T. Regge, "Gravity Theories With Asymptotically Flat Instantons," Nucl. Phys. B **195** (1982) 308. doi:10.1016/0550-3213(82)90402-3.
51. O. Chandia and J. Zanelli, "Topological invariants, instantons and chiral anomaly on spaces with torsion," Phys. Rev. D **55**, 7580 (1997) doi:10.1103/PhysRevD.55.7580 [hep-th/9702025].

Integrating Supergravity

Leonardo Castellani

Dipartimento di Scienze e Innovazione Tecnologica
Università del Piemonte Orientale, viale T. Michel 11, 15121 Alessandria, Italy

INFN, Sezione di Torino, via P. Giuria 1, 10125 Torino, Italy

Arnold–Regge Center, via P. Giuria 1, 10125 Torino, Italy

We review recent developments in integration on supermanifolds applied to supergravity theories. Poincaré duals in superintegration are used as a tool to relate component to superspace actions. These techniques help to clarify some issues in the group-geometric formulation of supersymmetric theories, initiated by Ne'eman and Regge in the late '70s, developed by the Torino collaborations in the '80s (see the review by Riccardo D'Auria in this volume) and used to this day in various applications.

1. Introduction

The idea of unifying gravity with gauge theories starting from their symmetry structure has an old and well-motivated history, culminating in the relatively recent AdS/CFT or gauge/gravity correspondences.

Even on the classical level a unified description of diffeomorphisms and gauge transformations has been proposed originally by Ne'eman and Regge in Ref. 1 and further developed in Refs. 2–5. Later applications and reviews can be found in Refs. 6–10, and in the contribution by Riccardo D'Auria in this volume.[11]

The setup is group-geometrical. The basic fields of the theory are the components of the vielbein one-form $\sigma^A = \sigma(z)^A{}_\Lambda dz^\Lambda$ on the manifold of a Lie group G, A being an index in the G Lie algebra, and z^Λ the coordinates of the group manifold. This vielbein satisfies the Cartan–Maurer (CM) equations.[a]

$$d\sigma^A + \frac{1}{2} C^A_{BC}\, \sigma^B \wedge \sigma^C = 0 \tag{1}$$

where C^A_{BC} are the structure constants of the G Lie algebra. The G vielbein $\sigma^A(z)$ has a fixed dependence on the coordinates z, and cannot therefore play the rôle of a dynamical

[a]For a summary on group manifold geometry see for ex.[11]

object. We consider instead a deformation of the group manifold vielbein not satisfying anymore the CM equations. The amount of deformation is measured by the curvature two-form:

$$R^A \equiv d\sigma^A + \frac{1}{2} C^A_{BC}\, \sigma^B \wedge \sigma^C. \tag{2}$$

This deformed vielbein can be interpreted as the vielbein of a "soft" group manifold $\tilde{G}$, diffeomorphic to the original rigid group manifold G.

For example consider $G =$ super-Poincaré group, and denote the vielbein on the $\tilde{G}$ manifold as $\sigma^A = (V^a, \omega^{ab}, \psi^\alpha)$. The index $A = (a, ab, \alpha)$ runs on the translations, Lorentz rotations and supersymmetries of the super-Poincaré Lie algebra. The $V^a, \omega^{ab}, \psi^\alpha$ components of the $\tilde{G}$ vielbein are identified with the vierbein, the spin connection and the gravitino of $d = 4$ simple supergravity. The curvature two-form defined as in (2) becomes

$$R^a = dV^a - \omega^{ab} \wedge V^c \eta_{bc} \tag{3}$$

$$R^{ab} = d\omega^{ab} - \omega^{ac} \wedge \omega^{bd} \eta_{cd} \tag{4}$$

$$\rho = d\psi - \frac{1}{4} \omega^{ab} \gamma_{ab} \psi \tag{5}$$

and one recognizes the familiar expressions for the torsion, the Lorentz curvature and the gravitino field strength. Bianchi identities are obtained by applying the exterior derivative to (3)–(5):

$$dR^a - \omega^a{}_b R^b + R^a{}_b V^b - i\bar{\psi}\gamma^a \rho = 0 \tag{6}$$

$$dR^{ab} - \omega^a{}_c R^{cb} + \omega^b{}_c R^{ca} = 0 \tag{7}$$

$$d\rho - \frac{1}{4} \omega^{ab} \gamma_{ab} \rho + \frac{1}{4} R^{ab} \gamma_{ab} \psi = 0 \tag{8}$$

(wedge product symbols omitted).

Tangent vectors dual to the vielbein σ^A, are denoted by t_B, so that $\sigma^A(t_B) = \delta^A_B$. Diffeomorphisms along tangent vectors $\varepsilon = \varepsilon^A t_A$ on $\tilde{G}$ are generated by the Lie derivative ℓ_ε. When applied to the $\tilde{G}$ vielbein, the variation under diffeomorphisms takes the suggestive form:

$$\ell_\varepsilon \sigma^A = d\varepsilon^A + C^A_{BC} \sigma^B \varepsilon^C + \iota_\varepsilon R^A \tag{9}$$

(where ι_ε is the contraction operator, see for ex. Ref. 11) and one recognizes on the right-hand side the G-covariant derivative of the infinitesimal parameter ε^A plus a curvature term. When the curvature term vanishes, i.e. when $\iota_\varepsilon R^A = 0$, the diffeomorphism takes the form of a *gauge transformation*, and the curvature is said to be *horizontal* along the t_A's entering the sum in $\varepsilon = \varepsilon^A t_A$.

Thus in group manifold geometry gauge transformations can be interpreted as *particular diffeomorphisms*, along the directions on which the curvatures are horizontal.

Supergravity Lagrangians in d spacetime dimensions are constructed using the (soft) group curvatures (2) and the vielbein, and are d-forms living in a $(d + m)$-dimensional (soft) group manifold.[b] To integrate them it is therefore necessary to multiply them by the Poincaré dual of the spacetime slice of $\tilde{G}$, so to obtain a $d + m$ top form. Integration of this top form on $\tilde{G}$ yields an action invariant under $\tilde{G}$ diffeomorphisms.

[b]The building rules for Lagrangian d-forms in the group-geometric framework can be found in Ref. 11.

For example the group manifold action of simple supergravity in $d = 4$ is the integral on the (4+10)-dimensional superPoincaré manifold:

$$S[V, \omega, \psi, \eta] = \int_{\tilde{G}} (R^{ab} V^c V^d \epsilon_{abcd} + 4\bar{\psi}\gamma_5\gamma_a\rho V^a)\ \eta_{M^4} \tag{10}$$

where all products between forms are wedge products, and η_{M^4} = Poincaré dual of M^4, a 10-form that localizes the Lagrangian on the spacetime submanifold M^4. The resulting action is invariant under diffeomorphisms on the (soft) superPoincaré manifold. In particular diffeomorphisms along fermionic directions are interpreted as local supersymmetry transformations.

Since the m-dimensional internal coordinates include Grassmann directions, Poincaré duals have to be extended to the case of supermanifolds. How this is done is recalled in the next Section (taken from Section 4 of Ref. 10) containing basic results in supermanifold integration,[c] and new developments concerning integral forms, discussed in Refs. 14–16.

2. Superintegration and integral forms

The construction of actions invariant under diffeomorphisms is solved "ab initio" in ordinary integration theory by *form* integration. The integral of a d-form

$$\omega^{(d)} = \omega_{[\mu_1 \cdots \mu_d]}(x)\ dx^{\mu_1} \wedge \cdots \wedge dx^{\mu_d} \tag{11}$$

on a d-dimensional manifold M^d is defined by

$$I = \int_{M^d} \omega^{(d)} \equiv \int_{M^d} \frac{1}{d!}\ \omega_{[\mu_1 \cdots \mu_d]}(x) \epsilon^{\mu_1 \cdots \mu_d}\ d^d x \tag{12}$$

i.e. by usual (Riemann–Lebesgue) integration on M^d of the function $\frac{1}{d!}\ \omega_{[\mu_1 \cdots \mu_d]}(x)\epsilon^{\mu_1 \cdots \mu_d}$, where $\epsilon^{\mu_1 \cdots \mu_d}$ is the Levi–Civita antisymmetric symbol in the coordinate basis, a tensor density of weight -1.

Under infinitesimal diffeomorphisms generated by the Lie derivative $\ell_\epsilon = \iota_\epsilon d + d\iota_\epsilon$ the form integral (12) transforms as

$$\delta I = \int_{M^d} \ell_\epsilon \omega^{(d)} = \int_{M^d} (\iota_\epsilon d + d\iota_\epsilon)\omega^{(d)} = 0 \tag{13}$$

since $d\omega^{(d)} = 0$ ($\omega^{(d)}$ is a top form) and $\int_{M^d} d(\iota_\epsilon \omega) = 0$ for appropriate boundary conditions. Thus we have checked invariance of the form integral under infinitesimal diff.s generated by the Lie derivative. Note that the existence of a top form, namely the fact that a d-form is closed on M^d, is crucial to ensure action invariance under diffeomorphisms.

Can we generalize form integration to *supermanifolds*, and use it to construct actions automatically invariant under superdiffeomorphisms? The answer to both questions is affirmative.

In analogy with the bosonic case, integration on forms living on supermanifolds is defined via integration of functions in superspace. Consider a function $\Phi(x, \theta)$, defined on a supermanifold $M^{d|m}$ with d bosonic coordinates x and m fermionic (anticommuting) coordinates θ^α. It is called a *superfield*, and can be expanded in the θ^α coordinates:

$$\Phi(x, \theta) = \phi(x) + \phi_{\alpha_1}(x)\theta^{\alpha_1} + \phi_{\alpha_1\alpha_2}(x)\theta^{\alpha_1}\theta^{\alpha_2} + \cdots + \phi_{\alpha_1 \cdots \alpha_m}(x)\theta^{\alpha_1} \cdots \theta^{\alpha_m}. \tag{14}$$

[c] See for ex. Ref. 12 for a recent review, or Ref. 13 for a textbook.

The functions $\phi_{\alpha_1\cdots\alpha_p}(x)$ are called *superfield components*, and have antisymmetrized indices due to the anticommuting θ's in the expansion (14). The integral of the superfield on $M^{d|m}$ is defined by Berezin integration:

$$\int_{M^{d|m}} \Phi(x,\theta)\; d^d x\; d^m\theta \equiv \int_{M^d} \frac{1}{m!}\phi_{\alpha_1\cdots\alpha_m}(x)\varepsilon^{\alpha_1\cdots\alpha_m}\; d^d x. \tag{15}$$

Only the highest component of Φ (corresponding to the maximal number of θ's) enters the integral on M^d.

Note the similarity between the two integrals (12) and (15). In fact we can define form integration in terms of Berezin integration. Consider the differentials dx in the d-form (11) as *anticommuting coordinates* $\xi^\mu = dx^\mu$, so that $\omega^{(d)}$ becomes a *function* of x and ξ:

$$\omega^{(d)}(x,\xi) = \omega_{[\mu_1\cdots\mu_d]}(x)\; \xi^{\mu_1}\cdots\xi^{\mu_d}. \tag{16}$$

Its Berezin integral on $M^{d|d}$ exactly yields the form integral (12). This is the key for a definition of superform integration on supermanifolds.

A natural generalization of a bosonic top form (11) is a $(d+m)$-superform:

$$\omega^{(d+m)}(x,\theta) = \omega_{[\mu_1\cdots\mu_d]\{\alpha_1\cdots\alpha_m\}}(x,\theta)\; dx^{\mu_1}\wedge\cdots\wedge dx^{\mu_d}\wedge d\theta^{\alpha_1}\wedge\cdots\wedge d\theta^{\alpha_m}. \tag{17}$$

Note that the $d\theta$ differentials *commute* (since the θ's are anticommuting), so that the indices α_i are symmetrized. For this reason $\omega^{(d+m)}$ cannot be a top form: a superform can have an arbitrary number of $d\theta$ differentials, and its exterior derivative does not vanish. Let's ignore for the moment this difficulty, and try to define a superform integral. Inspired by the observation in the preceding paragraph, we consider the superform $\omega^{(d+m)}(x,\theta)$ as a function of $x, \theta, dx, d\theta$, i.e. a function of the commuting variables $x, d\theta$ and anticommuting variables θ, dx. Its integral can be defined by Berezin integration on θ, dx, and usual Riemann–Lebesgue integration on $x, d\theta$. Here a second difficulty arises: the ordinary integration on the $u = d\theta$ coordinates produces integrals of the type

$$\int u^m\; d^m u \tag{18}$$

and there is no algorithmic way to assign a C-number to it. For the integral on the even variables $u = d\theta$ to make sense, the integrand must have compact support as a function of u. For this reason we consider functions of the $d\theta$'s which are *distributions* in $d\theta$ with support at the origin:

$$\omega(x,\theta,dx,d\theta) = \omega_{[\mu_1\cdots\mu_d]}(x,\theta)\; dx^{\mu_1}\cdots dx^{\mu_d}\delta(d\theta^1)\cdots\delta(d\theta^m). \tag{19}$$

These "functions" can be integrated on the supermanifold $M^{d+m|d+m}$ spanned by the $d+m$ bosonic variables $x, d\theta$ and $d+m$ fermionic variables dx, θ. The integral

$$\int_{M^{d+m|d+m}} \omega(x,\theta,dx,d\theta)\; d^d x\; d^m\theta\; d^d(dx)\; d^m(d\theta) \tag{20}$$

is defined by Berezin integration on the odd variables dx, θ and usual Riemann–Lebesgue integration on the even variables $x, d\theta$. Carrying out integration on the variables dx and $d\theta$ the integral becomes

$$\int_{M^{d|m}} \omega_{[\mu_1\cdots\mu_d]}(x,\theta)\epsilon^{\mu_1\cdots\mu_d}\; d^d x\; d^m\theta. \tag{21}$$

This integral can also be seen as an integral of the *form*:

$$\omega^{d|m} = \omega_{[\mu_1\cdots\mu_d]}(x,\theta)\delta(u^1)\cdots\delta(u^m)\, dx^{\mu_1}\wedge\cdots\wedge dx^{\mu_d}\wedge du^1\wedge\cdots\wedge du^m \tag{22}$$

where the even variables u are the differentials $d\theta$. Indeed, let us integrate this form with the recipe of considering it a function of x, θ, u and of the differentials dx,du, and then using Berezin and Riemann integration according to the odd or even grading of the variables. The result coincides with (21).

Thus the form $\omega^{d|m}$ can be integrated, even if it contains $d\theta$ differentials. We achieve this by confining the $d\theta$'s inside delta functions, and in this way overcome the first difficulty encountered with the superforms (17). But can $\omega^{d|m}$ overcome also the second difficulty, and be a *top form*? The answer is yes: the dx and du differentials are all anticommuting, so that their number in $\omega^{d|m}$ is already maximal, and multiplying it by $d\theta$ differentials gives zero because of the presence of the deltas. Therefore $d\omega^{d|m} = 0$, and $\omega^{d|m}$ is a *bona fide* top form. Since it can be integrated and it is a top form, $\omega^{d|m}$ is called an *integral top form.*

Finally, using the notation

$$\delta(u^1)\cdots\delta(u^m)\, du^1\wedge\cdots\wedge du^m \equiv \delta(u^1)\wedge\cdots\wedge\delta(u^m) \tag{23}$$

the integral top form can be rewritten (using $u = d\theta$):

$$\omega^{d|m} = \omega_{[\mu_1\cdots\mu_d]}(x,\theta)\, dx^{\mu_1}\wedge\cdots\wedge dx^{\mu_d}\wedge\delta(d\theta^1)\wedge\cdots\wedge\delta(d\theta^m) \tag{24}$$

or also

$$\omega^{d|m} = \omega_{[\mu_1\cdots\mu_d][\alpha_1\cdots\alpha_m]}(x,\theta)\, dx^{\mu_1}\wedge\cdots\wedge dx^{\mu_d}\wedge\delta(d\theta^{\alpha_1})\wedge\cdots\wedge\delta(d\theta^{\alpha_m}) \tag{25}$$

where indices α are antisymmetrized since the $\delta(d\theta^\alpha)$ anticommute, and

$$m!\; \omega_{[\mu_1\cdots\mu_d]} \equiv \omega_{[\mu_1\cdots\mu_d][1\cdots m]}. \tag{26}$$

In this notation μ and α indices play a similar role, and are both antisymmetrized. The numbers d, m are respectively called the *form number* and the *picture number*, and for integral top forms they coincide with the numbers of bosonic and fermionic dimensions of the supermanifold $M^{d|m}$.

We call "superforms" the forms of the kind (17), with dx and $d\theta$ differentials, without $\delta(d\theta)$'s. Thus superforms have a form number that counts the $dx, d\theta$ differentials, and zero picture number. For example the Lagrangian in (10) is a superform $L^{4|0}$.

3. Integration on submanifolds

Supergravity actions on supergroup manifolds $\tilde{G}$ can be written as integrals on the whole $\tilde{G}$ of a Lagrangian d-form multiplied by an appropriate Poincaré dual η_{M^d} of M^d, such that $L\wedge\eta_{M^d}$ becomes an integral top form. Let us see how this works for $N = 1$, $d = 4$ supergravity.

The supergravity Lagrangian in (10) is a (4|0) superform. The (soft) superPoincaré vielbein fields (V^a, ψ, ω^{ab}) depend on the superspace coordinates x^a, θ^α, with $\mu = 1, ..4$, $\alpha = 1, ..4.$ and on the 6 Lorentz coordinates y^{ab}. Only integral top forms of type (4|10) can be integrated on $\tilde{G}$. We need therefore a Poincaré dual of type (0|10), so that

$$L^{4|0}\wedge\eta^{0|10}_{M^4} \tag{27}$$

is an integral top form, i.e. of type (4|10). For this purpose we choose:

$$\eta^{0|10}_{M^4} = \varepsilon_{\alpha\beta\gamma\delta}\theta^\alpha\theta^\beta\theta^\gamma\theta^\delta \; \varepsilon_{\alpha'\beta'\gamma'\delta'}\delta(d\theta^{\alpha'}) \wedge \delta(d\theta^{\beta'}) \wedge \delta(d\theta^{\gamma'}) \wedge \delta(d\theta^{\delta'}) \\ \wedge \delta(y^{12})\delta(y^{13}) \cdots \delta(y^{34}) dy^{12} \wedge dy^{13} \wedge \cdots \wedge dy^{34} \tag{28}$$

and we find

$$\int_{\tilde{G}} L^{4|0} \wedge \eta^{0|10}_{M^4} = \int_{M^4} L^{4|0}(\theta = 0, d\theta = 0, y = 0, dy = 0). \tag{29}$$

We thus obtain the supergravity *spacetime* action of,[17,18] where all fields depend only on x-coordinates (the terms containing θ's and y are annihilated by the presence of the 4 θ's and the $\delta(y)$ in η) and have no "legs" $d\theta$ or dy because of the $\delta(d\theta)$ and dy differentials in η. Note that $\eta^{0|10}_{M^4}$ is closed and not exact.

Since multiplying by the Poincaré dual changes the picture number of the resulting form, η is also called a picture changing operator (PCO), a name borrowed from string theory and string field theory.

The Poincaré dual is by no means unique: we can orient the M^4 surface inside $\tilde{G}$ in many different ways. For example consider the PCO obtained by acting on η with an infinitesimal diffeomorphism in the θ directions:

$$\eta' = \eta + \ell_\epsilon \eta = \eta + d(\iota_\epsilon \eta). \tag{30}$$

This is still a PCO, being closed and not exact,[d] and dual to a submanifold diffeomorphic to the original M^4. Note also that the change in η is exact.

We can exploit the freedom in choosing η to relate different formulations of the same supergravity theory, cf. Section 5.

4. Symmetries

The action (10) with η given in (28) is the integral on $\tilde{G}$ of a top form: it is clearly invariant under diffeomorphisms on $\tilde{G}$. But what we are really interested in are the symmetries of the spacetime action as given on the r.h.s. of (29), where the variations are carried out only in the x-dependent fields in $L|_{\theta=0,d\theta=0,y=0,dy=0}$. The only symmetries guaranteed a priori are the 4-dimensional spacetime diffeomorphisms, the spacetime action being an integral of a 4-form on M^4.

Here resides most of the power of the group manifold formalism: if one considers the "mother" action (10) on $\tilde{G}$, the guaranteed symmetries are *all* the diff.s on $\tilde{G}$, generated by the Lie derivative ℓ_ε along the tangent vectors $\varepsilon = \varepsilon^A t_A$ of $\tilde{G}$. But how do these symmetries transfer to the spacetime action?

The variation of the group manifold action under diff.s generated by ℓ_ϵ is[e]

$$\delta S = \int_{\tilde{G}} \ell_\varepsilon (L \wedge \eta) = \int_{\tilde{G}} (\ell_\varepsilon L) \wedge \eta + L \wedge \ell_\varepsilon \eta = 0 \tag{31}$$

[d]Because η is closed and not exact, and d commutes with ℓ_ϵ.

[e]Recall $\ell_\varepsilon = \iota_\varepsilon d + d\iota_\varepsilon$ so that ℓ_ε(top form) $= d(\iota_\varepsilon$ top form).

modulo boundary terms. One has to vary the fields[f] in L as well as the submanifold embedded in $\tilde{G}$: the sum of these two variations gives zero[g] on the group manifold action S. But what we need in order to have a *spacetime* interpretation of all the symmetries of S, is really

$$\delta S = \int_{\tilde{G}} (\ell_\varepsilon L) \wedge \eta = 0. \tag{32}$$

If this holds, varying the fields ϕ inside L with the Lie derivative ℓ_ϵ as in (9), and then projecting on spacetime ($\theta = 0, d\theta = 0, y = 0, dy = 0$), yields spacetime variations

$$\delta\phi(\theta = d\theta = y = dy = 0) = \ell_\varepsilon \phi(x, \theta, y)|_{\theta=d\theta=y=dy=0} \tag{33}$$

that leave the spacetime action (29) invariant. We call them *spacetime invariances.* They originate from the diff. invariance of the group manifold action, and give rise to symmetries of the spacetime action (29) only when (32) holds. This happens if one of the following conditions is satisfied:

- the Lie derivative on η vanishes:

$$\ell_\varepsilon \eta = 0 \tag{34}$$

- the spacetime projection of the Lie derivative of L is exact:

$$(\ell_\varepsilon L)|_{\theta=d\theta=y=dy=0} = d\alpha. \tag{35}$$

In this case the variation (32)

$$\delta S = \int_{\tilde{G}} (\ell_\varepsilon L) \wedge \eta = \int_{M^4} (\ell_\varepsilon L)|_{\theta=d\theta=y=dy=0} \tag{36}$$

vanishes after integration by parts. The requirement (35) is equivalent to

$$(\iota_\varepsilon dL)|_{\theta=d\theta=y=dy=0} = d\alpha' \tag{37}$$

since $\lambda_\varepsilon = \iota_\varepsilon d + d\iota_\varepsilon$.

The Lagrangian L depends on the $\tilde{G}$-vielbein σ^A and its curvature R^A, so that also dL, after use of Bianchi identities, is expressed in terms of σ^A and R^A. Then condition (37) translates into a *condition on the contractions* $\iota_\varepsilon R^A$, i.e. a condition on the curvature components. In the example of $d = 4$ simple supergravity we find

$$dL = 2R^{ab}R^cV^d\varepsilon_{abcd} + 4\bar{\rho}\gamma_5\gamma_a\rho V^a - 4\bar{\psi}\gamma_5\gamma_a\rho R^a \tag{38}$$

after using the Bianchi identities (6)–(8).

Lorentz gauge transformations

It is immediate to see that if all curvatures are horizontal in the Lorentz directions (no "legs" along ω) then indeed $\iota_{\varepsilon^{ab}t_{ab}} dL = 0$, and Lorentz transformations are a spacetime

[f]Since ℓ_ε satisfies the Leibnitz rule, $\ell_\varepsilon L$ can be computed by varying in turn all fields inside L.
[g]In the following the vanishing of action variations will always be understood modulo boundary terms.

invariance of the supergravity action. This is essentially due to the absence of bare ω^{ab} in L. One can also verify that the horizontality constraints are compatible with the Bianchi identities (6)–(8).

The general diffeomorphism formula (9) yields the usual Lorentz transformations

$$\ell_{\varepsilon^{cd}t_{cd}} V^a = \varepsilon^a{}_b V^b \tag{39}$$

$$\ell_{\varepsilon^{cd}t_{cd}} \omega^{ab} = d\varepsilon^{ab} - \omega^a{}_c \varepsilon^{cb} + \omega^b{}_c \varepsilon^{ca} = \mathcal{D}\varepsilon^{ab} \tag{40}$$

$$\ell_{\varepsilon^{cd}t_{cd}} \psi = \frac{1}{4}\varepsilon^{ab}\gamma_{ab}\psi. \tag{41}$$

We can check directly the invariance of the action under these variations: indeed all curvatures and vierbeins appearing in (10) transform homogeneously, and only Lorentz invariant tensors appear in L.

Spacetime diffeomorphisms

Diff.s along tangent vectors ∂_μ dual to dx^μ are invariances of the spacetime action, since $\ell_{\varepsilon^\mu \partial_\mu} \eta_M = 0$ due to η_M containing only dy and $d\theta$ differentials, see Section 3.

Supersymmetry transformations

Diff.s along tangent vectors t_α dual to ψ^α are spacetime invariances provided $\iota_\epsilon dL|_{\theta=d\theta=0} = \textit{total derivative}$ with $\epsilon = \epsilon^\alpha t_\alpha$, that is to say

$$\iota_\epsilon dL = 2(\iota_\epsilon R^{ab})R^c V^d \varepsilon_{abcd} + 2R^{ab}(\iota_\epsilon R^c) V^d \varepsilon_{abcd} + 8\bar{\rho}\gamma_5\gamma_a(\iota_\epsilon\rho)V^a$$
$$-4\bar{\epsilon}\gamma_5\gamma_a\rho R^a - 4\bar{\psi}\gamma_5\gamma_a(\iota_\epsilon\rho)R^a - 4\bar{\psi}\gamma_5\gamma_a\rho(\iota_\epsilon R^a) = \textit{total derivative} \tag{42}$$

with $\theta = d\theta = 0$. This is a condition for the contractions on the curvatures, and it is satisfied by:

$$\iota_\epsilon R^a = 0 \tag{43}$$

$$\iota_\epsilon R^{ab} = -\varepsilon^{abef}\bar{\rho}_{ef}\gamma_5\gamma_g\epsilon V^g - \varepsilon^{efg[a}\bar{\rho}_{ef}\gamma_5\gamma_g\epsilon V^{b]} \equiv \bar{\theta}^{ab}_c \epsilon V^c \tag{44}$$

$$\iota_\epsilon \rho = 0. \tag{45}$$

Thus we have supersymmetry invariance of the spacetime action if the curvatures have the following parametrization on a basis of 2-forms:

$$R^a = R^a{}_{bc}\, V^b V^c \tag{46}$$

$$R^{ab} = R^{ab}{}_{cd} V^c V^d + \bar{\theta}^{ab}_c\, \psi\, V^c \tag{47}$$

$$\rho = \rho_{ab}\, V^a V^b \tag{48}$$

where we have taken into account also horizontality in the Lorentz directions. The conditions (43)–(45) are called "rheonomic conditions", and similarly (46)–(48) are called "rheonomic parametrizations" of the curvatures.

The diff.s along $\epsilon = \epsilon^\alpha t_\alpha$ (supersymmetry transformations) act on the fields according to the general formula (9), where the contractions on the curvatures are given in (43)–(45):

$$\ell_\epsilon V^a = i\bar{\epsilon}\gamma^a\psi \tag{49}$$

$$\ell_\epsilon \omega^{ab} = \bar{\theta}^{ab}_c \epsilon V^c \tag{50}$$

$$\ell_\epsilon \psi = \mathcal{D}\epsilon \equiv d\epsilon - \frac{1}{4}\omega^{ab}\gamma_{ab}\epsilon \tag{51}$$

with $\bar{\theta}^{ab}_c$ defined in (44).

As explained in,[11] the horizontality and rheonomic constraints on the curvatures are in fact an output of the equations of motion for the action (10). For a discussion on the action principle and the resulting field equations see again.[11]

The rheonomic constraints satisfy the Bianchi identities only on the shell of the field equations. This implies that the supersymmetry transformations (49)–(51) leave the supergravity spacetime action invariant, but do not close on a Lie superalgebra, unless use is made of the field equations. In this case we say that the supersymmetry algebra closes only on shell.

This situation can be cured by adding extra fields in the theory, called *auxiliary fields*, entering in the parametrization of the curvatures in such a way that the Bianchi identities do not imply propagation equations. The auxiliary fields are nondynamical fields, but their degrees of freedom (d.o.f.) are needed to ensure an equal number of off-shell d.o.f. of fermions and bosons. For $d = 4$ simple supergravity at least two sets of auxilary fields are known.[19–21]

5. A bridge between superspace and component actions

The possibility of different choices for the Poincaré dual offers a technique to relate component to superspace actions.

As discussed in Section 2, the group manifold action for a d-dimensional supergravity can be written as the superintegral:

$$S_{SG} = \int_{M^{d|m}} L^{d|0} \wedge \eta^{0|m} \tag{52}$$

where the Lagrangian $L^{d|0}$ is found by using the building rules recalled in.[11]

Suppose now that

$$dL^{d|0} = 0. \tag{53}$$

Then two Poincaré duals differing by a total derivative give rise to the same action when inserted into (52). As a consequence, the action (52) can be expressed in multiple ways, using different choices of η all in the same cohomology class. This observation has been used in Ref. 16 to relate component and superspace actions for the case of $d = 3$ supergravity.

6. Conclusions

The group-geometric approach to supergravity theories is emblematic of Tullio Regge's masterful ability to combine algebraic tools with geometry, producing in this way highly original results in physical applications. We have reviewed here some features of this approach, in the light of recent developments in superintegration theory.

References

1. Y. Ne'eman and T. Regge, "Gravity and Supergravity as Gauge Theories on a Group Manifold," Phys. Lett. **74B**, 54 (1978). doi:10.1016/0370-2693(78)90058-8.

2. A. D'Adda, R. D'Auria, P. Fré and T. Regge, "Geometrical Formulation of Supergravity Theories on Orthosymplectic Supergroup Manifolds," Riv. Nuovo Cim. **3N6**, 1 (1980). doi:10.1007/BF02724337.
3. R. D'Auria, P. Fré and T. Regge, "Graded Lie Algebra Cohomology and Supergravity," Riv. Nuovo Cim. **3N12**, 1 (1980). doi:10.1007/BF02905929.
4. T. Regge, "The Group Manifold Approach To Unified Gravity," Conf. Proc. C **8306271**, 933 (1983).
5. R. D'Auria and P. Fré, "Geometric Supergravity in d = 11 and Its Hidden Supergroup," Nucl. Phys. B **201**, 101 (1982) Erratum: [Nucl. Phys. B **206**, 496 (1982)]. doi:10.1016/0550-3213(82)90376-5, 10.1016/0550-3213(82)90281-4.
6. L. Castellani, R. D'Auria and P. Fré, "Supergravity and superstrings: A Geometric perspective. 3 Vol.s" Singapore, World Scientific (1991).
7. L. Castellani, R. D'Auria and P. Fré, *"Seven Lectures on the group manifold approach to supergravity and the spontaneous compactification of extra dimensions"*, Proc. XIX Winter School Karpacz 1983, ed. B. Milewski (World Scientific, Singapore).
8. L. Castellani, P. Fré and P. van Nieuwenhuizen, "A Review of the Group Manifold Approach and Its Application to Conformal Supergravity," Annals Phys. **136**, 398 (1981).
9. L. Castellani, "Group geometric methods in supergravity and superstring theories," Int. J. Mod. Phys. A **7**, 1583 (1992).
10. L. Castellani, "Supergravity in the group-geometric framework: a primer," Fortsch. Phys. **66**, 040014 (2018) doi:10.1002/prop.201800014 [arXiv:1802.03407 [hep-th]].
11. R. D'Auria, "Geometric Supergravity", this volume.
12. E. Witten, *Notes on Supermanifolds and Integration,* [arXiv:1209.2199 [hep-th]].
13. T. Voronov, *Geometric integration theory on supermanifolds.* Soviet Scientific Review, Section C: Mathematical Physics, 9, Part 1. Harwood Academic Publisher, Chur. 1991 (Second edition 2014).
14. L. Castellani, R. Catenacci and P. A. Grassi, "Supergravity Actions with Integral Forms," Nucl. Phys. B **889**, 419 (2014); [arXiv:1409.0192 [hep-th]].
15. L. Castellani, R. Catenacci and P. A. Grassi, "The Geometry of Supermanifolds and New Supersymmetric Actions," Nucl. Phys. B **899**, 112 (2015) [arXiv:1503.07886 [hep-th]].
16. L. Castellani, R. Catenacci and P. A. Grassi, "The Integral Form of Supergravity," JHEP **1610**, 049 (2016) [arXiv:1607.05193 [hep-th]].
17. D. Z. Freedman, P. van Nieuwenhuizen and S. Ferrara, "Progress Toward a Theory of Supergravity," Phys. Rev. D **13**, 3214 (1976).
18. S. Deser and B. Zumino, "Consistent Supergravity," Phys. Lett. B **62**, 335 (1976).
19. K. S. Stelle and P. C. West, "Minimal Auxiliary Fields for Supergravity," Phys. Lett. **74B**, 330 (1978). doi:10.1016/0370-2693(78)90669-X.
20. S. Ferrara and P. van Nieuwenhuizen, "The Auxiliary Fields of Supergravity," Phys. Lett. **74B**, 333 (1978). doi:10.1016/0370-2693(78)90670-6.
21. M. F. Sohnius and P. C. West, "An Alternative Minimal Off-Shell Version of N=1 Supergravity," Phys. Lett. **105B**, 353 (1981). doi:10.1016/0370-2693(81)90778-4.

RIVISTA DEL NUOVO CIMENTO VOL. 1, N. 5 1978

Gauge Theory of Gravity and Supergravity on a Group Manifold.

Y. NE'EMAN

Department of Physics and Astronomy, Tel Aviv University (*) - *Tel Aviv, Israel*
Center for Particle Theory, University of Texas (**) - *Austin, Tex.* 78712, *USA*

T. REGGE

School of Natural Sciences, The Institute for Advanced Study (***)
Princeton, N. J. 08540, *USA*
Istituto di Fisica dell'Università - 10125 *Torino, Italia*

(ricevuto il 13 Marzo 1978)

1. – Introduction.

In recent years, the intriguing properties of graded Lie algebras (GLA) have generated considerable interest [1-4]. In particular, it is widely felt that these structures may provide us with a new principle, unifying gravity with some of the fundamental interactions of particle theory. Such theories (and any gauge theory involving the action of a gauge group with a nontrivial action on space-time) are most naturally formulated by using the concept of

(*) Research supported in part by the U.S.-Israel Binational Science Foundation.
(**) Research supported in part by the U.S. Energy Research and Administration Grant E(40-1)3992.
(***) Research supported in part by NSF Grant 40768X.

(graded or ungraded) Grassmann algebras of forms in the context of the theory of Lie groups.

Much of the ground work for this kind of theory has been anticipated by CARTAN and is commonplace in the current mathematical literature. However, it is not as broadly known as it deserves among physicists. Part of our discussion is accordingly concerned with the problem of casting unified theories in the language of forms. In order to make the paper as self-sufficient as possible, we rederive known material of the theory of Lie groups, thus also displaying the consistency features of the method. Finally, existing theories are analysed and discussed in the light of the geometrical formalism.

2. – Differential geometry and Lie groups.

This is a vast subject of which we can only reproduce a few key definitions and results [5]. We also do not attempt to generalize systematically the theory to GLAs [6-7], but shall provide a few examples of interest.

Let M be a differentiable manifold with covering U_α and let x_α be co-ordinates on U_α. A p-form on U_α is an expression of the kind

$$\eta_\alpha = \sum_{\{A\}} \mathrm{d}x_\alpha^{A_1} \wedge \mathrm{d}x_\alpha^{A_2} \wedge \ldots \wedge \mathrm{d}x_\alpha^{A_p} E^{(\alpha)}_{A_1 \ldots A_p}(x_\alpha) , \qquad \{A\} = \{A_1, \ldots, A_p\} , \tag{2.1}$$

where (a, b represent the gradings of A, B when the variables are elements of a Grassmann algebra)

$$\mathrm{d}x_\alpha^A \wedge \mathrm{d}x_\alpha^B = -(-1)^{ab} \mathrm{d}x_\alpha^B \wedge \mathrm{d}x_\alpha^A$$

and the $E^{(\alpha)}_{\{A\}}$ are differentiable functions of the x_α. On $U_\alpha \cap U_\beta$ we identify the forms η_α and η_β provided that

$$\eta_\alpha = \sum_{\{A\}} \sum_{\{B\}} \mathrm{d}x_\alpha^{A_1} \frac{\partial x_\beta^{B_1}}{\partial x_\alpha^{A_1}} \wedge \mathrm{d}x_\alpha^{A_2} \frac{\partial x_\beta^{B_2}}{\partial x_\alpha^{A_2}} \wedge \ldots \wedge \mathrm{d}x_\alpha^{A_p} \frac{\partial x_\beta^{B_p}}{\partial x_\alpha^{A_p}} E^{(\beta)}_{B_1 \ldots B_p}(x_\beta(x_\alpha)) . \tag{2.2}$$

In this case, the set of all forms η_α defines a unique form η on M. In the sequel, unless specified, we drop the α-dependence.

The $\wedge$ product of differentials can be extended by linearity in an obvious way to generic forms and we have

$$\eta^{pa} \wedge \xi^{qb} = (-1)^{pq+ab} \xi^{qb} \wedge \eta^{pa} , \tag{2.3}$$

where η^{pa}, ξ^{qb} are p, q forms, respectively, and a, b are their GLA gradings.

The differential $\mathrm{d}\eta^p$ of a p-form is a $(p+1)$-form defined recursively by the properties

$$
(2.4)\qquad \begin{cases} \mathrm{d}(\mathrm{d}x^A) = 0\,, \qquad \mathrm{d}Q(x) = \sum\limits_A \mathrm{d}x^A \dfrac{\partial Q}{\partial x^A}\,, \\ \mathrm{d}(\eta^{pa}\wedge\xi^{qb}) = \mathrm{d}\eta^{pa}\wedge\xi^{qb} + (-1)^p\eta^{pa}\wedge\mathrm{d}\xi^{qb}\,. \end{cases}
$$

For any form, we always have $\mathrm{d}(\mathrm{d}\eta) \equiv 0$. We have adopted here ((2.2), (2.4)) the convention [7] of putting the differential (or the contravariant index) to the left of the derivative (or the covariant index), when summing over « dummy » indices. Note that the operator d behaves as a $p = 1, a = 0$ element in this product Grassmann manifold of differential forms.

Given manifolds M and N, a map λ, $M \xrightarrow{\lambda} N$, and a form ν^p on N, the « pull back » $\lambda^*\nu^p$ is a p-form on M, defined by the following procedure.

If ν^p is given by

$$
(2.5)\qquad \nu = \sum_{\{A\}} \mathrm{d}y^{A_1}\wedge\ldots\wedge\mathrm{d}y^{A_p}\, N_{A_1\ldots A_p}(y)\,, \qquad y\in N\,,
$$

and the map λ is realized by $y = \lambda(x)$, then we have

$$
(2.6)\qquad \lambda^*\nu = \sum_{\{B\}} \mathrm{d}x^{B_1}\frac{\partial y^{A_1}}{\partial x^{B_1}}\wedge\ldots\wedge\mathrm{d}x^{B_p}\frac{\partial y^{A_p}}{\partial x^{B_p}}\, N_{\{A\}}(\lambda(x))\,.
$$

It can be checked that

$$
(2.7)\qquad \begin{cases} \mathrm{d}\lambda^*\nu = \lambda^*\,\mathrm{d}\nu\,, \\ \lambda^*(\nu\wedge\sigma) = \lambda^*\nu\wedge\lambda^*\sigma\,. \end{cases}
$$

The d, $\wedge$ operations are thus co-ordinate independent and are preserved through maps. This property makes them ideal in treating covariant theories. Now we apply the algorithms of forms to Lie groups. (We refer the reader to ref. [2, 6, 7] for the effects of grading as in (2.4).)

Let G be a Lie group (or a formal group in the Berezin-Kats sense [8]) parametrized by $x^1 \ldots x^c$. We use in the sequel elements $x, y, z \ldots \in G$ whose parameters are $x^1 \ldots x^c$, $y^1 \ldots y^c$, etc. The element $x^A = 0$ denoted by e is the identity in G. The product on G is a map $G\cdot G \xrightarrow{\Lambda} G$ and we write

$$
z = xy \quad \text{instead of} \quad z = \Lambda(x, y)\,,
$$

satisfying the associativity condition

$$
t(xy) = (tx)y \qquad \forall t, x, y\,.
$$

In co-ordinates we have, for the product element z^A,

$$z^A = \Lambda^A(x^1 \dots x^c, y^1 \dots y^c), \tag{2.8}$$

yielding the differential forms

$$\mathrm{d}z^A = \mathrm{d}x^M\, V(x, y)^A_M + \mathrm{d}y^N\, W(x, y)^A_N. \tag{2.9}$$

In the sequel we use the $c\times 1$ row matrices

$$\mathrm{d}z = \{\mathrm{d}z^1 \dots \mathrm{d}z^c\}, \qquad \mathrm{d}x = \{\mathrm{d}x^1 \dots \mathrm{d}x^c\} \text{ etc.}$$

and consider V^A_M, W^A_N as $c\times c$ matrices. Briefly, then

$$\mathrm{d}z = \mathrm{d}x\, V + \mathrm{d}y\, W$$

Associativity implies $t(xy) = (tx)y$. This equation can be differentiated in all sets of variables and the results compared. One gets

$$\begin{cases} a) & V(t, x)\,V(tx, y) = V(t, xy), \\ b) & W(t, x)\,V(tx, y) = V(x, y)\,W(t, xy), \\ c) & W(tx, y) = W(x, y)\,W(t, xy). \end{cases} \tag{2.10}$$

As a consequence of (2.10) we have $V(x, e) = W(e, x) = 1$. The product on G gives two kinds of natural maps on G, the left and right translations

$$G \xrightarrow{l(a)} G, \qquad \text{where } l(a)(x) = ax,$$

$$G \xrightarrow{r(a)} G, \qquad \text{where } r(a)(x) = xa.$$

Consider now the ($c\times 1$ matrix) 1-forms

$$\begin{cases} \omega = \mathrm{d}x\, W(x^{-1}, x), \\ \pi = \mathrm{d}x\, V(x, x^{-1}). \end{cases} \tag{2.11}$$

By means of (2.10) it is easily checked that

$$l^*(a)\omega = \omega, \qquad r^*(a)\pi = \pi; \tag{2.12}$$

ω and π are, therefore, called Cartan's left and right invariant forms, respectively, or more briefly the LI and RI forms. By the properties of d and $\wedge$, any form obtained from $\omega(\pi)$ through these operations is also LI (RI). Therefore, if we expand

$$d\omega^A = \tfrac{1}{2} C^A_{\cdot BE}\omega^B \wedge \omega^E, \tag{2.13}$$

the $C^A_{\cdot BE} = - C^A_{\cdot EB}$ must be LI o-forms (that is functions) and are, therefore, constants, the structure constants of the Lie algebra of G. Similarly one has

$$d\pi^A = - \tfrac{1}{2} C^A_{\cdot BE}\pi^B \wedge \pi^E . \tag{2.14}$$

Equations (2.13), (2.14) are referred to as the Cartan-Maurer equations. Under right translations we have

$$r^*(a)\omega = dx\, V(x, a)\, W(a^{-1}x^{-1}, xa) = \tag{2.15}$$

$$= dx\, W(x^{-1}, x)\, W(a^{-1}, e)\, V(a^{-1}, a) = \omega\,\mathrm{ad}(a^{-1}),$$

where

$$\mathrm{ad}(a^{-1}) = W(a^{-1}, e)\, V(a^{-1}, a) = W(a^{-1}, e)\, V^{-1}(e, a^{-1}). \tag{2.16}$$

(In deriving (2.15) and (2.16), use has been made of (2.10).)

Similarly,

$$l^*(a)\pi = \pi V(e, a^{-1})\, W(a, a^{-1}) = \pi\,\mathrm{ad}(a). \tag{2.17}$$

By using (2.10) one further finds that

$$\mathrm{ad}(a)\,\mathrm{ad}(x) = \mathrm{ad}(ax) \tag{2.18}$$

so that ad(a) is a $c \times c$ representation of G, the adjoint representation. The matrix ad(a) is tied to the structure constants by useful identities.

Consider the map $G \times G \xrightarrow{\Lambda} G$. From the previous comments it is clear that $\Lambda^*\omega$, $\Lambda^*\pi$ also satisfy the Cartan-Maurer equations. If we use x, y, respectively, as co-ordinates on the first (second) copy of G, we can write $xy = \Lambda(x, y)$ and

$$\Lambda^*\omega = \omega_y + \omega_x\,\mathrm{ad}(y^{-1}) . \tag{2.19}$$

In (2.19), ω_y, ω_x are the Cartan LI forms written in the variables x, y, respectively. Moreover, x translates y on the left and does not appear in ω_y; y translates x on the right and appears in ad$(y^1)^-$ according to (2.15). By inserting (2.19) in (2.13), we find an identity relating 2-forms. Separate identification of the coefficients of $dx \wedge dx$, $dx \wedge dy$, $dy \wedge dy$ produces

$$\begin{cases} C^B_{\cdot FD}\,\mathrm{ad}(y^{-1})^{\cdot A}_B = C^A_{\cdot BE}\,\mathrm{ad}(y^{-1})^{\cdot B}_F\,\mathrm{ad}(y^{-1})^{\cdot E}_D , \\ d\,\mathrm{ad}(y^{-1})^{\cdot A}_B = C^A_{\cdot DE}\omega^D_y\,\mathrm{ad}(y^{-1})^{\cdot E}_B . \end{cases} \tag{2.20}$$

These identities are very useful in establishing the gauge covariance of field equations.

The map $a \to \mathrm{ad}(a^{-1})^T = \mathrm{cd}(a)$ is also a representation, the «coadjoint» representation of G, which is equivalent to the adjoint in a number of interesting cases (^{T} stands for the transposed matrix). Tangent vectors transform according to the coadjoint representation.

1-forms are essentially synonymous to covariant vector fields used by physicists. Contravariant fields appear in the mathematical literature as «tangent vectors» and are written in the form

$$U = u^M \frac{\partial}{\partial x^M}. \tag{2.21}$$

Given a 1-form $Q = \mathrm{d}x^M q_M$, we can define the scalar product $(U, Q) = u^M q_M$ as the «value» $Q(U)$ of Q on U. Clearly, $(\partial/\partial x^N)\,\mathrm{d}x^M = \delta^M_N$.

Given two tangent vectors T, U we have a natural bracket operation

$$[T, U] = \left[T^M \frac{\partial U^N}{\partial x^M} - U^M \frac{\partial T^N}{\partial x^M}\right] \frac{\partial}{\partial x^N}, \tag{2.22}$$

defining a Lie-algebra structure on tangent vectors. If we write

$$\omega^A = \omega_M^{\cdot A}\, \mathrm{d}x^M$$

and introduce $\Omega^M_{\cdot A}$ through

$$\omega_{M\cdot}^{\;A} \Delta_{A\cdot}^{\;N} = \delta^N_M \qquad \text{or} \qquad \Delta_B^{\cdot M} \omega_{M\cdot}^{\;A} = \delta^A_B, \tag{2.23}$$

we can then define tangent vectors D_A through

$$D_A = \Delta_A^{\cdot M} \frac{\partial}{\partial x^M}. \tag{2.24}$$

Clearly, $\omega^A(D_B) = \delta^A_{\cdot B}$. (Note that we use $A, B, \ldots$ for «Latin» or anholonomic indices and $M, N, \ldots$ for «Greek» or holonomic indices.)

After some algebraic manipulations one finds by using (2.23), (2.13)

$$[D_A, D_B] = C^E_{\cdot AB} D_E, \tag{2.25}$$

which shows that the D_A form a finite Lie algebra, *i.e.* by definition the Lie algebra $\mathscr{G}$ of G. In this role, the $C^E_{\cdot AB}$ reappear in a form more familiar to the physicist. However, (2.25) is fully equivalent in content to the Cartan-Maurer equations. From (2.23), (2.24) it is also clear that the D_A are invariant under left translations.

One can similarly introduce a right-invariant vector field S_B such that $\pi^A(S_B) = \delta^A_{\bullet B}$, and

$$[S_A, S_B] = -C^E_{\bullet AB} S_E \,. \tag{2.26}$$

The geometrical interpretation of D_A is clear when considering the identity $W(x^{-1}, x)\,W(x, e) = W(e, x) = 1$. From (2.11) and the second equation in (2.23) we have

$$D_A = W_A^{\bullet M}(x, e)\frac{\partial}{\partial x^M} \,. \tag{2.27}$$

But now, if $\psi(x)$ is a generic function on G, the expression

$$\mathrm{d}y^A D_A \psi(x) = \mathrm{d}y^A\, W_A^{\bullet M}(x, e)\frac{\partial}{\partial x^M}\psi(x) = \\ = \mathrm{d}y^A \frac{\partial x^M}{\partial y^A}\frac{\partial}{\partial x^M}\psi(x) = \mathrm{d}x^M\frac{\partial \psi}{\partial x^M} = \delta\psi \tag{2.28}$$

represents the infinitesimal change $\delta\psi$ of ψ under an infinitesimal translation to the right in the argument of ψ. In this sense, the D_A *generate right translations*. Similarly, the S_A generate left translations. Quite naturally then both the LI of D_A and the RI of S_A can be expressed as

$$[D_A, S_B] \equiv 0 \,. \tag{2.29}$$

3. – Examples of Lie groups of physical interest.

The (Euclidean) Poincaré group P is defined as the set of pairs $x = (\Xi, \boldsymbol{x})$, where $\Xi \in SO_4$, $\boldsymbol{x} \in \boldsymbol{R}^4$. The product rule is, by setting $t = (\Theta, \boldsymbol{t})$,

$$(Z, \boldsymbol{z}) = (\Xi\Theta, \Xi\boldsymbol{t} + \boldsymbol{x}) = (\Xi, \boldsymbol{x})(\Theta, \boldsymbol{t}) \,. \tag{3.1}$$

Thus the indices A, B etc. of sect. **2** run in the range 1, ..., 10. Because of the peculiar structure of P, we use $a = 1, \ldots, 4$ and the pairs (a, b) with $a > b$, instead of A. In this way, the differential form of (3.1), corresponding to (2.9), reads

$$\begin{cases} \mathrm{d}Z = \mathrm{d}\Xi\,\Theta + \Xi\,\mathrm{d}\Theta \,, \\ \mathrm{d}\boldsymbol{z} = \mathrm{d}\Xi\,\boldsymbol{t} + \mathrm{d}\boldsymbol{x} + \Xi\,\mathrm{d}\boldsymbol{t} \,. \end{cases} \tag{3.2}$$

The forms ω are obtained as in (2.11) by setting $\mathrm{d}\Xi, \mathrm{d}\boldsymbol{x} = 0$ in (3.2) and and then carrying out the replacements $(\Theta, \boldsymbol{t}) \to (\Xi, \boldsymbol{x})$ and $(\Xi, \boldsymbol{x}) \to (\Xi, \boldsymbol{x})^{-1}$.

The result is

$$(3.3)\qquad \begin{cases} \Omega = \Xi^{-1}\,\mathrm{d}\Xi\,, \\ \boldsymbol{o} = \Xi^{-1}\,\mathrm{d}\boldsymbol{x}\,. \end{cases}$$

Ω is thus a skew-symmetric matrix with elements $\omega^{ab} = -\,\omega^{ba}$. Componentwise we have

$$(3.4)\qquad \begin{cases} \omega^{ab} = \Xi^{ca}\,\mathrm{d}\Xi^{cb}\,, \\ o^{a} \;\; = \Xi^{ca}\,\mathrm{d}x^{c}\,. \end{cases}$$

The Cartan-Maurer equations (2.13) read

$$(3.5)\qquad \begin{cases} \mathrm{d}\omega^{ab} + \omega^{ac}\wedge\omega^{cb} = 0\,, \\ \mathrm{d}o^{a} \;\; + \omega^{ac}\wedge o^{c} \;\; = 0\,. \end{cases}$$

Similarly, the RI forms are (by putting $\mathrm{d}\Theta$, $\mathrm{d}t = 0$ and replacing $(\Theta, \boldsymbol{t})$ by $(\Xi, \boldsymbol{x})^{-1}$)

$$(3.6)\qquad \pi = \mathrm{d}\Xi\,\Xi^{-1}\,, \qquad \boldsymbol{p} = \mathrm{d}\boldsymbol{x} - \pi\boldsymbol{x}\,.$$

Tangent vectors appear as operators, linear in $\partial/\partial\Xi^{ab}$ and $\partial/\partial x^{a}$ and tangent to the defining variety of SO_4 with the normalization

$$\Xi^{ab}\,\Xi^{cb} = \delta^{ac}\,.$$

A convenient set of LI vectors is

$$(3.7)\qquad \begin{cases} D_{ab} = \Xi^{ca}\,\dfrac{\partial}{\partial\Xi^{cb}} - \Xi^{cb}\,\dfrac{\partial}{\partial\Xi^{ca}}\,, \\ D_{a} \;\; = \Xi^{ca}\,\dfrac{\partial}{\partial x^{c}}\,. \end{cases}$$

Clearly,

$$(3.8)\qquad \begin{cases} \Omega(D_{a}) \;\; = \boldsymbol{\omega}(D_{ab}) = 0\,, \\ \omega^{cf}(D_{ab}) = \delta^{ca}\delta^{fb} - \delta^{cb}\delta^{fa}\,, \\ \omega^{f}(D_{a}) \;\; = \delta^{f}{}_{a}\,. \end{cases}$$

For the RI vector fields we find

$$(3.9)\qquad \begin{cases} S_{ab} = x^{a}\,\dfrac{\partial}{\partial x^{b}} - x^{b}\,\dfrac{\partial}{\partial x^{a}} + \Xi^{af}\,\dfrac{\partial}{\partial\Xi^{bf}} - \Xi^{bf}\,\dfrac{\partial}{\partial\Xi^{af}}\,, \\ S_{a} \;\; = \dfrac{\partial}{\partial x^{a}}\,. \end{cases}$$

A general field $\psi(x)$ can be considered as an element of the Poincaré group with its x^M-dependence mentioned explicitly and its dependence on Ξ implicit and by fixing its « intrinsic » spin (except that we have to replace the Poincaré group P by its covering group). It is the usual convention (3.1) in defining the Poincaré group product that picks left translations (*i.e.* S_A) for co-ordinates and fields: $(\Lambda, \boldsymbol{a})(\Xi, \boldsymbol{x}) = (\Lambda\Xi, \Lambda\boldsymbol{x} + \boldsymbol{a})$, which, *e.g.*, for $\Xi = 0$, $\boldsymbol{x} = x^\mu$ and an infinitesimal P-transformation $\Lambda = \delta^{\mu\nu} + \varepsilon^{[\mu\nu]}$, $\boldsymbol{a} = \alpha^\mu$ yields $(1, x^\mu) \to (1, x^\mu + \varepsilon^{[\mu\nu]}x^\nu + \alpha^\mu)$. Had we used right translation, we would have found $(1, x^\mu) \to (1, x^\mu)$. The entire picture is thus RI and it is natural that the Hilbert-space considerations should lead to the (S_{ab}, S_a) set of (3.9) with « total » angular momentum S_{ab} including both that linear action on field components (« spin ») represented by the $\Xi^{[af}(\partial/\partial\Xi^{b]f})$ part, and a complementary « orbital » piece $x^{[a}(\partial/\partial x^{b]})$.

On the other hand, the procedure for « gauging » a symmetry implies the use of (co-ordinate invariant) forms, which thus have to be the LI forms ω^A (and their orthonormal LI algebra D_A). The LI set $\{D_{ab}, D_a\}$ is naturally factorized (up to a Lorentz transformation on the D_a) and will thus appear in the infinitesimal treatment of gauging [9, 10] with only D_{ab} giving rise to a gauge invariance of the Lagrangian. The action of D_a merges with general co-ordinate transformations on the x-space and a modified $\hat{D}_a$ can indeed be identified with the latter [10].

The formalism can be extended to supergroups. A most interesting example is provided by the graded Poincaré group (GP) [11-13]. Consider first a Grassmann algebra $A = A^+ \oplus A^-$, where A^+, A^- are, respectively, the even and odd terms in A. We can then form a set GP of triples $x = (\Xi, \boldsymbol{x}, \xi)$ with $x^a \in A^+$, $\xi^\alpha \in A^-$, $\Xi \in SO_4$. We consider $\boldsymbol{x}$ as an SO_4 vector, ξ as a Majorana spinor. We define a product

$$z = (Z, \boldsymbol{z}, \zeta) = xt = \left(\Xi\Theta, \Xi\boldsymbol{t} + \boldsymbol{x} + \frac{\bar{\xi}\gamma u(\Xi)\theta}{2}, u(\Xi)\theta + \xi\right) \tag{3.10}$$

with $t = (\Theta, \boldsymbol{t}, \theta)$.

This product is associative provided that

$$u^{-1}(\Xi)\gamma^a u(\Xi) = \Xi^{ab}\gamma^b \text{ or, in brief, } u^{-1}(\Xi)\gamma u(\Xi) = \Xi\gamma . \tag{3.11}$$

Upon differentiation of (3.10) we get

$$\begin{cases} dZ = d\Xi\,\Theta + \Xi\, d\Theta , \\ d\boldsymbol{z} = d\Xi\,\boldsymbol{t} + d\boldsymbol{x} + \frac{1}{2}\, d\bar{\xi}\,\gamma u(\Xi)\theta + \frac{1}{2}\bar{\xi}\gamma du(\Xi)\theta + \Xi\, d\boldsymbol{t} + \frac{1}{2}\bar{\xi}\gamma u(\Xi)\, d\theta , \\ d\zeta = du(\Xi)\theta + d\xi + u(\Xi)\, d\theta . \end{cases} \tag{3.12}$$

Repeating the procedure leading to ω we obtain

$$(3.13)\qquad \begin{cases} \Omega = \Xi^{-1}\,\mathrm{d}\Xi\,, \\ \boldsymbol{o} \;= \Xi^{-1}(\mathrm{d}\boldsymbol{x} - \frac{1}{2}\bar{\xi}\gamma\,\mathrm{d}\xi)\,, \\ \omega \;= u^{-1}(\Xi)\,\mathrm{d}\xi\,. \end{cases}$$

The corresponding RI forms are

$$(3.14)\qquad \begin{cases} \Pi = \mathrm{d}\Xi\,\Xi^{-1}\,, \\ \boldsymbol{p} \;= \mathrm{d}\boldsymbol{x} - \mathrm{d}\Xi\,\Xi^{-1}\boldsymbol{x} - \frac{1}{2}\mathrm{d}\bar{\xi}\,\gamma\xi - \frac{1}{4}\bar{\xi}\gamma\pi^{ab}\sigma^{ab}\,\xi\,, \\ \pi \;= \mathrm{d}\xi - \frac{1}{2}\pi^{ab}\sigma^{ab}\xi\,. \end{cases}$$

In deriving these results we used the identities (see the appendix)

$$(3.15)\qquad \begin{cases} \frac{1}{2}\pi^{ab}\sigma^{ab} = \mathrm{d}u(\Xi)u^{-1}(\Xi)\,, \\ \frac{1}{2}\omega^{ab}\sigma^{ab} = u^{-1}(\Xi)\,\mathrm{d}u(\Xi)\,. \end{cases}$$

Supersymmetric theories are frequently constructed [13, 14-16] in «superspace» $\mathbf{R}^{4/4}$. Elements of $\mathbf{R}^{4/4}$ form equivalence classes of GP under right multiplication by SO_4 elements. Therefore,

$$(\Xi,\, \boldsymbol{x},\, \xi) \simeq (\Xi\Theta,\, \boldsymbol{x},\, \xi)$$

and $\mathbf{R}^{4/4}$ is conveniently parametrized by (x, ξ) only. GP multiplication on the left-hand side by $(\Theta, \boldsymbol{t}, \theta)$, therefore, acts naturally on $\mathbf{R}^{4/4}$ and we have

$$(3.16)\qquad \begin{cases} x' = \Theta\boldsymbol{x} + \boldsymbol{t} + \bar{\theta}\gamma u(\Theta)\,\xi\,, \\ \xi' = u(\Theta)\,\xi + \theta\,. \end{cases}$$

GP thus becomes a generalization of the Euclidean group P on $\mathbf{R}^4$. Here the metric

$$(3.17)\qquad \mathrm{d}s^2 = \mathrm{d}x^a \otimes \mathrm{d}x^a = \boldsymbol{o}^a \otimes \boldsymbol{o}^a$$

is obviously left invariant on P and, therefore, P invariant on $\mathbf{R}^4$. Similarly, we have on $\mathbf{R}^{4/4}$, using (3.13),

$$(3.18)\qquad \mathrm{d}s^2 = \boldsymbol{o}^a \times \boldsymbol{o}^a + k\bar{\omega} \times \omega\,,$$

which is also GP invariant. Notice, however, that on odd superforms the product [7] $\wedge(\otimes)$ is commutative (anticommutative) and on even superforms

$\wedge(\otimes)$ is anticommutative (commutative) according to the usual convention (2.3). In this way we retrieve the well-known metric [17, 18] of the superspace « vacuum ». Also of great interest are the tangent vectors. The LI tangent vectors are given by

$$\left\{\begin{aligned} D_{ab} &= \Xi^{ca}\frac{\partial}{\partial \Xi^{cb}} - \Xi^{cb}\frac{\partial}{\partial \Xi^{ca}}\,, \\ D_a &= \Xi^{ca}\frac{\partial}{\partial x^c}\,, \\ D &= u^{-1}(\Xi)\left(\frac{\partial}{\partial \xi} - \frac{1}{2}\frac{\partial}{\partial x^m}\gamma^m\xi\right). \end{aligned}\right. \tag{3.19}$$

The RI tangent vectors are

$$\left\{\begin{aligned} S_{ab} &= x^a\frac{\partial}{\partial x^b} - x^b\frac{\partial}{\partial x^a} + \Xi^{af}\frac{\partial}{\partial \Xi^{bf}} - \Xi^{bf}\frac{\partial}{\partial \Xi^{af}} + \xi\sigma^{ab}\frac{\partial}{\partial \xi}\,, \\ S_a &= \frac{\partial}{\partial x^a}\,, \\ S &= \frac{\partial}{\partial \xi} + \frac{1}{2}\frac{\partial}{\partial x^m}\gamma^m\xi\,. \end{aligned}\right. \tag{3.20}$$

By neglecting the $\partial/\partial\Xi$ terms in S_A (spin), we obtain the infinitesimal generators of motions on $\boldsymbol{R}^{4/4}$. The D-components are related to the « covariant derivatives » of Salam and Strathdee [2]. The Maurer-Cartan equations appear as

$$\left\{\begin{aligned} &d\omega^{ab} + \omega^{ac}\wedge\omega^{cb} = 0\,, \\ &do^a + \omega^{ac}\wedge o^c + \tfrac{1}{2}\bar{\omega}\gamma^a\wedge\omega = 0\,, \\ &d\omega + \tfrac{1}{2}\omega^{ab}\sigma^{ab}\wedge\omega = 0\,. \end{aligned}\right. \tag{3.21}$$

In deriving (3.19) to (3.21), care must be taken to keep a consistent ordering of Fermi variables in order to avoid confusing results.

4. – Curvature and covariant derivatives.

Let G be endowed with a set ϱ^A of 1-forms (a « ϱ-bein » or « fundamental forms ») generalizing ω^A, but not subjected to the Maurer-Cartan equation. We define the *curvature* R^A as

$$R^A = d\varrho^A - \tfrac{1}{2}C^A_{\cdot BE}\,\varrho^B\wedge\varrho^E\,. \tag{4.1}$$

Hence ω^A has vanishing curvature. The covariant derivative of a set of p-forms η^A is given by (*)

$$(\mathrm{D}\eta)^A = \mathrm{d}\eta^A - C^A_{\cdot BE}\varrho^B \wedge \eta^E . \tag{4.2a}$$

It is convenient to introduce a similar definition of a covariant derivative of a « contravariant » set η_A:

$$(\mathrm{D}\eta)_A = \mathrm{d}\eta_A + C^E_{\cdot BA}\varrho^B \wedge \eta_E . \tag{4.2b}$$

If we vary ϱ^A in (4.1) by the infinitesimal $\delta\varrho^A$, we find

$$\delta R^A = (\mathrm{D}\delta\varrho)^A . \tag{4.3}$$

We also have

$$(\mathrm{D}^2\eta)^A = (\mathrm{D}(\mathrm{D}\eta))^A = - C^A_{\cdot BE} R^B \wedge \eta^E , \tag{4.4}$$

$$(\mathrm{D}R)^A = 0 \qquad \text{(Bianchi identities)}, \tag{4.5}$$

$$\mathrm{d}(\eta^A \zeta_A) = (\mathrm{D}\eta)^A \wedge \zeta_A + (-1)^p \eta^A \wedge (\mathrm{D}\zeta)_A . \tag{4.6}$$

We next consider two copies of G, $(G_x; G_y)$ with co-ordinates $(x; y)$, respectively, and a map λ: $G_x \xrightarrow{\lambda} G_y$. Let in general ω_x, ϱ_x, etc. (ω_y, ϱ_y, etc.) be forms on G_x (G_y), respectively. To any such map λ we define a *gauge transformation* $\Gamma(\lambda)$ as follows (see (2.19)):

$$\Gamma(\lambda)\varrho^A_x = \lambda^* \omega^A_y + \varrho^B_x \,\mathrm{ad}(\lambda^{-1}(x))^{\cdot A}_B . \tag{4.7}$$

(Recall that the dual map λ^* is defined in (2.6) and ad(a) in (2.16).)

It is often convenient to rewrite (4.7) for infinitesimal gauge transformations. In this case, we suppose that $\lambda(x)$ is given by $\lambda^M(x)$, where λ^M are infinitesimals. Using (2.11) and (2.20), we then find that the infinitesimal change is given by

$$\delta\varrho^A = \mathrm{D}\lambda^A , \tag{4.8}$$

where the r.h.s. is defined by (4.2a).

Using (4.1) and (2.20) one verifies that the curvature $\Gamma(\lambda)R^A_x$ calculated from $\Gamma(\lambda)\varrho^A_x$ is given by

$$\Gamma(\lambda) R^A_x = R^B_x \,\mathrm{ad}(\lambda^{-1}(x))^{\cdot A}_B . \tag{4.9}$$

(*) We write $\mathrm{D}^{(G)}$ for G only when the nature of the group has to be explicitly stated.

Similarly, by taking

$$(4.10a)\qquad \begin{cases} \Gamma(\lambda)\eta^A = \eta^B \operatorname{ad}(\lambda^{-1}(x))_B^{\cdot A}, \\ \Gamma(\lambda)\zeta_A = \zeta_B \operatorname{cd}(\lambda^{-1}(x))^B_{\cdot A} = \operatorname{ad}(\lambda(x))_A^{\cdot B}\zeta_B, \end{cases}$$

then

$$(4.10b)\qquad \begin{cases} (\mathrm{D}\Gamma(\lambda)\eta)^A = (\mathrm{D}\eta)^B \operatorname{ad}(\lambda^{-1}(x))_B^{\cdot A}, \\ (\mathrm{D}\Gamma(\lambda)\zeta)_A = (\mathrm{D}\zeta)_B \operatorname{cd}(\lambda^{-1}(x))^B_{\cdot A} = \operatorname{ad}(\lambda(x))_A^B(\mathrm{D}\zeta)_B. \end{cases}$$

Equations (4.8), (4.10) display the covariant nature of D and curvature.

Let $H \in G$ be a subgroup of G of dimension c'. Let $M = G/H$ be the corresponding homogeneous manifold. In the sequel, we often refer to the case $G = P$, $H = SO_4$, $M = \mathbf{R}^4$, or also $G = \mathrm{GP}$, $H = SO_4$, $M = \mathbf{R}^{4/4}$.

G can be considered as a principal bundle with fiber H, base space M. In the above-quoted case, this bundle is trivial, but already for $G = SO_3$, $H = SO_2$, $M = S^2$ one finds a nontrivial Hopf fibering, used in dealing with Dirac monopoles. Similar situations will arise in gravity and supergravity, and it is thus better to use from the start the general formal machinery developed for nontrivial bundles. In the following, we treat the conventional case of Bose-type manifolds and refer the reader to ref. [6-7] for the (Fermi) graded case, although we shall later develop in detail GP with a graded manifold.

Let U_α be a covering of M, p the projection $G \xrightarrow{p} M$. $p^{-1}(U_\alpha)$ is parametrized by elements (y, q_α, α), where $y \in M$, $q_\alpha \in H$. We give transition functions $\varphi_{\alpha\beta}(y) \in H$ on $U_\alpha \cap U_\beta$, such that, on $U_\alpha \cap U_\beta \cap U_\gamma$, $\varphi_{\alpha\beta}\varphi_{\beta\gamma} = \varphi_{\alpha\gamma}$. We then identify elements

$$(y, q_\alpha, \alpha) = (y, q_\beta, \beta),$$

where $q_\alpha = \varphi_{\alpha\beta} q_\beta$ and $y \in U_\alpha \cap U_\beta$. In this way, all sets $p^{-1}(U_\alpha)$ are « glued » together into G. On U_α we have a map χ_α:

$$p^{-1}(U_\alpha) \xrightarrow{\chi_\alpha} G,$$

$$y_\alpha(y, q_\alpha, \alpha) = q_\alpha.$$

Given the LI form ω^A on G, χ_α defines a form $\chi^*_\alpha\omega^A$ on $p^{-1}(U_\alpha)$. For each α we suppose given the form

$$(4.11)\qquad \varrho^A{}_\alpha = \chi^*_\alpha\omega^A + \tau^B{}_\alpha \operatorname{ad}(q_\alpha^{-1})_B^{\cdot A},$$

where $\tau^B{}_\alpha$ is a 1-form on M. If the set $\varrho^A{}_\alpha$ has to define a unique form ϱ^A on G independent of α, we must have the matching conditions

$$(4.12)\qquad \tau^A{}_\alpha = \varphi^*_{\beta\alpha}\omega^A + \tau^B{}_\beta \operatorname{ad}(\varphi^{-1}_{\beta\alpha})_B^{\cdot A}.$$

If (4.12) holds, ϱ^A is called a *factorized* c-bein on the pair (G, H). Let $\mathscr{G}$, $\mathscr{H}$ be the Lie algebras of G, H, respectively. We take a vector subspace $\mathscr{F}$ of $\mathscr{G}$ such that $\mathscr{G} = \mathscr{H} + \mathscr{F}$, $\mathscr{F} \cap \mathscr{H} = \emptyset$. Each element $a \in H$ induces a coadjoint transformation on $\mathscr{G}$ which leaves $\mathscr{H}$ invariant:

$$\mathrm{cd}(a)\mathscr{H} \subset \mathscr{H} .$$

If, however, $\mathscr{F}$ is such that $\mathrm{cd}(a)\mathscr{F} \subset \mathscr{F}$ as well, we say that $M = G/H$ is a *weakly reductive homogeneous manifold*. This happens in particular if H is compact, semi-simple and connected, or even discrete. If weak reducibility holds, $\mathrm{cd}(a)$ and $\mathrm{ad}(a)$ each reduce completely into two representations of H for $a \in H$. Correspondingly, we split the set of forms ω^A into sets ω_H, ω_F such that

$$\omega^H(\mathscr{F}) = 0 , \qquad \omega^F(\mathscr{H}) = 0 .$$

These sets are acted upon separately by the two representations in which $\mathrm{ad}(a)$ splits. Clearly we have also

$$[\mathscr{H}, \mathscr{F}] \subset \mathscr{F}$$

If, furthermore, $[\mathscr{F}, \mathscr{F}] \in \mathscr{H}$, M is a *symmetric* manifold. If weak reducibility holds, the set of forms ϱ_H with eqs. (4.11), (4.12) satisfies the conditions for a *Cartan connection* on G as a principal bundle. In this case the H-components of R^A represent *proper curvature* and the F-components are known as the *torsion*. If $[\mathscr{F}, \mathscr{F}] = 0$, then the curvature of ϱ_H as a Cartan connection coincides with the G-curvature (4.1). All these conditions are met when $G = P$, $H = SO_4$, $M = \mathbf{R}^4$, which can be identified as a structure with ordinary Riemannian geometry. In this case then (4.11) reduces to

$$(4.13)\qquad \begin{cases} \varrho^{ab} = (\Xi^{-1}\,\mathrm{d}\Xi)^{ab} + \tau^{cd}\,\Xi^{db}\,\Xi^{ca} , \\ \varrho^{a} \;\, = \Xi^{ca}\,\tau^{c} . \end{cases}$$

Clearly, then

$$(4.14)\qquad \begin{cases} R^{ab} = \Xi^{db}\,\Xi^{ca}\,\mathscr{R}^{cd} , \\ R^{a} \;\, = \Xi^{ca}\,\mathscr{R}^{c} , \end{cases}$$

where

$$(4.15)\qquad \begin{cases} \mathscr{R}^{cd} = \mathrm{d}\tau^{cd} + \tau^{ca}\wedge\tau^{ad} , \\ \mathscr{R}^{c} \;\, = \mathrm{d}\tau^{c} + \tau^{ca}\wedge\tau^{a} . \end{cases}$$

As we saw in (4.9),

$$(4.9a)\qquad R^A = \mathscr{R}^B_\alpha\, \mathrm{ad}(q_\alpha^{-1})_B^{\cdot A} .$$

In a Riemannian manifold, the condition $\mathscr{R}^a = 0$ is imposed and the metric tensor is given by

$$g_{\mu\nu}\, dy^\mu \otimes dy^\nu = \sum_c \tau^c \otimes \tau^c , \tag{4.16}$$

τ^c is then identified with the conventional vierbein, and $\mathscr{R}^c = 0$ allows one to calculate τ^{ca} from τ^a and $d\tau^a$ through a procedure which is essentially the known formula for the Christoffel symbols in terms of the metric tensor. Once τ^{ab} is known, the curvature $\mathscr{R}^{ab}$ calculated from (4.15) yields the conventional Riemann tensor.

Some generalized forms of conventional relativity have been proposed, where torsion does not necessarily vanish, according to the original ideas—long forgotten—of Cartan [9, 19-22]. These proposals, although far from any physical application [22], are nevertheless intringuing in that they remove in part the disparity of treatment between « internal » and « external » components of the ϱ^A, $\mathscr{R}^A$ multiplets and deal effectively with a Poincaré curvature rather than with a Lorentz curvature only, as used by EINSTEIN. To the extent that we might be tempted to give a more fundamental meaning to the group G itself rather than to a factor G/H, we should certainly move in that direction.

However, present theories with torsion do seem to have some drawback in this respect, since torsion does not propagate. While ordinary curvature does not vanish outside of matter, torsion does [22]. This disparity probably has its roots in the factorization hypothesis, which breaks from the very beginning the symmetry of the Poincaré group, since it factorizes out in a trivial way the Lorentz-group dependence. (We have used SO_4 in our calculation for simplicity, but all results obviously carry for $SO_{1,3}$ with minor modifications.) If we do not assume factorization, the dependence of the e-bein on the Lorentz variables must be dictated by the field equations and boundary conditions. Section **5** contains a discussion on how to attempt this task and on the difficulties encountered.

Now we review the Poincaré gauge transformations, before and after factorization. Equation (4.7) is our fundamental starting point, since it realizes a transformation in which the « y » copies of the group (the transformation parameters) are functions $y = \lambda(x)$ of the « x » copy variables (now used as the background manifold, an extension of space-time). For the P-gauge functions

$$U\colon\; U^{ab}(\Xi, x),\; u^a(\Xi, x) ,$$

we find the transformations

$$\begin{cases} (\varrho')^{ab} = (\Gamma(U)\varrho)^{ab} = (U^{-1})^{ac}\, dU^{cb} + (U^{-1})^{ac}\varrho^{cd}\, U^{db} , \\ (\varrho')^a \;= (\Gamma(U)\varrho)^a \;= (U^{-1})^{ac}\{\varrho^c + du^c + \varrho^{cd} u^d\} . \end{cases} \tag{4.17}$$

For infinitesimal transformations, $\varepsilon^{ab} = U^{ab} - \delta^{ab}$, $\varepsilon^a = u^a$, we get

$$(4.18)\qquad \begin{cases} \delta\varrho^{ab} = d\varepsilon^{ab} + \varrho^{ac}\varepsilon^{cb} - \varepsilon^{ac}\varrho^{cb} = D\varepsilon^{ab}\,, \\ \delta\varrho^{a} \;\;= d\varepsilon^{a} + \varrho^{ac}\varepsilon^{c} - \varepsilon^{ac}\varrho^{c} = D\varepsilon^{a}\,. \end{cases}$$

Note that D is here the Poincaré covariant derivative and includes the ε^{ac} action for ϱ^a. Holonomically, it entails derivatives and summations over all ten variables and differentials. However, we show in (5.12) that the full Poincaré gauge is too strong as a symmetry of the Einstein Lagrangian. Instead, factorization of the Ξ^{ab} variables according to (4.11), (4.12) reduces the gauge group to $H = SO_4$:

$$(4.19)\qquad \begin{cases} (\tau')^{ab} = (U^{-1})^{ac}\, dU^{cb} + (U^{-1})^{ac}\tau^{cd}\, U^{db}\,, \\ (\tau')^{a} \;\;= (U^{-1})^{ac}\tau^{c}\,. \end{cases}$$

Notice the similarity with (4.13), the factorization equation. The action of an infinitesimal ε^{ab} is the same as in (4.18),

$$(4.20)\qquad \begin{cases} \delta\tau^{ab} = d\varepsilon^{ab} + \tau^{ad}\varepsilon^{db} - \varepsilon^{ac}\tau^{cb} = D^{(H)}\varepsilon^{ab}\,, \\ \delta\tau^{a} \;\;= -\varepsilon^{ac}\tau^{c}\,. \end{cases}$$

We thus do not appeal to « gauge the translations », *i.e.* the u^a and ε^a gauge functions are not used.

A gauge theory expressed in terms of forms is automatically covariant, that is invariant under the group of general co-ordinate transformations (GCTG). An interesting further adaptation was developed originally by HEHL *et al.* [22] and by VON DER HEYDE [10], in the pursuance of the program begun by UTIYAMA, SCIAMA, KIBBLE [9] and aiming at a derivation of general relativity from gaugelike principles. The new equations are useful in connecting the GCTG action to that of the gauge group ($H \subset G$ in our above treatment). An infinitesimal element of GCTG is given by a tangent vector ε^M on G. We then have

$$(4.21)\qquad \delta x^M = \varepsilon^M$$

and

$$(4.22)\qquad \delta\varrho^B = \delta(dx^M \varrho_M^{\bullet B}) = dx^N \frac{\partial\varepsilon^M}{\partial x^N}\varrho_M^{\bullet B} + dx^M \varepsilon^N \frac{\partial\varrho_M^{\bullet B}}{\partial x^N} = $$
$$= dx^N\left[\frac{\partial\varepsilon^B}{\partial x^N} + \varepsilon^M\frac{\partial\varrho_N^{\bullet B}}{\partial x^M} - (-1)^{nm}\varepsilon^M\frac{\partial\varrho_M^{\bullet B}}{\partial x^N}\right],$$

where n, m are the gradings of the indices N, M for the case of a graded manifold and where we have defined an anholonomic co-ordinate variation

$$\varepsilon^B = \varepsilon^M \varrho_M^{\cdot B} . \tag{4.23}$$

The last line can be rewritten as

$$\delta\varrho^B = \mathrm{d}\varepsilon^B - 2(\varepsilon, \mathrm{d}\varrho^B) , \tag{4.24}$$

since

$$\mathrm{d}\varrho^B = -\frac{1}{2}(\mathrm{d}x^N \wedge \mathrm{d}x^M)\left(\frac{\partial\varrho_N^{\cdot B}}{\partial x^M} - (-1)^{nm}\frac{\partial\varrho_M^{\cdot B}}{\partial x^N}\right) \tag{4.25}$$

and the scalar product $(\varepsilon, \mathrm{d}\varrho^B)$ is a 1-form obtained by contracting ε^M with the second factor $\mathrm{d}x^M$ in $\mathrm{d}\varrho^B$. Using the G-covariant derivative $\mathrm{D}\varepsilon^B$ as in (4.2a), we find

$$\begin{cases} \delta\varrho^B = \mathrm{D}\varepsilon^B + C^B_{\cdot DE}\varrho^D \wedge \varepsilon^N \varrho_N^{\cdot E} - 2(\varepsilon, \mathrm{d}\varrho^B) = \\ \qquad\qquad = \mathrm{D}\varepsilon^B + (\varepsilon, -2\,\mathrm{d}\varrho^B + C^B_{\cdot DE}\varrho^D \wedge \varrho^E) , \\ \delta\varrho^B = \mathrm{D}\varepsilon^B - 2(\varepsilon, R^B) = \mathrm{D}\varepsilon^B - \varrho^E \varepsilon^F R_{EF}^{\cdot\cdot B} . \end{cases} \tag{4.26}$$

We refer to (4.26) as the anholonomized general co-ordinate transformation (AGCT). Whenever $R^B = 0$, the AGCT coincides with an ordinary gauge transformation (4.8).

Yet a third way of introducing vector fields on G consists in using the natural extension $\tilde{D}_B$ of the set D_B given in (2.27). We define the $\tilde{D}_B$ as in (2.23), (2.24):

$$\varrho^A(D_B) = \delta^A_{\cdot B} . \tag{4.27}$$

The commutators of the $\tilde{D}_B$ no longer close to form the Lie algebra $\mathscr{G}$ of G. In order to identify the necessary modifications, we start from the definition of curvature (4.1) in terms of the anholonomic components of R^A

$$R^A = \tfrac{1}{2}\varrho^B \wedge \varrho^E R_{BE}^{\cdot\cdot A} , \tag{4.28}$$

thus yielding

$$\mathrm{d}\varrho^A - \tfrac{1}{2}\varrho^B \wedge \varrho^E (C^A_{\cdot BE} + R_{BE}^{\cdot\cdot A}) = 0 . \tag{4.29}$$

This entails the (graded Lie bracket) commutation relations

$$[\tilde{D}_B, \tilde{D}_E] = (C^A_{\cdot BE} + R_{BE}^{\cdot\cdot A})\tilde{D}_A . \tag{4.30}$$

In case of factorization under a subgroup $H \subset G$, some important simplifications occur. In particular, $\tilde{D}_A = D_A$ for $A \in \{\mathscr{H}\}$, where { } denotes the range.

Indeed, for $F\in\{\mathscr{F}\}$

$$\varrho^F(D_A)=\omega^F(D_A')=0\,,$$

because τ_α^B in (4.11) contains only « horizontal » (*i.e.* $M=G/H$) differentials, orthogonal to D_A. Similarly, for $B\in\{\mathscr{H}\}$

$$\varrho^B(D_A)=\omega^B(D_A)=\delta^B_{\cdot A}\,.$$

Therefore, quite trivially, the commutation relations of the $\tilde{D}_A$, $A\in\{\mathscr{H}\}$, are the same as those of D_A. In addition, the commutator $[D_B,\tilde{D}_E]$ for $B\in\{\mathscr{H}\}$ is equal to the « flat » case for all E, since $R_{BE}^{\cdot\cdot A}=0$ whenever $B\in\{\mathscr{H}\}$, $\forall E$. Indeed, according to (4.9*a*), R^A has no vertical components since $\mathscr{R}^A$ is constructed on $M=G/H$ only.

The factorized *e*-bein for the pair (G,H) can be written (4.11) as

$$\text{(4.31)}\qquad \begin{cases} \varrho^A=\omega_H^A+\tau^B\text{ad}(q^{-1})_B^{\cdot A}\,, & A\in\{\mathscr{H}\}\,,\\ \varrho^A=\tau^B\,\text{ad}(q^{-1})_B^{\cdot A}\,, & A\in\{\mathscr{F}\}\,, \end{cases}$$

where, for simplicity, we neglect the index α and we write ω_H^A for $\chi_\alpha^*\omega^A$. As stated, $\tilde{D}_A=D_A$ for $A\in\{\mathscr{H}\}$. Now we compute explicitly $\tilde{D}_F$ for $F\in\{\mathscr{F}\}$. We use co-ordinates z^μ on $M=G/H$, this being the case if

$$\text{(4.32)}\qquad \begin{cases} T_C^{\cdot\mu}\tau_\mu^{\cdot B}=\delta_C^B\,,\\ T_C\;=\text{ad}(q)_C^{\cdot E}\,T_E^{\cdot\mu}\dfrac{\partial}{\partial z^\mu}\,, \end{cases}$$

so that

$$\text{(4.33)}\qquad \varrho^A(T_C)=\delta^A_{\cdot C}\,,\qquad A,C\in\{\mathscr{F}\}\,.$$

This suggests the form

$$\text{(4.34)}\qquad \begin{cases} \tilde{D}_C=\text{ad}(q)_C^{\cdot E}\,T_E^{\cdot\mu}\left(\dfrac{\partial}{\partial z^\mu}-\dfrac{1}{2}\sum\limits_{A\in\{\mathscr{H}\}}\tau_\mu^{\cdot A}S_A\right)=\Delta_C^{\cdot\mu}D_\mu^{(H)}\,, & C\in\{\mathscr{F}\}\,,\\ \tilde{D}_C=D_C\,, & C\in\{\mathscr{H}\}\,, \end{cases}$$

where S_A is the RI vector field on H. The quantity in the brackets corresponds to the conventional covariant derivative $D_\mu^{(H)}$ on M. Indeed, a short computation will show that (4.34) satisfies (4.27):

$$\varrho^A(\tilde{D}_B)=\delta^A_{\cdot B}\,,\qquad \forall A,B\,.$$

Now we apply the vector fields $\tilde{D}_A$ to x^M, as a specific realization of (4.21). This corresponds to

$$\varepsilon_A^{\cdot M} = \Delta_A^{\cdot M}\,, \qquad \varepsilon_A'^{\cdot B} = \delta_A^{\cdot B}$$

(by using (4.23)). Applying these fields to (4.26), we obtain

$$\delta_A \varrho^B = -\varrho^D (C^{\cdot B}_{\cdot DA} + R_{DA}^{\cdot\cdot B})\,, \tag{4.35}$$

which generalizes the global group transformation and exhibits the role of the $\tilde{D}_A$ in generating when «gauged» ordinary gauge transformations (4.8) for $A \in \{\mathscr{H}\}$ and AGCT for $A \in \{\mathscr{F}\}$ as in (4.26). Note, however, that the $\tilde{D}_A$ do not close to form a graded Lie algebra in a strict sense, unless $A \in \{\mathscr{H}\}$.

Specifying to P, we have the AGCT

$$\begin{cases} \delta\varrho^{ab} = \mathrm{D}\varepsilon^{ab} - 2(\varepsilon, R^{ab})\,, \\ \delta\varrho^{a} \;\, = \mathrm{D}\varepsilon^{a} - 2(\varepsilon, R^{a})\,, \end{cases} \tag{4.36}$$

which should be compared with (4.18).

However, as explained above, (4.2a) ensures that in the factorized case as given by (4.13) we have

$$R_{(cd)(ef)}^{\cdot\;\;\cdot\;\;(ab)} = R_{(cd)(ef)}^{\cdot\;\;\cdot\;\;a} = R_{(cd)e}^{\;\;\;\;(ab)} = R_{(cd)e}^{\;\;\;\;a} = 0\,, \tag{4.37}$$

so that (D is still $\mathrm{D}^{(P)}$)

$$\delta\varrho^{ab} = \mathrm{D}\varepsilon^{ab} - \varrho^c \varepsilon^d R_{cd}^{\cdot\cdot ab}\,, \tag{4.38a}$$

$$\delta\varrho^{a} \;\, = \mathrm{D}\varepsilon^{a} - 2(\varepsilon, R^{a})\,, \tag{4.38b}$$

so that ε^{ab} of SO_4 produces an ordinary gauge, whereas the translations ε^a are indeed replaced by an anholonomized co-ordinate transformation gauge. The corresponding Lie-algebra generators $\tilde{D}_{ab}$ and vector fields $\tilde{D}_a$ in the bundle of tangent frames are

$$\tilde{D}_{ab} = D_{ab}\,, \qquad \tilde{D}_a = \Xi^{ba} T_b^{\cdot\nu}(\partial_\nu - \tfrac{1}{2}\tau^{cd}(x) S_{cd}^{(H)})\,, \tag{4.39}$$

$$S_{cd}^{(H)} = \Xi^{cf}\frac{\partial}{\partial \Xi^{df}} - \Xi^{df}\frac{\partial}{\partial \Xi^{cf}}\,, \qquad T_b^{\cdot\nu}\tau_\nu^{\cdot a} = \delta_b^a \tag{4.40}$$

with the commutators

$$[D_{ab}, D_{cd}] = \delta_{ad} D_{bc} + \delta_{bc} D_{ad} - \delta_{ac} D_{bd} - \delta_{bd} D_{ac}\,, \tag{4.41a}$$

$$[D_{ab}, \tilde{D}_c] = \delta_{bc}\tilde{D}_a - \delta_{ac}\tilde{D}_b\,, \tag{4.41b}$$

$$[\tilde{D}_a, \tilde{D}_b] = R_{ab}^{\cdot\cdot cd} D_{cd} + R_{ab}^{\cdot\cdot c}\tilde{D}_c\,. \tag{4.41c}$$

For GP we have for (4.26) before factorization

$$(4.42a) \qquad \delta\varrho^{ab} = \mathrm{d}\varepsilon^{ab} + \varrho^{at}\wedge\varepsilon^{tb} - \varrho^{tb}\wedge\varepsilon^{at} - 2(\varepsilon, R^{ab}) = \mathrm{D}\varepsilon^{ab} - 2(\varepsilon, R^{ab}),$$

$$(4.42b) \qquad \delta\varrho^{a} = \mathrm{d}\varepsilon^{a} + \varrho^{at}\wedge\varepsilon^{t} - \varrho^{t}\wedge\varepsilon^{at} + \bar{\varrho}\gamma^{a}\varepsilon - 2(\varepsilon, R^{a}) = \mathrm{D}\varepsilon^{a} - 2(\varepsilon, R^{a}),$$

$$(4.42c) \qquad \delta\varrho = \mathrm{d}\varepsilon + \tfrac{1}{2}(\varrho^{ab}\sigma^{ab})\wedge\varepsilon - \tfrac{1}{2}\sigma^{ab}\varrho\wedge\varepsilon^{ab} - 2(\varepsilon, R) = \mathrm{D}\varepsilon - 2(\varepsilon, R)$$

and, after factorization of SO_4 (with $\mathrm{D} = D^{(\mathrm{GP})}$),

$$(4.43a) \qquad \delta\varrho^{ab} = \mathrm{D}\varepsilon^{ab} - \varrho^{c}\varepsilon^{d}R_{cd}^{\bullet\bullet ab} - \varrho^{c}\bar{\varepsilon}^{\alpha}R_{c\alpha}{}^{ab},$$

$$(4.43b) \qquad \delta\varrho^{a} = \mathrm{D}\varepsilon^{a} - 2(\varepsilon, R^{a}),$$

$$(4.43c) \qquad \delta\varrho = \mathrm{D}\varepsilon - 2(\varepsilon, R).$$

Aside from (4.41*a*), (4.41*b*), we have for the most general factorized case the commutators

$$(4.44a) \qquad [D_{ab}, \tilde{D}] = (\sigma_{ab})\tilde{D},$$

$$(4.44b) \qquad [\tilde{D}_{a}, \tilde{D}_{b}] = R_{ab}^{\bullet\bullet cd}D_{cd} + R_{ab}^{\bullet\bullet c}\tilde{D}_{c} + \bar{R}_{ab}\tilde{D},$$

$$(4.44c) \qquad C_{\alpha\beta}[\tilde{D}_{a}, \tilde{D}_{\beta}] = R_{a\alpha}^{\bullet\bullet cd}D_{cd} + R_{a\alpha}^{\bullet\bullet c}\tilde{D}_{c} + \bar{R}_{a\alpha}^{\bullet\bullet\gamma}D_{\gamma}$$

(note that, though β in $\tilde{D}_{\beta}$ is an « upper-spinor » index as in ξ^{β} and has to be lowered by $C_{\alpha\beta}$, we write it as a subindex like a or (ab))

$$(4.44d) \qquad \{\tilde{D}_{\alpha}, \tilde{D}_{\beta}\} = (\gamma^{a})_{\alpha\beta}\tilde{D}_{a} + R_{\alpha\beta}^{\bullet\bullet cd}D_{cd} + R_{\alpha\beta}^{\bullet\bullet c}\tilde{D}_{c} + \bar{R}_{\alpha\beta}^{\bullet\bullet\gamma}\tilde{D}_{\gamma}.$$

One last remark about factorization: had we chosen $H = G$, then factorization means trivially $\varrho^{A} = \omega^{A}$, that is empty space and a global symmetry only. If instead $H = e$, factorization imposes no conditions on ϱ^{A}. All D_{A}-commutators generate curvature terms, which is another way of visualizing the difficulties of providing for an invariant action, as discussed in sect. **5**.

5. – Lagrangians and field equations.

It is often argued in much of the recent work on unified theories that the final goal should be a purely « geometrical » theory. The definition of « geometrical » is at the moment not clearly stated and correspondingly there is room for disagreement as to whether many of the proposed theories really fit the criteria for geometry. In an attempt to narrow the discussion, we introduce in the sequel a number of interesting geometric criteria:

I) The « objects » or fields in the theory are c-beins ϱ^A on a convenient Lie group (or supergroup). Note that this generally implies a self-sourced theory with no explicit « matter » fields; alternatively, the geometric criteria characterize only the « generalized-vacuum » limit of the theory.

IIa) A Lagrangian density is formed from the c-beins by using the operations D and $\wedge$ only, plus contraction over indices.

IIb) In addition to IIa), we allow the use of the dual (*) of a form (enlarged definition).

IIIa) We consider a preferred subgroup $H \subset G$ and construct a theory which is gauge invariant under H, but not under the full gauged G.

IIIb) In addition to IIIa), we request factorization of the c-beins (reducing the « objects » to fields on the G/H variables) and vanishing of torsion in the G/H manifold. Alternatively, factorization may appear as the result of IIIa).

All theories of current interest meet I). Classical general relativity satisfies I), IIa), IIIb) and does not involve IIb). From the results of sect. 4, conditions I), IIIb) are obvious. IIa) follows from the (Trautman) action [21]

$$8A = \int_{R^4} R^{ab}\wedge\varrho^c\wedge\varrho^d\,\varepsilon_{abcd} = \int_{R^4} \mathscr{R}^{ab}\wedge\tau^c\wedge\tau^d\,\varepsilon_{abcd}\,, \tag{5.1}$$

which coincides with Einstein's Lagrangian. (Again, we denote anholonomic indices by Latin lower-case indices, using lower-case Greek indices for the space variables.) Variation of (5.1) proceeds as follows: we vary first τ^{ab}, thus obtaining

$$8A = \int_{R^4} \delta\mathscr{R}^{ab}\wedge\tau^c\wedge\tau^d\,\varepsilon_{abcd} = \int \mathrm{D}(\delta\tau)^{ab}\wedge\tau^c\wedge\tau^d\,\varepsilon_{abcd}\,, \tag{5.2}$$

where $\mathrm{D}(\delta\tau)^{ab}$ can be interpreted either as a G-covariant (*i.e.* Poincaré P-covariant) or as an H- (= Euclidean « Lorentz » SO_4) covariant differentiation (the latter is the « anholonomic » conventional one in general relativity) [9, 22]. In the last event, integration by parts and the antisymmetry in $(a \leftrightarrow b)$ yields the equation

$$\int \delta\tau^{ab}\wedge \mathrm{D}\tau^c\wedge\tau^d\,\varepsilon_{abcd} = 0\,. \tag{5.3}$$

Now, by definition, the H-covariant derivative $\mathrm{D}\tau^c$ is just the torsion $\mathscr{R}^c$. It follows

$$\int \delta\tau^{ab}\wedge\mathscr{R}^c\wedge\tau^d\,\varepsilon_{abcd} = 0\,. \tag{5.4}$$

Since $\delta\tau^{ab}$ is arbitrary, we get

$$(5.5)\qquad \begin{cases} \mathscr{R}^c \wedge \tau^d \varepsilon_{abcd} = 0\,, & \forall a, b\,, \\ \mathscr{R}^c \wedge \tau^d \qquad = \mathscr{R}^d \wedge \tau^c\,. \end{cases}$$

Now (5.5) implies $\mathscr{R}^c = 0$. Indeed, if we set

$$\mathscr{R}^c = u^c_{\cdot ab}\,\tau^a \wedge \tau^b \qquad \text{with } u^c_{\cdot ab} = -u^c_{\cdot ba}$$

and define $u^{eab} = \varepsilon^{efab} u^c_{\cdot ef}$, (5.5) implies

$$u^{cab} = u^{acb}\,, \qquad \text{but also } u^{cab} = -u^{cba}\,.$$

A few manipulations show that $u^{cab} \equiv 0 = u^c_{\cdot ab}$ and, therefore, $\mathscr{R}^c = R^c = 0$.

The variation in τ^a then yields

$$(5.6)\qquad (\mathscr{R}^{ab} \wedge \tau^f)\,\varepsilon_{abfc} = 0\,, \qquad \forall c\,.$$

If we set $dx^\mu \wedge dx^\nu \wedge dx^\lambda = \varepsilon^{\mu\nu\lambda\alpha}\, dv_\alpha$, $\mu, \nu, \lambda, \alpha = 1, \dots, 4$,

$$\mathscr{R}^{ab}_{\cdot\cdot\mu\nu} = \tau^a_{\cdot\varrho}\,\tau^b_{\cdot\sigma}\,\mathscr{R}^{\varrho\sigma}_{\cdot\cdot\mu\nu}\,,$$

(5.6) can be written as

$$\mathscr{R}^{\varrho\sigma}_{\cdot\cdot\mu\nu}\,\tau^a_{\cdot\varrho}\,\tau^b_{\cdot\sigma}\,\tau^f_{\cdot\lambda}\,\varepsilon_{abfc}\, dx^\mu \wedge dx^\nu \wedge dx^\lambda = 0\,,$$

by using

$$\tau^a_{\cdot\varrho}\,\tau^b_{\cdot\sigma}\,\tau^f_{\cdot\lambda}\,\varepsilon_{abfc} = \tau^{\cdot\beta}_c\,\varepsilon_{\beta\varrho\sigma\lambda}\,\Delta\,,$$

where $\Delta = \sqrt{g}$, and, if we delete the factors dv_α, $\tau^{\cdot\beta}_c$, we get

$$(5.7)\qquad \mathscr{R}^{\varrho\sigma}_{\cdot\cdot\mu\nu}\,\varepsilon_{\beta\varrho\sigma\lambda}\,\varepsilon^{\mu\nu\lambda\alpha} = 0$$

or

$$(5.8)\qquad \mathscr{R}^\alpha_{\cdot\beta} - \tfrac{1}{2}\mathscr{R}\delta^\alpha_{\cdot\beta} = 0\,,$$

which are Einstein's equations for empty space.

This method is clearly equivalent to Palatini's in that the vanishing of the torsion follows from the variational principle (and of course from factorization).

Tentatively, it would appear possible to assume (5.1) with nonfactorized c-beins (and with R^{ab} instead of $\mathscr{R}^{ab}$). One is then faced with two options, neither of which appears entirely satisfactory.

Since there is no factorization; there is no priviledged role of SO_4 (*i.e.* H) in P, and it would appear desirable to extend the integration to the entire group space (G) rather than to G/H only. The action then takes the form

$$8A = \int_P R^{ab} \wedge \varrho^c \wedge \varrho^d \varepsilon_{abcd} \wedge \nu \,, \tag{5.9}$$

where ν is a 6-form including the $d\Xi$ differentials (at least in the factorizable case). None of the obvious forms tried so far for ν seems to reproduce Einstein's equations. A second option is to consider

$$8A = \int_{\mathscr{M}^4} R^{ab} \wedge \varrho^c \wedge \varrho^d \varepsilon_{abcd} \,, \tag{5.10}$$

where $\mathscr{M}^4$ is a submanifold of G or rather a section over G, when this is taken as a bundle. This seems to be unnatural, for $\mathscr{M}^4$ was in the original intention the base space G/H. Furthermore, the very presence of $\mathscr{M}^4$ in the variational principle makes it a dynamical variable, thus also subject to variation. This implies the added complication that the equations describing the embedding of $\mathscr{M}^4$ in G should include arbitrary functions which must be considered as fields, in which case we would have strayed out of our conditions I) for a « geometric » theory.

If, however, we decide to overlook these difficulties and try (5.10), we find the field equations

$$\begin{cases} R^a \wedge \varrho^b - R^b \wedge \varrho^a = 0 \,, \\ R^{ab} \wedge \varrho^c \varepsilon_{abcd} \qquad = 0 \,, \end{cases} \tag{5.11}$$

resembling the factorized expressions (5.5), (5.6). Here, though lack of factorization inhibits direct deduction, heuristic arguments do yield $R^a = 0$. Interestingly enough, (5.11) which follows from the variation of the fields ϱ^{ab}, ϱ^a also suffices to guarantee that the variation in $\mathscr{M}^4$ vanishes trivially. In fact, the differential of the Lagrangian density

$$2R^{ab} \wedge R^c \wedge \varrho^d \varepsilon_{abcd}$$

vanishes as a consequence of (5.11).

This fact is not so strange. Any variation of $\mathscr{M}^4$ can be compensated by a change of co-ordinates in G; the embedding functions are thus unchanged, whereas ϱ^a, ϱ^{ab} do change. This of course is taken care of by the field equations (5.11). The option (5.10) thus appears to contain an automatic answer to the threat of « degeometrization » to which we alluded.

Returning to the further interpretation of the eqs. (5.11) of the unfactorized set, we remind ourselves that the factorized forms τ^{ab}, τ^a constructed through the gauges (4.11) satisfied $R^a = 0$ and eqs. (5.5). They thus also satisfy (5.11), since ϱ^{ab}, ϱ^a, R^a, R^{ab} can be reconstructed for the factorized case through (4.13), (4.14). Factorization thus supplies a solution for the unfactorized field equations (5.11). Now we can make the converse conjecture, namely that eqs. (5.11) uniquely yield the solution $\varrho^A \to \tau^A$ via (4.11), apart from a generic co-ordinate transformation on G. If this were true, factorization would be a result of (5.11). This would be consistent with our previous deduction from heuristic arguments that (5.11) implies $R^a = 0$, just as can be directly deduced from factorization. The same would now be true of (5.8), *i.e.* the Einstein equation. The theory, as extended here to forms on all of G, but with the action integral restricted to an arbitrary 4-dimensional submanifold, would still have the same physical content as Einstein's theory.

Actually, since the choice $\mathcal{M}^4$ has turned out to be irrelevant, we can choose $\mathcal{M}^4 \equiv \mathbf{R}^4$, *i.e.* identify the submanifold with «space-time». This then sets our Lagrangian (5.10) to be equal to the factorized (5.1), as a boundary condition

$$\varrho^A(\mathbf{R}^4) = \tau^A$$

with $\varrho^A(G)$ derivable from $\varrho^A(\mathbf{R}^4)$ through a transformation which is just the inverse of our factorization gauge (4.11). Conceivably, the unfactorized (5.10) may lead to new solutions (with torsion) in the presence of matter fields. However, in the present case, extending the forms to the entire P-manifold could be regarded just as a way of writing down simultaneously all gauge-related solutions (with appropriate co-ordinate transformations automatically completing the procedure).

The form (5.1) of the action brings us to another argument. Clearly, (5.1) is invariant under SO_4 (*i.e.* H) gauge transformations, but not under proper local Poincaré gauge transformations (that is including translations). It is instructive to see how this happens. By specializing to a translation-gauge transformation as appears in (4.18), we find

$$(5.12)\qquad \begin{cases} T(\lambda)\varrho^{ab} = \varrho^{ab}\,, \\ T(\lambda)\varrho^a \ = du^a + \varrho^a + \varrho^{ab}u^b = \varrho^a + \mathrm{D}u^a\,, \end{cases}$$

where u^a is the local translation and D is the SO_4 covariant derivative. Therefore, R^{ab} remains unchanged, but ϱ^a does get modified. The change in R^a can be computed directly:

$$(5.13)\qquad T(\lambda)R^a = \mathrm{D}T(\lambda)\varrho^a = \mathrm{D}\varrho^a + \mathrm{D}^2u^a = R^a + R^{ab}u^b\,.$$

Thus a torsionless e-bein acquires torsion under the action of a translation gauge. To no one's surprise, both the *Lagrangian and the field equation break the translational gauge symmetry.* We have seen in (4.24) how the invariance under general co-ordinate transformations (GCTG) in G/H can be represented as an AGCT translation gauge. Such a gauge corresponds to a translation generator obeying a new algebra with curvatures as structure functions, instead of structure constants. The Lagrangian obeys that AGCT gauge invariance, instead of (5.12). However, the curvatures themselves in (4.24)-(4.30) are determined by the Lagrangian through the equations of motion.

What is then the criterion by which we select a subgroup H (in this case SO_4), whose symmetry is not broken, out of the whole group G? Clearly, some part of the symmetry must be broken. A Lagrangian built out of curvatures only and *invariant under G is by definition a characteristic form* and yields only topological information, but no field equation.

MacDowell and Mansouri have recently [23] proposed a curious alternative to the customary Einstein Lagrangian. Their basic approach is to consider a $G = SO_5$, $H = SO_4$ theory (actually $SO_{3,2}$ and $SO_{3,1}$, respectively, in the physical case) and a factor space $S^4 = G/H$. One, therefore, has curvatures

$$R^{ab} = \mathrm{d}\varrho^{ab} + \varrho^{ac}\wedge\varrho^{cb} = -R^{ba} \tag{5.14}$$

with $a, b = 1, \ldots, 5$.

In order to compare this formalism with the previous one, we split the range by calling $\varrho^{a5} = \varepsilon\varrho^{a}$, where ε is a constant. We then have

$$\left\{\begin{aligned} R^{ab} &= \mathrm{d}\varrho^{ab} + \varrho^{ac}\wedge\varrho^{cb} - \varepsilon^2\varrho^{a}\wedge\varrho^{b}, \\ R^{a} &= \mathrm{d}\varrho^{a} + \varrho^{ac}\wedge\varrho^{c}, \end{aligned}\right. \qquad a, b = 1, \ldots, 4. \tag{5.15}$$

In the $\varepsilon \to 0$ limit, (5.15) reduces to (4.15). This limit corresponds to the Inönü-Wigner contraction of SO_5 into P. In ref. [23], however, the action integral is chosen just to be

$$8A = \frac{-1}{2\varepsilon^2}\int R^{ab}\wedge R^{cd}\varepsilon_{abcd5} = \frac{-1}{2\varepsilon^2}\int R^{ab}\wedge R^{cd}\varepsilon_{abcd}, \qquad a, b = 1, \ldots, 4. \tag{5.16}$$

Applying (5.15), we get

$$\begin{aligned} 8A = \frac{-1}{2\varepsilon^2}\int \bar{R}^{ab}\wedge\bar{R}^{cd}\varepsilon_{abcd} + \int \bar{R}^{ab}\wedge\varrho^{c}\wedge\varrho^{d}\varepsilon_{abcd} - \\ - \varepsilon^2\int \varrho^{a}\wedge\varrho^{b}\wedge\varrho^{c}\wedge\varrho^{d}\varepsilon_{abcd} = A_1 + A_2 + A_3, \end{aligned} \tag{5.17}$$

where

$$\bar{R}^{ab} = \mathrm{d}\varrho^{ab} + \varrho^{ac}\wedge\varrho^{cb}$$

is the « Poincaré » curvature. Now A_1 is a topological invariant and yields no contribution to the field equation. A_2 is the classical Einstein Lagrangian, A_3 is a cosmological term. In the limit $\varepsilon \to 0$, one retrieves the conventional theory and ε^{-1} is essentially the radius of the de Sitter space, *i.e.* the vacuum solution when cosmological terms are present.

In a way, the MacDowell-Mansouri formulation is more aesthetic in that only curvature components are used and the search for a Lagrangian is narrowed down to the choice of a suitable SO_5-breaking term.

A different way of looking at the issue is the following. Can we anticipate the form of a Poincaré-breaking Lagrangian that can be derived by contraction from an ansatz involving only curvature terms? In order to answer this question, it pays off to look at the procedure of contraction of a group in terms of the Cartan-Maurer equations. Contraction implies that one can choose the structure constants $C^A_{\bullet BE}(\varepsilon)$ as functions of a parameter ε and that the resulting groups G_ε are all isomorphic except at $\varepsilon = 0$, where the contracted group arises. We, therefore, have the Jacobi identities

$$C^A_{\bullet BG}(\varepsilon)\, C^B_{\bullet DE}(\varepsilon) + C^A_{\bullet BD}(\varepsilon)\, C^B_{\bullet EG}(\varepsilon) + C^A_{\bullet BE}(\varepsilon)\, C^B_{\bullet GD}(\varepsilon) = 0\,. \tag{5.18}$$

Differentiating, we find

$$\frac{\partial C^A_{\bullet BG}(\varepsilon)}{\partial\varepsilon}\, C^B_{\bullet DE}(\varepsilon) + C^A_{\bullet BG}(\varepsilon)\, \frac{\partial C^B_{\bullet DE}(\varepsilon)}{\partial\varepsilon} + \text{cycl} = 0\,. \tag{5.19}$$

Now the 2-form

$$\mu^A = \left.\frac{\partial C^A_{\bullet BG}}{\partial\varepsilon}\right|_{\varepsilon=0} \omega^B \wedge \omega^G$$

has vanishing covariant derivative at $\varepsilon = 0$ (with $\varrho^A = \omega^A$) as a consequence of (5.19). We have indeed

$$\begin{aligned}(\mathrm{D}\mu)^A = \mathrm{d}\mu^A - C^A_{\bullet BG}\omega^B\wedge\mu^G = \frac{1}{2}\,\frac{\partial C^A_{\bullet BG}}{\partial\varepsilon}\, C^B_{\bullet DE}\omega^D\wedge\omega^E\wedge\omega^G - \\ -\frac{1}{2}\,\frac{\partial C^A_{\bullet BG}}{\partial\varepsilon}\, C^G_{\bullet \Gamma \Xi}\omega^B\wedge\omega^D\wedge\omega^E - C^A_{\bullet BG}\omega^B\wedge\frac{\partial C^G_{\bullet ED}}{\partial\varepsilon}\,\omega^E\wedge\omega^D = \\ = \frac{\partial C^A_{\bullet BG}}{\partial\varepsilon}\, C^B_{\bullet DE} + C^A_{\bullet BG}\,\frac{\partial C^B_{\bullet DE}}{\partial\varepsilon}\,\omega^D\wedge\omega^E\wedge\omega^G = 0\,.\end{aligned} \tag{5.20}$$

We call any form η^A with vanishing covariant derivative a « pseudocurvature » form. Clearly

$$\zeta^A = C^A{}_{BD}\omega^B\wedge\omega^D \tag{5.21}$$

is a pseudocurvature corresponding to a trivial contraction of G, where $C^A{}_{BD}(\varepsilon) = f(\varepsilon) C^A{}_{BD}$ and $f(0) = 1$. Therefore, η^A is really an equivalence class modulo ζ^A (see the comment at the end of this section). These classes already appear in the Chevalley cohomology on Lie algebras [24]. The existence of a nontrivial class is related to a nontrivial contraction procedure. Indeed for P we have the nontrivial class $\mu^{ab} = \omega^a \wedge \omega^b$, $\mu^a = 0$, which corresponds to the contraction of SO_5 into P. On the contrary, SO_5 itself has no such class.

Thus, on SO_5 we have the MacDowell-Mansouri procedure but no pseudocurvature, while on P we have a pseudocurvature but the MacDowell-Mansouri procedure fails: $R^{ab}R^{cd}\varepsilon_{abcd}$ is not only Lorentz invariant, but Poincaré (gauge) invariant as well, and transforms away into a surface integral. The presence of pseudocurvature is, therefore, a warning that it is possible to construct a Lagrangian density directly with μ^A without « decontracting » first. Alternatively, μ^A is the infinitesimal difference between contracted and decontracted definitions of curvature. We should also note that we are interested more generally in all forms μ^A or ν_A possessing vanishing covariant derivatives.

On P we then have

$$\left\{\begin{aligned} (\mathrm{D}\mu)^{ab} &= \mathrm{d}\mu^{ab} + \omega^{at}\wedge\mu^{tb} - \omega^{tb}\wedge\mu^{at}\,, \\ (\mathrm{D}\mu)^{a} &= \mathrm{d}\mu^{a} + \omega^{at}\wedge\mu^{t} - \omega^{t}\wedge\mu^{at} \end{aligned}\right. \tag{5.22a}$$

and

$$\left\{\begin{aligned} (\mathrm{D}\nu)_{ab} &= \mathrm{d}\nu_{ab} + \omega^{bt}\wedge\nu_{at} - \omega^{at}\wedge\nu_{bt} + \omega^{b}\wedge\nu_{a} - \omega^{a}\wedge\nu_{b}\,, \\ (\mathrm{D}\nu)_{a} &= \mathrm{d}\nu_{a} + \omega^{at}\wedge\nu_{t}\,. \end{aligned}\right. \tag{5.22b}$$

If we find either μ^A or ν_A as a polynomial in ω^A, we can replace ω^A by ϱ^A in this polynomial. In this case, it is no longer true that $(\mathrm{D}\mu)^A = 0$, $(\mathrm{D}\nu)_A = 0$, but rather those derivatives become linear expressions in the curvature R^A. This provides an interesting argument when building Lagrangian densities. If we have (as in relativity) an action of the kind

$$A = \int R^A \nu_A\,, \tag{5.23a}$$

then variation in ϱ^A gives a contribution of the kind (symbolically)

$$\delta A = \int \left\{ (\mathrm{D}\delta\varrho)^A \nu_A + R^A\,\delta\varrho^B \frac{\delta\nu_A}{\delta\varrho^B} \right\} = 0\,.$$

Integrating the first term by parts we obtain

$$-\int \delta\varrho^A (\mathrm{D}\nu)_A + \int \delta\varrho^A R^B \frac{\delta\nu_B}{\delta\varrho^A} = 0\,,$$

or

$$-(\mathrm{D}\nu)_A + R^B \frac{\delta \nu_B}{\delta \varrho^A} = 0 . \tag{5.23b}$$

Equation (5.23) is identically satisfied for $\varrho^A = \omega^A$, since $R^B = 0$ and $(\mathrm{D}\nu)_A$ is linear in R^A. Therefore, $\varrho^A = \omega^A$ can be assumed as «empty space».

The above discussion would lead us to the conclusion that, if G is semi-simple, then the MacDowell-Mansouri procedure seems to be ideally suited for the construction of a Lagrangian. If, on the other hand, G is the result of a contraction, it has in a natural way a pseudocurvature form ν_A (we can take $\nu_{ab} = \varepsilon_{abcd}\varrho^c \wedge \varrho^d$, $\nu_a = 0$) which can be used in forming a Lagrangian. The G-symmetry in this Lagrangian is broken because pseudocurvature «remembers» which subgroup H in G is not affected by contraction (*e.g.*, SO_4, in going from SO_5 to P). The way in which this breaking occurs is dictated by the structure of the group G itself, rather than by some *ad hoc* choice of H. It should be emphasized that only the 2-form μ^A, with $(\mathrm{D}\mu)^A = 0$, can be related directly to the contraction procedure. A generic ps form (ps means pseudocurvature) with $(\mathrm{D}\zeta)^A$ or $(\mathrm{D}\nu)_A = 0$ is still useful in building Lagrangians, but its geometrical interpretation may be less straightforward, involving the use of various group invariants besides the group metric.

Finally, we remark that, if μ^A, ζ^A, ν_A, etc., are pseudocurvatures, then

$$\begin{cases} \lambda^A = C^A{}_{BD}\mu^B \wedge \zeta^D , \\ \lambda_A = C^B{}_{AD}\mu^D \wedge \nu_B \end{cases} \tag{5.24}$$

are also pseudocurvatures. This allows the construction of an algebra of pseudocurvatures of interest in trying to define a suitable Lagrangian formalism.

One final comment with respect to the possible use of the trivial-contraction ζ^A ps curvature (5.21). To apply it to an action such as (5.23*a*), we first have to lower the index:

$$\zeta_A = g_{AB}\zeta^B ,$$

where we have used the group metric (in a semi-simple group, its inverse g^{AB} also exists). If now we try the action

$$\int R^A \zeta_A ,$$

we discover that this is again a topological invariant and yields no equation of motion, for either semi-simple or contracted G. This is because ζ_A carries no information about a specific $H \subset G$, no symmetry breaking.

For our further discussion of supergravity, we include here the GP invariant derivatives of multiplets μ^A and ζ_A:

$$
(5.25) \quad \begin{cases} (\mathrm{D}\mu)^{ab} = \mathrm{d}\mu^{ab} + \varrho^{at}\wedge\mu^{tb} - \varrho^{tb}\wedge\mu^{at}\,, \\ (\mathrm{D}\mu)^{a} = \mathrm{d}\mu^{a} + \varrho^{at}\wedge\mu^{t} - \varrho^{t}\wedge\mu^{at} + \bar{\varrho}\gamma^a\mu\,, \\ \mathrm{D}\mu = \mathrm{d}\mu + \frac{1}{2}(\varrho^{ab}\sigma^{ab})\wedge\mu - \frac{1}{2}\sigma^{ab}\varrho\wedge\mu^{ab}\,, \end{cases}
$$

$$
(5.26) \quad \begin{cases} (\mathrm{D}\zeta)_{ab} = \mathrm{d}\zeta_{ab} - \varrho^{ta}\wedge\zeta_{tb} + \varrho^{tb}\wedge\zeta_{ta} + \frac{1}{2}\varrho^{T}(\sigma^{ab})^{T}(C\zeta) + \varrho^{b}\wedge\zeta_{a} - \varrho^{a}\wedge\zeta_{b}\,, \\ (\mathrm{D}\zeta)_{a} = \mathrm{d}\zeta_{a} - \varrho^{ta}\wedge\zeta_{b}\,, \\ \mathrm{D}\bar{\zeta} = \mathrm{d}\bar{\zeta} + \bar{\varrho}\gamma^a\zeta_a - \frac{1}{2}\varrho^{ab}\,\bar{\zeta}\sigma^{ab}\,. \end{cases}
$$

6. – Supersymmetric theories and the factorization of graded subgroups.

A consistent theory of supergravity has been formulated [25, 26] in a second-quantized formalism, by finding a set of local graded infinitesimal transformations on a set of fields including gravitation and leaving a Lagrangian invariant (after imposing a constraint equivalent to the vanishing of a generalized torsion). Various attempts have been made to find a geometrical formulation: a presentation of the theory in the context of a Palatini treatment of Einstein's relativity [26, 27], an attempt to derive it as a contraction limit in superspace [28], a partial description using a superspace with constant torsion in its physical vacuum [16] (see also ref. [7, 15]) and a generalization of the MacDowell-Mansouri treatment of gravity. Other works [32] have pursued the program for a generalized supersymmetry group (including internal degrees of freedom) and for the superconformal group.

In this article, we are interested in the supergroup GP and associated subgroups, *i.e.* the problem of picking $H \subset G$.

A) The first candidate is $H = SO_4$. In this case, $\mathrm{GP}/SO_4 = \mathbf{R}^{4/4}$, where $\mathbf{R}^{4/4}$ is the set of pairs $(\boldsymbol{x}, \xi)$ in a Grassmann algebra A, where $x^a \in A^+$, $\xi^\alpha \in A^-$ (see (3.16)), *i.e.* x^a is a commuting (ξ^α an anticommuting) variable. Notice, however, that, strictly speaking, each x^a contains more information than a real number. A itself is a real space of very high dimensionality, allowed by higher than $\mathbf{R}^4$, and containing all nilpotent elements of the form $A^- \otimes A^-$ etc. We refer to this space as «superspace». Theories using this form of H should have complete local Lorentz gauge invariance and complete covariance under general co-ordinate transformations $x' = f(x, \xi)$, $\xi' = g(x, \xi)$. In this class of theories we have the super-Riemannian ($V_{4,4}$ in a graded extension of the Schouten notation [22]) Arnowitt-Nath approach [14] and the alternative torsion ($U_{4,4}$) superspace of Wess-Zumino [16]. Theories based on this framework have nontrivial computational tasks to overcome and for this reason still have some loose ends in their final shape. In particular, the way in which ordinary general relativity is or should be contained in these theories has not been elucidated. In the $U_{4,4}$ version [16] the language of forms is used, but only field equations and no Lagrangian are given. In the $V_{4,4}$ theory [14], the standard Einstein tensor calculus is used in the graded version [29], up to the actual writing of a scalar curvature R. The resulting field equations have not yet been investigated in detail (even the listing of all fields has not been completed). A major stumbling block remains the need to use the Berezin integration formalism over the odd variables, a procedure which selects the term of the highest degree in those variables, with a resulting increase in the volume of computations.

B) A second choice is $H = SO_4 \otimes A^2 = GA_R$. This is a particular subgroup whose elements are the pairs

$$\begin{cases} (\Xi, 0, \xi_R)_{\mathrm{GP}} = (\Xi, \xi_R) \in GA_R\,, \\ \frac{1}{2}(1 + i\gamma_5)\xi_R = 0\,. \end{cases} \tag{6.1}$$

It is, therefore, a semi-direct product. A mirror image GA_L is obtained by considering

$$(\Xi, \xi_L),\ \tfrac{1}{2}(1 - i\gamma_5)\xi_L = 0 \tag{6.2}$$

instead of (6.1).

We have $GA_R \cap GA_L = SO_4$. This choice for H seems to have been overlooked. We remark, however, that GP is a contraction of $OSp_{1,4}$, which contains $OSp_{1,2C}$ as a subsupergroup. The contraction of $OSp_{1,2C}$ leads to GA_R. Those groups appear in the MacDowell-Mansouri version of supergravity [23], but it seems that their Lagrangian is not invariant under local GA_R gauge transformations. Moreover, their forms are written on $\mathbf{R}^4$ (or possibly our larger $\mathscr{A}^4_+$ since the same formalism would apply), and not on $\mathrm{GP}/\mathrm{GP}_R = \mathscr{M}^{4,2}_L$.

The elements of $\mathscr{M}^{4,2}_L$ are pairs

$$(\boldsymbol{x}, \xi_L),\ \tfrac{1}{2}(1 - i\gamma_5)\xi_L = 0\,, \qquad \boldsymbol{x} \in \mathscr{A}^4_+\,, \tag{6.3}$$

with an « analytic » 2-component ξ_L Majorana spinor projection.

C) We might also select the sub(graded)group $H = \mathrm{GP}_1$, defined as the set of triples

$$x = (\Xi, \boldsymbol{x}, \xi)\,,$$

where now $\xi \in A^-$, but $\boldsymbol{x} \in A^- \times A^-$, that is $\boldsymbol{x} \in A^+$ but is a nilpotent element. Obviously this is still a supergroup and also $\mathrm{GP}_1 \subset \mathrm{GP}$, but GP_1 is a proper sub(super)group of GP. By obvious considerations we have $\mathrm{GP}/\mathrm{GP}_1 = \mathbf{R}^4$, where this is now a purely ordinary manifold on the reals. This is thus the factorization appropriate for a (Euclidean) space-time theory of supergravity. We shall detail it at this stage as a heuristic model.

The GP_1 factorization has no counterpart in ordinary Lie groups since both GP_1 and GP have apparently the same Lie algebra. This seems a very interesting possibility for we would write forms on ordinary space, but the theory would still have local supersymmetric GP_1 invariance. The difference between GP_1 and GP lies only in the choice of the coefficient ring of the element of the Lie algebra.

The generic element

$$\alpha^{ij} S_{ij} + \alpha^i S_i + \bar{\alpha} S \tag{6.4}$$

will have the coefficients of the spinors S

$$\alpha, \bar{\alpha} \in A^- ,$$

those of the Lorentz generators S_{ij}

$$\alpha^{ij} \in A^+ ,$$

but the translations S_i have nilpotent coefficients in GP_1,

$$\alpha^i \in A^- \otimes A^- \subset A^+ , \qquad \alpha^i S_i \in GP_1 , \tag{6.5a}$$

as against either nilpotent or real ones in GP,

$$\alpha^i \in A^+ , \qquad \alpha^i S_i \in GP . \tag{6.5b}$$

Similarly, the space-time vector forms $\boldsymbol{o}$ take $A^+(A^- \otimes A^-)$ values on the Lie algebra $D_{GP}(D_{GP_1})$. Ω is always A^+-valued and ω has A^--values on the D of either group.

It is quite conceivable to write an A^+-valued form $\boldsymbol{o}$ as a sum

$$\boldsymbol{o} = \boldsymbol{o}_r + \boldsymbol{o}_1 ,$$

where $\boldsymbol{o}_r$ is the real-valued and $\boldsymbol{o}_1$ is the GP_1 nilpotent-valued part. We can thus also rewrite for GP the Cartan-Maurer equations (3.21) in the form

$$\left\{\begin{array}{l} d\omega^{ab} + \omega^{at}\wedge\omega^{tb} = 0 , \\ do^a_r + \omega^{at}\wedge o^t_r = 0 , \\ d\omega + \frac{1}{2}(\omega^{ab}\sigma^{ab})\omega = 0 , \\ do^a_1 + \omega^{at}\wedge o^t_1 + \frac{1}{2}\bar{\omega}\gamma^a\omega = 0 . \end{array}\right. \tag{6.6}$$

The GP_1 equations are obtained by putting

$$o_r = 0\,, \qquad o = o_1\,, \qquad o \in GP_1\,. \tag{6.7}$$

Thus (3.21) can be used for both GP and GP_1 provided we restrict the values of o suitably to nilpotent elements for GP_1. In a physical situation with curvature, these forms are replaced by a set ϱ^A with the same scheme of values.

We shall develop the GP_1 choice in more detail. It is particularly interesting if one looks for pseudocurvature forms.

We have no proof that the choice in the following is unique, but it does lead to a new «geometric» derivation of previously known results. By means of (3.21) it can be checked that the multiplet ζ_A given by

$$\begin{cases} \zeta_{ab} = \varepsilon_{abcd}\varrho^c \wedge \varrho^d\,, \\ \zeta_a \;= 0\,, \\ \xi_\alpha \;= -4i\bar{\varrho}\gamma^a\varrho^a\gamma_5 \quad \text{or} \quad \zeta^\alpha = -4i\gamma_5\gamma^a\varrho^a\varrho \end{cases} \tag{6.8}$$

has vanishing covariant derivative. We propose, therefore, the Lagrangian density

$$\begin{aligned} 8\mathscr{L} &= R^{ab}\wedge\zeta_{ab} + R^\alpha\wedge\xi_\alpha = R^{ab}\varrho^f\varrho^e\varepsilon_{abfe} + 4i\bar{R}\gamma_5\varrho^a\gamma^a\varrho = \\ &= R^{ab}\varrho^f\varrho^e\varepsilon_{abfe} + 2i\bar{R}\gamma_5\varrho^a\gamma^a\varrho - 2i\bar{\varrho}\gamma_5\varrho^a\gamma^a R + \text{surface terms}\,. \end{aligned} \tag{6.9}$$

Factorization implies that

$$\begin{cases} \varrho^{ab} = A^{ta}\tau^{ts}A^{sb} + A^{ta}\,\mathrm{d}A^{tb}\,, \\ \varrho^a \;= A^{ta}(\underline{D}a^t + \tau^t + \bar{\tau}\gamma^t\alpha - \frac{1}{2}\bar{\alpha}\gamma^t\,\underline{D}\alpha)\,, \\ \varrho \;\;= \mu(A^{-1})(\tau + \underline{D}\alpha)\,, \end{cases} \tag{6.10}$$

where $\underline{D}$ denotes the GR derivative,

$$\begin{cases} \underline{D}\alpha \;= \mathrm{d}\alpha + \frac{1}{2}\tau^{ab}\sigma^{ab}\alpha\,, \\ \underline{D}a^t = \mathrm{d}a^t + \tau^{ta}\wedge a^t\,, \end{cases} \tag{6.11}$$

$(\tau^{ts}, \tau^t, \tau)$ are forms on $\mathscr{R}^4$ and $(A, a, \alpha) \in GP_1$.

We set here

$$\begin{cases} \omega^{ab} = A^{ta}\,\mathrm{d}A^{tb}\,, \\ o \;\;= A^{-1}(\mathrm{d}a - \frac{1}{2}\bar{\alpha}\gamma\,\mathrm{d}\alpha)\,, \\ \omega \;\;= u(A^{-1})\,\mathrm{d}\alpha\,, \end{cases} \tag{6.12}$$

which satisfies (3.21), so that (6.10) indeed realizes (4.11). The corresponding factorization of the curvatures is given by

$$
(6.13)\qquad \begin{cases} R^{ab} = A^{ta}\mathscr{R}^{ts}A^{sb}\,, \\ R^{a} \;= A^{ta}(\mathscr{R}^{tb}a^{b} + \mathscr{R}^{t} + \bar{\mathscr{R}}\gamma^{t}\alpha - \frac{1}{4}\bar{\alpha}\gamma^{t}\mathscr{R}^{ef}\sigma^{ef}\alpha)\,, \\ R \;= u(A^{-1})(\mathscr{R} + \frac{1}{2}\mathscr{R}^{ab}\sigma^{ab}\alpha)\,, \\ \bar{R} \;= (\bar{\mathscr{R}} - \frac{1}{2}\bar{\alpha}\mathscr{R}^{ab}\sigma^{ab})u(A)\,. \end{cases}
$$

The factorized $\mathscr{R}^{A}$ are constructed from $(\tau^{ts}, \tau^{t}, \tau)$ by using the same formulae yielding R^{A} from $(\varrho^{ts}, \varrho^{t}, \varrho)$:

$$
(6.14)\qquad \begin{cases} \mathscr{R}^{ab} = \mathrm{d}\tau^{ab} + \tau^{ab}\wedge\tau^{tb}\,, \\ \mathscr{R}^{a} \;= \mathrm{d}\tau^{a} + \tau^{at}\wedge\tau^{t} + \frac{1}{2}\bar{\tau}\gamma^{a}\tau\,, \\ \mathscr{R} \;= \mathrm{d}\tau + \frac{1}{2}(\tau^{ab}\sigma^{ab})\tau = \underline{\mathrm{D}}\tau\,. \end{cases}
$$

We can now insert (6.13), (6.10) in (6.9). We denote by $\mathscr{L}_{\varrho}$ the Lagrangian density constructed as in (6.9) and by $\mathscr{L}_{\tau}$ the same with $\varrho^{A}\to\tau^{A}$. There is no A^{ta}, $\mathrm{d}A^{ta}$ dependence in $\mathscr{L}_{\tau}$ at all. It is, therefore, possible to choose $A^{ta} = \delta^{ta}$, $\mathrm{d}A^{ta} = 0$.

The final result appears to be rather complicated:

$$
\begin{aligned}
(6.15)\qquad -8\mathscr{L}_{\varrho} = {}& -8\mathscr{L}_{\tau} + 2\mathscr{R}^{ab}\mathscr{R}^{f}a^{e}\varepsilon_{abfe} - \mathscr{R}^{ab}\mathscr{R}^{et}a^{f}a^{t}\varepsilon_{abfe} + \\
& + 2i[(\bar{\mathscr{R}}\gamma_5\gamma^{f}\alpha) + (\bar{\alpha}\gamma_5\gamma^{f}\mathscr{R})]\mathscr{R}^{f} - 4i\bar{\mathscr{R}}\gamma_5\gamma^{f}\mathscr{R}a^{f} + \mathscr{R}^{ab}(\bar{\alpha}\gamma^{f}\underline{\mathrm{D}}\alpha)(\bar{\tau}\gamma^{e}\alpha)\varepsilon_{abfe} - \\
& - \tfrac{1}{4}\mathscr{R}^{ab}(\bar{\alpha}\gamma^{f}\underline{\mathrm{D}}\alpha)(\bar{\alpha}\gamma^{e}\underline{\mathrm{D}}\alpha)\varepsilon_{abfe} - i\bar{\mathscr{R}}\gamma_5(\bar{\alpha}\gamma^{f}\underline{\mathrm{D}}\alpha)\gamma^{f}\tau + i\bar{\tau}\gamma_5(\bar{\alpha}\gamma^{f}\underline{\mathrm{D}}\alpha)\gamma^{f}\mathscr{R} + \\
& + 2i\bar{\mathscr{R}}\gamma_5(\bar{\tau}\gamma^{f}\alpha)\gamma^{f}\underline{\mathrm{D}}\alpha - 2i\,\underline{\mathrm{D}}\bar{\alpha}\,\gamma_5(\bar{\tau}\gamma^{f}\alpha)\gamma^{f}\mathscr{R} + \tfrac{1}{2}\mathscr{R}^{ab}(\bar{\tau}\gamma_5\gamma^{f}\underline{\mathrm{D}}\alpha)\bar{\alpha}\gamma_5\gamma^{e}\alpha\varepsilon_{abfe}\,.
\end{aligned}
$$

The values of the action have to be real, *i.e.* in A^{+}. We constrain

$$
(6.16)\qquad \mathscr{R}\in A^{-}\,,\qquad \mathscr{R}^{f}\in(A^{-})^{2}\,,
$$

i.e. no real contribution to the factorized generalized torsion $\mathscr{R}^{f}$. This is a generalization of the

$$
\mathscr{R}^{f} = 0
$$

result of GR. Indeed, (6.16) implies that the only real allowed contribution to the GR torsion be given by $-\frac{1}{2}\bar{\tau}\gamma^{a}\tau$, as can be seen in (6.14). This can be considered as a boundary condition taken from the Kibble-Sciama U_4 version [22] of GR which fixes the contribution to (real) torsion so that it be proportional to the spin of the spinor fields present (here the τ_{μ} field with $J=\frac{3}{2}$). Since the original theory is postulated in GP, we have also allowed residual (nilpotent) contributions in addition in (6.16). Both conditions in (6.16) are satisfied by (6.13) provided $\alpha \in A^-$, $\boldsymbol{a} \in A^- \times A^-$ as required by GP_1.

Since the action is given by a d^4x integration, we have to define such an integral for functions with (nilpotent) values in $A^- \otimes A^-$. These are now products of either the $\alpha(x)$, $\boldsymbol{a}(x)$ gauge functions or of $(A^- \times A^-)$-valued forms:

$$I_N = \int \varphi_1(x) \dots \varphi_{2N}(x)\, d^4x \,. \tag{6.17}$$

We choose to define

$$I_N = 0 \qquad \text{for } N \geqslant 2 \,. \tag{6.18}$$

This choice ensures in (6.15) that

$$\int d^4x(\mathscr{L}_\varrho - \mathscr{L}_\tau) = 0 \tag{6.19}$$

and $\mathscr{L}_\tau$ as given by (6.15) with $\varrho \to \tau$ is GP_1 gauge supersymmetric in a natural way. This end product which is achieved without further assumptions by our choice of ζ_A shows a formal similarity with the supergravity theory of Freedman, van Nieuwenhuizen and Ferrara [23, 25, 26].

The relation to conventional supergravity is by no means trivial. It is of course obvious that all forms can be taken to have values in $[A^-]^0$ = reals, A^-, $[A^-]^2$ (= nilpotent even variables) only, because higher-order terms do not contribute to the action (an $[A^-]^3$-term in $\mathscr{L}$ must be paired to at least another A^--term since $\mathscr{L}$ is even, and this vanishes by (6.17)). Moreover, variations must be carried over consistently with (6.16). This we do by adding a term

$$\tfrac{1}{2}\Lambda^a \mathscr{R}^a \,, \qquad \Lambda^a \in A^- \otimes A^- , \tag{6.20}$$

to $\mathscr{L}$, where Λ^a is a Lagrange multiplier. Upon variation of τ, τ^a, τ^{ab} we find

$$\gamma^a \tau^a \mathscr{R} = 0 \qquad \text{(modulo } [A^-]^3) \,, \tag{6.21a}$$

$$\mathscr{R}^{ab} \tau^f \varepsilon_{abfe} - 2i\bar{\mathscr{R}}\gamma^5\gamma^e\tau + 2\, \underline{D}\Lambda^e = 0 \qquad \text{(modulo } [A^-]^4) \,, \tag{6.21b}$$

$$\mathscr{R}^f \tau^e \varepsilon_{abfe} = -\Lambda^a \tau^b + \Lambda^b \tau^a \qquad \text{(modulo } [A^-]^4) \,. \tag{6.21c}$$

Equations (6.21) are GP_1 invariant in this (6.16) « weak » sense. Within this definition we use (6.10) and (6.13) as the gauge transformations, viewing $\varrho = T_x(A, \boldsymbol{a}_1, \alpha)\tau$, $R = T_x(A, \boldsymbol{a}, \alpha)\mathscr{R}$. Moreover, since (6.21) is obviously SU_2 local gauge invariant, we may set $A^{ab} = \delta^{ab}$. The remaining gauge functions $a^b(x)$, $\alpha(x)$ with $x \in \mathbf{R}^4$ are of order $[A^-]^2$ and A^-, respectively. Within this choice, the only change in (6.21) which is not obviously of order $[A^-]^3$ is in (6.21'a)

$$\Delta = \tfrac{1}{2}\gamma^a \tau^a \mathscr{R}^{ef} \sigma^{ef} \alpha ,$$

but

$$\gamma^a \sigma^{ef} = \tfrac{1}{2}(\gamma^a \sigma^{ef} + \sigma^{ef}\gamma^a) + \tfrac{1}{2}(\gamma^a \sigma^{ef} - \sigma^{ef}\gamma^a) = \tfrac{1}{2}(i\gamma_5 \varepsilon^{aeft}\gamma^t + \delta^{ea}\gamma^f - \delta^{fa}\gamma^e) ,$$

yielding

$$\Delta = \tfrac{1}{4} i\gamma_5 \gamma^t \varepsilon^{aeft} \mathscr{R}^{ef} \tau^a \alpha + \tfrac{1}{2}\mathscr{R}^{at}\gamma^t \tau^a \alpha .$$

The first term is of order $[A^-]^3$ because of (6.21b) and the second is of order $[A^-]^3$ because of the Bianchi identity

$$(\mathrm{D}\mathscr{R})^\dagger = \mathscr{R}^{ta}\tau^a + \bar{\mathscr{R}}\gamma^t \tau = 0 . \tag{6.22}$$

Thus (6.21a) is weakly GP_1 gauge invariant. The gauge covariance of (6.21b), (6.21c) requires a change in Λ^a. If, according to (6.13) and (6.10),

$$\begin{cases} \mathscr{R}'^a = \mathscr{R}^a + \mathscr{R}^{at}a^t + \bar{\mathscr{R}}\gamma^a a - \tfrac{1}{4}\bar{\alpha}\gamma^a \mathscr{R}^{st}\sigma^{st}\alpha , \\ \mathscr{R}' \;\; = \mathscr{R} + \tfrac{1}{2}\mathscr{R}^{ab}\sigma^{ab}\alpha , \\ \tau' \;\; = \tau + \underline{\mathrm{D}}\alpha , \\ \tau^{b\prime} \; = \tau^b + \underline{\mathrm{D}}\alpha^b + \bar{\tau}\gamma^b\alpha - \tfrac{1}{2}\bar{\alpha}\gamma^b\,\underline{\mathrm{D}}\alpha , \end{cases} \tag{6.23}$$

then

$$\Lambda'^e = \Lambda^e + \frac{1}{2}\mathscr{R}^{ab}a^f \varepsilon_{abfe} - i\bar{\alpha}\gamma_5\gamma^e \mathscr{R} + \frac{i}{4}(\bar{\alpha}\sigma^{ab}\gamma_5\gamma^e\alpha)\mathscr{R}^{ab} \tag{6.24}$$

is a compensating variation for $\mathscr{R}^a$. That this is indeed so can be checked through a somewhat laborious procedure in which the identities and equations

$$\mathscr{R}^{ab}\tau^b = \mathscr{R}^{ab}\tau^f \varepsilon_{abfe} = 0 \qquad \text{modulo } [A^-]^2 \tag{6.25}$$

(but not modulo $[A^-]^4$) are repeatedly used. Clearly, Λ^a in (6.21) can be deduced from $\mathscr{R}^a$ through some linear duality operation, which shows that the relation between these two sets of forms is involutory. If $\Lambda^a = 0$, then $\mathscr{R}^a = 0$ and (6.21) reduces to conventional supergravity.

However, this identification is formal, since the latter theory [23, 25, 26] does not restrict the (odd-grading) spinor gauge functions by conditions such as (6.17).

7. – Supergravity.

Conventional supergravity rests instead on an action with a Lagrangian density defined as a 4-form, integrated on $\mathcal{M}^4$, a submanifold of GP, in analogy to unfactorized general relativity (5.10):

$$A = \int_{\mathcal{M}^4} R^A \wedge \zeta_A = \tfrac{1}{8}\int_{\mathcal{M}^4} (R^{ab}\wedge\zeta_{ab} + R^\alpha\wedge\zeta_\alpha)\,. \tag{7.1}$$

On a generic $\mathcal{M}^4$, the variational equations are [25-27]

$$R^a = 0\,, \tag{7.2}$$

$$R^{ab}\varrho^f\varepsilon_{abfe} - 2i\bar{R}\gamma_5\gamma_e\varrho = 0\,, \tag{7.3}$$

$$\gamma^a\varrho^a R = 0\,. \tag{7.4}$$

The equations of motion impose severe restrictions on the components $R_{BE}^{\cdot\cdot A}$. Before we derive these results, it is quite interesting to analyse the relationship between a generic solution to eqs. (7.2)-(7.4) and its restriction to a particular $\mathcal{M}^4 \in$ GP.

In analogy to the case of $\mathcal{R}^4 \subset P$, we claim that the entire physics is already contained in any single $\mathcal{M}^4$ and that the conventional supersymmetry transformations [25] of supergravity relate the fields on $\mathcal{M}^4$ to the fields on any other subvariety $\mathcal{M}^{4'}$. Therefore, solving the equations of motion (7.2)-(7.4) on GP is somewhat tantamount to an exponentiation of all ordinary and supersymmetric transformations, just as solving the unfactorized Einstein equations (5.11) amounts to carrying out all possible SO_4 gauge transformations. Here, however, our Lagrangian has no supersymmetric gauge transformation in the standard sense defined by (4.7), and there is no simplifying factorization of the odd Grassmann variables. The integration of supersymmetries is, therefore, just as difficult as integrating the equations of motion. In this sense, the ξ^α should be regarded as fermionic « times ».

This conclusion is strengthened by a close analysis of the commutation relations of $\tilde{D}_A$ in GP, in view of equations (7.2)-(7.4) which imply the vanishing of all except a handful of the anholonomic components $R_{BE}^{\cdot\cdot a}$.

First, (7.2) asserts that

$$R_{BE}^{\cdot\cdot a} \equiv 0\,, \qquad \forall B, E\,. \tag{7.5a}$$

We next observe that (7.4) implies the vanishing of the 3-form $\gamma^a\varrho^a R$. Contracting (7.4) with the tangent vectors $(\tilde{D}_c, \tilde{D}_a, \tilde{D}_\beta)$ and using (4.27)

$$\varrho^a(\tilde{D}_\alpha) = 0\,,$$

we obtain (with γ^e acting on the unwritten spinor index of R)

$$\gamma^e R_{\alpha\beta} = 0 .$$

It follows that

$$R_{\alpha\beta} = 0 . \tag{7.5b}$$

We may contract (7.4) next with $(\tilde{D}_e, \tilde{D}_f, \tilde{D}_\alpha)$ and obtain

$$\gamma^e R_{f\alpha} - \gamma^f R_{e\alpha} = 0 .$$

Taking $e \neq f$ and multiplying by $\gamma^e\gamma^f$ yields

$$\gamma^f R_{f\alpha} + \gamma^e R_{e\alpha} = 0$$

and, for $f \neq e \neq s$,

$$-\gamma^s R_{s\alpha} = \gamma^f R_{f\alpha} = -\gamma^e R_{e\alpha} = \gamma^s R_{s\alpha} = 0 ,$$

so that

$$R_{s\alpha} = 0 . \tag{7.5c}$$

This proof holds if we replace the spinor index α in $\tilde{D}_\alpha$ with a «vertical» index $A =: (ab)$. Thus

$$R_{(ab)c} = R_{(ab)\alpha} = R_{(ab)(cd)} = 0 . \tag{7.5d}$$

We note that only R_{cd} survives.

Turning to (7.3), contracting with $(\tilde{D}_e, \tilde{D}_\alpha, \tilde{D}_\beta)$ and using (7.5c) yields

$$R^{\bullet\bullet ab}_{\alpha\beta}\, \varepsilon_{abce} = 0 ,$$

so that

$$R^{\bullet\bullet ab}_{\alpha\beta} = 0 . \tag{7.5e}$$

Similarly,

$$R^{\bullet\ \bullet\ ab}_{(cd)(ef)} = R^{\bullet\ \bullet ab}_{(cd)\alpha} = 0 \tag{7.5f}$$

and contraction of (7.3) with $(\tilde{D}_m, \tilde{D}_p, \tilde{D}_{st})$ leads to

$$R^{\bullet\ \bullet ab}_{(st)m} = 0 . \tag{7.5g}$$

We thus find that all curvature components involving an SO_4 index (cd) vanish, as would be expected from an SO_4-factorized theory. The Ξ^{cd} variables appear factorized in the expressions for the curvature, at least. To that extent, we may regard this fact as consistent with the hypothesis of factorization of SO_4 appearing as a result of the equations of motion. Indeed, we surmise as for general relativity that this does occur here too. Curvature is in any case restricted to $\mathbf{R}^{4/4} = \mathrm{GP}/H$, *i.e.* to superspace. The commutation relations of the $\tilde{D}_A$ are thus given by (4.41*a*)-(4.41*b*) and (4.44*a*)-(4.44*d*).

Contraction of (7.3) with $(\tilde{D}_m, \tilde{D}_p, \tilde{D}_\alpha)$ leads to

$$R_{\alpha m}^{\bullet\bullet ab}\,\varepsilon_{abpe} - R_{\alpha p}^{\bullet\bullet ab}\,\varepsilon_{abme} - 4i\bar{R}_{mp}{}^{\dot{\beta}}(\gamma_5\gamma_e)_\alpha^\beta = 0 \tag{7.6}$$

(the dot on the first β in the last term denotes a « lower spinor »), which leads to a nontrivial relation between the spinor curvature $\bar{R}_{mp}$ and the spinor components of the ordinary curvature R^{ab}. Relation (7.6) is the same as eq. (11*c*) of ref. [30], in which $\bar{R}_{mp}$ is denoted $\bar{H}_{mp}$ and $R_{\alpha m}^{\bullet\bullet ab}$ is essentially Σ_{mab}, the somewhat unwieldy expression given for the variation of the relativity connection field, in the original papers on supergravity [25-27]. That this is indeed $R_{\alpha m}^{\bullet\bullet ab}$ can be seen from (4.43*a*).

Finally, we need to compute

$$R_{cd}^{\bullet\bullet ab} = 2\tilde{\Delta}_d^{\bullet N}\tilde{\Delta}_c^{\bullet M}R_{MN}^{\bullet\bullet ab}\,. \tag{7.7}$$

A possibile source of confusion arises in trying to calculate anholonomic components by using the inverses of the conventional vierbeins instead of the inverse 14×14 matrix $\tilde{\Delta}_A^{\bullet M}$ (though the summation over M, N is now reduced to vector-vector and vector-spinor MN sets only). This explains the occurrence of Ω_{abcd} in ref. [30] as distinct from R_{abcd}. It is the Ω_{abcd} of ref. [30] which coincides with the present $R_{cd}{}^{ab}$, while R_{abcd} of ref. [30] is an auxiliary quantity of no physical and geometrical content.

In matching these components with those of ref. [30], care should be taken in setting up a correspondence between our geometrical operators $\tilde{D}_A$ (and, in general, any vector field on GP) and the set of second-quantized operators $\mathscr{S}(x)$, $\mathscr{H}(x)$, $\mathscr{T}^{ab}(x)$ of ref. [30]. The latter can be written as a GP multiplet $\mathscr{H}_A(x)$ with the commutation relations

$$[\mathscr{H}_B(x), \mathscr{H}_E(x')] = \delta(x, x')(C^A_{\bullet BE} + R_{BE}^{\bullet\bullet A})\,\mathscr{H}_A(x) \tag{7.8}$$

in analogy to our (4.41*a*), (4.41*b*), (4.44*a*)-(4.44*d*) set. Note that with the vanishing of components in (7.5), (4.44*b*)-(4.44*d*) now become

$$[\tilde{D}_a, \tilde{D}_b] = R_{ab}^{\bullet\bullet cd}D_{cd} + \bar{R}_{ab}\tilde{D}\,, \tag{7.9a}$$

$$C_{\alpha\beta}[\tilde{D}_a, \tilde{D}_\beta] = R_{a\alpha}^{\bullet\bullet cd}D_{cd}\,, \tag{7.9b}$$

$$\{\tilde{D}_\alpha, \tilde{D}_\beta\} = (\gamma^a)_{\alpha\beta}\tilde{D}_a\,, \tag{7.9c}$$

and (4.43) consist in (the D are GP-covariant derivatives)

$$
\left\{
\begin{aligned}
\delta\varrho^{ab} &= \mathrm{D}\varepsilon^{ab} - \varrho^c \varepsilon^d R_{cd}^{\bullet\bullet ab} - \varrho^c \bar{\varepsilon}^{\dot{\alpha}} R_{c\alpha}{}^{ab}, \\
\delta\varrho^a &= \mathrm{D}\varepsilon^a, \\
\delta\varrho &= \mathrm{D}\varepsilon - \varrho^c \varepsilon^d R_{cd},
\end{aligned}
\right. \tag{7.10}
$$

which are the conventional transformations of supergravity (see (4.42*a*)-(4.42*c*) for the variations contained in $\mathrm{D}\varepsilon^a$ and $\mathrm{D}\varepsilon$).

Supergravity, however, has meaning on a 4-dimensional manifold $\mathscr{M}^4$ only which we may take to be $\xi^\alpha = 0$. (4.43) or (7.10) are then relating the restriction of the forms ϱ^A on $\mathscr{M}^4$ ($\xi^\alpha = 0$) before AGCT to the corresponding restrinction on $\mathscr{M}^{4\prime}$ ($\xi'^\alpha = 0$) after AGCT; clearly any covariant theory admits formulae (4.43), but so far only in supergravity the $R_{BE}^{\bullet\bullet A}$ components are all functionals of the restriction $\varrho^A|_{\mathscr{M}^4}$ and (7.10) is a true transformation. In this sense supergravity is the only known supersymmetric theory with structural group GP.

8. – Conclusions.

The following facts have emerged from the foregoing discussion.

1) It is possible and indeed convenient to develop the formalism of gravitational theories as gauge theories on a group manifold. The groups are P or GP for gravity and supergravity, but further enlargements should follow the same method.

When this is done, the connection forms and the vierbeins appear as a single c-bein on G (c being the dimension of G), which also play the role of connections on G in computing covariant derivatives and curvatures. In particular, ordinary curvatures and torsions are naturally unified as expected in Einstein-Cartan theories.

2) An action is formed under the assumption that it reduces to the Einstein action in the appropriate cases and be compatible with the vanishing of curvature (« flat » empty space). The notion of pseudocurvature is introduced and related to Chevalley cohomology and to the Segal-Wigner-Inönü theory of group contraction.

Pseudocurvatures provide an interesting way of achieving the same ends as the MacDowell-Mansouri [23] procedure, but with methods which are purely intrinsic to G. The method reproduces conventional supergravity [25-26].

3) In all these theories, the action is strictly gauge invariant under SO_4 gauges only. Moreover, the Lagrangian density is always a 4-form and is integrated over an arbitrary $\mathscr{M}^4$ variety in G. However, physics is seen to be completely determined by what happens on a single $\mathscr{M}^4$. The transfer of information from any $\mathscr{M}^4$ to any other $\mathscr{M}^{4'} \subset G$ implies either ordinary SO_4-gauged symmetry or supersymmetry as an AGCT gauge. In a way, restricting a GP SO_4-factorized theory to $\boldsymbol{R}^{4/4}$ includes as partial $\boldsymbol{R}^4$ slices all possible supersymmetry-related conventional supergravity theories. $\boldsymbol{R}^4 \to \boldsymbol{R}^{4/4}$ « determinism » has been realized elsewhere [31].

4) Factorization is seen to be a key simplifying feature in these theories. We conjecture that, if a Lagrangian is gauge invariant under any $H \subset G$, then it is H-factorizable as a consequence of the equations of motion. In this sense, factorization should not be an independent postulate.

It is possible to give a heuristic proof of the factorization hypothesis for solutions infinitesimally close to a factorized one. All these solutions can be reduced to factorized ones by an infinitesimal co-ordinate transformation on G. In the large, however, it may be that there are discrete families of factorized solutions with the same boundary conditions and still topologically distinct. This interesting possibility remains an open question.

5) The procedure can be in principle extended to other groups. It would also be desirable to extend the theory so as to include « matter » distributions whose energy-momentum, supersymmetry and spin density tensors are non-trivial functions on G.

* * *

The authors would like to thank Prof. J. PRENTKI and the Theory Division of CERN for their hospitality during part of this work. They would also like to thank Prof. J. A. WHEELER and the Center for Theoretical Physics at the University of Texas, Austin, for the hospitality to one of us (TR) during part of the continuation of this work.

APPENDIX

The (Euclidean) Dirac algebra.

Metric (1, 1, 1, 1):

(A.1) $$\gamma^a\gamma^b + \gamma^b\gamma^a = 2\delta^{ab},$$

(A.2) $$\sigma^{ab} = \tfrac{1}{4}(\gamma^a\gamma^b - \gamma^b\gamma^a),$$

(A.3) $$\gamma^0 = -i\gamma^4 = -(\gamma^0)^{-1}\,,$$

(A.4) $$\gamma_5 = \gamma^0\gamma^1\gamma^2\gamma^3 = -(\gamma_5)^{-1}\,,$$

(A.5) $$C\gamma^a = -\gamma^{aT}C\,,$$

(A.6) $$A\gamma^\mu = -\gamma^{\mu\dagger}A \qquad (\mu = 0, 1, 2, 3)$$

With the choice

(A.7) $$A = -C = \gamma^4$$

we have

(A.8) $$(\gamma^a)^\dagger = (\gamma^a)^{-1} = \gamma^a\,,$$

(A.9) $$C^T = -C\,,$$

(A.10) $$(C\gamma^a)^T = C\gamma^a\,,$$

(A.11) $$(C\sigma^{ab})^T = C\sigma^{ab}\,,$$

(A.12) $$(C\gamma_5)^T = -C\gamma_5\,,$$

(A.13) $$(C\gamma_5\gamma^a)^T = -C\gamma_5\gamma^a\,,$$

(A.14) $$C^2 = 1\,,$$

and the adjoint (= charge conjugate) Majorana spinor is defined as

(A.15) $$\bar\psi = (C\psi)^T\,.$$

Commutators.

(A.16) $$[\sigma^{ab}, \sigma^{cd}] = \delta^{ad}\sigma^{bc} - \delta^{ac}\sigma^{bd} + \delta^{bc}\sigma^{ad} - \delta^{bd}\sigma^{ac}\,,$$

(A.17) $$[\gamma^a, \sigma^{cd}] = \delta^{ac}\gamma^d - \delta^{ad}\gamma^c\,.$$

The generators

(A.18) $$S_{ab}(x) = x^a\frac{\partial}{\partial x^b} - x^b\frac{\partial}{\partial x^a}\,,$$

(A.19) $$S_{ab}(\xi) = \frac{\partial}{\partial\xi^\alpha}\,\sigma^{ab,\alpha}{}_{\beta}\,\xi^\beta\,,$$

(A.20) $$[S_{ab}(x), S_{ab}(\xi)] = 0$$

obey the same algebra as (A.16)

(A.21) $$[S_{ab}, S_{cd}] = \delta_{ad}S_{bc} - \delta_{ac}S_{bd} + \delta_{bc}S_{ad} - \delta_{bd}S_{ac}\,.$$

Fierz transformations utilized in sect. 6:

(A.22) $$(\bar\omega\gamma_e\chi)\,\varepsilon^{abfe} = i(\bar\chi\sigma^{ab}\gamma_5\gamma^f\omega) - i(\bar\omega\sigma^{ab}\gamma_5\gamma^f\chi)$$

for the case of $\{\chi, \omega\} = 0$, and

$$(\bar{\omega}\gamma_e\omega)\,\varepsilon^{abfe} = -2i(\bar{\omega}\sigma^{ab}\gamma_5\gamma^f\omega) \tag{A.23}$$

for $[\omega, \omega] = 0$.

REFERENCES

[1] See, for example, B. ZUMINO: *Proceedings of the XVII International Conference on High-Energy Physics*, edited by J. R. SMITH (Chilton, Didcot, 1974), p. 254.
[2] A. SALAM and J. STRATHDEE: *Phys. Rev. D*, **11**, 521 (1975).
[3] L. CORWIN, Y. NE'EMAN and S. STERNBERG: *Rev. Mod. Phys.*, **47**, 573 (1975).
[4] S. FERRARA: *Riv. Nuovo Cimento*, **6**, 105 (1976); P. FAYET and S. FERRARA: *Phys. Rep.*, **32** C, 249 (1977).
[5] For a previous exposition of the theory for physicists, see B. S. DEWITT: *Relativity groups and topology*, in *Proceedings of the Les Houches 1963 Seminar*, edited by C. DEWITT and B. DEWITT (New York, N. Y., 1964).
[6] F. A. BEREZIN and D. A. LEITES: *Dokl. Akad. Nauk SSSR*, **224**, (1975); (English translation: *Sov. Math. Dokl.*, **16**, 1218 (1975)).
[7] B. ZUMINO: *Proceedings of the Conference on Gauge Theories and Modern Field Theory, Northeastern University, Boston, 1975*, edited by R. ARNOWITT and P. NATH (Cambridge, Mass., 1976), p. 255.
[8] F. A. BEREZIN and G. I. KATZ: *Mat. Sb. (SSSR)*, **82**, 343 (1970) (English translation: **11**, 311 (1970)).
[9] T. W. B. KIBBLE: *Journ. Math. Phys.*, **2**, 212 (1961); D. W. SCIAMA: in *Recent Developments in General Relativity* (Oxford, 1962), p. 415.
[10] P. VON DER HEYDE: *Phys. Lett.*, **58** A, 141 (1976).
[11] YU. A. GOLFAND and E. P. LIKHTMAN: *JETP Lett.*, **13**, 452 (1971).
[12] J. WESS and B. ZUMINO: *Nucl. Phys.*, **70** B, 39 (1974).
[13] A. SALAM and J. STRATHDEE: *Nucl. Phys.*, **76** B, 477 (1974).
[14] P. NATH and R. ARNOWITT: *Phys. Lett.*, **56** B, 177 (1975).
[15] V. P. AKULOV, D. V. VOLKOV and V. A. SOROKA: *JETP Lett.*, **22**, 396 (1975) (English translation, p. 187).
[16] J. WESS and B. ZUMINO: *Phys. Lett.*, **66** B, 361 (1977).
[17] G. WOO: *Lett. Nuovo Cimento*, **13**, 546 (1975).
[18] P. P. SRIVASTAVA: *Lett. Nuovo Cimento*, **13**, 161, 657 (1975).
[19] E. CARTAN: *Compt. Rend.*, **174**, 593 (1922); *Ann. Ecole Norm. Sup.*, **40**, 325 (1923); **41**, 1 (1924); **42**, 17 (1925).
[20] F. W. HEHL: Thesis, Technical University Clausthol (1970); *Phys. Lett.*, **36** A, 225 (1971).
[21] A. TRAUTMAN: *Bull. Acad. Pol. Sci., Ser. Sci. Math. Astron. Phys.*, **20**, 185, 503, 895 (1972); **21**, 345 (1973); *Symp. Math.*, **12** (Bologna, 1973), p. 30.
[22] F. W. HEHL, P. VON DER HEYDE, G. D. KERLICK and J. M. NESTER: *Rev. Mod. Phys.*, **48**, 393 (1976).
[23] S. W. MACDOWELL and F. MANSOURI: *Phys. Rev. Lett.*, **38**, 739 (1977).
[24] C. CHEVALLEY and S. EILENBERG: *Trans. Amer. Math. Soc.*, **63**, 85 (1948).
[25] D. Z. FREEDMAN, P. VON NIEUWENHUIZEN and S. FERRARA: *Phys. Rev. D*, **13**, 3214 (1976).

[26] S. DESER and B. ZUMINO: *Phys. Lett.*, **62** B, 335 (1976).
[27] D. Z. FREEDMAN and P. VAN NIEUWENHUIZEN: *Phys. Rev. D*, **14**, 912 (1976).
[28] P. NATH and R. ARNOWITT: *Phys. Lett.*, **65** B, 73 (1976).
[29] R. ARNOWITT, P. NATH and B. ZUMINO: *Phys. Lett.*, **56** B, 81 (1975).
[30] C. TEITELBOIM: *Phys. Rev. Lett.*, **38**, 1106 (1977).
[31] L. BRINK, M. GELL-MANN, Y. NE'EMAN, P. RAMOND and J. SCHWARZ: *Superspace Lecture Notes*, Aspen Center for Physics, *1977 Supergravity Seminar* (unpublished).
[32] A. H. CHAMSEDDINE and P. C. WEST: *Nucl. Phys.*, **129** B, 39 (1977).

Direttore responsabile: CARLO CASTAGNOLI

Stampato in Bologna dalla Tipografia Compositori coi tipi della Tipografia Monograf
Questo fascicolo è stato licenziato dai torchi il 24-VII-1978

Volume 74B, number 1, 2 PHYSICS LETTERS 27 March 1978

GRAVITY AND SUPERGRAVITY AS GAUGE THEORIES ON A GROUP MANIFOLD

Yuval NE'EMAN
Department of Physics and Astronomy, Tel Aviv University [1], *Tel Aviv, Israel*
and Center for Particle Theory, University of Texas [2], *Austin, TX 78712, USA*

and

Tullio REGGE
School of Natural Sciences, The Institute for Advanced Study [3], *Princeton, NJ 08540, USA*
and Istituto di Fisica, Universita di Torino, I-10125 Torino, Italy

Received 18 January 1978

We construct generalizations of gravity, including supergravity, by writing the theory on the group manifold (Poincaré for gravity, the graded-Poincaré group for supergravity). The action involves forms over the group, restricted to a 4-dimensional submanifold. The equations of motion produce a Lorentz gauge in gravity and supergravity, and an additional anholonomic supersymmetric coordinate transformation which reduces to the "local supersymmetry" of supergravity.

In this letter we propose a method for the construction and discussion of generalizations of conventional Einstein gravity theory, including recent theories of supergravity [1].

The idea is to write such a theory on a group manifold. Let G be (a non semi-simple) Lie group of dimension c (e.g. the Poincaré group P for gravity, the graded Poincaré [2] formal-group [3] GP for supergravity). On G we give a set of c forms ρ^A and define G-curvature as the 2-forms

$$R^A = \mathrm{d}\rho^A - \tfrac{1}{2}\rho^B \wedge \rho^E C_{BE}^{\cdot\cdot A}\,, \tag{1}$$

where $C_{BE}^{\cdot\cdot A} = -(-1)^{be} C_{EB}^{\cdot\cdot A}$ are the (graded, with b, e the gradings [4] of B, E) structure constants of the Lie algebra $\mathcal{G}$ of G. For the case $\rho^A = \omega^A$, where ω^A are the left invariant Cartan forms on G, $R^A = 0$ and eq. (1) realizes the Cartan–Maurer equations. We refer to this case as the "flat" geometry.

We regard the ρ^A as the "objects" of the theory, to be treated as the potentials of a Yang–Mills G-gauge theory, and the R^A are the corresponding field-strenghts. Given any G-multiplet η^A (η_A) we define G-covariant derivatives as,

$$(\mathrm{D}\eta)^A = \mathrm{d}\eta^A - \rho^B \wedge \eta^E C_{BE}^{\cdot\cdot A}\,,$$
$$(\mathrm{D}\eta)_A = \mathrm{d}\eta_A + \rho^B \wedge \eta_E C_{BA}^{\cdot\cdot E}\,. \tag{2}$$

Gravitational theories follow from an action of the form

$$A = \int_{\mathcal{M}^4} R^A \wedge \zeta_A\,, \tag{3}$$

where $\mathcal{M}^4$ is any 4-dimensional submanifold of G, and $\zeta_A = \zeta_A(\rho)$ is a 2-form constructed as a quadratic polynomial in the ρ^A. A should be stationary with respect to all variations of ρ^A and of $\mathcal{M}^4$. This last condition turns out, however, to be trivially satisfied by virtue of the general covariance of the theory. The theory should admit "flat" space $\rho^A = \omega^A$ as a solution. The field equations can be symbolically written as

$$R^A \wedge \delta\zeta_A/\delta\rho^B - \mathrm{D}\zeta_B = 0\,, \tag{4}$$

[1] Research supported in part by the US–Israel Binational Science Foundation.
[2] Research supported in part by the U.S. Energy Research and Development Administration Grant e(40-1) 3992.
[3] Research supported in part by NSF Grant 40768X.

Volume 74B, number 1,2 PHYSICS LETTERS 27 March 1978

which is satisfied by $\rho^A = \omega^A$, $R^A = 0$ only if

$$D\zeta_A(\omega) = 0\,. \tag{5}$$

A form ζ_A satisfying eq. (5) will be designated as a pseudo-curvature.

Clearly, the action (3) is not invariant under all G-gauge transformations. For example, Einstein gravity [5], which can be written as in eq. (3),

$$\zeta_{ab} = \epsilon_{abcd}\rho^c \wedge \rho^d,\; \zeta_a = 0,\; A = \tfrac{1}{8}\int_{\mathcal{M}^4} R^{ab} \wedge \zeta_{ab}\,, \tag{6}$$

is only SO(3, 1) gauge-invariant. Therefore, the assumption of a pseudo-curvature ζ_A breaks the G-symmetry. Pseudo-curvatures are tied to a non-trivial Segal–Inönu–Wigner contraction procedure from a semi-simple group G' to G; in the above case, from SO(3, 2) to P. In fact, if $C_{BE}^{\;\cdot\;\cdot A}(\epsilon)$ are the structure constants of G', and $C_{BE}^{\;\cdot\;\cdot A}(0) = C_{BE}^{\;\cdot\;\cdot A}$ those of G,

$$\left.\frac{\partial C_{BE}^{\;\cdot\;\cdot A}(\epsilon)}{\partial \epsilon}\right|_{\epsilon=0} \rho^B \wedge \rho^E = \zeta^A\,, \tag{7}$$

is a pseudo-curvature. Pseudo-curvatures are related to the Chevalley cohomologies on groups, and to the MacDowell–Mansouri derivations [6] of gravity theories.

The action (3) for supergravity is given by selecting ζ_A as:

$$\zeta_{ab} = \epsilon_{abcd}\,\rho^c \wedge \rho^d\,, \quad \zeta_a = 0\,,$$
$$\bar{\zeta}_\alpha = -4\mathrm{i}\bar{\rho}\gamma^a\rho^a\gamma_5 \quad \text{or} \quad \zeta^\alpha = -4\mathrm{i}\gamma_5\gamma^a\rho^a\rho\,. \tag{8}$$

We assume the validity of the field equations (4) throughout G. Clearly, if A is invariant under a gauge subgroup $H \subset G$ (i.e. SO(3, 1) for both P and BP), a solution on G can be derived from its boundary value on $Z = G/H$ through the so-called factorization hypothesis,

$$\rho^A = \omega_H^A + \tau^B \operatorname{ad}(q^{-1})_B^{\;\cdot A}\,, \tag{9}$$

where, locally, G has coordinates (z, q), $z \in Z$, $q \in H$, τ^B are the (factorized) forms on Z, ω_H^A the restrictions of the Cartan forms ω^A to H. We conjecture that in general ρ^A is determined by its boundary values on any $\mathcal{M}^4$, apart from a generic coordinate transformation on G. This can be proved to hold if a solution of eq. (4) is sufficiently close to a factorized one. Globally, one cannot exclude the possible existence of "twisted" topologically inequivalent factorized solutions sharing the same boundary conditions on $\mathcal{M}^4$.

On P (GP), eq. (9) becomes,

$$\rho^{ab} = (\Xi^{-1}\,\mathrm{d}\Xi)^{ab} + \tau^{cd}\Xi^{db}\Xi^{ca}\,,$$
$$\rho^a = \Xi^{ca}\tau^c \quad (\rho = u(\Xi^{-1})\tau)\,, \tag{10}$$

where $\Xi^{ab} \in H$, $\Xi^{ab}\Xi^{cb} = \delta^{ac}$ (we disregard the Minkowski metric). Z is $\mathbf{R}^4$ in P, and $\mathbf{R}^{4/4}$ ("superspace", with 4 Bose and 4 Fermi dimensions [7]) in GP. For P and Einstein's gravity, Z coincides with $\mathcal{M}^4$ so that eq. (9) connects solutions on possible choices of $\mathcal{M}^4$ as discussed. For GP, $\mathcal{M}^4$ is still a submanifold of $\mathbf{R}^{4/4}$. In principle, eqs. (4), (5) should determine the extension of the form τ^A to $\mathbf{R}^{4/4}$ from their restriction on $\mathcal{M}^4$. However, this is not achieved through a gauge transformation as it is for the Ξ^{ab} variables on SO(3, 1). In order to compute the change in the ρ^A while moving infinitesimally from a surface $\mathcal{M}^4$ to $\mathcal{M}^{4'}$, we utilize a generalization to an arbitrary group G of a set of transformations defined by Von der Heyde [8] and by Hehl and collaborators [9]. Carrying out an infinitesimal coordinate change $x^M \to x^M + \epsilon^M$ (M, N are holonomic indices; A, B are anholonomic) we find for the change in ρ^A:

$$\delta\rho^A = D\epsilon^A - 2(\epsilon, R^A)\,, \tag{11}$$

where (this is an "anholonomized general coordinate transformation", AGCT)

$$\epsilon^A = \epsilon^M \rho_M^{\;A}\,, \quad \rho^A = \mathrm{d}x^M \rho_M^{\;A}\,,$$

and

$$R^A = \tfrac{1}{2}\,\mathrm{d}x^M \wedge \mathrm{d}x^N R_{MN}^{\;\;\;A}\,,$$
$$(\epsilon, R^A) = \tfrac{1}{2}\mathrm{d}x^M\epsilon^N R_{MN}^{\;\;\;A} = \tfrac{1}{2}\rho^B\epsilon^C R_{BC}^{\;\;\;A}\,. \tag{12}$$

The ϵ^A are related to the lapse and shift functions of the standard canonical formalism of general relativity. For the index A in the Majorana range, eq. (11) yields the "local supersymmetry" transformations of supergravity. However, the exact comparison with supergravity requires the repeated use of field equations (4), (5). For GP, the field equations are,

$$R^a = 0\,, \tag{13a}$$

$$R^{ab}\rho^f\epsilon_{abfe} - 2\mathrm{i}\bar{R}\gamma_5\gamma_e\rho = 0\,, \tag{13b}$$

$$\gamma^a\rho^a R = 0\,. \tag{13c}$$

Volume 74B, number 1,2 PHYSICS LETTERS 27 March 1978

Direct calculation of the anholonomic components $R_{BC}{}^{A}$ from eq. (13) yields

$$R_{BC}{}^{a}=0\,,\quad R_{\alpha\beta}{}^{A}=0\,,\quad R_{(ab)B}{}^{A}=0\,, \tag{14}$$

where a is a (translation) vector index, (ab) a skew-symmetric tensor value (for an SO(3, 1) direction) component, α, β are Majorana spinor indices for the odd generators. The surviving components can be seen to coincide with those of ref. [10], after some algebra involving eq. (13b). Inserting these components in eq. (11) reproduces the conventional supergravity transformations

$$\delta\rho^{ab}=\mathrm{D}\epsilon^{ab}-\rho^{c}\epsilon^{d}R_{cd}{}^{ab}-\rho^{c}\bar{\epsilon}^{\dot\alpha}R_{c\alpha}{}^{ab}\,,$$
$$\delta\rho^{a}=\mathrm{D}\epsilon^{a}\,,\quad \delta\rho=\mathrm{D}\epsilon-\rho^{c}\epsilon^{d}R_{cd}\,, \tag{15}$$

where

$$\mathrm{D}\epsilon^{ab}=\mathrm{d}\epsilon^{ab}+\rho^{at}\wedge\epsilon^{tb}-\rho^{tb}\wedge\epsilon^{at}\,,$$
$$\mathrm{D}\epsilon^{a}=\mathrm{d}\epsilon^{a}+\rho^{at}\wedge\epsilon^{t}-\rho^{t}\wedge\epsilon^{at}+\bar{\rho}\gamma^{a}\epsilon\,, \tag{16}$$
$$\mathrm{D}\epsilon=\mathrm{d}\epsilon+\tfrac{1}{2}(\rho^{ab}\sigma^{ab})\wedge\epsilon-\tfrac{1}{2}\sigma^{ab}\rho\wedge\epsilon^{ab}\,.$$

In this sense, the exponentiation of eqs. (11) starting from an arbitrary $\mathcal{M}^4$ is "condensed" into forms ρ^A on $\mathbf{R}^{4/4}$, and the $\mathbf{R}^{4/4}$ theory can be viewed as the collection of supersymmetric-related theories on $\mathcal{M}^4$. Formula (11) therefore solves in principle the problem of constructing supersymmetric transformations for any theory. Clearly, the field equations impose very stringent conditions on $R_{BE}{}^{A}$, and supergravity is certainly unique in yielding these restrictions. However, any covariant theory on GP or $\mathbf{R}^{4/4}$ is compatible with eq. (11), with the specific form of $R_{BE}{}^{A}$ depending on the field equations. The values of these curvature components may then be much more complicated and there is no a priori guarantee that $R_{BE}{}^{A}$ are functionals of the restrictions of ρ^A on $\mathcal{M}^4$ only, as needed in order that eq. (11) be an effective transformation. In the present formulation, the choice of an action is highly restricted by eq. (5). In all supersymmetry theories, we expect the $\mathcal{M}^4$ to remain an even 4-dimensional manifold. This avoids the use of the Berezin integration [3] on the variables, whose formal difficulties have impeded the development of viable 8-dimensional actions [11,12]. From our point of view, the odd variables are just a shorthand for a collection of Fermi fields needed to specify $\mathcal{M}^4\in G$.

An extended exposition of these ideas will be published elsewhere.

References

[1] D.Z. Freedman, P. van Nieuwenhuizen and S. Ferrara, Phys. Rev. D13 (1976) 3214;
S. Deser and B. Zumino, Phys. Lett. 62B (1976) 335.

[2] Yu.A. Golfand and E.P. Likhtman, JETP Lett. 13 (1971) 452;
J. Wess and B. Zumino, Nucl. Phys. B70 (1974) 39.

[3] F.A. Berezin and G.I. Katz, Mat. Sb. 82 (1970) 124; English transl. 11 (1970) 311.

[4] L. Corwin, Y. Ne'eman and S. Sternberg, Rev. Mod. Phys. 47 (1975) 573.

[5] A. Trautman, Bull. Acad. Pol. Sci. Ser. Sci. Mat. Astron. Phys. 20 (1972) 185, 503, 895; 21 (1973) 345.

[6] S.W. MacDowell and F. Mansouri, Phys. Rev. Lett. 38 (1977) 739.

[7] A. Salam and J. Strathdee, Nucl. Phys. B76 (1974) 477.

[8] P. Von der Heyde, Phys. Lett. 58A (1976) 141.

[9] F.W. Hehl, P. Von der Heyde, G.D. Kerlick, J.M. Nester, Rev. Mod. Phys. 48 (1976) 393.

[10] C. Teitelboim, Phys. Rev. Lett. 38 (1977) 1106.

[11] J. Wess and B. Zumino, Phys. Lett. 66B (1977) 361.

[12] L. Brink, M. Gell-Mann, Y. Ne'eman, P. Ramond and J. Schwarz, Superspace Lecture Notes, Supergravity Seminar (Aspen Center for Physics, 1977), unpublished.

PART VII

2+1 Quantum Gravity

2+1 Dimensional Gravity

J. E. Nelson

Dipartimento di Fisica Sezione Teorica,
Università degli Studi di Torino, and
I.N.F.N. Sezione di Torino,
via Pietro Giuria 1, 10125 Torino, Italy

It gives me great pleasure to review some of the joint work by Tullio Regge and myself. We worked intensely on 2+1-dimensional gravity from 1989 for about five years, and published 16 articles. I will present and review two of our early articles, highlighting what I believe are the most important results, some of them really surprising, and discuss later developments.

Introduction

2+1-dimensional gravity is still a fascinating theory. One can study many issues of the four-dimensional theory without its technical complications, and the only other lower dimensional theory, two-dimensional gravity, has an action which is a topological invariant. It has a Chern–Simons formulation,[1–3] and the curvature is constant, proportional to the cosmological constant (positive, negative, or zero). There are no local degrees of freedom, only global topological ones, and only when there are non trivial topologies. However, it does have local exact solutions,[4] and admits three-dimensional black holes.[5]

Often considered as a "toy model" without physical relevance, 2+1-dimensional gravity has, nevertheless, interesting connections to other theories, e.g. in the context of the Ads-CFT correspondence,[6] pure Chern–Simons theory on a manifold with boundary induces, for certain boundary conditions, the Wess–Zumino–Witten model[7] on the boundary.[8] For different boundary conditions, it induces Liouville theory at infinity.[9] Three-dimensional Chern–Simons theory can also arise as a string theory.[10]

Tullio Regge always liked discrete physics, preferably with very few degrees of freedom, so when I showed him the 1989 Witten preprint on 2+1-dimensional gravity[3] which describes a theory with a finite number of only global topological degrees of freedom, he was very enthusiastic. We worked intensely on 2+1-dimensional gravity from then on, for about five years, and published 16 articles (see e.g. Refs. 11–19). Here I review and complement two of our early articles[11,12] which set the scene for many others. I will highlight some important results, e.g. the emergence of braid group relations, and the quantum group $SU(2)_q$, and discuss later developments.

In both articles we studied the first order form of 3-dimensional (2 space, 1 time) de Sitter/anti de Sitter (AdS) gravity, where the gauge group is $SO(3,1)$ for cosmological constant $\Lambda > 0$, and $SO(2,2)$ for $\Lambda < 0$, and spacetime is $\mathbb{R} \times \Sigma_g$, Σ_g being a compact Riemann surface of genus g. In a previous article[13] we had studied the Poincare case i.e. with $\Lambda = 0$, where the gauge group is $ISO(2,1)$. We later showed[14] that this can be obtained in the $\Lambda \to 0$ limit.

In Section 1 I discuss the first article,[11] in which we constructed the algebra of observables for the torus $\mathbb{T}^2$, $g = 1$, and studied some of its representations. There are connections to knot theory, and to quantum groups. The quantum algebra of observables is the starting point for the second article[12] which I discuss in Section 2. Here we studied the symmetries of the quantum algebra of observables and how to generate them, implemented for any genus and at the quantum level. In Section 3 I discuss further developments, in particular the algebraic structures which have been rediscovered in the mathematical and mathematical physics literature.

This text is intended as a companion to articles.[11,12] It cannot be understood without having both of these to hand. Please note that in both articles the notation is unconventional. We used square brackets [,] to denote Poisson brackets, and round brackets (,) to denote commutators. I will refer to equations in both articles as equation (), but equations in this text simply as ().

1. Homotopy groups and 2 + 1 dimensional quantum De Sitter gravity, J. E. Nelson, T. Regge and F. Zertuche, Nucl. Phys. B339 (1990) 516

In this article our construction used AdS connections ω^{ab} defined on page 517, and their Poisson brackets Eq. (1.4). These Poisson brackets and the AdS constraints $R^{ab} = d\omega^{ab} - \omega^{ac} \wedge \omega_c{}^b = 0$ can be seen most clearly from the (2+1)-dimensional spacetime splitting $\mathbb{R} \times \Sigma$ of the Einstein action

$$I_{Ein} = \int d^3x \sqrt{-{}^{(3)}g}\, ({}^{(3)}R - 2\Lambda) = \frac{\alpha}{4} \int dt \int d^2x \, \epsilon^{ij} \epsilon_{abcd} \, (\omega^{cd}{}_j \, \dot{\omega}^{ab}{}_i - \omega^{ab}{}_0 R^{cd}{}_{ij}). \quad (1)$$

In Section 2 we formally solved the constraints $R^{ab} = 0$ (Einstein's equations) by the expression

$$d\psi^{ab} = \omega^{ac} \, \psi_c{}^b \quad (2)$$

where ψ^{ab} is an SO$(3,1)$ or SO$(2,2)$-valued zero-form. We argued that (2), when restricted to a path γ on Σ_g, will lead to a set of ψ^{ab} which depend only on the homotopy class of γ and its starting and end points. Joining these points (closing the path γ) means that the new ψ^{ab}, now denoted $\psi^{ab}(\gamma)$ depends only on the homotopy class of the closed path γ, and the base point. I.e. these $\psi^{ab}(\gamma)$ furnish a representation of the fundamental group

π_1 of Σ_g, whose presentation is given in Eqs. (1.5) and (1.6). We also argued that, from Eq. (1.4), that the single elements of the 4×4 matrices $\psi(\gamma)$ and $\psi(\sigma)$, where $\gamma, \sigma \in \pi_1$ will have zero Poisson brackets iff they are obtained from integration of (2) over paths with zero intersection. We proved this in Section 3. This was the first hint of a quantum group structure though it differed from that of e.g. Ref. 20 where elements of the same matrix may not commute but elements from different matrices do, the exact opposite of what we found.

We calculated the Poisson brackets of the matrices $\psi(\gamma)$ and $\psi(\sigma)$ by expanding them along infinitesimal paths, see Eq. (3.2). The result is given in Eqs. (3.4), and the r.h.s. indeed depends on the intersection number s. It is valid only when ρ, σ have a single intersection i.e. $s = \pm 1$. Clearly the sign of s depends on the orientation of the paths ρ, σ.

The spinor version of Section 4 is more useful, where the spinor equivalent of $\psi(\gamma)$ are 2×2 matrices $S^{\pm}$. They are related through Eq. (4.3). Their Poisson brackets are given in Eqs. (4.8). Details can be found in Appendix A. The $\pm$ refer to the two copies of the $S^{\pm}$. For $SO(2,2)$ $(k = -1, \Lambda < 0)$ the $S^{\pm}$ are real and independent, i.e. $S^{\pm} \in SL(2,\mathbb{R})$, whereas for $SO(3,1)$ $(k = 1, \Lambda > 0)$ they are complex conjugates, i.e. $S^{\pm} \in SL(2,\mathbb{C})$.

Equations (4.8), though messy, are highly suggestive. In the first, the r.h.s. depends on the intersection number s. Note also that ρ_1, σ_1 denote the segments of the paths ρ, σ before the intersection, and ρ_3, σ_3 denote the segments after (the paths are divided as $\rho = \rho_3\rho_2\rho_1$ and similarly σ). The infinitesimal segments with a single intersection are ρ_2, σ_2.

Viewed in this way, if e.g. ρ is the path top left to bottom right and σ is the path bottom left to top right (with both start and endpoints identified), then the first of Eqs. (4.8) can be represented graphically as the classical bracket between the paths ρ, σ, shown in Figure 1.

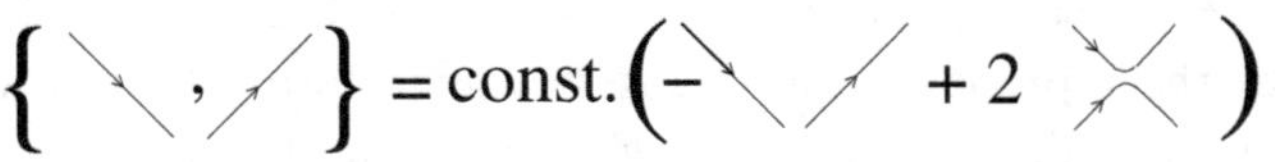

Figure 1. The Poisson bracket of the paths ρ, σ.

This was a surprise, but we had seen similar relationships in knot theory. In fact, we later realised that Figure 1 is highly reminiscent of the Kauffman polynomial[21] for classical knots and links, and that it had already appeared in.[22] We extended the relationship to knot theory in[15,16] where we constructed, for genus 2, a group of automorphisms of $\pi_1(\Sigma)$ and a group of canonical transformations on the classical algebra of observables. Its symmetries are those of the braid group for 6 $(=2g+2)$ threads.

In Section 5 taking traces, Eq. (5.1), ensures gauge invariance, where a gauge transformation on a path ρ is $\rho \to \nu\rho\nu^{-1}$ for any open path ν, i.e. a shift of the base point. For 2+1-dimensional gravity this is equivalent to a diffeomorphism (coordinate transformation). The traces Eq. (5.1) is clearly gauge invariant. The result is the classical non-linear algebra, Eq. (5.2). There is another version, Eq. (5.4), which Regge and I did not study. It is linear, but subject to non-linear constraints (Eq. (5.3)).[a] In Eq. (5.2) u, v refer to two oriented

[a]Roger Picken and I have studied this second version, and it has led us to a new quantum group structure, some interesting properties in quantum geometry, and a quantisation of the Goldman bracket.[23]

closed paths (cycles) with a single intersection on Σ_g, and there are three independent traces $c(u), c(v), c(uv)$. Intersection ± 1 is sufficient for the torus $g = 1$, and the rest of[11] discusses, for the + algebra, the consequences of Eq. (5.2), its quantisation, and a preliminary study of its representations.

The three traces in Eq. (5.2) are denoted, for the + algebra, $x = c^+(u), y = c^+(v), z = c^+(uv)$, and their Poisson brackets are given in Eq. (6.1). They have a remarkable triangular, or cyclical structure, and were easy to quantise in Section 6. Firstly define the commutators as $i\hbar$ times the r.h.s. of the classical brackets Eq. (6.1) and replace the product xy with $\frac{1}{2}(xy + yx)$. The result is the commutator, Eq. (6.2), which I repeat

$$(x, y) = xy - yx = e^{i\theta}\ (z - xy) \tag{3}$$

and cyclical permutations of x, y, z. The parameter $e^{i\theta}$ depends on Λ, $\hbar$, and the intersection number s. The $-$ algebra is identical to (3) but with $\theta \to -\theta$. Equation (6.2) is rearranged more symmetrically in Eq. (6.3). This was the second surprise. In fact Eq. (6.3) is referred to as the cyclical presentation of $SU(2)_q$, with $q = e^{i\theta}$.[24,25]

We also found some preliminary isometries of the algebra Eq. (6.3), e.g. Eq. (6.6) corresponds to the Dehn twist known as $S : U \to V^{-1}, V \to U$. We later extended this discussion to any genus in.[18]

We found an important central element, Eq. (6.4) (actually two, one for each $\pm$ algebra). In the classical limit $\theta \to 0$ it can be expressed, using Eq. (5.3), as

$$F^2 = 1 + 2xyz - x^2 - y^2 - z^2 = tr(\ \mathbb{I}S(U)\ S(V)\ S(U^{-1})\ S(V^{-1})\ ) \tag{4}$$

and therefore $F^2 = 0$ can be considered as an application of the group identity $UVU^{-1}V^{-1} = 1$ to the representations S. It is worth noting that, considering the two $\pm$ algebras, there are actually six traces $x^\pm, y^\pm, z^\pm$, so the theory is overprescribed. But there are two central elements $(F^2)^\pm$, leaving four independent variables, which characterise the two degrees of freedom we wished to describe. Further, setting $F^2 = constant$ (4) defines a cubic surface, and we used that to classify the various solutions, showing that only the unbounded (hyperbolic) solutions correspond to an $SO(2,2)$, $\Lambda < 0$ theory, whereas the bounded (parabolic) solutions correspond to the unphysical $SO(4)$. In fact it was later shown in[26] that, by considering both $\pm$ algebras, of the nine possibilities (hyperbolic, elliptic, parabolic for each algebra), only the "hyperbolic-hyperbolic" solutions are physical. For solutions with elliptic or parabolic holonomies, spacetime still has the topology $\mathbb{R} \times \mathbb{T}^2$, but the toroidal slices are not spacelike.

In Sections 7 and 8 we discussed the bounded and unbounded representations for $\Lambda < 0$, and those for $\Lambda > 0$, respectively.

The unphysical bounded representations corresponding to $SO(4)$ are a direct generalisations of the theory of angular momentum. Using raising and lowering operators $K^\pm$, Eq. (7.1), and with $z = sin\mu$, $\mu^\dagger = \mu$, $|z| < 1$ then $z, K^\pm$ satisfy the commutators of $SU(2)_q$. The quantum Casimir F^2, Eq. (6.4), is written in terms of these variables and finite-dimensional representations are constructed. There are a number of results, all incomplete, but one is worth noting, i.e. when $\frac{\pi}{\theta} = q$ is an integer. Then, from the Weyl representations Eq. (7.15), the products A^q, B^q, C^q of the unitary operators A, B, C lie in the center of the algebra. This is typical of quantum algebras.[27]

Unbounded representations are obtained by letting μ be non-hermitian. With $z^\dagger = z$ it follows that $\mu^\dagger + \mu = \pi$, and $\mu = \frac{\pi}{2} + i\nu$, ν real, with conjugate ζ. The Weyl representation has hermitian operators A, B, C.

The de Sitter ($\Lambda > 0$) algebra of Section 9 is similar, but with two sets of complex conjugate traces, and θ replaced by $-i\theta$. The limit $\theta \to 0$ corresponds to the linear representations of $SO(3,1)$.[28]

2. 2 + 1 quantum gravity, J. E. Nelson and T. Regge, Phys. Lett. B272 (1991) 213

In this short letter[12] we streamlined many things, the presentation is pretty tight, and, as we said in the abstract "Here all of the previous results are implemented for any genus and at the quantum level". It can be considered as a prelude to[19] where we studied the representations of the $g = 2$ theory in some detail.

We extended the quantum algebra Eq. (6.3) of Ref. 11 or (3) to paths with more than one intersection, for $\Lambda < 0$ (and setting $s = 1$). To this end we used a different representation of π_1 of (Σ_g) for genus g. It has generators t_i, $i = 1 \cdots 2g + 2$, related to the $U_i, V_i, i = 1...g$ in Eq. (6), and they satisfy the three identities in Eq. (7). The t_i run around the perimeter of the polygon, Figure 1 of,[12] but there are also diagonals (compound paths) like e.g. d_{37}, which is homotopic to $t_3 t_4 t_5 t_6$. One of the advantages of using the the t_i is that the number of intersections is at most ± 2.[17] Two paths, either perimeter or compound, that "touch" on the border of the polygon have intersection number ± 1, those that do not touch anywhere have zero intersection, and those that cross inside the polygon have intersection ± 2. The quantum traces corresponding to the paths d_{ij} are denoted $a_{ij} = c(d_{ij})$. For intersection ± 1 the commutator (3) applies, and for zero intersection the commutator is zero. Many of the commutators of traces corresponding to paths with intersection ± 2 were obtained by using the programme Mathematica, the others by repeated use of the quantum Jacobi identity $(A,(B,C)) + (B,(C,A)) + (C,(A,B)) = 0$. These are reported in Eqs. (12) and (13), where $K = e^{i\theta}$.

The quantum algebra Eq. (6.3) is highly overprescribed, e.g. for $g = 2$ there are 15 independent a_{ij}, but in,[18] we showed that for the classical algebra, Eq. (5.2) of,[11] by using a set of rank and trace identities, that there are precisely $6g - 6$ independent traces, the number of independent moduli on a Riemann surface of genus g.

We also found a set of generators for the quantum mapping class group. In analogy with the quantisation for $g = 1$ in[11] we parametrised one trace, a_{jk}, for j, k fixed by $a_{jk} = \frac{\cos\psi}{\cos\frac{\theta}{2}}$ (the analogue of $z = sin\mu$) where, for hyperbolic solutions, ψ is not hermitian, and therefore functions of ψ are not periodic. We looked for functions of ψ that generate, through their commutators, transformations on all the other a_{ij} that have non-zero commutators with this a_{jk}, by starting with a triangular structure (the analogue of x, y, z). For the triangular structure the intersections are ± 1 but were easily extended to intersections ± 2. The result is that the quantum counterpart of the classical Dehn twists is generated by conjugation with Gaussians of ψ, see Eqs. (17)–(19), by using the BCH formula

$$e^A B e^{-A} = B + [A, B] + \frac{[A, [A, B]]}{2!} + \dots . \tag{5}$$

3. Later developments

In a later article[19] we studied the representations of the $g = 2$ theory in some detail. We were able to reduce the 15 independent traces to 6 $(6g - 6)$ by using choosing three commuting traces, parametrised by three angles, and their conjugate momenta. This is highly non-unique, but we were also able to obtain formulae for the remaining 12 traces.

Waelbrook and Zertuche have considered extensions to conformal gravity, supergravity, and introduced a time variable (one of the traces).[29]

With other collaborators I have studied the relationship between this approach and the second order metric formalism of Moncrief and others,[30] and the linear version of the algebra, Eq. (5.4) of Ref. 11.

Steve Carlip and I compared the second order ADM metric formalism to the first order holonomy formalism for the torus, showing that they can be considered as "Schrödinger" and "Heisenberg" pictures, respectively. With a hyperbolic representation of the algebra (3) the holonomy parameters are related to the torus modulus and momentum through a time-dependent canonical transformation. The quantisation of the two classically equivalent formulations differs by terms of order $O(\hbar^3)$, negligible for small $|\Lambda|$. Further, the quantum modular group splits the holonomy representation Hilbert space into physically equivalent orthogonal "fundamental regions" that are interchanged by modular transformations.[31]

Vincent Moncrief and I used linear combinations of constants of motion to satisfy the AdS algebra so$(2, 2)$ in either ADM or holonomy variables. Quantisation is straightforward in terms of the holonomy parameters, and the modular group is generated by these conserved quantities. On inclusion of the (time-independent part of) the Hamiltonian three new global constants are derived, and the quantum algebra extends to the conformal algebra so$(2, 3)$.[32]

Roger Picken and I have studied the linear algebra Eq. (5.4) of Ref. 11, which Regge and I did not study. It is linear, but subject to non-linear constraints (Eq. (5.3)). It has led us to a new quantum group structure, some interesting properties in quantum geometry, a quantisation of the Goldman bracket,[22] and a theory of intersecting loops on surfaces.[23]

The algebraic structures that we found, both classical and quantum, have been rediscovered and developed in the mathematical and mathematical physics literature, in particular their symmetries and invariants. I give just a few examples.

For $g = 1$ the classical cubic Casimir (4) is the Markov polynomial,[33] and the elements x, y, z have been used as elements of 3×3 Stokes matrices.[34]

For $g = 2$ we constructed an explicit representation of the mapping class group acting on $\pi_1(\Sigma)$.[15,16] It was reviewed in Ref. 35.

For arbitrary g our set of mapping class group induced observables corresponds to the centraliser of a non-trivial mapping class group.[36] The quantum algebra, Eqs. (13) of Ref. 12 is isomorphic to the non-standard q-deformed algebra $U_q'(so_n)$, $n = 2g + 2$,[37] and can be obtained by quantising the coordinates of Teichmuller space.[38] Further, the quantum algebra has been the subject of 2 UK EPSRC research grants.[39] It also appears in tropical geometry.[40]

Final remarks

In this first-order approach, the constraints have been solved exactly. There is no Hamiltonian, and no time development. One can think of this formalism as a "Heisenberg picture" time-independent description of the entire spacetime, or as initial data for some choice of time. It is related, for the torus, to the time dependent ADM theory.[31]

We only studied the + algebra but a physical theory would need both $\pm$ copies of the quantum algebra Eq. (3). Indeed, when Carlip and I made the connection to the ADM theory, for the torus,[31] we used a hyperbolic representation of the three traces x, y, z which satisfy (3), with parameters $r^{\pm}_{1}, ^{\pm}_{2}$. The unique action of (2+1)-dimensional gravity can then be written as $\int \alpha(r_1^- dr_2^- - r_1^+ dr_2^+)$.

Tullio Regge had clearly been thinking about the relationship between discrete geometry, topology, and higher dimensions for a long time. He often talked about his dream of a "swiss cheese universe",[41] the torus, or doughnut ($g = 1$) and a pretzel ($g = 2$) being low genus examples. His idea was that a lower-dimensional system with very high genus g could mimic higher dimensions. For example, a discrete scaffolding of a three-dimensional space with "fattened pipes" can be considered as an approximation to a high genus two-dimensional surface Σ_g. On Σ_g there would be a set of closed loops (cycles), each cycle circling either a hole or a pipe. In this scenario the curvature would be constant on Σ_g, and the cycles should satisfy certain identities, e.g. Eq. (1.6) of Ref. 11 or Eq. (7) of Ref. 12.

This idea was considered in the context of loop quantum gravity.[42] There is a noncanonical graded Poisson algebra of classical observables, which are nonlocal and relate to loops in a three-manifold Σ. The quantum theory involves a deformation of this algebra.

More recently a discretised and complex Chern–Simons theory was quantised, and led to a four-dimensional model with non-zero Λ.[43]

Leonid Chekhov, Regge and I made a preliminary attempt to study the large genus limit. As a first step, we expressed the observables in terms of Penner–Fock coordinates, and used a graph description with a discrete, finite number of components. By considering continuous non-discrete components in a scaling limit some elements of the algebra become commuting, the remaining 2 continuous elements satisfying Poisson brackets for string coordinates. All of the observables can be expressed in terms of these.[44]

It was a great pleasure and privilege to have known, and worked with Tullio Regge, the most brilliant scientist I ever met. We shared moments of enlightment, and the pure joy of discovery. Thank you, Tullio.

Acknowledgments

This work was supported by INFN - Istituto Nazionale di Fisica Nucleare - Iniziative Specifiche GSS, and MIUR - Ministero dell'Istruzione, Università e della Ricerca Scientifica e Tecnologica. I am grateful to the University of Turin for one year of teaching leave, and to Roger Picken for his help in preparing the figure for the braid group relation Figure 1.

References

1. S. Deser, R. Jackiw and G. 't Hooft, Ann. Phys. **152** (1984) 220.
2. A. Achúcarro and P. K. Townsend, Phys. Lett. **B180** (1986) 89.
3. E. Witten, Nucl. Phys. **B 311** (1989) 46.
4. A. A. Garcia-Diaz, *Exact Solutions in Three-Dimensional Gravity*, (Cambridge University Press, 2017).
5. M. Banados, C. Teitelboim, J. Zanelli, Phys.Rev.Lett. **69** (1992) 1849.
6. J. Maldacena, Adv. Theor. Math. Phys. **2** (1998) 231.
7. J. Wess and B. Zumino, Phys. Lett. **B37** (1971) 95; E. Witten, Comm. Math. Phys. **92** (1984) 455.
8. E. Witten, Comm. Math. Phys. **121** (1989) 351.
9. O. Coussaert, M. Henneaux and P. van Driel, Class. Qu. Grav. **12** (1995) 2961.
10. E. Witten, Prog.Math. **133** (1995) 637.
11. J. E. Nelson, T. Regge and F. Zertuche, Nucl. Phys.**B339** (1990) 516.
12. J. E. Nelson and T. Regge, Phys. Lett. **B272** (1991) 213.
13. J. E. Nelson and T. Regge, Nucl. Phys. **B328** (1989) 190.
14. J. E. Nelson and T. Regge, in *Mechanics, Analysis and Geometry: 200 years after Lagrange*, M. Francaviglia (ed.), (Elsevier, 1991).
15. J. E. Nelson and T. Regge, Comm. Math. Phys. **141** (1991) 211.
16. J. E. Nelson and T. Regge, Int. J. Mod. Phys. **B6** (1992) 1847.
17. J. E. Nelson and T. Regge, Int. J. Mod. Phys. **B6** (1992) 2091.
18. J. E. Nelson and T. Regge, Comm. Math. Phys. **155** (1993) 561.
19. J. E. Nelson and T. Regge, Phys. Rev. **D50** (1994) 5125.
20. L. D. Faddeev, N. Yu and L. A. Takhtajan, *Quantisation of Lie Groups and Lie Algebras*, LOMI-preprint E-14-87.
21. L. Kauffman, Trans. Am. Math. Soc. **318** 2 (1990) 417.
22. Goldman W. M., Invariant functions on Lie groups and Hamiltonian flows of surface group representations, Invent. Math. **85** (1986) 263.
23. J. E. Nelson and R. F. Picken, Int. J. Mod. Phys. **A 24** Issue: 15 (20 June 2009) 2839; Gen. Rel. Grav. **43** (2011) 777; Adv. Theor. Math. Phys. **18** 3 (2014) 709.
24. D. B. Fairlie, J. Phys. **A 23** 5 (1990) L183.
25. S. Majid, Int. J. Mod. Phys. **A51** (1990).
26. K. Ezawa, *Reduced Phase Space of the First Order Einstein Gravity on* $\mathbb{R}\times T^2$, Osaka preprint OU-HET-185 (1993).
27. W. A. Schnizer, Comm. Math. Phys. **163** 2 (1994) 293.
28. M. A. Naimark, *Les representations lineare du groupe de Lorentz*, (Dunod, Paris 1982).
29. H. Waelbroeck and F. Zertuche, Phys. Lett. **B254** (1991) 424; Mod. Phys. Lett.**A7** (1992) 2715; Phys.Rev. **D50** (1994) 4966.
30. V. Moncrief, J. Math.Phys. **30** (1989) 2907; A. Hosoya and K. Nakao, Class. Qu. Grav. **7** (1990) 163, Prog. Theor. Phys. **84** (1990) 739; Y. Fujiwara and J. Soda, Prog. Theor. Phys. **83** (1990) 733.
31. S. Carlip and J. E. Nelson, Phys. Rev. **D 51** 10 (1995) 5643: Phys. Lett. **B 324** (1994) 299.

32. J. E. Nelson and V. Moncrief, Int. J. Mod. Phys. D **6**, 5 (1997) 545.
33. A. I. Molev, Rev. Math. Phys. **20** (2008) 173.
34. B. Dubrovin, in *Integrable systems and quantum groups* (M. Francaviglia, S. Greco, Eds.), Lect. Notes. Math. **1620** (Springer, 1996) 120.
35. J. E. Nelson, Zentralblatt fur Mathematik 821.57009 (1995).
36. R. A. Baadhio, Nucl. Phys. **B441** (1995) 383.
37. A. M. Gavrilik, Proc. Inst. of Mathematics of NAS of Ukraine, **30**, 2, (2000) 304.
38. L. Chekhov and V. V. Fock, Czech. J. Phys. **50** (2000) 1201.
39. M. Mazzocco *Quantum Frobenius manifolds, Nelson-Regge algebra and Riemann-Hilbert problem*, EP/D071895/1, EP/D071895/2 (2006-2012).
40. L. Chekhov and M. Mazzocco, in *Tropical Geometry and Integrable Systems* (Am. Math. Soc., 2012) 85.
41. T. Regge, in P. P. Delsanto and A. W.Saenz (eds.) *New perspectives on problems in classical and quantum physics*, Part 1, (Gordon and Breach, 1998) 361.
42. C. Rovelli and L. Smolin, Nucl.Phys. **B331** (1990) 80.
43. H. M. Haggard, M. Han, W. Kaminski and A. Riello, Phys. Lett. **B752** (2016) 258.
44. L. Chekhov, J. E. Nelson and T. Regge, Lett.Math.Phys. **78** 1, (2006) 17.

Nuclear Physics B339 (1990) 516–532
North-Holland

HOMOTOPY GROUPS AND (2 + 1)-DIMENSIONAL QUANTUM DE SITTER GRAVITY

J.E. NELSON

Istituto Nazionale di Fisica Nucleare, Sezione di Torino, Via Pietro Giuria 1, Torino 10125, Italy

T. REGGE

Dipartimento di Fisica Teorica dell'Universita' di Torino, Via Pietro Giuria 1, Torino 10125, Italy

F. ZERTUCHE

Fellow of Universidad National Autonoma de Mexico and of SISSA, Strada Costiera 11, Miramare, Trieste, Italy

Received 2 January 1990

The integrated CCR (canonical commutation relations) for (2 + 1)-dimensional de Sitter/anti-de Sitter gravity in first-order formalism are used to derive generalised spinor representations Ψ: $\pi_1(\Sigma) \to SL(2,\mathbb{C})/SL(2,\mathbb{R}) \otimes SL(2,\mathbb{R})$ of the fundamental group of the initial data surface Σ^2. The matrix elements of $\Psi(U), \Psi(V)$, with $U \neq V$ are in general non-commuting according to a set of formal rules which are related to the theory of quantum groups. The Poincaré theory appears as a limiting case.

1. Introduction

In ref. [1] we have analysed Poincaré (2 + 1)-dimensional quantum gravity and derived the complete set of gauge-invariant physical observables which are associated with global integrated connections having a direct topological significance.

In this paper we derive commutation relations for the integrated connections and the associated quantum representation Ψ: $\pi_1 \to G$, where $G = SO(3,1)$ or $SO(2,2)$ (and their corresponding spinor groups) of the homotopy group $\pi_1(\Sigma^2)$ of the initial data surface Σ^2. As in ref. [1] the physically relevant dynamical quantities are the gauge-invariant central functions on π_1, i.e. functions $a(x)$ with the property $a(x) = a(y^{-1}xy)$, $x, y \in \pi_1$.

The field variables are the de Sitter spin connections $\omega^{ab} = \omega^{ab}{}_{\alpha} dx^{\alpha}$ with $a, b = 0,1,2,3$, $\alpha = 0,1,2\ldots$. The tangent Minkowski metric η_{ab} has signature $(-1,1,1,k)$, $k = +1$ for de Sitter, $k = -1$ for anti-de Sitter space. The riemannian curvature is

$$R^{ab} = d\omega^{ab} - \omega^{at} \wedge \omega_t{}^b . \tag{1.1}$$

0550-3213/90/\$03.50 © Elsevier Science Publishers B.V.
(North-Holland)

In terms of these fields the lagrangian density is given by the Chern–Simons three-form

$$\alpha\left(\mathrm{d}\omega^{ab} - \tfrac{2}{3}\omega^{a}{}_{t} \wedge \omega^{tb}\right) \wedge \omega^{cd}\epsilon_{abcd}/8, \tag{1.2}$$

and is equal to the scalar curvature plus cosmological term,

$$-\tfrac{1}{2}\left(\mathrm{d}\omega^{\mathscr{a}\mathscr{b}} - \omega^{\mathscr{a}}{}_{\mathscr{d}} \wedge \omega^{\mathscr{d}\mathscr{b}} + \Lambda e^{\mathscr{a}} \wedge e^{\mathscr{b}}\right) e^{\mathscr{c}} \epsilon_{\mathscr{a}\mathscr{b}\mathscr{c}}, \tag{1.3}$$

where $\mathscr{a}, \mathscr{b}, \mathscr{c}, \mathscr{d} = 0,1,2$ only, provided the triads $e^{\mathscr{a}} = \alpha\omega^{\mathscr{a}3}$, $\epsilon_{\mathscr{a}\mathscr{b}\mathscr{c}3} = -\epsilon_{\mathscr{a}\mathscr{b}\mathscr{c}}$ and the cosmological constant $\Lambda = \frac{1}{3}k\alpha^{-2}$. In the sequel $\sqrt{k}$ means unambiguously $+1$ for $k=1$ and $+i$ for $k=-1$. The action (1.2) and (1.3) leads to the standard CCR:

$$\left[\omega^{ab}{}_{i}(x), \omega^{cd}{}_{j}(y)\right] = k\alpha^{-1}\epsilon_{ij}\epsilon^{abcd}\delta^{2}(x-y), \tag{1.4}$$

or equivalently,

$$\left[e^{\mathscr{a}}{}_{i}(x), \omega^{\mathscr{b}\mathscr{c}}{}_{j}(y)\right] = -\epsilon_{ij}\epsilon^{\mathscr{a}\mathscr{b}\mathscr{c}}\delta^{2}(x-y),$$

$$\left[e^{\mathscr{a}}{}_{i}(x), e^{\mathscr{b}}{}_{j}(y)\right] = \left[\omega^{\mathscr{a}\mathscr{b}}{}_{i}(x), \omega^{\mathscr{c}\mathscr{d}}{}_{j}(y)\right] = 0,$$

where $x, y \in \Sigma^2$ are generic points on the $x^0 = \text{const.}$ surface Σ^2; $i, j = 1,2$ are spatial vector indices on Σ^2 and $\epsilon_{ij} = -\epsilon_{ji}$, $\epsilon_{12} = 1$. The variational equations derived from (1.2) are $R^{ab} = 0$ and imply that space-time is everywhere locally de Sitter. We have $\alpha R^{\mathscr{a}3} = R^{\mathscr{a}} = \mathrm{d}e^{\mathscr{a}} - \omega^{\mathscr{a}\mathscr{b}} \wedge e_{\mathscr{b}}$, called the torsion.

As in ref. [1] we assume that Σ^2 is a compact surface of genus g with no a priori complex structure assumed. Space-time M^3 is then given by $\Sigma^2 \times \mathbb{R}^1$, where time runs along $\mathbb{R}^1$. We recall that the spinor group of SO(2, 2) is SL(2, $\mathbb{R}$) ⊗ SL(2, $\mathbb{R}$), that of SO(3, 1) is SL(2, $\mathbb{C}$). In both cases the number of real parameters is six.

The fundamental group π_1 of a surface Σ^2 of genus g admits a presentation with the generators

$$U_1, V_1, U_2, V_2, \ldots, U_g, V_g, \tag{1.5}$$

and the single relator

$$U_1V_1U_1^{-1}V_1^{-1}U_2V_2U_2^{-1}V_2^{-1} \ldots U_gV_gU_g^{-1}V_g^{-1} = 1. \tag{1.6}$$

In the sequel we shall deal with 4×4 representations $\Psi(\pi_1)$ of this group on SO(2, 2) and SO(3, 1). We set $C_n = U_nV_nU_n^{-1}V_n^{-1}$, $n = 1, \ldots, g$.

By adjoining the relators $C_n = 1$ to the single relator (1.6) we would obtain the abelianization of π_1 which is isomorphic to the first homology group $\mathrm{H}_1(\Sigma^2, \mathbb{Z})$. This group is a $2g$-dimensional module with coefficients in the ring of integers.

In sect. 2 we define the integrated connections with appropriate boundary conditions. In sect. 3 we discuss quantum commutators among matrix elements of

$\Psi(x), \Psi(y)$ where x, y are intersecting paths in Σ^2. Appropriate conventions for G and its spinor groups are established in sect. 4. The central functions of Ψ and their quantum commutators are discussed in sect. 5. The anti-de Sitter algebra is introduced and discussed in sect. 6. In sect. 7 we discuss the bounded representations of the SO(4) algebra, in sect. 8 those of the SO(2,2) algebra and in sect. 9 those of the SO(3,1) de Sitter algebra. Appendix A contains a number of essential formulae.

2. The integrated connection

In this section we shall derive explicit formulas for the CCR of the components of the integrated connection Ψ. Consider two generic points A, B in M^3 and a path γ joining them. Let γ be parametrized as $x(t)$, with $x(0) = A$, $x(1) = B$. On γ we consider the differential equation

$$d\Psi/dt = \omega_t \Psi, \tag{2.1}$$

where $\omega_t = \omega_\alpha T^\alpha$ and T is the tangent vector along γ, subject to the boundary condition $\Psi(0) = 1$. The value of the solution at $t = 1$ will depend on the homotopy class of γ only and will be denoted by $\Psi(B, A, \gamma)$ or simply by $\Psi(\gamma)$. The generic solution to eq. (2.1) is then given by $\Psi(B) = \Psi(B, A, \gamma)\Psi(0)$. Consider now a point C, a second path γ' with endpoints B, C and the solution $\Psi(C, B, \gamma')$. We have then $\Psi(C, A, \gamma'\gamma) = \Psi(C, B, \gamma')\Psi(B, A, \gamma)$ where $\gamma'\gamma$ is the path going from A to C obtained by joining the paths γ', γ at B.

The element $\Psi(B, A, \gamma)$ applied to a multiplet belonging to a 4×4 representation of G at A gives the Levi-Civita transported multiplet at B. Setting B = A we obtain the subset $\Psi(B, B, \{\gamma\}) = \Psi(B, W)$, where $W = \{\gamma\}$ is the homotopy class of γ and $W \in \pi_1$. The function $\Psi(B, W)$ provides us with a representation of π_1.

So far we have dealt with commuting purely classical fields and the paths γ, γ' are generic in M^3. We now restrict ourselves to Σ^2 and consider quantized fields. Since $M^3 \simeq \Sigma^2 \times \mathbb{R}^1$, the fundamental groups $\pi_1(\Sigma^2)$ and $\pi_1(M^3)$ are isomorphic and will be denoted by the common symbol π_1. It is then clear from (1.3) that the single elements of matrices like $\Psi(B_1, U)$ and $\Psi(B_2, V)$ corresponding to generic elements $U, V \in \pi_1$ will commute iff they can be obtained by integrating over paths which do not intersect. It will be therefore convenient at first to consider commutation relations involving different base points B_1, B_2 in order to avoid intersecting at the common base point. In the remaining part of the paper we derive a set of recursive rules which allows us to deal directly with the integrated commutation relation.

The ensuing structure reminds us of the quantum groups considered by Faddeev et al. [4] but differs from them in several important aspects. In a quantum group there is non-commutativity only among matrix elements of the same group element, here it occurs instead among matrix elements of $\Psi(U_n)$ and $\Psi(V_n)$ which behave like conjugate pairs.

In the sequel we distinguish clearly among the commutator of two matrices $(\Psi,\Psi')_a{}^b = \Psi_a{}^t\Psi_t'{}^b - \Psi_a'{}^t\Psi_t{}^b$ for which we use the round brackets, and the Poisson bracket of their elements $[\Psi_{ab},\Psi'_{cd}]$, $a,b,c,d=1,\ldots,4$ for which we use square brackets. By $[\Psi,\Psi']$ we mean the four index array of the above Poisson bracket. Similarly we proceed for generic quantum matrices.

3. Intersecting paths

We consider now the generic case of two oriented paths ρ,σ in Σ^2 joining the pairs of points A, B and C, D, respectively, and the corresponding integrated connections $\Psi(\mathrm{B},\mathrm{A},\rho)$ and $\Psi(\mathrm{D},\mathrm{C},\sigma)$. If ρ and σ have no points in common then $[\Psi(\rho),\Psi(\sigma)]=0$.

Let ρ and σ be parametrized by t and u respectively, with $0\leqslant t\leqslant 1$, $0\leqslant u\leqslant 1$, with A, B labelled by 0 and 1 on ρ, and C, D labelled by 0 and 1 on σ. We assume they have a single intersection point P labelled by $t=t^0$ on ρ and by $u=u^0$ on σ.

Next we divide ρ into $\rho_1=\{0\leqslant t\leqslant t^0-\epsilon\}$, $\rho_2=\{t^0-\epsilon\leqslant t\leqslant t^0+\epsilon\}$, $\rho_3=\{t^0+\epsilon\leqslant t\leqslant 1\}$ and correspondingly σ into $\sigma_1=\{0\leqslant u\leqslant u^0-\epsilon\}$, $\sigma_2=\{u^0-\epsilon\leqslant u\leqslant u^0+\epsilon\}$, $\sigma_3=\{u^0+\epsilon\leqslant u\leqslant 1\}$, for $\epsilon\ll 1$. We obtain in this way

$$\Psi(\mathrm{B},\mathrm{A},\rho)=\Psi(\rho_3)\Psi(\rho_2)\Psi(\rho_1),\qquad \Psi(\mathrm{D},\mathrm{C},\sigma)=\Psi(\sigma_3)\Psi(\sigma_2)\Psi(\sigma_1). \tag{3.1}$$

In this decomposition the only intersecting paths are ρ_2 and σ_2. It follows that $[\Psi(\rho_i),\Psi(\sigma_k)]=0$ for i or $k=1,3$.

On the other hand, $\Psi(\rho_2)$ and $\Psi(\sigma_2)$ are infinitesimally close to the identity and can be written at first order in ϵ from (2.1) as

$$\Psi(\rho_2)=1+\int_{t^0-\epsilon}^{t^0+\epsilon}\mathrm{d}t\,\omega_j(x(t))\,\mathrm{d}x^j(t)/\mathrm{d}t, \tag{3.2}$$

and a similar expression can be obtained by replacing ρ_2,t,t^0 with σ_2,u,u^0. The vectors $\mathrm{d}x^i/\mathrm{d}t,\mathrm{d}x^j/\mathrm{d}u$ are tangent to ρ_2,σ_2 and therefore to Σ^2. We see immediately from (1.4) that

$$\left[\Psi^{ab}(\rho_2),\Psi^{cd}(\sigma_2)\right]=k\alpha^{-1}\int_{t^0-\epsilon}^{t^0+\epsilon}\mathrm{d}t\int_{u^0-\epsilon}^{u^0+\epsilon}\mathrm{d}u\,J\epsilon^{abcd}\delta^2(x(t)-x(u)),$$

where the sign $s(\rho,\sigma)$ of the Jacobian $J=(\mathrm{d}x^j/\mathrm{d}t)(\mathrm{d}x^i/\mathrm{d}u)\epsilon_{ji}$ defines the orientation of the intersection of ρ and σ. Clearly,

$$J\delta^2(x(t)-x(u)) = -s\delta(t-t^0)\delta(u-u^0), \tag{3.3}$$

where we set briefly $s=-s(\rho,\sigma)=s(\sigma,\rho)$ and therefore

$$[\Psi^{ab}(\rho_2),\Psi^{cd}(\sigma_2)] = -k\alpha^{-1}\epsilon^{abcd}s,$$

$$[\Psi^{ab}(\sigma_2),\Psi^{cd}(\rho_2)] = k\alpha^{-1}\epsilon^{abcd}s. \tag{3.4}$$

We may now assemble the Poisson bracket for ρ,σ by using eqs. (3.1). The resulting formula is not informative and we shall instead proceed by assembling similar formulas for the spinor groups.

Eqs. (3.4) are valid only when the paths ρ,σ have a single intersection. If they have no intersection then the Poisson brackets are trivially zero. If they have $m>1$ intersections then one can express one of the paths, say ρ, as $\rho=\rho_1\rho_2\rho_3\cdots\rho_m$, where each ρ_i, $i=1,\dots,m$, has a single intersection and eqs. (3.4) can be applied. Alternatively, one can give a direct geometrical derivation extending the one we have given here for $m=1$.

4. Quantum spinor representations

We now assemble the spinor quantum representations of $\pi_1(\Sigma^2)$. Define

$$\Delta(x)=\Delta_i(x)\mathrm{d}x^i=\tfrac{1}{4}\omega^{ab}(x)\gamma_{ab}. \tag{4.1}$$

Therefore, from $R^{ab}=0$, it follows that $\mathrm{d}\Delta-\Delta^2=0$. From (1.4) we have

$$[\Delta_i(x),\Delta_j(y)] = k\epsilon_{ij}\epsilon^{abcd}\gamma_{ab}\otimes\gamma_{cd}\delta^2(x-y)/16\alpha$$

$$= -k\epsilon_{ij}\gamma_5\gamma_{ab}\otimes\gamma^{ab}\delta^2(x-y)/8\alpha. \tag{4.2}$$

(For γ-matrix conventions and identities see appendix A.) The spinor equivalent $S(\mathrm{B},\mathrm{A},\sigma)$ of $\Psi(\mathrm{B},\mathrm{A},\sigma)$ of sect. 2 is given by

$$\Psi(\mathrm{B},\mathrm{A},\sigma)^{ab}\gamma_b = S(\mathrm{B},\mathrm{A},\sigma)^{-1}\gamma^a S(\mathrm{B},\mathrm{A},\sigma) \tag{4.3}$$

and satisfies

$$\mathrm{d}S(x)=\Delta(x)S(x), \tag{4.4}$$

where $S(x)$ denotes $S(\mathrm{B},\mathrm{A},\sigma)$, and x are the coordinates of the endpoint B on Σ^2. The 4×4 representation $\Delta(x)$, $S(x)$ can be decomposed into 2×2 irreducible parts by using the projectors $p_\pm$ (see appendix A) as follows:

$$\Delta(x) = \Delta^+(x) \otimes p_+ + \Delta^-(x) \otimes p_- ,$$

$$S(x) = S^+(x) \otimes p_+ + S^-(x) \otimes p_- , \tag{4.5}$$

so that

$$\mathrm{d}S^\pm(x) = \Delta^\pm(x) S^\pm(x) . \tag{4.6}$$

It follows from eqs. (4.2) and (4.5) that the 2×2 representations have the Poisson brackets

$$\left[\Delta_i^\pm(x), \Delta_j^\pm(y)\right] = \pm i/(2\alpha\sqrt{k})\epsilon_{ij}\sigma^m \otimes \sigma^m \delta^2(x-y),$$

$$\left[\Delta_i^+(x), \Delta_j^-(y)\right] = 0. \tag{4.7}$$

For the integrated connections $S^\pm(\rho)$ we repeat the derivation of sect. 3 by replacing Ψ with S and ω with Δ and find (see appendix A)

$$\left[S^\pm(\rho)_\alpha{}^\beta, S^\pm(\sigma)_\gamma{}^\tau\right] = \pm i(s/2\alpha\sqrt{k})$$

$$\times\left(-S^\pm(\rho)_\alpha{}^\beta S^\pm(\sigma)_\gamma{}^\tau + 2S^\pm(\rho_3\sigma_1)_\alpha{}^\tau S^\pm(\sigma_3\rho_1)_\gamma{}^\beta\right),$$

$$\left[S^+(\rho)_\alpha{}^\beta, S^-(\sigma)_\gamma{}^\tau\right] = 0. \tag{4.8}$$

5. The algebra of traces

We define now for a generic closed path τ the traces

$$c^\pm(\tau) = \tfrac{1}{2} S_\alpha^{\pm\alpha}(\tau) . \tag{5.1}$$

Some comments are in order.

(i) Traces are gauge-invariant quantities as expressed in $c^\pm(\tau) = c^\pm(\nu\tau\nu^{-1})$ for any open path ν. This corresponds to a shift of base point along ν. Their Dirac and Poisson brackets coincide.

(ii) In taking the trace of both sides of eq. (4.8) we must assume that the base points B, B′ of ρ and σ are different to avoid ambiguous contributions from intersections at the common base point. We denote $\delta = \sigma_1^{-1}\rho_1$ the open path from B to B′. In the r.h.s. of eq. (4.8) we encounter the trace

$$c(\sigma_1\sigma_3\rho_1\rho_3) = c(\sigma_3\rho_1\rho_3\sigma_1) = c\left(\sigma\sigma_1^{-1}\rho_1\rho_3\sigma_1\right) = c(\sigma\rho') ,$$

where $\sigma = \sigma_3\sigma_1$, $\rho' = \delta\rho\delta^{-1}$ are now both based in B′. This means that σ, ρ'

identify elements u, v of the homotopy group $\pi_1(\Sigma^2, \mathrm{B}')$ based on B′. We write then $c(u), c(v), c(uv)$ in place of $c(\sigma), c(\rho'), c(\sigma\rho')$ etc. for every group element. By comment (i) we have $c(v) = c(\rho) = c(\rho')$. With these remarks we take the trace (4.8) and find

$$[c^{\pm}(v), c^{\pm}(u)] = \pm is/(2\alpha\sqrt{k})(-c^{\pm}(v)c^{\pm}(u) + c^{\pm}(uv)),$$
$$[c^{+}(v), c^{-}(u)] = 0. \tag{5.2}$$

(iii) For 2×2 matrices the following identity holds:

$$c^{\pm}(u)c^{\pm}(v) = \tfrac{1}{2}(c^{\pm}(vu) + c^{\pm}(vu^{-1})), \tag{5.3}$$

and therefore eq. (5.2) can be also written as

$$[c^{\pm}(v), c^{\pm}(u)] = \pm is/(4\alpha\sqrt{k})(c^{\pm}(uv) - c^{\pm}(uv^{-1})). \tag{5.4}$$

Under this form the algebra of traces is given an infinite Lie algebra $\mathscr{L}$ structure subject to the non-linear constraints (5.3). It can be checked that these constraints generate an ideal in the enveloping algebra of $\mathscr{L}$.

The general theory of representations of Lie algebras subject to non-linear constraints of the form (5.3) is yet to be developed. Here we take a different attitude. We remark that by using eq. (5.3) repeatedly we can compute recursively all the traces $c^{\pm}(w)$, $w \in \pi_1$, starting from a finite set of traces. For instance all traces on the subgroup generated by u, v can be computed from $c^{\pm}(u)$, $c^{\pm}(v), c^{\pm}(uv)$. It is therefore reasonable to assume that we can derive a representation for $\mathscr{L}$ once we know an appropriate representation for the above set of traces.

In the case $k = 1$, i.e. $\mathrm{G} = \mathrm{SO}(3,1)$ the traces $c^{+}(w)$ and $c^{-}(w)$ are mutually complex conjugate; if $k = -1$, i.e. $\mathrm{G} = \mathrm{SO}(2,2)$ then both traces $c^{\pm}(w)$ are real and independent observables. These statements are consistent with the conjugacy properties of (5.2).

We first examine the quantisation of the $k = -1$ case.

6. The anti-de Sitter algebra

We discuss first the algebra of the $c^{+}(w)$ and from now on omit the plus sign where there is no ambiguity. That of $c^{-}(w)$ will be later obtained by trivial changes of sign.

There is no a priori bound on the trace of an SL(2, $\mathbb{R}$) matrix. For simplicity we set $c(u) = x$, $c(v) = y$, $c(uv) = c(vu) = z$ and consider eq. (5.2):

$$[x, y] = -(s/2\alpha)(z - xy), \tag{6.1}$$

and cyclical permutations of x, y, z. We set $xy - yx = (x, y) = i\hbar[x, y]$ and symmetrise the xy product. The symmetrised commutator is therefore

$$(x, y) = -iu\left(z - \tfrac{1}{2}(xy + yx)\right), \tag{6.2}$$

where $u = s\hbar/2\alpha = -2\tan(\frac{1}{2}\theta)$. $|\theta| < \pi$, or alternatively,

$$e^{\frac{1}{2}i\theta}\, xy - e^{-\frac{1}{2}i\theta}\, yx = i \sin\theta z \text{ and cyclical}, \tag{6.3}$$

where in (6.3) we have replaced x, y, z with $x/\cos(\frac{1}{2}\theta)$, etc. This rescaling is irrelevant in the $\hbar \to 0$ limit as it corresponds to changes of the order of $\hbar^2$. The algebra (6.3) is not a Lie algebra but it is related to the Lie algebra of the quantum SU(2) groups [4, 6]. It admits the central element

$$F^2 = \cos^2\tfrac{1}{2}\theta + 2e^{i\theta/2}\, xyz - e^{i\theta}(x^2 + z^2) - e^{-i\theta}\, y^2. \tag{6.4}$$

Notwithstanding its appearance note that

(i) F^2 is hermitian;

(ii) F^2 is cyclically symmetric in x, y, z;

(iii) F^2 plays the same role for the algebra (6.3) as the Casimir operator in the theory of angular momentum;

(iv) The algebra (6.3) and F^2 are invariant under the substitutions

$$x \to -x, \qquad y \to -y, \qquad z \to z \tag{6.5}$$

and

$$x \to y, \qquad y \to x, \qquad z \to -z + (xy + yx)/\cos(\tfrac{1}{2}\theta). \tag{6.6}$$

Both (6.5) and (6.6) generate an infinite discrete group of automorphisms, to be related later to the Dehn or mapping class group of automorphisms of the homotopy group of the Riemann surface of genus g. It is of the utmost importance that (6.5) and (6.6) should be implemented at the quantum level. Before we proceed to analyse the quantum theory we look closely into the classical limit $\theta \to 0$ since this is relevant to the discussion of the irreducible representations of the algebra (6.3).

In this limit F^2 reduces to $1 + 2xyz - x^2 - y^2 - z^2$. The cubic surface $F^2 = \Phi =$ const. admits the following classification:

(1) $\Phi > 1$. The surface is four-sheeted and all sheets map into each other by the substitutions (6.5). Writing

$$F^2 = (1 - x^2)(1 - y^2) - (z - xy)^2,$$

we see that if $|x|, |y| < 1$ then $F^2 < 1$; therefore at most one of the variables x, y, z

can lie the range $-1, 1$. In this case, since F^2 is cyclically symmetric, we may suppose $|x| > 1$, $|z| > 1$, $|y| < 1$. It follows $F^2 < 0$ and therefore all three variables x, y, z must be hyperbolic.

(2) $1 > \Phi > 0$. The surface is then five-sheeted. Four of these five sheets (case 2a) correspond to those already discussed in case (1). Writing

$$1 - z^2 = \Phi - \phi(x, y),$$

where $\phi(x, y) = x^2 - 2zxy + y^2$ which is positive definite for $|z| < 1$, it follows that $1 - z^2 < \Phi$, $|z| < \sqrt{(1-\Phi)}$. By symmetry a new sheet [case (2b)] arises with $|x|, |y|, |z| < \sqrt{(1-\Phi)}$, leading to the limiting case of the usual angular momentum theory when $x, y, z \ll 1$, because then $\Phi \sim 1 - x^2 - y^2 - z^2 < 0$.

(3) $\Phi = 0$. The surface acquires four conical points in which the central sheet joins the four external ones. Case (3) can be parametrized by putting $x = \cos\phi$, $y = \cos\psi$, $z = \cos(\phi + \psi)$ for the central sheet [(case (3a)] and $x = \cosh(\phi)$, $y = \cosh(\psi)$, $z = \cosh(\phi + \psi)$ with changes of sign indices by (6.5) for the external ones [(case (3b)].

(4) $\Phi < 0$. The central sheet merges with the external ones into a unique unbounded sheet.

Consider now elements u, v with intersection number $s = 1$ and a representation $S(u), S(v)$ on $SL(2, \mathbb{R})$. If the rotation axis of an element $S \in SL(2, \mathbb{R})$ is timelike (spacelike) then the rotation angle is circular (hyperbolic). We use, as in ref. [1], the pseudo-Pauli real matrices

$$\tau^0 = i\sigma^2, \qquad \tau^1 = \sigma^3, \qquad \tau^2 = \sigma^1,$$

$$\tau^i\tau^k + \tau^k\tau^i = 2\eta^{ik}, \qquad i, k = 0, 1, 2, \qquad \eta^{ik} = (-1, 1, 1).$$

Suppose $|x|$, $|y| < 1$, the elements $S(u), S(v) \in SL(2, \mathbb{R})$ can be written as

$$S(u) = x + \sqrt{(1-x^2)}\,(M_j\tau^j),$$

$$S(v) = y + \sqrt{(1-y^2)}\,(N_j\tau^j),$$

with $M_i M_k \eta^{ik} = -1$, $N_i N_k \eta^{ik} = -1$. $\frac{1}{2}\operatorname{Tr} S(uv)$ is then

$$z = c(uv) = xy + \sqrt{(1-x^2)(1-y^2)}\,(M \cdot N),$$

with $(M \cdot N) = M_i N_k \eta^{ik} = (z - xy)/\sqrt{(1-x^2)(1-y^2)}$.

Since $|(M \cdot N)| > 1$ it follows that $0 > 1 - (N \cdot M)^2 = F^2/(1-x^2)(1-y^2)$, i.e. $\Phi < 0$, and therefore corresponds to case (4).

If G = SO(4) and $S(u), S(v) \in$ SU(2) then we would have $1 > \Phi > 0$ [(case (2a)] but this does not correspond to a de Sitter/anti-de Sitter theory. If $|x|, |y| > 1$ then M_a, N_a are spacelike and $\Phi > 0$ [cases (1) and (2b)]. The limiting case 3a corresponds to $S(u), S(v)$ having the same axis and being restricted to an abelian subgroup of G.

In summary, unbounded open sheets, whatever the value of Φ, correspond to $k = -1$, G = SO(2, 2) configurations. The bounded sheets [(cases (2a) and (3a)] correspond to G = SO(4). We conjecture that the classical limit of the irreducible quantum representations of the algebra (6.3) is patterned after the sheets of the $F^2 = \Phi$ surface. In particular, a representation where $|x|, |y|, |z| < 1$ is not acceptable since it leads to an SO(4) theory. Also we conjecture that the SO(3, 1) representations correspond to the complex manifold F^2 = complex const.

It is nevertheless interesting to investigate the bounded representations of the algebra (6.3) even if they do not have a direct physical significance, because they are the direct generalisations of the theory of the angular momentum.

7. Quantum operators

We look for hermitian and bounded representations of the algebra (6.3). Since $|z| < 1$ we set $z = \sin\mu$. In analogy with the angular momentum theory we define raising and lowering operators

$$K^{\pm} = e^{i\theta/2} x \pm iy\, e^{\pm i\mu}, \tag{7.1}$$

satisfying the commutation relations

$$(\mu, K^{\pm}) = \pm\theta K^{\pm}, \qquad (K^+, K^-) = \sin\theta \sin 2\mu. \qquad (7.2), (7.3)$$

Eqs. (7.2) and (7.3) are identical, apart from trivial scaling by constant factors, to the commutation relations of the infinitesimal algebra of quantum groups [6], a deformation of the Lie algebra of SU(2) defined on the universal enveloping algebra. Moreover, we have the relations

$$(\cos\mu K^+)^\dagger = \cos\mu K^-, \qquad \mu^\dagger = \mu, \tag{7.4}$$

$$F^2 = \cos^2\tfrac{1}{2}\theta \cos^2\mu + \sin^2\tfrac{1}{2}\theta \sin^2\mu - \tfrac{1}{2}(K^+K^- + K^-K^+). \tag{7.5}$$

With these operators we can construct finite-dimensional representations of the algebra (6.3). Let $|p\rangle$ be an eigenfunction of μ:

$$\mu|p\rangle = p\theta|p\rangle. \tag{7.6}$$

We then have

$$\mu K^{\pm}|p\rangle = (p \pm 1)\theta K^{\pm}|p\rangle, \tag{7.7}$$

provided $K^{\pm}|p\rangle$ is not null. For some constants α_p, β_p we then have

$$K^{-}|p\rangle = \alpha_p|p-1\rangle, \qquad K^{+}|p\rangle = \beta_{p+1}|p+1\rangle. \tag{7.8}$$

Applying eq. (7.3) to $|p\rangle$ we find

$$\alpha_p\beta_p - \alpha_{p+1}\beta_{p+1} = \sin\theta \sin 2p\theta. \tag{7.9}$$

Eq. (7.9) is a finite difference equation which can be integrated once we know the boundary conditions. These follow from fixing the upper and lower bounds for p in the spectrum of μ. Suppose that $a \leqslant p \leqslant b$ with $\alpha_a = \beta_{b+1} = 0$. From the upper boundary condition we get

$$\alpha_p\beta_p = \sin(p+b)\theta \sin(b-p+1)\theta. \tag{7.10}$$

The lower boundary condition is satisfied if

$$\sin(a+b)\theta \sin(b+1-a)\theta = 0. \tag{7.11}$$

The dimension q of the representation is then $q = b+1-a$. The hermiticity condition (7.4) yields

$$\beta_p = \alpha_p^* \cos(p-1)\theta/\cos p\theta, \tag{7.12}$$

$$|\alpha_p|^2 = \frac{\cos p\theta}{\cos(p-1)\theta} \sin(p+b)\theta \sin(b-p+1)\theta. \tag{7.13}$$

Representations of type A, B are characterised by $\sin(a+b)\theta = 0$ and $\sin(b-a+1)\theta = 0$, respectively. Moreover, in the range $a \leqslant p \leqslant b$ we must have $|\alpha_p|^2 > 0$. A detailed analysis of these representations is beyond the scope of this paper; it overlaps with results coming from the theory of quantum groups [4, 6] and we discuss a few cases of interest only. By interchanging x and y we can always obtain $\theta > 0$.

We set $w(p) = \cos(2p-1)\theta$, $w_0 = -\cos\theta$, $w_1 = \cos(2b+1)\theta = \cos(2a-1)\theta$ and from eq. (7.13) we have either $w(p) < \min(w_0, w_1)$ or $w(p) > \max(w_0, w_1)$. Interesting cases are

Type A: all the conditions are satisfied if $q-1 < \pi/\theta$ and $q = 2b+1$. We obtain in this way an algebra which is the direct generalisation of the usual SU(2) representations;

Type B: all the conditions are satisfied if $2b < q < 2b+1$ and $(b-a+1) = \pi/\theta = q$. It follows that θ must be quantised. This representation has no analogue in the SU(2) theory.

For both types we have

$$F^2 = \cos^2(b+\tfrac{1}{2})\theta \tag{7.14}$$

(see eqs. (6.4) and (7.5)). That is $0 \leqslant F^2 \leqslant 1$ which corresponds to cases (2) and (3) of sect. 6. In no way our discussion is exhaustive or complete and preliminary numerical investigations indicate that the subset of values of θ in the range $0, \pi$ for which $|\alpha_p|^2$, as defined in eq. (7.13), is positive is a very complicated function of a and b, possibly involving fractals.

The coefficients α_p, β_p completely determine the operators x, y, z. Clearly, eq. (7.13) determines the absolute value of α_p only. The phase can be fixed by performing a unitary transformation $O \to SOS^{-1}$ where S is a function of μ. Besides type A and B there are other interesting representations (type C), corresponding in the classical limit to case (3a) of sect. 6. In type C no state is ever annihilated by the raising and lowering operators $K^{\pm}$, and it is best to proceed by defining directly the operators x, y, z without reference to the previous analysis. Indeed we set:

$$x = \tfrac{1}{2}(A + A^{-1}), \qquad y = \tfrac{1}{2}(B + B^{-1}), \qquad z = \tfrac{1}{2}(C + C^{-1}). \tag{7.15}$$

If A, B, C are unitary operators satisfying

$$AB = BA\,e^{i\theta}, \qquad BC = CB\,e^{i\theta}, \qquad CA = AC\,e^{i\theta}, \tag{7.16}$$

$$ABC = BCA = CAB = e^{i\theta/2}, \tag{7.17}$$

x, y, z generate the algebra (6.3). Eqs. (7.16) and (7.17) define a well-known case of the Weyl algebra and we may satisfy them by setting

$$A = e^{i(\theta x + \phi)}, \qquad B = e^{-i(p+\psi)}, \qquad C = e^{i(p - \theta x - \phi + \psi)}$$

where

$$xp - px = (x, p) = i.$$

Here ψ and ϕ are arbitrary constant phases. Note that the exponents in A, B, C add to zero and that $F^2 = 0$. This construction admits a natural extension to any genus g as follows.

Let H be the linear space spanned by x^i, p_i, $i = 1, \ldots, g$ with $(x^i, p_j) = i\delta^i{}_j$ and λ: $\pi_1 \to$ H with $\lambda(uv) = \lambda(u) + \lambda(v)$, $-i\theta s(u,v) = (\lambda(u), \lambda(v))$, $\lambda(u) = \lambda(u)^\dagger$. The quantum trace is then defined by $c(u) = \cos\lambda(u)$. This class of representations is however geometrically quite trivial because it implies that, apart from quantum fluctuations, all rotation axes for all $w \in \pi_1$ actually coincide.

If $\theta = \pi/q$ (type B), where q is an integer then the representation has dimension q and A^q, B^q, C^q lie in the center of the algebra. In this case representations with different ϕ and ψ are generally inequivalent. If q is not a prime then the corresponding representation may be reducible. This analysis of the representations of the algebra (6.3) is however incomplete and will be the subject of a further study. It should be clear that types A, B, C all correspond to $|z| \leqslant 1$, i.e. bounded representations valid for G = SO(4) only. The more physically relevant representations correspond to $|z|$ being unbounded and are discussed in sect. 8.

8. The unbounded representations

A heuristic discussion of the representations of the algebra (6.3) can be based on the following ansatz. We set

$$K^+ = f(\mu)e^{i\zeta}, \qquad K^- = f(\pi - \mu)e^{-i\zeta}, \tag{8.1), (8.2}$$

where $f(\mu)$ has period 2π and for a generic $g(\mu)$:

$$(\zeta, \mu) = i\theta, \quad \text{i.e.} \quad \zeta = i\theta\partial/\partial\mu, \qquad e^{-i\zeta} g(\mu) = g(\mu + \theta)e^{-i\zeta} \tag{8.3}$$

and as before $z = \sin\mu$. Without loss of generality we assume $\cos\mu \geqslant 0$. Eqs. (7.2), (7.3) and (7.5) are then satisfied if

$$f(\mu)f(\pi - \mu + \theta) = \cos(\mu - \theta)\cos\mu - F^2. \tag{8.4}$$

In the following if $g(\mu)$ is a generic analytic complex function of μ we define $g^*(\mu) = (g(\mu^\dagger))^\dagger$. For types A, B, C of sect. 7, where $\mu = \mu^\dagger$ and $\zeta = \zeta^\dagger$, we have in addition from eq. (7.4)

$$f^*(\pi - \mu)\cos\mu = f(\mu + \theta)\cos(\mu + \theta). \tag{8.5}$$

Unbounded representations are obtained by relaxing the hermiticity of μ and ζ but maintaining $z^\dagger = z$ and eqs. (8.1)–(8.5). Accordingly,

$$\mu^\dagger = \pi - \mu, \qquad \zeta^\dagger = -\zeta \tag{8.6}$$

and therefore,

$$(\cos\mu K^\pm)^\dagger = -\cos\mu K^\pm, \tag{8.7}$$

which is again consistent with eq. (8.5).

Bounded representations are characterized by discrete values of z and of μ and by the relation

$$|f(\mu)|^2 = (\cos^2(\mu - \tfrac{1}{2}\theta) - F^2)/\cos\mu, \tag{8.8}$$

which follows from eqs. (8.4) and (8.5). This relation leads to eq. (7.13) if z is discrete, but it is stronger if $z = \sin\mu$ ranges in the interval $(-1, 1)$. In this case $\cos\mu$ vanishes at the ends of the interval only and is otherwise positive. It follows that

$$\cos^2(\mu - \tfrac{1}{2}\theta) - F^2 > 0 \tag{8.9}$$

must hold if the bounded range must be admitted. This means that

$$F^2 < 1, \tag{8.10}$$

which corresponds to cases 2–4 of sect. 6.

Representations of case 2, i.e. satisfying eqs. (8.9) and (8.10) and $F^2 > 0$ have the ordinary angular momentum algebra as a limiting case when $\theta \to 0$, correspond to the SU(2) gauge theory and the range of allowed values of z is disconnected from $|z| \geqslant 1$.

In the hyperbolic range (i.e. SO(2,2) gauge) the conditions (8.8)–(8.10) are not required and we may set $\mu = \frac{1}{2}\pi + i\nu$, $\zeta = \theta\partial/\partial\nu$, the unbounded operators $e^{\pm i\zeta}$ are hermitian and defined over an everywhere dense subset of states in the Hilbert space of square integrable functions of ν. It is easy to check however that z and x, y as defined by eq. (7.1) map odd (even) functions of ν into odd (even) functions of ν and that the corresponding representation of the algebra (6.3) is therefore reducible.

Accordingly, we restrict ourselves to the Hilbert subspaces of wave functions $\psi(\mu)$ which are periodic and of definite parity, i.e. $\psi(\mu) = \pm\psi(\pi - \mu)$ and $\psi(\mu + 2n\pi) = \psi(\mu)$. These representations correspond to cases (1) and (2) of sect. 6 and are directly relevant to our discussion.

Case (4) of sect. 6 corresponds here to $F^2 < 0$, whereby eq. (8.9) is always satisfied and the circular range joins the hyperbolic range at $|z| = 1$. The natural Hilbert space is now that of the functions of z which are square integrable over the range $-\infty < z < +\infty$. This space can be described alternatively as the space of functions of μ which are defined in the union of the real line and of the lines $\mu = n\pi + \frac{1}{2}\pi + i\nu$ (ν real) in the complex plane of μ and have period 2π in μ. The operators $K^{\pm}$ are then defined over the everywhere dense subset of states in this Hilbert space represented by functions of μ which are holomorphic in the strips $|\mu - n\pi - \frac{1}{2}\pi| \leqslant \theta$. This peculiar class of representations covers both the circular and the hyperbolic range.

Case (3b), in analogy to case (3a) already discussed at the end of sect. 7, can be extended to G = SO(2,2) theory. To this purpose we define as in eq. (7.15)

$$x = \tfrac{1}{2}(A + A^{-1}), \qquad y = \tfrac{1}{2}(B + B^{-1}), \qquad z = \tfrac{1}{2}(C + C^{-1}),$$

but now A, B, C are hermitian and given by

$$A = e^{\theta x + \phi}, \qquad B = e^{p + \psi}, \qquad C = e^{-\theta x - p - \psi - \phi}, \tag{8.11}$$

where ψ, ϕ are real and, as before, $(x,p)=i$. This representation has the outer sheet $|z|, |x|, |y| > 1$ of the cubic surface $\Phi = 0$ as a classical limit. The other three sheets follow by replacing e^{ψ}, e^{ϕ} with $\pm e^{\psi}, \pm e^{\phi}$.

9. The de Sitter algebra

As pointed out at the end of sect. 5, if $G = SO(3,1)$ we obtain two sets of complex conjugate traces $c^+(w), c^-(w)$. Setting $x = c^+(u)$, $y = c^+(v)$, $z = c^+(uv)$ we obtain the commutation relations

$$(x,y) = -(\hbar s/2\alpha)\left(z - \tfrac{1}{2}(xy + yx)\right), \tag{9.1}$$

and the ones which follow by cyclical permutations of x, y, z. Proceeding analogously to sect. 6 we set $s\hbar/4\alpha = -\tanh(\theta/2)$, thus obtaining

$$e^{\theta/2}xy - e^{-\theta/2}yx = \sinh(\theta)z \quad \text{(and cyclical)}. \tag{9.2}$$

Eqs. (9.2) must be supplemented by the conjugate set and by the normality condition

$$(O^+, O) = 0, \tag{9.3}$$

where O^+, O are any one of the operators x^+, y^+, z^+ and x, y, z respectively. F^2 is then defined as

$$F^2 = \cosh^2(\tfrac{1}{2}\theta) + 2e^{\theta/2}xyz - e^{-\theta}(x^2 + z^2) - e^{\theta}y^2, \tag{9.4}$$

and is not a hermitian operator. Apart from this, eqs. (9.2) and (9.4) can be obtained from (6.3) and (6.4) by replacing θ with $-i\theta$, and F^2 satisfies the conditions (ii)–(iv) of sect. 6. This algebra deserves a detailed discussion and we limit ourselves to a heuristic treatment. We set $z = \sin\mu$ where the spectrum of μ is now the whole complex plane. As before we also define

$$K^{\pm} = e^{\theta/2}x \pm iy\, e^{\pm i\mu}. \tag{9.5}$$

We introduce the operator ζ satisfying

$$(\zeta, \mu) = \theta, \qquad \text{i.e.} \quad \zeta = \theta\partial/\partial\mu, \tag{9.6}$$

and as in sect. 8 we assume a wave function $\psi(\mu)$ with definite parity and periodic, i.e. $\psi(\pi - \mu) = \pm\psi(\mu)$ and $\psi(\mu + 2n\pi) = \psi(\mu)$, with μ ranging now over the whole complex plane $\mathbb{C}$. $\psi(\mu)$ is not holomorphic but belongs to $L^2(\mathbb{C})$. This suggests defining the operators $K^{\pm}$ as in sect. 8 where now $f(\mu) = f(\mu + 2\pi n)$ satisfies eq. (8.4) and θ is replaced by $-i\theta$. Besides these operators we introduce their hermitian conjugate, notice however that now $(K^+)^\dagger$ is no longer related to K^- by a relation of the kind (8.7). In this scheme the algebra O (O^+) is then generated by μ, ζ (μ^+, ζ^+) and O, O^+ satisfy (9.3). This construction reduces in the limit $\theta \to 0$ to the one given by Naimark [5] for the linear representations of SO(3, 1).

The de Sitter algebra admits also a case (3) similar to the ones discussed in sects. 7 and 8. We assume the same anzatz (7.15) but now x is a complex variable and $p = -i\partial/\partial x$.

The discussion of the algebras (6.3), (9.2) and their representations carried out in this paper is clearly preliminary and orientative and should be followed by a more rigorous approach. It is however evident from our discussion that these algebras generalise in a nontrivial way the Lie algebras of the SU(2), SL(2, ℝ), SL(2, ℂ) groups and that their structure raises a number of interesting questions, among which the precise relationship between our approach and that of quantum groups [4, 6].

Appendix A

$$
\begin{aligned}
&e^{\bullet}{}_{\alpha} e^{\triangle}{}_{\beta} e^{\circ}{}_{\gamma} \epsilon_{\bullet\triangle\circ} = g^{1/2} \epsilon_{\alpha\beta\gamma}, \\
&\epsilon^{\alpha\beta\gamma} \epsilon_{\sigma\tau\gamma} = \delta^{\alpha}{}_{\tau} \delta^{\beta}{}_{\sigma} - \delta^{\alpha}{}_{\sigma} \delta^{\beta}{}_{\tau}, \\
&\epsilon_{012} = 1 = -\epsilon^{012}, \qquad e^{\bullet}{}_{\alpha} e^{\triangle}{}_{\beta} \epsilon_{\bullet\triangle\circ} = \epsilon_{\alpha\beta\gamma} e_{\circ}{}^{\gamma} g^{1/2}, \\
&e_{\circ}{}^{\gamma} = g^{\gamma\beta} \eta_{\circ\bullet} e^{\bullet}{}_{\beta}, \qquad e_{\circ}{}^{\alpha} e^{\bullet}{}_{\alpha} = \delta^{\bullet}{}_{\circ}.
\end{aligned}
\tag{A.1}
$$

The γ-matrices are defined by

$$
\gamma_a \gamma_b + \gamma_b \gamma_a = 2 g_{ab}, \qquad \gamma_{ab} = \tfrac{1}{2}(\gamma_a, \gamma_b),
$$

and $\gamma_5 = k \gamma_0 \gamma_1 \gamma_2 \gamma_3$ satisfies

$$
\gamma_5^2 = -k, \qquad -\tfrac{1}{2} \epsilon^{abcd} \gamma_a \gamma_b = \gamma_5 \gamma^c \gamma^d. \tag{A.2}
$$

In terms of the Pauli matrices the representation is $\gamma_0 = i\sigma_2 \otimes \sigma_3$, $\gamma_1 = \sigma_3 \otimes \sigma_3$, $\gamma_2 = -1 \otimes \sigma_1$, $\gamma_3 = -\sqrt{k}\, \sigma_1 \otimes \sigma_3$, $\gamma_5 = i\sqrt{k}\, 1 \otimes \sigma_2$. The projectors

$$
P_{\pm} = \pm\left(\gamma_5 \pm i\sqrt{k}\, 1\right) / \left(2i\sqrt{k}\right)
$$

satisfy $P_{\pm}^2 = P_{\pm}$, $P_+ P_- = P_- P_+ = 0$. In this representation,

$$
P_{\pm} = 1 \otimes p_{\pm}, \qquad p_{\pm} = \tfrac{1}{2}(1 \pm \sigma_2). \tag{A.3}
$$

It follows that

$$\gamma^{ab} \otimes \gamma_{ab} = -4(\sigma^i \otimes p_+) \otimes (\sigma^i \otimes p_+) - 4(\sigma^i \otimes p_-) \otimes (\sigma^i \otimes p_-), \quad \text{(A.4)}$$

where

$$\sigma^i{}_\alpha{}^\beta \sigma^i{}_\gamma{}^\tau = -\delta_\alpha{}^\beta \delta_\gamma{}^\tau + 2\delta_\alpha{}^\tau \delta_\gamma{}^\beta. \quad \text{(A.5)}$$

References

[1] J.E. Nelson and T. Regge, Nucl. Phys. B328 (1989) 190
[2] E. Witten, Nucl. Phys. B311 (1988/89) 46;
S. Deser, R. Jackiw and G. 't Hooft, Ann. Phys. (N.Y.) 152 (1984) 220;
S. Deser and R. Jackiw, Ann. Phys. (N.Y.) 153 (1984) 405
[3] H.M. Farkas and I. Kra, Riemann surfaces (Springer-Verlag, Berlin, 1980)
[4] L.D. Faddeev, N.Yu. Reshestikhin and L.A. Takhtajan, Quantisation of Lie groups and Lie algebras, LOMI-preprint E-14-87 and references therein
[5] M.A. Naimark, Les representations lineaires du groupe de Lorentz (Dunod, Paris, 1982)
[6] V.G. Drinfeld, *in* Proc. Int. Congress of Mathematicians (Univ. of California Press, Berkeley, 1986) p. 798;
S. Woronowicz, Publ. Res. Inst. Math. Sci. (Kyoto University) 23 (1987) 117; Commun. Math. Phys. 111 (1987) 613; Invent. Math. 93 (1988) 35;
M. Jimbo, Lett. Math. Phys. 10 (1985) 63; 11 (1986) 247; Commun. Math. Phys. 102 (1987) 537;
L.C. Biedenharn, A q-boson realization of the quantum group $SU_q(2)$ and the theory of q-tensor operators, invited paper at the Clausthal Summer Workshop on Mathematical Physics, Clausthal-Zellerfeld, 1989

Physics Letters B 272 (1991) 213–216
North-Holland

PHYSICS LETTERS B

2+1 quantum gravity

J.E. Nelson and T. Regge

Dipartimento di Fisica Teorica dell'Università di Torino, via Pietro Giuria 1, I-10125 Turin, Italy

Received 16 September 1991

In previous work we analysed the algebra of observables of the initial data surface Σ^2 for 2+1 de Sitter gravity, and constructed for genus 2 the group of autmorphisms H of the homotype group π_1 induced by the mapping class group. The group H induces a group D of canonical transformations on the algebra of observables which is related to the braid group. Here all of the previous results are implemented for any genus and at the quantum level.

In refs. [1–3] the algebra of observables A for classical gravity in 2+1 dimensions was explicitly defined and calculated for genus $g=1$, 2 and we discussed the role and the relationships among the mapping-class group, the braid group [4] and their representations in terms of canonical transformations on A. The lagrangian density for pure gravity with negative cosmological constant Λ in 2+1 dimensions is the Chern–Simons three-form [2]

$$\tfrac{1}{8}\alpha(\mathrm{d}\omega^{AB}-\tfrac{2}{3}\omega^{a}{}_{T}\wedge\omega^{TB})\wedge\omega^{CD}\epsilon_{ABCD}\,, \tag{1}$$

with A, $B=0, 1, 2, 3$, $\eta_{AB}=(-1, 1, 1, -1)$, $\Lambda=-\frac{1}{3}\alpha^{-2}$, $\omega^{AB}=\omega^{BA}$. The action (1) leads to the Poisson brackets

$$[\omega^{AB}{}_{i}(x), \omega^{CD}{}_{j}(y)]=\alpha^{-1}\epsilon_{ij}\epsilon^{ABCD}\delta^{2}(x-y)\,, \tag{2}$$

$i, j=1, 2$, $x, y\in\Sigma^2$ and the field equations

$$R^{AB}=\mathrm{d}\omega^{AB}-\omega^{AT}\wedge\omega_{T}{}^{B}=0 \tag{3}$$

implying that space–time is locally anti-de Sitter. The integrated connections yields the representation $\psi: \pi_1(\Sigma^2, \mathrm{B})\to \mathrm{G}$, where the gauge group G is here the spinor group $\mathrm{SL}(2, \mathbb{R})\otimes\mathrm{SL}(2, \mathbb{R})$ of $\mathrm{SO}(2, 2)$. The algebra A of observables is generated by the traces $c^{\pm}(a)=c^{\pm}(a^{-1})=\frac{1}{2}\,\mathrm{tr}[S^{\pm}(a)]$, $a\in\pi_1$ where $S^{\pm}$ is the partial map into the upper (lower) spinor group. We discuss here the + sign only, the lower spinor case being trivially related to the upper.

The fundamental group π_1 of a surface Σ^2 of genus g [5] admits a presentation with the generators

$$U_1, V_1, U_2, V_2, \ldots, U_g, V_g \tag{4}$$

and the single relator

$$\mathrm{Rel}=U_1V_1U^{-1}V_1^{-1}U_2V_2U_2^{-1}V_2^{-1}\ldots$$
$$\times U_gV_gU_g^{-1}V_g^{-1}=1\,. \tag{5}$$

This presentation is replaced here by an equivalent one which is more suitable for our purpose. We set

$$t_1=V_g^{-1}U_1U_2\ldots U_{g-1}\,,$$
$$t_{2i}=(U_iU_{i+1}\ldots U_{g-1})^{-1}V_{i-1}^{-1}U_iV_iU_{i+1}U_{i+2}\ldots U_{g-1}\,,$$
$$t_{2i+1}=(U_iU_{i+1}\ldots U_{g-1})^{-1}U_{i+1}U_{i+2}\ldots U_{g-1}\,,$$
$$t_{2g}=V_{g-1}^{-1}U_g\,,$$
$$t_{2g+1}=V_g\,,$$
$$t_{2g+2}=U_g^{-1}\,. \tag{6}$$

One can show that

$$t_1t_2\ldots t_{2g+2}=V_g^{-1}\,\mathrm{Rel}\,V_g=1\,,$$
$$t_1t_3\ldots t_{2g+1}\equiv 1\,,$$
$$t_2t_4\ldots t_{2g+2}\equiv 1 \tag{7}$$

and that a presentation of π_1 is given by the generators $t_1\ldots t_{2g+2}$ and the relators (7). Let λ be the abelianisation map $\pi_1\to \mathrm{H}_1(\Sigma_2)$, where H_1 is the first homology group of Σ_2. Let $\mathrm{int}(\alpha, \beta)$, $\alpha, \beta\in\mathrm{H}_1$ denote the skew bilinear intersection form on H_1 and let $\langle a,$

Volume 272, number 3,4 PHYSICS LETTERS B 5 December 1991

$b\rangle = \mathrm{int}(\lambda(a), \lambda(b))$ where $a, b \in \pi_1$. We have then $\langle U_i, U_j\rangle = \langle V_i, V_j\rangle = 0$, $\langle U_i, V_j\rangle = \delta_{ij}$ and therefore $\langle t_i, t_j\rangle = \delta_{i,j-1} - \delta_{i,j+1}$ where here and in the sequel and by virtue of (7) lower indices are understood as mod $2g+2$. All these results can be displayed graphically on a polygon (see fig. 1) of $2g+2$ vertices $P_1 \ldots P_{2g+2}$ arranged anticlockwise where the generator t_i, $i = 1, \ldots, 2g+2$ is represented by the side from P_{i+1} to P_i.

The diagonal from P_k to P_i is denoted by $d_{ik} = t_i t_{i+1} \ldots t_{k-1}$ with $d_{ii} = 1$, $d_{i,i+1} = t_i$, $d_{ik} d_{km} = d_{im}$. The first relator in (7) is then the clockwise perimeter of the polygon. One can show then that:

(a) $\langle d_{lk}, d_{mj}\rangle = 0$ if d_{lk}, d_{mj} have no common point, i.e., if the interval l to k contains either both j, m or neither.

(b) $\langle d_{mj}, d_{jk}\rangle = 1$ (-1) if m, j, k run anticlockwise (clockwise).

(c) $\langle d_{kj}, d_{ml}\rangle = 2$ (-2) if k, m, j, l run anticlockwise (clockwise), d_{kj} and d_{ml} intersect inside the polygon and the interval i to k contains either m or l.

Cases (a), (b), (c) exhaust all possibilities and this presentation has an explicit cyclical symmetry of order $2g+2$ extending the results in ref. [3] for $g = 1$, 2. Given $a, b \in \pi_1$ such that $\langle a, b\rangle = 0$ [case (a)] from refs. [1–3] we have the Poisson brackets (PB) of the normalised traces $[c(a), c(b)] = 0$ and correspondingly we set the commutators $(c(a), c(b)) = 0$ in the quantum theory.

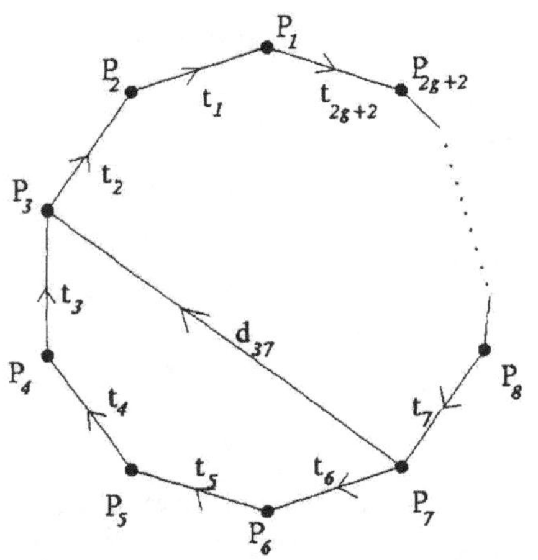

Fig. 1. $2g+2$ polygon representing a Riemann surface of genus g.

If instead $\langle a, b\rangle = 1$ [case (b)] then $\langle b, b^{-1}a^{-1}\rangle = 1$, $\langle b^{-1}a^{-1}, a\rangle = 1$ and we obtain the following set of PB:

$$[c(a), c(b)] = 1/2\alpha\,[-c(a)c(b) + c((ab)^{-1})], \tag{8}$$

and cyclical permutations in a, b, $(ab)^{-1}$.

Since $S(a)$, $S(b) \in \mathrm{SL}(2, \mathbb{R})$ the classical identity holds:

$$c(a)c(b) = \tfrac{1}{2}[c(ab) + c(ab^{-1})], \tag{9}$$

and therefore (8) can be rewritten as

$$[c(a), c(b)] = 1/4\alpha\,[c(ab) - c(ab^{-1})]. \tag{10}$$

Under (10) traces are given an infinite Lie algebra L structure subject to non-linear constraints (9) which generate an ideal I in the enveloping algebra of L. By repeated used of (9) we can compute recursively all the $c(a)$, $a \in \pi_1$, starting from the set of traces $a_{ij} = a_{ji} = c(d_{ij})$. For case (b) a set of quantum commutators having (8) as classical limit and satisfying the Jacobi identities is given by

$$\begin{aligned}(c(a), c(b)) &= (K-1)[c(ab) - c(a)c(b)] \\ &= (1 - 1/K)[c(ab) - c(b)c(a)] \\ &= (1-K)[c(ab^{-1}) - c(b)c(a)] \\ &= (1/K - 1)[c(ab^{-1}) - c(a)c(b)]\,,\end{aligned} \tag{11}$$

where $K = (4\alpha - \mathrm{i}\hbar)/(4\alpha + \mathrm{i}\hbar) = \Phi^2 = \exp(\mathrm{i}\theta)$ and the quantum counterpart of (9) is given by

$$c(ab) + c(ab^{-1}) = c(a)c(b) + c(b)c(a)\,.$$

Case (c) leads to commutators which are completely determined from those of the cases (a), (b) by repeated use of the Jacobi identities. For instance, if m, j, l, k run anticlockwise $\langle d_{jk}, d_{lm}\rangle = 2$:

$$(a_{jk}, a_{lm}) = (K - 1/K)(a_{jl}a_{mk} - a_{kl}a_{jm})\,, \tag{12}$$

but also

$$\begin{aligned}&(a_{mk}, a_{jl}) = (a_{mj}, a_{kl}) = 0\,, \\ &(a_{jk}, a_{kl}) = (1 - 1/K)(a_{jl} - a_{kl}a_{jk})\,, \\ &(a_{jk}, a_{km}) = (1/K - 1)(a_{jm} - a_{jk}a_{km})\,, \\ &(a_{jk}, a_{lm} - (K+1)a_{kl}a_{km}) = 0\,.\end{aligned} \tag{13}$$

The quantum algebra defined by the above brackets cannot be identified immediately with the algebra of observables A since the variables a_{ij} are not independent already at the classical level (see refs. [1–3]). Indeed generic elements u, $v \in SL(2, \mathbb{R})$ can be considered as vectors in $\mathbb{R}^4$ endowed with the scalar product $u \cdot v = \frac{1}{2}\,\mathrm{tr}(uv^{-1})$. Write then $d_{ik} = d_{in}d_{nk} = d_{in}\,(d_{kn}^{-1})$ where n is some fixed vertex. We see that the rank of the (classical matrix A of elements a_{ik} cannot exceed 4 and this manifests itself as a set of polynomial constraints on the a_{ik}. Other constraints follow from the relators but are more complicated to express (see ref. [3] for details). Furthermore, the a_{ij} are not invariant under those components of the groups of diffeomorphisms which are not connected to the identity and generate Dehn twists.

Consider now the case (b) triple d_{mj}, d_{jk}, $d_{km} \in \pi_1$, where m, j, k run anticlockwise and set $a_{jk} = \cos\psi / \cos(\frac{1}{2}\theta)$. ψ is hermitian (antihermitian) if the invariant axis of $S(d_{jk})$ in SO(2, 1) is timelike (spacelike). We define raising and lowering operators as

$$M^{\pm} = a_{km} - \exp(\tfrac{1}{2}\mathrm{i}\theta \pm \mathrm{i}\psi)\, a_{mj}\,, \tag{14}$$

so that for any function $F(\psi)$

$$M^{\pm} F(\psi \pm \theta) = F(\psi) M^{\pm}\,. \tag{15}$$

Strictly speaking in order to define $M^{\pm}$ we need to consider extended representations of A which include the algebra $A_1 \supset A$ of functions of ψ not restricted to be even in ψ as they are for a_{jk}. These representations are in general reducible if restricted to A. Let $F(\psi)$ satisfy the condition

$$F(\psi) = F(\psi \pm \theta)\exp(\pm \mathrm{i}\psi + \tfrac{1}{2}\mathrm{i}\theta)\,, \tag{16}$$

then we have

$$F(\psi) a_{kl} F(\psi)^{-1} = (1+K) a_{jk} a_{kl} - K a_{jl}\,,$$

$$F(\psi) a_{mj} F(\psi)^{-1} = (1+K) a_{jk} a_{mj} - K a_{km}\,,$$

$$F(\psi) a_{km} F(\psi)^{-1} = a_{mj}\,. \tag{17}$$

The action of $F(\psi)$ is trivial for case (a) and can be extended to case (c) since from the last of (13) it follows that

$$F(\psi) a_{lm} F(\psi)^{-1} = a_{lm} - (1+K) a_{kl} a_{km}$$

$$- (K^2+K) a_{lj} a_{mj} + (1+K)^2 a_{jk} a_{kl} a_{mj}, \tag{18}$$

and $F(\psi)$ induces an automorphism $D(a)$ on A which is the quantum counterpart of the classical Dehn twist [3]. A solution of (16) is given by

$$F(\psi) = \exp(-\mathrm{i}\psi^2/2\theta)\,, \tag{19}$$

but then it is not periodic since

$$G(\psi) = F(\psi + 2\pi) F(\psi)^{-1}$$

$$= \exp(-2\mathrm{i}\pi\psi/\theta - 2\mathrm{i}\pi^2/\theta)\,, \tag{20}$$

and in order to represent it we need to enlarge the algebra A_1 further to an algebra A_2 which includes also non-periodic functions of ψ. The Dehn twists would then appear as automorphisms which are inner in A_2 but not in A. In many ways they resemble the discrete symmetries of the Lorentz group which require doubling of the spinor components. $G(\psi)$ lies in the center of A_2 since $(G(\psi), M^{\pm}) = 0$.

As already pointed out in refs. [1–3] the case where θ/π is rational is very interesting as can already be seen when all $S(a)$ have a common axis, i.e., all matrices $S(a)$ are of the form

$$S(a) = \begin{vmatrix} e(a) & 0 \\ 0 & e(a)^{-1} \end{vmatrix}.$$

The quantum trace is then defined as $c(a) = \frac{1}{2}[e(a) + e(a)^{-1}]/\cos\frac{1}{2}\theta$. The operators $e(a)$ and the c number Φ obey the following rules:

(1) $e(a)^{\dagger} = e(a)^{-1}$;

(2) $e(1) = 1$;

(3) $e(a)\Phi = \Phi e(a)$;

(4) Let $\langle a, b\rangle = p$, then $e(ab) = \Phi^{p}\, e(a)\, e(b) = \Phi^{-p} e(b)\, e(a)$.

We call E the group generated by all $e(a)$ and Φ. Rules (2)–(4) imply (11)–(13) and that E is a Weyl–Heisenberg group.

An explicit representation $T(\mathrm{E})$ for E is then given by $T(e(a)) = \exp[\mathrm{i}\psi(a)]$ where $\psi(ab) = \psi(a) + \psi(b)$ and $(\psi(a), \psi(b)) = \mathrm{i}\theta\langle a, b\rangle$. If $m\theta = 2p\pi$, then, for all $a \in \pi_1$, $e(a)^m = G(\psi(a))^{-p}$, $e(a)^m$ lies in the center of E and $T(\mathrm{E})$ splits into subrepresentations where all $T(e(a)^m)$ are multiples of the identity but on which we cannot in general represent the Dehn twists. The structure of $T(\mathrm{E})$ is most easily displayed by using the standard presentation (4)–(5), whereby

$$(\psi(U_i), \psi(U_j)) = 0, \quad (\psi(V_i), \psi(V_j)) = 0\,,$$

$$(\psi(U_i), \psi(V_j)) = \mathrm{i}\theta\delta_{ij}\,, \tag{21}$$

whence we see that $T(\mathrm{E}) = \otimes_i T(\mathrm{E}_i)$ where $T(\mathrm{E}_i)$ is generated by $\psi(U_i)$, $\psi(V_i)$. In the above quoted case where $\theta=2p\pi/m$ each $T(E_i)$ admits an irreducible representation of dimension m. Clearly the homology group H_1 rather than the homology group is represented faithfully on E and the mapping class group acts through $Sp(2g, \mathbb{Z})$ on the linear span of the $\psi(a)$. It is conceivable that the representations and the Casimir operators already discussed for $g=1$, 2 can be extended to generic g [6].

References

[1] J.E. Nelson and T. Regge, Nucl. Phys. B 328 (1989) 190.
[2] J.E. Nelson, T. Regge and F. Zertuche, Nucl. Phys. B 339 (1990) 516.
[3] J.E. Nelson and T. Regge, 2+1 gravity for genus > 1, preprint DFTT 37/90, Commun. Math. Phys., to appear.
[4] J.S. Birman, Braids, links and the mapping class group, Annals of Mathematics Studies (Princeton U.P., Princeton, N.J. 1975).
[5] H.M. Farkas and I. Kra, Riemann surfaces, (Springer, Berlin, 1980).
[6] J.E. Nelson and T. Regge, in preparation.

PART VIII

Methods of Algebraic Geometry

Introduction to the paper "Duality Group for Calabi–Yau Two-Moduli Space"

Anna Ceresole[1] and Riccardo D'Auria[2,1]

[1]*Istituto Nazionale di Fisica Nucleare (INFN), Sezione di Torino, Italy*
[2]*DISAT, Politecnico di Torino, Corso Duca degli Abruzzi 24, I-10129 Turin, Italy*

This contribution, with no pretence at originality or completeness, is conceived to put in perspective the paper "Duality group for Calabi–Yau two-moduli space"[1] reported in this volume, where Tullio Regge provided key insight and a new method to solve a specific instance of an open and quite debated problem: the computation of the target space duality group acting on the moduli space $\mathcal{M}$ of string theory. This is the space of deformations by truly marginal operators of the underlying superconformal field theory, and is known to classify inequivalent string vacua. The knowledge of its local and global properties is essential towards the full comprehension of the symmetries and the physical couplings of the low energy effective lagrangian obtained by compactification down to four dimensions.

Duality symmetries, that are ubiquitous in physics,[2] have played in the early '90s a major rôle in frontier high energy physics, within supersymmetric Yang–Mills theories, string theory and supergravity.[3–5] Here, we describe in some detail the algebro-geometric and field-theory methods that were developed to extract information on the non-perturbative regime of string and supergravity theories, along the personal viewpoint of.[6–9]

Historically, the geometric interpretation of the six extra dimensions of space-time via a Kähler manifold of vanishing first Chern Class, or Calabi–Yau manifold,[10] marked a first important step toward understanding the structure of supergravity Lagrangians at low energy. Since then, the approach to these issues has been deeply changed by the discovery of D-branes, of the rich variety of flux vacua, the gauging procedure of supergravity symmetries, and the dualities relating perturbative and non-perturbative regimes of a given theory.[11] All these new ingredients are strictly intertwined: for example, some of the fluxes are sourced by the D-brane RR-charges and moreover, their presence triggers a gauging of the original theory. This behaviour was first seen in the Freund–Rubin solution of the $D = 11$ supergravity, the theory which is now believed to provide a low energy formulation of M-theory.[12] When fluxes and branes are present, one is led to consider compactifications on more general backgrounds than 4-dimensional $N = 1$ or $N = 2$ supergravities, in particular focusing on internal manifolds which, differently from the Calabi–Yau, are not Ricci flat.

Notwithstanding this, the original treatment of duality symmetry for (*ungauged*) supergravities through compactification on Calabi–Yau Ricci flat 3-folds remains a very important starting point towards more complex situations. The low energy fields in four or higher-dimensional supergravities are still studied today by an expansion of the original fields along the same basis of forms which defines the Calabi–Yau reduction.

The computation of the duality group for string compactifications on Calabi–Yau manifolds with more than a single modulus was an unsolved problem at the time. We were well aware that Regge was not following the latest developments of string theory, in spite of having planted its seeds many years back, in the context of the dual models for hadrons (see Veneziano's contribution in this volume). However, we proposed it to him because of his profound knowledge of algebraic geometry. Actually, during the years at IAS he had previous worked on the monodromy properties of Feynman diagrams, which lead to tools still useful today in the study of scattering amplitudes (see the contribution of G. Passarino in this volume), and gave him the right perspective. His incredible ability to enter immediately into the core of a physical topic and to translate it into precise mathematical terms, quickly led to an elegant solution. The final outcome was quite interesting also because it contradicted the common naive expectations about the general structure of the duality group associated to the periods of a given algebraic curve, adding new insight and an alternative method to the analysis of Picard–Fuchs equations.[6] We apologise that lack of space forces us to omit many references, but we point the reader to those contained in the original paper that we comment here.

1. Duality symmetries in supergravity

Many important developments in string theory and supergravity as its low energy description, obtained towards the end of the last century, where tied to the concept of *duality*. In the simplest case, T-*duality* relates string compactifications on a torus of radius R to those on a torus of radius α'/R. More generally, T-duality transformations are described in toroidal compactifications by the group O(d, d; R). It was later argued that string and supergravity theories enjoy S-*duality*, that relates the perturbative behaviour of a theory in a weak coupling regime, $g^2 << 1$ to its behaviour at strong coupling, $g^{-2} << 1$, generally described by a different perturbative theory.

Duality between weak and strong coupling was already known since many decades in physics, both in condensed matter (Ising model) as in field theory, famous examples being the equivalence between sine-Gordon and Thirring model in two space-time dimensions. In four space-time dimensions, an outstanding example is the Dirac $\mathbf{Z}_2$ duality between electric and magnetic field strengths in Maxwell electromagnetism and its extension to a SL(2, $\mathbf{Z}$) in the presence of a θ-angle. Duality is welcome to a theoretical physicist, as the strongly coupled phase is the hardest to examine and a weak-strong duality allows us to study it via the much easier weakly coupled formulation.

T- and S-duality symmetries appear in supergravity as isometries of the Lagrangian σ model. In particular, when the number of supersymmetries N is greater than two, $2 < N \leq 8$, the scalar fields are coordinates of a σ-model whose geometry is constrained to a be a coset manifold G/H where H is the maximal compact subgroup of G. The isometry Lie group G comprises both T- and S-duality and is called U-*duality* group.[4]

To set the stage, along,[19] we focus on the generic structure of the bosonic sector of an ungauged supergravity theory, coupled to an arbitrary number n of vector multiplets. This leads to a total number $n+g$ of U(1) gauge fields, where g denotes the number of gauge fields in the gravitational multiplet. Correspondingly, there are $2(n+g)$ electric and magnetic field strengths (F^Λ, G_Λ). In $N = 2$ supergravity, which is for us the most relevant, we have just one graviphoton, $g = 1$, and $2n + 2$ gauge fields. The relevant bosonic $D = 4, N = 2$

supergravity Lagrangian, limited to the Einstein and the kinetic terms of scalar and vector fields, reads:

$$\mathcal{L} = \det e \left[-\frac{R}{2} + \frac{1}{2} g_{ij} \partial_\mu \phi^i \partial^\mu \phi^j + i \left(\bar{\mathcal{N}}_{\Lambda\Sigma} F^{\Lambda(-)}_{\mu\nu} F^{\Sigma(-)\mu\nu} - \mathcal{N}_{\Lambda\Sigma} F^{\Lambda(+)}_{\mu\nu} F^{\Sigma(+)\mu\nu} \right) \right] + \dots, \quad (1)$$

where $\det e$ is the vierbein determinant, $\Lambda = 0, 1, \dots, n$ and $F^{\Lambda(\pm)}$ denote the self-dual and anti-selfdual part of F^Λ. $\mathcal{N}_{(\Lambda\Sigma)}(\phi) = R_{\Lambda\Sigma}(\phi) + \mathrm{i} I_{\Lambda\Sigma}(\phi)$ is a *field dependent matrix of coupling constants* whose real part $\mathcal{R}_{\Lambda\Sigma}$ plays the role of a generalized θ-term of ordinary field theory. The scalar fields for $N = 2$ are complex, and they parameterize a very specific type of Kähler manifold to be defined below. Therefore it is usual to write the kinetic term of the scalars in complex coordinates $z, \bar{z}$ as $g_{\alpha\bar{\beta}} \partial_\mu z^\alpha \partial^\mu \bar{z}^{\bar{\beta}}$.

In ungauged supergravities considered at the time, the matter fields are neutral with respect to the $U(1)^{n+1}$ gauge group of the gauge fields A^Λ_μ, and no minimal coupling term appears in the Lagrangian. The structure of the bosonic kinetic term remains formally the same in the extension to higher N supersymmetries.

The $n + 1$ equations of motion and the $n + 1$ Bianchi identities result in a set of $2n + 2$ equations invariant under the *symplectic subgroup* $\mathrm{Sp(2n+2,R)}$ of the general linear transformations of $\mathrm{GL(2n+2,R)}$. In principle, any transformation under $\mathrm{GL(2n+2,R)}$ would leave invariant this linear set. However, the restriction to the symplectic group arises from the requirement that the coupling matrix $\mathcal{N}_{\Lambda\Sigma}$ remains symmetric in the indices $(\Lambda\,\Sigma)$ under a general linear transformation.[16] Besides acting as a group G of isometries on the scalar fields, $\mathrm{Sp(2n+2,R)}$ acts linearly in a symplectic $(2n+2)$-dimensional representation D^Λ_Σ of G on the $(2n+2)$ *electromagnetic* field strengths $H^\Lambda = (F^\Lambda, G_\Lambda)$, with $\Lambda = 1, \dots 2n+2$. Here F^Λ and G_Λ[a] denote the electric and their magnetic-dual field strengths respectively and the index Λ, $(\Lambda = 1, \dots n+1)$, includes both the vectors of the vector multiplets and the graviphoton of the gravitational multiplet. The symplectic transformation acting on the $2(n+1)$-dimensional vector $H^\Lambda = (F^\Lambda, G_\Lambda))$ is in fact a duality between electric and magnetic configurations, generalizing the $\mathbf{Z_2}$ duality of pure electromagnetism. This relation between the isometry group G and the symplectic transformations is a consequence of the supersymmetry of the Lagrangian: since scalar and vector fields are in the same supersymmetry multiplet, in order to preserve supersymmetry, an isometry on the scalar fields must be accompanied by a suitable transformation of the electromagnetic fields.

As an $\mathrm{Sp(2n+2,R)}$ transformation mixes electric and magnetic field strengths, an *electric-magnetic duality symmetry* holds for the on-shell configurations, so that when duality symmetry is present, we can in principle obtain important informations on the non perturbative aspects of the theory. It must be kept in mind, however, that this duality *is not a symmetry of the Lagrangian*, but only of the equations of motion: that is, it is only realized for the *on-shell* configurations. This result is just an extension of the Dirac duality exchanging Bianchi identities and equations of motion in Maxwell electromagnetism.

The action of the symplectic transformations on the Lagrangian can be easily ascertained

[a]Because of the non standard coupling of the scalar fields to the gauge fields, the magnetic field strength is not simply ${}^*F^A$, but it is given by $\frac{1}{2}\frac{\partial \mathcal{L}}{\partial F_{\mu\nu}} = \mathrm{Re(z,\bar{z})F_{\mu\nu} - Im(z,\bar{z})^*F_{\mu\nu}}$.

writing the generic Sp(2n + 2, R) matrix as

$$S = \begin{pmatrix} A & B \\ C & D \end{pmatrix} \tag{2}$$

where A, B, C, D are $(n+1) \times (n+1)$ block matrices obeying the relations

$$A^t\, C = C^t\, A; \qquad B^t\, D = D^t\, B; \qquad A^t D - C^t\, B = 1. \tag{3}$$

All terms in the Lagrangian result to be invariant, *except* the kinetic term of the vectors. More precisely, the transformations where $B \neq 0$ change in an essential way the vector kinetic term, and therefore the Lagrangian, so that we are carried to a new Lagrangian describing the perturbative behaviour of another theory. In particular, when $B \neq 0$, the coupling constant undergoes the fractional transformation

$$\mathcal{N}'(z', \bar{z}') = (C + D\mathcal{N}(z, \bar{z}))(A + B\mathcal{N}(z, \bar{z}))^{-1}. \tag{4}$$

As the inversion of the coupling constant $\mathcal{N}_{(\Lambda\Sigma)} \to -1/\mathcal{N}^{(\Lambda\Sigma)}$ belongs to these transformations, we are led to the conclusion that *the presence of the U-duality group G implies a weak-strong duality symmetry* and the transformed Lagrangian encodes the strong coupling regime of the original one. When $B = 0$ and $C = 0$, the Lagrangian is invariant, while when $B = 0$ and $C \neq 0$ it varies with a total derivative. These latter transformations describe the *classical* symmetries of the theory.[19]

At the *quantum* level, the duality group G must become *discrete*, since the Dirac–Schwinger–Zwanziger quantization of electric and magnetic charges implies that they span a lattice: therefore the group must act on H^Λ as a a discrete Sp$(2n+2, \mathbf{Z})$ transformation. This makes it quite difficult in general to find the *quantum duality group*. Once the discrete duality group has been determined, the space of the *physical vacua* of the theory is obtained by quotienting by it the moduli space.

Physically, a point of the moduli space fixes the vacuum expectation value of the corresponding scalar field in the four-dimensional Lagrangian. As the low energy physical parameters (σ-model metric, Yukawa couplings, etc.) are functions of the moduli, it is of the outmost importance to know, besides its local properties, also its *global structure*. This allows us to identify the set of possible vacua of the theory, eliminating the redundancy due to duality. Furthermore, since the duality group is a modular group of the quotient space, the physical low energy parameters of the Lagrangian like the metric $g_{\alpha\bar{\beta}}$ and the Yukawa couplings $W_{\alpha\beta\gamma}$ must transform as modular forms under duality.

2. Moduli spaces and their monodromies

In the early '90s, special attention was given to four-dimensional supergravities obtained as 10D compactification breaking the supersymmetry to $N = 1$, since the chirality property of the $N = 1$ supergravity makes it a potential candidate to describe physics of the real world. The issue was to arrive to the determination of the physical parameters of the $N = 1$ low energy lagrangian, such as the Yukawa couplings and the σ-model metric.

Technically, these could be obtained either in terms of the *marginal deformations* of the superconformal field theory (SCFT), (namely zero-genus string σ-model), or equivalently, in terms of the *deformations* of the Calabi–Yau 3-fold. In both approaches, such deformations are exactly the *moduli.* In the effective Lagrangian the moduli appear as *massless neutral scalar fields* parameterizing a σ-model and having a vanishing potential.

In the SCFT approach, the moduli are directly related to the correlation functions of a two-dimensional SCFT, which usually are quite difficult to compute. More insight can be obtained via the Landau–Ginzburg superpotential and its *Chiral Ring,*[13] or, what is equivalent, using a geometric description in terms of *Calabi–Yau* compactification of ten-dimensional $N = 1$ (heterotic) or $N = 2$ supergravity. In the following we shall focus on the *geometrical Calabi–Yau approach,* where most of the results concerning discrete duality have been obtained.[14]

From the four-dimensional point of view, the *local* geometry of the moduli space, encoded in the metric of the σ-model, is completely fixed by supersymmetry: in 4D and $N = 1$, the σ-model is described by a Kähler–Hodge manifold, while in $N = 2$ one has the more restricted structure of a *Special Kähler manifold,* to be defined in the coming sections. On the other hand, in 4D compactification of the ten-dimensional Heterotic or Type IIB theories on a Calabi–Yau 3-fold, both the corresponding $N = 1$ or $N = 2$ scalar geometry must exhibit the same Special Kähler geometry of the $N = 2$, $D = 4$ supergravity. Indeed the massless scalar fields in 4D are by definition the *moduli* of the same Calabi–Yau 3-fold, and therefore they share the same Special Kähler geometry. Therefore as the metric of the σ-model of the $D = 4$, $N = 2$ Lagrangian is obtained from compactification from ten dimensions on a Calabi–Yau, it must coincide with the metric of the moduli space of the CY manifold itself.

Considering also the global geometry of the moduli space, we can relate the two formulations of a duality symmetric theory through the action of the modular group. This can be done if we determine the *monodromy* group arising from the singularities of the moduli space, that is of these points where the local description in terms of the original objects ceases to be valid in the chosen chart of the perturbative expansion.
We shall show that the analysis of the monodromies can be done in terms of a set of differential equations named Picard–Fuchs equations, as was actually done for just one modulus. However, already in the case of two moduli it become very involved and it is in this context that the alternative method elaborated by Regge shows its power.

3. $N = 2$ Super–Yang–Mills theories

In the same period of CY compactifications, many studies were focussed on the electric-magnetic duality of $N = 2$ and $N = 4$ supersymmetric YM theories, which was linked to dilaton-axion duality in superstring theories. In particular, $N = 2$ YM theories are much more interesting than $N = 4$, since they display both perturbative and non perturbative phenomena.

An important advance was made in 1994 by Seiberg and Witten.[15] They studied the vacuum structure of the $D = 4$, $N = 2$ supersymmetric Yang–Mills theory with gauge group SU(2) and they determined the quantum exact non-perturbative behaviour of the low energy effective Lagrangian. The theory was found to exhibit an *electromagnetic duality,* with its strongly coupled behaviour described by a weakly coupled theory of monopoles. This triggered a remarkable progress in the understanding of non perturbative properties of supersymmetric field theories, but also in the strings and supergravity domain.

For example, it is by now well established that the various perturbatively defined string/supergravity theories are just non perturbative equivalent descriptions of one fundamental, but as yet largely unknown theory, called M-theory, whose low energy Lagrangian is described by $11D$ supergravity.[12]

It is very likely that the beautiful analysis given in[15] was motivated by the previously described results in the string and supergravity framework. In fact, the electromagnetic duality of the supersymmetric Yang–Mills theory is essentially determined by the monodromy group in the complex moduli space, exactly as in supergravity. More precisely, this duality is determined by a monodromy group Γ which is a subgroup of $Sp(2r, \mathbf{Z})$, where r is the rank of the Yang–Mills group G. When $r = 1$ ($G = SU(2)$), Seiberg and Witten identified Γ as the group $\Gamma(2) \subset SL(2, \mathbf{Z}) \simeq \mathrm{Sp}(\mathbf{2}, \mathbf{Z})$ where $\Gamma(2)$ is the discrete group of matrices that are equivalent to the identity modulo 2. It should be no surprise that the computation of these monodromies is strongly reminiscent of the analogous analysis of Calabi–Yau moduli space which we shall report here in some detail, showing that the monodromy group Γ, is a discrete subgroup of $Sp(2n + 2, \mathbf{Z})$ and is related to the three-form cohomology of the CY manifold.
Moreover, the monodromy analysis for the $N = 2$ SYM moduli space is strictly analogous to the one in $N = 2$ supergravity.[8] In both supergravity and SYM we need a restricted form of the Kähler geometry called respectively *Special Kähler Geometry* (SKG in the following) and *Rigid Special Kähler Geometry*, the latter being a particular limit of SKG. These can be related to the Picard–Fuchs differential equations associated to the moduli space. The existence of an holomorphic function $F(z)$, called *prepotential*, characteristic of these geometries will be essential to obtain exact results for the duality group and to connect the perturbative and non perturbative regimes.

4. Calabi–Yau 3-folds and special Kähler geometry

In order to set the stage for the computation of the duality groups, we give a short resume of the Calabi–Yau and Special Kähler geometry in $N = 2$ supergravity. We shall restrict the discussion to moduli spaces of one complex dimensions, since in the general case the differential equations (Picard–Fuchs equations) are quite complicated.

4.1. *Calabi–Yau manifolds*

An n-dimensional Calabi–Yau manifold (CY in the following) can be defined as a compact Kähler manifold whose first Chern class vanishes. This implies that the holonomy is SU(n) instead of U(n), and that the metric is Ricci flat, $R_{i\bar{\jmath}} = 0$. We shall mostly refer to CY of complex dimensions $n = 3$.

In algebraic geometry, the moduli space of a Kähler manifold is parametrised by deformations of its *Kähler class* and of its *complex structure*, while the moduli space of the CY can be viewed as the space of deformations of the Ricci flat metric. If g_{ij} (i=1,...,6) and $g_{ij}+\delta g_{ij}$ are both Ricci flat, one can show that $\delta g_{\alpha\bar{\beta}}$ and $\delta g_{\alpha\beta}$, $(\alpha,\bar{\alpha}=1,\ldots,3)$ satisfy the Lichnerowitz equation. Thus we can associate to $\delta g_{\alpha\bar{\beta}}$ the *harmonic* (1,1)-form

$$\omega_{(1,1)} = \delta g_{\alpha\bar{\beta}}(z;y)dy^{\alpha}\wedge d\bar{y}^{\bar{\beta}} \tag{5}$$

describing the Kähler class deformations, and the *harmonic* (2,1)-form

$$\phi_{\alpha}^{(2,1)} = \Omega_{\alpha\beta}^{(3,0)\bar{\gamma}}(z;y)g_{\bar{\lambda}\bar{\gamma}}(z;y)dy^{\alpha}\wedge dy^{\beta}\wedge d\bar{y}^{\bar{\lambda}} \tag{6}$$

for the complex structure ones. Here z^i denote the moduli, y^i are coordinates of the CY manifold and $\Omega_{\alpha\beta}^{\bar{\gamma}}\equiv\Omega_{\alpha\beta\gamma}g^{\gamma\bar{\gamma}}$ is its unique (3,0)-form. The holomorphic 3-form Ω, together with the (2,1)-forms $\phi_{\alpha}^{(2,1)}$ and their complex conjugates, define the Dolbeault–de Rham cohomology ring of order three. From these definitions (and form the low energy effective theory or SCFT), the moduli space for the CY deformations results in a product space,

$$\mathcal{M}(t,z) = \mathcal{M}_{(1,1)}(t)\times\mathcal{M}_{(2,1)}(z)\,. \tag{7}$$

The factors for (1,1)- and (2,1)- are associated to the scalars in the vector multiplets t^i and in the hypermultiplets z^{α}, respectively. The suffixes (1,1) and (2,1) indicate the (complex) dimensions of $\mathcal{M}_{(1,1)}$ is $h_{(1,1)}$ (the dimension of the cohomology group of $(1,1)$-forms on the given CY manifold), and analogously $h_{(2,1)}$ denotes the dimension of $\mathcal{M}_{(2,1)}$ (corresponding to the dimension of the $(2,1)$-forms of the same CY).

In general, one refers to only one of these spaces, namely to the complex structure moduli space $\mathcal{M}_{(2,1)}$. This is because the physical quantities associated to the complex structure moduli z^{α} like the σ-model metric, the Yukawa couplings etc., are *quantum exact* at the string σ-model level in virtue of some powerful non renormalization theorems: they include all the perturbative and non perturbative (world sheet instantons) corrections to the σ-model tree level result. On the other hand, the metric and couplings for the Kähler class t^i are corrected both at the 1-loop perturbative level and, non perturbatively, by world-sheet instantons.

Interestingly, it was first supposed in SCFT and later discovered in Calabi–Yau geometry, that there exists a duality symmetry between couples of Calabi–Yau spaces: for each CY with homology numbers (1,1) and (2,1) there exists a mirror CY whose homology numbers are exchanged: $(1,1)\to(2,1)$ and $(2,1)\to(1,1)$. This correspondence is named *mirror symmetry.*[17] Mirror symmetry enables us to obtain *quantum* exact results also for the space of (1,1) moduli, in that to compute the geometry of the moduli space of the Kähler class $\mathcal{M}_{(1,1)}$ of a given CY, one just computes the geometry of the moduli space of the complex structures $\mathcal{M}_{(2,1)}$ in its mirror.

Concerning $\mathcal{M}_{(2,1)}$, the metric can be given by the Weyl–Petersson metric

$$g_{\alpha\bar{\beta}}(z,\bar{z}) = \frac{\int_{\mathcal{M}}\phi_{\alpha}\wedge\phi_{\bar{\beta}}}{\int_{\mathcal{M}}\Omega\wedge\bar{\Omega}} = \partial_{\alpha}\partial_{\bar{\beta}}K(z,\bar{z}), \tag{8}$$

where $K(z,\bar{z})$ is the Kähler potential. This metric coincides with the σ-model metric appearing in the kinetic term of the complex structure moduli fields $z(x)$ in the Lagrangian

$$\mathcal{L} = \ldots + g_{\alpha\bar{\beta}}(z,\bar{z})\partial_{\mu}z^{\alpha}\partial_{\nu}\bar{z}^{\bar{\beta}}G^{\mu\nu} + \ldots. \tag{9}$$

The properties of the moduli space of the CY 3-fold are encoded in the *periods.* Starting from $\Omega_{(3,0)}(z)$, they are defined as

$$X^A(z) = \int_{\gamma_A} \Omega^{(3,0)}(z;y); \qquad F_A(z) = \int_{\gamma^B} \Omega^{(3,0)}(z;y) \qquad (A,B = 0,1,\dots n), \tag{10}$$

where γ_A, γ^B are the basis of the third homology group of the CY manifold, satisfying $\gamma_A \cap \gamma^B = \delta_A^B$, $\gamma_A \cap \gamma_B = \gamma^A \cap \gamma^B = 0$. Using in these definitions the relation

$$\int_{\mathcal{M}} \Omega \wedge \frac{\partial \Omega}{\partial X^A} = 0\,, \tag{11}$$

one finds that

$$X^A \partial_B F_A - F_B = 0 \rightarrow X^A F_A = 2F(X), \tag{12}$$

so that $F(X^A(z))$ is a *holomorphic homogeneous function of degree two* and $F_A = \frac{\partial F}{\partial X^A}$, and the periods are not independent. Moreover from (8) one finds that the Kähler potential can be written in terms of the periods as

$$K = -\log \mathrm{i}(X^A \bar{F}_A - \bar{X}^A F_A)\,, \tag{13}$$

which is invariant under symplectic transformations acting on the basis cycles γ_A, γ^B with symplectic metric. This particular Kähler potential describes what is called *Special Kähler geometry* .

Finally, in order to make contact with the physical parameters of the effective $N = 2$ supergravity Lagrangian, we mention that the Yukawa couplings can be given as

$$W_{\alpha\beta\gamma}(z) = \int_{\mathcal{M}} \Omega \wedge \partial_\alpha \partial_\beta \partial_\gamma \Omega = \partial_\alpha X^A \partial_\beta X^B \partial_\gamma X^C \frac{\partial^3 F}{\partial X^A \, \partial X^B \, \partial X^C}. \tag{14}$$

Note that, since the $\mathcal{M}_{(2,1)}$ manifold has complex dimension n, the periods X^A, $(A = 0, 1, \dots, n)$ can be chosen as a set of *projective* coordinates. As affine coordinates we can take $t^a = \frac{X^a}{X^0}$, $(a = 1, \dots, n)$, named *special coordinates.* This reflects, in particular, that the periods F_A are not independent from the X^A, in agreement with our previous remark. Using the t^a coordinates and defining the function $\mathcal{F}(t) = (X^0)^2 F$, the Kähler potential in special coordinates becomes

$$K(t,\bar{t}) = -\log \mathrm{i}\left[2(\mathcal{F} - \bar{\mathcal{F}} - (\partial_a \mathcal{F} + \partial_{\bar{a}} \mathcal{F})(t^a - \bar{t^{a}})\right]\,, \tag{15}$$

while the Yukawa couplings (22) take the simple form $W_{abc}(t) = \partial_a \partial_b \partial_c \mathcal{F}$.

It is known that the periods (X^A, F_A) of a holomorphic d-form of a d-dimensional algebraic curve in $CP(d+1)$ satisfy a system of differential equations of Fuchsian type, called Picard–Fuchs equations (PFE), which can be derived from the equation of the algebraic curve defining the CY.[7] In our case, one starts from the curve of the CY 3-fold in CP(4), $\mathcal{W}(y, \psi_i) = 0$, where ψ_i are the moduli, and the procedure allows us to write the Picard–Fuchs system of linear differential equations. Solving the Picard–Fuchs equations we find the periods X^A, F_A, and then the Kähler potential and the Yukawa couplings from Eqs. (13) and (14). Moreover, what is more important for our purposes, the monodromy of the Picard–Fuchs equations together with the symmetries of the defining polynomial $\mathcal{W}(y, \psi_i) = 0$ of the CY-threefold, allow us to reconstruct the full *duality group* of the moduli space.

4.2. *Special Kähler geometry*

Historically Special Kähler geometry[20] was first defined in $N = 2$ effective Lagrangians as constraints required by $N = 2$ supersymmetry on the Kähler geometry, before Calabi–Yau compactifications were recognized as the right way to reproduce $N = 1$ and $N = 2$ theories from ten dimensions. As already remarked, even if originally arising in the $N = 2$ theory, the Special Kähler constraint must hold also in $N = 1$, when the effective Lagrangian is obtained from ten dimensions by compactification on a Calabi–Yau 3-fold.

Firstly, one defines a *Kähler–Hodge* manifold, relevant for the $N = 1$ effective Lagrangian, as a Kähler manifold such that the Kähler 2-form coincides with the first Chern class: there is a U(1) bundle $L(\mathcal{M}_n)$ such that $c_1(L) = \mathrm{i} g_{\alpha\bar{\beta}} dz^\alpha \wedge d\bar{z}^{\bar{\beta}}$.

A Special Kähler is a restricted Kähler–Hodge manifold. Among many equivalent definitions, the most convenient for our purposes is to define a Kähler–Hodge manifold as *Special* if the following set of differential relations is satisfied:

$$D_\alpha \begin{pmatrix} V \\ U_\beta \\ \bar{U}_{\bar{\beta}} \\ \bar{V} \end{pmatrix} = \begin{pmatrix} U_\alpha \\ -\mathrm{i}\, C_{\alpha\beta\gamma} g^{\gamma\bar{\gamma}} \bar{U}_{\bar{\gamma}} \\ g_{\alpha\bar{\beta}} \bar{V} \\ 0 \end{pmatrix}, \tag{16}$$

where $D_\alpha = \frac{D}{\partial z^\alpha}$ is the Kähler and coordinate covariant derivative, $V(z, \bar{z})$ is a section of a complex line bundle and α, β run on $1, \ldots, n$, n being the (complex) dimensions of the Kähler manifold parameterized by the complex coordinates (moduli) z^α. The covariant derivative D_α on a generic vector field ϕ_β, is defined by

$$\begin{aligned} D_\alpha \phi_\beta &= \left(\partial_\alpha \phi_\beta + \frac{p}{2} \partial_\alpha K\right) - \Gamma^\gamma_{\alpha\beta} \cdot \phi_\gamma, \\ D_{\bar{\alpha}} \phi_\beta &= \left(\partial_{\bar{\alpha}} \phi_\beta + \frac{\bar{p}}{2} \partial_{\bar{\alpha}} K\right) \phi_\beta \end{aligned} \tag{17}$$

where $p, (\bar{p})$ is the Kähler weight of $\phi_\beta(\bar{\phi}_{\bar{\beta}})$, respectively. For the $2n+2$-dimensional vector of Eq. (16) $p = 1/2$. Note that we may always convert a covariantly holomorphic section $\psi(z, \bar{z})$ into a holomorphic section $\tilde{\psi}(z)$: that is $D_{\bar{\alpha}} \psi = 0 \to \tilde{\partial}_{\bar{\alpha}} \psi = 0$, where $\tilde{\psi} = e^{\bar{p}K/2} \psi$ has Kähler weight $(p - \bar{p}, 0)$.

The integrability of (16) gives a restriction on the Riemann tensor

$$R^\delta_{\alpha\bar{\beta}\gamma} = g_{\alpha\bar{\beta}}\, \delta^\delta_\gamma + g_{\gamma\bar{\beta}}\, \delta^\delta_\alpha - C_{\alpha\gamma\epsilon} g^{\epsilon\bar{\epsilon}}\, C_{\bar{\beta}\bar{\delta}\bar{\epsilon}} g^{\delta\bar{\delta}}, \tag{18}$$

and also requires the $C_{\alpha\beta\gamma}$ tensor to be completely symmetric in its three indices, covariantly holomorphic and curl-free on each of them, namely $D_{\bar{\alpha}} C_{\beta\gamma\delta} = 0$, $D_{[\alpha}\, C_{\beta]\gamma\delta} = 0$.

The system (16) can be written in a more compact form as

$$(\mathbf{1}\partial_\alpha - \mathcal{A}_\alpha)\, \mathbf{U} = \mathbf{0}, \tag{19}$$

where $\mathbf{U}$ is the vector $(V, U_\alpha, \bar{U}_{\bar{\alpha}}, \bar{V})^t$, and

$$\mathcal{A}_\alpha = \begin{pmatrix} -\partial_\alpha K & \delta^\beta_\alpha & 0 & 0 \\ 0 & -\delta^\beta_\gamma \partial_\alpha K + \Gamma^\gamma_{\alpha\beta} & -\mathrm{i} C_{\alpha\beta\gamma} g^{\gamma\bar{\gamma}} & 0 \\ 0 & 0 & 0 & g_{\alpha\bar{\beta}} \\ 0 & 0 & 0 & 0 \end{pmatrix}. \tag{20}$$

In the low energy $N = 2$ four-dimensional theory, the defining equations of Special geometry (16) are solved by a $2n+2$-dimensional *holomorphic* vector (X^A, F_A), which transforms as an Sp$(2n+2)$ vector. Thus, all geometric quantities can be expressed in terms of these holomorphic *prepotentials* (X^A, F_A). By direct computation (for example using Bianchi identities in superspace), one can show that the relations (16) are identically satisfied by a Kähler potential given by

$$K = -\log \mathrm{i}(\mathrm{X}^{\mathrm{A}}(\mathrm{z})\bar{\mathrm{F}}_{\mathrm{A}}(\bar{\mathrm{z}}) - \bar{\mathrm{X}}^{\mathrm{A}}(\bar{\mathrm{z}})\,\mathrm{F}_{\mathrm{A}}(\mathrm{z}))\,, \tag{21}$$

which is invariant under symplectic transformations. Moreover, there exists a holomorphic function $F(X^A)$, homogeneous of degree two, such that $F_A = \frac{\partial F}{\partial Z^A}$ and that the Yukawa couplings $W_{\alpha\beta\gamma}(z)$ are also holomorphic and given by

$$W_{\alpha\beta\gamma}(z) = \partial_\alpha X^A \partial_\beta X^B \partial_\gamma X^C \frac{\partial^3 F}{\partial X^A\, \partial X^B\, \partial X^C}. \tag{22}$$

Comparing Eqs. (21) and (22) with Eqs. (13) and (14), one is led to identify the Special geometry sections (X^A, F_A) with the periods of the Calabi–Yau 3-fold. Then, the moduli space has the Special geometry structure, and the Yukawa couplings coincide with the tensor $W_{\alpha\beta\gamma} = e^K C_{\alpha\beta\gamma}$.

5. Picard–Fuchs equations and duality group

5.1. *Derivation of the Picard–Fuchs equations*

Here we recall the procedure to derive the Picard–Fuchs equations from the knowledge of a given algebraic curve, defined by the vanishing of a homogeneous polynomial $W(y; z^\alpha)$ in $CP(d+1)$. Consider the unique holomorphic d-form, which always exists on a CY-d-fold, in the integral representation:

$$\Omega(z) = \oint_\gamma \frac{\omega(y^i)}{W(y^i; z)}\,, \tag{23}$$

where ω is the volume element

$$\omega = \sum (-1)^k y^k dy^1 \wedge \cdots \wedge \hat{dy}^k \wedge \cdots \wedge y^{d+2}, \tag{24}$$

(the hat on the differential means that it must be omitted) and γ is a small 1-dimensional cycle winding the hypersurface $W(y; z) = 0$. One may represent an element of the Hodge cohomology ring $\bigoplus_0^k H^{(d-q,q)}$ as

$$\Omega_A(z) = \oint_\gamma \frac{p_A(y)}{W^{k+1}}\omega, \qquad A = 0, \ldots, \dim H^d, \tag{25}$$

where p_A ($p_0 = 1, \Omega_0 = \Omega$) are homogeneous polynomials of degree $\alpha = k\nu$, ν being the degree of W. The polynomials p_A span a basis in the ring of all polynomials *modulo* the gradient of W. Then the Ω_A span a basis of H^d.

Given these definitions, the period matrix of W is

$$V_A^B = \oint_{\Gamma_B} \frac{p_A(y)}{W^{k+1}}\omega, \tag{26}$$

where Γ_A are a basis of homology cycles of $H_{d+1}(CP(d+1) - W)$.

The differential equations obeyed by the periods V_A^B can be obtained by differentiation with respect to the moduli z^α. Their explicit form is obtained by decomposing the polynomials obtained by differentiation in terms of the basis p_A modulo $\partial_\alpha W$, and further using that the d-form

$$\oint_{\Gamma_B} \frac{1}{W^{k+1}}[kY^i\partial_i W - W\partial_i Y^i]\omega \tag{27}$$

is exact if the Y^i are homogeneous polynomials of degree $k\nu - (d+1)$. In this way one gets the system of Picard–Fuchs equations:

$$\left[\delta_A^B \frac{\partial}{\partial z^\alpha} - (A_\alpha)_A^B\right] V_B^{\hat{C}}(z) = 0, \tag{28}$$

where A_α is a connection matrix. If we are considering a CY 3-fold and n moduli, the capital indices A,B,...run from 1 to $2n+2$, the dimension of the cohomology group $H^{(3)}$, and α runs from 1 to n, the (complex) dimension of the moduli space $\mathcal{M}_{(2,1)}$. For each index $\hat{C}$ the solution of (28) is a column vector, and there are $2n+2$ linearly independent solutions which span the matrix $V_B^{\hat{C}}(z)$.

As a simple example, we may take the cubic torus ($d=1$ Calabi–Yau), where we have

$$W(y^i; z^\alpha) = \frac{1}{3}\left(y_1^3 + y_2^3 + y_3^3\right) - \psi\, y_1 y_2 y_3. \tag{29}$$

We define $\omega_1 = \int_\Gamma \frac{1}{W}\omega$, with the volume 2-form given in (24). Then the periods are $\frac{\partial \omega_1}{\partial \psi} = \omega_2 = \int_\Gamma \frac{y_1 y_2 y_3}{W}\omega$ and $\frac{\partial \omega_2}{\partial \psi} = \omega_3 = \int_\Gamma \frac{x^2 y^2 z^2}{W}\omega$.

Using the identity $(1-\psi)^3 y_1^2 y_2^2 y_3^2 = y_1 y_3^2\left[\psi^2 y_2 \partial_3 W + \psi y_3 \partial_1 W + y_1 \partial_2 W\right]$ and integrating by parts yields

$$(1-\alpha^3)\frac{\partial \omega_2}{\partial \psi} = \psi \int \frac{y_3}{W^2}\omega + 2\psi^2 \int \frac{y_1 y_2 y_3}{W^2}\omega. \tag{30}$$

Using $y_3 = y_3\partial_3 W + \psi y_1 y_2 y_3$ in the first term of the right-hand side and again integrating by parts gives the following Picard–Fuchs equation:

$$\frac{d}{d\psi}\begin{pmatrix}\omega_0\\ \omega_1\end{pmatrix} = \begin{pmatrix} 0 & 1 \\ \frac{\psi}{1-\psi^3} & \frac{3\psi^2}{1-\psi^3}\end{pmatrix}, \tag{31}$$

where ω_0, ω_1 are the periods of the torus. The system (31) can also be transformed into the second order differential equation

$$\left(\frac{d^2}{d\psi^2} - \frac{3\psi^2}{1-\psi^3}\frac{d}{d\psi} - \frac{\psi}{1-\psi^3}\right)\omega_0 = 0. \tag{32}$$

The next example is given by the family of Calabi–Yau 3-folds described by a quintic polynomial in $CP(4)$ with a complex modulus deformation. This is very important, because it was the first one modulus deformation of a CY manifold for which the Picard–Fuchs equation was explicitly solved and the duality group determined. The defining equation in $CP(4)$ is

$$\mathcal{W} = y_1^5 + y_2^5 + y_3^5 + y_4^5 + y_5^5 - \psi y_1^5 y_2^5 y_3^5 y_4^5 y_5^5, \tag{33}$$

where the one modulus deformation is associated to the marginal operator $p = y_1 y_2 y_3 y_4 y_5 y$. A basis for the ring $\mathcal{R}$ is given by $(1, p, p^2, p^3)$, which is actually a subring of all the possible 101 complex structure deformations of the quintic. Since the dimension of the ring is $r=4$ we have a four-dimensional cohomology subring of $H^{(3)}(W)$, and the matrix $\mathcal{A}_A^B$ is a (4×4) matrix in Eq. (28). We shall give the associated PF equation and duality group for these two cases in the next section.

5.2. *The monodromy and duality group*

To uncover the *global geometry* of moduli space, we observe that the *duality group* Γ is related to the modular group of the moduli space. Once we know the defining polynomial $\mathcal{W}$ of the algebraic curve (the CY 3-fold in our case), Γ can be determined by computing the singularity locus of of the curve $\mathcal{W}$ and the monodromy group Γ_M of the embedding space CP(4).
Note that in general the monodromy group is not the complete duality group, as one should also take into account the symmetry group Γ_W of the defining polynomial $\mathcal{W}$ (which is a much easier task).

Thus, the monodromy group of the moduli space coincides with that of the *Picard–Fuchs equations* for the periods (X^A, F_A) of the CY curve $\mathcal{W} = 0$. For one modulus, such the monodromy is defined by the behaviour of a basis of solutions $(f_1(z), f_2(z), \ldots, f_n(z))$ of a $n - th$-order differential equation at a point z_1, under analytic continuation along a closed loop around a singularity z_0. This brings to a new solution at z which must be expressible as a linear combination of the basis:

$$(f_1(z), f_2(z), \ldots, f_n(z)) \to \left(\hat{f}_1(z), \hat{f}_2(z), \ldots, \hat{f}_n(z)\right) = (f_1(z), f_2(z), \ldots, f_n(z))\, A(z), \tag{34}$$

where the $n \times n$ non-singular matrix $A(z)$ defines the monodromy around z_1. If the equation has r singular points we obtain r monodromy matrices $A(z_1), A(z_2), \ldots, A(z_r)$ and if we compose closed loops around z_i and z_j, there corresponds a matrix $(A(z_1)\, A(z_j))$. More generally, $A(z_1), A(z_2), \ldots, A(z_r)$ generate a group, the *monodromy group* Γ_M of the differential equation. Moreover $\mathbf{1}$ is the loop contractible to a point and A^{-1} is the loop obtained running z in the opposite direction. This procedure was first applied in the seminal paper of Candelas and de la Ossa,[17] for a Calabi–Yau 3-fold described by the quintic homogeneous polynomial W in CP(4) given in (33). They were able to solve exactly the Picard–Fuchs equations and thus to find the monodromy and the duality group of the 1-dimensional moduli space of $\mathcal{W}$, given in (33).

The computation involves several steps and for simplicity we shall display them here for a one dimensional Calabi–Yau, namely a *torus*, which in essence is analogous to the quintic equation analysed in Ref. 17.

The Picard–Fuchs equation for the cubic torus given in Eq. (31) as a system of two ordinary differential equations, or alternatively by the second order differential Eq. (32). This exhibits as singular regular points $\psi = 1, \alpha, \alpha^2$, and $\psi = \infty$, where $\alpha = \exp 2\pi i/3$. It is sufficient to compute the monodromy matrix around $\psi = 1$ since the matrices T_1, T_2 around α, α^2 can be obtained from T_0 by conjugation with the matrix $\mathcal{A}$ representing $\psi \to \alpha\psi$, namely

$$\begin{pmatrix} U_1 \\ U_2 \end{pmatrix} \to \begin{pmatrix} 1 & 0 \\ 0 & \alpha \end{pmatrix} \begin{pmatrix} U_1 \\ U_2 \end{pmatrix} \tag{35}$$

and we can rescale the basis to have $\det \mathcal{A} = 1$ since we are interested only in projective representations:

$$\mathcal{A} = \begin{pmatrix} \alpha^{-1/2} & 0 \\ 0 & \alpha^{1/2} \end{pmatrix}. \tag{36}$$

Using the matrix $\mathcal{A}$ we have

$$T_1 = \mathcal{A}\, T_0\, \mathcal{A}^{-1}; \qquad T_2 = \mathcal{A}^2\, T_0\, \mathcal{A}^{-2}. \tag{37}$$

Furthermore a loop enclosing $1\alpha, \alpha^2, \infty$ is contractible, so that

$$T_\infty T_2 T_1 T_0 = 1 \to T_\infty = (T_2 T_1 T_0)^{-1}. \tag{38}$$

To compute the monodromy around $\psi = 1$ it is convenient to use in Eq. (32) the variable $z = \psi^3$. Applying the method previously described and nicely summarised in,[22] one obtains:

$$\left(9z(1-z)\frac{d^2}{dz^2} - (6-15z)\frac{d}{dz} - 1\right)\omega = 0. \tag{39}$$

This is a hypergeometric functions of parameters $(1/3, 1/3, 2/3)$ with two solutions around $z = \psi = 0$ given by

$$U_1 = \frac{\Gamma(1/3)}{\Gamma(2/3)} F(1/3, 1/3, 2/3; \psi^3) \tag{40}$$

$$U_2 = \frac{\Gamma(2/3)}{\Gamma(4/3)} \psi F(2/3, 2/3, 4/3; \psi^3). \tag{41}$$

This Fuchsian equation has monodromy generators around $z = \psi^3 = 0$ and $z = \psi^3 = 1$. The monodromy generator T_0 around $z = 0$ can be computed using the properties of the hypergeometric function. By analytic continuing around $z = 1$, the two solution acquire a logarithmic factor $log(1-z)$ due to the coincidence of the roots of the indicial equation at $z = \psi^3 = 1$. Using finally the Kummer relations for hypergeometric functions and performing a loop around $z = 1$ the monodromy around is easily computed. One finds

$$T_0 = \begin{pmatrix} 1 - \mathrm{i}\sqrt{3} & \mathrm{i}\sqrt{3} \\ -\mathrm{i}\sqrt{3} & 1 + \mathrm{i}\sqrt{3} \end{pmatrix} \tag{42}$$

and from Eqs. (37) and (36) we also get

$$T_1 = \begin{pmatrix} 1 - \mathrm{i}\sqrt{3} & \alpha^{-1}\mathrm{i}\sqrt{3} \\ -\alpha\mathrm{i}\sqrt{3} & 1 + \mathrm{i}\sqrt{3} \end{pmatrix} \qquad T_2 = \begin{pmatrix} 1 - \mathrm{i}\sqrt{3} & \alpha^{-2}\mathrm{i}\sqrt{3} \\ -\alpha^2\mathrm{i}\sqrt{3} & 1 + \mathrm{i}\sqrt{3} \end{pmatrix}. \tag{43}$$

We now observe that the modular or duality group for the torus is known a priori to be $\mathrm{SL}(2; \mathbf{Z})$, and therefore the monodromy group must be one of its subgroups. Since the matrices T_i are not valued in $\mathbf{Z}$, there must exist a change of basis giving matrices with integer entries. One finds that in an appropriately chosen basis the monodromy matrices take the form:

$$T_0 = \begin{pmatrix} 1 & 3 \\ 0 & 1 \end{pmatrix}; \qquad T_1 = \begin{pmatrix} -5 & 12 \\ -3 & 7 \end{pmatrix}; \qquad T_2 = \begin{pmatrix} -2 & 3 \\ -3 & 4 \end{pmatrix}. \tag{44}$$

One recognizes that the monodromy group Γ_M is isomorphic to $\Gamma(3)$, the group of matrices equivalent to the identity modulo 3 which is a normal subgroup of the modular group: a well known result in mathematics. The quotient group is a symmetry group Γ_W of the defining Eq. (31), generated by the transformations $\psi \to \alpha\psi$ and $\psi \to \frac{\psi+2}{\psi-1}$. Therefore $\Gamma/\Gamma_M = \Gamma_W$.

The computation of the duality group for the torus just sketched is intended to be propaedeutic to the analogous one for the 1-modulus deformed Calabi–Yau 3-fold described by the quintic Eq. (33).

The corresponding 4th order Picard–Fuchs equation is

$$\left(\frac{d^4}{d\psi^4}-\frac{10\psi^4}{1-\psi^5}\frac{d^3}{d\psi^3}-\frac{25\psi^3}{1-\psi^5}\frac{d^2}{d\psi^2}-\frac{15\psi^2}{1-\psi^5}\frac{d}{d\psi}-\frac{\psi}{1-\psi^5}\right)V=0\,. \tag{45}$$

According to our previous discussion, the solutions of this equation represent the periods of the holomorphic uniquely defined 3-form $\Omega(z)$ in (10). Since the periods are not functionally independent, (12), as a counterpart there is a functional relation satisfied by the coefficients of the differential Eq. (45):

$$W_3 = a_1 - \frac{da_2}{d\psi} - \frac{1}{2}\frac{d^2 a_3}{d\psi^2} = 0, \tag{46}$$

where we have denoted with a_k the coefficients of $\frac{d^k}{d\psi^k}$. This condition is an invariant statement since, under ψ reparametrization, W_3 transforms as a covariant tensor of order three. For many moduli this statement generalizes to the property that the connection A_α of the associated linear system is gauge equivalent to a Sp(2n+2) connection. In our case, in the gauge where A_α is valued in Sp(4) ($n = 2$, since we have one electric and one magnetic field strength) we have

$$\hat{V} = \begin{pmatrix} V_1 & V_2 & V_3 & V_4 \\ x & x & x & x \\ x & x & x & x \\ x & x & x & x \end{pmatrix} \in \mathrm{Sp}(4), \tag{47}$$

where (V_1, V_2, V_3, V_4) is (X^0, X^1, F_1, F_0). To compute the duality group Γ, we observe that that the group Γ_W of the quintic equation $W = 0$ (33) is invariant under $\mathcal{A} : \psi \to \alpha\psi$ with $\alpha = e^{\frac{2\pi i}{5}}$, since it can be undone by rescaling one of the coordinates y_i.

The monodromy group Γ_M of the Picard–Fuchs equation will be represented by 4×4 matrices on the four periods V_1, V_2, V_3, V_4 solution of (45) and the same will also be true for Γ_W since the transformation $\mathcal{A} : \psi \to \alpha\psi$ leaves the differential operator invariant and therefore can only induce linear transformation on the solutions V_1, V_2, V_3, V_4. Moreover, by using the gauge (47) both Γ_W and Γ_M can be represented by Sp(4) matrices.

Let us sketch the procedure to compute Γ_M given in Ref. 17. The differential equation has regular singular points at $\psi = \alpha^k$, $k = (0, 1, 2, 3, 4)$ and $\psi = \infty$. The corresponding monodromy matrices are denoted as T_k. As for the torus, if we call $\mathcal{A}$ the 4×4 matrix representing $\psi \to \alpha\psi$, it is sufficient to compute the monodromy around $\psi = 1$ owing to the relations:

$$T_k = \mathcal{A}^{-1} T_0 \mathcal{A}; \qquad T_\infty T_4 T_3 T_2 T_1 T_0 = 1. \tag{48}$$

Again as for the torus, it is convenient to set $z = \psi^{-5}$ which transforms the quintic equation as follows:

$$\left(\frac{d^4}{dz^4} - \frac{2(4z-3)}{z(1-z)}\frac{d^3}{dz^3} - \frac{72z-35}{5z^2(1-z)}\frac{d^2}{dz^2} - \frac{24z-5}{5z^3(1-z)}\frac{d}{dz} - \frac{24}{625z^3(1-z)}\right)V(z) = 0, \tag{49}$$

which has singular points at $z = 0, 1, \infty$. A solution at $z = 0$; $(\psi = \infty)$ is given by the generalized hypergeometric function

$$\omega_0 = {}_4F_3\left(\frac{1}{5}, \frac{2}{5}, \frac{3}{5}, \frac{4}{5}; 1, 1, 1; \psi^{-5}\right). \tag{50}$$

Using known properties of the hypergeometric functions, this solution around $\psi = \infty$ can be continued around $\psi = 0$. Moreover, since $\mathcal{A} : \psi \to \alpha\psi$ leaves invariant the differential operator, we can construct, using the notation of,[17] five other solutions:

$$\omega_j(\psi) = \omega_0(\alpha^j \psi)\,, \tag{51}$$

which can be shown to satisfy $\sum_j \omega_j(\psi) = 0$. On the basis $(\omega_2, \omega_1, \omega_0, \omega_4)$ one can then compute the monodromy generator $\mathcal{A}$. Again using analytic continuation around $z = 1$ where the differential equation has a double root, one may also compute in the same basis the T_0 generator. Finally one performs a change of basis on the $(\omega_2, \omega_1, \omega_0, \omega_4)$ to obtain a symplectic basis, so that the generator are given by Sp(4) matrices and one obtains:

$$\mathcal{A} = \begin{pmatrix} 1 & -1 & -5 & 3 \\ 0 & 1 & 8 & 5 \\ 1 & -1 & -4 & 3 \\ 0 & 0 & 1 & 1 \end{pmatrix}, \qquad T_0 = \begin{pmatrix} 1 & 0 & 0 & 0 \\ 0 & 1 & 0 & 0 \\ -1 & 0 & 1 & 0 \\ 0 & 0 & 0 & 1 \end{pmatrix}, \tag{52}$$

which act on the right of the vector $(V_1, V_2, V_3, V_4) = (X^A, F_A), (A = 0, 1)$.

6. Seiberg and Witten approach to $N = 2$ SYM

In this section we shall apply the above methods to the computation of the duality group in $N = 2$ supersymmetric Yang–Mills theories, introduced in section 4. We start by reviewing in a nutshell the approach used by Seiberg and Witten in Ref. 15 to uncover the exact quantum moduli space for the $N = 2$ supersymmetric Yang–Mills theory. This short resumè is partly based on Ref. 18, while we refer to the original paper for all details and for a thorough discussion of the particle spectrum.

The $D = 4$, $N = 2$ Yang–Mills theory with gauge group SU(2) is, like supergravity, an effective theory, containing a gauge vector multiplet (A_μ, λ, ϕ). The scalar potential in the Lagrangian is

$$V(\phi) = Tr[\phi, \phi^\dagger], \tag{53}$$

and it vanishes when ϕ belongs to the U(1) subgroup of SU(2), namely $\phi = a\,\sigma^3$. Hence the complex parameter a is a *modulus* describing a 1-dimensional manifold $\mathcal{M}$, namely the complex plane. It is convenient to redefine the modulus as $u = 2a^2$. Classically, this theory becomes singular when $u = 0$ since at this point the $W^\pm$ vector bosons, which where integrated out as massive excitations at the scale of the Higgs vacuum expectation value, would become massless. The bosonic Lagrangian reads

$$\mathcal{L}_{YM} = \mathrm{Im}(\tau)\mathrm{F}_{\mu\nu}\mathrm{F}^{\mu\nu} + \mathrm{Re}(\tau)\mathrm{F}_{\mu\nu}{}^*\mathrm{F}^{\mu\nu} + \mathrm{Im}(\tau)\partial_\mu \mathrm{a}\,\partial^\mu \bar{\mathrm{a}} + \ldots \tag{54}$$

where $\tau = \frac{\theta(a)}{\pi} + \mathrm{i}\frac{8\pi}{g^2(a)}$. An important point is that supersymmetry fixes τ to be the second derivative of an *holomorphic* function $\mathcal{F}(a)$, namely $\tau = d^2\mathcal{F}/da^2$.
At the quantum level, the classical (complex) coupling constant $\frac{\theta(a)}{\pi} + \mathrm{i}\frac{8\pi}{g^2(a)}$ acquires two type of corrections: one is a perturbative correction at 1-loop, and the other is non perturbative being given by an infinite sum of instanton corrections. The holomorphic function $\mathcal{F}(a)$ is given by

$$\mathcal{F} = \frac{1}{2}\tau_0\, a^2 + \mathrm{i}a^2 \log\frac{a^2}{\Lambda^2} + \frac{1}{2\pi\mathrm{i}}\sum_1^\infty c_n \left(\frac{\Lambda}{a}\right)^{4n}, \tag{55}$$

where Λ is the dynamically generated scale at the quantum level.

Using physical arguments, Seiberg and Witten proposed that in the quantum corrected theory, the moduli space must acquire two new singularities at $u = \pm\Lambda^2$, while the $u = 0$ singularity disappears. Since in the classical case $u = 0$ is the point where the $W^{\pm}$ bosons have vanishing mass, we must then expect that at the new singularity points $u = \pm\Lambda^2$ *new particle excitations* must become *massless.* Taking into account the 't Hooft–Polyakov monopole solution, they concluded that these massless excitations should be *monopoles.* Moreover, using semi classical considerations, one can recognize that the monopoles are BPS states, that is they saturate the bound $m^2 \geq |Z|^2$, where Z is the central charge of the $N = 2$ supersymmetry algebra and that the BPS monopoles must belong to the hypermultiplet sector. Writing the central charge as

$$Z = e\, a + g\, a_D, \tag{56}$$

one can consider a_D as the scalar field dual to a, thus belonging to the magnetic vector multiplet $(A_{\mu D}, a_D)$. In the neighborhood of the singularities $u = \pm\Lambda^2$, physics will be described in terms of the dual vector multiplet. Introducing the quantum corrected holomorphic dual function $\mathcal{F}_D(a_D)$, such that $a_D = \frac{d\mathcal{F}}{\partial a}$, one can see that when $a_D \to 0$ the theory is weakly coupled. One then shows that, when the electric charge e is zero, the monopole central charge, and hence the mass, vanishes as $a_D \to 0$, that is when we reach $u = \pm\Lambda^2$. Thus, in the neighborhood of the point $u = \infty$ we have a theory which at weak coupling is well described by the Lagrangian (54). Moreover, it is completely fixed by the holomorphic function $\mathcal{F}(a)$, where all the 1-loop quantum corrections and the infinite sum of instanton corrections, are well defined in terms of the modulus a. However this description ceases to be valid outside the radius of convergence of the instanton sum and, near the two singularities $u = \pm\Lambda^2$. There, we are in the strong coupling regime where the natural variable to use is the magnetic dual a_D and a holomorphic function F_D. This means that $\mathcal{F}$ and $\mathcal{F}_D$ are not functions but *sections*, and to fix completely the theory we must patch together the local behaviour in the different neighborhoods, to obtain a consistent global description in the whole moduli space $\mathcal{M}$. Since the 1-loop term correction to F^a is logarithmic, a loop around the singular points would give rise to a monodromy matrix T acting on the two sections a and a_D:

$$\begin{pmatrix} a_D \\ a \end{pmatrix}' = T \begin{pmatrix} a_D \\ a \end{pmatrix}. \tag{57}$$

Setting for symplicity $\Lambda^2 = 1$, we have three monodromy matrices $(T_\infty; T_{\pm 1})$ in correspondence of the three singularities $(u = \infty; u = \pm 1)$. From the asymptotic behaviour around infinity one finds

$$\begin{pmatrix} a_D(u) \\ a(u) \end{pmatrix} \simeq \begin{pmatrix} \frac{i}{\pi}\sqrt{2u}\,\log u \\ \sqrt{u}/2 \end{pmatrix} \tag{58}$$

from which we find that a loop around $u = \infty$ gives

$$\begin{pmatrix} -1 & -2 \\ 0 & -1 \end{pmatrix}. \tag{59}$$

Near $u = 1$ we have

$$\begin{pmatrix} a_D(u) \\ a(u) \end{pmatrix} \simeq \begin{pmatrix} c_0(u-1) \\ a_0 + \frac{i}{\pi} a_d \log a_D \end{pmatrix} \tag{60}$$

giving the monodromy generator

$$T_1 = \begin{pmatrix} -1 & 2 \\ -2 & 3 \end{pmatrix}, \tag{61}$$

and since a loop encircling the three singularities is contractible to a point, the third matrix is found from the vanishing cycle relation $T_{\Lambda^2} T_{-\Lambda^2} T_\infty = 1$, namely

$$T_{-1} = \begin{pmatrix} -1 & 0 \\ -1 & 3 \end{pmatrix}. \tag{62}$$

The group generated by these matrices is the group $\Gamma_2 \subset Sp(2, \mathbf{Z})$, that is the group of matrices equivalent to the identity modulo 2.

6.1. *Duality in $N = 2$ Super–Yang–Mills theories*

In this section we show that the results of Seiberg and Witten for the $N = 2$ SYM theory can be equivalently obtained by the approach based on the Picard–Fuchs, as for Calabi–Yau moduli space. In fact the computation of the duality group becomes quite efficient if one recognizes the $N = 2$ Yang–Mills theory as a rigid limit of $N = 2$ supergravity.

We consider generic $N = 2$ SYM theory with n vector multiplets, with SU(n + 1) gauge group. (A^A, X^A), with $(A = 1, \ldots, n)$ are the gauge and scalar fields of the vector multiplets (the gaugini do not play any role here).

For generic flat directions of the Yang–Mills scalar potential, the gauge group is broken to $\mathrm{U}(1)^n$ and the scalar fields become *moduli* parameterizing a complex manifold:

$$V(X) = 0 \rightarrow Tr[X^A, X_A] = 0 \rightarrow \begin{pmatrix} X_1 & 0 & 0 \\ 0 & \ddots & 0 \\ 0 & 0 & X_n \end{pmatrix} \tag{63}$$

with $\sum X^A = 0$. There is a residual symmetry under the Weyl group: $X^A \rightarrow -X^A$. The scalar fields X^A describe a Kählerian sigma model

$$G_{A\bar{B}} \partial X^A \partial \bar{X}^B \tag{64}$$

with metric $G_{A\bar{B}} = \mathrm{i}\partial_A \partial_{\bar{B}} \left(F_C \bar{X}^C - \bar{F}_C X^C \right)$. These coordinates X^A are the analogue of the "special coordinates" of Calabi–Yau moduli spaces since here the number of variables n equals the dimension of $\mathcal{M}_n$. To promote the metric from special to generic coordinates z^α, $(\alpha = 1, \ldots, n)$, we introduce n holomorphic functions $X^A(z)$.

The general discussion for many moduli is based on several observations. First, in rigid supersymmetry the scalar fields X^A span a pure Kähler rather than a Kähler–Hodge geometry. Second, in rigid $N = 2$ there is a restriction on the Kähler geometry, similarly to that in $N = 2$ supergravity, which gives rise to the *Rigid Special geometry* (RSKG). Specifically, if one introduces the $2n$-dimensional holomorphic vector $V = (X^A, F_A)$, the supersymmetry constraints on the $N = 2$ YM theory imply the identities

$$D_\alpha V = V_\alpha \tag{65}$$

$$D_\alpha V_\beta = -\mathrm{i} W_{\alpha\beta\gamma} V^\gamma; \qquad V^\gamma = g^{\gamma\bar{\gamma}} V_{\bar{\gamma}} \tag{66}$$

$$D_\alpha \bar{V}_{\bar{\gamma}} = 0, \tag{67}$$

where D_α is the covariant derivative similar to (16), but without the Kähler connection $\partial_\alpha K$, as the geometry is restricted to a Kähler rather than a Kähler–Hodge manifold, without the extra U(1) bundle.
The integrability conditions of these equations gives the curvature

$$R_{\alpha\bar{\beta}\gamma\bar{\delta}} = -W_{\alpha\beta\sigma}W_{\bar{\beta}\bar{\delta}\bar{\sigma}}G^{\sigma\bar{\sigma}}, \tag{68}$$

which, in comparison with the supergravity case, misses the term $g_{\alpha\bar{\beta}}\,\delta^\delta_\gamma + g_{\gamma\bar{\beta}}\,\delta^\delta_\alpha$ as a result of the absence of the Kähler connection in the covariant derivative. The solution of the defining Eq. (65) can be given in terms of the holomorphic vector $(X^A(z), F_A(z))$, and by a Kähler potential of the form

$$K(z,\bar{z}) = -i(F_A\bar{X}^A - \bar{F}_A X^A)\,. \tag{69}$$

One then shows that the prepotentials F_A are related to the X^A and can be derived from a holomorphic function $F(X)$ as $F_A = \frac{\partial F}{\partial X^A}$, as in supergravity. Analogously, the Yukawa couplings in (68) are

$$W_{\alpha\beta\gamma} = \partial_\alpha X^A \partial_\beta X^B \partial_\gamma X^C \partial_A \partial_B \partial_C F\ , \tag{70}$$

so that in the special coordinates X^A they become $W_{ABC} = \partial_A\partial_B\partial_C F(X)$.
The metric in special coordinates X^A, reads

$$G_{AB} = \partial_A \partial_{\bar{B}}\, K = -2\mathrm{Im}F_{AB}\,, \tag{71}$$

while in a generic frame it is

$$G_{\alpha\bar{\beta}} = \partial_\alpha\partial_\beta K(z,\bar{z}) = \mathrm{i}\left(\partial_\alpha F_A\,\partial_{\bar{\beta}}\bar{X}^A - \partial_\alpha X_A\,\partial_{\bar{\beta}}\bar{F}^A\right)\,. \tag{72}$$

The $N = 2$ SYM Lagrangian in this set up also contains, similarly to $N = 2$ supergravity 1, the field dependent coupling matrix $\mathcal{N}_{AB}$, which can be computed from the two entries of Eq. (91) as $\mathcal{N}_{AB} = e^C_\alpha F_{AC} e^\alpha_{\bar{B}} = F_{AB}$. The general result then is

$$L_{YM} = \mathrm{ImF_{AB}}\partial_\mu \mathrm{X^A}\partial^\mu \mathrm{\bar{X}^B} + \mathrm{ImF_{AB}F^A_{\mu\nu}F^{B\mu\nu}} + \mathrm{ReF_{AB}F^A \wedge F^B} - \mathrm{V(X)} + \ldots\,. \tag{73}$$

The same result can also be derived using the CY geometry. We start from a 1-fold, namely a (complex) one-dimensional curve in $CP(2)$, that is a Riemann surface. If ω is a 1-form of the $H^{(1)}$ cohomology, we define the periods as for the CY 3-fold:

$$\oint_{\gamma_A} \omega = X^A(u); \qquad \oint_{\gamma_A} \omega = F_A(u), \tag{74}$$

where γ_A, γ^A are a basis of one-dimensional cycles. The Kähler potential and the metric are obtained as

$$K(u,\bar{u}) = \mathrm{i}\oint_{\mathcal{M}} \omega \wedge \bar{\omega} = \mathrm{i}(X^A\bar{F}_A - \bar{X}^A\,F_A), \tag{75}$$

$$g_{\alpha\bar{\beta}} = \mathrm{i}\oint_{\mathcal{M}} \partial_\alpha\omega \wedge \partial_{\bar{\beta}}\bar{\omega} = \mathrm{Im}F_{AB}\,\partial_\alpha X^A \partial_{\bar{\beta}} X^B > 0 \tag{76}$$

and the *period matrix* $\mathcal{N}_{AB}$ playing the role of a field dependent coupling constant of the scalars to the vector field strengths (see Eq. (73)) is

$$\mathcal{N}_{AB} = \oint_{\gamma^A} \partial_A\omega = F_{AB}, \tag{77}$$

reproducing the results obtained by Rigid Special geometry.

The important issue now is to have an equation that allows us to compute the periods. Like in supergravity, we perform a gauge transformation on the system (65) to transform it into a purely holomorphic form, *i.e.* a system of equations where the covariant derivative D_α is replaced by its purely holomorphic part $\hat{D}_\alpha$, with a purely holomorphic Christoffel connection. Using the flat holomorphic vielbein

$$e_\alpha^A = \partial_\alpha X^A, \tag{78}$$

which is a $n \times n$ matrix, we have

$$\hat{\Gamma}^\gamma_{\alpha\beta}(z) = (\partial_\beta e^a_\alpha)\, e_a^{-1\gamma}, \tag{79}$$

whose associated curvature vanishes. Moreover V^γ is replaced by a holomorphic vector

$$V^\alpha = (0, e^{-1\alpha}_{A}). \tag{80}$$

Since the first equation is left invariant by constant translations $V \to V + c$, it is possible to consider (V, V_α, V^α) as a $(2n+1)$-dimensional vector, namely:

$$V = (1, X^A, F_A) \tag{81}$$

$$V_\alpha = (0, e^A_\alpha, e^B_\alpha F_{AB}) \tag{82}$$

$$V^\alpha = (0, 0, e^{-1\alpha}_{A}). \tag{83}$$

In terms of the $(2n+1) \times (2n+1)$ solution matrix $\mathcal{V} = \left(V, V_\beta, V^\beta\right)^t$ the holomorphic system can be written as

$$\mathcal{D}_\alpha \mathcal{V} = 0\ , \tag{84}$$

with

$$\mathcal{D}_\alpha = \partial_\alpha - \mathcal{A}_\alpha \tag{85}$$

and the flat connection $\mathcal{A}_\alpha$ is given by

$$\mathcal{A}_\alpha = \begin{pmatrix} 0 & \delta^\gamma_\alpha & 0 \\ 0 & \hat{\Gamma}^\gamma_{\alpha\beta}(z) & (W_\alpha)_{\beta\gamma}(z) \\ 0 & 0 & -\hat{\Gamma}^\beta_{\alpha\gamma}(z) \end{pmatrix}. \tag{86}$$

Note that there is a $ISp(2n)$ acting on the right-hand side of V (or $\mathcal{V}$) represented as $\begin{pmatrix} 1 & 0 \\ C & M \end{pmatrix}$ with $M \subset \mathrm{Sp}(2n)$. This follows from the fact that the submatrix

$$\begin{pmatrix} \hat{\Gamma}^\gamma_{\alpha\beta} & W_{\alpha\beta\gamma} \\ 0 & -\hat{\Gamma}^\beta_{\alpha\gamma} \end{pmatrix} \tag{87}$$

is valued in the Lie algebra of $Sp(2n)$ with respect to the metric

$$Q = \begin{pmatrix} 0 & -1_n \\ 1_n & 0 \end{pmatrix}. \tag{88}$$

In special coordinates, $e^A_\alpha = \delta^A_\alpha$, $\hat{\Gamma} = 0$ so that the connection $\mathcal{A}_\alpha$ reduces to

$$\mathcal{A}_\alpha = \begin{pmatrix} 0 & \delta^\gamma_\alpha & 0 \\ 0 & 0 & W_{\alpha\beta\gamma}(z) \\ 0 & 0 & 0 \end{pmatrix} \tag{89}$$

which is nilpotent of degree three ($(\mathcal{A}_\alpha)^3 = 0$).

Equation (84) can be rewritten as a system of third order differential equations for the upper component of $\mathcal{V}$,

$$\hat{D}_\alpha (W^{-1\hat{\gamma}})^{\sigma\beta} \hat{D}_{\hat{\gamma}} \partial_\beta V = 0 \ , \tag{90}$$

(where $\hat{\gamma}$ is a priori not summed over). Since the first equation in (80) can be used to just start with V_α, it can be reduced to a second order differential equation for $V_\beta = \partial_\beta V$. We can then delete the first entry in (83)

$$V_\alpha = \left(e^A_\alpha, e^B_\alpha F_{AB}\right) \tag{91}$$

$$V^\alpha = \left(0, e_A^{-1\alpha}\right), \tag{92}$$

and write the connection as

$$\mathcal{A}_\alpha = \begin{pmatrix} \hat{\Gamma}^\gamma_{\alpha\beta} & W_{\alpha\beta\gamma} \\ 0 & -\hat{\Gamma}^\beta_{\alpha\gamma} \end{pmatrix} . \tag{93}$$

This reduces in special coordinates to

$$\mathcal{A}_\alpha = \begin{pmatrix} 0 & (W_\alpha)_{\beta\gamma} \\ 0 & 0 \end{pmatrix} \tag{94}$$

which is then nilpotent of degree two, and $W_{\alpha\beta\gamma}$ is an n-dimensional abelian subalgebra.

At this point we can connect this to Seiberg and Witten's analysis. First of all, we restrict ourselves to one modulus corresponding to the gauge group SU(2), so that we can omit the index A in the prepotentials: (X, F). Then, we have to identify the SW modulus a and its dual a_D, with the two entries of Eq. (91) which reduce to $(e, e\frac{d^2F}{dX^2})$, while Eq. (92) becomes $(0, e^{-1})$. Note that the coordinates (a, a_D) of SW are "special coordinates", while a generally covariant description contains the general modulus z.

Now for one modulus ($n = 1$), Eq. (90) becomes

$$(\hat{D}W^{-1}\hat{D}\hat{\partial})V = 0 \ , \tag{95}$$

and setting $U = \partial V$ one finds

$$(\partial + \hat{\Gamma})W^{-1}(\partial - \hat{\Gamma})U = 0 \ . \tag{96}$$

This yields a second order equation

$$\partial^2 U + a_1 \partial U + a_0 U = 0 \tag{97}$$

with

$$a_1 = -\partial \log W \tag{98}$$

$$a_0 = \partial \log W\hat{\Gamma} - \partial\hat{\Gamma} + \hat{\Gamma}^2; \qquad \hat{\Gamma} = \partial \log e \, . \tag{99}$$

Knowing a_1, a_0, one can compute the Yukawa coupling $W \equiv W_{111}$ and e, ($e = 1$ in special coordinates). Therefore the general solution in the one parameter case is

$$U = \partial_z V = (\partial_z X, \partial_{zz} F) = (e, e\frac{d^2F}{dX^2}) \tag{100}$$

where the last expression is given in special coordinates and e is the vielbein component that here plays the role of a rescaling factor. Taking the ratio of the two solutions one recovers $\tau = \frac{\partial^2 F}{\partial X^2}$. Using the τ variable we see that the general metric Eq. (72) becomes

$$G_{z\bar{z}} = |e(z)|^2 \mathrm{Im}\tau = |e(z)|^2 \frac{\partial^2 F}{\partial X^2} \equiv G_{A\bar{B}} \tag{101}$$

so that the $N = 2$ supersymmetric $N = 2$, Yang–Mils Lagrangian takes the form used in Ref. 15 in special coordinates. If we denote $\frac{d^2F}{dX^2} = \tau$ (73) becomes

$$L_{YM} = \mathrm{Im}\,\tau\partial_\mu \mathrm{X}\partial^\mu \bar{\mathrm{X}} + \mathrm{Im}\tau\,\mathrm{F}_{\mu\nu}\mathrm{F}^{\mu\nu} + \mathrm{Re}\tau\mathrm{F}_{\mu\nu} * \mathrm{F}^{\mu\nu} - \mathrm{V(X)} + \ldots \tag{102}$$

which is identical to the Lagrangian used in Ref. 15, namely Eq. (54), with the substitution,[b] $a \to X$.

$$\mathcal{L}_{(SYM)} = \mathrm{Im}\tau\,(\partial_\mu \mathrm{a}\partial^\mu \mathrm{a} + \mathrm{F}_{\mu\nu}\mathrm{F}^{\mu\nu}) + \mathrm{Re}\tau\mathrm{F}_{\mu\nu}{}^*\mathrm{F}^{\mu\nu}. \tag{103}$$

As an illustration, let us derive the differential equation for the particular one parameter case of Seiberg and Witten. Consider the family E_u of genus one Riemann surfaces

$$y^2 = (x+1)(x-1)(x-u)\ , \tag{104}$$

which, in homogeneous coordinates ($x \to \frac{x}{z}, y \to \frac{y}{z}$) can be described by the vanishing of the homogeneous polynomial $\mathcal{W}$ in CP^2,

$$\mathcal{W}(x,y,z) = -zy^2 + x(x^2 - z^2) - u\ z(x^2 - z^2). \tag{105}$$

For convenience, we change variables to ($x \to x + z, z \to x - z$), obtaining

$$\mathcal{W} = -(x-z)y^2 + xz(x-z) - u\ xz(x-z)\ . \tag{106}$$

The differential equation associated to (106) can now be derived using the standard techniques described in subsection 6.1. Define the integrals

$$U_0 = \int \frac{\omega}{\mathcal{W}}; \qquad U_1 \equiv \frac{dU_0}{du} = \int \frac{\omega}{\mathcal{W}^2} xz(x-z) \tag{107}$$

where ω is a volume form, which form a basis of the cohomology $H^1(E_u)$. By differentiating under the integral sign and using the "vanishing relations" $\frac{\partial\mathcal{W}}{\partial x} = \frac{\partial\mathcal{W}}{\partial y} = \frac{\partial\mathcal{W}}{\partial z} = 0$, one can show that the vector (U_0, U_1) satisfies a regular, singular matrix differential equation

$$\frac{d}{du}\begin{pmatrix} U_0 \\ U_1 \end{pmatrix} = \begin{pmatrix} 0 & 1 \\ \frac{2u}{(1-u^2)} & \frac{1}{4(1-u^2)} \end{pmatrix}\begin{pmatrix} U_0 \\ U_1 \end{pmatrix}\ , \tag{108}$$

which is equivalent to the second order differential equation

$$\frac{d^2U_0}{du^2} - \frac{2u}{(1-u^2)}\frac{dU_0}{du} - \frac{1}{4(1-u^2)}U_0 = 0\ . \tag{109}$$

Comparing (109) with (98) and (99) one can read out that

$$W = \frac{1}{(u^2-1)}$$

and that $\hat{\Gamma}$ satisfies

$$-\frac{2u}{u^2-1}\hat{\Gamma} - \partial\hat{\Gamma} + \hat{\Gamma}^2 = \frac{1}{4(u^2-1)}\ . \tag{110}$$

[b]While in $N = 2$ supergravity the metric of the scalars and the (field dependent) coupling of the vectors were different objects, here they coincide. For many moduli, the computation of the coupling matrix $\mathcal{N}_{AB}$ differs in the rigid and supergravity cases. When the graviphoton sits among the vector fields, the number of projective coordinates is $n+1$ and differs from the number n of the moduli. Restricting from the local to the rigid case, the coupling matrix is given by $\partial_\alpha F_A(\partial\alpha X^A)^{-1} = F_{AC}e^C_\alpha e^\alpha_B = F_{AB} = \tau$. Therefore in the rigid limit metric and vector couplings become the same and the coefficient of the generalized θ term, $\mathrm{Re}\tau$.

Writing $\hat{\Gamma} = \partial \log e$, then (110) coincides with Eq. (109) for U_0, in agreement with the fact that, according to (100), e is one of the solutions of (109).

As a check, we may compute the asymptotic behaviour of the solutions $(U_0^{(1)}, U_0^{(2)})$ of this Fuchsian equation around the singular points $u = 1, -1, \infty$. We find

$$u \to \pm 1 \quad (U_0^{(1)}, U_0^{(2)}) \approx (c_1, log(u \mp 1)) \tag{111}$$

$$u \to \infty \quad (U_0^{(1)}, U_0^{(2)}) \approx (u^{-1/2}, u^{-1/2} logu). \tag{112}$$

Recalling that $U_0^{(1)} = \partial V^{(1)}, U_0^{(2)} = \partial V^{(2)}$ and the third period $V^{(3)} = c$, one finds

$$u \to \pm 1 \quad (V^{(1)}, V^{(2)} + V^{(3)}) \approx (u \mp 1, c + c'(u \mp 1) log(u \mp 1)) \tag{113}$$

$$u \to \infty \quad (V^{(1)}, V^{(2)} + V^{(3)}) \approx (u^{1/2}, u^{1/2} logu) \tag{114}$$

in agreement with the behaviour of the periods (a, a_D) of Ref. 15.

The change of variables $u \to 1 - 2z$, puts Eq. (109) into the form of an hypergeometric equation of parameters $(\frac{1}{2}, \frac{1}{2}, 1)$ so that $U_0^1 = {}_2F_1\left[\frac{1}{2}, \frac{1}{2}, 1; \frac{1-u}{2}\right]$. Using the standard relations among hypergeometric functions, one can reconstruct the monodromy matrices of the periods in a symplectic basis thus retrieving the same generators of the monodromy group Γ_2 given in Ref. 15.

7. Conclusions

We have described as concisely as possible the mathematical methods to determine the local and global geometry of moduli spaces in string/supergravities compactified on a Calabi–Yau 3-fold. The main point is that, in principle, all the low energy couplings can be computed by the methods of algebraic geometry in conjunction with supersymmetry. It is important to stress that the invariance and monodromy of the Picard–Fuchs equations allow the determination of the target space duality which, in particular, connects the physics from small to large volume (T-duality) and the weak and strong coupling regime of the same theory albeit with a different Lagrangian.

These considerations should put into context the paper[1] of this volume. The examples we have presented here only refer to deformations by a *single (complex) modulus*, and they are tractable without excessive difficulty. However, as we consider moduli spaces of higher dimensionality (more then one modulus), the Picard–Fuchs equation becomes a fourth-order *system* of *partial* differential equations, and the explicit solution becomes very difficult to find.

For this reason it is very important to recognize that the alternative approach proposed by Tullio Regge in the interesting collaboration that yielded the paper included in this volume, even if related, is different from the use of the Picard–Fuchs equations. However, it opens the way to the search for the duality group for *two moduli Calabi–Yau deformations* and possibly also to a higher number of moduli.

References

1. A. Ceresole, R. D'Auria and T. Regge, Nucl. Phys. B **414** (1994) 517, [hep-th/9307151].

2. E. Castellani, "Dualities and Intertheoretic Relations". In: Suárez M., Dorato M., Rédei M. (eds.) EPSA Philosophical Issues in the Sciences. Springer, Dordrecht
3. M. K. Gaillard and B. Zumino, Nucl. Phys. B **193** (1981) 221.
4. C. M. Hull and P. K. Townsend, Nucl. Phys. B **438** (1995) 109 [hep-th/9410167].
5. J. Polchinski, Stud. Hist. Phil. Sci. B **59** (2017) 6 [arXiv:1412.5704 [hep-th]].
6. A. Ceresole, R. D'Auria, S. Ferrara, W. Lerche and J. Louis, Int. J. Mod. Phys. A **8** (1993) 79 [hep-th/9204035].
7. A. Ceresole, R. D'Auria, S. Ferrara, W. Lerche, J. Louis and T. Regge, AMS/IP Stud. Adv. Math. **1** (1996) 281.
8. A. Ceresole, R. D'Auria and S. Ferrara, Phys. Lett. B **339** (1994) 71 [hep-th/9408036].
9. M. Billo, A. Ceresole, R. D'Auria, S. Ferrara, P. Fre, T. Regge, P. Soriani and A. Van Proeyen, Class. Quant. Grav. **13** (1996) 831 doi:10.1088/0264-9381/13/5/007 [hep-th/9506075].
10. P. Candelas, G. Horowitz, A. Strominger and E. Witten, Nucl. Phys. **B268** (1985) 46. See also M. Green, J. Schwarz and E. Witten, *Superstring Theory*, Cambridge University Press (1987).
11. J. Polchinski, "String Theory", Cambridge Monographs on Mathematical Physics String Theory, Cambridge University Press (2005).
12. K. Becker, M. Becker and J. H. Schwarz, "String theory and M-theory: A modern introduction," Cambridge University Press (2006).
13. W. Lerche, C. Vafa and N. P. Warner, Nucl. Phys. B **324** (1989) 427. doi:10.1016/0550-3213(89)90474-4.
14. L. Andrianopoli, M. Bertolini, A. Ceresole, R. D'Auria, S. Ferrara, P. Fre and T. Magri, J. Geom. Phys. **23** (1997) 111 [hep-th/9605032].
15. N. Seiberg and E. Witten, Nucl. Phys. B **426** (1994) 19 Erratum: [Nucl. Phys. B **430** (1994) 485] [hep-th/9407087].
16. A. Ceresole, R. D'Auria and S. Ferrara, Nucl. Phys. Proc. Suppl. **46** (1996) 67 [hep-th/9509160].
17. P. Candelas, X. C. De La Ossa, P. S. Green and L. Parkes, Nucl. Phys. B **359** (1991) 21.
18. W. Lerche, Nucl. Phys. Proc. Suppl. **55B** (1997) 83 [Fortsch. Phys. **45** (1997) 293] [hep-th/9611190].
19. A. Ceresole, R. D'Auria, S. Ferrara and A. Van Proeyen, Nucl. Phys. B **444** (1995) 92 [hep-th/9502072].
20. A. Strominger, Commun. Math. Phys **133** (1990) 163; L. Castellani, R. D'Auria and S. Ferrara, Phys. Lett. **B241** (1990) 57, Class. Quantum Grav. **1** (1990) 317.
21. A. Ceresole, Theor. Math. Phys. **98** (1994) 243 [Teor. Mat. Fiz. **98** (1994) 358].
22. W. Lerche, D. J. Smit and N. P. Warner, Nucl. Phys. B **372** (1992) 87 [hep-th/9108013].

Nuclear Physics B414 (1994) 517–537
North-Holland

NUCLEAR PHYSICS B

Duality group for Calabi–Yau two-moduli space *

A. Ceresole [a,c,1], R. D'Auria [a,c] and T. Regge [b,c]

[a] *Dipartimento di Fisica, Politecnico di Torino, Corso Duca Degli Abruzzi 24, I-10129 Turin, Italy*
[b] *Dipartimento di Fisica Teorica, Università di Torino, Via P. Giuria 1, I-10125 Turin, Italy*
[c] *INFN, Sezione di Torino, Turin, Italy*

Received 30 July 1993
Accepted for publication 24 August 1993

We present an efficient method for computing the duality group Γ of the moduli space $\mathscr{M}$ for strings compactified on a Calabi–Yau manifold described by a two-moduli deformation of the quintic polynomial immersed in $\mathbb{CP}(4)$, $\mathscr{W} = \frac{1}{5}(y_1^5 + \ldots + y_5^5) - a y_4^3 y_5^2 - b y_4^2 y_5^3$. We show that Γ is given by a three-dimensional representation of a central extension of B_5, the braid group on five strands.

1. Introduction

Generic string models are believed to exhibit target space duality [1], which is a discrete symmetry acting on the moduli space of the underlying superconformal field theory (SCFT), or of the relevant compactified six-dimensional manifold. Duality symmetry, whose origin is strictly related to the fact that the string is a one-dimensional object, leaves invariant the spectrum and supposedly the interactions, and quite generally describes quantum symmetries of the low energy effective action. Such symmetry has been explicitly found in simple models like toroidal compactifications and their orbifolds, where it generalizes the $R \to 1/R$ symmetry of the bosonic string compactified on a circle. The most celebrated example is the effective lagrangian obtained by compactification of the heterotic string on a six-torus in the large radius limit [2]. The Kähler class untwisted modulus corresponding to the volume size $t = 2(R^2 + i\sqrt{g})$ parametrizes the homogeneous space $SU(1, 1)/U(1)$, so that this sector of the theory is invariant under $PSL(2, \mathbb{R})$. However, when other (twisted) sectors are introduced and/or quantum corrections computed, the residual symmetry is given by the duality group $SL(2, \mathbb{Z})$ with generators $t \to t + 1$ and $t \to -1/t$. The determination of the duality symmetry group Γ for Calabi–Yau compactifications, or in broader sense for (2, 2) $c = 9$ SCFT, is in general a very difficult problem, so far solved only for few examples in

* Supported in part by MURST.
1 E-mail address: ceresole@to.infn.it.

0550-3213/94/$07.00
SSDI 0550-3213(93)E0435-3

which the moduli space is one-dimensional. In particular, in a seminal paper [3], Candelas et al. treated in detail the class of Calabi–Yau three-folds immersed in $\mathbb{CP}(4)$ given by the one-parameter deformation of the quintic polynomial $\mathscr{W}_0$,

$$\mathscr{W}=\mathscr{W}_0-\psi y_1 y_2 y_3 y_4 y_5, \quad \mathscr{W}_0=\tfrac{1}{5}\left(y_1^5+y_2^5+y_3^5+y_4^5+y_5^5\right), \tag{1.1}$$

where y_i are homogeneous coordinates on $\mathbb{CP}(4)$. Using the monodromy properties of the solutions of the Picard–Fuchs equations, these authors were able to reconstruct the full duality group of the one-dimensional space parametrized by ψ. It was found that Γ has two generators, $\{A, T\}$ of order 5 and ∞ respectively, and their representation in terms of 4×4 matrices acting on the periods was given. After that, several other examples of one-dimensional deformations of other Calabi–Yau manifolds defined by polynomials in weighted projective space have been developed along the same lines [4].

The general computation of the duality group for more than one modulus has not been successfully attempted so far, because of the mathematical complexity that is encountered in going from the one-dimensional case to the higher dimensional ones. However, a two-moduli case is presently under investigation [5]. Moreover, it has been observed [6] that for any number of moduli, the translational symmetry on the special variables $t^a \to t^a + n^a$, $n^a \in \mathbb{Z}$, is always a subgroup of the duality group.

In this paper we present an efficient method for determining the duality group for a two-moduli deformation of the quintic polynomial $\mathscr{W}_0$,

$$\mathscr{W}=\mathscr{W}_0-a y_4^3 y_5^2-b y_4^2 y_5^3. \tag{1.2}$$

This example of a Calabi–Yau manifold is a subspace of the general 101-dimensional deformation space of $\mathscr{W}_0$ which gives rise to zero Yukawa couplings for the associated effective lagrangian. Because of that, if taken by itself, it constitutes more of a toy model as far as the low energy lagrangian is concerned. However, it provides a two-moduli example for which the duality group can be determined completely, displaying the power of some general techniques of algebraic geometry which were developed and applied to the study of the monodromy groups of Feynman integrals many years ago [7,8].

Our result for the duality group is surprisingly simple: Γ is given by an U(1, 2) valued (projective) representation of B_5, the braid group on five strands. The representation acts on the periods associated to the uniquely defined holomorphic three-form Ω which always exists on a Calabi–Yau manifold. In terms of the defining polynomial $\mathscr{W}$, the fundamental period is defined by the integral representation [16,17]

$$\omega_0(a, b)=\oint_\gamma \frac{\omega}{\mathscr{W}(y; a, b)}, \tag{1.3}$$

where ω is the volume element

$$\omega = \sum(-1)^i y_i \, \mathrm{d}y_1 \wedge \ldots \wedge \mathrm{d}\hat{y}_i \wedge \ldots \wedge \mathrm{d}y_5 \tag{1.4}$$

(the hat means that the corresponding differential must be omitted), and γ is an element of the basis for the homology cycles of $H_{(4)}(\mathbb{CP}(4) - \mathscr{W}; \mathbb{Z})$. There are as many independent integrals ω_0^I as there are elements of the basis, $\gamma^I \subset H_{(4)}(\mathbb{CP}(4) - \mathscr{W}; \mathbb{Z})$.

Quite generally, if L^{N-1} is the singularity locus of an algebraic variety $\mathscr{W}$ parametrized by N moduli, the monodromy group Γ acting on the periods of $\mathscr{W}$ is given by a representation of the fundamental group π_1 of the embedding space $\mathbb{CP}(N)$.

The computation of the homotopy group π_1 is based on the use of the following two theorems [9,10].

Theorem 1 (Picard–Severi). Let L^{N-1} be the $N-1$ complex dimensional singular locus of a given algebraic variety. If the (complex) projective line $\mathbb{CP}(1) \subset \mathbb{CP}(N)$ is generic with respect to L^{N-1} (i.e., it avoids all singular points of L^{N-1}), then we have the isomorphism

$$\pi_1\left(\mathbb{CP}(1) - \left[\mathbb{CP}(1) \cap L^{N-1}\right]; B\right)/\mathrm{G} \approx \pi_1\left(\mathbb{CP}(N) - L^{N-1}; B\right), \tag{1.5}$$

where B is the base point and G is an invariant subgroup of $\pi_1(\mathbb{CP}(1) - [\mathbb{CP}(1) \cap L^{N-1}]; B)$.

In other words, $\pi_1(\mathbb{CP}(N) - L^{N-1}; B)$ is obtained from $\pi_1(\mathbb{CP}(1) - [\mathbb{CP}(1) \cap L^{N-1}]; B)$ by adding the identities $\rho = \mathbb{1} \forall \rho \in \mathrm{G}$.

A method for deriving such identities has been provided by van Kampen [11], and we shall use it in our particular case to obtain the isomorphism of eq. (1.5). As the singular locus of the algebraic variety $\mathscr{W}$ is given by an equation of the form $L(a, b) = 0$, we are interested in the case $N = 2$.

The second theorem, due to Zariski [10], allows to understand that, as far as the identification of the fundamental group π_1 is concerned, the case of more than two moduli can be essentially reduced to the $N = 2$ case, so that the computation of π_1 in general will not be more difficult than the one under study. However, for several variables it is in general much harder to find the singular locus and the behaviour of the algebraic variety in its neighbourhood, and therefore the determination of the monodromy group can be more involved.

Theorem 2 (Zariski). If the complex projective plane $\mathbb{CP}(2)$ is generic with respect to L^{N-1} and if $B \in [\mathbb{CP}(2) - \mathbb{CP}(2) \cap L^{N-1}]$, then the map

$$\pi_1\left(\mathbb{CP}(2) - \left[\mathbb{CP}(2) \cap L^{N-1}\right]; B\right) \to \pi_1\left(\mathbb{CP}(N) - L^{(N-1)}; B\right) \tag{1.6}$$

is an isomorphism.

We see that, in virtue of Zariski's theorem, the study of the homotopy group of $\mathbb{CP}(N) - L^{N-1}$ is reduced to the study of the homotopy of the complement of the one-dimensional curve $L^1 = \mathbb{CP}(2) \cap L^{N-1}$ on a generic two-dimensional section. Since the singular locus of the variety (1.2) is already one-dimensional ($N = 2$), theorem 1 is sufficient for our present purposes.

Before proceeding to the actual determination of the duality group, we recall what is the local geometry associated to the moduli space $\mathcal{M}$ of $\mathcal{W}$. The two-moduli family of Calabi–Yau deformations given by eq. (1.2) was first introduced in ref. [14] as a particular tensor product example of five copies of minimal models with $k = 3$ and $c = \frac{9}{5}$. It was observed that in such model there are restrictions given by charge conservation which enforce the condition $W_{\alpha\beta\gamma} = 0$ for the Yukawa couplings. Because of that, the constraint of Special Geometry [12],

$$R_{\alpha\bar{\beta}\gamma\bar{\delta}} = g_{\alpha\bar{\beta}}g_{\gamma\bar{\delta}} + g_{\alpha\bar{\delta}}g_{\gamma\bar{\beta}} - e^{2K}W_{\alpha\gamma\rho}W_{\bar{\beta}\bar{\delta}\bar{\sigma}}g^{\rho\bar{\sigma}}, \tag{1.7}$$

reduces to

$$R_{\alpha\bar{\beta}\gamma\bar{\delta}} = g_{\alpha\bar{\beta}}g_{\gamma\bar{\delta}} + g_{\alpha\bar{\delta}}g_{\gamma\bar{\beta}}, \quad \alpha, \bar{\beta}, \gamma, \bar{\delta} = 1, 2. \tag{1.8}$$

Thus, the local geometry of $\mathcal{M}$ is given by a two-dimensional Kähler manifold with constant curvature, which, according to the classification of ref. [15], corresponds to the coset space $U(1, 2)/[U(1) \otimes U(2)]$. The global structure of such moduli space is determined by modding out the discrete isometries given by the duality group Γ.

In principle, for the two-moduli case, one would expect a six-dimensional representation of the modular group. In fact, the dimension of the $H^{(3)}$ cohomology group is given by $2h_{21} + 2$, where $h_{21} = \dim H^{(2,1)}$, the number of complex structure moduli of the Calabi–Yau manifold. It turns out, however, that our representation is only three-dimensional, as this specific model is singular due to the vanishing of the Yukawa couplings $W_{\alpha\beta\gamma}$. In such case, the Picard–Fuchs equations for the periods of $\mathcal{W}$ are of second order rather than fourth order [13]. The six-dimensional representation, which in the symplectic basis is valued in $Sp(6, \mathbb{Z})$, splits into two three-dimensional representations of $U(1, 2)$ according to the embedding $6 = 3 + \bar{3}$ of $U(1, 2)$ in $Sp(6)$. Later, we shall verify explicitly that the three-dimensional representation of B_5 is actually valued in $U(1, 2)$.

2. The fundamental group of $\mathcal{W}(y; a, b)$

In this section we compute the fundamental group $\pi_1(\mathbb{CP}(2) - L; B)$ of the algebraic variety $\mathcal{W} = 0$. The first step consists of finding the singular locus L of eq.

(1.2), which is given by solving simultaneously the equations

$$\frac{\partial \mathscr{W}}{\partial y_i} = 0, \quad i = 1,\ldots,5. \tag{2.1}$$

A straightforward computation yields the one-dimensional complex curve

$$L(a, b) = 108(a^5 + b^5) - 80a^3b^3 - 165a^2b^2 + 30ab - 1 = 0, \tag{2.2}$$

which represents the locus of the complex points of the original variety $\mathscr{W} = 0$ where two or more of the roots coincide.

For the derivation of the van Kampen relations among the homotopy generators around the various branches of $L(a, b)$, it is important to know where $L(a, b)$ itself has multiple points. These are found by solving the equations $L(a, b) = \partial L/\partial a = \partial L/\partial b = 0$, which give the location of the multiple roots

$$(a, b) = (\rho^k, \rho^{-k}), \quad (a, b) = -\tfrac{1}{4}(\rho^k, \rho^{-k}), \quad k = 0,\ldots,4, \tag{2.3}$$

where $\rho = e^{2\pi i/5}$. The first set of roots in (2.3) corresponds to nodes with two distinct complex conjugate tangents (which are isolated points for the real section of (2.2) represented by real values of a and b). The second set instead represents (second order) cusps, since the hessian $\partial^2 L/\partial a\, \partial b$ is degenerate at these points.

We may obtain a more elegant and geometrically intuitive representation of the curve $L(a, b) = 0$ by choosing new coordinates (p, q) such that the real section corresponding to real values of p and q exhibits the previous singular points in the real (p, q) plane. It is sufficient to set

$$\begin{cases} a = p + iq, \\ b = p - iq, \end{cases} \tag{2.4}$$

and we find

$$\begin{aligned} L(p, q) = {} & -1 + 30(p^2 + q^2) - 165(p^4 + q^4) - 80(p^6 + q^6) + 216p^5 \\ & - 330p^2q^2 - 2160p^3q^2 - 240(p^4q^2 + p^2q^4) + 1080pq^4. \end{aligned} \tag{2.5}$$

In the real plane of (p, q) the multiple points from eq. (2.3) take now the real values

$$\begin{aligned} (p, q) &= (\cos(\tfrac{2}{5}\pi k), \sin(\tfrac{2}{5}\pi k)), \\ (p, q) &= -\tfrac{1}{4}(\cos(\tfrac{2}{5}\pi k), \sin(\tfrac{2}{5}\pi k)), \end{aligned} \tag{2.6}$$

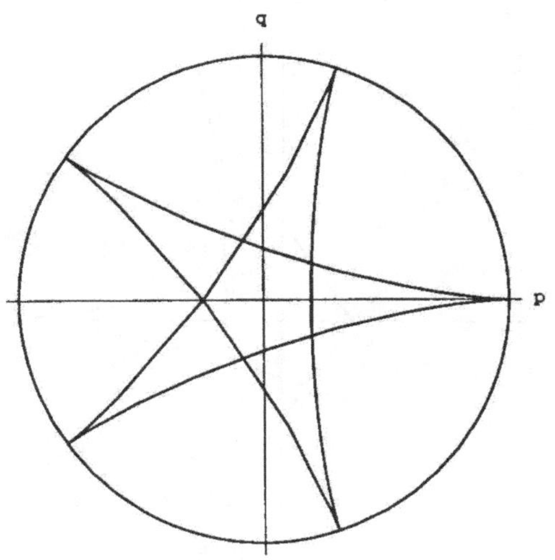

Fig. 1.

respectively. Actually, the curve $L(p, q) = 0$ can be put in a parametric form by setting

$$\begin{cases} p = \frac{1}{5}(3 \cos 2t + 2 \cos 3t), \\ q = \frac{1}{5}(3 \sin 2t - 2 \sin 3t), \quad 0 \leqslant t \leqslant 2\pi, \end{cases} \tag{2.7}$$

and can be recognized as a pentacuspidal hypocycloid (the curve described by a point of a circle of radius $R = \frac{1}{5}$ rolling inside a circle of $R = 1$), whose graph is represented in fig. 1.

According to the Picard–Severi theorem, we now take a generic line C through the base point $B = (0, 0)$ which intersects the real branch of the hypocycloid in four (finite) points p_i (fig. 2). To each of such representative points, we attach a generator of $\pi_1(\mathbb{C} - (\mathbb{C} \cap L); B)$ by constructing a loop which leaves B along a straight line, makes an infinitesimal loop around p_i counterclockwise in the complex plane $\mathbb{C}$ containing the real line, and comes back to B again in the opposite direction. If the straight line encounters a real branch of L, it will be taken to undercross it by describing a small semicircle in the complex plane. If by varying the angular coefficient of the straight line no critical point of L is encountered, the corresponding points give rise to equivalent loops. Generally inequivalent loops are obtained if two straight lines intersect L along points belonging to two different real branches of L emanating from the critical points. We thus obtain 15 loops, five of which go around the sides of the pentagonal figure described by L and ten going around the branches emanating from the cusps (five of these are shown in fig. 3).

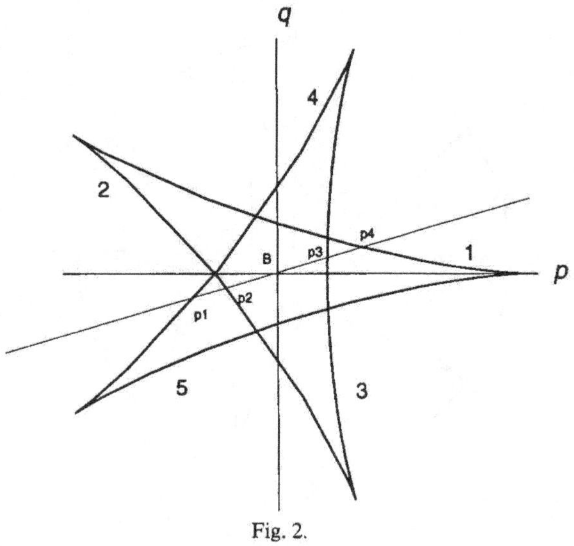

Fig. 2.

We now quote the van Kampen relations [11] between loops which, added to the free group generated by the above-mentioned 15 generators, make it isomorphic to $\pi_1(\mathbb{CP}(2) - L;\ B)$.

For the nodes corresponding to transversal intersections of the real branches of L, we have (see fig. 4)

$$\alpha\beta = \beta\alpha, \quad \alpha = \alpha', \quad \beta = \beta', \tag{2.8}$$

i.e., we can slide the representative loops across the node without any change, and

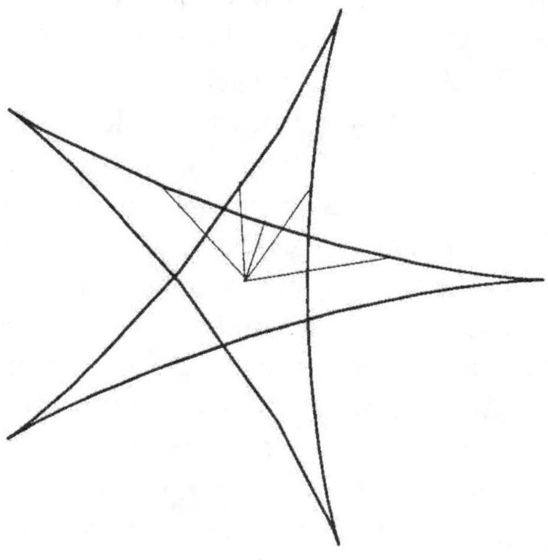

Fig. 3.

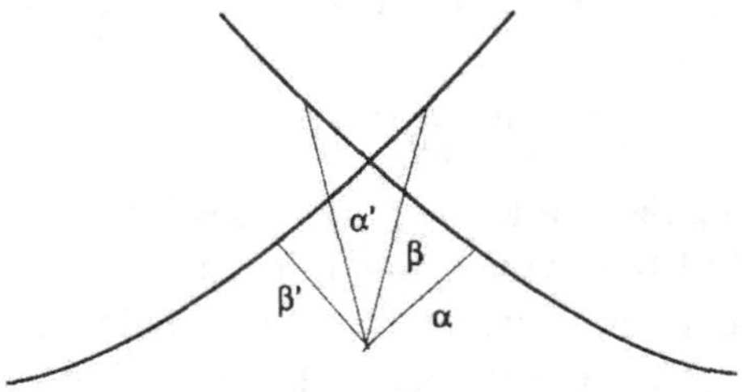

Fig. 4.

the loops around two branches commute. In this way, the 15 generators reduce to five independent ones.

For each cusp we have the relation (see fig. 5)

$$\alpha\beta\alpha = \beta\alpha\beta. \tag{2.9}$$

Let us enumerate in increasing order the five branches of the hypocycloid described successively by the point of the small circle rolling inside the big circle (fig. 2). Then, denoting by α_i, $i = 1,\dots,5$, the loops winding around the five branches of fig. 1, we have the following set of relations among the generators:

$$\begin{aligned}
&\alpha_1\alpha_3 = \alpha_3\alpha_1, \quad \alpha_1\alpha_2\alpha_1 = \alpha_2\alpha_1\alpha_2,\\
&\alpha_1\alpha_4 = \alpha_4\alpha_1, \quad \alpha_2\alpha_3\alpha_2 = \alpha_3\alpha_2\alpha_3,\\
&\alpha_2\alpha_4 = \alpha_4\alpha_2, \quad \alpha_3\alpha_4\alpha_3 = \alpha_4\alpha_3\alpha_4,\\
&\alpha_2\alpha_5 = \alpha_5\alpha_2, \quad \alpha_4\alpha_5\alpha_4 = \alpha_5\alpha_4\alpha_5,\\
&\alpha_3\alpha_5 = \alpha_5\alpha_3, \quad \alpha_5\alpha_1\alpha_5 = \alpha_1\alpha_5\alpha_1.
\end{aligned} \tag{2.10}$$

We note that the subset of relations (2.10) not involving α_5 coincide with the defining relations of the braid group B_5, with four generators α_i, $i = 1,\dots,4$,

$$\begin{cases} \alpha_i\alpha_j = \alpha_j\alpha_i, & |i-j| \geqslant 2,\\ \alpha_i\alpha_{i+1}\alpha_i = \alpha_{i+1}\alpha_i\alpha_{i+1}, & i = 1,\dots,3. \end{cases} \tag{2.11}$$

However, it may be easily verified that the element of B_5 exchanging the first and

Fig. 5.

fifth strand can be written in terms of the four generators α_i by the word

$$\alpha_5 = (\alpha_4\alpha_3\alpha_2)\alpha_1(\alpha_4\alpha_3\alpha_2)^{-1}. \tag{2.12}$$

With a little effort, using the van Kampen relations for $\alpha_1, \ldots, \alpha_4$, one checks that the extra relations of (2.10) involving α_5 are indeed verified. Therefore, we come to the conclusion that, if the relation (2.12) also holds among the monodromy generators α_i $(i = 1, \ldots, 5)$ of the hypocycloid, then $\pi_1(\mathbb{C}(2) - L;\ B)$ is isomorphic to B_5.

The reason why we do not find eq. (2.12) among the van Kampen relations is that we have not considered the critical point of L at ∞ and the associated generator. Rather than studying such critical point, we shall give evidence in the sequel that eq. (2.12) must be satisfied by the monodromy generators, so that indeed B_5 coincides with the fundamental group associated to $\mathscr{W}$.

3. Behaviour of the periods around the singular curve

Till now we have only considered the abstract presentation of the fundamental group in terms of its generators α_i. To obtain an explicit realization on the periods of $\mathscr{W}$, it is necessary to consider their leading behaviour in the neighbourhood of the singularity locus $L(a, b)$. Let us evaluate the integral defined in (1.3) on a suitable contour. Setting $y_3 = 1$, we may rewrite it in the following way:

$$\omega_0 = 5\oint \mathrm{d}y_2\mathrm{d}y_4\mathrm{d}y_5 \oint \frac{\mathrm{d}y_1}{y_1^5 + f(a, b, y_2, y_4, y_5)}, \tag{3.1}$$

where

$$f = y_2^5 + y_4^5 + y_5^5 + 1 - 5ay_4^3y_5^2 - 5by_4^2y_5^3. \tag{3.2}$$

Performing the last integration on the cycle $|y_1| = \text{const.}$ (posing $y_1^5 = t$), gives immediately $\frac{2}{5}\pi i(-f)^{4/5}$, so that we may write

$$\omega_0 = 2\pi i \oint \frac{\mathrm{d}y_2\mathrm{d}y_4\mathrm{d}y_5}{\left(y_2^5 + y_4^5 + y_5^5 + 1 - 5ay_4^3y_5^2 - 5by_4^2y_5^3\right)^{4/5}}$$

$$= 2\pi i \oint \mathrm{d}y_4\mathrm{d}y_5 \oint \frac{\mathrm{d}y_2}{\left[y_2^5 + g(a, b, y_4, y_5)\right]^{4/5}}, \tag{3.3}$$

where

$$g = y_4^5 + y_5^5 + 1 - 5ay_4^3y_5^2 - 5by_4^2y_5^3. \tag{3.4}$$

Again, with $y_2 = g^{1/5}u$, the last integral on the cycle $|y_2| = $ const. yields

$$g^{-3/5}\oint \frac{du}{(u^5+1)^{4/5}}, \tag{3.5}$$

which is a number independent of a and b, so that we arrive at

$$\omega_0 = \text{const.} \times \oint \frac{dy_4 dy_5}{g^{3/5}}. \tag{3.6}$$

After the further change of variables

$$y_4 = \sigma\tau^{1/2}, \quad y_5 = \sigma\tau^{-1/2}, \tag{3.7}$$

we find

$$\omega_0 = \text{const.} \times \oint \frac{d\tau}{\tau} \oint \frac{\sigma\, d\sigma}{[1+\sigma^5 h(\tau)]^{3/5}}, \tag{3.8}$$

where $h(\tau) = \tau^{5/2} + \tau^{-5/2} - 5a\tau^{1/2} - 5b\tau^{-1/2}$. Setting at last $\xi = \sigma h^{1/5}$, the σ cyclic integral gives $h^{2/5}$ times a purely numerical integral. Therefore we obtain that the periods ω_0 can be written as a simple one-dimensional integral

$$\omega_{ij} = \text{const.} \times \oint_{\gamma_{ij}} \frac{d\tau}{(\tau^5 - 5a\tau^3 - 5b\tau^2 + 1)^{2/5}} \tag{3.9}$$

along a set of suitably chosen contours γ_{ij}. The quintic polynomial $p(\tau) = \tau^5 - 5a\tau^3 - 5b\tau^2 + 1$ has five roots τ_i, and the possible singularities of the integral (3.9) arise as pinching singularities due to the coincidence of two roots encircled by a figure-eight contour (see fig. 6).

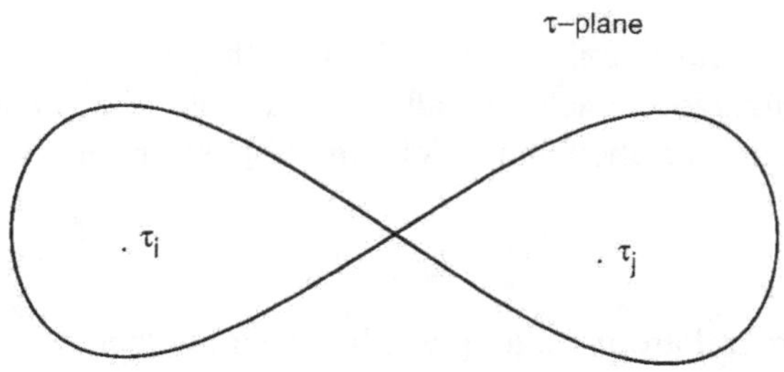

Fig. 6.

One can check that the singular locus of the singularities of $P(\tau)$ is given again by the curve $L(a, b) = 0$ of eq. (2.2) by computing the resultant between $P(\tau)$ and $\mathrm{d}P/\mathrm{d}\tau$.

At this point, we may conclude that the monodromy group is exactly B_5. In fact we note that the braid group B_n can be identified with the fundamental group of the space of all unordered sets of n distinct complex numbers τ_i, $i = 1, \ldots, n$. More precisely, $B_n = \pi_1(\mathbb{C}^n/S_n \setminus S/S_n)$, where S_n is the permutation group on five elements and S is the union of the hyperplanes $\tau_i = \tau_j$ $\forall (i, j)$ [18]. Identifying the numbers τ_i with the roots of a polynomial $P_n(\tau)$ we see that in our case the monodromy group of the periods ω_0 must be a subgroup of B_5, since $P(\tau)$ is a quintic polynomial. On the other hand, B_5 contains the element given by (2.12) exchanging the strands one and five. Since

$$\alpha_5 = \alpha_4 \alpha_3 \alpha_2 \alpha_1 \alpha_2^{-1} \alpha_3^{-1} \alpha_4^{-1} \tag{3.10}$$

satisfies the relations (2.10) involving α_5, the group whose presentation is given by (2.10) must actually coincide with B_5.

There is yet another independent argument leading to the same conclusions. Given a general polynomial of fifth degree, we may always fix three of its five coefficients by a Möbius transformation to arbitrary values. As $P(\tau)$ contains only two parameters, it is in fact a gauge fixed form of a generic quintic polynomial, and therefore the associated subgroup of B_5 is really B_5 itself.

One can show that the set of all possible figure-eight contours γ_{ij} encircling a couple of roots can be expressed linearly in terms of three of them, thus confirming that the number of linearly independent periods of $\mathscr{W}$ is indeed equal to three. To see this, let us denote by $\gamma_{i,i+1}$ the figure-eight contour encircling two consecutive roots τ_i, τ_{i+1} (loops encircling non-consecutive roots are easily written as products of the $\gamma_{i,i+1}$'s, e.g. $\gamma_{i,i+2} = \gamma_{i,i+1} \circ \gamma_{i+1,i+2}$, etc.), and by $\omega_{i,i+1}$ the corresponding period. Only three of them are independent, since they satisfy the following two relations:

$$\sum_{k=1}^{5} \omega_{k,k+1} = 0, \quad \sum_{k=1}^{5} e^{-ik4\pi/5} \omega_{k,k+1} = 0. \tag{3.11}$$

The second relation easily follows from the fact that $\gamma_{12} \circ \gamma_{23} \circ \gamma_{34} \circ \gamma_{45} \circ \gamma_{51}$ is homotopic to a single loop encircling all the five roots of $P(\tau)$, and the integral is regular at ∞. The first relation can be obtained by observing that

$$\omega_{i,i+1} = I_{i+1} - I_i, \tag{3.12}$$

where I_i is the integral around a loop winding counterclockwise around the simple root τ_i. Therefore, the representation of B_5 on the $\omega_{i,i+1}$ is three-dimensional, thus confirming the observation made in the introduction that the vanishing of the Yukawa couplings reduces the six-dimensional Sp(6, $\mathbb{Z}$) valued representation of the periods into the $3+\bar{3}$ representation which will be later shown to belong to U(1, 2).

The behaviour of $\omega_{ij}(a, b)$ around the $L(a, b)=0$ singularity can now be obtained by expanding $P(\tau)$ in the neighbourhood of a point τ_0 where $P(\tau)$ vanishes together with its first derivative:

$$P(\tau)=(\tau-\tau_0)^2+H(\tau, a, b),$$

$$\lim_{\tau\to\tau_0} H(\tau, a, b)=L(a, b). \tag{3.13}$$

If we now set $\tau-\tau_0=H^{1/2}\eta$, then we find

$$\omega_{ij}=\text{const.}\times L^{1/10}\oint\frac{\mathrm{d}\eta}{(\eta^2+1)^{3/5}}\equiv\text{const.}\times L^{1/10}(a, b), \tag{3.14}$$

thus finding that, upon performing a loop around a branch of the hypocycloid, the integral acquires a phase $\mathrm{e}^{i\pi/5}\equiv z$.

An alternative way of reaching the same conclusion is to consider the effect of analytic continuation of $\omega_{i,i+1}$ around the loop $\alpha_i\in B_5$ in the (a, b) plane,

$$\omega_{i,i+1}\xrightarrow{\alpha_i}\mathrm{e}^{i\pi/5}\omega_{i,i+1}. \tag{3.15}$$

Indeed, when the path α_i winds around the ith branch of the hypocycloid, the roots $\tau_i(a, b)$, $\tau_{i+1}(a, b)$ of the polynomial $P(\tau)$ are exchanged. In the τ plane, $\gamma_{i,i+1}$ gets deformed as in fig. 7. The final result is a new circuit followed in opposite direction where the base point B is on a different Riemann sheet. This gives the same integral as before except for a phase $-(\mathrm{e}^{2\pi i})^{-2/5}=z$.

The correctness of this result can also be ascertained by studying the leading behaviour of the period ω_0 from the Picard–Fuchs equations obeyed by it. Using

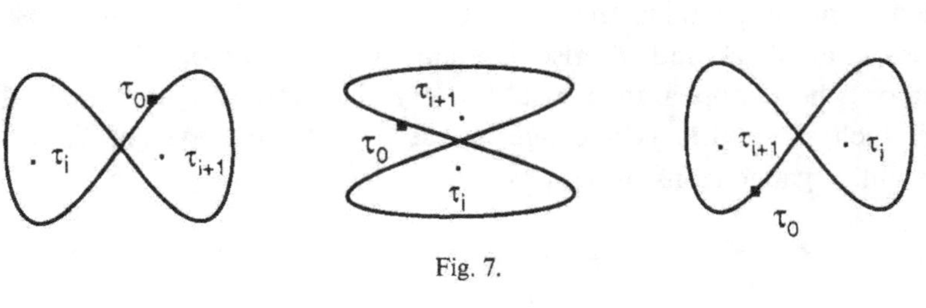

Fig. 7.

the methods of refs. [16,17], one derives the following set of three partial differential equations:

$$\frac{\partial^2\omega_0}{\partial a^2}=\frac{1}{1-4ab}$$

$$\times\left(6b^2\frac{\partial^2\omega_0}{\partial a\,\partial b}+3a\frac{\partial^2\omega_0}{\partial b^2}+12b\frac{1-2ab}{1-4ab}\frac{\partial\omega_0}{\partial a}+\frac{12a^2}{1-4ab}\frac{\partial\omega_0}{\partial b}\right),$$

$$\frac{\partial^2\omega_0}{\partial b^2}=\frac{1}{1-4ab}$$

$$\times\left(6a^2\frac{\partial^2\omega_0}{\partial a\,\partial b}+3b\frac{\partial^2\omega_0}{\partial a^2}+12a\frac{1-2ab}{1-4ab}\frac{\partial\omega_0}{\partial b}+\frac{12b^2}{1-4ab}\frac{\partial\omega_0}{\partial a}\right),$$

$$\frac{\partial^2\omega_0}{\partial a\,\partial b}=\frac{1}{1-13ab}$$

$$\times\left(6a^2\frac{\partial^2\omega_0}{\partial a^2}+6b^2\frac{\partial^2\omega_0}{\partial b^2}+11a\frac{\partial\omega_0}{\partial a}+11b\frac{\partial\omega_0}{\partial b}+\omega_0\right). \tag{3.16}$$

The study of the singularities of eqs. (3.16) is better achieved by writing the associated linear system. Setting

$$\omega_1=\frac{\partial\omega_0}{\partial a}=\oint_\Gamma\frac{y_4^3y_5^2}{\mathscr{W}^2}\omega,\quad \omega_2=\frac{\partial\omega_0}{\partial b}=\oint_\Gamma\frac{y_4^2y_5^3}{\mathscr{W}^2}\omega, \tag{3.17}$$

and by elimination of the mixed derivatives in (3.16), we find

$$\frac{\partial}{\partial a}\boldsymbol{\Pi}=A(a,b)\boldsymbol{\Pi},\quad \frac{\partial}{\partial b}\boldsymbol{\Pi}=B(a,b)\boldsymbol{\Pi},\quad \boldsymbol{\Pi}=\begin{pmatrix}\omega_0\\ \omega_1\\ \omega_2\end{pmatrix}, \tag{3.18}$$

where the 3×3 matrices $A(a,b)$ and $B(a,b)$ are rational functions of a and b whose denominator contains the singular locus $L(a,b)$. Actually, most of the matrix elements of A and B also contain an extra factor of $1-4ab$ in the denominator whose appearance would imply the extra singular locus $4ab=1$. However, such singularity is a gauge artifact, as the two systems in (3.18) are covariant under gauge transformations

$$A\to\mathscr{N}^{-1}A\mathscr{N}-\mathscr{N}^{-1}\partial\mathscr{N} \tag{3.19}$$

(and similarly for B), where $\mathcal{N}$ belongs to the Borel subgroup of lower triangular matrices of GL(3, $\mathbb{C}$). The examination of the Picard–Fuchs system in the neighbourhood of the variable $v = ab = \frac{1}{4}$ gives in fact perfectly regular solutions in the new basis

$$\tilde{\omega}_0 \equiv \omega_0,$$

$$\tilde{\omega}_1 = \omega_1 + 8a^5\omega_2,$$

$$\tilde{\omega}_2 = 2(64a^5 - 5)\omega_1 - (1 + 32a^5)\omega_2. \tag{3.20}$$

The same conclusion is found by replacing the two linear systems in (3.18) by two third order ordinary differential equations in a and b for ω_0, with coefficients depending on the other variable, where the singularity $ab = \frac{1}{4}$ is absent. Furthermore, the fuchsian analysis of the linear system (3.18) or of the two third order differential equations in the neighbourhood of $L(a, b) = 0$ gives again the behaviour $L^{1/10}$ around the singularity, thus confirming our previous analysis.

4. The monodromy generators

In this section we determine the representation $\mathcal{R}(\alpha_i) \subset \mathrm{GL}(3, \mathbb{C})$ of the fundamental group acting on V, the vector space spanned by three independent periods. The computation is made by extending the representation of the group B_5 to a representation on the group ring over the complex field $\mathbb{C}$. Noting that α_i acts on the period $\omega_{i,i+1}$ as an analytic continuation around the ith branch of $L(a, b)$, the corresponding discontinuities are given by

$$\begin{aligned} &[\mathcal{R}(\alpha_i) - 1]\,\omega_{i,i+1} = (z-1)\,\omega_{i,i+1}, \\ &[\mathcal{R}(\alpha_i) - 1]\,\omega_{j,j+1} = 0, \qquad i \neq j, j+1 \end{aligned} \tag{4.1}$$

Hence we introduce the (Picard–Lefschetz) discontinuity operators [18]

$$\mathcal{R}(\alpha_i) = (z-1)u_i + 1, \tag{4.2}$$

where u_i are one-dimensional projection operators obeying $u_i^2 = u_i$. The first set of relations (2.10) imply

$$u_i u_j = u_j u_i, \tag{4.3}$$

where i, j are non-contiguous indices and we are considering 1, 2, 3, 4, 5 cyclicly ordered so that 1 and 5 are contiguous. By right multiplication with u_j and left multiplication with u_i, we get

$$u_i u_j = u_j u_i u_j \equiv \lambda u_j,$$

$$u_i u_j = u_i u_j u_i \equiv \lambda u_i, \tag{4.4}$$

where we have used the fact that since the u_i are one-dimensional projection operators, for any operator $\mathscr{O}$,

$$u_i \mathscr{O} u_i = \lambda_{\mathscr{O}} u_i. \tag{4.5}$$

Eqs. (4.4) then imply

$$u_i u_j = u_j u_i = 0, \tag{4.6}$$

which is a relation much stronger than (4.3). Indeed, eq. (4.6) has an intuitive meaning since e.g. u_1, u_3 correspond to projection of the integral (3.9) around the disconnected figure-eight circuits winding the roots τ_1, τ_2 and τ_3, τ_4 respectively. From the second set of (2.10) we get

$$(z-1)^2 u_i u_{i+1} u_i + z u_i = (z-1)^2 u_{i+1} u_i u_{i+1} + z u_{i+1}. \tag{4.7}$$

Again, since the u_i are one-dimensional projection operators, we have

$$u_i u_{i+1} u_i = \rho u_i,$$

$$u_{i+1} u_i u_{i+1} = \sigma u_{i+1}. \tag{4.8}$$

Multiplying the two equations in (4.8) by u_{i+1} on the right and u_i on the left respectively, we find $\rho = \sigma$, so that (4.7) gives

$$u_i\left[(z-1)^2 \rho + z\right] = u_{i+1}\left[(z-1)^2 \rho + z\right] \tag{4.9}$$

or

$$\rho = z/(z-1)^2. \tag{4.10}$$

Using the relation (3.10) in the form $\alpha_5 \alpha_1 \alpha_3 \alpha_2 = \alpha_4 \alpha_3 \alpha_2 \alpha_1$, we find from (4.2)

$$\begin{aligned}(z-1)^3 u_5 u_4 u_3 u_2 + (z-1)^2 u_5 u_4 u_3 + (z-1) u_5 u_4 + u_5 \\ = (z-1)^3 u_4 u_3 u_2 u_1 + (z-1)^2 u_3 u_2 u_1 + (z-1) u_2 u_1 + u_1.\end{aligned} \tag{4.11}$$

By right multiplication with u_2 and left multiplication with u_5, using (4.8)–(4.10), we get

$$(z-1) z^2 u_5 u_5 u_4 u_3 u_2 = u_5 u_1 u_2, \tag{4.12}$$

where the final form of the coefficient on the LHS of (4.12) is obtained by using the relations obeyed by the tenth roots of unity $z = e^{i\pi/5}$ (e.g. $1 - z + z^2 - z^3 + z^4 \equiv 0$). Obviously, equations analogous to (4.12) are also obeyed by similar products of u_i's with indices cyclically permuted. Such relations allow us to replace a product of four contiguous u_i operators in decreasing order with (a coefficient times) the product of three contiguous u_i's in increasing order, the first and the last factors being the same in both expressions.

We are now ready to construct the explicit representation for the α_i's. We select three arbitrary linearly independent basis vectors ω_{51}, ω_{12}, ω_{45} defined as the eigenvectors of u_5, u_1, u_4 corresponding to eigenvalue one,

$$\begin{aligned} u_5\omega_{51} &= \omega_{51} \equiv \Psi_{51}, \\ u_1\omega_{51} &= \frac{1}{z-1}\omega_{12} \equiv \Psi_{12}, \\ u_4\omega_{51} &= \frac{1-z}{z}\omega_{45} \equiv \Psi_{45}, \end{aligned} \tag{4.13}$$

where the factors in front of ω_{ij} have been chosen in such a way that the action of the cyclic permutation Z of $\mathbb{Z}_5 \subset B_5$ gives $Z\omega_{i,i+1} = \omega_{i+1,i+2}$ with no extra phase. The application of u_i, $i = 1, \dots, 5$, to any basis vector gives a linear combination of them, namely

$$u_i u_k \Psi_{51} = p_i \Psi_{51} + q_i \Psi_{12} + r_i \Psi_{45}, \quad k = 5,\, 1,\, 4. \tag{4.14}$$

It is now easy to compute the coefficients p_i, q_i, r_i by repeated use of the formulae (4.8)–(4.11). For instance, if we take $i = 3$, then

$$\begin{aligned} u_3\Psi_{51} &= u_3 u_5 \Psi_{51} = 0, \\ u_3\Psi_{45} &= u_3 u_4 \Psi_{51} = p\Psi_{51} + q u_1 \Psi_{51} + r u_4 \Psi_{51}. \end{aligned} \tag{4.15}$$

On the other hand,

$$u_3 u_4 \Psi_{51} = u_3 u_4 u_5 \Psi_{51}, \tag{4.16}$$

and, using (4.12),

$$u_3 u_4 u_5 \Psi_{51} = (z-1) z^2 u_3 u_2 u_1 u_5 \Psi_{51} = p\Psi_{51} + q u_1 \Psi_{51} + r u_4 \Psi_{51}. \tag{4.17}$$

Multiplying the last two sides by u_2 on the left, we obtain

$$-(z-1) z^2 \frac{z}{(z-1)^2} u_2 u_1 \Psi_{51} = q u_2 u_1 \Psi_{51} \quad \to q = -\frac{z^3}{z-1}. \tag{4.18}$$

Applying now u_3 on the left of (4.15), we find

$$u_3 u_4 \Psi_{51} = r u_3 u_4 \Psi_{51} \quad \rightarrow r = 1. \tag{4.19}$$

Finally, acting with u_4 in (4.15), we have

$$u_4 u_3 u_4 \Psi_{51} = (p+r) u_4 \Psi_{51}$$

$$\rightarrow -\frac{z}{(z-1)^2} u_4 \Psi_{51} = (p+r) u_4 \Psi_{51} \quad \rightarrow p = \frac{1-z^3}{(z-1)^3}. \tag{4.20}$$

In the same way one can compute the coefficients for the action of any other projection operator u_i. The final result for the monodromy operator α_i on the basis $\{\omega_{45}, \omega_{51}, \omega_{12}\}$ is

$$\mathscr{R}(\alpha_1) = \begin{pmatrix} 1 & 0 & 0 \\ 0 & 1 & 1 \\ 0 & 0 & z \end{pmatrix}, \quad \mathscr{R}(\alpha_2) = \begin{pmatrix} 1 & 0 & 0 \\ 0 & 1 & 0 \\ z^4 & -1+z-z^2 & z \end{pmatrix},$$

$$\mathscr{R}(\alpha_3) = \begin{pmatrix} z & z^2(1-z) & z^2 \\ 0 & 1 & 0 \\ 0 & 0 & 1 \end{pmatrix}, \quad \mathscr{R}(\alpha_4) = \begin{pmatrix} z & 0 & 0 \\ -z & 1 & 0 \\ 0 & 0 & 1 \end{pmatrix},$$

$$\mathscr{R}(\alpha_5) = \begin{pmatrix} 1 & 1 & 0 \\ 0 & z & 0 \\ 0 & -z & 1 \end{pmatrix}. \tag{4.21}$$

From eqs. (4.21) we may verify that indeed α_5 satisfies the relation (2.12). Furthermore, we can use (4.21) to compute the monodromy generator around ∞, which we have disregarded until now. It can be shown that the generator α_∞ can be written by the following word:

$$\alpha_\infty = \alpha_4 \alpha_2 \alpha_3 \alpha_5 \alpha_1 \alpha_5, \tag{4.22}$$

and we obtain

$$\mathscr{R}(\alpha_\infty) = \begin{pmatrix} z^2 & 1+z^2 & z(1+z^2) \\ -z^2 & -1-z^2 & -z^3 \\ -1 & -1 & -1-z \end{pmatrix}. \tag{4.23}$$

We note that the eigenvalues of $\mathscr{R}(\alpha_\infty)$ are $\{-1, -1, -z\}$ thus showing the presence of a singularity at ∞ with critical exponent $-\frac{2}{5}$. This can also be confirmed by the behaviour by the integral (3.9) for large values of a and b. We find in this case

$$\omega_0 \overset{a,b\to\infty}{\sim} \oint \frac{d\tau}{(5a\tau^3+5b\tau^2)^{2/5}}, \tag{4.24}$$

and by the rescaling $a \to \lambda\xi$, $b \to \lambda\eta$, we find

$$\omega_0 \sim \oint \frac{d\tau}{(\xi\tau^3+\eta\tau^2)^{-2/5}}\lambda^{-2/5} = \text{const.} \times \lambda^{-2/5}, \tag{4.25}$$

thus confirming the critical behaviour computed from (4.23).

5. The duality group

It is known that the full duality group of the moduli space is given not only by the monodromy group of the periods, Γ_M, but also by the symmetry group of the defining polynomial $\mathscr{W}$, $\Gamma_{\mathscr{W}}$. We now want to show that the symmetries of the defining polynomial $\mathscr{W}(y;\ a,\ b)=0$ give at most a central extension for the monodromy group B_5 acting on the three-dimensional basis of the periods.

It is easily seen that the transformations leaving invariant $\mathscr{W}$ are given by

$$\begin{cases} a \to \rho a, \\ b \to \rho^{-1} b, \end{cases} \tag{5.1}$$

with $\rho^5 = 1$, as they can be undone by the linear coordinate transformation

$$\begin{cases} y_4 \to \rho y_4, \\ y_5 \to \rho^{-1} y_5. \end{cases} \tag{5.2}$$

Since there is apparently no other action with this property, we conclude that the duality group of the superpotential is given by $\mathbb{Z}_5$.

In order to find the representation U of the transformations (5.1) on the periods, we observe that on any integral, say ω_{51}, the transformation (5.1) can be compensated in the integrand by the map

$$\tau \to \rho^3 \tau. \tag{5.3}$$

On the other hand, choosing $\rho = e^{4\pi i/5}$, the transformations (5.1) on the $(p,\ q)$ real plane correspond to a rotation of an angle $\frac{4}{5}\pi$, mapping the fifth branch of the hypocycloid into the first, so that γ_{51} is mapped into γ_{12}. Taking into account that $d\tau \to \rho^3\, d\tau$, we find

$$U:\quad \omega_{51} \to e^{2\pi i/5}\omega_{12}, \tag{5.4}$$

and analogous relations for cyclically permuted indices. We now observe that the transformation $\omega_{51} \to \omega_{12}$ is realized by the monodromy operator

$$Z = \mathscr{R}(\alpha_3\alpha_2\alpha_1\alpha_5) = \begin{pmatrix} 0 & 1 & 0 \\ 0 & 0 & 1 \\ z^4 & -1+z-z^2 & z-1 \end{pmatrix}, \tag{5.5}$$

which corresponds to a generator of the cyclic subgroup $\mathbb{Z}_5 \subset B_5$. It follows that

$$UZ = e^{2\pi i/5}\mathbb{1} \tag{5.6}$$

on any period $\omega_{i,i+1}$ and therefore also on the selected basis $\{\omega_{45}, \omega_{51}, \omega_{12}\}$. Thus we conclude that, unless there is some element of B_5 represented by UZ, the U transformation gives a central extension of the braid group B_5. The above central extension gives the full duality group of the moduli space of the Calabi–Yau manifold.

Notice that our result differs from the previously studied one-dimensional examples, where the full duality group Γ was given by the semidirect product $\Gamma_M \ltimes \Gamma_{\mathscr{W}}$ of the monodromy group of the periods and the symmetry group of $\mathscr{W}$, as suggested in ref. [17]. We further remark that, since the moduli space is two-dimensional, we may take as coordinates the ratios $t_1 = \omega_{12}/\omega_{51}$, $t_2 = \omega_{45}/\omega_{51}$, which correspond to a linear combination of the "special" variables of Special Geometry. Hence, on t_1, t_2, the action of the full duality group is given by a faithful projective three-dimensional representation of B_5.

Recalling that the full symmetry of the moduli space is given by modding out by Γ the local moduli space, we obtain that the geometry of $\mathscr{M}$ is given by

$$\mathscr{M} = \frac{U(1,2)}{U(1) \otimes U(2)}/\hat{B}_5, \tag{5.7}$$

where $\hat{B}_5$ is the previously introduced central extension of B_5.

6. Conclusions

Some comments are in order. We have found a three-dimensional representation of the monodromy group for the three fundamental periods ω_{45}, ω_{51}, ω_{12} given by the three integrals associated to independent loops of the integral (3.9), or, equivalently, to the top solution of the system of differential equations (3.18). We know that B_5 must act as a group of discrete isometries on the local moduli space U(1, 2)/[U(1) ⊗ U(2)] and therefore our matrices should belong to the U(1, 2) group. In fact, it can be shown that the matrices (4.21) satisfy

$$\mathscr{R}(a_i^\dagger) g \mathscr{R}(a_i) = g, \tag{6.1}$$

where g is the metric given by

$$g = \begin{pmatrix} 1 & -z^2 & 1-z^3 \\ z^3 & 1-z+z^4 & -z^2 \\ 1+z^2 & z^3 & 1 \end{pmatrix}. \tag{6.2}$$

Since g has one positive and two negative eigenvalues, indeed $\alpha_i \in$ U(1, 2). As we have already remarked, there must exist a canonical basis for the $H_{(3)}$ homology of the Calabi–Yau, where the direct sum of the 3 and $\bar{3}$ representations of U(1, 2) given by (4.21) and their complex conjugate take values in Sp(6, $\mathbb{Z}$), six being the dimension of $H_{(3)}$. We leave the determination of this change of basis to forthcoming work.

Let us summarize our results. Starting with the family of curves given in eq. (1.2) we have been able to compute exactly the duality group of the periods associated to $\mathscr{W}(y;\, a,\, b)$ by means of some very efficient and powerful techniques of algebraic geometry, without resorting to the explicit computation of the periods of e.g. via solutions of the Picard–Fuchs equations. Our method is in principle applicable also to more complicated situations where more moduli are present and/or Yukawa couplings are non-vanishing. In fact, the computation of the fundamental group $\pi_1(\mathbb{CP}(N) - L^{N-1};\, B)$ is always possible in virtue of the fundamental theorems of Picard–Severi and Zariski, together with the van Kampen relations. The actual construction of the monodromy group relies however also on the knowledge of the behaviour of the periods around the singular locus of the defining polynomial. In our case, this computation was derived from the study of the one-dimensional integral, which simplifies the actual task. It is clear that in general one cannot expect that the periods can always be reduced to such one-dimensional integrals, and the examination of the leading singularity can be more involved. Still, it is important to realize that, as mentioned in sect. 3, the analysis of the singularities can always be done in a systematic way from the linear system of Picard–Fuchs equations using standard techniques of fuchsian analysis.

We hope to extend these techniques for the computation of the duality group to more complicated examples in future publications.

It is a pleasure to thank G. Ponzano for many illuminating discussions during the early stages of this work. We also thank P. Candelas, X. De La Ossa and S. Ferrara for many interesting comments.

References

[1] K. Kikkawa and M. Yamasaki, Phys. Lett. B149 (1985) 357;
N. Sakai and L. Senda, Prog. Theor. Phys. 75 (1986) 692;
V.P. Nair, A. Shapere, A. Strominger and F. Wilczek, Nucl. Phys. B287 (1987) 402;
A. Giveon, E. Rabinovici and G. Veneziano, Nucl. Phys. B322 (1989) 167;
A. Shapere and F. Wilczek, Nucl. Phys. B320 (1989) 167;
M. Dine, P. Huet and N. Seiberg, Nucl. Phys. B322 (1989) 301;
J. Molera and B. Ovrut, Phys. Rev. D40 (1989) 1150;
J. Lauer, J. Maas and H.P. Nilles, Phys. Lett. B226 (1989) 251; Nucl. Phys. B351 (1991) 353;
W. Lerche, D. Lüst and N.P. Warner, Phys. Lett. B231 (1989) 418;
M. Duff, Nucl. Phys. B335 (1990) 610;
A. Giveon and M. Porrati, Phys. Lett. B246 (1990) 54; Nucl. Phys. B355 (1991) 422;
Giveon, N. Malkin and E. Rabinovici, Phys. Lett. B238 (1990) 57;
J. Erler, D. Jungnickel and H.P. Nilles, Max Planck Institute preprint MPI-Ph/91-90;
S. Ferrara, D. Lüst, A. Shapere and S. Theisen, Phys. Lett. B233 (1989) 147;
J. Schwarz, Caltech preprint CALT-68-1581 (1990); Phys. Lett. B272 (1991) 239; *in* Strings: Stony Brook 1991 (World Scientific, Singapore, 1991);
J. Erler, D. Jungnickel and H.P. Nilles, Max Planck Institute preprint MPI-Ph/91-81

[2] E. Witten, Phys. Lett. B155 (1985) 151

[3] P. Candelas, X.C. de la Ossa, P.S. Green and L. Parkes, Phys. Lett. B258 (1991) 118; Nucl. Phys. B359 (1991) 21

[4] D. Morrison, Essays on mirror manifolds, ed. S.T. Yau (International Press, 1992);
A. Font, Nucl. Phys. B391 (1993) 358;
A. Klemm and S. Theisen, Nucl. Phys. B389 (1993) 153

[5] P. Candelas, private communication; Talk held at Eloisatron Project, Erice, December 1992;
P. Candelas, X. De La Ossa, A. Font, S. Katz and D.R. Morrison, Mirror symmetry for two parameter models-I, CERN-TH.6884/93

[6] R. D'Auria and S. Ferrara, String quantum symmetries from Picard–Fuchs equations and their monodromy, CERN-TH.6777/93, POLFIS-TH.24/93

[7] T. Regge, The fundamental group of Poincaré and the analytic properties of Feynman relativistic amplitudes, Nobel Symposium Series VIII (1968);
G. Ponzano and T. Regge, *in* Proc. Varna Conf., 1968

[8] G. Ponzano, T. Regge, E.R. Speer and M.J. Westwater, Commun. Math. Phys. 15 (1969) 83; 18 (1970) 1

[9] See O. Zariski, Algebraic surfaces, second ed. (Springer, Berlin, 1971)

[10] O. Zariski, Ann. Math. 38 (1937) 131

[11] E.R. van Kampen, Amer. J. Math. 55 (1933) 255

[12] A. Strominger, Commun. Math. Phys. 133 (1990) 163;
L. Castellani, R. D'Auria and S. Ferrara, Phys. Lett. B241 (1990) 57; Class. Quantum Grav. 1 (1990) 317

[13] A. Ceresole, R. D'Auria, S. Ferrara, W. Lerche and J. Louis, Intern. J. Mod. Phys. A8 (1993) 79

[14] L.J. Dixon, V.S. Kaplunovsky and J. Louis, Nucl. Phys. B329 (1990) 27

[15] E. Cremmer and A. Van Proyen, Class. Quantum Grav. 2 (1985) 445

[16] A. Cadavid and S. Ferrara, Phys. Lett. B267 (1991) 193

[17] W. Lerche, D. Smit and N. Warner, Nucl. Phys. B372 (1992) 87

[18] V.I. Arnold, S.M. Gusein-Zade and A.N. Varchenko, Singularities of differentiable maps, Vol. II (Birkhauser, Basel)

PART IX

Popularization of Science

Regge Multimedial Science Popularizer

Piero Bianucci

La Stampa

Tullio Regge wrote 394 articles for the Italian daily newspapers. Twenty-one appeared in the "Gazzetta del Popolo", the newspaper of his debut, which ceased publication on July 30th, 1981. 346 articles followed in "La Stampa" (of which 123 in the "Tuttoscienze" supplement) and 27 in "la Repubblica". For thirty years, from 1979 to 2009, Tullio Regge's signature was diligently present, and it only became less frequent when he became a Member of the European Parliament (1989–1994). The collaboration with "la Repubblica" is a brief parenthesis at the beginning of the millennium: it coincided with a column he was given by the monthly magazine "Le Scienze", belonging to the same editorial group as the Roman newspaper. Regge used that column above all to confront polemically controversial topics: the defense of GMOs, the criticism of the "precautionary principle" of the Greens, the fight against cigarettes and the tobacco multinationals, the denunciation of pseudoscience, and equal opportunities for the disabled. Then he returned to "La Stampa".

He wrote quickly, as spontaneously as he spoke. This was his secret, however transparent and unconscious. Each piece was made up of simple words and his phrases were short, rhythmic, and almost jazzy, a style he had absorbed from English, perhaps from the scientific supplement of the "New York Times", where Walter Sullivan, one of the best American science reporters, wrote. Obviously the scientific terms were the same, and he did not always have the space to explain unusual concepts such as symmetry breaking, the quantum void, or gravitational singularities. But Regge minimised the technicalities and courageously simplified complex concepts, at the cost of exposing himself to criticism from his more supercilious colleagues.

It is immodest to say this, but I am responsible for the birth of the journalist Regge. I had met him on the occasion of an interview as he passed through Italy when he was still at the Institute for Advanced Studies in Princeton, New Jersey. For the past ten years I had been curating the cultural page of the "Gazzetta del Popolo": I proposed to him to write current scientific articles, a request he granted me when in 1979, after receiving the prestigious Einstein Medal, he returned permanently to Turin. It amused him very much. "It's his new toy," his wife Rosanna Cester smiled with indulgence.

Fun was the only incentive that could drag him into such a widespread activity, but soon the diffusion of science became a growing commitment for Regge. Contemporary to the first articles in the "Gazzetta del Popolo" were the very crowded conferences held at the Cultural Union of Turin on the occasion of the centennial of Einstein's birth. The conferences allowed a much more relaxed language than the articles: thousands of people followed one of his evenings at the Palasport while the screen showed 80 slides of planets, stars, nebulae, and galaxies that Regge had brought back from the United States. Then the first images from the "Pioneer", "Viking" and "Voyager" probes, launched in the solar system, were circulating. The laser pointer looked like a bazooka and it was a technological marvel. At the Pellerina Park during the summer, the town council organised the "Green Points": one evening Regge attracted ten thousand people to his talk about the fundamentals of computer science, another time showing the Moon and Jupiter with a camera set in the focus of an amateur telescope.

Around the middle of the 1980s, popularisation of science in newspapers, squares and theatres found a natural outlet in radio and television programs that I curated at the time at the Rai Turin office. Directed by Bruno Gambarotta, four series of 12 episodes each were created for the Department of School and Education, dedicated to astronomy (*Towards the infinite and back*, with among the guests Margherita Hack, Mario Girolamo Fracastoro, Giovanni Gòdoli, Marcello Coradini, Francesco Bertola, Giancarlo Setti), to particle physics (*Journey inside the atom*, with Primo Levi, Gian Carlo Wick, Carlo Rubbia, Chen Ning Yang, Nicola Cabibbo, Toraldo di Francia, Remo Ruffini), to the concept of symmetry in physics, mathematics, crystallography, chemistry, biology, art, literature, music (*Behind the mirror*, with the composer Luciano Berio, the anglist Claudio Gorlier, the mathematician Alberto Conte, the physicist Vittorio De Alfaro, the neuroscientist Ruggero Pierantoni and the biologist Pier Carlo Marchisio) and life sciences (*Good cells, bad cells*, a series curated by Marchisio himself). They were programs of extreme sobriety: 28-minute episodes, three chairs, basic set, a few still images, some short films. But the School Education Department allowed for in-depth discourses that would be unthinkable on current TV and Bruno Gambarotta succeeded in devising elegant acronyms, with music by Ludovico Einaudi, who was thirty at the time. Unfortunately, the broadcast was programmed with little care. The episode with the Nobel Prize winner in physics, Chen Ning Yang, was interrupted to broadcast to united networks the President of the Republic's New Year's speech, and the second half remained unpublished. However the didactic result was appreciated and still now some occasional rerun can be seen on the Rai School Channel.

In 1987 Regge had even more opportunities to practice the popularisation of science with the birth of the annual *Experimenta* exhibition, promoted by the Piedmont Region, and of the cycles of spectacularised conferences *Giovedì Scienza* at the Colosseo Theatre in Turin. The first edition of *Experimenta* introduced in Italy the teaching method of the Exploratorium of San Francisco founded in 1967 by Frank Oppenheimer, brother of Robert (the physicist who had led the atomic bomb project and then the Institute for Advanced Study at Princeton). The novelty consisted in involving the visitor directly in the experiments according to the rule "learning by doing", a reclamation of the hands as extensions of the brain, and of the whole body in the case of immersive experiments. The interactivity in science education inaugurated at *Experimenta* was a great success, and this led the Region to organise 19 other annual editions on the most diverse themes: Regge presided over the scientific committee for a long time. *Giovedí Scienza* has been filling the 1400 seats of the Colosseo Theatre for 33 years, and from 2013 it also has a streaming broadcast. The conferences have exceeded 400 thousand overall views. Since 1987, 500 conferences have been staged by 400 lecturers, including several Nobel prizewinners.

In the 1990s the personal computer came into common usage and the Internet began to spread, even more so when at CERN the www hypertext protocol was developed by Tim Berners Lee. Initially this was intended for physicists at CERN, and then was made available to the internet community at large. For some time Regge had been programming small home computers to produce drawings with allusive, ironic, playful titles. The popularity of personal computers offered Regge a new dissemination tool that allowed the material interactivity of *Experimenta* to be transferred to the screen, in a sort of virtual laboratory: the CD-ROM. With the collaboration of Federico Tibone, a physicist who had graduated from Turin with Regge and worked in England on the JET program for controlled nuclear fusion [see his contribution to this volume], the CD-ROM *Tullio Regge explains Einstein's relativity* was born, and it extended into tens of thousands of copies with "La Stampa-Tuttoscienze". This success was repeated with *In the space between the stars, From Galileo Galilei to the Galileo probe* (summary of the most recent astronomical discoveries curated in collaboration with the present writer), *Close to the music*, with Luciano Berio, and *The arrow of time*, with Federico Tibone as co-author.

In parallel, a publishing activity was born. The articles written for the "Gazzetta del Popolo" formed the first science popularisation book by Tullio Regge, *Chronicles of the Universe*, published in 1981 by Paolo Boringhieri (now Bollati Boringhieri). Many successful books followed with various publishers, including *The Wonders of the Real* (1987), *Infinity* (1994), *Science and the Environment* (1996, fruit of the experience as member of the European Parliament, written with Maurizio Pallante), the chapter dedicated to cosmology in the *Atlas of the Universe* by Piero Bianucci and Walter Ferreri (Utet, 1997), *The Endless Universe* (1999), *Letter to the young on science* (2004 — a passionate warning against anti-science and pseudoscience), and *The infinite search* (2012, autobiography written with Stefano Sandrelli). A case in itself is *Dialogue* (1984), a conversation between Tullio Regge and Primo Levi organised by Ernesto Ferrero; it first appeared in the Olivetti editions of the Community and then Einaudi published many reprints: an extraordinary autobiographical testimony of two protagonists of the second half of the twentieth century.

It is interesting to take a closer look at Regge's journalistic production. The debut in the "Gazzetta del Popolo" took place on the "third page" with a series of articles on Einstein and relativity. The "third page" (now replaced by cultural inserts variously located in the daily newspapers) was reserved for humanistic culture. Science was excluded, subordinated to the humanistic culture of Crocian origin. The articles of Regge on the "third page" therefore marked a strong discontinuity in Italian journalism and represent a challenge to the dominant culture.

Most of the articles are linked to current scientific topics but there are also incursions into habits and social commitment. Two articles trespass on science fiction, anticipating a narrative trend that was expressed in the collection of stories *Don't be afraid* (Ed. La Stampa, 1999). The first, *Perpetual Motion*, appeared in the "Gazzetta del Popolo" on August 5th 1979. It has thermodynamics as its theme but is written in narrative form: it is the story of a certain Santeddu, convinced to have invented perpetual motion, who clashes with an eminent professor of physics. He is ultimately proven to be right and subtracts the university chair from his adversary: pages written on an ironic and paradoxical register, where the author himself is recognised in the role of the expelled professor, and the other is one of the "misunderstood geniuses" who sometimes challenge "official science".

The second "anomalous" article, *The library of Borges puts into crisis the universe*, published October 24, 1981 in Tuttolibri, is a curious commentary on the famous story by Borges — *The Library of Babel* of 1941. The parodic register is immediately evident in the style "Á la Borges", perpetual motion returns, the "Newton–Einstein–Santeddu theory of gravitation" is cited, and a manuscript from the Upper Middle Ages signed by Eco also appears. Taking the indications of the Argentine writer to the letter, Regge wondered if the Library could really exist and calculated the number of books that could be obtained by combining the 25 signs of the alphabet in all possible ways. The books described by Borges have 450 pages each of 40 lines. They therefore contain 656 thousand signs. All the possible books, in the great majority devoid of sense because they consist of random sequences of letters, are therefore 25 raised to 656 thousand, a number of about 900 thousand digits. Regge observed that, in cubic centimetres, the volume of the current universe is represented by a mere 85 digits; the library would not be in the universe, unless it is open, obliged to expand to infinity. But perhaps there is no need for this hypothesis: the mass of books required by Borges's story would be so large that the library would collapse into a black hole. With its symbolic meanings, the outcome of the calculation deeply affected the Argentine writer, who responded with a long and equally paradoxical interview published by "Tuttolibri" on 12 December 1981.

How understandable is Regge's journalistic writing? Let us try an objective answer. In the United States around 1950 the Austrian Rudolf Flesch had devised a formula to measure the readability of an expository text. Taking up this idea, in 1988 at the University of Rome "La Sapienza", a group of linguists and educators led by the historian of the Italian language, Tullio De Mauro, elaborated the Gulpease comprehensibility index. This index, unlike that of Flesch, which only takes into account the brevity of words and sentences, has as its background the *Basic vocabulary* compiled by De Mauro, having as reference the frequency of use of the 250 thousand Italian words. The words considered "basic" are 7050. Among these, 2000 represent 86 percent of written and spoken language, according to their frequency of use, and should be understandable to all. The others are subdivided into words "of high usage" (around 2300) or of "high availability", again depending on their frequency. The Gulpease algorithm discriminates between these groups and takes syntactic complexity into account: a parataxic text with short side-by-side sentences is favoured, while many subordinate clauses drastically lower the index. This allows us to evaluate not only the agility of writing but also — albeit approximately — the information transmitted by a text, and to calibrate the measure on the reader's cultural level.

The Gulpease Index ranges from 0 to 100. A value below 40 identifies texts accessible only to those with university studies; the intermediate value between the two extremes distinguishes texts within reach of those with a high school diploma. Writings with Gulpease Index under 60 are difficult for those with a compulsory education diploma. Those who attended only primary school are already in difficulty with an index below 80. That said, we measured a sample of Tullio Regge articles, ranging from physics, mathematics, to current affairs and habits, according to the Gulpease Index. Here are the results.

On March 24, 1989, newspapers around the world reported a resounding fact: two electrochemists had obtained nuclear fusion at room temperature, paving the way for clean and almost unlimited energy. Regge commented on the news with technical detachment but in the ironic final joke he put forward a prophetic doubt: "I would have delayed the announcement till the beginning of next month". That is, on 1 April, precisely "April Fools' Day" , a classic of the Anglo-Saxon press that Regge had imported into Italy with the complicity of the editor of "La Stampa". The article on cold fusion has an index 50: a reading suitable for those with higher education, difficult for those who have middle schools, almost incomprehensible for those who have only primary school. Yet, of the 550 words of the article, 419 fall within the basic vocabulary, 59 are "of high usage" and 3 of "high availability". The most difficult sentence has index 39. The closing sentence has 64.

Mathematical entities and physical reality is an article written for the monthly magazine "Le Scienze". It concerns the eternal problem of whether numbers have ontological consistency (such as Platonic ideas) or are mere conventions that sometimes represent the laws of nature. Here the joke is already in the first sentence: "As is well known, the great problems that beset humanity are often resolved after a libation" (index 59: fault of the italian words *sovente* and *libagione*). The overall index is 50 again. Rather severe. But no sentence falls below 37, many are above 60, the theme is philosophical as well as scientific, and we are on a periodical magazine of aristocratic dissemination.

In August 1985 many were falling into ruin by gambling with the Lotto on 34, a "latecomer" number. Regge explains that gamblers unfortunately ignore the calculation of probabilities and denounces the iniquity of the State gambling. Out of a total of 667 words, 609 fall within the basic vocabulary. The Gulpease Index rises to 55.

On 11 October 1986, Regge remembers in "La Stampa" one of his masters, the Russian physicist Gleb Wataghin, who has just died at the age of 86. In a total 404 words, 328 are part of the 2000 basic vocabulary. The index is 56.

On November 20, 2002, Regge deals with the increasingly frequent discoveries of exoplanets and the possibility that some of them are habitable: in search of life in space consists of 542 words, of which 465, that is 89%, included in the Basic Vocabulary. The Gulpease Index is 52.

December 3, 2002 marks the anti-smoking article *Why the no-global are silent on Big Tobacco* ("la Repubblica"). There are 641 words, 561 belong to the basic vocabulary, the index is 52. *Venice and the disabled* is another article from 2002: 471 words, 393 included in the basic vocabulary, index 54.

The most day-to-day topics (the Lotto game, smoking, architectural barriers) have higher comprehensibility indices than strictly scientific topics. But there is something strange: for Regge's articles, the Gulpease indexes always appear too severe with respect to the impression one gets when reading. It is as if the algorithm finds abstruse texts that instead are easily accessible even to the less educated reader. The explanation is simple. Algorithms do not measure empathy, and above all, they do not have a sense of humor.

1. List of all the articles by Tullio Regge in daily newspapers

1.1. *"La Gazzetta del Popolo"*

(1) 10 January 1979, p. 3 "A man, an equation".
(2) 18 January 1979, p. 3 "Universe day zero".
(3) 6 February 1979, p. 3 "On the spaceship of the future".
(4) 15 February 1979, p. 3 "Space Odyssey".
(5) 17 April 1979, p. 3 "Those poor "negro" scientists of culture".
(6) 17 May 1979, p. 3 "After Harrisburg".
(7) 22 July 1979, p. 5 "Children of the sun star".
(8) 5 August 1979, p. 5 "Perpetual motion".
(9) 28 August 1979, p. 3 "But those twins betray Einstein".
(10) 21 September 1979, p. 3 "But does God play dice?"
(11) 26 September 1979, p. 3 "We children of the big bang".
(12) 19 January 1980, p. 3 "The physicist who discovered Newton's mistake".
(13) 8 February 1980, p. 3 "The woodworm of the cosmos is called entropy".
(14) 11 March 1980 p. 3 "Does God Play Dice?"
(15) 22 March 1980 p. 3 "Inside the atom".
(16) 28 March 1980, p. 3 "Nuclear power plants? Here's how they work".
(17) 15 October 1980, p. 3 "Democritus was right (but not really in everything)".
(18) 22 October 1980, p. 3 "Who saw a quark?"
(19) 24 January 1981, p. 14 "Nuclear fusion, today the new Prometheus dreams of it".
(20) 29 April 1981, p. 14 "The other universe (non-existent) of antimatter".
(21) 13 June 1981, p. 16 "Alien brother if you are there hit a beat".

1.2. *"La Stampa" and "La Stampa–TuttoScienze"*

(1) 28/10/1981 Neutrino hunting alive or dead to understand the universe.
(2) 9/12/81 Proton: all spying on his death.

(3) 12/16/81 How Einstein got the Nobel.
(4) 20/1/82 Vital equations. The debate on the anthropic principle.
(5) 3/2/82 Carbon dioxide is growing. Will we have oranges in Scandinavia?
(6) 17/2/82 Will the wandering asteroid become a colony?
(7) 10/3/82 The patience to resist the charm of the black hole.
(8) 28/4/82 The vice of discovering Einstein's mistakes.
(9) 1/5/82 Aristotle dixit but the scientist does not listen to it.
(10) 3/5/82 Italy changes from the old emigration to America.
(11) 5/5/82 Einstein–Bohr. The very mysterious quanta.
(12) 10/5/82 The megacity project is causing debate and controversy.
(13) 12/5/82 The only certain thing is an incomplete truth.
(14) 17/5/82 Italy changes from the mass race towards educational qualifications to the search for practical outlets.
(15) 19/5/82 A syllabary of quantum mechanics: waves or corpuscles of light a story that does not end.
(16) 26/5/82 A fossil from space.
(17) 23/6/82 The universe is full of memories.
(18) 21/7/82 The mysteries of a water molecule.
(19) 23/7/82 A war cry, neutral science.
(20) 8/9/82 How does life come from inert molecules?
(21) 15/9/82 The discoveries of biochemistry seen by a great physicist.
(22) 22/9/82 The assembly line of life.
(23) 29/9/82 The biochemistry seen from the physicist.
(24) 11/17/82 Many atomic surprises if water is weighed.
(25) 8/12/82 Difficult to find another land.
(26) 15/12/82 How the double helix of DNA works.
(27) 24/12/82 A scientist and a writer, hopes and fears.
(28) 12/1/83 A cosmic spinning top the fastest pulsar.
(29) 23/1/83 The experiment of the physicist Rubbia in Geneva.
(30) 16/2/83 Not even 3C273 can disprove Einstein.
(31) 9/3/83 Old age and death of a star.
(32) 6/4/83 With nuclear fusion man will imitate the stars.
(33) 14/7/83 At the international meeting GR10 in Padua Don Giovanni and relativity.
(34) 9/19/83 The earth vibrates like a drum to the rhythm of gravitational waves.
(35) 11/15/83 New hypotheses of cheerful scientists. The universe will not end.
(36) 12/21/83 Quarks, with six bricks the universe is built.
(37) 1/18/84 Slow down time. Einstein's miracle verified in the atom.
(38) 2/2/84 I don't believe in astrology because I know astronomy.
(39) 4/18/84 Kapitsa's long journey through the planet of absolute zero.
(40) 3/5/84 Many mysterious twists in the first second of the Big Bang.
(41) 6/6/84 The most round spheres ever built to confirm Einstein's theory.
(42) 8/8/84 Energy explorers.
(43) 3/10/84 From span to absolute precision.
(44) 18/10/84 Antimatter captured - Rubbia Carlo Nobel Prize.

(45) 31/10/84 Paul Dirac prophet of antimatter.
(46) 3/11/84 book by the famous German scientist Heisenberg.
(47) 4/12/84 Primo Levi - Regge Tullio, dialogue on science.
(48) 12/19/84 The year of the Italian Nobel Prize.
(49) 9/1/85 True: the quasars come from the Big Bang.
(50) 1/23/85 India remembers the great Romanujan Srinivasa.
(51) 2/27/85 How many dimensions for the secrets of the new physics.
(52) 3/4/85 The cosmos at the crossroads: hot death or cold death.
(53) 21/4/85 Primo Levi explores the thought of others.
(54) 8/5/85 Where is the glue that joins the galaxies?
(55) 9/7/85 In search of absolute security.
(56) 8/18/85 Science and humanistic culture against the fantalatino.
(57) 8/23/85 The frenzy of the lot, the calculation of probabilities.
(58) 18/9/85 Einstein also updates.
(59) 1/10/85 America far away.
(60) 11/26/85 Regge Tullio and the controversy over the plant.
(61) 12/18/85 Balance of a year of science, at the end of the atom a string.
(62) 12/28/85 The physicist Regge Tullio struggling with the machine.
(63) 10/1/86 Conference with scientists and philosophers.
(64) 23/1/86 Pro and against the new plants of Enel in Trino.
(65) 5/2/86 Within a year we will know if in nature there is a fifth force.
(66) 2/25/86 Prof. Regge and the ecological initiative.
(67) 27/2/86 Seminars and lessons: Einstein in Naples.
(68) 5/3/86 For the nuclear fusion Italy is at a crossroads.
(69) 3/14/86 Giotto sent photos and data up to 577 kilometers.
(70) 3/4/86 A baron in Rome at the chair factory.
(71) 4/30/86 The teaching warning of the terrible accident.
(72) 10/5/86 There are more lasting threats than radioactivity.
(73) 8/6/86 The cherry on the ice cream.
(74) 6/15/86 At school Darwin or the Bible?
(75) 28/6/86 The physicist also intervenes in the open controversy.
(76) 6/7/86 When science faces the paranormal world.
(77) 11/7/86 Rol replies to Regge. The magician is not needed.
(78) 7/16/86 Causes and consequences. A report from scientists.
(79) 8/17/86 After 10 years adventurous return to campus.
(80) 8/28/86 Reflections on Chernobyl and nuclear safety.
(81) 10/9/86 Higgs teaches that nothingness is complicated.
(82) 2/10/86 Surprise in Turin in the mathematical library.
(83) 11/10/86 Wataghin, pioneer of the cosmos, died at 86 years old.
(84) 10/16/86 Levi Montalcini Rita and other scientists.
(85) 10/29/86 Not so brilliant students look for the professor.
(86) 6/11/86 That multitude of asteroids speaks of space.
(87) 11/20/86 The universe could end with a leap into the void.
(88) 3/12/86 Physicists make a small Big Bang.

(89) 12/16/86 Einstein virtue of a creator.
(90) 23/12/86 Boredom at the faculty council. Sadistics in quotation marks.
(91) 18/1/87 The controversy Rubbia Colombo, neutral scientist not to be found.
(92) 18/2/87 Science is neutral, scientists no.
(93) 20/3/87 At 94 he was one of the fathers of quantum theory (De Broglie).
(94) 21/3/87 Skies forbidden to the theologian.
(95) 9/4/87 Energy Superconductors, frontiers of the 2000s.
(96) 14/4/87 An intellectual between two cultures. An artist for science.
(97) 15/4/87 Maxi electric powerplants irreconcilable with the environment.
(98) 29/4/87 In subatomic democracy all electrons are equal.
(99) 3/6/87 Superconductivity at room temperature.
(100) 10/6/87 The more the physical research advances the more the concept of elementary particle becomes vague.
(101) 2/7/87 Mountain View. Discovered by NASA, the journey of almost 1000 years of the phantom planet.
(102) 10/7/87 Nuclear power and human error in Chernobyl.
(103) 5/8/87 Galloni, Ruberti and the university: between the two litigants I choose Totò
(104) 7/10/87 The scientific and philosophical debate on quantum mechanics.
(105) 23/10/87 For the referendum on the nuclear question, Regge: I will vote in this way.
(106) 11/26/87 Between USSR astronauts in the city of stars.
(107) 12/23/87 No more fines for speeding but for dangerous momentum.
(108) 6/1/88 A concept of dynamics analyzed by the gyroscope.
(109) 9/1/88 Physics Fiction, Regge Tullio and the spirits of Skey Malcom. Those ghosts called UWO.
(110) 23/1/88 The views of Saturday, shadows of the Big Bang.
(111) 26/1/88 Meeting with Chung W Kim witness of the bomb.
(112) 7/3/88 Jewish culture and Europe.
(113) 1/4/88 Magnetic nuclear resonance enters the kitchen.
(114) 6/4/88 Harsh controversy in the United States for Alvarez's theory of dinosaurs.
(115) 27/4/88 The point on a fundamental property of the physical world.
(116) 5/31/88 Sun effect in search of the beautiful summer.
(117) 6/29/88 Stephen Hawking's bold theories to describe the dawn of the cosmos.
(118) 23/7/88 Needless alarmism about environmental radiation.
(119) 9/13/88 Congress in Perth: scientific news and threats to the environment.
(120) 3/11/88 Moral and ecology between Australia and Thaiti.
(121) 11/24/88 Tullio Regge on the discovery announced in the USA.
(122) 7/1/89 The selected works of the great physicist come out.
(123) 21/1/89 Against the barbarians a slight barrier for the handicapped.
(124) 4/2/89 The greenhouse effect, the ozone hole and the rains.
(125) 3/3/89 The environment also pays the price.
(126) 4/3/89 The negative effects of the Enel monopoly.
(127) 8/3/89 The future is not worth the Russian roulette.
(128) 3/16/89 The problem of architectural barriers remains.

(129) 22/3/89 The scale of minimum and enormous physical quantities.
(130) 3/23/89 Cold nuclear fusion. The sensational announcement of two researchers.
(131) 3/29/89 Moral and physical considerations about electricity.
(132) 1/4/89 New hypotheses on the origin of the universe (fossil neutrinos).
(133) 2/4/89 Neutrini fossili joke of Regge.
(134) 9/4/89 Atomic risk in the abyss. Nuclear waterloo.
(135) 11/4/89 Cold fusion does not solve today's problems.
(136) 12/4/89 Technical progress did not act in the same way.
(137) 18/4/89 Disabled USL and missed promises.
(138) 4/5/89 The smog for heating, the condominium is not green.
(139) 25/5/89 The spinal unit. Inefficiency does not exist.
(140) 6/26/89 A turning point, a challenge. Culture and image.
(141) 6/28/89 Lep was born at the Cern in Geneva, hunting the Higgs particle.
(142) 5/7/89 Pollution and engines, diesel and the Sumerian.
(143) 20/7/89 Energy crisis and missed opportunities.
(144) 3/8/89 Better a radio telescope. I wouldn't go to Mars.
(145) 8/19/89 Electricity and heating, heat should not be wasted.
(146) 8/30/89 Technology, first, save siliceous gel.
(147) 3/10/89 Fairbank died. He was a hero of normal physics.
(148) 12/10/89 Regge replies to Pera. Frankenstein does not live here.
(149) 8/11/89 A disabled person in Strasbourg.
(150) 9/11/89 The rights of young children, genes in crisis when examining life.
(151) 10/11/89 Cinderella science.
(152) 6/12/89 Secrets of mental calculation.
(153) 10/7/90 Poland suffering from coal.
(154) 8/25/90 The energy crisis makes our choices more difficult.
(155) 26/10/90 Regge and the evil that killed Bernstein: damn smokers.
(156) 7/11/90 If nature is left-handed.
(157) 12/12/90 Nuclear energy, endless puzzle.
(158) 12/27/90 Energy crisis. A course in thermodynamics.
(159) 12/28/90 Here are the 5 characters of the year.
(160) 9/1/91 A scale for atomic risks.
(161) 1/30/91 Nuclear risk is still in the east.
(162) 8/2/91 Heating: too much waste in the name of the law.
(163) 20/2/91 The museums where science makes the show.
(164) 20/3/91 Energy and eco-generation: there is a treasure in hot water.
(165) 5/6/91 The involuntary censor.
(166) 10/23/91 Birthday parties, upcoming surprises.
(167) 10/30/91 Energy policies.
(168) 11/11/91 The nuclear fusion experiment. The remote future of our energy.
(169) 11/12/91 Killer vice.
(170) 12/29/91 Flag down, USSR goodbye.
(171) 20/1/92 European premiere, Italy launches the policy of slaps in Strasbourg.
(172) 22/1/92 Dear Alberoni, you're wrong, scientific rationality is neither dry nor amoral.

(173) 1/4/92 Uranium corals, environment killer.
(174) 2/4/92 There are no killer corals that have absorbed uranium.
(175) 4/22/92 He researched energy and the atom, goodbye to the physicist Wick, master and gentleman.
(176) 29/7/92 Nuclear plants to the east. Nuclear power is at risk.
(177) 8/19/92 And if we all became vegetarians?
(178) 8/22/92 B Series animals.
(179) 9/9/92 Relativity Congress for Einstein, a difficult marriage.
(180) 24/12/92 Hoax. The bitter welcome of bureaucracy.
(181) 12/29/92 Woody Allen, who cares about scandals?
(182) 12/1/93 A decree is not enough to change habits.
(183) 25/2/93 The EEC has commissioned Tullio Regge to study extraterrestrials.
(184) 3/3/93 UFO. The European community wants the truth.
(185) 24/3/93 And if the child is not perfect?
(186) 5/5/93 I don't believe it. Regge replies to the Israeli physicist who claims to have invented the time machine.
(187) 14/8/93 Bohr the physicist of the quanta.
(188) 22/9/93 Physics? Video games will teach it to you.
(189) 3/10/93 The handicapped tell about love.
(190) 10/11/93 My friend Eduardo Caianiello.
(191) 12/31/93 The only flaw of Poggiolini.
(192) 2/1/94 Enemies of the rituals.
(193) 3/1/94 He skimmed the Nobel. The physicist Occhialini is dead.
(194) 15/1/94 Diaries from Brussels.
(195) 12/2/94 False sightings to trigger panic.
(196) 3/20/94 Suffering and conscience.
(197) 12/4/94 How nice, it's mathematical.
(198) 04/27/94 Theory confirmed, found the needle in the haystack.
(199) 7/5/94 A 300-year-old puzzle.
(200) 29/6/94 Pure or applied research?
(201) 20/7/94 The night of the landing on the memories.
(202) 4/8/94 NEW FORM OF LIFE. That computer virus is watching us.
(203) 8/28/94 Regge accuses Arrabal's esotericism.
(204) 30/8/94 Brescia: parks are born without light pollution.
(205) 11/9/94 Ecology friendly to the economy.
(206) 9/13/94 Origins of the cosmos. Encrypted messages from the stars. The endless race to the whys of the cosmos.
(207) 23/9/94 Magnetic and inimitable personality.
(208) 11/21/94 Scandal in England, the psychosis of the wrong mother.
(209) 11/30/94 Is the cosmos younger?
(210) 14/12/94 The Nobel Nash on the border between genius and madness.
(211) 21/12/94 Nuclear waste, plutonium, instructions for use.
(212) 12/31/94 Berliner's art violated with a mobile phone.
(213) 14/1/95 But Rai must not kill Leonardo.

(214) 21/1/95 Regge comments on the discovery of the thirteenth zodiac sign.
(215) 1/3/95 From Einstein to anti-espionage.
(216) 8/3/95 Now the physics is at the top. End of the journey at the center of matter.
(217) 12/4/95 The cigarette is still the great killer.
(218) 24/5/95 Paranormal science. The last dance with the three-legged table.
(219) 5/7/95 The universe is growing but there are those who doubt it.
(220) 12/7/95 The children of Chernobyl. A report 9 years after the accident.
(221) 2/8/95 World Relativity Conference. An indestructible theory.
(222) 8/23/95 The colors revealed, as they are formed at the atomic level.
(223) 8/25/95 A letter from Tullio Regge.
(224) 8/30/95 The limits of truth. What will remain of Popper?
(225) 20/9/95 The Roswell UFO. A virtual autopsy?
(226) 18/10/95 Death of Rudolf Peierls. He launched the atomic alarm.
(227) 6/12/95 The discoveries made thanks to serendipity. Genius by chance.
(228) 9/12/96 Einstein was a man too.
(229) 31/12/95 The faces of 95.
(230) 17/1/96 The stages of discovery (from an idea of ??Dirac, the antimatter)
(231) 1/30/96 Regge criticizes the abuses, Scalfaro approves.
(232) 3/27/96 Classical culture, scientific culture.
(233) 3/5/96 Regge responds to Vattimo: technological progress is not a metaphor.
(234) 3/10/96 Research, charm and sweat.
(235) 10/24/96 An offer of peace between religion and science.
(236) 12/12/96 A debate between risk and environment. Less hypocrisies.
(237) 12/21/96 He was an astrophysicist and popularizer, Sagan died.
(238) 12/29/96 Karadzic the most fatal lies.
(239) 19/2/97 Between Latin and research. Prejudices about science.
(240) 5/3/97 The atomic laser was born.
(241) 6/3/97 Many evils could be avoided, the key word is prevention.
(242) 11/4/97 The discoverer Hale, useless to study physics, I am unemployed, the comet does not provide work.
(243) 16/4/97 There is not only the anti-scientific interpretation of Heidegger. (Regge responds to Vattimo)
(244) 4/17/97 Do I condemn Hitler or Stalin?
(245) 12/6/97 Research into energy and reason.
(246) 5/7/97 Space exploration needs scientific success and great funding.
(247) 2/10/97 3001 the odyssey returns and the dead resurrect.
(248) 31/12/97 The faces of 97.
(249) 4/2/98 Teleportation, is the dream approaching?
(250) 1/4/98 Einstein marks another point. Italian research confirms the gravitomagnetic force.
(251) 20/5/98 Why I reject the special supervised label. Alt to bioinquisition.
(252) 19/9/98 The errors of the monopoly.
(253) 11/25/98 Heisenberg beats Heidegger.
(254) 11/28/98 The nuclear issue is still reopened.

(255) 12/16/98 Cloning the hunt of moralists for fake Frankensteins.
(256) 12/19/98 The universe is getting bigger and growing at enormous speed.
(257) 11/1/99 Whoever denies cloning does not think about the ill.
(258) 27/1/99 Nuclear fusion, the future. How to turn on the sun in the laboratory.
(259) 3/3/99 Russian power plants, the sword of Damocles.
(260) 4/3/99 Regge remembers. Chalkboard companions.
(261) 3/17/99 (Regge replies to Grillo's attack on the Novartis industry) Dear Grillo, anathemas are not enough to feed the world.
(262) 12/5/99 Exceptional experiment at Harvard University. Pedalling at the speed of light.
(263) 5/25/99 It was the Torquemada of the prophets.
(264) 21/7/99 Waiting with terror for the year 2000. Against the new year.
(265) 8/8/99 Eclypses, incomparable show of the moon over the sun.
(266) 29/9/99 Between logic and physics. If the computer idles.
(267) 3/11/99 As a boy you studied piano ...
(268) 12/11/99 (TORINOSETTE) The scientist is having fun (Regge debuts in narrative).
(269) 11/22/99 Cosmos endless puzzle.
(270) 24/11/99 Controversy between Regge and the Greens. Biotechnology challenge.
(271) 12/22/99 Regge responds to Celli on transgenic foods.
(272) 12/29/99 From the atom to the borders of the cosmos.
(273) 1/1/2000 The fate of the universe.
(274) 23/2/2000 The green dodecalogue on biotechnology.
(275) 5/19/2000 Time and eternity, the unresolved doubt.
(276) 5/31/2000 After the agreement on disarmament signed at the UN, atomic peace is scary.
(277) 11/10/2000 Invention returns to the forefront.
(278) 11/29/2000 We have technicians and experience to proceed with waste treatment. The antinuclear demagogue.
(279) 16/12/2000 They are the children of light.
(280) 9/2/2001 Freedom of research, science will save the ragout.
(281) 3/15/2001 Europe to the states: do not prevent experiments.
(282) 9/11/2003 Teller, father of the atomic force, died the last Martian.
(283) 9/16/2003 The failure of the Cancun conference. The GMO is not the devil.
(284) 19/9/2003 A myth denied by history and news. Nature and its sibyls.
(285) 26/9/2003 How many ways to kill the earth.
(286) 29/9/2003 Nuclear power is not the devil.
(287) 2/10/2003 Dear Mattioli, I have the right to be an apostate.
(288) 10/21/2003 Too much light in the city, let us wonder about the sky
(289) 10/24/2003 Technology and internet at the service of disability. (Torinosette)
(290) 10/29/2003 An English research and Italian fears, do not fire on GMOs.
(291) 6/11/2003 The right scenario for nuclear power is repeated.
(292) 9/11/2003 The great artist was also a scholar of numbers and proportions.
(293) 11/11/2003 We need an agreement on uranium, the nightmare of terrorists with the atomic bomb.

(294) 11/15/2003 Fears of atomic disasters or environmental contamination are groundless.
(295) 11/19/2003 Special six hundred years of the University of Turin.
(296) 26/11/2003 The light bulb has revolutionized the world.
(297) 11/28/2003 Exaggerated alarm.
(298) 4/12/2003 The prejudices against cremation.
(299) 12/16/2003 How to curb tobacco's damage, more taxes and less smoke.
(300) 5/1/2004 The scientist, that's why the mission is important.
(301) 6/1/2004 The Kyoto agreements and nuclear policy.
(302) 9/1/2004 Stalingrad and Stalin (on Torino sette).
(303) 10/1/2004 Many risks few advantages. But is it really worth it?
(304) 21/1/2004 The attack on the famous physicist. Hawking's black hole. Enemy of the animals.
(305) 4/2/2004 The great characters of science. That evening with Gian Carlo Wick.
(306) 10/2/2004 Household appliances on fire, the strange fires of Messina.
(307) 13/2/2004 Spare parts for the body.
(308) 28/2/2004 The decree on nuclear waste is confused and incomplete.
(309) 9/3/2004 Ecotourism damages the environment, enemy of the animals.
(310) 13/3/2004 The military affected by Hodgkin's lymphoma.
(311) 19/4/2004 The evolution of computer viruses, Darwin in the PC.
(312) 29/4/2004 Doubts and nightmares about climate, the greenhouse effect.
(313) 4/5/2004 Prestigious scientific award to 4 young people of Acqui, the boys of the stars.
(314) 11/5/2004 Greenpeace's assault on GMOs. A return to the Middle Ages.
(315) 5/13/2004 Too easy to think of green men.
(316) 23/6/2004 A century after birth.
(317) 9/29/2004 The legendary discoverer of the incompleteness theorem.
(318) 30/9/2004 From OGM to cloning, when the antiscientific syndrome hits.
(319) 1/10/2004 With the signing of the Kremlin the protocol will enter into force.
(320) 2/10/2004 Young people, I urge you to science.
(321) 10/27/2004 There is good in GMOs too.
(322) 10/11/2004 Energy alarm.
(323) 11/22/2004 Rice modified with x-rays.
(324) 21/1/2005 A choice without alternatives.
(325) 20/2/2005 An extraordinary portrait in the new edition of the 1984 dialogue between the chemist and the physicist.
(326) 4/4/2005 A strong no on stem cells.
(327) 6/4/2005 A Europe for science and vice versa.
(328) 16/5/2005 The difficult relationship between science and philosophy.
(329) 5/27/2005 Regge: why I say yes.
(330) 29/7/2005 The resources of the human brain.
(331) 10/31/2005 The alternative is a difficult turning point, in Italian homes the waste is high.
(332) 10/1/2006 Games and disabilities, the first challenge in Turin against the step.

(333) 18/10/2006 Salvation is to convert to GMOs.
(334) 22/2/2006 Neutrino, mystery solved.
(335) 12/3/2006 The subtle pleasure of becoming a torchbearer in the crowd.
(336) 04/26/2006 A hi-tech sanctuary in front of the Andes captures cosmic rays.
(337) 5/31/2006 Outside the Middle Ages.
(338) 6/28/2006 Scientists and green, talk to each other now.
(339) 12/9/2006 The infinite race in the microscopic.
(340) 10/25/2006 The deception of catastrophism.
(341) 25/2/2007 A nice but useless gesture.
(342) 14/6/2007 It's Croce's fault.
(343) 25/7/2007 Cigarettes are back to attack.
(344) 9/9/2007 An ancient wall of distrust.
(345) 7/10/2007 The salami and the throat.
(346) 2/4/2008 Energy dossier Analysis :enough fears, solutions exist.
(347) 5/6/2008 Return to prejudice
(348) 26/2/2009 Pros and cons of splitting an atom.

1.3. *April's fool jokes*

(1) 1/4/88 Magnetic nuclear resonance enters the kitchen (La Stampa)
(2) 1/4/89 New hypotheses on the origins of the universe (fossil neutrinos) (La Stampa)
(3) 1/4/92 Killer uranium corals in the environment (La Stampa) DISCLAIM: 2/4/92 There are no killer corals that have absorbed uranium. (The print)
(4) 1/4/98 Einstein scores another point. (TuttoScienze)
(5) 1/4/2001 Telepathy moves things. Here are the proofs. (The Republic)

1.4. *la Repubblica*

(1) Do not lock yourself in ivory towers 7/16/2000
(2) Telepathy moves things. Here is the proof. 1/4/2001
(3) Nostalgia for the Middle Ages 8/4/2001
(4) "Zero risk" progress is only a beautiful utopia. 04/17/2001
(5) Here is the music of the universe. 04/30/2001
(6) Table radiation. 10/5/2001
(7) Too many alarms about the future. Science is not evil. 05/26/2001
(8) Environment, many words but no choice. 06/15/2001
(9) The color of the universe. 01/11/2002
(10) If the sun becomes a dwarf. 5/2/2002
(11) Venice and the disabled 16/2/2002
(12) The blinders on the environment 11/4/2002
(13) Great fault of the Latinists 17/4/2002
(14) Too many prejudices about nuclear energy. 14/5/2002
(15) Transgenic foods and Italian autarchy. 06/04/2002
(16) All the risks of environmentalism. 07/05/2002
(17) In search of lost stars. 11/7/2002

(18) The temptation to dissolve the CNR. 07/29/2002
(19) If light loses speed. 09/08/2002
(20) When fires save forests. 08/27/2002
(21) From tobacco a hope for the third world. 09/28/2002
(22) The emigrants of science. 9/10/2002
(23) In search of life in space. 20/11/2002
(24) But why are the non-global silent on big tobacco? 3/12/2002
(25) INFM: so a resource of the country is dispersed. 1/2/2002
(26) That map of the sky that reveals the mysteries of the Big Bang 5/3/2003
(27) All research numbers 3/6/2003

Regge's Cartoon Physics

Federico Tibone

By the mid 1990s, CD-ROM drives, sound cards, speakers and increased RAM had turned the average personal computer into a digital multimedia machine. This opened the way to the development of applications that could involve moving images, sound and hypertextual links. High-quality video was still out of the question; but software-driven animations, sound effects and spoken commentary could be used together to produce a new kind of interactive experience. In particular, one could now think of using computer applications to aid the teaching of physics or, more generally, to popularize scientific ideas. The theory of relativity seemed like an ideal testbed to try out these new techniques, for at least two reasons. On the one hand, relativity is part of the curriculum in science-oriented high schools, but it is nigh impossible to experiment with it in a school laboratory. Einstein's *Gedankenexperimenten*, however, can be effectively presented in an idealized, simulated world. In addition, the general public's fascination with relativity means that, even outside formal education, there is always a market for new ways to present Einstein's ideas.

Basic simulations of special relativity

Friends who worked with Tullio Regge said he might be keen on this sort of project and it would be worth discussing it with him. So I went to see him at home. Three minutes into my pitch for a computer simulation of special relativity, Regge raised his hand and said: "Let me show you something". So I followed his wheelchair to the office, where he proceeded to open a folder on his computer, double-clicked a file, and there it was — he had already done it. In his spare time, Regge had learned to program in BASIC and produced a set of interactive hypertexts, as exemplified by Figure 1. Using the keyboard, one could navigate through a number of screens describing a relativistic effect, then set the values of a few parameters and watch the resulting animation. The approach was straightforward: scale down the value of c, so that a beam of light can be visualized as it moves across the computer screen, then explore what happens when objects move at a sizeable, freely variable fraction of c.

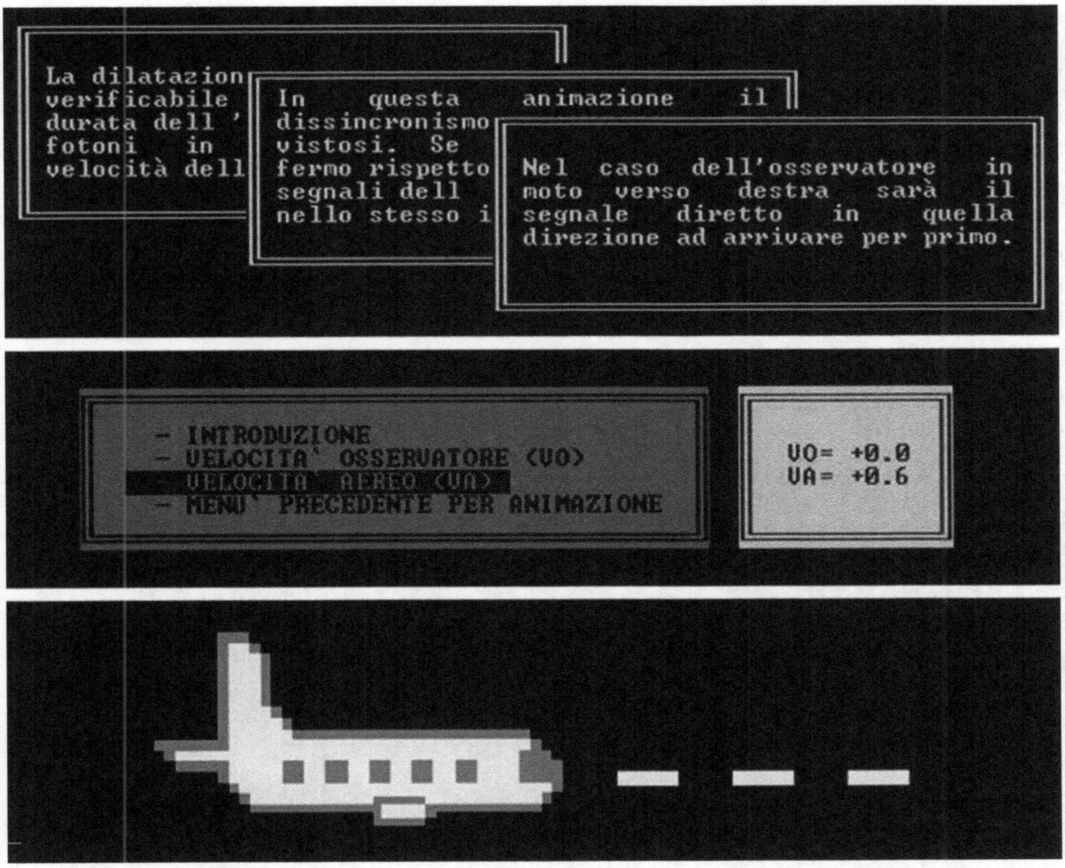

Figure 1. Regge's BASIC simulation of special relativity (1991).

Regge's simulations mostly used simple geometrical objects. You would see, for instance, how a square became more and more rectangular, contracting in the direction of relative motion, as its speed increased towards the speed of light. An undergraduate student might usefully get the message of these little programs, but we agreed that something friendlier was needed for younger students, and for a wider audience of non-specialists. So a more ambitious plan was drawn: the application would be an interactive cartoon, to be drawn by professionals and programmed by a software house, using the best scripting tools available at the time. It would be stored on a CD-ROM; this was the only practical way to distribute hundreds of megabytes of software, at a time when Internet connection was through 56 kbps modems. All interactions within the program would rely on the mouse, both to adjust physical parameters, by dragging graphical handles, and to start the corresponding animations. A click on other buttons would give access to animated explanations of the various relativistic effects, with commentary spoken by actors' voices.

An unusual co-production

It was not difficult to find a publisher prepared to invest in the project and capable of giving it a wide distribution. In fact, thanks to the appeal of Regge's authorship, two publishers agreed to co-produce the CD-ROM: Editrice La Stampa, a national daily newspaper, would sell it at newsagents, while Zanichelli editore, producer of the best science textbooks in Italy, would market it in high schools. Six months' work led to the finished product, a program so rich in content that it would take a committed user several hours to explore it all. It included a detailed chronology of Einstein's life and related historical events, a screen-saver based on Einstein's quotations and Regge's voice answering some frequently asked questions about relativity. But the real novelty was represented by the interactive animations. Time dilation, for example, was demonstrated using a cartoon version of "Einstein's clock", the universal time-keeper, here shaped like an hourglass (Figure 2).

Figure 2. "Einstein's clock" in Regge's relativity CD-ROM (© La Stampa/Zanichelli editore 1997).

A beam of light oscillates up and down, reflected by mirrors at the top and bottom of the hourglass, with a bing sound effect at each reflection. Two identical clocks are shown on the screen and the user can set either clock in motion. When one does, it becomes apparent — both visually and audibly — that the moving clock ticks more slowly than the stationary one, the effect becoming more dramatic as one increases the relative speed of the moving clock. In addition, at any time one can click on the moving clock to see things from its point of view: now the other clock is seen moving in the opposite direction, and one can sees that its time slows down, thus highlighting the perfect symmetry between inertial frames.

Regge's creativity surfaces in the invention of an object inspired by his longstanding passion for the works of Salvador Dalí, especially the celebrated Montres molles. "Dalí's clock" (Figure 3) is remarkable in that it displays, in a single animated object, three separate effects predicted by special relativity. When observed in motion, the hands take longer to go around the face (time dilation) and the clock is deformed along the direction of motion (Lorentz–FitzGerald contraction); at the same time, its hands also show that rigid bodies do not exist in a relativistic world. The hands of the moving clock are deformed, since each of their points has a different velocity component along the direction of motion, corresponding to a different dilation of time. The deformation varies in time, while each hand rotates at relativistic speed; as a result, Dalí's clock in motion really looks like a surreal, dynamical "soft clock".

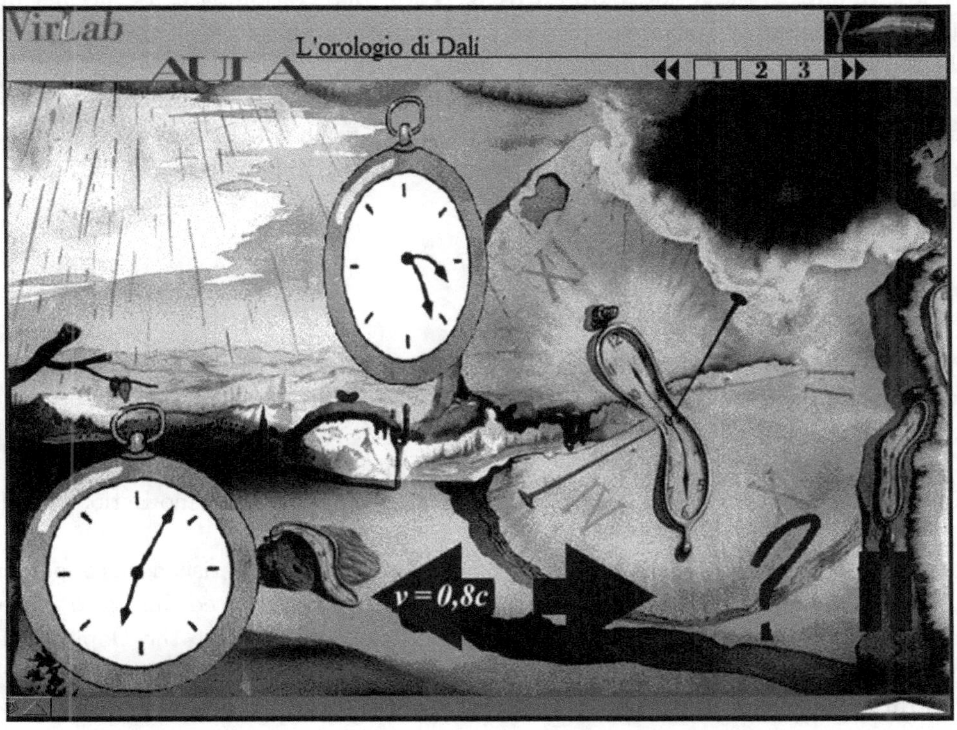

Figure 3. "Dalí's clock" in Regge's relativity CD-ROM (© La Stampa/Zanichelli editore 1997).

Norwegian woods

The relativity CD-ROM was remarkably successful. In June 1997, when it was first published, 35,000 copies were sold by newsagents in a week. For most buyers, it would be the first experience of a multimedia computer program. In the following school year the CD-ROM was also well received by teachers, so much so that a special edition was soon commissioned by Zanichelli. The program was expanded to include curricular subjects, such as Lorentz transformations and Minkowski space-time, and additional interactive problems for use in the classroom or at home.

A digital application devoted to science was a novelty in this early Internet era, and Regge was invited to give several public presentations. Unexpectedly, one such invitation came from the Italian Cultural Institute in Oslo, Norway. As soon as he heard the news, Regge happily told his wife: "Rosanna, we are off to Hessdalen!". He had read about the mysterious lights observed in that remote Norwegian valley and, ever curious about unexplained phenomena, was keen to go and have a look for himself. So it was that one February afternoon, after a presentation at Trondheim University, we were driven through the snow-clad country. At dusk, dressed up like an Arctic explorer, Regge was helped off his wheelchair and onto the back of a snowmobile. There followed an adventurous 20-minute ride, climbing over hill after hill in fresh snow, following invisibile tracks through birch woods known only to the reckless local drivers. At the observation base, a container in the middle of nowhere, a researcher was retrieving tapes with the most recent recordings — this was before webcams made remote measurements possible. Flattered to meet the famous theoretician, he gave a tour of the equipment, played recordings of the lights and showed results of spectral analyses. The famous lights eluded us that night, and the science left Regge rather unimpressed. For him, however, the snowmobile ride alone had been well worth the trip.

Berio's Canons for Tullio Regge

Following the success of the relativity CD-ROM, the publishers asked for the format to be expanded into a series on various science topics. So the following years saw the production of three more programs. One was devoted to astronomy, with Piero Bianucci as co-author, another dealt with entropy and the arrow of time. Most exciting for Regge, however, was the final CD-ROM in the series, scripted with his longtime friend Luciano Berio, who chose the title *Vicino alla musica* (Figure 4).

Regge, who played the piano and had a deep passion for music, cherished this opportunity to do work together with the Maestro. In the program, Regge used animated cartoons to explain the physical properties of sound, from Mersenne's laws to the Doppler effect, while Berio reviewed intervals and scales used across the centuries and the musical styles, demonstrating his points with dozens of excerpts from classical and contemporary works.

A special bonus emerged from the collaboration when Berio composed for the occasion six short canons of increasing complexity, all starting with the first five notes C-D-E-F-G, and dedicated them to Regge. Berio's *Canoni in do re mi fa sol* were then used to explore the border between music and noise, exploiting the computer's ability to synthesize sounds

Figure 4. Regge's computer drawing for the cover of Vicino alla musica (1999).

and randomize numbers. In the CD-ROM, one could play the canons selecting different instruments or noises for each voice, or even ask the program to randomly assign a different timbre to every single note.

Legacy in form of an app

Twenty years after its first publication, Regge's cartoon relativity has been re-packaged as *RelativitApp*, an app for tablets and smartphones. Regge's touch is still well visible in this modernized version of the program. One of his favourites was the story of the fall into the black hole, that sums up general relativity. An intrepid astronaut decides to explore a black hole, while his girlfriend remains on the spaceship and observes the scene from a safe distance (Figure 5).

From *her* viewpoint, he appears to descend ever more slowly, as time dilates with the increasing gravitational field, until at the event horizon his image freezes, turning redder and redder until it finally vanishes into the infrared. All the while, the user can listen in to the radio communications between the two astronauts. In the frame of the spaceship her voice sounds normal, while his transmitted voice becomes lower and lower in pitch, due to the red shift, reminding one of HAL 9000 being switched off in *2001: A Space Odyssey.*

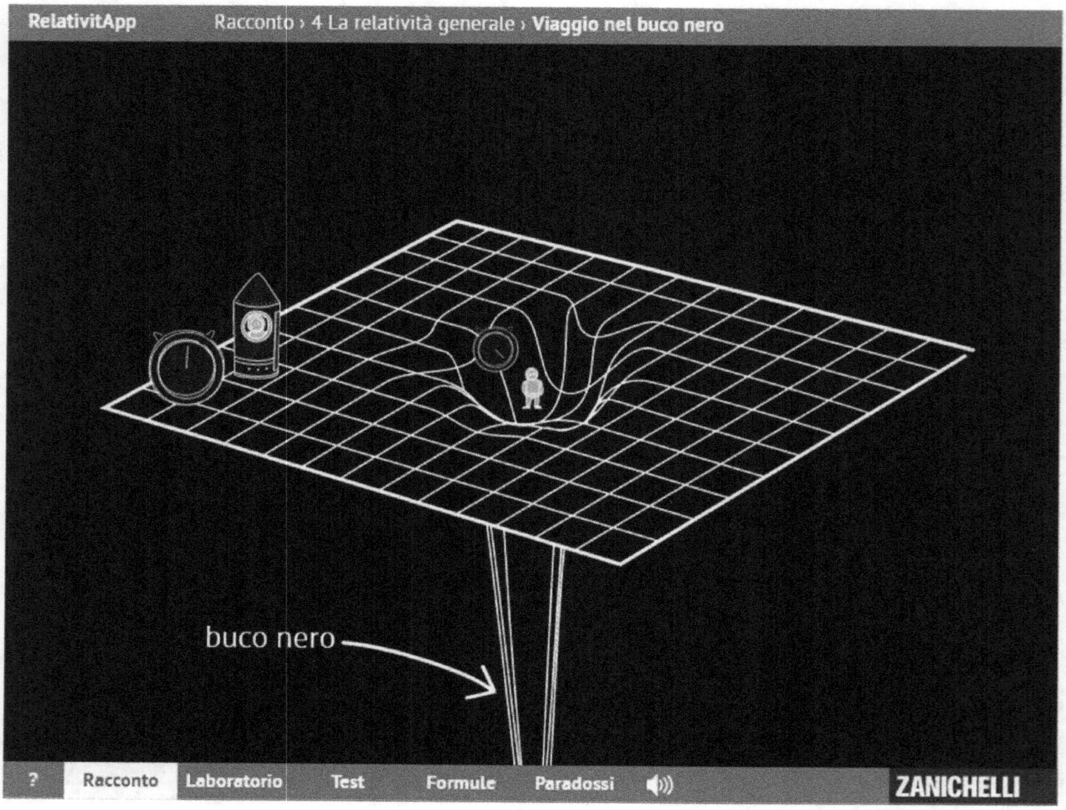

Figure 5. The fall into the black hole, from RelativitApp (© Zanichelli editore 2014).

But what does *he* perceive, as he looks up and talks via radio to his girlfriend? While descending ever faster, the explorer sees the sky shrinking to a porthole, due to the gravitational bending of light falling into the black hole; at the same time, all the stars above appear bluer and bluer to him. And as he radios the spaceship, his voice sounds normal, while the girl's transmitted voice gets higher and higher in pitch, due to the blue shift of the radio waves. For the poor chap, however, all is over very quickly: in a fraction of a second he is spaghettified by the hole's tidal forces.

An unhappy ending, you could say. But Regge would probably argue that the unique spectacle witnessed by the dying astronaut may have been worth it.

Web reference:
RelativitApp can be browsed at http://online.scuola.zanichelli.it/relativita/demo/

Tullio Regge: His Imprint on Science Popularization in Italy

G. Peruzzi

Department of Physics and Astronomy, University of Padova
Via Marzolo 8, I-35131 Padova - giulio.peruzzi@unipd.it

I met Tullio Regge in the late 1990s, when we started working together on various projects related to the popularization of relativity and quantum mechanics. I thus had the occasion to discuss with him not only about physics but also about other topics, ranging from politics to art and literature. By interacting with him I have had the opportunity to appreciate the great scientist, the cultivated man, his curiosity, his sense of humour and his capacity to look at questions in culture, politics, and society from unexpected or unconventional points of view. Here I try to outline the Regge's original approach to the science popularization, in the light of his wide cultural interests.

1. Regge and science popularization: significant milestones and background

After graduating in physics at the University of Turin in 1952, and after various experiences abroad culminated in 1965 when he was made a member of the Institute for Advanced Study, Regge permanently came back to Italy in 1979. In the same year Regge started being involved in science popularization.

As Ubaldo Richard and Gleb Wataghin, in different ways, played a crucial role in the choice of Regge to study physics instead of engineering, the involvement of Regge in science popularization was first of all influenced by two persons: Piero Bianucci, with whom Regge collaborated since 1979,[a] and Primo Levi with whom Regge published a *Dialogue* in 1984.

As for his political activity, the more important experience for Regge was his election as a member of the European Parliament, as an independent in the list of the Italian Communist Party for the parliamentary term 1989–1994; on this occasion, he worked in the Committee on Energy, Research and Technology.

[a]Let us recall the original events of science popularization they realized in Turin, like "Scienza al Parco", "Experimenta", "GiovedìScienza", and the articles Regge wrote for the Turin newspaper "Gazzetta del Popolo" and "Tuttoscienze", the scientific supplement founded by Bianucci in 1981 of the Turin newspaper "La Stampa".

Political involvement and a growing sensibility for the responsibility of scientists in science popularization are the two main interconnected features inspiring his original approach. His approach is not only intended for laymen or the "man in the street"[b] but also directed to the members of national and international public institutions and to the business community (the political and economical leadership). This is why, alongside books (such as "Infinito. Viaggio ai limiti dell'universo", 1994), articles in "Tuttoscienze" and his column in the Italian journal *Le Scienze* (initially called "Tulllio Regge ci scrive" and later "L'opinione di Tullio Regge"), since 1990 Regge participated with Piero Angela in the Italian Committee for the Investigation of Claims of the Paranormal (CICAP), and he committed himself in the European Parliament and in various forums (for example, since 1997, he was a member of the States General of Piedmont).

As noted previously, the commitment of Regge to science popularization, that plays a key role in the development of the field in Italy, started first of all from his awareness (at that time rare) of scientists' responsibility toward society. As we can read in his column "Tullio Regge ci scrive" in *Le Scienze* on February 1998:[1]

> "The scientific world, launched into a race, has neglected the laymen and has left a dangerous void in its wake, opened to the incursion of 'demolition guys' [magicians, pseudosciences and millennium fears]. Many of my colleagues are convinced that the public support of research is an obvious fact, which we can take for granted also for the next century. I wish they were right.
>
> We need to move in due time, and regain the space we have lost: science must go on not only to satisfy our thirst for knowledge, but also to be prepared us to the very serious crises facing mankind. Otherwise there will be a rude awakening."

For this reason too, the topics selected by Regge for his popular papers are not only related to current issues of science, but also connected "to the very serious crises facing the mankind", such as pollution, climate change, genetically modified organisms, and energy issues. He sometimes addresses ordinary people and other times the political and economical leadership. For example, in his column "Tullio Regge ci scrive" in *Le Scienze* on April 1998,[2] he writes: "For more than one year I have been cooperating with the States General of Piedmont as a consultant on scientific and environmental regional issues, and I have been dealing with problems which are very relevant also at a national level, not to say at a planetary level. One of these is the air pollution caused by cars' exhaust fumes." After a review of precise data and a critical review of some of the proposed solutions, Regge observes that in Italy there are too many legal norms, and makes clear the costs of this normative hypertrophy for companies and society. Note that this final part of the article begins, in his style, with the words: "Italy is a Republic based on norms".

[b]This last lexeme is translated by Regge with the Italian word "pedone", instead of the analogue Italian lexeme "uomo della strada": a further way to mark his originality.

2. Some thoughts on physics

Obviously, what Regge considers as peculiar characters of physical research have a fundamental influence on his personal attitude on science popularization. The kind of work we have done together dealt with issues of history of physics, that naturally compelled us to deal with some epistemological questions.

One day I asked Regge what he thought about the existence of a fundamental level of reality and about the so called "Theories of Everything". "I conjecture — he answered me — that, in fact, there is no ultimate level. Physics research will continue forever (and stop only when sadly governments will stop funding research), because every sound theory at some level of energy or length scale will be overcome by a new theory at a more fine-grained level, at higher energy or, equivalently, lower length scale. Instead of the existence of an ultimate level, I consider more likely the existence of an infinite sequence of more and more fine-grained levels, like an onion in which there is no final skin, or like a Russian doll (also known as matryoshka) composed by infinite Russian nesting dolls, one smaller than the other."

These ideas are reccurrent in Regge's speeches and articles which are not only intended for laymen but also for the scientific community (see for example the article "Some thoughts on the future of unified models", *Journal of Physics A: Mathematical and Theoretical*, 2007[4]).

On another occasion, we were talking about theory construction in modern physics. The starting point of our conversation was a letter that Einstein had sent to his friend Solovine on May 7, 1952. Regge was actually sympathetic to Einstein's point of view. It is worth quoting the following passage from Einstein's letter:

> As for the epistemological question, you completely misunderstood me; I probably expressed myself badly. I see the matter schematically in this way:

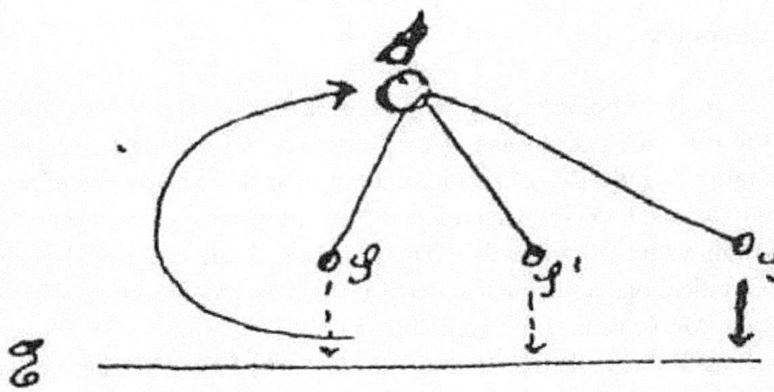

Figure 1. Einstein's sketch reproduced from his letter to Solovine, May 7, 1952.

> (1) The E's (immediate experiences) are our data.

(2) The axioms from which we draw our conclusions are indicated by A. Psychologically the A's depend on the E's. But there is no logical route leading from the E's to the A's, but only an intuitive connection (psychological), which is always likely to be "reconsidered" ("auf Widerruf").

(3) Logically, specific statements S, S', S" are deduced from A; these statements can lay claim to exactness.

(4) The S's are connected to the E's (verification through experience). Closer examination shows that this procedure also belongs to the extralogical (intuitive sphere), for the relation between the notions show up in S and the immediate experiences is not logical in nature. But the relation between S's and E's is (pragmatically) much less uncertain than the relation between the A's and the E's. (Take the notion 'dog' and the corresponding immediate experiences.) If such a relationship could not be set up with a high degree of certainty (though it may be beyond the reach of logic), logical machinery would have no value in the 'comprehension of reality' (example: theology).

What this all boils down to is the eternally problematical connection between the world of ideas and that which can be experienced (immediate experiences of the senses).

That the axioms (or principles) of a scientific theory are always to be reconsidered, and that the connection between axioms and experimental data is only "intuitive" are two aspects of Einstein's epistemology that have inspired Regge's approach to science and his very personal and fruitful activity in science popularization. In particular, the fact that the connection between axioms and experimental data are not logical in nature, offers to the scientist a freedom space, the space for "a free play with concepts".[5] This is an important ingredient of scientific thought, emphasizing the role of scientific imagination, in which Regge excels both as a scientist and as a science popularizer.

3. Science and society

Another feature of Regge's point of view and style in science popularization is due to his awareness that the ethos of science has something to do with democracy: his passion for scientific dissemination is indeed based on his convinction that the degree of scientific acculturation is related to social well-being and even to participative democracy. Although we had no occasion to enter into the details of this question, I am sure that Regge would have shared some of the reflections and conclusions of Robert Merton about this issue. Therefore it is worth sketching Merton's original contribution.

On October 1942, Robert Merton, one of the fathers of the sociology of science, published an article entitled "A note on science and democracy".[6] It was a troubling period, during the Second World War. At the beginning of his paper (c.f. p. 115) Merton observes: "Incipient and actual attacks upon the integrity of science have led scientists to recognize their dependence on particular types of social structure. Manifestos and pronouncements by associations of scientists are devoted to the relations of science and society. An institution under attack must reexamine its foundations, restate its objectives, seek out its rationale. Crisis invites self-appraisal". Merton examines different aspects of the question and concludes: "Although detailed monographs assembling the needed comparative materials are few and scattered, they provide some basis for the provisional assumption that science is afforded opportunity for development in a democratic order which is integrated with the ethos of science". Then Merton enumerates and analyses what the main pillars of the ethos of science are:

> "*Universalism.* [...] The acceptance or rejection of claims entering the lists of science is not to depend on the personal or social attributes of their protagonist; his race, nationality, religion, class and personal qualities are as such irrelevant (c.f. p. 118). [...]
>
> *Communism.* 'Communism', in the non-technical and extended sense of common ownership of goods, is a second integral element of the scientific ethos (c.f. p. 121). [...] The institutional conception of science as part of the public domain is linked with the imperative for communication of findings (c.f. p. 122). [...] The communism of the scientific ethos is incompatible with the definition of technology as 'private property' in a capitalistic economy (c.f. p. 123). [...]
>
> *Disinterestedness.* Science, as is the case with the professions in general, includes disinterestedness as a basic institutional element. Disinterestedness is not to be equated with altruism nor interested action with egoism. Such equivalences confuse institutional and motivational levels of analysis. A passion for knowledge, idle curiosity, altruistic concern with the benefit of humanity and a host of other special motives have been attributed to the scientist. The quest for distinctive motives appears to have been misdirected. It is rather a distinctive pattern of institutional control of a wide range of motives which characterizes the behavior of scientists (c.f. p. 124). [...]
>
> *Organized scepticism.* Organized scepticism is variously interrelated with the other elements of the scientific ethos. It is both a methodologic and an institutional mandate. The suspension of judgment until 'the facts are at hand' and the detached scrutiny of beliefs in terms of empirical and logical criteria have periodically involved science in conflict with other institutions. [...] This appears to be the source of revolts against the so-called intrusion of science into other spheres. Such resistance on the part of organized religion has become less significant as compared with that of economic and political groups. The opposition may exist quite apart from the introduction of specific scientific discoveries which appear to invalidate particular dogmas of church, economy or state. It is rather a diffuse, frequently vague, apprehension that scepticism threatens the current distribution of power. Conflict becomes accentuated whenever science extends its research to new areas toward which there are institutionalized attitudes or whenever other institutions extend their area of control. In modern totalitarian society,

> anti-rationalism and the centralization of institutional control both serve to limit the scope provided for scientific activity (c.f. p. 126)."

All these pillars of the ethos of science are resonant with Regge's approach to science popularization, like Pasteur's assessment that "science has no country, the scientist should have one" (c.f. Address at the Inauguration of the Pasteur Institute on 14 Nov, 1888). However, one can remark that modern science was born in a context in which 'scientists' are not freely and quietly dedicated to their researches, but, as Paolo Rossi observes:[7] "In those decades, Europe did not saw only witch trials and the work of Inquisition Tribunals (for which our thoughts go out to Giordano Bruno and Galileo). Europe witnessed the Thirty Years' War, and was crossed far and wide by armies of mercenaries committing all sorts of atrocities. Moreover the plague cutted in half the inhabitants of towns like Milan, Seville, Naples, London. Yet, only in an ideal Republic, which aspired to be independent from the fightings, conflicts and miseries of the world, there could emerge such an amazing statement — due to Francis Bacon — according to which a science carried out for the glory and the power of his own country is morally less noble than a science serving the whole of humanity".[7]

Coming back to Italy, a way of better understanding the context in which Regge's involvement in science popularization developed is to consider what happened in Italy after the Second World War. Let us quote in this respect what Enrico Bellone claimed in his plenary lecture at the Congress of the Italian Physical Society, held in Palermo on 6–11 October 2000: "During some decisive years in the Italian history the country has taken the blind path of modernization without innovation: blind, because grounded on the hypothesis that it would have been possible to modernize Italy without leveraging on curiosity driven research and cutting edge technologies".[8]

Going further back in time, if we look at the events after the unification of Italy (completed in 1871), we can observe that from the years 1860s and up to the first decades of the 20th century the political leadership could not neglect the presence of scientists, because scientists at that time formed a fundamental part of the political leading class. The exchange between the German historian Theodor Mommsen and the Minister of Finance Quintino Sella is well known. It took place in 1871 shortly after the annexation of Rome to Italy. Mommsen addressed the following question to Sella: "But what do you intend to do in Rome? This is what makes the rest of us uneasy. One does not remain in Rome without some cosmopolitan purpose. What do you intend to do?"[9] Sella, who was a professor at the (now known as) Turin Polytechnic before starting his political career, replied: "Yes, we have a cosmopolitan plan: Science".[10]

Though at the beginning of the 20th century, the process started in the last decades of the 19th century slowed down, the importance of the dialogue between science, politics and society was reaffirmed and kept alive, and the project of financing and promoting the development of science was pursued, thanks to the work of people like Vito Volterra, Orso Mario Corbino and Antonio Garbasso (at the same time scientists and politicians, like Quintino Sella).

However, after then the First World War, fascism and the Second World War interrupted these efforts. It seemed difficult to continue the dialogue between science and society, as the Italian society was struggling with survival problems, and science was impoverished by infamous racial laws and wounded its dignity by its acquiescence to fascism. On the contrary, in the beginning of the 1950s, the country lived an unexpected spring, with a dialogue between science and society enhanced by new entrepreneurial dynamics. These are the years of Adriano Olivetti and Mario Tchou, Domenico Marotta, Enrico Mattei, Felice Ippolito, and Giulio Natta, as well as the years of the "Centro Informazioni Studi ed Esperienze" (CISE), founded by Giuseppe Bolla, Carlo Salvetti, Giorgio Salvini, Mario Silvestri, in which various Italian industries took parts (for example, FIAT, Montecatini, Adriatic Society of Electricity — SADE, Pirelli and Falck).

This period met a quick end around 1970. Since then, the divergence between science, on the one side, and politics, entrepreneurship, and society, on the other side, has increased. The reasons for this process are various. The following three are surely the most fundamental ones:

1. The backwardness of the ruling class, lacking scientific and technological acculturation;
2. The specificity of Italian business fabric, mostly made up of small and medium-sized enterprises, which has progressively lost ground in the areas of research and development;
3. The persistence of some "images of science" produced "by people who have no idea about what is actually the day-to-day scientific research activity",[8] and what the education and training for research is.

Let us briefly comment only the third of these reasons. As Bellone observed in the lecture at the Congress of Palermo mentioned above, these spurious images of science are all convergent in a sole thesis: "Science is potentially dangerous, but it could be useful if controlled by politics". We can roughly identify four ideological stances, very different one from the other but all convergent in this fake image of science. (1) Influential left wing parts — even when they do not agree with marxist models in which science is reduced to one of the superstructures determined by the economic structure or base — consider that science is not neutral to social needs. Scientific research can thus give useful results only if governed by a political selection of social relevant needs. However, establishing a strict connection between scientific research and the satisfaction of social relevant needs has worrying implications for the development of science, which is by its very nature free and, firstly, curiosity driven. (2) In the Catholic area, it is often pointed out that scientific and technological practices are at the origin of materialistic conceptions, which can undermine the primacy of the human being, hence the need for an extra-scientific, political control of science in order to avoid a risks of this kind. (3) Among right wing intellectuals it is common to find an image of science as a dangerous instrument in the hand of the masses, with the obvious consequence of pointing out once again the primacy of politics with respect to science. (4) Finally, in the context of the so-called "secular thinking" we can find the belief that the Western world is becoming dehumanized under the influence of the so-called "techno-science", the rationality of which must be sharply criticised for this reason.

This image of science, emerging from the convergence of the above ideological stances, has become more and more popular. The construction and diffusion of this spurious image is accompanied, often in contradictory forms, by a critical assessment of scientific rationality and technology. The growing fear for science and technology is an aspect used by the different ideological positions for building political consensus. The result is that the decline of science, in Italy and abroad, implies a decline of democracy.

Tullio Regge was one of the first to be well aware of these trends, and his action in science popularization was a conscious effort to intervene against them. His commitment as a scientist to science popularization and democracy has been a strong stimulus for the Italian society and has left its indelible imprint on Italian scientific community.

References

1. T. Regge, "Maghi, pseudoscienze e paure di fine millennio", Le Scienze, n. 354, 7 (February 1998).
2. T. Regge, "Sopravvivere in una Repubblica fondata sulle norme", Le Scienze, n. 356, 7 (April 1998).
3. A. Einstein, *Letters to Solovine 1906-1955*, (New York: Philosophical Library, 1987).
4. T. Regge, "Some thoughts on the future of unified models", J. Phys. A: Math. Theor. **40**, 3205-3206 (2007).
5. A. Einstein, Autobiographical Notes, in P. A. Schlipp, *Albert Einstein: Philosopher-Scientist*, 295 (New York: Tudor Publishing Co., second edition, 1951).
6. R. Merton, "A Note on Science and Democracy", J. Legal & Pol. Soc. **1**, 115-126 (1942).
7. P. Rossi, *La nascita della scienza moderna in Europa*, (Bari: Laterza, 1997).
8. E. Bellone, "Storia delle scienze e cultura", Il Nuovo Saggiatore **16**, n. 5/6, 21-23 (2000).
9. F. Chabod, *Italian Foreign Policy: The Statecraft of the Founders, 1870-1896* (Princeton: Princeton University Press, 1996), 155.
10. Q. Sella, *Discorsi Parlamentari*, Vol. I (Roma : Tip. della Camera dei deputati, 1887), 292.

PART X

Reminiscences

Some Reminiscences about Tullio Regge

Sergio Ferrara

Theoretical Physics Department, CERN CH-1211 Geneva 23, Switzerland
INFN - Laboratori Nazionali di Frascati Via Enrico Fermi 40, I-00044 Frascati, Italy
Department of Physics and Astronomy and Mani L. Bhaumik Institute for Theoretical Physics, U.C.L.A., Los Angeles CA 90095-1547, USA

It is for me a pleasure and an honour to pay a small tribute to Tullio Regge, in the occasion of a collected volume organised by his friends, colleagues, students and collaborators.

Tullio Regge, one of the most brilliant theoretical physicists of the post war period, left a unique legacy of deep and fundamental contributions in theoretical particle physics, where physical ideas and sophisticated mathematical tools were used as basic ingredients to formulate new theories. Among them probably the most famous is the one about Regge poles, and their basic use in hadron physics and in bootstrap ideas for scattering amplitudes in string theory starting with the duality formulated by G. Veneziano.

The other major achievement is the so-called Regge calculus, a very original formulation of quantum gravity based on triangulations of space-time manifolds, a technique able to both regularise gravity but also give a non-perturbative definition, as the case of lattice gauge Yang–Mills theories.

It is not the scope of this note to enumerate all important contributions of Regge's research but to spend few words on the occasions I had to meet Tullio and to discuss the overlapping research interests. In the early seventies (1973) supersymmetric field theories were formulated by Julius Wess and Bruno Zumino, inspired by the so-called super gauge symmetries of dual resonance models with fermionic degrees of freedom. The original Wess–Zumino model involved only spin 1/2 and spin zero particles but they were soon generalised, in 1974, to include spin one and non-abelian Yang–Mills interactions by Zumino and the author and also by A. Salam and J. Strathdee. Regge was fascinated by this new symmetry of Lagrangian field theory but he was still waiting for the inclusion of gravity in the game. Supergravity was formulated two years later (by D. Freedman, P. Van Nieuwenhuizen and the author) and extended to a first order formalism by Stanley Deser and Zumino. Tullio Regge and Yuval Ne'eman introduced a new formulation of supergravity, based on the concept of supergroup (group in the case of gravity) and later called rheonomic approach. The nice aspect of this approach is that the language of forms is used from the very beginning and the standard space-time Lagrangian involves forms over the group restricted to a four-dimensional slice. The Regge formulation of supergravity was continued and expanded by his group of collaborators in Turin, among them P. Fré, R. D'Auria and later A. Ceresole and L. Castellani. The Regge approach to supergravity is particularly useful in the case of N-extended supergravity (theories with N supersymmetries) where the geometry of the supergroup manifold is merged with the so-called duality symmetries, R symmetries and sigma models target spaces, involving coset spaces or spaces with exceptional holonomies. This is also the case for higher dimensional theories such as eleven dimensional supergravity, which can be formulated in alternative ways in the rheonomic approach, as well as in black hole dynamics. The Turin group soon became an outstanding group in all subsequent developments of supergravity and superstring theories (the latter also formulated in Turin under the leadership of Sergio Fubini, a senior colleague of mine at CERN). A comprehensive review of the group manifold approach and its application to supergravity in higher dimensions can be found in a three-volume book by L. Castellani, R. D'Auria and P. Fré: "Supergravity and Superstrings: A Geometric perspective", published by World Scientific (1991).

As a personal recollection my interaction with Tullio started when he made a long visit at CERN in 1983 and we had many discussions on the emerging rich structure of supergravity; his interest was already in geometric aspects and the way supergravity intervened to restrict the geometry. It is curious that, even though I talked to him on several subjects related to supergravity in the many visits I made to Turin in subsequent years, due to my frequent collaborations with colleagues from Turin University and "Polytechnic" Institute, I happened to have two collaborations with him only in 1996. The first was a long paper published in "Classical and Quantum Gravity (n. 13, (1996), 831) with a large group also including Antoine Van Proeyen from Leuven and Billo', Ceresole, D'Auria, Fré, and Soriani from Turin. The subject was on the search of non-perturbative dualities in $N = 2$ Yang–Mills coupled to supergravity and their superstring origin, on compacification on Calabi–Yau threefolds. The second paper was actually the contribution to a volume on mirror symmetry based on a workshop organised by S.T. Yau at the MSRI in Berkeley (Mathematical Science Research Institute). I gave an invited talk at the workshop and the subsequent proceeding contribution also included A. Ceresole, R. D'Auria, W. Lerche (CERN), and J. Louis (Munich). The content of this contribution was on the role of Picard–Fuchs equations in special geometry and target space duality. The year 1996 was the 20th anniversary of supergravity and Tullio was especially impressed and interested on the prominent role that algebraic geometry was playing in the Kaluza–Klein approach to supergravity and in its connection to 10D superstrings and 11D M theory. In later years until he passed away in 2014 I rarely happened to meet Tullio but I was glad to have participated to the inaugural ceremony of the Arnold–Regge Center in Turin, in February 2017, whose director is my friend and colleague Prof. Pietro Fré.

Tullio Regge from a Son's Perspective

Daniele Regge

University of Turin and Candiolo Cancer Institute, Turin

When I was twelve, I used to play chess with my father in our house on 45 Veblen Circle in Princeton, New Jersey. He was upset when I won a game in 6 moves using a variation of the Benoni Opening. That was probably the only time I felt as if I was doing something at his level. The rest of the time I simply thought he was a genius and indeed it was not always easy for me to cope with it. When I was in my first year of high school, he tried to teach me exponential calculus, but I failed to follow his reasoning. From then on, he thought I was just normal and let me choose my life. So, in the end I decided I would study medicine, a choice I was not convinced of at the beginning. But looking back I believe it was a great idea.

My sisters and I had fun moments with my father and mother, as for example the time we traveled from Seattle to Denver through beautiful landscapes of national parks including Craters of the Moon in Idaho, or Yellowstone where we had close encounters with black bears, or sliding down the hillside with sleds at Princeton Battlefield State Park bordering our garden or traveling to Mexico to look at a total solar eclipse for his 60th birthday. We had the opportunity to meet interesting people, including young fellows attending the Institute for Advanced Studies, some that would later become renowned scientists.

However, my parents spent much time traveling for work and we were frequently left to babysitters or sent to aunts and grandparents. My father and mother had good connections with physicists in Russia and Israel where they went several times and I remember them also visiting Brazil, other south American countries and going on a trip round the world. My father made it a challenge to learn idioms from the places he visited in order to be able to give speeches in the local language. He eventually was able to speak eight languages fluently. Even when we were together, gathering for lunch for example, he was usually trying to solve some complicated formulas and would not talk to us and was annoyed when we were too noisy. To get inspiration he would spend time playing our light brownish half tail piano, usually before dinner.

My father at times was childish. He loved playing jokes like stealing food from our plates during lunch or tying knots on the back of the kitchen apron of Maria, our faithful domestic worker. He also liked to write unlikely stories on April Fool's Day on the local newspapers. The idea behind it was to trick the reader into believing a very improbable story by giving him a very scientifically sound explanation. For example, in one story he wrote that Finnish and Russian researchers found proof of the existence of telepathy. The scientists noted that a drop of helium at very low temperatures resonated when humans expressed strong emotions. In another short story he tried to convince readers that proof was found on pictures taken by Mars Global Explorer of the past existence of giant worms on the surface of Mars. Due to the realistic description and the reputation of my father as a scientific popularizer many people thought the stories were true.

When I took my decision to study medicine, I did not think my father would be interested in my profession. However, he was curious about everything and eventually his beliefs were able to make a significant impact also on people's health. First and most importantly he contributed to the reduction of the number of cigarette deaths in Italy. He gained consciousness on the dangers of smoking when some of his best friends died from respiratory tract cancer. He was particularly upset when in 1966 Robert Oppenheimer, whose pipe was always lit, was diagnosed with throat cancer; he would die a year later. Lung cancer was a rare disease in the 19th century with only 140 cases being reported in the literature by 1900.[1] Unfortunately, in the 1920s lung cancer incidence dramatically increased. It was not until the 1950s that population studies showed convincingly that tobacco smoking was the leading cause of respiratory tract cancer.[2] Nevertheless, the tobacco industry refused to admit the harms of smoking until the new millennium and today lung cancer kills approximately 1.5 million people per year globally.

My father was actively battling cigarette smoke starting from the 1990s. At the time I had just taken my position as chief radiologist at Candiolo Cancer Institute in Torino. He asked me to provide data on lung cancer and cigarette smoking to write opinion articles on popular Italian newspapers such as La Stampa and La Repubblica, where he enjoyed a wide audience and freedom of expression. His writings would focus on the harms of tobacco, criticizing the ambiguity of politicians that were very keen on accusing the overrated harms of GMO's while being silent on tobacco smoking.[3] In 2001 he proposed to approach health environmental risks from a probabilistic perspective, suggesting that while "zero health risks" do not exist one should focus first on prevention of most damaging substances. He therefore proposed that the Italian government should adopt primary prevention on cigarette smoking and to increase taxes on cigarettes. Just a few years later Australia and New Zealand would decide to increase the cost of a pack of Marlboro to over 20 dollars, which is three times more costly than in Italy. In 2001 he was invited by the Minister of Health to be part of a working group were he clearly stated his point of view. Two years later the Italian parliament passes a revolutionary law for those times, the Legge Sirchia, that prohibited the use of tobacco in public places. Probably as a consequence of those restrictions from 2003 to 2014 the number of smokers in Italy decreased from 23.8 % to 19.5 %.[4] As a testimony of the passion he put in the battle against cigarette smoke my father later published a series of pictures with his computer showing a skull with a cowboy hat smoking a cigarette. In the background were some of his favorite cinema actors and Oppenheimer smoking a pipe, all of whom died of cancer (Fig. 1).

In 1949, at the age of 18, my father did not pass the medical assessment to enter the

Figure 1.

military service as he could not lift his arms above his head. He was later diagnosed with facioscapulohumeral muscular dystrophy, a rare genetic disease characterized by progressive muscle weakness. By the age of 40 he used walking sticks; in 1990 after a serious fall in the Strasbourg hemicycle, he decided to move with a wheelchair. My father did not show any distress for his physical inability, on the opposite I have always had the impression that his condition allowed him to travel freely with his imagination. Nevertheless, he was conscious of the millions of handicapped people around the world that did not have the freedom to move independently because of architectural barriers. In Italy he was appointed president of the Association of Prevention and Research of Handicap; in Europe he fought battles to promote new laws to reduce architectural barriers. My father also represented barriers in his art. Several of his computer paintings show steep and improbable stairs in the style of Escher's Relativity (Fig. 2).

Throughout my medical career my father would ask me to update him on new discoveries on cancer diagnosis and treatment, which I was always very happy to provide. When he eventually stopped, I knew he had lost interest in life.

Figure 2.

References

1. Proctor RN. The history of the discovery of the cigarette-lung cancer link: evidentiary traditions, corporate denial, global toll. Tob Control. 2012 Mar; 21(2):87-91. doi: 10.1136/tobaccocontrol-2011-050338. Review. Erratum in: Tob Control. 2013 Jan; 22(1):62. PubMed PMID: 22345227.
2. Hammond E.C., Horn D. The relationship between human smoking habits and death rates: a follow-up study of 187,766 men. J Am Med Assoc. 1954 Aug 7; 155(15):1316-28. PubMed PMID: 13174399.
3. https://www.fondazionebassetti.org/0due/docs/biotech/regge-precauzione.htm
4. http://www.governo.it/sites/governo.it/files/air_tabacchi.pdf

PART XI

Full Bibliography of Scientific Works

From InSPIRE -HEP: `http://inspirehep.net/search?ln=en&ln=en&p=find+a+regge&of=hb&action_search=Search&sf=earliestdate&so=a&rm=&rg=250&sc=0`

References

1. T. Regge and M. Verde, "Scattering problems with spin-orbit couplings," Nuovo Cim. **10** (1953) 997. doi:10.1007/BF02773097
2. T. Regge, "Tensor force and intermediate coupling in p-shell nuclei," Nuovo Cim. **11** (1954) 285. doi:10.1007/BF02781395
3. B. Bosco and T. Regge, "On the polarization of high energy nucleons scattered by a nuclear field," Nuovo Cim. **12** (1954) 285. doi:10.1007/BF02781184
4. F. Mandl and T. Regge, "Reaction $p + p \to \pi^+ + d$ with Polarized Protons," Phys. Rev. **99** (1955) 1478. doi:10.1103/PhysRev.99.1478
5. T. Regge, "Angular correlations between decay planes of tau+ mesons," Nuovo Cim. **4** (1956) 1615. doi:10.1007/BF02746401
6. T. Regge, "On the Properties of spin-2 particles," Nuovo Cim. **5** (1957) 325 Erratum: [Nuovo Cim. **6** (1957) 1233]. doi:10.1007/BF02855242
7. T. Regge and J. A. Wheeler, "Stability of a Schwarzschild singularity," Phys. Rev. **108** (1957) 1063. doi:10.1103/PhysRev.108.1063
8. T. Regge, "Gravitational fields and quantum mechanics," Nuovo Cim. **7** (1958) 215. doi:10.1007/BF02744199
9. T. Regge, "On the Analytic behaviour of the eigenvalue of the S-matrix in the complex plane of the energy," Nuovo Cim. **9** (1958) 295. doi:10.1007/BF02724933
10. T. Regge, "Symmetry properties of Clebsch-Gordan's coefficients," Nuovo Cim. **10** (1958) 544. doi:10.1007/BF02859841
11. T. Regge, "Analytic properties of the scattering matrix," Nuovo Cim. **8** (1958) 671. doi:10.1007/BF02815247
12. T. Regge, "Construction of potentials from resonance parameters," Nuovo Cim. **9** (1958) 491. doi:10.1007/BF02725104
13. T. Regge, "Introduction to complex orbital momenta," Nuovo Cim. **14** (1959) 951. doi:10.1007/BF02728177
14. T. Regge, "Symmetry properties of Racah's coefficients," Nuovo Cim. **11** (1959) 116. doi:10.1007/BF02724914
15. T. Regge, "Bound states, shadow states and Mandelstam representation," Nuovo Cim. **18** (1960) 947. doi:10.1007/BF02733035
16. T. Regge, "General Relativity Without Coordinates," Nuovo Cim. **19** (1961) 558. doi:10.1007/BF02733251
17. T. Regge, "Topics on complex angular momenta," Proceedings of the 11th International Conference on High-energy Physics (ICHEP 62), 4-11 Jul 1962. Geneva, Switzerland, CNUM: C62-07-04.
18. A. Bottino, A. M. Longoni and T. Regge, "Potential scattering for complex energy and angular momentum," Nuovo Cim. **23** (1962) 954. doi:10.1007/BF02731254
19. T. Regge, "Mathematical Theory of Potential Scattering," Proceedings of the IAEA Seminar on Theoretical Physics, Palazzino Miramare, Trieste, Itally (1962), CNUM: C62-07-16.
20. T. Regge, "Recent progress in potential scattering theories," Lect. Theor. Phys. **6** (1964) 373.
21. V. de Alfaro and T. Regge, "Potential Scattering," North Holland Publishing Company (1965).
22. T. Regge, "The fundamental group of poincare and the analytic properties of feynman relativistic amplitudes," Conf. Proc. C 680519 (1968) 285.
23. T. Regge, "Algebraic Topology Methods in the Theory of Feynman Relativistic Amplitudes," Proceedings of Battelle rencontres - Seattle (1967), C67-07-016.

24. G. Ponzano, T. Regge, E. R. Speer and M. J. Westwater, "The monodromy rings of a class of self-energy graphs," Commun. Math. Phys. **15** (1969) 83. doi:10.1007/BF01645374
25. G. Ponzano, T. Regge, E. R. Speer and M. J. Westwater, "The monodromy rings of one loop feynman integrals," Commun. Math. Phys. **18** (1970) 1. doi:10.1007/BF01649638
26. A. Molinari and T. Regge, "Phonon excitations in liquid He II," Phys. Rev. Lett. **26** (1971) 1531. doi:10.1103/PhysRevLett.26.1531
27. T. Regge, "Veneziano Amplitudes, Angular Momentum Coefficients And Related Symmetries," In *Varenna 1972, Proceedings, Developments In High Energy Physics*, New York 1972, 111-116.
28. T. Regge, "S Matrix And Feynman Integrals. (talk)," In *J R Klauder, Magic Without Magic*, San Francisco 1972, 161-170.
29. T. Regge, E. R. Speer and M. J. Westwater, "The monodromy rings of the necklace graphs," Fortsch. Phys. **20** (1972) 365. doi:10.1002/prop.19720200603
30. A. Molinari and T. Regge, "Rotational motion at large angular momentum in even-even deformed nuclei," Phys. Lett. **41B** (1972) 93. doi:10.1016/0370-2693(72)90434-0
31. B. L. Hu and T. Regge, "Perturbations on the Mixmaster Universe," Phys. Rev. Lett. **29** (1972) 1616. doi:10.1103/PhysRevLett.29.1616
32. T. Regge and C. Teitelboim, "Improved Hamiltonian for general relativity," Phys. Lett. **53B** (1974) 101. doi:10.1016/0370-2693(74)90355-4
33. A. J. Hanson and T. Regge, "The Relativistic Spherical Top," Annals Phys. **87** (1974) 498. doi:10.1016/0003-4916(74)90046-3
34. T. Regge and C. Teitelboim, "Role of Surface Integrals in the Hamiltonian Formulation of General Relativity," Annals Phys. **88** (1974) 286. doi:10.1016/0003-4916(74)90404-7
35. R. A. Broglia, A. Molinari, G. Pollarolo and T. Regge, "On the back-bending in the rotational motion," Phys. Lett. **50B** (1974) 295. doi:10.1016/0370-2693(74)90672-8
36. M. Rasetti and T. Regge, "Vortices in He II, Current Algebras and Quantum Knots," Print-75-0207 (IAS,PRINCETON).
37. T. Regge and C. Teitelboim, "General Relativity a La String: A Progress Report," Conference: C75-07-07.2, pp. 77-87.
38. A. Molinari and T. Regge, "On the Density Oscillations in Finite Systems," PRINT-75-0392 (IAS,PRINCETON).
39. R. A. Broglia, A. Molinari, G. Pollarolo and T. Regge, "Band hybridization in rotational motion," Phys. Lett. **57B** (1975) 113. doi:10.1016/0370-2693(75)90002-7
40. A. J. Hanson, T. Regge and C. Teitelboim, "Constrained Hamiltonian Systems," RX-748, PRINT-75-0141 (IAS,PRINCETON).
41. F. Lund, M. Rasetti and T. Regge, "Statistical Mechanics Models and the Modular Group," Commun. Math. Phys. **51** (1976) 15. doi:10.1007/BF01609049
42. R. A. Broglia, A. Molinari and T. Regge, "On the density oscillations in finite systems," Annals Phys. **97** (1976) 289. doi:10.1016/0003-4916(76)90039-7
43. F. Lund and T. Regge, "Unified Approach to Strings and Vortices with Soliton Solutions," Phys. Rev. D **14** (1976) 1524. doi:10.1103/PhysRevD.14.1524
44. G. Barucchi and T. Regge, "Conformal Properties of a Class of Exactly Solvable n Body Problems in Space Dimension One," J. Math. Phys. **18** (1977) 1149. doi:10.1063/1.523384
45. T. Regge, "Old Problems and New Hopes in S Matrix Theory," Publ. Res. Inst. Math. Sci. Kyoto **12** (1977) 367. doi:10.2977/prims/1195196616
46. F. Lund, M. Rasetti and T. Regge, "Ising and Dimer Models on the Lobachevsky Plane," Teor. Mat. Fiz. **33** (1977) 246. doi:10.1007/BF01036598
47. T. Regge, "The Gauge Group and Geometry of Supergravity," Conference: C78-09-11, pp. 477-479 Proceedings, Austin Texas 1978.
48. Y. Ne'eman and T. Regge, "Gravity and Supergravity as Gauge Theories on a Group Manifold," Phys. Lett. **74B** (1978) 54. doi:10.1016/0370-2693(78)90058-8

49. Y. Ne'eman and T. Regge, "Gauge Theory of Gravity and Supergravity on a Group Manifold," Riv. Nuovo Cim. **1N5** (1978) 1. doi:10.1007/BF02724472
50. Y. Ne'eman, T. Regge and J. Thierry-Mieg, "Ghost-Fields, BRS and Extended Supergravity as Applications of Gauge Geometry," In *Ruffini, R. (ed.) et al.: Matter particled* 301-303, 19th International Conference on High-Energy Physics (ICHEP 78).
51. A. J. Hanson and T. Regge, "Torsion And Quantum Gravity," LBL-8240, Conference: C78-09-11 (Austin Group Theor.1978:354), p. 354.
52. A. D'Adda, R. D'Auria, P. Fre and T. Regge, "Gauge Theories Of Supergravity On Orthosymplectic Supergroup Manifolds.," 3rd Johns Hopkins Workshop on Current Problems in High-Energy Particle Theory (1979), Florence, Italy, CNUM: C79-05-30.
53. F. Gliozzi, T. Regge and M. A. Virasoro, "Gauge Fields As Phonon Excitations In A Condensate Of Dual Strings," Phys. Lett. **81B** (1979) 178. doi:10.1016/0370-2693(79)90517-3
54. A. D'Adda, R. D'Auria, P. Fre and T. Regge, "Geometrical Formulation Of Supergravity As A Theory On A Supergroup Manifold.," Supergravity Workshop (1979), Stony Brook, New York, CNUM: C79-09-27.
55. R. D'Auria, T. Regge and S. Sciuto, "A General Scheme for Bidimensional Models With Associate Linear Set," Phys. Lett. **89B** (1980) 363. doi:10.1016/0370-2693(80)90143-4
56. R. D'Auria, T. Regge and S. Sciuto, "Group Theoretical Construction of Two-dimensional Models With Infinite Set of Conservation Laws," Nucl. Phys. B **171** (1980) 167. doi:10.1016/0550-3213(80)90366-1
57. R. D'Auria, P. Fre and T. Regge, "Group Manifold Approach To Supergravity And The Coupling Of Matter Fields.," 4th Workshop on Current Problems in High-Energy Particle Theory (1980) Bad Honnef, Germany, CNUM: C80-06-02.
58. T. Regge, "Unified Gravity Theories.," International School of Physics 'Enrico Fermi': From Nuclei to Particles, 23 Jun-5 Jul 1980. Varenna, Italy, CNUM: C80-06-23.4.
59. A. D'Adda, R. D'Auria, P. Fre and T. Regge, "Geometrical Formulation of Supergravity Theories on Orthosymplectic Supergroup Manifolds," Riv. Nuovo Cim. **3N6** (1980) 1. doi:10.1007/BF02724337
60. R. D'Auria, P. Fre and T. Regge, "Supergravity and Cohomology Theory: Progress and Problems in $D = 5$," Europhysics Study Conference on Unification of the Fundamental Interactions (1980), Erice, Italy, CNUM: C80-03-17.
61. R. D'Auria, P. Fre and T. Regge, "Coupling of Matter to Supergravity From the Rheonomic Symmetry Point of View and the Origin of the Auxiliary Fields," Nucl. Phys. B **176** (1980) 243. doi:10.1016/0550-3213(80)90075-9
62. M. Rasetti and T. Regge, "Dimer Covering and Ising Model on Lattices Homogeneous Under the Icosahedral Group," Riv. Nuovo Cim. **4N1** (1981) 1. doi:10.1007/BF02740736
63. R. D'Auria, P. Fre and T. Regge, "Graded Lie Algebra Cohomology and Supergravity," Riv. Nuovo Cim. **3N12** (1980) 1. doi:10.1007/BF02905929
64. T. Regge, "Gauge Theories of Unification," Print-80-0579 (TEL AVIV).
65. R. D'Auria, T. Regge and P. Fre, "The Supergroup Manifold Formulation of the Wess-Zumino Multiplet Revisited: The Supercurrent and the Supergravity Auxiliary Fields," Nucl. Phys. B **188** (1981) 342. doi:10.1016/0550-3213(81)90260-1
66. R. D'Auria, E. Maina, T. Regge and P. Fre, "Geometrical First Order Supergravity in Five Space-time Dimensions," Annals Phys. **135** (1981) 237. doi:10.1016/0003-4916(81)90155-X
67. T. Regge, "Group Manifold Approach In Unified Gravity Theories," In Trieste 1981, Proceedings, Differential Geometric Methods In Theoretical Physics, 25-31C81-06-30, pp. 25-31.
68. R. D'Auria, P. Fre and T. Regge, "Recent Developments In The Group Manifold Approach," Europhysics Study Conference: Unification of Fundamental Interactions II: 6-14 Oct 1981. Erice, Italy, CNUM: C81-10-06.
69. R. D'Auria and T. Regge, "Gravity Theories With Asymptotically Flat Instantons," Nucl. Phys. B **195** (1982) 308. doi:10.1016/0550-3213(82)90402-3

70. R. d'Auria, P. Fre and T. Regge, "Group Manifold Approach To Gravity And Supergravity Theories," IC-81-54, C81-04-22-2, Conference: C81-04-22 (Trieste Supergrav.School 1981:0421), p. 0421.
71. R. D'Auria, P. Fre, E. Maina and T. Regge, "A New Group Theoretical Technique for the Analysis of Bianchi Identities and Its Application to the Auxiliary Field Problem of $D = 5$ Supergravity," Annals Phys. **139** (1982) 93. doi:10.1016/0003-4916(82)90007-0
72. T. Regge, "Unified Gravity Theories," International Conference on Unified Theories and their Experimental Tests, 16-18 Mar 1982. Venice, Italy, CNUM: C82-03-16.
73. P. Fre and T. Regge, "Group Manifold Approach And D = 11 Supergravity.," Annecy Meeting on Supersymmetry and Supergravity at LAPP, 10-12 Jan 1983. Annecy-le-Vieux, France, CNUM: C83-01-10.1.
74. R. D'Auria, P. Fre and T. Regge, "Consistent Supergravity in Six-dimensions Without Action Invariance," Phys. Lett. **128B** (1983) 44. doi:10.1016/0370-2693(83)90070-9
75. M. Rasetti and T. Regge, "Quantum Vortices," CERN-TH-3643.
76. T. Regge, "An Elementary Course On General Relativity," CERN-83-09, CERN-YELLOW-83-09.
77. T. Regge, "The Group Manifold Approach To Unified Gravity," Conf. Proc. C **8306271** (1983) 933.
78. M. Rasetti and T. Regge, "Quantum Vortices And Diff (r3)," 2nd International Colloquium on Group Theoretical Methods in Physics (GROUP 12), 5-10 Sep 1983. Trieste, Italy, CNUM: C83-09-05.1.
79. T. Regge, "Free Differential Algebras and Relativistic Theories," Fundam. Theor. Phys. **9** (1984) 433. doi:10.1007/978-94-009-6469-3_26
80. T. Regge, "A Science Fiction Scenario," Presented at the Old and New Problems in Fundamental Physics Conference: C84-10-25.2, pp. 181-183 Proceedings.
81. J. E. Nelson and T. Regge, "Exterior Canonical Formalism For Unified Gravity," First Torino Meeting on Superunification and Extra Dimensions, 22-28 Sep 1985. Turin, Italy, CNUM: C85-09-22.1.
82. J. E. Nelson and T. Regge, "Covariant Canonical Formalism For Gravity Theories."
83. J. E. Nelson and T. Regge, "Covariant Canonical Formalism For Gravity And Supergravity," 4th Marcel Grossmann Meeting on the Recent Developments of General Relativity, 17-21 Jun 1985. Rome, Italy, CNUM: C85-06-17.1.
84. T. Regge, "Lie Groups And Gravity Theories," Pramana **25** (1985) 349. doi:10.1007/BF02846760
85. A. D'Adda, J. E. Nelson and T. Regge, "Covariant Canonical Formalism for the Group Manifold," Annals Phys. **165** (1985) 384. doi:10.1016/0003-4916(85)90302-1
86. J. E. Nelson and T. Regge, "Covariant Canonical Formalism for Gravity," Annals Phys. **166** (1986) 234. doi:10.1016/0003-4916(86)90057-6
87. A. Lerda, J. E. Nelson and T. Regge, "The Group Manifold Hamiltonian for Supergravity," Phys. Lett. **161B** (1985) 297. doi:10.1016/0370-2693(85)90765-8
88. A. Lerda, J. E. Nelson and T. Regge, "Covariant Canonical Formalism for Supergravity," Phys. Lett. **161B** (1985) 294. doi:10.1016/0370-2693(85)90764-6
89. T. Regge, "Comments On The Canonical Formalism For General Relativity," VII National Meeting on General Relativity and Gravitational Physics, 3-6 Sep 1986. Rapallo (Genoa), Italy, CNUM: C86-09-03.1.
90. A. Prastaro and T. Regge, "The Group Structure Of Supergravity," Ann. Inst. H. Poincare Phys. Theor. **44** (1986) 39.
91. T. Regge, "Gravity On A Lattice," In *Sant Feliu De Guixols 1984, Proceedings, Supersymmetry, Supergravity and Related Topics*, 109-122.
92. T. Regge, "On Broken Symmetries and Gravity," Phys. Rept. **137** (1986) 31. doi:10.1016/0370-1573(86)90068-2

93. T. Regge, "An Elementary Course On General Relativity," RX-1149. Seminar at the Univ. of Valencia in 1985.
94. A. Lerda, J. E. Nelson and T. Regge, "Covariant Canonical Formalism For Polynomial Supergravity In Any Dimension," Int. J. Mod. Phys. A **2** (1987) 1643. doi:10.1142/S0217751X87000855
95. T. Regge, "On Strings And J. Douglas Variational Principle," IN *COMO 1987, PROCEEDINGS, DIFFERENTIAL GEOMETRICAL METHODS IN THEORETICAL PHYSICS* 145-148.
96. D. Zawischa, U. Regge and R. Stapel, "Effective interaction and the staggering of nuclear charge radii," Phys. Lett. B **185** (1987) 299. doi:10.1016/0370-2693(87)91003-3
97. T. Regge, "Discrete Yang-mills Theories," In *Gotsman, E. (Ed.), Tauber, G. (Ed.): From Su(3) To Gravity*, 395-407.
98. J. E. Nelson and T. Regge, "Lorentz invariance and the group manifold approach," Nuovo Cim. B **100** (1987) 405.
99. J. E. Nelson and T. Regge, "Group Manifold Derivation Of Canonical Theories," 8th Italian Conference on General Relativity and Gravitational Physics, 30 Aug-3 Sep, 1988. Trento, Italy, CNUM: C88-08-30.
100. J. E. Nelson and T. Regge, "Canonical theories from the group manifold," International Meeting on Geometrical and Algebraic Aspects of Nonlinear Field Theory, 23-28 May 1988. Amalfi, Italy, CNUM: C88-05-23.2.
101. J. E. Nelson and T. Regge, "Lorentz invariance in tangent space," 23-28 May 1988. Amalfi, Italy, CNUM: C88-05-23.2.
102. J. E. Nelson and T. Regge, "Group manifold derivation of canonical theories," Int. J. Mod. Phys. A **4** (1989) 2021. doi:10.1142/S0217751X89000820
103. J. E. Nelson and T. Regge, "Canonical normal coordinates," Gen. Rel. Grav. **21** (1989) 645. doi:10.1007/BF00760625
104. J. E. Nelson and T. Regge, "Normal coordinates in canonical form," 8th Italian Conference on General Relativity and Gravitational Physics, CNUM: C88-08-30 (1988).
105. J. E. Nelson and T. Regge, "Canonical (2+1)-dimensional quantum gravity," 13th Johns Hopkins Workshop on Current Problems in Particle Theory: Knots, Topology and Quantum Field Theory, CNUM: C89-06-14.1.
106. J. E. Nelson and T. Regge, "Homotopy Groups and (2+1)-dimensional Quantum Gravity," Nucl. Phys. B **328** (1989) 190. doi:10.1016/0550-3213(89)90099-0
107. J. E. Nelson, T. Regge and F. Zertuche, "Homotopy groups and (2+1)-dimensional quantum de Sitter gravity," Nucl. Phys. B **339** (1990) 516. doi:10.1016/0550-3213(90)90359-L
108. J. E. Nelson and T. Regge, "SU(2)-q and (2+1) quantum gravity," Proceedings CNUM: C90-03-01.1.
109. J. E. Nelson and T. Regge, "2+1 quantum gravity up to 2 handles," Proceedings CNUM: C90-09-25.1.
110. J. E. Nelson and T. Regge, "(2+1) quantum gravity," Phys. Lett. B **272** (1991) 213. doi:10.1016/0370-2693(91)91822-D
111. J. E. Nelson and T. Regge, "(2+1) gravity for genus ¿ 1," Commun. Math. Phys. **141** (1991) 211. doi:10.1007/BF02100010
112. J. E. Nelson and T. Regge, "(2+1)-dimensional quantum gravity," Int. J. Mod. Phys. B **6** (1992) 2091. doi:10.1142/S0217979292001031
113. J. E. Nelson and T. Regge, "The Mapping class group for genus 2," Int. J. Mod. Phys. B **6** (1992) 1847. doi:10.1142/S021797929200089X
114. J. E. Nelson and T. Regge, "2+1 Dimensional Quantum Gravity," Conference: C90-03-01.1, pp. 173-181.
115. J. E. Nelson and T. Regge, "(2+1) gravity for high genus," Class. Quant. Grav. **9** (1992) S187. doi:10.1088/0264-9381/9/S/012

116. J. E. Nelson and T. Regge, "Invariants of (2+1)-quantum gravity," Commun. Math. Phys. **155** (1993) 561. doi:10.1007/BF02096728 [hep-th/9209031].
117. T. Regge, "2+1 quantum gravity," Prepared for 10th Italian Conference on General Relativity a Conference: C92-09-01, pp. 399-405.
118. A. Ceresole, R. D'Auria and T. Regge, "Duality group for Calabi-Yau 2 moduli space," Nucl. Phys. B **414** (1994) 517. doi:10.1016/0550-3213(94)90439-1 [hep-th/9307151].
119. J. E. Nelson and T. Regge, "Quantization of (2+1) gravity for genus 2," Phys. Rev. D **50** (1994) 5125. doi:10.1103/PhysRevD.50.5125 [gr-qc/9311029].
120. A. Ceresole, R. D'Auria, S. Ferrara, W. Lerche, J. Louis and T. Regge, "Picard-Fuchs equations, special geometry and target space duality," AMS/IP Stud. Adv. Math. **1** (1996) 281.
121. J. E. Nelson and T. Regge, "Quantum gravity and quantum groups," In *Francaviglia, M. (ed.): Mechanics, analysis and geometry* 503-512.
122. T. Regge, "Invariants of 2+1 quantum gravity," In *Varenna 1994, Quantum groups and their applications in physics* 397-405.
123. M. Billo, A. Ceresole, R. D'Auria, S. Ferrara, P. Fre, T. Regge, P. Soriani and A. Van Proeyen, "A Search for nonperturbative dualities of local N=2 Yang-Mills theories from Calabi-Yau threefolds," Class. Quant. Grav. **13** (1996) 831. doi:10.1088/0264-9381/13/5/007 [hep-th/9506075].
124. T. Regge and R. Zecchina, "Exact solution of the Ising model on group lattices of genus g ¿ 1," J. Math. Phys. **37** (1996) 2796. doi:10.1063/1.531690 [cond-mat/9603072].
125. T. Regge, "The Swiss cheese hypothesis," In *Delsanto, P.P. (ed.) et al.: New perspectives on problems in classical and quantum physics* 361-366.
126. T. Regge and R. Zecchina, "Combinatorial and topological approach to the 3-D Ising model," J. Phys. A **33** (2000) 741. doi:10.1088/0305-4470/33/4/308 [cond-mat/9909168].
127. T. Regge and R. M. Williams, "Discrete structures in gravity," J. Math. Phys. **41** (2000) 3964. doi:10.1063/1.533333 [gr-qc/0012035].
128. L. Chekhov, J. E. Nelson and T. Regge, "Extension of geodesic algebras to continuous genus," Lett. Math. Phys. **78** (2006) 17. doi:10.1007/s11005-006-0112-4 [gr-qc/0510119].
129. T. Regge, "Some thoughts on the future of unified models," J. Phys. A **40** (2007) 3205. doi:10.1088/1751-8113/40/12/S19
130. T. Regge, "Short Autobiography."
131. T. Regge, "Introduction à l'étude du groupe des rotations."
132. T. Regge and C. Teitelboim, "General Relativity à la string: a progress report," arXiv:1612.05256 [hep-th].

1. Outreach and Teaching works

(1) Relatività, in Enciclopedia, vol. XI, Einaudi, Torino, 1980, pp. 820-889.
(2) Sette lezioni su Einstein, with E. Amaldi, V. de Alfaro, D. Galletto, L. Geymonat, F. Pacini, M. Rasetti, Stampatori, Torino, 1980.
(3) Cronache dell'Universo, Boringhieri, Torino, 1981.
(4) Spazio, tempo, relatività, Loescher, Torino, 1981.
(5) Gravitazione e cosmologia, in Enciclopedia del Novecento, vol. VI, Istituto dell'Enciclopedia Italiana, 1982, pp. 90-112.
(6) Dialogo, con Primo Levi, Edizioni di Comunità, Milano, 1984.
(7) Le meraviglie del reale, La Stampa, Torino, 1987.
(8) Gli eredi di Prometeo. L'energia nel futuro, La Stampa, Torino, 1993.
(9) Europa Universitas. Tre saggi sull'impresa scientifica europea, con G. Giorello e S. Veca, Feltrinelli, Milano, 1993.

(10) Infinito. Viaggio ai limiti dell'universo, Mondadori, Milano, 1994.
(11) Scienza e ambiente. Un dialogo, con M. Pallante, Bollati Boringhieri, Torino, 1996.
(12) Non abbiate paura. Racconti di fantascienza, La Stampa, Torino, 1999.
(13) L'universo senza fine. Breve storia del Tutto: passato e futuro del cosmo, Mondadori, Milano, 1999.
(14) Spazio, tempo e universo, con G. Peruzzi, UTET Libreria, Torino, 2003.
(15) Lettera ai giovani sulla scienza, Rizzoli, Milano, 2004.
(16) L'infinito cercare. Autobiografia di un curioso, con S. Sandrelli, Einaudi, Torino, 2012.

2. CD-Rom

- La relatività di Einstein, La Stampa-TuttoScienze, in collaboration with Zanichelli Publishers (1997).
- Nello spazio tra le stelle, with P. Bianucci, La Stampa-TuttoScienze, in collaboration with Zanichelli Publishers (1998).
- La freccia del tempo, with F. Tibone, La Stampa-TuttoScienze, in collaboration with Zanichelli Publishers (1999).
- Vicino alla musica, with L. Berio, La Stampa-TuttoScienze, in collaboration with Zanichelli Publishers (1999).

About the Editors

Leonardo Castellani, born in Fribourg (Switzerland) in 1954, is Professor of Theoretical Physics at the University of East Piedmont. He obtained the Ph.D in Theoretical Physics at the State University of New York at Stony Brook, USA, in 1981, and worked as postdoctoral Fellow at Caltech, and as Senior Fellow at CERN. His research interests are in supergravity, superstrings and noncommutative geometry. His outreach activities include entry-level books (for Springer and Corriere della Sera) on quantum mechanics and teleportation.

Anna Ceresole is an Italian Research Director in Theoretical Physics at the National Institute for Nuclear Physics (INFN), working in the Turin University division. After graduating from there in 1984, she received her PhD from SUNY at Stony Brook (USA) in 1989, she was a postdoc at CALTECH till 1992, and worked as an assistant professor at the Polytechnic of Turin till 2003, when she joined the Turin Division of INFN, where presently she coordinates the Theory group. Her research studies focus on algebraic and geometric aspects of quantum field theory, conformal invariant theories, supergravity and superstrings. She has been active within several EU research networks and since 2017 she is the national PI of the INFN special project *Gauge Theories, Supergravity and String Theory.* She sits in the managing board of the Arnold Regge Center.

Riccardo D'Auria is a Professor Emeritus in Theoretical Physics at the Polytechnic University in Turin, where he became an INFN researcher in 1967 and continued as an assistant and then associate professor till 1985. He moved to Padua University as a full professor and then moved back to Turin Polytechnic, where he was Director of the Department of Physics until 1999. He paid regular scientific visits to CERN, at the Universities of Princeton, Stony Brook and UCLA, where he delivered special courses to PhD students in Theoretical Physics. He is the author of about 200 publications in international journals, with 20 articles being co-authored with T. Regge. His publication with L. Castellani and P. Fré is a three-volume book *Supergravity and Superstrings: A Geometric Perspective*, (World Scientific, Singapore, 1990) and, together with M. Trigiante, the book *From Special Relativity to Feynman Diagrams* (Springer).

Pietro G. Fré, born in Alessandria, Italy, in 1952, is a theoretical physicist, active in the fields of supergravity, superstring theory, cosmology and geometrical field theories. He graduated from the University of Torino in 1974 where he is presently Full Professor of Theoretical Physics. Previously, he was Coordinator of the Elementary Particle Sector at SISSA (Trieste). He was post-doctoral fellow at Caltech and at CERN and he was visiting scientist in several Universities and Laboratories in Europe and in the USA. From 2009 to 2017, he served as Scientific Counselor of the Italian Embassy in Moscow. He is (co)-author of various scientific monographs: in particular *Gravity, a Geometrical Course* (Springer 2012), and he is active also in the field of history of science and its popularization. He is Director of the Arnold-Regge Center for Algebra, Geometry and Theoretical Physics.

www.ingramcontent.com/pod-product-compliance
Lightning Source LLC
LaVergne TN
LVHW080751300926
849162LV00008BA/1791